WHALES, WHALING, AND OCEAN ECOSYSTEMS

*The publisher gratefully acknowledges the generous
contribution to this book provided by
the Gordon and Betty Moore Fund in Environmental Studies.*

*Financial support for the development of this volume was
provided by the Pew Fellows Program in Marine
Conservation, the National Marine Fisheries Service,
and the U.S. Geological Survey.*

*The illustrations preceding each chapter were drawn by
Kristen Carlson through the Mills Endowment to the
Center for Ocean Health, University of California, Santa Cruz.*

WHALES, WHALING, AND OCEAN ECOSYSTEMS

Edited by

JAMES A. ESTES

DOUGLAS P. DEMASTER

DANIEL F. DOAK

TERRIE M. WILLIAMS

ROBERT L. BROWNELL, JR.

UNIVERSITY OF CALIFORNIA PRESS
Berkeley   Los Angeles   London

University of California Press, one of the most distinguished university presses in the United States, enriches lives around the world by advancing scholarship in the humanities, social sciences, and natural sciences. Its activities are supported by the UC Press Foundation and by philanthropic contributions from individuals and institutions. For more information, visit *www.ucpress.edu*.

University of California Press
Berkeley and Los Angeles, California

University of California Press, Ltd.
London, England

Library of Congress Cataloging-in-Publication Data

Whales, whaling, and ocean ecosystems / J.A. Estes ... [et al.].
    p. cm.
    Includes bibliographical references and index.
    ISBN-13: 978-0-520-24884-7 (cloth : alk. paper)
    ISBN-10: 0-520-24884-8 (cloth : alk. paper) 1.   Whaling—Environmental aspects. 2.   Marine ecology. 3.   Whales—Ecology.
I. Estes, J. A. (James A.), 1945-
    SH381.W453 2007
    333.95′95—dc22
                                                            2006013240

Manufactured in the United States of America
10   09   08   07   06
10   9   8   7   6   5   4   3   2   1

The paper used in this publication meets the minimum requirements of ANSI/NISO Z39.48-1992 (R 1997) *(Permanence of Paper)*.

Cover photograph: Predator-prey interactions between killer whales and baleen whales, and how such behavioral interactions may have been altered by modern industrial whaling, has emerged as an intriguing and controversial topic of research. Detail of "The Greenland whale." © The New Bedford Whaling Museum.

CONTENTS

LIST OF CONTRIBUTORS

JOHN P.Y. ARNOULD Deakin University, Burwood, Victoria, Australia

LISA T. BALLANCE Southwest Fisheries Science Center, NMFS, La Jolla, California

LANCE G. BARRETT-LENNARD Vancouver Aquarium Marine Science Center, Vancouver, British Columbia, Canada

JOEL BERGER Wildlife Conservation Society, Teton Valley, Idaho

BODIL A. BLUHM University of Alaska, Fairbanks, Alaska

TREVOR A. BRANCH University of Washington, Seattle, Washington

DANIEL W. BROMLEY University of Wisconsin, Madison, Wisconsin

ROBERT L. BROWNELL, JR. Southwest Fisheries Science Center, NMFS, La Jolla, California

PHILLIP J. CLAPHAM Alaska Fisheries Science Center, NMFS, Seattle Washington

DANIEL P. COSTA University of California, Santa Cruz, California

KENNETH O. COYLE University of Alaska, Fairbanks, Alaska

DONALD A. CROLL University of California, Santa Cruz, California

ERIC M. DANNER University of California, Santa Cruz, California

DOUGLAS P. DEMASTER Alaska Fisheries Science Center, NMFS, Seattle, Washington

DANIEL F. DOAK University of California, Santa Cruz, California

C. JOSH DONLAN Cornell University, Ithaca, New York

TIMOTHY E. ESSINGTON University of Washington, Seattle, Washington

JAMES A. ESTES U.S. Geological Survey, University of California, Santa Cruz, California

KARIN A. FORNEY Southwest Fisheries Science Center, NMFS, Santa Cruz, California

KATHY A. HEISE University of British Columbia, Vancouver, British Columbia, Canada

ROGER P. HEWITT Southwest Fisheries Science Center, NMFS, La Jolla, California

RAYMOND C. HIGHSMITH University of Alaska, Fairbanks, Alaska

GEORGE L. HUNT, JR. University of California, Irvine, California

JEREMY B.C. JACKSON University of California, San Diego, California

PETER KAREIVA Santa Clara University, Santa Clara, California

MATTHEW J. KAUFFMAN University of Wyoming, Laramie, Wyoming

BRENDA KONAR University of Alaska, Fairbanks, Alaska

RAPHAEL KUDELA University of California, Santa Cruz, California

DAVID R. LINDBERG University of California, Berkeley, California

JASON S. LINK Northeast Fisheries Science Center, NMFS, Woods Hole, Massachusetts

HEIKE K. LOTZE Dalhousie University, Halifax, Nova Scotia, Canada

MARC MANGEL University of California, Santa Cruz, California

PAUL S. MARTIN University of Arizona, Tucson, Arizona

RANSOM A. MEYERS Dalhousie University, Halifax, Nova Scotia, Canada

CASEY O'CONNOR Santa Clara University, Santa Clara, California

MICHAEL K. ORBACH Duke University, Beaufort, North Carolina

ROBERT T. PAINE University of Washington, Seattle, Washington

STEPHEN R. PALUMBI Stanford University, Pacific Grove, California

BETE PFISTER Alaska Fisheries Science Center, NMFS, Seattle, Washington

JOHN F. PIATT Alaska Science Center, U.S. Geological Survey, Anchorage, Alaska

ROBERT L. PITMAN Southwest Fisheries Science Center, NMFS, La Jolla, California

NICHOLAS D. PYENSON University of California, Berkeley, California

RANDALL R. REEVES Okapi Wildlife Associates, Hudson, Quebec, Canada

GARY W. ROEMER New Mexico State University, Las Cruces, New Mexico

JOE ROMAN University of Vermont, Burlington, Vermont

DONALD B. SINIFF University of Minnesota, St. Paul, Minnesota

CRAIG R. SMITH University of Hawaii at Manoa, Honolulu, Hawaii

TIM D. SMITH Northeast Fisheries Science Center, NMFS, Woods Hole, Massachusetts

ALAN M. SPRINGER University of Alaska, Fairbanks, Alaska

BERNIE R. TERSHY University of California, Santa Cruz, California

WAYNE Z. TRIVELPIECE Southwest Fisheries Science Center, NMFS, La Jolla, California

GUS B. VAN VLIET Auke Bay, Alaska

PAUL R. WADE Alaska Fisheries Science Center, NMFS, Seattle, Washington

MICHAEL J. WEISE University of California, Santa Cruz, California

HAL WHITEHEAD Dalhousie University, Halifax, Nova Scotia, Canada

TERRIE M. WILLIAMS University of California, Santa Cruz, California

NICHOLAS WOLF University of California, Santa Cruz, California

BORIS WORM Dalhousie University, Halifax, Nova Scotia, Canada

CHRISTOPHER YUAN-FARRELL Santa Clara University, Santa Clara, California

LIST OF TABLES

LIST OF FIGURES

PREFACE

The idea behind this book was born from two related but heretofore largely unconnected realities. One is a growing understanding of the powerful and diverse pathways by which high-trophic-level consumers influence ecosystem structure and function. The other is that most of the great whales were greatly reduced by commercial whaling. Drawing these truths together in our minds, we could easily imagine that ocean ecosystems were profoundly influenced by the loss of the great whales and could be profoundly influenced by the pattern of their future recovery. However, astonishingly little scientific work has attempted to flesh out these speculations about past or current ocean ecosystems.

Recognizing our ignorance about the role of the great whales in ocean ecosystems, we decided to convene a symposium in April 2003 on whaling and whale ecology. This volume is the product of that symposium. Our goal, in both the symposium and this volume, was to examine the ecological roles of whales, past and present, from the broadest set of viewpoints possible. We then hoped, perhaps naïvely, to develop a unified synopsis and synthesis from the conclusions. We realized at the outset that this would be a challenging task. Nature is difficult to observe on the high seas, great whales are especially cryptic, and no one had bothered to record how the oceans may have changed as the whales were being depleted. Although whales have figured prominently in the history of the oceans and the growth of human civilization, depressingly little is really known about their natural history, general biology, and ecology. Despite the best efforts of many dedicated people, scientists remain far apart on even the most basic of questions, such as how many whales are there now, and how much do they need to eat?

The paucity of concrete information about whale ecology means that our group, like others before us, was left using retrospection, analogy with other systems, and broad ecological theory to squeeze inferences and conclusions out of the few hard data that are available on whales. Because of these difficulties, we invited scientists with diverse backgrounds, perspectives, and opinions to think and write about the ecological effects of whales and whaling. By our choice, many of these people were not experts on whales, or even experts on ocean science. Rather, they were creative thinkers who could provide novel approaches to difficult questions. In the beginning, our hope was that this eclectic group would think deeply and that their interactions would add new insights into the ecology of whales. In the end, we found deeply ingrained differences in the approach to scientific investigation that proved difficult to overcome and offered little common ground for progress.

Given our ignorance about something so elemental as whale abundance, its little wonder that the perspectives of our different contributors on a vastly more difficult problem—understanding the roles of these astonishing creatures in ocean food web dynamics—would prove contentious. As the editors of this volume, our views on the science were deeply divided on most issues of substance, and feelings ran strong in a number of cases. The most contentious issues were also the simplest and most fundamental things one would like to know about the great whales: How many existed before commercial whaling, and how many live in the oceans today? These numbers are essential for any theoretical evaluation of how whales and whaling influenced ocean ecosystems. Another issue of major disagreement was the degree to which one can discern the top-down effects of whales and whaling from the bottom-up effects of shifting oceanographic patterns on the dynamics of ocean food webs. Given the historical nature of the problem and the intractability of the animals, neither side could be proven unequivocally. This impasse led to an even more fundamental philosophical difference: Is it irresponsible for us as scientists to frame and publish theories that cannot be definitively proven at the moment, or should we publish only those hypotheses and explanations that have achieved consensus approval within a community of scientists and policy makers? With no consensus on these issues, we have let the contributors to this volume state their own opinions, so that you, the reader, can clearly see the arguments from all sides and draw your own conclusions.

Whether we have succeeded or failed in furthering humankind's understanding of the ecological consequences of whales and whaling is probably a matter of debate. Some will find the results enlightening, whereas others will no doubt think we have done more harm than good by including speculation along with hard facts in the contributions. Regardless of where one stands on the science of whale ecology, the indisputable point remains that great whales were once considerably more abundant than they are now. In the end we are united on one front—a hope that the content of this volume stimulates others to explore further the role of great whales in ocean ecosystems and to consider the related question of how the way we manage them will influence our oceans' future.

D. F. Doak, J. A. Estes, and *T. M. Williams,*
Santa Cruz, California
R. L. Brownell, Jr., Pacific Grove, California
D. P. DeMaster, Seattle, Washington

ONE

Introduction

JAMES A. ESTES

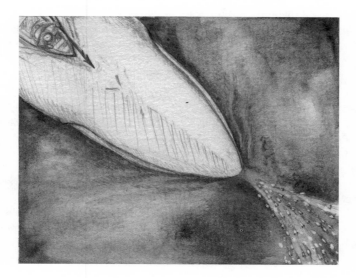

Overharvesting has led to severe reductions in the abundance and range of nearly every large vertebrate species that humans have ever found worth pursuing. These megafaunal reductions, dating in some cases from first contact with early peoples (Martin 1973), are widely known. In contrast, remarkably little is known about the ecological consequences of megafaunal extirpations. Whales and whaling are part of that legacy. Most people know that large whales have been depleted, but little thought has been given to how the depletions may have influenced ocean ecosystems. This volume is an exploration of those influences.

My own interest in the ecological effects of whaling has a complex and serendipitous history, beginning with a view of species interactions strongly colored by first-hand observations of the dramatic and far-reaching influence of sea otters on kelp forest ecosystems (Estes and Palmisano 1974; Duggins et al. 1989; Estes et al. 2004). Sea otters prey on herbivorous sea urchins, thus "protecting" the kelp forest from destructive overgrazing by unregulated sea urchin populations. The differences between shallow reef systems with and without sea otters are every bit as dramatic and far-reaching as those that exist between clear-cuts and old growth forests on the land. I had long thought that the sea otter–kelp forest story was an unusual or even unique case, but have now come to realize that many other species of

large vertebrates exert similarly important ecological influences on their associated ecosystems and that today's world is a vastly different place because of what we have done to them. Accounts of the influences of elephants in Africa (Owen-Smith 1988); wolves in North America (McLaren and Peterson 1994; Ripple and Larsen 2000; Berger et al. 2001); coyotes in southern California (Crooks and Soulé 1999); fishes in North American lakes (Carpenter and Kitchell 1993) and rivers (Power 1985); large carnivores in Venezuela (Terborgh et al. 2001); and what Janzen and Martin (1982) termed "neotropical anachronisms"—dysfunctional ecosystems resulting from early Holocene extinctions of the New World megafauna—provide compelling evidence for significant food web effects by numerous large vertebrates in a diversity of ecosystems. This view was recently reinforced by the realization that coastal marine ecosystems worldwide have collapsed following historical overfishing (Jackson et al. 2001). The belief that whaling left an important imprint on ocean ecosystems was easy to embrace.

That belief, however, was founded far more on principle and analogy than it was on empirical evidence. My real entrée into the ecology of whales and whaling was set off by a seemingly unrelated event—the collapse of sea otters in southwest Alaska. In truth, the possibility that the sea otter's welfare was in any way related to whaling never dawned on me until recently. But the search for an explanation of the sea otter decline led my colleagues and me to increased predation by killer whales as the likely cause (Estes et al. 1998), although at the time we didn't understand why this happened. However, we knew that various pinnipeds in the North Pacific Ocean and southern Bering Sea had also declined in the years preceding the sea otter collapse and thus surmised that a dietary switch by some of the pinniped-eating killer whales may have caused them to eat more sea otters. In the search for an ultimate cause, we therefore presumed that factors responsible for the pinniped declines were also responsible for the sea otter collapse. Like most others at that time, we believed that the pinniped declines had been driven by nutritional limitation, the purported consequence of ocean regime shifts and/or competition with fisheries (Alaska Sea Grant 1993; National Research Council 1996). However, that belief changed from acceptance to suspicion to doubt as a number of inconsistencies and uncertainties with the nutritional limitation hypothesis became apparent (National Research Council 2003). This growing doubt, coupled with evidence that killer whales had caused the sea otter decline, made it easy to imagine that predation by killer whales was responsible for the pinniped declines as well. Demographic and energetic analyses of killer whales and their prey indicated that this possibility was imminently feasible, and in the admittedly complex ecological milieu of the North Pacific Ocean and southern Bering Sea it seemed the most parsimonious explanation. As we assembled additional information, two remarkable patterns emerged—first, that the coastal marine mammal declines began in earnest following the collapse of the last phase of industrial whaling in the North Pacific Ocean; second, that the various populations of pinnipeds and sea otters declined sequentially, one following the next in a seemingly well-ordered manner. These patterns led us to surmise that whaling was likely an important driver of the megafaunal collapse, the proposed mechanism being a dietary shift by killer whales from great whales to other, smaller marine mammal species after the great whales had become sufficiently rare (Springer et al. 2003). This hypothesis, though admittedly simplistic and immensely controversial, stimulated my interest in the connection between whales and ocean food webs.

The question of how whales and whaling influenced ocean food webs is a much broader one, both from the standpoints of process and geography. The idea of a book to consider these larger issues arose from discussions with Dan Doak and Terrie Williams. We recognized that any such effort would require people with expertise on the great whales. Bob Brownell and Doug DeMaster thus joined us. A little further thought led us to identify three main pathways by which the great whales, and their demise due to whaling, may have influenced ocean food webs. One such pathway was as prey for other predators, along the lines of the hypothesis summarized in the preceding paragraphs—a sort of bottom-up effect with indirect food web consequences. Another pathway for the great whales was as consumers—a top-down effect in the traditional sense. The third potentially important food web pathway for the great whales was as detritus—effects on the flux of carbon and other nutrients via scavengers and other detritivores. These food web pathways provide a roadmap for where and how to look for the influences of whales and whaling on ocean ecosystems. The big question, of course, is whether or not any or all of these imagined pathways are important. At one extreme, the great whales may be little more than passengers in ocean ecosystems largely under the control of other processes. At the other extreme is the possibility that one or more of these pathways drive the structure and function of ocean ecosystems in significant ways. Given the enormous number of whales that inhabited the world's oceans before the whalers took them, the diversity of habitats they occupied and prey they consumed, and their large body sizes and high metabolic rates, it is easy to imagine that their losses were important ecologically.

Imagining and knowing, however, are very different things. The problem before us is to evaluate the potential effects of whales and whaling on ocean ecosystems in rigorous and compelling ways, and the challenges of this task are substantial. For one, the events of interest are behind us. We have relatively little information on ocean ecosystems from earlier periods when whales were abundant. This difficulty is compounded by the facts that estimates of abundance for many of the whale populations are poorly known and that the ocean environment is highly dynamic. Climate regime shifts have important effects on production, temperature, and the distribution and abundance of species (Mantua and Hare 2002; Chavez et al. 2003). El Niño–Southern Oscillation (ENSO) events, which have been widely recognized and carefully studied only during the past several decades, exert strong influences on ocean ecosystems over even shorter time periods (Diaz and Pulwarty 1994). Furthermore, open ocean ecology seems to have focused almost exclusively on bottom-up forcing processes. Although no reasonable scientist could possibly believe that bottom-up forcing does not influence the dynamics of ocean ecosystems, the focus on this perspective of food web dynamics and population regulation has relegated species at higher trophic levels to an implicit status of passengers (as opposed to drivers) in ocean ecosystem dynamics. Finally, large whales are not the only organisms to have been removed in excess from the world oceans. Immense numbers of predatory fishes also have been exploited, substantially reducing many populations before, during, and after the whaling era (Pauly et al. 1998, 2002). The largely unknown food web effects of these fisheries, while potentially of great importance, confound our efforts to understand the effects of whales and whaling on ocean ecosystems.

The news is not all bad. There are reasons to hope that significant progress will be made in understanding the ecological consequences of whales and whaling. Ocean ecosystems have been perturbed by the removal of large whales. A great experiment was thus done, and if this experiment did create significant change, records of that change surely exist. The trick is finding them. Such records might be discovered in anoxic basin sediment cores, isotopic analyses, or any number of historical databases looked at with the question of whaling in mind. Another useful feature of the problem is that the effects of whaling were replicated at different times and places. This spatio-temporal variation in the demise of whale populations offers further opportunity for analyses. Finally, none of the great whales have been hunted to global extinction. With protection, most populations that have been monitored have started to recover (Best 1993), and some may have fully recovered (e.g., the eastern North Pacific gray whale). Thus there is the potential for recovery of not only the whales but of their food web interactions, and the opportunity to watch this happen in real time. A more powerful instrument of learning is difficult to imagine.

Understanding the effects of whales and whaling on ocean ecosystems is a complex problem. The unraveling of what one might know and learn requires people with diverse interests and experience. We have attempted to assemble an appropriately eclectic group to write this book. Some of the authors are experts on the biology and natural history of the great whales; their knowledge is also essential to the reconstruction of what happened during the era of industrial whaling. Other authors, while perhaps knowing little about whales, were invited because of their expertise in such diverse areas as history, economics, policy, physiology, demography, genetics, paleontology, and interaction web dynamics. Still others were invited because of their knowledge of other ecosystems in which either the perspective of process differs from that of ocean ecologists or the evidence for ecological roles of large vertebrates is clearer.

The volume is divided into five sections. The first (Background) provides a backdrop by reviewing the theory and evidence for food web processes and summarizing what is known about the history and ecological role of large consumers in other ecosystems. The second section (Whales and Whaling) presents a variety of relevant information on the natural history of whales and on the consequences of whaling to the whales themselves, including several accounts focusing specifically on killer whales and killer whale-large whale relationships. The third section (Process and Theory) examines how and why food web interactions involving great whales might occur. Relevant aspects of their morphology and physiology as well as general assessments of their potential roles as predators, prey, and detritus are presented in this section. The fourth section (Case Studies) includes a variety of more specific accounts of the effects of whales and whaling in various ocean ecosystems. This is necessarily the book's most diverse and unstructured section, because we are asking the question retrospectively; the evidence has not been gathered in a systematic manner; and the participating scientists have widely varying opinions and perspectives on the nature of the problem and the meaning of the data. Some chapters focus on species, others on regions, and still others on parts of ecosystems. Some chapters are strictly empirical, whereas others are more synthetic or theoretical. Whaling was a human endeavor, ultimately driven by human needs and human behavior. The book's fifth and final section (Social Context) thus considers whaling from the perspectives of economics, policy, and law. The concluding chapter, by Peter Kareiva, Christopher Yuan-Farrell, and Casey O'Connor, is a retrospective view of what the other authors have written—how the question of whales and whaling has been addressed to this point, how it might be approached in the future, and how the various issues surrounding whales and ocean ecosystems compare with other problems in applied ecology and conservation biology.

This book reflects the collective wisdom of a group of people with a remarkable range of knowledge and perspective. My particular hope is that our efforts will stimulate others to think about how different today's oceans might be if the great whale fauna were still intact. In an increasingly dysfunctional world of nature, in which food web dynamics remain poorly known and grossly underappreciated, my greater hope is that our efforts will serve as a model for thinking about what conservationists and natural resource managers must do to restore and maintain ecologically effective populations of highly interactive species (Soulé et al. 2003)—one of the twenty-first century's most pressing needs and greatest challenges.

Acknowledgments

I thank Doug Demaster and Terrie Williams for comments on the manuscript.

Literature Cited

Alaska Sea Grant. 1993. *Is it food?: Addressing marine mammal and seabird declines.* Workshop Summary, Alaska Sea Grant Report 93-01. Fairbanks: University of Alaska.

Berger, J., P. B. Stacey, L. Bellis, and M. P. Johnson. 2001. A mammalian predator-prey imbalance: Grizzly bear and wolf extinction affect avian neotropical migrants. *Ecological Applications* 11: 967–980.

Best, P. B. 1993. Increase rates in severely depleted stocks of baleen whales. *ICES Journal of Marine Science* 50: 169–186.

Carpenter, S. R. and J. F. Kitchell. 1993. *The trophic cascade in lakes.* Cambridge, UK: Cambridge University Press.

Chavez, F. P., J. Ryan, S.E. Lluch-Cota, and M. Ñiquen C. 2003. From anchovies to sardines and back: multidecadal change in the Pacific Ocean. *Science* 299: 217–221.

Crooks, K. R. and M. E. Soulé. 1999. Mesopredator release and avifaunal extinctions in a fragmented system. *Nature* 400: 563–566.

Diaz, H. F. and R. S. Pulwarty. 1994. An analysis of the time scales of variability in centuries-long ENSO-sensitive records. *Climatic Change* 26: 317–342.

Duggins, D. O., C. A. Simenstad, and J. A. Estes. 1989. Magnification of secondary production by kelp detritus in coastal marine ecosystems. *Science* 245: 170–173.

Estes, J. A. and J. F. Palmisano. 1974. Sea otters: their role in structuring nearshore communities. *Science* 185: 1058–1060.

Estes, J. A., E. M. Danner, D. F. Doak, B. Konar, A. M. Springer, P. D. Steinberg, M. T. Tinker, and T. M. Williams. 2004. Complex trophic interactions in kelp forest ecosystems. *Bulletin of Marine Science* 74(3): 621–638.

Estes, J. A., M. T. Tinker, T. M. Williams, and D. F. Doak. 1998. Killer whale predation on sea otters linking oceanic and nearshore ecosystems. *Science* 282: 473–476.

Jackson, J. B. C., M. X. Kirby, W. H. Berger, K. A. Bjorndal, L. W. Botsford, B. J. Bourque, R. Bradbury, R. Cooke, J. A. Estes, T. P. Hughes, S. Kidwell, C. B. Lange, H. S. Lenihan, J. M. Pandolfi, C. H. Peterson, R. S. Steneck, M. J. Tegner, and R. Warner. 2001. Historical overfishing and the recent collapse of coastal ecosystems. *Science* 293: 629–638.

Janzen, D. H. and P. S. Martin. 1982. Neotropical anachronisms: the fruits the gomphotheres ate. *Science* 215: 19–27.

Mantua, N. J. and S. R. Hare. 2002. The Pacific decadal oscillation. *Journal of Oceanography* 58: 35–44.

Martin, P. S. 1973. The discovery of America. *Science* 179: 969–974.

McLaren, B. E. and R. O. Peterson. 1994. Wolves, moose and tree rings on Isle Royale. *Science* 266: 1555–1558.

National Research Council. 1996. *The Bering Sea ecosystem*. Washington, D.C.: National Academy Press.

———. 2003. *Decline of the Steller sea lion in Alaskan waters*. Washington, D.C.: National Academy Press.

Owen-Smith, N. 1988. *Megaherbivores: the influence of very large body size on ecology*. Cambridge, U.K., and New York: Cambridge University Press.

Pauly, D., V. Christensen, J. Dalsgaard, R. Froese, and F. Torres, Jr. 1998. Fishing down marine food webs. *Science* 279: 860–863.

Pauly, D., V. Christensen, S. Guénette, T. J. Pitcher, U. R. Sumaila, C. J. Walters, R. Watson, and D. Zeller. 2002. Towards sustainability in world fisheries. *Nature* 418: 689–695.

Power, M. E. 1985. Grazing minnows, piscivorous bass, and stream algae: dynamics of a strong interaction. *Ecology* 66: 1448–1456.

Ripple, W. J. and E. J. Larsen. 2000. Historic aspen recruitment, elk, and wolves in northern Yellowstone National Park, U.S.A. *Biological Conservation* 95: 361–370.

Soulé, M. E., J. A. Estes, J. Berger, and C. Martinez del Rio. 2003. Ecological effectiveness: conservation goals for interactive species. *Conservation Biology* 17: 1238–1250.

Springer, A. M., J. A. Estes, G. B. van Vliet, T. M. Williams, D. F. Doak, E. M. Danner, K. A. Forney, and B. Pfister. 2003. Sequential megafaunal collapse in the North Pacific Ocean: an ongoing legacy of industrial whaling? *Proceedings of the National Academy of Science* 100: 12223–12228.

Terborgh, J., L. Lopez, V. P. Nuñez, M. Rao, G. Shahabuddin, G. Orihuela, M. Riveros, R. Ascanio, G. H. Adler, T. D. Lambert, and L. Balbas. 2001. Ecological meltdown in predator-free forest fragments. *Science* 294: 1923–1926.

BACKGROUND

TWO

Whales, Interaction Webs, and Zero-Sum Ecology

ROBERT T. PAINE

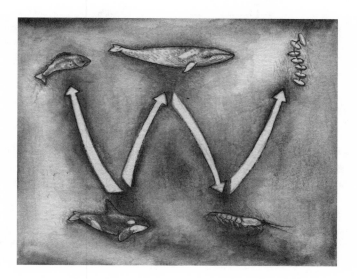

Food webs are inescapable consequences of any multispecies study in which interactions are assumed to exist. The nexus can be pictured as links between species (e.g., Elton 1927) or as entries in a predator by prey matrix (Cohen et al. 1993). Both procedures promote the view that all ecosystems are characterized by clusters of interacting species. Both have encouraged compilations of increasingly complete trophic descriptions and the development of quantitative theory. Neither, however, confronts the issue of what constitutes a legitimate link (Paine 1988); neither can incorporate the consequences of dynamical alteration of predator (or prey) abundances or deal effectively with trophic cascades or indirect effects. Thus one challenge confronting contributors to this volume is the extent to which, or even whether, food webs provide an appropriate context for unraveling the anthropogenically forced changes in whales, including killer whales (*Orcinus orca*), their interrelationships, and the derived implication for associated species.

A second challenge is simply the spatial vastness (Levin 1992) of the ecological stage on which whale demography and interactions are carried out. This bears obvious implications for the amount, completeness, and quality of the data and the degree to which "scaling up" is permissible. Manipulative experiments, equivalent to those that have proven so revealing on rocky shores and even more so in freshwater ecosystems, are clearly impossible. Buried here, but of critical

importance, is the "changing baseline" perspective (Pauly 1995, Jackson et al. 2001): Species abundances have changed, and therefore the ecological context, but by how much?

This essay begins with a brief summary of experimental studies that identify the importance of employing interaction webs as a format for further discussion of whales and ocean ecosystems. The concept, while not novel, was developed by Paine (1980) as "functional" webs; Menge (1995) provided the more appropriate term, *interaction web*. My motivation is threefold:

1. Such studies convincingly demonstrate that species do interact and that some subset of these interactions bear substantial consequences for many associated species.

2. The studies also reveal the panoply of interpretative horrors facing all dynamic community analysis: Individual species will have different, and varying, per capita impacts; nonlinear interactions are rampant; and indirect effects are commonplace.

3. The preceding two points raise another question: Are oceanic assemblages so fundamentally different from terrestrial, lentic, and shallow-water marine ones (perhaps because of an ecological dilution due to their spatial vastness) that different organizational rules apply?

I next develop a crucial aspect of my argument that interaction webs provide a legitimate and useful framework. I call this aspect "zero-sum ecology." It invokes a mass balance equilibrium, implying that carbon is not being meaningfully sequestered from or released to global ecosystems over time spans appropriate to current whale ecology. It differs from Hubbell's (2001) similar perspective by focusing on energy rather than individual organisms. That is, the global cycling of organic carbon is more or less in balance, and thus all photosynthetically fixed carbon is returned to the global pool via bacterial or eukaryote respiration. Hairston et al. (1960) developed the same theme. Its primary implication is that removal of substantial biomass from one component of an ecosystem should be reflected in significant changes elsewhere, identified perhaps as increased (or decreased) biomass and population growth rates, alteration of diets as the spectrum of prey shifts, or changes in spatial distribution. Interaction webs are intended to portray these dynamics qualitatively and fit comfortably with multispecies models such as that of May et al. (1979).

The terminal section discusses a varied set of studies that collectively suggest that whales, including *O. orca*, at oceanic spatial scales could have played roles analogous to those demonstrated for consumers of secondary production in much smaller, experimentally tractable systems. Acceptance or denial of their relevance is at the crux of the question: Do whales and their interspecific interactions matter, or how might they, or could their consequences have been anticipated or predicted under an onslaught of anthropogenic forcing? The concluding paragraphs argue for an open-mindedness in addressing this question. Frankly, I do not *know* whether whales mattered (ecologically, not esthetically), but their large mass, physiology (homeothermy), and diminished numbers, even at characteristically huge spatial scales, implies that suggestion of significant roles in the ocean's economy should not be summarily dismissed or ignored. Resolution surely will involve an interplay between compilation and analysis of historical information (e.g., whaling records); modeling (e.g., using EcoSim/Ecopath; see Walters et al. 1997); newer data on demographic trends, density, diet, and so forth; and, equally, the degree to which analogy with data-rich exploited fish and shark populations proves relevant.

Interaction Webs

Charles Darwin was an insightful experimentalist, and many of his tinkerings produced striking results, although the resolving power of such interventions in the organization of "nature" was unrecognized or underappreciated in his time. One kind of controlled manipulation is represented by Darwin's (1859: 55) grass clipping exercise or Paine (1966). Such studies identify phenomena such as changes in species richness, distribution pattern, or even production, and their results are often broadly repeatable despite minimal appreciation of the root mechanisms. Another kind of study involves manipulation of some variable such as density

manipulation or specific nutrient inputs, with the goal of a much more precise understanding of how that segment of a system functions. Both kinds of study provide a basis for prediction, the former qualitative, the latter quantitative. Both also imply that species are dynamically linked and that changes in some species' density, prey or nutrient availability, or system trophic structure are highly likely to be reflected in changes elsewhere in the ecosystem.

These relationships constitute the domain of interaction webs. As identified earlier, such webs differ from the more descriptive linkage patterns and energy flow webs because they focus on the change subsequent to some manipulation rather than a fixed, seemingly immutable pattern. No standardized graphic protocols have been developed, and none are attempted here. On the other hand, an increasing number of review articles attest to a recognition that understanding the complexities of multispecies relationships is both a vital necessity and the handmaid of successful ecosystem management. Interaction webs provide the matrix in which such understanding can be developed.

An early review of experimentally induced alteration in assemblage structure (Paine 1980) introduced the term *trophic cascade* and provided a coarse taxonomy of food webs. That perspective was encouraged by a number of seminal studies, some of which described dramatic assemblage changes after an invasion (Brooks and Dodson 1965; Zaret and Paine 1973) or recovery of an apex predator (Estes and Palmisano 1974). Supplementing these results were manipulative experiments in which species of high trophic status were removed, excluded, or added (Paine 1966, Sutherland 1974, Power et al. 1985). Other studies employed experimental ponds (Hall et al. 1970) or their smaller cousins, "cattle tanks" (Morin 1983), and even whole lakes (Hassler et al. 1951, Schindler 1974). The foregoing references are but a small fraction of studies identifying the consequences of nutrient alteration, consumers jumbling the consequences of competitive interactions, or apex predators influencing whole community structure.

By 1990 this conceptual framework, already hinted at by Forbes (1887) and clearly visible in the work of Brooks and Dodson (1965), had been deeply explored in freshwater ecosystems (Carpenter et al. 1985, Carpenter and Kitchell 1993). The ecologically polarizing jargon of "top-down" (predator control) and "bottom-up" (production control) developed rapidly. An ecumenical review by Power (1992) established the obvious—that both forces exist, and it is their relative importance that should be evaluated. A majority of recent reviews concentrate on trophic cascades, a top-down forcing phenomenon and one easily produced in experimentally tractable assemblages and equally visible in large-geographic-scale, heavily fished systems. For instance, Sala et al. (1998), Fogarty and Murawski (1998), and Pinnegar et al. (2000) expand on fisheries' impacts in marine shallow-water, rocky-surface systems. Pace et al. (1999) identify cascades as widespread in a diversity of systems ranging from insect guts to open oceans; Shurin et al.

(2002), in an examination of 102 field experiments, found predation effects strongest in freshwater and marine benthic webs and weakest in marine plankton and terrestrial assemblages. Duffy (2002), Schmitz (2003), and Van Bael et al. (2003) have continued to develop an appreciation of the ubiquity, but not the generality, of top-down influences, as have Banse (2002) and Goericke (2002) for blue-water systems.

What is the relevance of small-scale, generally short-duration studies for whales, the biologists invested in their study, and even biological oceanographers investigating ecological events at spatial scales ranging from hundreds to thousands of square kilometers? With the exception of Estes's research (e.g., Estes and Palmisano 1974; Estes et al. 1998), we generally do not know. Those studies, conducted along the shallow shoreline of western North America, basically trace a predator-induced cascade from higher trophic levels to benthic algae attached to rocky surfaces, on which the relatively simple process of interference competition for space predominates. Is an understanding of such a dynamic transferable to open-water assemblages in which (presumably) exploitation competition occurs, in addition to the impacts of consumption by higher trophic levels? Again, we do not know. However, it appears shortsighted to reject top-down influences dogmatically, given their undisputed presence in some, though perhaps not all, other ecosystems.

Zero-Sum Ecology

Hairston et al. (1960) based their seminal paper in part on a presumption that all photosynthetically fixed carbon was utilized. That is, natural gas, coal, and oil were not being deposited; globally, carbon fixed equaled carbon respired. In the short run, this appears to be correct, although increasing atmospheric CO_2 and uncertainty about carbon sources and sinks clouds the issue. The same might be said for methane ices or clathrates, CH_4 trapped in ice under known conditions of ambient pressure and temperature: There seems to be a balance between carbon in the sedimentary reservoirs and carbon fluxes (Kvenvolden 1998). Carbon atoms bound in rock (e.g., $CaCO_3$) are probably immaterial at the time scales considered here.

If these assumptions are correct, they have profound implications for interaction webs. What goes in must come out; a major alteration of living biomass and its maintenance requirements in one sector of a food web will surely be reflected by changes elsewhere. Because the rudimentary natural history outlining relationships known for more accessible systems is lacking, or minimized, for environments inhabited by whales, the prediction of and search for consequences have been hampered.

An analogy developed by Robert MacArthur catches the sense of the situation nicely. If a shelf is filled to capacity with books, every withdrawal permits an addition. On the other hand, on a shelf characterized by numerous gaps, removal (= extraction) or addition should be of small,

undetectable, or no consequence. A zero-sum ecological perspective implies that removals at the magnitude of suspected whale extractions must have had effects. Without knowing what these might have been, and in the absence of any serious attempt to document specific resultant changes, we must resort to inference or conjecture. The following section develops arguments that changes within oceanic ecosystems must have occurred.

Consequential Interactions in Large, Open Ecosystems

We know that organisms interact, both from direct observation and as demonstrated by controlled manipulation at small spatial scales. As the spatial domain increases, however, our knowledge base diminishes accordingly—to the point, perhaps, where interactions at the community level (that is, beyond the obvious acts of feeding or being eaten) become uncertain and obscure. Various lines of evidence suggest that consequential interactions or interrelationships do exist in pelagic ecosystems. The examples that follow are hardly exhaustive; rather they indicate the kinds of natural or imposed phenomena that are capable of altering biological assemblages at large spatial scales. The absence of information on trophic responses is better viewed as lost opportunity rather than absence of effect. The majority of these examples have been categorized within the framework proposed by Bender et al. (1984) into pulse perturbations, which are relatively instantaneous impacts, and press experiments, in which the perturbation is maintained. Plagues and massive diebacks provide examples of the former, commercial fisheries of the latter.

Pulse Perturbations

EEL GRASS DISAPPEARANCE. Between 1930 and 1933, a "wasting disease" caused the disappearances of about 90% of the eel grass (Zostera marina) in coastal waters of the north Atlantic (Short et al. 1987). Zostera is a major source of detritus and a food for birds. Its precipitous decline decimated populations of migratory waterfowl, led to loss of a commercial scallop fishery, impacted a lagoon's invertebrate assemblage, and generated the first documented extinction of a marine invertebrate: the limpet Lottia alveus (Carlton et al. 1991; Stauffer 1937).

DIADEMA DIEBACK. The catastrophic collapse of populations of the sea urchin Diadema antillarum provides a second example. Prior to 1983 this species was ubiquitous on coral reefs in the Caribbean basin. Within little more than a year, an epidemic of unknown cause had killed from 93% to nearly 100% of these urchins within a 3.5 million km^2 region (Lessios 1988). Community effects ranged from increases in benthic algal percent cover, increases in the rate of fish herbivory, decreases in bioerosion, reduced coral recruitment, and numerical increases in other urchin species, implicating previously unsuspected interspecific competition (Lessios

1988). Generalizations based on the demise of this single species are confounded by hurricane damage and overexploitation of large herbivorous fishes (Hughes 1994; Paine et al. 1998). Nonetheless, the lesson seems certain: The ecological consequences of mass, nearly instantaneous, mortality of an important grazer ramified throughout this ecosystem, and many of the resultant population shifts could have been predicted correctly a priori.

Catastrophic mortality events are not uncommon in marine near-shore ecosystems. The list of taxa involved ranges from corals and sponges to dolphins and seals. To the extent that single species are involved, these precipitous declines provide rare opportunities to probe the role a species plays in community organization. Do they occur in noncoastal oceans? We do not know. However, pulse perturbations at large spatial scales appear to retain the salient hallmarks characterizing small-scale experimental studies: dramatic community alteration, shifts in trophic structure, and important indirect consequences.

Press Perturbations

Sustained industrial fishing operations in the world's coastal and central oceans can be considered press perturbations. Evidence for their effects is seen in the growing evidence for overexploitation of apex predators (Myers and Worm 2003) and an increasingly accepted metaphor of marine food webs being "fished down" (Pauly et al. 1998). The following examples only hint at the magnitude and complexity potentially induced when the density of apex predators is altered.

SHALLOW-WATER, BENTHIC SYSTEMS: SEA OTTERS. Trophic cascades occur when the addition or deletion of some higher-level consumer leads to major shifts in species composition at lower levels. One could consider keystone species effects (e.g., Paine 1966) as a muted cascade, although only two "trophic levels" were involved. A much clearer example, involving three and possibly four levels, has been developed by Estes et al. (1998). By the time (1911) sea otter exploitation was terminated, the species was locally extinct throughout much of its original range. The existence of a few local populations, however, led to the development of two contrasting states, providing the comparisons detailed in the classic study by Estes and Palmisano (1974). Although considering these regions as either two-level (without otters) or three-level (with otters) is a gross simplification, it has enabled a striking range of studies in this experimentally intractable system. For instance, Simenstad et al. (1978) illustrate that two- and three-state systems (excluding humans as a trophic level) characterize prehistoric Aleut middens and probably resulted from local overexploitation of sea otters. Duggins et al. (1989), by comparing islands with and without sea otters, quantified a range of indirect effects: At islands with otters, and therefore robust kelp populations, mussels and barnacles grow roughly twice as fast as they do at otterless islands. Stable carbon isotope analyses identified detrital kelp as the supplemental energy source. Estes and Duggins (1995) have demonstrated the geographic ubiquity of ecological transformation from a two- to three-level system as otter populations recovered and, in the process, promoted kelp bed development. Finally, Estes et al. (1998) have shown that the recent entry of killer whales into this food web, which added a fourth trophic level (comparable to the prehistoric Aleut impact), has generated the compositional shifts anticipated at all three linked lower levels.

I believe that otters at high population densities represent a sustained press perturbation just as, in their absence, sea urchins do. Furthermore, the results seem generalizable over thousands of kilometers of shoreline. Thus, while a trophic cascade and positive indirect effects on species not eaten by otters (barnacles) are clearly identified, it remains basically a shallow-water, benthic study system whose applicability to truly pelagic systems is open to question.

DEEP-WATER, BENTHIC SYSTEMS: ATLANTIC COD-SHRIMP INTERACTIONS. Commercial fisheries generate a wealth of data on fish abundances, often expressed as time series. Severely depleted Atlantic cod (Gadus morhua) stocks and those of their commercially valuable prey, the shrimp Pandalus borealis, have been subjected to meta-analysis by Worm and Myers (2003). These species represent a natural predator-prey coupling: Eight of nine analyses of mixed stocks in different geographic regions revealed significantly inverse abundances. When all north Atlantic regions are combined, cod catch statistics are also inversely related to catches of two large crustaceans (snow crabs and American lobsters). Such dominating top-down influences, expressed at a large geographic scale, are certain to have cascading effects on other benthic species (Worm and Myers 2003). Witman and Sebens's (1992) comparison of Western Atlantic seamounts supports this opinion. At offshore sites with abundant cod, crabs were scarce and brittle stars were subjected to significantly greater predation rates. The reverse pattern characterizes coastal sites with fishery-depleted cod populations. Since these invertebrates are themselves important predators, cascading impacts at still lower trophic levels, while difficult to measure, should have occurred.

Overexploitation of high-trophic-status predators such as cod can be anticipated to induce changes in community structures. The Worm and Myers (2003) analysis provides one striking example. The observation and small-scale experiments of Witman and Sebens (1992) substantiate that opinion. Nonetheless, these relationships, even if in deep water, are still firmly anchored to a benthic component, again calling into question their relevance to purely pelagic interactions. A variety of approaches to the latter are explored next.

OCEANIC, TOTALLY PELAGIC SYSTEMS: SHARKS, TUNA. Industrialized fishing occurs in the world's open oceans on almost unimaginable scales: purse seines two to three hundred meters in depth, baited lines in excess of 50 kilometers in length. They have extracted a toll on apex consumers (tuna, sharks) and have often been characterized by a substantial bycatch. Because of public concern about impacts on such charismatic species as albatross, turtles, and dolphins, as well as for fundamental management reasons, a substantial database exists for some fisheries. For instance, Essington et al. (2002) and Schindler et al. (2002) employed bioenergetic models (Kitchell et al. 1977) that balance mean (population) metabolism and prey consumption to estimate the consequences of fishing effects on blue sharks and small tuna. One goal was to evaluate the efficiency of different fishing strategies. Both analyses employed what can be considered a hybrid food web modeling approach. Both concluded that commercial fisheries have strong effects on the character of trophic linkages in pelagic webs, that these effects may alter predator life history traits and have an impact on lower trophic level prey as well. Such a result is to be anticipated if "zero-sum" energy balances apply in the open ocean. Lack of both interest and a persuasive conceptual framework, and little or no funding for studies of tangential consequences, have constrained understanding of even minimal ecosystem effects of these press perturbations.

Similar interpretations involving a very different approach have been obtained by Stevens et al. (2000), who examined the effects of fishing on sharks and their allies employing the modeling technique EcoSim. For instance, their Table 1 summarizes the status of 17 stocks, of which 12 have collapsed or are in decline. The models compared the consequences of shark removal in three different ecosystems. In general, there were numerous surprises with respect to population trends, and many of the outcomes were not as predictable as anticipated. It seems impossible to know whether such trends reflect "reality" or whether, when "surprises" occur, they are due to unrealistic parameter estimates or incomplete knowledge of that particular food web.

Although none of the foregoing studies included whales, it is certain that species do not exist in an ecological vacuum, and therefore, interactions must be present. When the impact of diminished apex predator mass was evaluated in a food web context, population density of other food web members was found to change. Direct and indirect effects were implicated in these multi–trophic level simulations. In an earlier study employing linked differential equations in which whales were a key component, May et al. (1979) identified realistic population trajectories dependent on "treatment." That is, whether whales and/or krill were protected or exploited had identifiable consequences for seals and penguins. A variety of models, in fact, may provide the surest way to generate testable predictions in these large, open systems. As the quality and trophic extent of parameter estimates increases, the role of whales at even historic population levels might be evaluated. If modeling is to be effective, however, many of the attributes of interactions will have to be first identified and then quantified.

Other Kinds of Interactions

Nature is highly variable—a condition confounded by anthropogenic forcing—so strict categorization of influences is difficult if not inappropriate. In addition to pulse and press perturbations attributed to commercial overfishing, at least one other category producing ecological change exists. All species vary naturally in abundance in space through time. When these variations are haphazard (stochastic) in magnitude, timing, or place, they are frustratingly useless in an analytical sense. However, a few marine species vary in highly predictable ways. Shiomoto et al. (1997) have cautiously described the consequences of biannual variation in pink salmon catch in the subarctic North Pacific. This variation of slightly more than an order of magnitude translates in years of high salmon abundance to both fewer herbivorous zooplankton and a reduction of their predators (carnivorous macro-zooplankton), and increased concentrations of chlorophyll a, an index of phytoplankton abundance and primary production. When salmon were scarce, the pattern reversed. The implication of top-down dynamical consequence is inescapable.

Other quasi-cyclic variations exist—Pacific decadal oscillations, El Niño, and La Niña especially—but they are much too general in effect to provide the unambiguous, species-specific signals that most interaction webs, qualitative or quantitative, require.

Conclusions

"Do or did whales matter?" is our primary issue. If global carbon flux is in equilibrium, at least to the extent that coal, oil, gas, and clathrates are not being deposited (which would signal no surplus of production over consumption), reducing the biomass of large-bodied consumers must have had effects on other elements of the food web. The experimental evidence for cascades (and ripples) from all kinds of ecosystems with a benthic component is unequivocal: Top-down influences are often important. Natural history detail, direct observation, and small-scale experimentation all substantiate this conclusion. Additional support comes from less tractable, larger systems, such as the Great Lakes (Madenjian et al. 2002) and the Northeast Pacific (Estes et al. 1998). A "weight of evidence" argument (National Research Council 2003) would support the view that whales did matter and that systematic exploitation, reducing their biomass and changing their spatial distribution, must have left an ecosystem imprint.

Sadly, there is little direct proof that this conjecture is correct, in part because of the spatial vastness of the whales' domain and in part to the near-total absence of trophic linkage detail and even taxonomy of the food web's membership. Probably the cleanest signal will come from predicting indirect consequences and then evaluating their robustness, much in the spirit of May et al. (1979) and Springer et al. (2003). However, it is almost certain that any confidence in the predicted consequences of whale removal will be compromised by both uncertainty about historic population sizes (Roman and Palumbi 2003) and concurrent overexploitation of other apex consumers. A more optimistic vision is that most whale stocks will recover eventually to sustainable levels. In that eventuality, again assuming zero-sum ecosystem energetics and a sufficiently strong ecological signal, resurrection of these apex consumers will provide an important probe for understanding the organization of oceanic ecosystems at global scales.

Acknowledgments

I want to thank Jim Estes for inviting my contribution to this volume and especially for allowing me, in my terms, to try to "convert the heathen." Financial support came from the Andrew W. Mellon Foundation and the Pew Fellows Program in Marine Conservation. The Makah Indian Nation, by sanctioning research on Tatoosh Island, has allowed me the freedom to develop my perspective; I remain continuingly grateful.

Literature Cited

Banse, K. 2002. Steemann Nielsen and the zooplankton. *Hydrobiologia* 480: 15–28.

Bender, E. A., T. J. Case, and M. E. Gilpen. 1984. Perturbation experiments in community ecology: theory and practice. *Ecology* 65: 1–13.

Brooks, J. L. and S. I. Dodson. 1965. Predation, body size, and composition of plankton. *Science* 150: 28–35.

Carlton, J. T., G. J. Vermeij, D. R. Lindberg, D. A. Carlton, and E. C. Dudley. 1991. The first historical extinction of a marine invertebrate in an ocean basin: the demise of the eelgrass limpet *Lottia alveus*. *Biological Bulletin* 180: 72–80.

Carpenter, S. R., J. F. Kitchell, and J. R. Hodgson. 1985. Cascading trophic interactions and lake productivity. *BioScience* 35: 634–639.

Carpenter, S. R., and J. F. Kitchell, eds. 1993. *The trophic cascade in lakes*. Cambridge, UK: Cambridge University Press.

Cohen, J. E., R. A. Beaver, S. H. Cousins, D. L. DeAngelis, L. Goldwasser, K. L. Heong, R. D. Holt, A. J. Kohn, J. H. Lawton, N. Martinez, R. O'Malley, L. M. Page, B. C. Patten, S. L. Pimm, G. A. Polis, M. Rejmanek, T. W. Schoener, K. Schoenly, W. G. Sprules, J. M. Teal, R. E. Ulanowicz, P. H. Warren, H. M. Wilbur, and P. Yodzis. 1993. Improving food webs. *Ecology* 74: 252–258.

Darwin, C. 1859. *The origin of species by means of natural selection*. New York: Random House.

Duffy, J. E. 2002. Biodiversity and ecosystem function: the consumer connection. *Oikos* 99: 201–219.

Duggins, D. O., C. A. Simenstad, and J. A. Estes. 1989. Magnification of secondary production by kelp detritus in coastal marine ecosystems. *Science* 245: 170–173.

Elton, C. 1927. *Animal ecology*. New York: Macmillan.

Essington, T. E., D. E. Schindler, R. J. Olson, J. F. Kitchell, C. Boggs, and R. Hilborn. 2002. Alternative fisheries and the predation rate of yellow fin tuna in the Eastern Pacific Ocean. *Ecological Applications* 12: 724–734.

Estes, J. A. and D. O. Duggins. 1995. Sea otters and kelp forests in Alaska: generality and variation in a community ecological paradigm. *Ecological Monographs* 65: 75–100.

Estes, J. A. and J. F. Palmisano. 1974. Sea otters: their role in structuring nearshore communities. *Science* 185: 1058–1060.

Estes, J. A., M. T. Tinker, T. M. Williams, and D. F. Doak. 1998. Killer whale predation on sea otters linking oceanic and nearshore ecosystems. *Science* 282: 473–476.

Fogarty, M. J. and S. A. Murawski. 1998. Large-scale disturbance and the structure of marine systems: fisheries impacts on Georges Bank. *Ecological Applications* 8 (Supplemental): 175–192.

Forbes, S. A. 1887. The lake as a microcosm. *Bulletin of the Peoria Scientific Association* 1877: 77–87. Reprinted (1925) in the *Bulletin of the Illinois Natural History Survey* 15: 537–550.

Goericke, R. 2002. Top-down control of phytoplankton biomass and community structure in the monsoonal Arabian Sea. *Limnology and Oceanography* 47: 1307–1323.

Hairston, N. G., F. E. Smith, and L. B. Slobodkin. 1960. Community structure, population control, and competition. *American Naturalist* 94: 421–425.

Hall, J. D., W. E. Cooper, and E. E. Werner. 1970. An experimental approach to the production dynamics and structure of freshwater animal communities. *Limnology and Oceanography* 15: 839–928.

Hassler, A. D., O. M. Brynildson, and W. T. Helm. 1951. Improving conditions for fish in brown-water lakes by alkalization. *Journal of Wildlife Management* 15: 347–352.

Hubbell, S. P. 2001. *The unified neutral theory of biodiversity and biogeography*. Princeton, NJ: Princeton University Press.

Hughes, T. P. 1994. Catastrophes, phase shifts, and large-scale degradation of a Caribbean coral reef. *Science* 265: 1547–1551.

Jackson, J. B. C., M. X. Kirby, W. H. Berger, K. A. Bjorndal, L. W. Botsford, B. J. Bourque, R. Bradbury, R. Cooke, J. A. Estes, T. P. Hughes, S. Kidwell, C. B. Lange, H. S. Lenihan, J. M. Pandolfi, C. H. Peterson, R. S. Steneck, M. J. Tegner, and R. Warner. 2001. Historical overfishing and the recent collapse of coastal ecosystems. *Science* 293: 629–638.

Kitchell, J. F., D. J. Stewart, and D. Weininger. 1977. Application of a bioenergetics model to yellow perch (*Perca flavescens*) and walleye (*Stizostedion vitreum vitreum*). *Journal of the Fisheries Research Board of Canada* 34: 1922–1935.

Kvenvolden, K. A. 1998. A primer on the geological occurrence of gas hydrates, in *Gas Hydrates: Relevance to World Margin Stability and Climatic Change*. J. P. Henriet and J. Mienert, eds. Special Publication 137. London: Geological Society of London, pp. 9–30.

Lessios, H. A. 1988. Mass mortality of *Diadema antillarum* in the Caribbean: What have we learned? *Annual Review of Ecology and Systematics* 19: 371–393.

Levin, S. A. 1992. The problem of pattern and scale in ecology. *Ecology* 73: 1943–1967.

Madenjian, C. P., G. L. Fahnenstiel, T. H. Johengen, T. F. Nalepa, H. A. Vanderploeg, G. W. Fleischer, P. J. Schneeberger, D. M. Benjamin, E. B. Smith, J. R. Bence, E. S. Rutherford, D. S. Lavis, D. M. Robertson, D. J. Jude, and M. P. Ebener. 2002. Dynamics of the Lake Michigan food web, 1970–2000. *Canadian Journal of Fishery and Aquatic Science* 59: 736–753.

May, R. M., J. R. Beddington, C. W. Clark, S. J. Holt, and R. M. Laws. 1979. Management of multi-species fisheries. *Science* 205: 267–277.

Menge, B. A. 1995. Indirect effects in marine rocky intertidal interaction webs: patterns and importance. *Ecological Monographs* 65: 21–74.

Morin, P. J. 1981. Predatory salamanders reverse the outcome of competition among three species of anuran tadpoles. *Science* 212: 1284–1286.

Myers, R. A. and B. Worm. 2003. Rapid worldwide depletion of predatory fish communities. *Nature* 423: 280–283.

National Research Council. 2003. *Decline of the Steller sea lion in Alaskan waters.* Washington, DC: National Academy Press.

Pace, M. L., J. J. Cole, S. R. Carpenter, and J. F. Kitchell. 1999. Trophic cascades revealed in diverse ecosystems. *Trends in Ecology and Evolution* 14: 483–488.

Paine, R. T. 1966. Food web complexity and species diversity. *American Naturalist* 100: 65–75.

———. 1980. Food webs: linkage, interaction strength, and community infrastructure. *Journal of Animal Ecology* 49: 667–685.

———. 1988. Food webs: road maps of interactions or grist for theoretical development? *Ecology* 69: 1648–1654.

Paine, R. T., M. J. Tegner, and E. A. Johnson. 1998. Compounded perturbations yield ecological surprises. *Ecosystems* 1: 535–545.

Pauly, D. 1995. Anecdotes and the shifting baseline syndrome of fisheries. *Trends in Ecology and Evolution* 10: 430.

Pauly, D., V. Christensen, J. Dalsgaard, R. Froese, and F. Torres, Jr. 1998. Fishing down marine food webs. *Science* 279: 860–863.

Pinnegar, J. K., N. V. C. Polunin, P. Francour, F. Badalamenti, R. Chemello, M. -L. Harmelin-Vivien, B. Hereu, M. Milazzo, M. Zabala, G. D'Anna, and C. Pipitone. 2000. Trophic cascades in benthic marine ecosystems: lessons for fisheries and protected-area management. *Environmental Conservation* 27: 179–200.

Power, M. E. 1992. Top-down and bottom-up forces in food webs: Do plants have primacy? *Ecology* 73: 733–746.

Power, M. E., W. J. Matthews, and S. A. Stewart. 1985. Grazing minnows, piscivorous bass, and stream algae: dynamics of a strong interaction. *Ecology* 66: 1448–1456.

Roman, R. and S. R. Palumbi. 2003. Whales before whaling in the North Atlantic. *Science* 301: 508–510.

Sala, E., C. F. Boudouresque, and M. Harmelin-Vivien. 1998. Fishing, trophic cascades, and the structure of algal assemblages: evaluation of an old but untested paradigm. *Oikos* 82: 425–439.

Schindler, D. E., T. E. Essington, J. E. Kitchell, C. Boggs, and R. Hilborn. 2002. Sharks and tunas: fisheries impacts on predators with contrasting life histories. *Ecological Applications* 12: 735–748.

Schindler, D. W. 1974. Eutrophication and recovery in experimental lakes: implications for lake management. *Science* 184: 897–898.

Schmitz, O. J. 2003. Top predator control of plant biodiversity in an old field ecosystem. *Ecology Letters* 6: 156–163.

Shiomoto, A., K. Tadokoro, K. Nagasawa, and I. Ishida. 1997. Trophic relations in the subarctic North Pacific ecosystem: possible feeding effects from pink salmon. *Marine Ecology Progress Series* 150: 75–85.

Short, F. T., L. K. Muehlstein, and D. Porter. 1987. Eel grass wasting diseases: cause and recurrence of a marine epidemic. *Biological Bulletin* 173: 557–562.

Shurin, J. B., E. T. Borer, E. W. Seabloom, K. Anderson, C. A. Blanchette, B. Broitman, S. D. Cooper, and B. S. Halpern. 2002. A cross-ecosystem comparison of the strength of trophic cascades. *Ecology Letters* 5: 785–791.

Simenstad, C. A., J. A. Estes, and K. W. Kenyon. 1978. Aleuts, sea otters, and alternate stable-state communities. *Science* 200: 403–411.

Springer, A. M., J. A. Estes, G. B. van Vliet, T. M. Williams, D. F. Doak, E. M. Danner, K. A. Forney, and B. Pfister. 2003. Sequential megafaunal collapse in the North Pacific Ocean. *Proceedings of the National Academy of Science* 100: 12223–12228.

Stauffer, R. C. 1937. Changes in the invertebrate community of a lagoon after disappearance of the eel grass. *Ecology* 18: 427–431.

Stevens, J. D., R. Bonfil, N. K. Dulvy, and P. A. Walker. 2000. The effects of fishing on sharks, rays, and chimaeras (chondrichthyans), and the implications for marine ecosystems. *ICES Journal of Marine Science* 57: 476–494.

Sutherland, J. P. 1974. Multiple stable points in natural communities. *American Naturalist* 108: 859–873.

Van Bael, S. A., J. D. Brawn, and S. K. Robinson. 2003. Birds defend trees from herbivores in a neotropical forest. *Proceedings of the National Academy of Science* 100: 8304–8307.

Walters, C., V. Christensen, and D. Pauly. 1997. Structuring dynamic models of exploited ecosystems from trophic mass-balance assessments. *Reviews in Fish Biology and Fisheries* 7: 39–172.

Witman, J. D. and K. P. Sebens. 1992. Regional variation in fish predation intensity: a historical perspective in the Gulf of Maine. *Oecologia* 90: 305–315.

Worm, B. and R. A. Myers. 2003. Meta-analysis of cod-shrimp interactions reveals top-down control in oceanic food webs. *Ecology* 84: 162–173.

Zaret, T. M. and R. T. Paine. 1973. Species introduction in a tropical lake. *Science* 182: 449–455.

THREE

Lessons from Land

Present and Past Signs of Ecological Decay and the Overture to Earth's Sixth Mass Extinction

C. JOSH DONLAN, PAUL S. MARTIN, AND GARY W. ROEMER

We are currently experiencing the sixth major extinction event in the world's history (Thomas et al. 2004). This event is more pervasive than the previous five and is overwhelmingly human-driven. Nevertheless, distinguishing between the proximate and ultimate causes of extinction is often difficult (Caughley 1994). Alongside these extinctions, we have recently witnessed a number of complex species interactions that have restructured entire ecosystems and contributed to the decline of biodiversity. These interactions often involve apex predators, suggesting that species of high trophic status play important roles in ecosystem function (Estes 1995; Terborgh et al. 1999). This putative role of predators as key players maintaining biodiversity, combined with their widespread decline (e.g., Laliberte and Ripple 2004), presents a situation with high stakes for the conservation of biodiversity.

This volume addresses the question of how the removal of whales in various fisheries influenced the workings of modern oceans. Despite the recent and dramatic nature of these events, their consequences are both controversial and poorly documented. Here, we offer a comparative view of other systems in which food web reorganization has apparently followed anthropogenic disturbances to key vertebrates. Joining evidence from a series of observations in terrestrial ecosystems with coastal marine case studies

(Jackson, Chapter 4 of this volume; Springer et al., Chapter 19 of this volume), we argue that prehistoric, historic, and present-day reductions in some vertebrate populations may have been initially triggered by human action. We will further argue that these reductions subsequently distorted ecological dynamics, leading to ecosystem simplification or decay. We first turn to deep history in an effort to examine terrestrial vertebrate extinctions in the late Pleistocene and Holocene; we discuss potential mechanisms and present day conservation implications of losing our Pleistocene fauna. We then discuss historic and contemporary examples of strong species interactions and ecosystem decay, starting on land and moving to coastal seas. Some of these interactions are associated with the overexploitation of key species; all are related to human impacts. An emerging synthesis that terrestrial biodiversity is often strongly influenced by species interactions of high trophic status and how human action can perturb such interactions—from 50,000 years ago to the present—offers a bold new view for marine environs. Within the coastal realm, the evidence is reasonably strong for both the importance of top-down forcing (Paine 1966, 2002) and the impacts of human overharvesting (Jackson et al. 2001; Hjermann et al. 2004).

We suggest that, taken in aggregate, the observed ecological dynamics in a variety of prehistoric, historic, and present-day settings are of paramount importance in understanding the current biodiversity crisis and strengthening the case that high-trophic-level consumers play vital roles in structuring ecosystems (Soulé et al. 2003). What the land provides us better than any other class of ecosystem is a view of the past and of how now-extinct creatures shaped the life histories of extant species through what must have been the selective forces of strong species interactions. This is one potential window for better understanding the historical ecology of the sea.

Lessons from Land's Past

The Demise of Our Pleistocene Heritage

Looking to the past—*near time*, or the last 50,000 years—we come upon a cluster of remarkable extinctions that represent a prophecy for the mass extinction unfolding today. The Pleistocene extinctions, along with potential mechanisms, offer insights into contemporary terrestrial and marine ecosystems and their dynamics. Understanding the import of the Pleistocene extinctions and the ecological and evolutionary interactions that were lost with them (Janzen and Martin 1982) are of paramount importance for conservation (Flannery 1995; Donlan and Martin 2004; Foreman 2004).

In near time, biogeographers recognize over 100 extinctions of large continental vertebrates (>45 kg), including a host of large mammals and flightless birds, some giant lizards, terrestrial crocodiles, and giant tortoises. Most megafaunal extinctions occurred in the Americas and Australia, with smaller numbers of vertebrates going extinct in Madagascar and New Zealand. North America alone lost 31 genera of large terrestrial mammals. South America lost even more (Martin 2002). American losses included elephants, ground sloths, glyptodonts, equids, camelids, cervids, tayassuids, giant bears, saber-tooth cats, and two endemic mammalian orders restricted to South America, the Litopterna and Notoungulata. Calibrated radiocarbon dating places these losses at or slightly later than 13,000 YBP (Martin 2002). Over 30,000 years earlier, Australia lost a series of giant marsupials, along with giant lizards such as *Megalania* (450 kg or more) and terrestrial crocodiles. The cause of these near-time extinctions have been debated for decades (Martin and Wright 1967; Martin and Klein 1984; MacPhee 1999), and the search for an answer has unfailingly provoked controversy among anthropologists, archaeologists, ecologists, geographers, and vertebrate paleontologists, not to mention natural historians in general and the public at large.

Two scenarios vie for preeminence among those searching for the cause of the Pleistocene extinctions: (1) lethal changes in climate and (2) rampant predation, or *overkill*, owing to the initial invasion and spread of *Homo sapiens*. Mounting evidence now points toward a human cultural model (Alroy 2001; Fiedel and Haynes 2003; Lyons et al. 2004). Perhaps

the most powerful support for the overkill hypothesis is the time-transgressive nature of extinction events coincident with the arrival of humans, starting in Australia ca. 46,000 YBP, then North and South America ca. 13,000 YBP, then the islands of Oceania ca. 3000 YBP, Madagascar ca. 2400 YBP, and lastly New Zealand less than 500 years ago (Steadman 1995; Martin and Steadman 1999; Roberts et al. 2001; Worthy and Holdaway 2002). With such a pattern, a climate-driven scenario is problematic.

Even when the near time extinctions are viewed on a continent-to-continent perspective, new information on Quaternary climate elucidates additional problems with climate-driven scenarios. It is clear that climate played little, if any role, in the Australian extinctions ca. 46,000 YBP. Rather, the arrival and expansion of humans appear to be the likely culprit (Flannery 1999). Those who favor a climatic explanation for extinction in the Americas commonly turn to the Younger Dryas (YD) cold snap, recorded in thickness, dustiness, and other features of ice cores from Greenland. This cold snap was followed by an abrupt warming event at 11,570 YBP, more than 1,000 years after the disappearance of the megafauna. Discordant timing acquits a warming event from any causal primacy in the American Pleistocene extinctions. Those favoring a climate model in the Americas must then link the YD with megafaunal extinction. There is no such link. Changes during the ca. 1,000-year-long YD were no more abrupt or cold than the 24 climatic oscillations over the previous 100,000 years (Alley and Clark 1999; Alley 2000). The plant fossil record from thousands of fossil packrat middens in western North America have documented that range shifts rather than extinction were the norm (Betancourt et al. 1990), and only a single plant extinction is known from the late Pleistocene (Jackson and Weng 1999). Large mammals are more widely distributed and mobile than are plants and small mammals. Considering these advantages, combined with their general diets (Davis et al. 1984; Hanson 1987), American megafauna would likely not have been challenged by shifts in plant distributions—they would have moved, not gone extinct. Further, evidence from Beringia suggest a different relationship between Pleistocene herbivores and plant communities: The late Pleistocene shift from a grass dominated steppe to a vegetation mosaic dominated by mosses appears not to be driven by climate but rather by the loss of large mammalian grazers (Zimov et al. 1995). Perhaps top-down forcing via strong species interactions was more important than environmental forcing in some Pleistocene ecosystems.

Short of a natural catastrophe, it seems unlikely that a climatic crisis sufficient to force an extraordinary loss of Pleistocene mammals from such large regions could have escaped detection in the wealth of proxy climatic data now available within near time. With the exception of a few clinging to a vague climate model (Grayson and Meltzer 2003), more and more researchers are convinced of a cultural mechanism driving the Pleistocene extinctions (Martin and Steadman 1999; Miller et al. 1999; Alroy 2001; Worthy and Holdaway

2002; Fiedel and Haynes 2003; Kerr 2003; Lyons et al. 2004). Now, it appears, "The right question probably isn't whether people were involved, but how?" (O'Connell 2000).

Despite mounting evidence for the role of humans in the demise of Pleistocene megafauna, the details of the original American overkill model (Mosimann and Martin 1975) and a more recent model (Alroy 2001) are likely unrealistic, for a number of important reasons (Fiedel and Haynes 2003). First, human populations during the Clovis era were likely an order of magnitude less abundant than previously modeled; probably no more than 50,000 people resided in North America ca. 13,000 YBP (Haynes 2002). Most megafaunal extinctions took place within 400 years of human arrival, suggesting that a few people over a short time period had to be responsible for the extinctions (Fiedel and Haynes 2003). Other factors include the actual behavior of the Clovis people; evidence suggests that humans probably did not spread in a wavelike fashion across the continent, as once envisioned (Anderson 1990; Dincauze 1993). Given this new information, how could a small human population wipe out a continent full of large mammals?

Perhaps the American continent was not as full as once suspected. Although estimating Pleistocene densities of large mammals is difficult, these behemoths, such as mammoths and giant ground sloths, were not predator-free upon human arrival. They would have been vulnerable to a suite of predators, including giant short-faced bears *(Arctodus simus)*—the most powerful predator of Pleistocene North America (Kurtén and Anderson 1980). Bones of *Arctodus* were associated with mammoths at Huntington Canyon, Utah, and Mammoth Hot Springs, South Dakota. Bones of young mastodons in Friesenhan Cave, Texas, lay with those of dirk-tooth cats *(Megantereon hesperus)*, suggesting another predator-prey relationship rarely revealed by the fossil record (Kurtén and Anderson 1980). A broken Beringian lion tooth *(Panthera leo atrox)* stuck in the muzzle of an extinct Alaskan bison *(Bison priscus)* provides another example, reminiscent of African lions using a muzzle bite to choke African bovids to death (Guthrie 1990).

The famous tar pits of Rancho la Brea depict a Pleistocene ecosystem in which carnivores completely utilized prey carcasses and suffered an abundance of broken teeth in the process, suggesting intense interspecific competition among carnivores for what may have been limited prey (Van Valkenburgh and Hertel 1993). Tens of thousands of bones and teeth of saber tooth cats *(Smilodon fatalis)* have been excavated from La Brea, representing at least 1,200 individuals (Quammen 2003; Kurtén and Anderson 1980). Remains of dire wolves *(Canis dirus)* are also abundant. Potential prey would have included camels, bison, horses, ground sloths, and other large herbivores. From these associations, along with current observations of large-mammal predator-prey dynamics from the African Serengeti (Sinclair et al. 2003), we can be relatively certain a large array of predators preyed upon the Pleistocene herbivores when humans arrived over the Bering Land Bridge.

Perhaps the first Americans lent a helping hand to these Pleistocene predators, triggering a cascade of extinctions (Janzen 1983). Contemporary and historic examples in systems reveal how apex predators can trigger wholesale food web changes that include the precipitous decline of prey species (Estes et al. 1998; Roemer et al. 2002). In some cases, relatively few predators triggered the declines. Could Pleistocene hunters have played a similar role (Janzen 1983; Kay 2002)?

Two ecological scenarios might have been operating. First, nomadic hunters could have moved into a new area, depleted the stock of large herbivores, and subsequently moved to another area, leaving the diverse and abundant assemblage of native predators to drive the reduced herbivore populations to extinction, with overexploitation ultimately leading to the extinction of native predators as well (Janzen 1983). Alternatively, the loss of strong megaherbivore-plant interactions through targeted hunting by humans, could have resulted in the loss of nutrient-rich and spatially diverse vegetation, an effect that has been observed contemporaneously in Africa with elephants (Owen-Smith 1987, 1988). The loss of these "keystone herbivores" could have triggered a series of ecological events leading to wholesale extinction (Owen-Smith 1987). However, certain evidence does not support this second scenario. Seventeen North and South American genera went extinct between 11,400 and 10,800 RCBP (radiocarbon years before present), with proboscideans clustering toward the end of this period and other smaller herbivores (e.g., *Equus* and *Camelops*) disappearing earlier (Fiedel and Haynes 2003, and references therein). One would expect the opposite pattern with the keystone herbivore hypothesis (Owen-Smith 1987). Nonetheless, Pleistocene herbivores have often been viewed as being regulated simply and exclusively from the bottom up, a view inconsistent with key paleoecological evidence and contemporary examples (Van Valkenburgh and Hertel 1993; Terborgh et al. 1999; Kay 2002; Sinclair et al. 2003).

Long-term studies in the Serengeti support the premise that most mammalian herbivores are regulated by predation (Sinclair et al. 2003). Herbivore populations appear to be regulated by the diversity of both predators and prey, and by the body size of the herbivore relative to other herbivores and predators in the community. Large predators not only feed on large prey but also affect smaller prey species. Consequently, smaller prey are eaten by a diversity of predators of varying body size, from small to large (Sinclair et al. 2003). During the Pleistocene in North America, both predators and herbivores were larger and more diverse than they are in the Serengeti today (Van Valkenburgh and Hertel 1993; Martin 2002; Sinclair et al. 2003), and similar processes could have operated as long as the diversity of body sizes existed.

If North American Pleistocene herbivores were regulated from the top down by predators, the presence of a highly interactive novel predator, humans, could have triggered a cascade of extinctions (Figure 3.1). This may have been the case a few thousand years ago in Oceania, with the arrival of

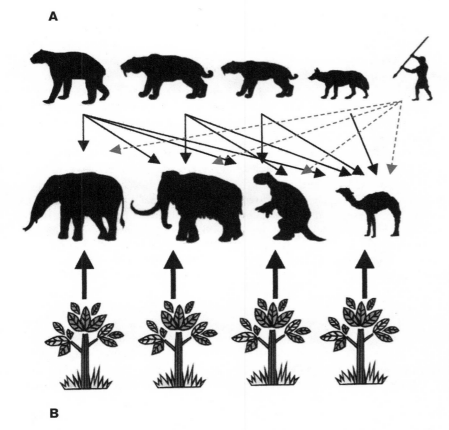

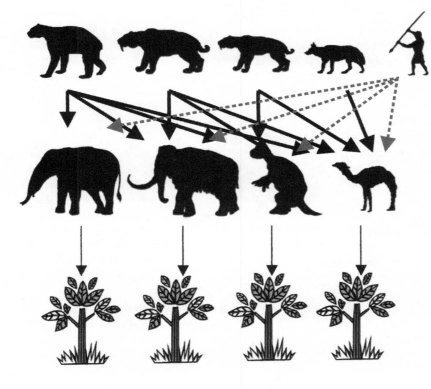

FIGURE 3.1. Hypothesized trophic relations between Pleistocene megafauna, humans, and primary production in North America. (A) Prior views suggested that the trophic web was bottom-up driven: Plants controlled herbivore populations, which in turn influenced the abundance of carnivores. Predators, including humans, had little influence over herbivore populations. (B) More plausible is a top-down view whereby large carnivores, from the formidable short-faced bear to the dire wolf, had an equally important effect on megaherbivore population dynamics. Once humans reached North America, they depleted megaherbivores, in turn reducing the prey available, causing declines in their predators—triggering wholesale ecosystem collapse.

humans to islands containing pygmy elephants *(Stegodon trigonocephalus florensis* and *S. sompoensis)* and Komodo dragons *(Varanus komodoensis,* Diamond 1987). Prior to the arrival of humans to the Wallacea islands (~4000 BC), pygmy stegodonts were present on a number of islands, along with the Komodo dragon (Auffenberg 1981). While the exact mechanism remains elusive, Diamond (1987) suggests that the Komodo dragon evolved to prey on pygmy elephants, and that the arrival of a novel hunter, humans, caused not only the extinction of the pygmy elephants but a decline of their specialized predator as well. The largest living lizard is now confined to just five islands, feeding mainly on livestock and other introduced prey (Auffenberg 1981).

Although uncertainty will always be present with these near-time extinctions, it is incontestable that the late Pleistocene extinctions were a unique event in the history of life (Alroy 1999; Barnosky et al. 2004). And these extinctions abruptly ended a multitude of species interactions—leaving many species anachronistic in their landscape.

The Ghosts of Evolution's Past: Anachronisms on the Landscape

The ecologies of certain large-seeded plants and animals are incomplete when not viewed through the lens of the Pleistocene (Janzen and Martin 1982; Janzen 1986), taking into account the loss of an entire suite of plant-herbivore and predator-prey interactions ca. 13,000 YBP. Many of the prominent, large perennial plants of the Chihuahuan desert *(Opuntia, Yucca, Acacia, Prosopis)* are in at least partial ecological and evolutionary disequilibria with the loss of their Pleistocene grazers, browsers, and seed dispersers (Janzen 1986). Tropical palms *(Scheelea, Bactris),* nitrogen-fixing legumes *(Acacia, Hymenaea, Prosopis),* and other Central and North American trees *(Crescentia, Asimina, Maclura)* were likely dispersed by a suite of Pleistocene large herbivores, including gomphotheres, ground sloths, and horses (Janzen and Martin 1982; Barlow 2000). Plant-herbivore studies in Africa, as well as observations of European horses and cattle in Central America, support this hypothesis (Janzen 1981; Yumoto et al. 1995; Barlow 2000, and references therein).

American pronghorn *(Antilocapra americana)* provide an animal example: four million years of directional selection to avoid swift predators such as the American cheetah *(Miracinonyx)* came abruptly to an end for the pronghorn in the late Pleistocene (Byers 1997). These "ghosts of predators past" are reminders of just how fast the American cheetah was. Pronghorn, with speeds of 100 km/hr, are second only to the African cheetah *(Acinonyx jubatus,* Lindstedt et al. 1991). With such evolutionary forces now absent, one must wonder whether the antelope is slowing.

These ecological and evolutionary losses and their ecological and conservation implications have only recently been appreciated, and largely for terrestrial systems (Janzen and Martin 1982; Martin 1999; Burney et al. 2002; Steadman and

Martin 2003; Donlan and Martin 2004). For example, the current distributions of some extreme anachronistic plants (Osage orange [*Maclura pomifera*] and *Crescentia alata)* with megaherbivore dispersal syndromes have been severely reduced compared to their distributions in the Pleistocene (Gentry 1983; Schambach 2000). Of seven *Maclura* species present in North America in the Pleistocene, a single species now survives. Could such taxa be on their way out because of the loss of important species interactions?

Finally, could the extinctions of the late Pleistocene and Holocene serve as an overture to the current mass extinction event underway? We now turn to some contemporary examples of ecosystem decay in terrestrial systems, in an effort to elucidate possible similarities between human-driven biodiversity loss in the Pleistocene and the removal or addition of species from high trophic levels in contemporary time.

Lessons from Land's Present

Several contemporary studies of terrestrial ecosystems have revealed linkages between human action, strong interspecific interactions, intersystem connectivity, and trophic reorganization that have led to ecosystem simplification or degradation. The removal or addition of large vertebrates appears to be the proximate driver in these dynamics. Examples include the ecological collapse in tropical forest fragments bereft of predators (Terborgh et al. 1997; Terborgh et al. 2001); substantially altered plant-herbivore-ecosystem dynamics following the loss of predators from the Greater Yellowstone Ecosystem (Berger 1999; Ripple and Larsen 2000; Berger et al. 2001; Ripple et al. 2001); loss of an apex carnivore from habitat fragments in an urbanized landscape that released a subsidized predator (domestic/feral cats, *Felis catus),* which then caused declines in avian diversity (Soulé et al. 1988; Crooks and Soulé 1999); and the reorganization of a food web on the California Channel Islands that caused the decline of the island fox *(Urocyon littoralis)* because of heightened predation by golden eagles *(Aquila chrysaetos),* a community shift ultimately triggered by the presence of an exotic species (Roemer et al. 2001; Roemer et al. 2002). These examples contribute to the mounting evidence supporting the importance of apex predators in maintaining biodiversity and point to the complex interaction web pathways through which these effects are manifested.

Ecological Meltdown at Lago Guri

The creation of one of the world's largest hydroelectric dams in the Caroni Valley of Venezuela provides insight into how the loss of apex carnivores can lead to ecosystem decay. In 1986 Lago Guri reservoir reached its highest levels, flooding 4,300 km^2 of tropical forest and forming a series of newly isolated "island" fragments (0.1 to 350 ha in size). Because many of these fragments were too small to maintain viable populations and too far from the mainland to be recolonized, they lost their apex predators, such as jaguars *(Panthera onca),* mountain lions *(Puma concolor),* and harpy eagles *(Harpia harpyja).* Since 1993, John Terborgh and colleagues have documented the resulting ecological decay of these now predator-free islands (Terborgh et al. 1997; Terborgh et al. 1999; Rao et al. 2001; Terborgh et al. 2001). On the smaller islands (1 to 10 ha), seed predators (rodents) and generalist foliovores (iguanas [*Iguana iguana*], howler monkeys [*Alouatta seniculus*], and leaf-cutter ants [*Atta* spp. and *Acromyrmex* sp.]) experienced ecological release that resulted in an increase in population densities by one to two orders of magnitude. The hyperabundant consumers in turn unleashed a trophic cascade, causing significant changes in the plant community. Seedling and sapling densities and the recruitment of certain canopy trees were severely reduced. Lago Guri herbivores, released from top-down regulation, appear to be transforming the once species-rich forest into a simpler, peculiar collection of unpalatable plants (Rao et al. 2001; Terborgh et al. 2001; also see Donlan et al. 2002 for another experimental example).

The dynamics at Lago Guri elucidate a second important lesson: The proximate mechanism causing biodiversity decline is often uniquely determined by idiosyncratic patterns of species occurrence (Terborgh et al. 1997). For instance, on one island, olive capuchin monkeys *(Cebus olivaceus)* became hyperabundant, decimating bird populations by raiding nests. Meanwhile, on similarly sized islands nearby that lacked this mesopredator, bird populations persisted. On other islands, ecological decay involved not olive capuchins but rather leaf-cutter ants or agoutis *(Dasyprocta aguti).* Thus, while the ultimate cause of biodiversity decline at Lago Guri appears to be the loss of top predators, the proximate mechanisms by which this decline is achieved are often complex and unpredictable. These observations suggest that predicting extinctions will prove difficult.

Predator-Prey Disequilibria in North America

The loss of predators in temperate ecosystems of North America has produced similar ecological effects. In the Greater Yellowstone Ecosystem, and throughout the western United States, poaching and predator control programs in the late 1800s and early 1900s resulted in a regionwide reduction of apex predators, including the grizzly bear *(Ursus arctos),* gray wolf *(Canis lupus),* and mountain lion (Laliberte and Ripple 2004). The eradication of predators set off a series of ecological ripples, in some cases with far-reaching consequences. In Yellowstone National Park (YNP), herbivory by elk *(Cervus elaphus)* essentially halted aspen *(Populus tremuloides)* recruitment starting in the 1920s, coinciding with the eradication of wolves from the region, once a significant source of elk mortality (Ripple and Larsen 2000; Ripple et al. 2001). Elk populations in the northern Greater Yellowstone Ecosystem have skyrocketed since the cessation of artificial control programs, from approximately 4,500 in the late 1960s to 20,000 by 1995 (Soulé et al. 2003). Heavy elk browsing on willows *(Salix* spp.) precipitated landscape change, including the near-disappearance of beaver wetlands. The beaver wetland ecosystem promotes willow establishment by raising the water table and enhancing productivity (Naiman et al. 1986). The loss of beaver *(Castor canadensis)* and its associated ecosystem has also been observed in Rocky Mountain National Park; here, too, these changes appear linked to the loss of large carnivores (Berger et al. 2001; Soulé et al. 2003). Beaver losses have further triggered biological and physical changes, including a 60% decline in willow stands, a lowering of the water table, increased erosion, and streambed channelization.

A similar dynamic has been documented in the southern Greater Yellowstone Ecosystem, where moose *(Alces alces)* have become hyperabundant following large predator reductions (Berger et al. 2001). In the region of Grand Teton National Park a comparison between public lands outside the park (where humans hunt moose) and the park proper (where hunting is prohibited and moose densities are five times higher) demonstrated that moose overbrowse riparian habitats in complex ways. Like elk herbivory, moose herbivory substantially altered the distribution and abundance of willow, in this case with cascading impacts on the diversity and abundance of nesting migrant songbirds. Bird species richness was reduced by 50% within the park proper in comparison to surrounding areas where moose densities were lower. Mitigating for the past management action of removing predators in this National Park will prove challenging.

Coyotes, Habitat Fragmentation, and Mesopredator Release

With a decline in wolves across North America, a smaller canid, the coyote *(Canis latrans),* prospered in areas previously occupied by wolves, expanding their range from the western United States across the continent (Laliberte and Ripple 2004). In some systems coyotes became the apex predator, creating effects that trickled through lower trophic levels, ultimately influencing biodiversity. For example, in the highly urbanized area of southern California, natural sage-scrub habitats have become increasingly fragmented, resulting in patches of varying size and ecological history (Soulé et al. 1988; Bolger et al. 1991). In fragments juxtaposed to urban areas, several native (grey fox, striped skunk, and raccoon) and exotic (opossum and domestic cats) predators are present. Some patches are large enough to support coyotes, which directly kill or exclude the smaller mesopredators (Crooks and Soulé 1999). Mesopredators become abundant in patches where coyotes are absent, and this increase reduces avian diversity. The abundance of certain scrub-breeding birds is lower in coyote-free fragments, often less than 10 individuals in the smaller fragments (Bolger et al. 1991). Such low population sizes result in higher local extinction rates, with perhaps as many as 75 extirpations over the past 100 years in this southern Californian ecosystem (Bolger et al. 1991). Crooks and Soulé (1999) conclude that presence or absence of coyotes, subsequent mesopredator activity, and

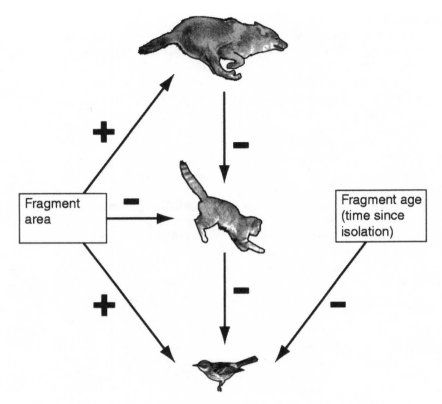

FIGURE 3.2. Species interactions between apex predators and fragmentation. Coyotes limit the distribution of mesopredators such as feral cats. In the absence of coyotes, mesopredators reduce the species and numbers of scrub-breeding birds in coastal California habitats. The ecological histories of the habitat fragments interact with the presence or absence of an apex predator to structure the ecological community. Effect of interactions (positive or negative) is indicated with plus or minus. Reproduced from Crooks and Soulé 1999.

fragmentation effects interact to structure these ecological communities (Figure 3.2).

Turning Predators into Prey: Trophic Reorganization of an Island Ecosystem

The recent arrival of a novel apex predator on the California Channel Islands further demonstrates the ability of apex predators to change ecosystems swiftly. In the mid-1990s, island fox *(Urocyon littoralis)* populations in Channel Islands National Park (CINP) declined rapidly (Roemer et al. 2002). By 1999, less than 200 foxes were known to be alive on the three northern Channel Islands, where just 6 years earlier an estimated 3,600 occurred (Roemer et al. 1994). Disease was initially suspected, but further investigation identified the presence of an exotic species, the feral pig *(Sus scrofa),* as the primary driver of the declines (Roemer et al. 2000; Roemer et al. 2001; Roemer et al. 2002). Pigs, by acting as an abundant prey, enabled mainland golden eagles to colonize the northern Channel Islands. Eagles, in turn, preyed upon the unwary fox as well, causing its decline. Pigs, with their high fecundity,

larger body size, and more nocturnal habit, could cope demographically with heightened levels of predation: Eagles fed mainly on small piglets, and when piglets reached a certain size (~10 kg), they became immune to eagle predation. In contrast, foxes were far more susceptible to eagle predation because of their low fecundity, their small adult body size (~2 kg), and the fact that they are often active during the day. This interaction, a form of apparent competition (Holt 1977), led to an asymmetrical effect on these two species. Eagle predation had little effect on the exotic pig but drove the endemic fox toward extinction (Figure 3.3).

The presence of pigs, and subsequently golden eagles, further triggered a reorganization of the island food web. Historically, island foxes were the largest terrestrial carnivore and were competitively dominant to the island spotted skunk *(Spilogale gracilis amphiala,* Crooks and Van Vuren 1995; Roemer et al. 2002). Prior to the arrival of eagles on Santa Cruz Island, foxes were captured 35 times more frequently than skunks (Roemer et al. 2002). As foxes declined, skunks were released from competition and increased dramatically (skunk capture success increased 17-fold; Figure 3.3).

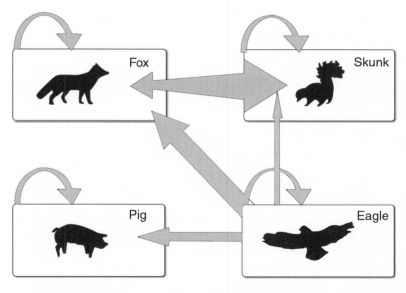

B

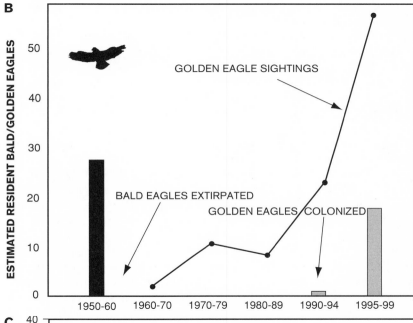

GOLDEN EAGLE SIGHTINGS

BALD EAGLES EXTIRPATED

GOLDEN EAGLES COLONIZED

ESTIMATED RESIDENT BALD/GOLDEN EAGLES

50

40

30

20

10

0

1950-60 1960-70 1970-79 1980-89 1990-94 1995-99

C

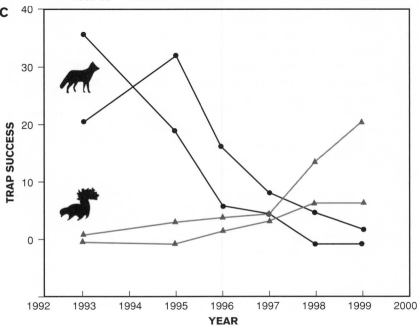

TRAP SUCCESS

40

30

20

10

0

1992 1993 1994 1995 1996 1997 1998 1999 2000

YEAR

FIGURE 3.3. Interspecific interactions and trophic reorganization on the California Channel Islands. (A) Schematic representation of inter-specific interactions among native, exotic, and colonizing vertebrates. Native island foxes were competitively superior to native island spotted skunks. Pigs, once introduced, enabled golden eagles to colonize the islands. Eagles then impacted the unwary fox, driving it toward extinction. (B) Bald eagles may have competed for nest sites on the Channel Islands. After bald eagles were extirpated by direct persecution and organochlorine con-taminants, golden eagles could freely colonize the islands. (C) After fox numbers declined owing to golden-eagle predation, skunks were released from fox competition, and their numbers dramatically increased. Trap success data are from two widely spaced trapping grids on Santa Cruz Island. Modified from Roemer et al. 2002.

Although pigs were linked to the decline in foxes on the northern Channel Islands, the ultimate cause of this complex interaction may have been a result of historic, human-induced perturbations to the islands, adjacent mainland, and to the surrounding marine environment (Roemer et al. 2001). European agricultural practices, together with overgrazing by introduced herbivores, reduced vegetative cover on the islands and probably increased the vulnerability of foxes to avian predators. Environmental contamination of the marine environment with DDT led to the extirpation of the bald eagle *(Haliaeetus leucocephalus)* from the Channel Islands by 1960 (Figure 3.3; Kiff 1980). Unlike golden eagles, which are terrestrial predators, bald eagles are primarily piscivorous and forage over marine environments; they are also territorial and aggressive toward other raptors, and thus they may have competed with golden eagles for nest sites (Roemer et al. 2001). Finally, increased urbanization along the southern California coast reduced golden eagle habitat, possibly displacing them to new hunting grounds on the islands (Harlow and Bloom 1989). Although speculative, this complex series of anthropogenic disturbances could have facilitated golden eagle colonization of the islands and their subsequent effects on island foxes.

Looking Seaward

Compared with terrestrial ecosystems, the oceans suffered few extinctions during the Pleistocene and Holocene (Martin 2002). The Caribbean monk seal *(Monacus tropicalis)* and Steller's sea cow *(Hydrodamalis gigas)* are the rare exceptions. Nonetheless, ecological extinctions *(sensu* Estes et al. 1989) in marine environments are widespread and appear driven mainly by overharvesting (Jackson et al. 2001). Many of these events were precipitated by the loss of highly interactive species, whose decline elicited wholesale ecosystem simplification and degradation (Estes et al. 1989; Soulé et al. 2003). Examples include sea otters *(Enhydra lutris)* in the northern Pacific (Estes et al. 1998; Springer et al. 2003), predatory and herbivorous fishes on coral reefs (Hughes 1994; Pandolfi et al. 2003), and green turtles *(Chelonia mydas)* and dugongs *(Dugong dugon)* in tropical sea grass communities (Jackson et al. 2001).

The kelp forests of the Northern Pacific offer the most compelling example of the pervasive effects caused by ecological extinction. This ecosystem is inhabited by a multitude of strongly interacting species, including kelps, strongylocentrotid sea urchins, and sea otters. Historically, kelp forests were abundant as sea otters preyed on sea urchins, which prevented the sea urchins from overgrazing the kelp (Estes and Palmisano 1974). Subsequent to human occupation of the western Aleutian archipelago (~2500 YBP), sea otter harvesting by aboriginal Aleuts triggered community shifts (Simenstad et al. 1978). Fur traders exacerbated this community transformation in the 1800s by hunting the sea otter to near extinction. Kelp forests declined or disappeared, grazed away by sea urchins released from sea otter predation. With

legal protection, sea otters and their kelp forest ecosystems recovered. This sequence of historic events provides strong empirical support for the importance of a highly interactive predator in the maintenance of an entire ecosystem at a regional scale (Estes and Duggins 1995).

The ecology of the Northern Pacific also provides a principal example of how human perturbation can trigger cascading events across ecosystems, in this case from the open ocean to coastal kelp forests. Populations of seals, sea lions, and, most recently, sea otters collapsed sequentially throughout the Northern Pacific during the latter decades of the twentieth century. Killer whales *(Orcinus orca)* appear to be responsible for the declines, but human overfishing of large whales during the mid-twentieth century may have ultimately triggered this chain of events (Springer et al. 2003). The advent of large-scale industrial whaling in the North Pacific following World War II drove the decline of the great whales in this region. Springer et al. (2003) hypothesized that as the large whales grew scarce, killer whales switched from feeding on large whales to feeding on smaller marine mammals, effectively "fishing-down" the marine food web. In sum, the onset of industrial whaling, and the subsequent ecological extinction of the great whales, appear to have triggered an unprecedented ecological chain reaction that ultimately caused the deforestation of coastal kelp forests. (Figure 3.4, Estes et al. 1998; Springer et al. 2003). Remarkably, few killer whales may have been responsible for the sequence of events; six individuals preying exclusively on sea otters could have driven the decline in otters throughout the Aleutian archipelago (Estes et al. 1998). These ecological extinctions and their wide-ranging effects on other species may be more common than previously appreciated in other coastal marine ecosystems (Jackson et al. 2001).

Many coastal ecosystems were degraded long before ecologists began to study them (Jackson 1997). These degradations appear to have been caused by the ecological extinction of highly interactive species through overfishing. Paleoecological, archeological, and historical data lend credence to this view (Jackson et al. 2001). For example, coral reef ecosystems began to decay centuries ago, long before the recent outbreaks of coral disease and bleaching (e.g., Harvell et al. 1999). Such proximate drivers were preceded by declines in large predatory and herbivorous fishes caused by human overharvesting. Hence, overfishing may ultimately best explain the long-term, global declines in coral reefs (Pandolfi et al. 2003). Similarly, recent mass mortality of seagrass beds is often associated with increases in sedimentation, turbidity, or disease (Hall et al. 1999; Abal et al. 2001). Yet the ecological extinction of large vertebrate herbivores through overfishing may have increased the ecosystem's vulnerability to these agents of change. For example, historic estimates for the endangered green turtles in the Caribbean alone were as high as 33 million individuals, and there may have been over 100,000 dugongs among the seagrass beds in Moreton Bay, Australia (Jackson et al. 2001). Jackson et al. (2001) have argued that the increased seagrass abundance that resulted

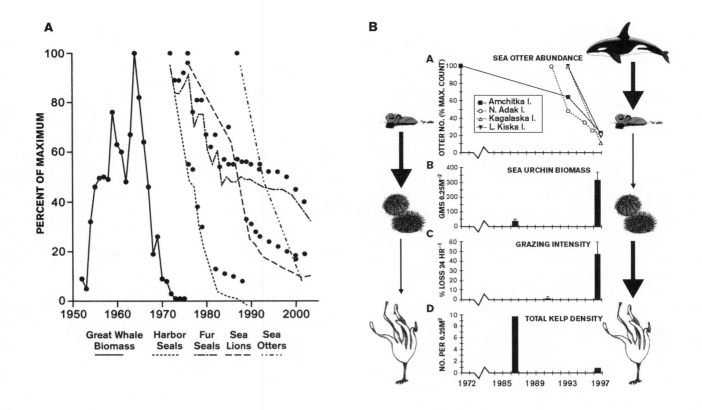

FIGURE 3.4. Ecological events in the northern Pacific Ocean following industrial whaling and the decline of the great whales. (A) The sequential collapse of marine mammals (shown as percent of maximum) likely caused by killer whales "fishing down" the food web from the open ocean to the coastal waters. (B) Subsequent cascading events in coastal waters of the Aleutian Islands, Alaska, where sea otters declined, sea urchins increased along with grazing intensity, and the density of kelp plummeted. Arrows represent strong and weak species interactions with and without killer whale predation. Reproduced from Springer et al. 2003 and Estes et al. 1998.

from the loss of these abundant herbivores greatly increased their vulnerability to the spread of disease.

These marine scenarios highlight that functional extinction of species precedes ultimate extinction and that the decline of highly interactive marine species appears to hasten the dissolution of ecological interactions, which then leads to ecological decay. In most coastal marine systems, overexploitation of highly interactive species appears to have set the stage for the cascade of events that followed. By this view, such perturbations may reduce resilience to further disturbance, allowing proximate forces such as increased turbidity, pollution, and disease to act as the *coup de grace* furthering ecosystem degradation.

Conclusion

The ecological decay witnessed in the terrestrial systems discussed are all linked to changes in the abundance of highly interactive predators—loss of natives in some cases and addition of exotics in others. In all cases, however, biodiversity loss resulted from complex species interactions. As seen on the islands of Lago Guri, such events and their ecological consequences are difficult to predict, both in timing and the precise pathways of change. On the Channel Islands, feral pigs were introduced over 100 years ago, and bald eagles were extirpated roughly 40 years before colonization by golden eagles. It is not apparent why this community shift took decades to occur. Looking seaward to the northern Pacific, we see hints of similar spatial and temporal unpredictability, where large-scale whaling appears to have triggered a series of population declines, via an apex predator, with widespread ecosystem consequences. Both the killer whales of the northern Pacific and the golden eagles of California demonstrate how human-induced perturbations can cascade across ecosystems at the regional scale.

Ecological impacts resulting from the loss or gain of highly interactive species can also be swift and often triggered by very few individuals. As few as six killer whales could have caused the precipitous decline in sea otters in the Aleutian archipelago (Estes et al. 1998; Williams et al. 2004), and as few as seven golden eagles may have driven the island fox declines (Roemer et al. 2001). Few, fast, and fickle may be common themes associated with human-induced effects on ecosystems as mediated by changes in the abundance, presence, or behavior of apex predators. Yet while these immediate short-term effects may be difficult to predict, the longer-term changes—ecosystem simplification and biodiversity loss—triggered by human perturbation are increasingly clear. Ecological simplification and biodiversity loss are not only of the present; rather, they have been going on for the last 50,000 years. As a consequence, many species characters, and the ecological landscapes those species inhabit, are anachronistic.

These terrestrial examples from New World ecosystems, raise an important question about the oceans: To what degree are the life histories and current patterns of distribution and abundance of marine organisms the result of the recent mass reductions of fishes, invertebrates, and whales and the concomitant loss of ecological interactions that have come at the hand of human exploitation of these key marine consumers? Hardly anyone, it seems, has thought about such possibilities. If our view of the land and the coastal oceans (Jackson, Chapter 4 of this volume) are any indication of the processes operating in the open sea, it is difficult to imagine that whales and whaling did not have profound influences on marine biodiversity.

Acknowledgments

We thank Jim Estes for the invitation to contribute to this volume, and Jim Estes and Dan Doak for improving an earlier version of this manuscript.

Literature Cited

Abal, E.-G., W.-C. Dennison, and P.-F. Greenfield. 2001. Managing the Brisbane River and Moreton Bay: an integrated research/management program to reduce impacts on an Australian estuary. *Water Science and Technology* 43: 57–70.

Alley, R. B. 2000. *The two-mile time machine: ice cores, abrupt climate change and our future*. New Brunswick, NJ: Princeton University Press.

Alley, R. B. and P. U. Clark. 1999. The deglaciation of the northern hemisphere: a global perspective. *Annual Reviews of Earth and Planetary Sciences* 27: 149–182.

Alroy, J. 1999. Putting North America's end-Pleistocene megafaunal extinction in context: large scale analyses of spatial patterns, extinction rates, and size distributions, in *Extinctions in near time: causes, contexts, and consequences*. R. D. E. MacPhee, ed. New York: Plenum, pp. 105–143.

———. 2001. A multispecies overkill simulation of the end-Pleistocene megafaunal mass extinction. *Science* 292: 1893–1896.

Anderson, D. G. 1990. The Paleoindian colonization of eastern North America: a view from the Southeastern United States, in *Early Paleoindian economies of eastern North America*. K. B. Tankersley and B. L. Isaac, eds. Research in Economic Anthropology, Supplement 5. Greenwich, CT: JAI Press, pp. 163–216.

Auffenberg, W. 1981. *The behavioral ecology of the Komodo monitor*. Gainesville: University of Florida Press.

Barlow, C. C. 2000. *The ghosts of evolution: nonsensical fruit, missing partners, and other ecological anachronisms*. New York: Basic Books.

Barnosky, A. D., C. J. Bell, S. D. Emslie, H. T. Goodwin, J. I. Mead, C. A. Repenning, E. Scott, and A. B. Shabel. 2004. Exceptional record of mid-Pleistocene vertebrates helps differentiate climatic from anthropogenic ecosystem perturbations. *Proceedings of the National Academy of Sciences* 101: 9297–9302.

Berger, J. 1999. Anthropogenic extinction of top carnivores and interspecific animal behaviour: Implications of the rapid decoupling of a web involving wolves, bears, moose and ravens. *Proceedings of the Royal Society Biological Sciences Series B* 266: 2260–2267.

Berger, J., P. B. Stacey, L. Bellis, and M. P. Johnson. 2001. A mammalian predatory-prey imbalance: Grizzly bear and wolf extinction affect avian neotropical migrants. *Ecological Applications* 11: 947–960.

Betancourt, J. L., T. R. Van Devender, and P. S. Martin. 1990. *Packrat middens: the last 40,000 years of biotic change*. Tucson: University of Arizona Press.

Bolger, D. T., A. C. Alberts, and M. E. Soule. 1991. Occurrence patterns of bird species in habitat fragments sampling extinction and nested species subsets. *American Naturalist* 137: 155–166.

Burney, D. A., D. W. Steadman, and P. S. Martin. 2002. Evolution's second chance. *Wild Earth* (Summer): 12–15.

Byers, J. A. 1997. *American pronghorn: social adaptations and the ghosts of predators past*. Chicago: University of Chicago Press.

Caughley, G. 1994. Directions in conservation biology. *Journal of Animal Ecology* 63: 215–244.

Crooks, K. R. and M. E. Soulé. 1999. Mesopredator release and avifaunal extinctions in a fragmented system. *Nature* 400: 563–566.

Crooks, K. R. and D. Van Vuren. 1995. Resource utilization by two insular endemic mammalian carnivores, the island fox and island spotted skunk. *Oecologia* (Berlin) 104: 301–307.

Davis, O. K., L. D. Agenbroad, P. S. Martin, and J. L. Mead. 1984. The Pleistocene dung blanket of Bechan Cave, Utah, in *Contributions in Quaternary vertebrate paleontology: a volume in memorial of John E. Guilday*. H. H. Genoways and M. R. Dawson, eds. Pittsburgh: Carnegie Museum of Natural History, pp. 267–282.

Diamond, J. M. 1987. Did Komodo dragons evolve to eat pygmy elephants? *Nature* 326: 832.

Dincauze, D. F. 1993. Pioneering in the Pleistocene: large Paleoindian sites in the Northeast in *Archaeology of Eastern North America: papers in honor of Stephen Williams*. J. B. Stoltman, ed. Archaeological Report 25. Jackson: Mississippi Department of Archives and History, pp. 43–60.

Donlan, C. J. and P. S. Martin. 2004. Role of ecological history in invasive species management and conservation. *Conservation Biology* 18: 267–269.

Donlan, C. J., B. R. Tershy, and D. A. Croll. 2002. Islands and introduced herbivores: conservation action as ecosystem experimentation. *Journal of Applied Ecology* 39: 235–246.

Estes, J. A. 1995. Top-level carnivores and ecosystem effects: questions and approaches, in *Linking species and ecosystems*. C. G. Jones and J. H. Lawton, eds. New York: Chapman and Hall, pp. 151–158.

Estes, J. A. and D. O. Duggins. 1995. Sea otters and kelp forests in Alaska: generality and variation in a community ecological paradigm. *Ecological Monographs* 65: 75–100.

Estes, J. A., D. O. Duggins, and G. B. Rathbun. 1989. The ecology of extinctions in kelp forest communities. *Conservation Biology* 3: 252–264.

Estes, J. A., and J. F. Palmisano. 1974. Sea otters: their role in structuring nearshore communities. *Science* 185: 1058–1060.

Estes, J. A., M. T. Tinker, T. M. Williams, and D. F. Doak. 1998. Killer whale predation on sea otters linking oceanic and nearshore ecosystems. *Science* 282: 473–476.

Fiedel, S. and G. Haynes. 2003. A premature burial: comments on Grayson and Meltzer's "Requiem for North American overkill." *Journal of Archaeological Science* 31: 109–120.

Flannery, T. F. 1995. *The future eaters: an ecological history of the Australasian lands and people.* New York: Braziller.

———. 1999. Debating extinction. *Science* 283: 182–183.

Foreman, D. 2004. *Rewilding North America: a vision for conservation in the 21st century.* Washington, DC: Island Press.

Gentry, A. H. 1983. Dispersal and distrubution of Bignoniaceae. *Sonderbände des Naturwissenschaftlichen Vereins in Hamburg* 7: 187–199.

Grayson, D. K. and D. J. Meltzer. 2003. A requiem for North American overkill. *Journal of Archaeological Science* 30: 585–593.

Guthrie, R. D. 1990. *Frozen fauna of the Mammoth Steppe: the story of Blue Babe.* Chicago: University of Chicago Press.

Hall, M.-O., M.-J. Durako, J.-W. Fourqurean, and J.-C. Zieman. 1999. Decadal changes in seagrass distribution and abundance in Florida Bay. *Estuaries* 22: 445–459.

Hanson, R. M. 1987. Shasta ground sloth food habits. *Paleobiology* 4: 302–319.

Harlow, D. L. and P. H. Bloom. 1989. Buteos and the golden eagle, in *Proceedings of the Western Raptor Management Symposium and Workshop.* B. G. Pendleton, editor. Washington, DC: National Wildlife Federation, pp. 102–110.

Harvell, C. D., K. Kim, J. M. Burkholder, R. R. Colwell, P. R. Epstein, D. J. Grimes, E. E. Hofmann, E. K. Lipp, A. D. M. E. Osterhaus, R. M. Overstreet, J. W. Porter, G. W. Smith, and G. R. Vasta. 1999. Emerging marine diseases: climate links and anthropogenic factors. *Science* 285: 1505–1510.

Haynes, G. 2002. *The early settlement of North America: the Clovis era.* Cambridge, UK: Cambridge University Press.

Hjermann, D. O., G. Ottersen, and N. C. Stenseth. 2004. Competition among fishermen and fish causes the collapse of Barents Sea capelin. *Proceedings of the National Academy of Sciences* 101: 11679–11684.

Holt, R. D. 1977. Predation, apparent competition, and the structure of prey communities. *Theoretical Population Biology* 12: 197–229.

Hughes, T. P. 1994. Catastrophes, phase shifts, and large-scale degradation of a Caribbean coral reef. *Science* 265: 1547–1551.

Jackson, J. B. C. 1997. Reefs since Columbus. *Coral Reefs* 16 (Supplement): S23–S32.

Jackson, J. B. C., M. X. Kirby, W. H. Berger, K. A. Bjorndal, L. W. Botsford, B. J. Bourque, R. H. Bradbury, R. Cooke, J. Erlandson, J. A. Estes, T. P. Hughes, S. Kidwell, C. B. Lange, H. S. Lenihan, J. M. Pondolfi, C. H. Peterson, R. S. Steneck, M. J. Tegner, and R. R. Warner. 2001. Historical overfishing and the recent collapse of coastal ecosystems. *Science* 293: 629–638.

Jackson, S. T. and C. Weng. 1999. Late Quaternary extinction of a tree species in eastern North America. *Proceedings of the National Academy of Sciences* 96: 13847–13852.

Janzen, D. H. 1981. Guanacaste tree seed-swallowing by Costa Rican horses. *Ecology* 62: 587–592.

———. 1983. The Pleistocene hunters had help. *The American Naturalist* 121: 598–599.

———. 1986. Chihuahuan desert Nopaleras: defaunated big mammal vegetation. *Annual Review of Ecology and Systematics* 17: 595–636.

Janzen, D. H. and P. S. Martin. 1982. Neotropical anachronisms: the fruits the gomphotheres ate. *Science* 215: 19–27.

Kay, C. E. 2002. False gods, ecological myths, and biological reality, in *Wilderness and political ecology: aboriginal influences and the original state of nature.* C. Kay and R. T. Simmons, eds. Salt Lake City: University of Utah Press, pp. 238–261.

Kerr, M. 2003. Megafauna died from the big kill, not big chill. *Science* 300: 885.

Kiff, L. F. 1980. Historical changes in resident populations of California Islands raptors, in *The California Islands: proceedings of a multidisciplinary symposium.* D. M. Power, ed. Santa Barbara, CA: Santa Barbara Museum of Natural History, pp. 651–673.

Kurtén, B. and E. Anderson. 1980. *Pleistocene mammals of North America.* New York: Columbia University Press.

Laliberte, A.-S., and W.-J. Ripple. 2004. Range contractions of North American carnivores and ungulates. *Bioscience* 54: 123–138.

Lindstedt, S. L., J. F. Hokanson, D. J. Wells, S. D. Swain, H. Hoppeler, and V. Navarro. 1991. Running energetics in the pronghorn antelope. *Nature* 353: 748–750.

Lyons, S. K., F. A. Smith, and J. H. Brown. 2004. Of mice, mastodons, and men: human-mediated extinctions on four continents. *Evolutionary Ecology Research* 6: 339–358.

MacPhee, R. D. E., ed. 1999. *Extinctions in near time: causes, contexts, and consequences.* New York: Kluwer Academic.

Martin, P. S. 1999. Deep history and a wilder West, in *Ecology of Sonoran Desert plants and plant communities.* R. H. Robichaux, ed. Tucson: University of Arizona Press, pp. 255–290.

———. 2002. Prehistoric extinctions: in the shadow of man, in *Wilderness and political ecology: aboriginal influences and the original state of nature.* C. Kay and R. T. Simmons, eds. Salt Lake City: University of Utah Press, pp. 1–27.

Martin, P. S. and R. G. Klein. 1984. *Quaternary extinctions: a prehistoric revolution.* Tucson: University of Arizona Press.

Martin, P. S. and D. W. Steadman. 1999. Prehistoric extinctions on islands and continents, in *Extinctions in near time: causes, contexts, and consequences.* R. D. E. MacPhee, ed. New York: Kluwer Academic, pp. 17–53.

Martin, P. S. and H. E. Wright, eds. 1967. *Pleistocene extinctions: the search for a cause.* New Haven, CT: Yale University Press.

Miller, G. H., J. W. Magee, B. J. Johnson, M. L. Fogel, N. A. Spooner, M. T. McCulloch, and L. K. Ayliffe. 1999. Pleistocene extinction of *Genyornis newtoni*: human impact on Australian megafauna. *Science* 283: 205–208.

Mosimann, J. E., and P. S. Martin. 1975. Simulating overkill by Paleoindians. *American Scientist* 63: 304–313.

Naiman, R. J., J. M. Milillo, and J. M. Hobbie. 1986. Ecosystem alteration of boreal forest streams by beaver *(Castor canadensis)*. *Ecology* 67: 1254–1369.

O'Connell, J. F. 2000. An emu hunt, in *Australian archeaeologist: collected papers in honor of Jim Allen.* A. Anderson and T. Murray, eds. Canberra: Coombs Academic Publishing, pp. 172–181.

Owen-Smith, N. 1987. Pleistocene extinctions: the pivotal role of megaherbivores. *Paleobiology* 13: 351–362.

Owen-Smith, R. N. 1988. *Megaherbivores: the influence of very large body size on ecology.* Cambridge, UK: Cambridge University Press.

Paine, R. T. 1966. Food web complexity and species diversity. *The American Naturalist* 100: 65–75.

———. 2002. Trophic control of production in a rocky intertidal community. *Science* 296: 736–739.

Pandolfi, J. M., R. H. Bradbury, E. Sala, T. P. Hughes, K. A. Bjorndal, R. G. Cooke, D. McArdle, L. McClenachan, M. J. H. Newman, G. Peredes, R. R. Warner, and J. B. C. Jackson. 2003. Global trajectories of the long-term decline of coral reef ecosystems. *Science* 301: 955–958.

Quammen, D. 2003. *Monster of God: the man-eating predator in the jungles of history and the mind.* New York: W. W. Norton.

Rao, M., J. Terborgh, and P. Nunez. 2001. Increased herbivory in forest isolates: implications for plant community structure and composition. *Conservation Biology* 15: 624–653.

Ripple, W. J. and E. J. Larsen. 2000. Historic aspen recruitment, elk, and wolves in northern Yellowstone National Park, USA. *Biological Conservation* 95: 361–370.

Ripple, W. J., E. J. Larsen, R. A. Renkin, and D. W. Smith. 2001. Trophic cascades among wolves, elk and aspen on Yellowstone National Park's northern range. *Biological Conservation* 102: 227–234.

Roberts, R. G., T. F. Flannery, L. K. Ayliffe, H. Yoshida, J. M. Olley, G. J. Prideaux, G. M. Laslett, A. Baynes, M. A. Smith, R. Jones, and B. L. Smith. 2001. New ages for the last Australian megafauna: continent-wide extinction about 46,000 years ago. *Science* 292: 1888–1892.

Roemer, G. W., T. J. Coonan, D. K. Garcelon, J. Bascompte, and L. Laughrin. 2001. Feral pigs facilitate hyperpredation by golden eagles and indirectly cause the decline of the island fox. *Animal Conservation* 4: 307–318.

Roemer, G. W., T. J. Coonan, D. K. Garcelon, C. H. Starbird, and J. W. McCall. 2000. Spatial and temporal variation in the seroprevalence of canine heartworm antigen in the island fox. *Journal of Wildlife Diseases* 36: 723–728.

Roemer, G. W., C. J. Donlan, and F. Courchamp. 2002. Golden eagles, feral pigs and insular carnivores: how exotic species turn native predators into prey. *Proceedings of the National Academy of Sciences* 99: 791–796.

Roemer, G. W., D. K. Garcelon, T. J. Coonan, and C. Schwemm. 1994. The use of capture-recapture methods for estimating, monitoring, and conserving island fox populations, in *The Fourth California Islands Symposium: update on the status of resources.* W. L. Halvorsen and G. J. Maender, eds. Santa Barbara, CA: Santa Barbara Museum of Natural History, pp. 387–400.

Schambach, F. F. 2000. Spiroan traders, the Sanders site, and the Plains interaction sphere. *Plains Anthropologist* 45: 7–33.

Simenstad, C. A., J. A. Estes, and K. W. Kenyon. 1978. Aleuts, sea otters, and alternate stable state communities. *Science* 200: 403–411.

Sinclair, A. R. E., S. Mduma, and J. S. Brashares. 2003. The patterns of predation in a diverse predator-prey system. *Nature* 425: 288–290.

Soulé, M. E., D. T. Bolger, A. C. Alberts, R. M. Sauvajot, J. Wright, M. Sorice, and S. Hill. 1988. Reconstructed dynamics of rapid extinctions of chaparral-requiring birds in urban habitat islands. *Conservation Biology* 2: 75–92.

Soulé, M. E., J. A. Estes, J. Berger, and C. Martinez del Rio. 2003. Ecological effectiveness: conservation goals for interactive species. *Conservation Biology* 17: 1238–1250.

Springer, A. M., J. A. Estes, G. B. van Vliet, T. M. Williams, D. F. Doak, E. M. Danner, K. A. Forney, and B. Pfister. 2003. Sequential megafaunal collapse in the North Pacific Ocean: An ongoing legacy of industrial whaling? *Proceedings of the National Academy of Sciences:* 12223–12228.

Steadman, D. W. 1995. Prehistoric extinctions of Pacific island birds: biodiversity meets zooarchaeology. *Science* 267: 1123–1131.

Steadman, D. W. and P. S. Martin. 2003. The late Quaternary extinction and future resurrection of birds on Pacific islands. *Earth-Science Reviews* 61: 133–147.

Terborgh, J., J. A. Estes, P. Paquet, K. Ralls, D. Boyd-Heger, B. J. Miller, and R. F. Noss. 1999. The role of top carnivores in regulating terrestrial ecosystems, in *Continental conservation: scientific foundations of regional reserve networks.* M. E. Soulé and J. Terborgh, eds. Washington, DC: Island Press, pp. 39–64.

Terborgh, J., L. Lopez, V. P. Nunez, M. Rao, G. Shahabuddin, G. Orihuela, M. Riveros, R. Ascanio, G. H. Adler, T. D. Lambert, and L. Balbas. 2001. Ecological meltdown in predator-free forest fragments. *Science* 294: 1923–1926.

Terborgh, J., L. Lopez, J. Tello, D. Yu, and A. R. Bruni. 1997. Transitory states in relaxing bridge islands, in *Tropical forest remnants: ecology, management, and conservation of fragmented communities.* W. F. Laurance and R. O. Bierregaard, eds. Chicago: University of Chicago Press, pp. 256–274.

Thomas, J. A., M. G. Telfer, D. B. Roy, C. D. Preston, J. J. D. Greenwood, J. Asher, R. Fox, R. T. Clarke, and J. H. Lawton. 2004. Comparative losses of British butterflies, birds, and plants and the global extinction crisis. *Science* 303: 1879–1881.

Van Valkenburgh, B. and F. Hertel. 1993. Tough times at La Brea: tooth breakage in large carnivores of the late Pleistocene. *Science* 261: 456–459.

Williams, T. M., J. A. Estes, D. F. Doak, and A. M. Springer. 2004. Killer appetites: assessing the role of predators in ecological communities. *Ecology* 85: 3373–3384.

Worthy, T. H. and R. N. Holdaway 2002. *The lost world of the Moa: prehistoric life of New Zealand.* Bloomington: Indiana University Press.

Yumoto, T., T. Maruhashi, J. Yamagiwa, and N. Mwanza. 1995. Seed-dispersal by elephants in a tropical rain forest in Kahuzi-Biega National Park, Zaire. *Biotropica* 27: 526–530.

Zimov, S. A., V. I. Chuprynin, A. P. Oreshko, F. S. Chapin III, J. F. Reynolds, and M. C. Chapin. 1995. Steppe-tundra transition: a herbivore-driven biome shift at the end of the Pleistocene. *American Naturalist* 146: 765–794.

FOUR

When Ecological Pyramids Were Upside Down

JEREMY B.C. JACKSON

The systematic destruction of the great whales was a stupendous act of modern ecological folly that rivals the extirpation of the once vast herds of American bison in efficiency, speed, and last-minute second thoughts that marginally spared the species (Isenberg 2000; Clapham and Baker 2002; Kareiva et al., Chapter 30 of this volume; Pfister and DeMaster, Chapter 10 of this volume). It is therefore remarkable that so little serious attention has been paid until quite recently to the ecological consequences of the relatively sudden removal of whales in terms of their past roles both as keystone consumers in food chains and as once-superabundant prey and carrion (Katona and Whitehead 1988; Smith and Baco 2003; Springer et al. 2003). The bison were replaced by cattle, which rivaled them in size and abundance and competed with them for graze even while the bison were in precipitous decline. Without either the bison or the cattle, the western short-grass prairies would have been invaded by a host of different plant and animal species, and the ecology of the plains would have even more drastically changed. Whales, however, are the largest animals on the planet, and there was nothing even vaguely comparable to replace them when they were gone. Whales were also enormously abundant. How abundant is a question more controversial than it should be (Whitehead 2002; Roman and Palumbi 2003; Springer et al. 2003), but the total for all species combined was certainly many millions,

so the ecological consequences of the removal of so many behemoths must have been profound.

In the absence of sufficient knowledge for whales, we can nevertheless ask what have been the ecological consequences of the removal of other once-abundant very large animals, on the land or in the ocean, to help provide a framework for thinking about the whales and the questions they raise for marine conservation and management now and in the future. The most striking example concerns the once great mammoths, giant sloths, saber-toothed cats, and other large animals that roamed the earth during the Ice Ages and whose extinction, notwithstanding the stubbornness of a few die-hard anthropologists and paleontologists, had more than a little to do with hunting by people (Flannery 1994; Flannery and Roberts 1999; Martin and Steadman 1999; Barnosky et al. 2004; Vermeij 2004; Donlan et al., Chapter 3 of this volume). Megafauna, which are arbitrarily defined as animals greater than 44 kg (100 pounds), suffered disproportionately in these extinctions, especially in such regions as the Americas, Australia, and Madagascar, where humans arrived comparatively recently. By roughly ten thousand years ago, only a few thousand years after people first arrived, North America had lost 33 out of 46 genera of megafaunal mammals, and South America had lost 50 out of approximately 60 such genera (Barnosky et al. 2004). Australia lost 14 of 16 genera of megafaunal mammals about forty-five thousand years ago, which coincides roughly with the time that people first arrived in Australia (Diamond 2001; C. Roberts et al. 2001; Barnosky et al. 2004). In contrast, Eurasia lost only nine out of 25 megafaunal mammal genera, and Africa lost only eight out of 44 (Barnosky et al. 2004). Extinction rates for small mammals were vastly lower everywhere (Barnosky et al. 2004).

Increasingly detailed analyses of terminal distributions of megafauna in relation to climate and vegetation (Stuart et al. 2004) confuse proximate and ultimate factors and ignore the realities of (1) the dramatic global diaspora of modern *Homo sapiens* (Barnosky et al. 2004); (2) the massive loss of predators and accompanying disruption of food webs (Donlan et al., Chapter 3, this volume); and (3) the earlier, often repeated, survival of the same species during the many previous episodes of similarly if not exactly identically severe and rapid climate change during the past million years (Barnosky et al. 2004). The major difference, of course, is that all the previous climatic crises lacked modern *Homo sapiens*. Indeed, ever since animals invaded the land from the sea some four hundred million years ago, animals larger than 44 kilograms have dominated the ecology of terrestrial ecosystems (Owen-Smith 1988; Benton 1990; Janis and Damuth 1990). The only exceptions are the comparatively brief interludes of one to a few million years following the greatest mass extinctions, especially the devastating events at the end of the Permian and Cretaceous, which wiped the large-animal slate nearly clean (Erwin 2001). The severe cold snaps of the latest Pleistocene were a trifle by comparison.

The ecological consequences of the loss of all these megafauna are not entirely understood, but there is abundant evidence of dramatic ecological changes in the Americas and Australia (Janzen and Martin 1982; Flannery 1994; Flannery and Roberts 1999; Donlan et al., Chapter 3 of this volume). Food webs were severely disrupted by the loss of almost all of the top predators and megafaunal herbivores, and patterns of vegetation changed greatly in response. The vast herds of bison in North America were partly an artifact of the elimination of all other large herbivores that had competed with bison for graze. Seed dispersal and distribution patterns of Neotropical plants with large, armored seeds, whose germination depended on consumption by large frugivores, were greatly altered by the extinction of 15 genera of large herbivores; a pattern that was apparently reversed at the eleventh hour by the arrival of goats, cattle, and pigs four centuries ago (Janzen and Martin 1982). And perhaps most spectacularly, the elimination of large herbivores in Australia resulted in the accumulation of vast amounts of uneaten vegetation, which was vulnerable to wildfires, which in turn transformed much of the vegetation to arid scrub, controlled more by fires than by herbivores (Flannery 1994).

These huge changes in the land are the simple (and many people would say the highly desirable) outcome of the deliberate, unrelenting efforts of humanity to replace large, dangerous animals with cattle, sheep, pigs, goats, and dogs and to replace insufficiently productive forests with range lands and agriculture (Diamond 1997). Exploitation acts to reverse ecological succession at the price of species diversity and structural complexity, but with the overriding benefit (from the point of view of *H. sapiens*) of greatly increased productivity (Margalef 1968). The consequence is more than six billion of us and counting. Such perspective, however, unfortunately rarely extends to our modern understanding of the oceans, where ignorance of large animal ecology is particularly acute, and because everything beneath the surface is out of sight and out of mind. Indeed, there is very little collective public wisdom and common sense about the oceans, to the extent that otherwise intelligent scientists refused to believe that the oceans were exhaustible until well into the twentieth century (Cushing 1988). More remarkably, hunting and gathering are still the dominant mode of obtaining food from the sea—increasingly high-tech, nevertheless Stone Age in mentality (Pauly et al. 2002). Even today, people who should know better deny that our fisheries are unsustainable and on the brink of near-total collapse (Schiermeier 2002). And at gatherings such as the symposium that gave rise to this volume, academic scientists find themselves arguing about what abundance of great whales may or may not have been possible in the past based on ecological principles and data derived from heavily exploited ecosystems that are only a pale shadow of what they were before.

Historical ecology, coupled with detailed study of the few quasi-pristine ecosystems remaining in the sea, provides a much needed antidote to the canonical wisdom of modern marine ecology, which is based almost exclusively on the study of severely degraded ecosystems. History provides us with many a story, and, much as in the phenomenon of small-world networks (Buchanan 2002), the stories converge on a remarkably common outcome (Jackson 2001; Jackson et al. 2001; Pandolfi et al. 2003, 2005; Lotze et al. 2006). The quasi-pristine ecosystems are time machines that enable us to examine the underlying structure and mechanisms behind the stories (Polovina 1984; Jackson and Sala 2001; Friedlander and DeMartini 2002) and thereby contribute much needed credibility to ever-skeptical ecologists who deny what they cannot see for themselves. This brief essay addresses some of the more obvious historical insights for the past ecological roles of large marine animals and their implications for how ocean ecosystems used to be.

Time Machines on the Land

Almost every summer for the past twelve years I have visited the cavern of Peche Merle in the heart of the vast limestone plateaux that were the Ice Age hunting grounds of people in central France fifteen thousand years ago (Lartigaut 1993). Some fifty steps down the narrow cement staircase, dim electric lamps briefly illuminate sections of the walls and ceilings of the cavern as visitors make their way along the guided tour. The lights reveal extraordinary murals of mammoths, oryx, bison, horses, and other animals painted by Ice Age

Michelangelos, who somehow made their way down to the ink-black depths of the cavern with their precious pigments and charcoal to render hidden masterpieces for reasons we shall never fully know. Each summer, as my understanding of the regional dialect has slowly improved, I catch new bits of information and see animals in places where I had not seen them before. The resulting mental collector's curve tells a consistent story. These people lived in a world dominated by large animals that were the focus of their practical and spiritual lives. Hundreds of similar caverns scattered across southern Europe, including the most famous but now-closed artistic pinnacles of Lascaux and Altamira, tell the same story. Most of the approximately fifty species of vertebrates depicted by the Ice Age artists (Roussot 1997) weighed 44 kilograms or more, and a dozen or so weighed more than 500 kilograms (Bernosky et al. 2004). Judging from the debris at kill sites and kitchen middens, many of these big animals were important prey (Bernosky et al. 2004).

Back up the steps at Peche Merle, the largest animals in Western Europe today, except in zoos and rare alpine refuges, are cows. European ecologists almost exclusively study "wild" animals such as voles, mice, lemmings, songbirds, grouse, and beetles in remnant landscapes of hedgerows, heaths, and downs (Elton 1927, 1966). American and Australian ecologists have had a little more to work with, studying remnant or artificially exploded populations of mountain sheep, moose, deer, wolves, big cats, bison, kangaroos, and bear in much larger remnant forests, prairies, and deserts (Deevey 1947; Hutchinson 1978; McLaren and Peterson 1994; Berger et al. 2001a; Ripple and Beschta 2004). Nevertheless, perusal of modern ecology textbooks, or any issue of the leading ecological journals since their beginnings nearly a century ago, tells a similar story of an inordinate fondness for songbirds, rodents, and insects, all too commonly studied in extremely tiny plots over only a few years (Kareiva and Anderson 1988).

The only reasonably intact terrestrial ecosystems with abundant large animals that have been intensively studied by ecologists are the East African plains and woodlands—wild animal parks such as the Serengeti, which constitute virtual time machines that, until recently, took us back almost to Peche Merle time with their extraordinary abundance and biomass of elephants, wildebeest, buffalos, rhinoceroses, hippopotamuses, giraffes, zebras, hyenas, lions, and crocodiles (Sinclair and Arcese 1995; Kruuk 1972; Schaller 1972). Why so few megafauna in Africa did not go extinct earlier as on other tropical continents is unknown, but it presumably has much to do with their extended behavioral coevolution with modern *Homo sapiens* over the last 150,000 years (Berger et al. 2001b).

Of course, most of the animal species in the Serengeti are smaller than these megafauna, and to my knowledge there are no good numbers on the weight frequency distribution of biomass per hectare of all animals from elephants and wildebeest to termites and ants (Sinclair et al. 2003). Suffice it to say that the proportional biomass of animals heavier than 44 kilograms is vastly greater there than anywhere else, especially compared with the canonical European countryside studied by Charles Elton (1927, 1966), who was the first to think rigorously about the ecology of animal size. Moreover, the ecological roles of the largest species in the Serengeti, such as elephants and wildebeest, are paramount in controlling the distribution, abundance, three-dimensional structure, species composition, and productivity of the vegetation that, in turn, affects all other animal species, big and small alike (Owen-Smith 1988; Sinclair and Arcese 1995; McNaughton and Banyikwa 1995). Even today, with the rapid encroachment of more and more people, poachers, and cattle, wildebeest modulate the ecological rhythm of the Serengeti in ways I doubt most ecologists could have imagined without direct observations and experiments.

Time Machines in the Sea

The basic principles of marine ecology arose from the study of microscopic phytoplankton and zooplankton of the open sea and of small coastal animals and plants in environments such as the rocky intertidal, mud flats, estuaries, and coral reefs. Most of these ecosystems had long since lost their once great abundance of whales, walruses, sea cows, seals, dolphins, sea turtles, sharks, rays, and large fish. These large animals persist in the public consciousness in ancient Persian stone friezes in the Louvre, books such as *Moby-Dick*, films such as *Free Willy* and *Jaws*, and myriad stuffed animals for children. But despite occasional mention in elementary textbooks (Nybakken and Bertness 2005) and in the context of conservation, large animals are nearly absent from the pages of advanced textbooks and scientific journals of marine ecology and biological oceanography, which are instead dedicated to microbes, dinoflagellates, diatoms, zooplankton, and sardines in the water column and barnacles, mussels, sea urchins, starfish, and damselfish on the sea floor (Parsons et al. 1984; Sale 1991; Valiela 1995; Bertness et al. 2001).

It is revealing to explore how the study of such degraded ecosystems has affected ecological thought about the trophic structure and ecological pyramids of biomass and production in the sea. Even before the revolution of ecological energetics (Lindeman 1942), it was well understood that the number of individuals always decreases and individual size generally increases as one moves up the food chain from producers to lower- and then higher-order consumers (Elton 1927; Odum 1953). It was also well known that pyramids of biomass were most commonly right side up, although they could be at least partially inverted if the primary producers were small and reproduced much more rapidly than their consumers, as is commonly the case for some aquatic phytoplankton and zooplankton (Odum 1959).

However, almost all of the data for biomass pyramids in the oceans were, and still are, for primary producers and consumers, with little or no information about higher-level consumers, which are typically lumped together as "fish" or "nekton" that cannot be studied with plankton nets. The importance of such sampling problems cannot be overemphasized, because they reinforce the schism between biological oceanographers, interested in mechanisms of primary production and consumption toward the bottom of the food chain and the recently discovered microbial loop (Parsons et al. 1984; Valiela 1995; Azam 1998), and ecologists and fisheries biologists, interested primarily in population dynamics, species interactions, and community structure of consumers closer to the top of the food chain, which is the bulk of the curriculum in marine biology (Levinton 2001; Nybakken and Bertness 2005) and fisheries (Jennings et al. 2001).

As a result, there are almost no data for the biomass of all the different trophic levels of marine food webs in the same place. Even Hardy's (1924) classic food web of the North Sea is topped by "adult herring," and the few studies that attempt to put it all together (see Barnes and Hughes 1999) were compiled in the 1980s and 1990s, when most of the big animals were gone. For example, the excellent detailed study of Chesapeake Bay in summer (Baird et al. 1995) shows an ecosystem dominated by zooplankton, bacteria, scavengers and deposit feeders, and phytoplankton—in that order, with fish in trace amounts. Yet this is the ecosystem that, three centuries ago, was choked in summer by menhaden, shad, enormous sturgeon, bluefish, striped bass, salmon, sharks, and whales (Wharton 1957). Thus it is hardly surprising that empirically based ecological pyramids for marine ecosystems are almost always right side up, with most of the animal biomass represented by small invertebrates, and apex predators represented by sardines, anchovies, snails, or crabs. Compilations are pitifully few, but the ratios of free-living vertebrate to invertebrate biomass are now sometimes precariously close to zero (Alinao et al. 1993; Baird et al. 1995; Jackson and Sala 2001).

Experimental investigations of these remnant ecosystems have clearly demonstrated the near-universal importance of biological interactions among species for both top-down and bottom-up determinants of distribution, abundance, and community composition (Worm et al. 2002; Paine, Chapter 2 of this volume), the absence of large animals notwithstanding. Such studies have also helped fundamentally to elucidate a more holistic landscape vision of ecosystems (Paine and Levin 1981) that incorporates the characteristic regime of perturbations and productivity of any particular environment into a dynamic concept of the "baseline" community (Jackson 1991; Dayton et al. 1998), rather than the narrower, small-scale perspective based on the study of a few isolated quadrats (Connell 1978). But in the oceans there are, so far, no well-protected, well-studied, quasi-pristine wild animal parks like the Serengeti to elucidate the incompleteness of our ecological vision in the absence of large animals.

The resulting ignorance of marine large animal ecology distorts basic ecological understanding and lowers our sights for conservation and management (Dayton et al. 1998; Jackson and Sala 2001). The recent burst of historical ecological studies notwithstanding, fisheries managers are reluctant to accept the implications for their woefully inadequate underestimates

of pristine stocks (Pauly 1995; Pauly et al. 2002); assessments of the ecological health of estuaries tend to ignore the lost top-down potential of filter feeders such as oysters and menhaden to help control eutrophication in comparison to simpler measurements of nutrient loading (Boesch 2001); and even advocates of marine reserves and other conservation strategies fail to understand the full potential of the policies they propose (C. Roberts et al. 2001). And now our ignorance is enshrined in textbook dogma as fundamental truth: "The numbers of individuals and total biomass decrease at successive trophic levels *because the amount of available energy decreases*" (Thurman and Trujillo 2004, p. 421, italics mine). So much for the insights of the great pioneers in ecosystem ecology.

In the face of such conviction I admit to having wondered whether my own historical reconstructions of coastal ecosystems (Jackson et al. 2001) could possibly be correct. But all doubts vanished during my first dive on the fore-reef at Palmyra Atoll three years ago. Large sharks and fishes of all kinds, both carnivores and herbivores, were spectacularly abundant, and sea urchins and other free living invertebrates large enough to see were spectacularly rare. The ratio of free-living vertebrate to invertebrate biomass seemed closer to infinity than to zero, and the rocky reef surface was almost entirely carpeted with living stony corals and cementlike coralline algae, except for countless deep gashes, the feeding scars of parrotfish and wrasses, everywhere apparent in the cracks and crevices of the reef as well as the open surface above. Fleshy algae, sponges, and soft corals were rarer than on any reef I had seen before. Most creatures less hardy than cement were in the stomachs of the fish, and much of the cement was as well.

My impressions at Palmyra might be dismissed as the delusions of a wishful historical ecologist, but for the fundamental paper by Friedlander and DeMartini (2002), which describes the trophic structure, species composition, average fish size, and biomass of the remote and lightly fished coral reef communities in the northwestern Hawaiian Islands (NWHI) compared with the reefs of the heavily fished main Hawaiian Islands (MHI). Grand mean fish standing stock was 260 percent greater in the NWHI than the MHI. Most spectacularly, large apex predators costituted 54 percent of the total fish biomass in the NWHI in contrast to only 3 percent in the MHI, whereas herbivorous fishes constituted only 18 percent of the total fish biomass in the NWHI in contrast to 55 percent in the MHI. The paper presents no data for free-living invertebrates such as lobsters, crabs, and sea urchins to compare with the fishes, but, at least for the fish community, the NWHI trophic pyramid is truly upside down.

Additional evidence from French Frigate Shoals in the NWHI (Polovina 1984) provides a vertebrate-to-invertebrate biomass ratio of 10 percent. This can be contrasted with a ratio of 2 percent at heavily fished Bolinao in the Philippines (Alinao et al. 1993, which, in the absence of comparable data from the MHI, represents the only data available). Moreover, many of the megafauna at French Frigate Shoals, including sea turtles, monk seals, and sharks, have been heavily fished in the past, either in the NWHI or as highly migratory species from surrounding waters, and the sea turtles and monk seals are red-listed as endangered species. Thus the pristine ratio of vertebrate to invertebrate biomass was almost certainly 20 percent or more, and I suspect the same is true today for the fore reef at Palmyra. This suggests a total animal trophic pyramid of biomass with a big bulge at the top, regardless of the biomass of small cryptic and planktonic invertebrates at lower trophic levels. If this were indeed the case, then the distribution of biomass among different trophic levels of marine food webs was strikingly similar to the predictions of Hairston et al. (1960) in their classic paper. Further studies of the entire animal food web throughout the archipelago, as well as the more detailed fishing history of the individual islands, are clearly needed to resolve the topology of the web. But regardless of these details to come, the trophic structure of the NWHI is not close to textbook right side up.

The implications of these differences between the NWHI and MHI are even more apparent when we consider the food web structure and per capita interaction strengths (*sensu* Paine 1980, 1992) among the different species for about one thousand square kilometers of Caribbean coral reef in the U.S. Virgin Islands (Bascompte et al. 2005). This is the largest food web ever examined quantitatively, with 249 species (or combined trophic species) and 3,313 interactions. It comprises all pelagic and benthic communities in the region down to 100 meters depth, including detritus, four groups of primary producers, 35 invertebrate taxa, 208 fish and shark species, sea turtles, and sea birds (Randall 1967; Opitz 1996). Of greatest importance here, the dietary data used to calculate interaction strengths were obtained from a great variety of habitats in the 1950s, when fish were still moderately abundant before the modern collapse of Caribbean reef communities in the 1970s and 1980s (Hughes 1994; Knowlton 2001; Gardener et al. 2003; Pandolfi et al. 2003). What we really want, of course, is comparable food web data for Palmyra Atoll or from the northwest Hawaiian Islands, but while we await those data, Randall's and Opitz's compilations are sufficient to demonstrate just how wrong we can be about the natural state of affairs without good data on large animals.

The Caribbean web is dominated by a matrix of weak interactions interspersed by few strong interactions, and the frequency distribution of per capita interaction strengths approximates a log normal distribution ranging over an astounding seven orders of magnitude (Bascompte et al. 2005). All of this matches previous studies (Paine 1992; Fagan and Hurd 1994; Raffaelli and Hall 1995; Wootton 1997; Goldwasser and Roughgarden 1997), but Bascompte and his colleagues broke new ground by dissecting how strong and weak interactions are distributed among the basic building blocks of the web, defined as *tritrophic food chains* (TFCs), in which a top predator P eats a consumer C, which in turn consumes a resource R. Co-occurrence of strong interactions on two consecutive levels of a TFC may set off destabilizing *trophic cascades,* whereby reduction in the predator (e.g.,

shark, barracuda) results in increased abundance of the consumer (e.g., grouper, snapper) and decrease in the resource (e.g., parrotfish, lobster) (Paine 1980; Carpenter and Kitchell 1993; Pace et al. 1999; Pinnegar et al. 2000; Shurin et al. 2002). In contrast, the occurrence of *omnivory*, in which P consumes both C and R, may stabilize food web dynamics (Fagan 1997; McCann and Hastings 1997; McCann et al. 1998).

For the Caribbean web, Bascompte et al. (2005) found that co-occurrence of two strong interactions happened less frequently among all possible TFCs than expected by chance, whereas omnivorous links were more frequent than expected. Thus the web is structured such that destabilizing trophic cascades should be buffered by omnivory, and simulation analyses support that inference. Nevertheless, the simulations also show that the presence of paired strong interactions is potentially highly destabilizing, and this effect is particularly likely in the face of human exploitation, because fishing selectively targets a highly biased sample of species belonging to the highest tropic levels (Pauly et al. 1998; Myers and Worm 2003). For example, just ten species of heavily fished sharks account for 48 percent of the predators in the strongly interacting TFCs in the Caribbean web, and most of these same TFCs include resource species at the base, such as parrotfishes and other herbivores, that are important grazers of macroalgae. This is just one more compelling piece of evidence that trophic cascades, caused by the extreme overfishing of sharks (Baum et al. 2003) and other top predatory fishes, contributed to the demise of reef corals and the rise of seaweeds on Caribbean reefs (Hughes 1994; Knowlton 2001; Gardener et al. 2003). "The community-wide impacts of fishing are stronger than expected because fishing preferentially targets species whose removal can destabilize the web" (Bascompte et al. 2005).

These results also reveal just how wildly unpredictable the consequences of removal of specific top predators might be, and they reveal the folly of assuming that things can be made right again by the selective culling of undesired species to restore the stocks of desirable fisheries species. Yet this is exactly what has been proposed by the Japanese Government with regard to whales as the culprits for decreased fish stocks (Struck 2001) and by the Canadian Government with regard to harp seals versus fisheries (Lavigne 1995; Buchanan 2002). The absurdity of such arguments is obvious in the context of extreme global overfishing in the entire global ocean (Hutchings and Myers 1994; Myers et al. 1996; Myers et al. 1997; Jackson et al. 2001; Pauly et al. 2002; Baum et al. 2003; Myers and Worm 2003) and the unpredictable consequences of culling (Paine et al. 1998; Scheffer et al. 2001). Nevertheless, the Government of Canada embarked on the massive slaughter of half a million harp seals per year with no increase in the cod. The lack of response of the cod is hardly surprising because the seals interact with at least 150 other species in the food web besides cod (Lavigne 1995), and which species might or might not benefit from the removal of seals is highly uncertain at best (Yodzis 1988). Moreover, as we have already seen for the Caribbean web,

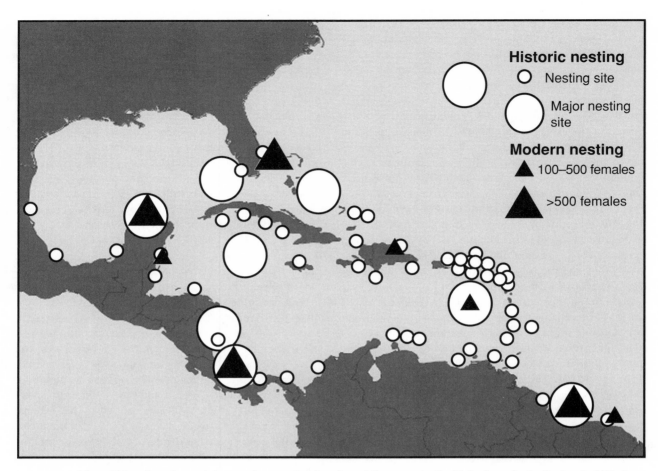

Historic nesting
○ Nesting site
○ Major nesting site

Modern nesting
▲ 100–500 females
▲ >500 females

FIGURE 4.1. Map of historic and remaining modern nesting beaches of the green turtle *Chelonia mydas* in the tropical Western Atlantic for the past 1,000 years (McClenachan et al., 2006). Modern nesting beaches with fewer than 100 nesting beaches are not shown.

uncertainty and instability of the web increase even more if the species removed belongs to many strongly interacting TFCs, as do large sharks, whales, harp seals, and cod (Solé and Montoya 2000; Bascompte et al. 2005). Everything we know about the historical trajectories of decline of marine ecosystems due to overfishing and the structure and function of food webs suggests that the culling of large predators to increase more desirable fish stocks is bound to fail.

Carrying Capacities Past and Present

Green turtles and sea cows were the wildebeest and bison of the tropical coastal oceans, widely reported in fantastic numbers by early explorers and naturalists (Jackson 1997). But how big a number is "fantastic," and could the seagrasses and algae they fed upon have really sustained such extraordinary grazing? My students and I revisited this question for Caribbean green turtles with the assistance of 163 archeological and historical references that provided basic data on the former geographic distribution of nesting beaches and abundance (McClenachan et al., in press). We found evidence for nine major and 50 minor nesting beaches throughout the tropical western Atlantic (Figure 4.1).

Seventeen of those beaches were eliminated by intensive hunting, and only four remain with more than 500 nesting females per season.

The most important remaining beach is at Tortuguero in Costa Rica, where persistent and dedicated protection has seen the number of nesting females rise to 26,100 in 1999 (based on data in Troeng and Rankin 2005). We used this number and our nesting beach data to calculate an estimate of historic adult green turtle abundance for the entire region of about 1.8 million. But this number is absurdly low, because early Jamaican hunters killed well over one million nesting females at Grand Cayman alone in the seventeenth and eighteenth centuries as food for slaves (Jackson 1997; McClenachan et al., 2006), not to mention the enormous depredations there by buccaneers and Cuban fishermen, who somehow neglected to report their catch to the British authorities in Jamaica. Nevertheless, calculations based on the Jamaican data for Grand Cayman, one of the nine major historical beaches, gives an historic abundance of six and one half million adults, which translates to 58.5 million for the nine major nesting beaches alone. Assuming very conservatively that the remaining 50 minor beaches supported only ten percent of the Cayman population gives a total of

91 million adult green turtles for the entire tropical western Atlantic in the late seventeenth century. That abundance contrasts with the current population estimates of much less than half a million (the data for the recent are remarkably vague!) and is more than the estimated biomass of all the bison in North America before the arrival of the horse and the rifle (Isenberg 2000).

This huge number is still well within the calculated carrying capacity of turtles based on the modern regional distribution and productivity of turtlegrass, which constitutes more than 85 percent of their food. Nevertheless, the turtles had an enormous impact on the turtlegrass (McClenachan et al., 2006). Adult green turtles were twice as large before we hunted them to near-extinction than they are today, and they consumed somewhere between 43 to 86 percent of the standing crop and 22 to 45 percent of the total annual turtlegrass production (the ranges reflect different-sized animals used in the calculations, based on the estimate of 91 million adults). The turtles grazed throughout the entire regional distribution of five million hectares of seagrass meadows, where they maintained large, seemingly bare grazing patches of heavily cropped turtlegrass grazed down to the sediment surface, at the junction of the rhizomes and the blades (Thayer et al. 1982, 1984). Just as with the wildebeest in the Serengeti (McNaughton and Banyikwa 1995), intensive turtle grazing increases the production of the turtlegrass in the patches until the energy of the plants is exhausted and the turtles move on to potentially greener pastures (Thayer et al. 1982, 1984). Sea urchin grazing also increases turtlegrass productivity, although in apparently smaller amounts (Valentine et al. 1997).

Turtlegrass meadows today are strikingly different from when green turtles reigned supreme (Jackson 1997, 2001; McClenachan et al., in press). Historic populations of green turtles consumed and efficiently metabolized at least one-third of the total production of the turtlegrass, with the help of abundant microbial symbionts that break down cellulose in the turtles' guts (Thayer et al. 1982, 1984). Now, consumption by surviving grazers, including fish, sea urchins, and various smaller invertebrates, may be low or high, apparently due to the extent of overfishing among areas studied (Thayer et al. 1982, 1984; Valentine and Heck 1999; Kirsch et al. 2002). However, grazing fishes and sea urchins only partially metabolize the seagrass they consume, because they lack microbial symbionts to break down cellulose in any abundance (Thayer et al. 1984). Today turtlegrass grows up to 50 cm tall, in contrast to historic heights of 10 cm, and the long blades break off, drift about, and slowly decompose on the bottom in particles too large for efficient microbial attack. Flux of carbon and nitrogen to the sediments is increased by an order of magnitude or more and is buried and lost forever from the grazing food chain that supports the fish we eat—not to mention the unimaginable numbers of tiger sharks once sustained by the turtles as prey. Extirpation of green turtles shook Caribbean ecosystems from top to bottom.

There is a similar story for the eleven million hawksbill turtles that once consumed 39 to 83 percent of the standing crop and total annual sponge production throughout the tropical western Atlantic, with obvious implications for what was natural on coral reefs in terms of sponges versus corals and the species composition of sponge and coral communities (McClenachan et al., 2006). To return to the topic of seals and fishes, an estimated half a million or more adult West Indian monk seals consumed the equivalent of all of the pitiful remnant biomass of fish in Caribbean coastal environments (McClenachan et al., 2006). If the monk seals were not already extinct, one could well imagine yet another misguided program to cull the seals to increase the fish. Yet, just as in Canada, the fishes would not recover, because the real problem with the fish is that *we* ate them and are still eating them, and this time the seals are not even around to conveniently take the blame.

Less-than-Zero-Sum Ecology?

So great is the influence of Popperian hypothesis testing and falsification in ecology that the most careful and detailed historical descriptions are commonly dismissed as untestable anecdotes. Ecologists accept the once-great abundance of North American bison, because their slaughter was documented in great detail (Isenberg 2000), and the time machine of the Serengeti makes it vaguely imaginable. Ditto the billions of passenger pigeons that blackened the skies, because they disappeared barely a century ago and supported one of the most spectacular and well-documented episodes of gluttony in history (Price 1999). But these are the rare exceptions, and usually we deny the once-great existence of anything we ate more than a century ago (Mowat 1996).

What is really scary about this unbridled antihistorical determinism is that it flies in the face of everything we have learned about the "methodical dismantling of the world's ecosystems" (Buchanan 2002) and the common trajectories of change that emerge both from the theoretical study of small-world networks (Solé and Montoya 2000; Buchanan 2002; Bascompte et al. 2005) and from detailed historical ecological reconstructions of change (Jackson et al. 2001; Pandolfi et al. 2003; Lotze et al. 2006). Details differ from system to system and place to place, but the details matter surprisingly little for the generality of the conclusions and their profound implications for conservation and management (Pandolfi et al. 2005). Perhaps this attitude will change with wider acceptance of Bayesian inference based on weight of evidence, but by then, I suspect, the last remnants of large wild animals will be gone.

If the carbon fixed by photosynthesis is entirely utilized by organisms within an ecosystem, then the surplus generated by the demise of one group of consumers as a result of exploitation should be taken up by another unexploited group. This is the basic premise of zero-sum ecology (Hubbell 2001; Paine, Chapter 2 of this volume), which greatly simplifies calculations of ecosystemwide historical impacts of species, such as whales, whose abundance was very greatly reduced prior to ecological investigation. But the sea turtles and the monk seals discussed here hint at a far darker image of a "less-than-zero-sum ecology," whereby a large proportion of global ocean production is lost from the grazing food chain to the netherworld of burial in sediments. In such cases, the present biomass of consumers may be only a small fraction of the historical biomass, and projections based on present values are bound to be grossly incorrect. Similarly, though for mostly different reasons, the increasing global emergence of eutrophic "dead zones," inhabited by tiny flagellates and microbes, which are key components of the microbial loop (Azam 1998) and make better meals for jellyfish than for copepods and fish (Diaz 2001; Parsons and Lalli 2002), also results in the deposit of vast quantities of unconsumed organic matter for burial on the sea floor (Rabalais et al. 2002). All of this carbon sinking may be good for global carbon credits against greenhouse gas emissions, but it is potentially catastrophic for the future supply of food from the sea, which ultimately depends on the grazing food chain, whether for wild fish or aquaculture.

In summary, marine ecosystems were once dominated by large animals, and ecological pyramids of animals were at least partially upside down. Removal of large animals necessarily destabilized food webs, with the resultant further loss of productivity that decreases the potential for recovery of the same species of large animals whose loss set off this chain reaction of events. We need to keep all of these things very clearly in mind when we make strong statements based on ecological calculations and insights derived from degraded, remnant ecosystems about what may or may not have been possible for the abundance and ecological significance of great whales.

Acknowledgments

Heike Lotze, Loren McClenachan, Marah Newman, Andy Rosenberg, Enric Sala, George Sugihara, and my colleagues in the historical ecology and ecosystem function working groups at the National Center for Ecological Analysis and Synthesis (NCEAS) have contributed greatly to my thinking about these questions. This work was supported by NCEAS, the History of Marine Animal Populations (HMAP) program, and the William E. and Mary B. Ritter Chair of the Scripps Institution of Oceanography.

Literature Cited

Alinao, P. M., L. T. McManus, C. L. Nanola, M. D. Fortes, G. C. Trono, and G. C. Jacinto. 1993. Initial parameter estimations of a coral reef flat ecosystem in Bolinao, Panganistan, northwestern Philippines, in *Trophic models of aquatic ecosystems*. V. Christensen and D. Pauly, eds. ICLARM Conference Proceedings 26. Manila: International Center for Living Aquatic Resources Management, pp. 252–258.

Azam, F. 1998. Microbial control of oceanic carbon flux: The plot thickens. *Science* 280: 694–696.

Baird, D., R. E. Ulanowicz, and W. R. Boynton. 1995. Seasonal nitrogen dynamics in the Chesapeake Bay, a network approach. *Estuarine and Coastal Shelf Science* 41: 37–62.

Barnes, R. S. K. and R. N. Hughes. 1999. *An introduction to marine ecology*, 3rd edition. Oxford: Blackwell Science.

Barnosky, A. D., P. L. Koch, R. S. Feranec, S. L. Wing, and A. B. Shabel. 2004. Assessing the causes of Late Pleistocene extinctions on the continents. *Science* 306: 70–75.

Bascompte, J., C. J. Melián, and E. Sala. 2005. Interaction strength combinations and the overfishing of a marine food web. *Proceedings of the National Academy of Sciences* 102: 5443–5447.

Baum, J. K., R. A. Myers, D. G. Kehler, B. Worm, J. Harley, and P. A. Doherty. 2003. Collapse and conservation of shark populations in the northwest Atlantic. *Science* 299: 389–392.

Benton, M.J. 1990. Reptiles, in *Evolutionary Trends*. K. J. McNamara, ed. Tucson: University of Arizona Press, pp. 279–300.

Berger, J., P. B. Stacey, L. Bellis, and M. P. Johnson. 2001a. A mammalian predator-prey imbalance: Grizzly bear and wolf extinction affect avian neotropical migrants. *Ecological Applications* 11: 967–980.

Berger, J., J. E. Swenson, and I. -L. Persson. 2001b. Recolonizing carnivores and naïve prey: conservation lessons from Pleistocene extinctions. *Science* 291: 1036–1039.

Bertness, M. D., S. D. Gaines, and M. E. Hay, eds. 2001. *Marine community ecology.* Sunderland, MA: Sinauer Associates.

Boesch, D. F. 2001. Measuring the health of the Chesapeake Bay: Toward integration and prediction. *Environmental Research Section A* 82: 134–142

Buchanan, M. 2002. *Small worlds and the groundbreaking theory of networks.* New York: W. W. Norton.

Carpenter, S. R. and J. F. Kitchell. 1993. *The trophic cascade in lakes.* Cambridge, UK: Cambridge University Press.

Clapham, P. J. and C. S. Baker. 2002. Whaling, modern, in *Encyclopedia of marine mammals*. W. F. Perrin, B. Wursig, and J. G. M. Thewissen, eds. San Diego: Academic Press, pp. 1328–1332.

Connell, J. H. 1978. Diversity in tropical rain forests and coral reefs. *Science* 199: 1302–1310.

Cushing, D. H. 1988. *The provident sea.* Cambridge, UK: Cambridge University Press.

Dayton, P. K., M. J. Tegner, P. B. Edwards, and K. L. Riser. 1998. Sliding baselines, ghosts, and reduced expectations in kelp forest communities. *Ecological Applications* 8: 309–322.

Deevey, E. S. 1947. Life tables for natural populations of animals. *Quarterly Review of Biology* 22: 283–314.

Diamond, J. 1997. *Guns, germs, and steel: the fates of human societies.* New York: W. W. Norton.

Diamond, J. M. 2001. Australia's lost giants. *Nature* 411: 755–757.

Diaz, R. J. 2001. Overview of hypoxia around the world. *Journal of Environmental Quality* 30: 275–281.

Elton, C. 1927. *Animal ecology.* London: Sidgwick and Jackson.

Elton, C. S. 1966. *The pattern of animal communities.* London: Methuen.

Erwin, D. H. 2001. Lessons from the past: biotic recoveries from mass extinctions. *Proceedings of the National Academy of Sciences* 98: 5399–5403.

Fagan, W. F. 1997. Omnivory as a stabilizing feature of natural communities. *American Naturalist* 150: 554–567.

Fagan, W. F. and L. E. Hurd. 1994. Hatch density variation of a generalist arthropod predator: population consequenes and community impact. *Ecology* 75: 2022–2032.

Flannery, T. 1994. *The future eaters: an ecological history of the Australasian lands and people.* New York: Grove Press.

Flannery, T. F. and R. G. Roberts. 1999. Late Quaternary extinctions in Australia. in *Extinctions in near time: causes, contexts, and consequences.* MacPhee, R. D. E., ed. New York: Kluwer/Plenum, pp. 239–255.

Friedlander, A. M. and E. E. DeMartini. 2002. Contrasts in density, size, and biomass of reef fishes between the northwestern and main Hawaiian islands: the effects of fishing down apex predators. *Marine Ecology Progress Series* 230: 253–264.

Gardener, T. A., I. Cote, J. A. Gill, A. Grant, and A. R. Watkinson. 2003. Long-term region-wide declines in Caribbean corals. *Science* 301: 958–960.

Goldwasser, L. and J. Roughgarden. 1997. Sampling effects and the estimation of food-web properties. *Ecology* 78: 41–54.

Hairston, N. G., F. E. Smith, and L. Slobodkin. 1960. Community structure, population control, and competition. *American Naturalist* 94: 421–425.

Hardy, A. C. 1924. The herring in relation to its animate environment. Part I. The food and feeding habits of the herring with special reference to the east coast of England. *Fishery Investigations (London)* Series 2, 7: 1–53.

Hubbell, S. P. 2001. *The unified neutral theory of biodiversity and biogeography.* Princeton, NJ: Princeton University Press.

Hughes, T. P. 1994. Catastrophes, phase shifts, and large-scale degradation of a Caribbean coral reef. *Science* 265: 1547–1551.

Hutchings, J. A. and R. A. Myers. 1994. What can be learned from the collapse of a renewable resource? Atlantic cod, *Gadus morhua,* of Newfoundland and Labrador. *Canadian Journal of Fisheries and Aquatic Science* 51: 2126–2146.

Hutchinson, G. E. 1978. *An introduction to population ecology.* New Haven, CT: Yale University Press.

Isenberg, A. C. 2000. *The destruction of the bison.* Cambridge, UK: Cambridge University Press.

Jackson, J. B. C. 1991. Adaptation and diversity of reef corals. *BioScience* 41: 475–482.

———. 1997. Reefs since Columbus. *Coral Reefs* 16 (Suppl.): S23–S32.

———. 2001. What was natural in the coastal oceans? *Proceedings of the National Academy of Sciences* 98: 5411–5418.

Jackson, J. B. C., M. X. Kirby, W. H. Berger, K. A. Bjorndal, L. W. Botsford, B. J. Bourque, R. H. Bradbury, R. Cooke, J. Erlandson, J. A. Estes, T. P. Hughes, S. Kidwell, C. B. Lange, H. S. Lenihan, J. M. Pondolfi, C. H. Peterson, R. S. Steneck, M. J. Tegner, and R. R. Warner. 2001. Historical overfishing and the recent collapse of coastal ecosystems. *Science* 293: 629–638.

Jackson, J. B. C. and E. Sala. 2001. Unnatural oceans. *Scientia Marina* 65 (Suppl. 2): 273–281.

Janis, C. M. and J. Damuth. 1990. Evolutionary trends, in *Evolutionary trends.* K. J. McNamara, ed. Tucson: University of Arizona Press, pp. 301–345.

Janzen, D. H. and P. S. Martin. 1982. Neotropical anachronisms: the fruits the gomphotheres ate. *Science* 215: 19–27.

Jennings, S., M. J. Kaiser, and J. D. Reynolds. 2001. *Marine fisheries ecology.* Oxford: Blackwell Science.

Kareiva, P. and M. Andersen. 1988. Spatial aspects of species interactions: the wedding of models and experiments, in *Community ecology.* A. Hastings, ed. New York: Springer Verlag, pp. 38–54.

Katona, S. and H. Whitehead. 1988. Are cetaceans ecologically important? *Oceanography and Marine Biology Annual Reviews* 26: 553–568.

Kirsch, K. D., J. F. Valentine, and K. L. Heck, Jr. 2002. Parrotfish grazing on turtlegrass *Thalassia testudinum*: evidence for the importance of seagrass consumption in food web dynamics of the Florida Keys National Marine Sanctuary. *Marine Ecology Progress Series* 227: 71–85.

Knowlton, N. 2001. The future of coral reefs. *Proceedings of the National Academy of Sciences* 98: 5419–5425.

Kruuk, H. 1972. *The spotted hyena.* Chicago: University of Chicago Press.

Lartigaut, J., ed. 1993. *Histoire du Quercy.* Toulouse: Editions Privat.

Lavigne, D. M. 1995. Seals and fisheries, science and politics. Paper presented at the 11th Biennial Conference on the Biology of Marine Mammals, Orlando, Florida, 1995. Available at www.imma.org/orlando.pdf.

Levinton, J. S. 2001. *Marine biology: function, biodiversity, ecology.* New York: Oxford University Press.

Lindeman, R. L. 1942. The trophic-dynamic aspect of ecology. *Ecology* 23: 399–418.

Lotze, H. K., H. S. Lenihan, B. J. Bourque, R. H. Bradbury, R. G. Cooke, M. C. Kay, S. M. Kidwell, M. X. Kirby, C. H. Peterson, and J. B. C. Jackson. 2006. Depletion, degradation, and recovery potential of estuaries and costal seas. *Science* 312: 1806–1809.

Lotze, H. K. and I. Milewski. 2004. Two centuries of multiple human impacts and successive changes in a North Atlantic food web. *Ecological Applications* 14: 1428–1447.

Margalef, R. 1968. *Perspectives in ecological theory.* Chicago: University of Chicago Press.

Martin, P. S. and D. W. Steadman. 1999. Prehistoric extinctions on islands and continents, in *Extinctions in near time: causes, contexts, and consequences.* R. D. E. MacPhee, ed. New York: Kluwer/Plenum, pp. 17–55.

McCann, K. and A. Hastings. 1997. Re-evaluation of the omnivory-stability relationship in food webs. *Proceedings of the Royal Society of London, Series B, Biological Sciences* 264: 1249–1254.

McCann, K., A. Hastings, and G. Huxel. 1998. Weak trophic interactions and the balance of nature. *Nature* 395: 794–798.

McClenachan, L., J. B. C. Jackson, and M. J. H. Newman. 2006. Conservation implications of historic sea turtle nesting beach loss. *Frontiers in Ecology and the Environment* 4: 290–296.

McLaren, B. E. and R. O. Peterson. 1994. Wolves, moose, and tree rings on Isle Royale. *Science* 266: 1555–1558.

McNaughton, S. J. and F. F. Banyikwa. 1995. Plant communities and herbivory, in *Serengeti II: dynamics, management, and conservation of an ecosystem.* A. R. E. Sinclair and P. Arcese, eds. Chicago: University of Chicago Press, pp. 49–70.

Mowat, F. 1996. *Sea of slaughter.* Shelburne, VT: Chapters Publishing.

Myers, R. A., N. J. Barrowman, J. M. Hoenig, and Z. Qu. 1996. The collapse of cod in eastern Canada: the evidence from the tagging data. *ICES Journal of Marine Science* 53: 629–640.

Myers, R. A., J. A. Hutchings, and N. J. Barrowman. 1997. Why do fish stocks collapse? The example of cod in Atlantic Canada. *Ecological Applications* 7: 91–106.

Myers, R. A. and B. Worm. 2003. Rapid worldwide depletion of predatory fish communities. *Nature* 423: 280–283.

Nybakken, J. W. 2001. *Marine biology: an ecological approach.* San Francisco: Addison Wesley Longman.

Nybakken, J. W. and M. D. Bertness. 2005. *Marine biology: an ecological approach,* 6th edition. San Francisco: Pearson Benjamin Cummings.

Odum, E. P. 1953. *Fundamentals of Ecology.* Philadelphia: W. B. Saunders Company.

Odum, E. P. 1959. *Fundamentals of Ecology,* 2nd edition. Philadelphia: W. B. Saunders Company.

Opitz, S. 1996. *Trophic interactions in Caribbean coral reefs.* ICLARM Technical Reports 43. Makati City, Philippines: International Center for Living Aquatic Resources Management.

Owen-Smith, R. N. 1988. *Megaherbivores: the influence of very large body size on ecology.* Cambridge, UK: Cambridge University Press.

Pace, M. L., J. J. Cole, S. R. Carpenter, and J. F. Kitchell. 1999. Trophic cascades revealed in diverse ecosystems. *Trends in Ecology and Evolution* 14: 483–488.

Paine, R. T. 1980. Food webs: Linkage, interaction strengths, and community. *Journal of Animal Ecology* 49: 669–685.

———. 1992. Food-web analysis through field measurement of per capita interaction strength. *Nature* 355: 73–75.

Paine, R. T. and S. A. Levin. 1981. Intertidal landscapes: disturbance and the dynamics of pattern. *Ecological Monographs* 51:145–178.

Paine, R.T., M.J. Tegner, and E.A. Johnson. 1998. Compounded perturbations yield ecological surprises. *Ecosystems* 1: 535–545.

Pandolfi, J. M., R. H. Bradbury, E. Sala, T. P. Hughes, K. A. Bjorndal, R. G. Cooke, D. McArdle, L. McClenachan, M. J. H. Newman, G. Peredes, R. R. Warner, and J. B. C. Jackson. 2003. Global trajectories of the long-term decline of coral reef ecosystems. *Science* 301: 955–958.

Pandolfi, J.M., J.B.C. Jackson, N. Baron, R.H. Bradbury, H. Guzman, T. P. Hughes, F. Micheli, J. Ogden, H. Possingham, C. V. Kappel, and E. Sala. 2005. Are US coral reefs on the slippery slope to slime? *Science* 307: 1725–1726.

Parsons, T. R. and C. M. Lalli. 2002. Jellyfish population explosions: Revisiting a hypothesis of possible causes. *La Mer* 40: 111–121.

Parsons, T.R., M. Takahashi, and B. Hargrave. 1984. *Biological oceanographic processes,* 3rd edition. Oxford: Pergamon Press.

Pauly, D. 1995. Anecdotes and the shifting baseline syndrome of fisheries. *Trends in Ecology and Evolution* 10: 430.

Pauly, D., V. Christensen, J. Dalsgaard, R. Froese, and F. Torres, Jr. 1998. Fishing down marine food webs. *Science* 279: 860–863.

Pauly, D., V. Christensen, S. Guénette, T. J. Pitcher, U. R. Sumaila, C. J. Walters, R. Watson, and D. Zeller. 2002. Towards sustainability in world fisheries. *Nature* 418: 689–695.

Pinnegar, J. K., N. V. C. Polunin, P. Francour, F. Badalamenti, R. Chemello, M. -L. Harmelin-Vivien, B. Hereu, M. Milazzo, M. Zabala, G. D'Anna, and C. Pipitone. 2000. Trophic cascades in benthic marine ecosystems: lessons for fisheries and protected-area management. *Environmental Conservation* 27: 179–200.

Polovina, J. J. 1984. Model of a coral reef ecosystem. I. The ECOPATH model and its application to French Frigate Shoals. *Coral Reefs* 3: 1–11.

Price, J. 1999. *Flight maps: Adventures with nature in modern America.* New York: Basic Books.

Rabalais, N. N., R. E. Turner, and W. J. Wiseman, Jr. 2002. Gulf of Mexico hypoxia, a. k. a. "The Dead Zone." *Annual Reviews of Ecology and Systematics* 33: 235–263.

Raffaelli, D. and S. Hall. 1995. Assessing the relative importance of trophic links in food webs, in *Food webs, integration of patterns, and dynamics.* G. Polis and K. Winemiller, eds. New York: Chapman and Hall, pp. 185–191.

Randall, J. E. 1967. Food habits of reef fishes of the West Indies. *Studies in Tropical Oceanography* 5: 665–847.

Ripple, W. J. and R. L. Beschta. 2004. Wolves and the ecology of fear: Can predation risk structure ecosystems? *BioScience* 54: 755–766.

Roberts, C. M., J. A. Bohnsack, F. Gell, J. P. Hawkins, and R. Goodridge. 2001. Effects of marine reserves on adjacent fisheries. *Science* 294: 1920–1923.

Roberts, R. G., T. F. Flannery, L. K. Ayliffe, H. Yoshida, J. M. Olley, G. J. Prideaux, G. M. Laslett, A. Baynes, M. A. Smith, R. Jones, and B. L. Smith. 2001. New ages for the last Australian megafauna: continent-wide extinction about 46,000 years ago. *Science* 292: 1888–1892.

Roman, J. and S. R. Palumbi. 2003. Whales before whaling in the North Atlantic. *Science* 301: 508–510.

Roussot, A. 1997. *L'art préhistorique.* Bordeaux: Sudouest.

Sale, P.F., ed. 1991. *The ecology of fishes.* San Diego: Academic Press.

Schaller, G. 1972. *The Serengeti lion: a study of predator-prey relations.* Chicago: University of Chicago Press.

Scheffer, M., S. Carpenter, J. A. Foley, C. Folke, and B. Walker. 2001. Catastrophic shifts in ecosystems. *Nature* 413: 591–596.

Schiermeier, Q. 2002. How many more fish in the sea? *Nature* 419: 662–665.

Shurin, J.B., E. T. Borer, E. W. Seabloom, K. Anderson, C. A. Blanchette, B. Broitman, S. D. Cooper, and B. S. Halpern. 2002. A cross-ecosystem comparison of the strength of trophic cascades. *Ecology Letters* 5: 785–791.

Sinclair, A. R. E. and P. Arcese. 1995. *Serengeti II: dynamics, management, and conservation of an ecosystem.* Chicago: University of Chicago Press.

Sinclair, A. R. E., S. Mduma, and J. S. Brashares. 2003. Patterns of predation in a diverse predator-prey system. *Nature* 425: 288–290.

Smith, C. R. and A. R. Baco. 2003. Ecology of whale falls at the deep-sea floor. Oceanography and Marine Biology: an Annual Review 41: 311–354.

Solé, R. and J. Montoya, 2000. Complexity and fragility in ecological networks. Working paper 00-11-060, Santa Fe Institute, Santa Fe, New Mexico. Available at www.santafe.edu/sfi/publications/00wplist.html.

Springer, A. M., J. A. Estes, G.B. van Vliet, T. M. Williams, D.F. Doak, E. M. Danner, K. A. Forney, and B. Pfister. 2003. Sequential megafaunal collapse in the North Pacific Ocean: An ongoing legacy of industrial whaling? *Proceedings of the National Academy of Sciences* 100:12223–12228.

Struck, D. 2001. Japan blames whales for lower fish catch. *International Herald Tribune* (July 28–29).

Stuart, A. J., P. A. Kosintsev, T. F. G. Higham, and A. M. Lister. 2004. Pleistocene to Holocene extinction dynamics in giant deer and wooly mammoth. *Nature* 431: 684–689.

Thayer, G. W., D. W. Engel, and K. A. Bjorndal. 1982. Evidence for the short-circuiting of the detritus cycle of seagrass beds by the green turtle, *Chelonia mydas. Journal of Experimental Marine Biology and Ecology* 62: 173–183.

Thayer, G. W., K. A. Bjorndal, J. C. Ogden, S. L. Williams, and J. C. Zieman, 1984. Role of larger herbivores in seagrass communities. *Estuaries* 7: 351–376.

Thurman, H. V. and A. P. Trujillo. 2004. *Introductory oceanography,* 10th edition. Upper Saddle River, NJ: Pearson Prentice Hall.

Troeng, S. and E. Rankin. 2005. Long term conservation efforts contribute to positive green turtle Chelonia mydas nesting trend at Tortuguero, Costa Rica. *Biological Conservation* 121: 111–116.

Valentine, J. F., K. L. Heck, Jr., J. Busby, and D. Webb. 1997. Experimental evidence that herbivory increases shoot density and productivity in a subtropical turtlegrass (*Thalassia testudinum*) meadow. *Oecologia* 112: 193–200.

Valentine, J. F. and K. L. Heck, Jr. 1999. Seagrass herbivory: evidence for the continued grazing of marine grasses. *Marine Ecology Progress Series* 176: 291–302.

Valiela, I. 1995. *Marine ecological processes,* 2nd ed. New York: Springer-Verlag.

Vermeij, G. J. 2004. *Nature: an economic history.* Princeton, NJ: Princeton University Press.

Wharton, J. 1957. *The bounty of Chesapeake.* Jamestown, VA: Virginia Anniversary Celebration Corporation, Jamestown 390th Anniversary Historical Booklet 13.

Whitehead, H, 2002. Estimates of the current global population size and historical trajectory for sperm whales. *Marine Ecology Progress Series* 242: 295–304.

Wootton, J. T. 1997. Estimates and tests of per capita interaction strength: diet, abundance, and impact of intertidally foraging birds. *Ecological Monographs* 67: 45–64.

Worm, B., H. K. Lotze, H. Hillebrand, and U. Sommer. 2002. Consumer versus resource control of species diversity and ecosystem functioning. *Nature* 417: 848–851.

Yodzis, P. 1988. The indeterminacy of ecological interactions, as perceived through perturbation experiments. *Ecology* 69: 508–515.

FIVE

Pelagic Ecosystem Response to a Century of Commercial Fishing and Whaling

TIMOTHY E. ESSINGTON

The dominant role of humans as structuring agents in ecosystems is now widely acknowledged (Vitousek et al. 1997). Human activity has led to widespread alteration of marine coastal ecosystems through changes in biogeochemical cycling of nutrients, carbon and contaminants, and hydrologic regimes and through the direct and indirect effects accompanying the targeted removal of marine life. Jackson et al. (2001) conclude that human exploitation and concomitant changes in food web structure is primary among the anthropogenic alterations of coastal ecosystems. These anthropogenic alterations of community and ecosystem structure are now well documented for a diversity of ecosystems, including coral reef (McCanahan et al. 1994), estuaries (Kiddon et al. 2003), continental shelf (Fogarty and Murawski 1998) and semienclosed seas (Caddy 1993).

Nearly all of the aforementioned examples derive from ecosystems in close geographic proximity to land, where humans might be expected to exert their greatest influences. To date, there has been no documented restructuring of pelagic "blue water" ecosystems induced by human activity. This absence may be partly due to logistic limitations that make human influences much more difficult to detect; the enormous spatial extent of these ecosystems preclude all but the most basic monitoring efforts. There is rising concern, however,

that these ecosystems have not been spared the influence of human activity. The rise of industrial fishing and whaling in the latter half of the last century brought unprecedented changes in the efficiency with which high-seas organisms were exploited. Across the globe, whaling operations have killed nearly two million baleen and toothed whales since 1950, according to the UN Food and Agricultural Organization (FAO), rapidly depleting whale populations (Tillman 1977; Whitehead et al. 1997; Read and Wade 2000; Whitehead 2002). Industrial fishing targeting large pelagic predators removed roughly one-quarter million tonnes of large tunas (albacore, bluefin, bigeye, and yellowfin), billfish (marlins, sailfish, and swordfish), and pelagic sharks (threshers, mako, and blue shark) in 1950, whereas catches now exceed two million tonnes annually (FAO; Figure 5.1). Some have hypothesized that the apex predator guild has been depleted by as much as 90% since the advent of modern industrial fishing (Myers and Worm 2003), although other more detailed stock assessments suggest more modest declines (Hampton 2000; Cox et al. 2002b). Regardless of the magnitude of decline, it is clear that the combination of whaling and fishing has dramatically diminished the abundance of large, high-trophic-level animals in these ecosystems.

Given the potentially important role of apex predators as structuring agents in ecosystems, some have speculated that

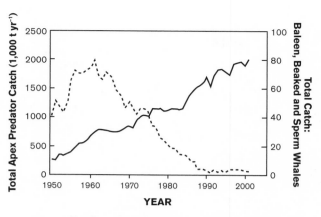

FIGURE 5.1. Total catches of apex fish predators (solid line) and large cetaceans (dashed line) in the Pacific Ocean, 1950–2001. Apex fish predators include all billfish, sharks, and tunas except skipjack and blackfin tuna.

whaling and fishing might have dramatically altered pelagic food webs (Myers and Worm 2003). This chapter attempts to address this hypothesis by using a food web model to characterize contemporary food web structure and to reconstruct historical food web structure prior to the advent of industrial whaling and fishing. The strategy for this modeling exercise was to specify the contemporary state of the food web in terms of biomass, productivities, consumption rates, food habits, and fisheries and to use this information to initialize a dynamic food web model. The dynamic food web model was then used to simulate the changes in food web structure that might be expected in a food web without whaling and fishing, in the hope of gaining a glimpse of what the pre-exploitation food web might have looked like. This type of "back to the future" exercise cannot capture the profound changes that accompany catastrophic shifts in ecosystem reorganization (Peterson 2000; Scheffer et al. 2001), and for that reason, interpretation of these models is contingent on the assumptions of no radical change in ecosystem structuring variables. This exercise can, however, point to whether there have been dramatic changes in the energy flow throughout the food web and, if so, what might be the consequences.

Methods

The region considered here, the Central North Pacific (CNP), consists of the tropical and subtropical waters lying north of the equator extending to 40° N and ranging from 130° E to 150° W. This region supports a substantial fishery whose landings are dominated by purse-seine-caught skipjack tuna (*Katsuwanus pelamis*), from the western Pacific Ocean, but also include long-line-captured bigeye tuna *(Thunnus obesus),* yellowfin tuna *(Thunnus albacares),* albacore *(Thunnus alalunga),* and bluefin tuna *(Thunnus thunnus).* Baleen and toothed whales also use this region as feeding and breeding grounds.

A food web model for the CNP was previously developed for purposes other than reconstruction of historical food web

structure (Cox et al. 2002a). I used this existing model as starting point and modified the model structure so that I could more readily compare gross changes in food webs caused by fishing and whaling. For that reason, biomass pools specified in earlier models were aggregated into distinct functional feeding groups. These groups were defined on the basis of similarities in food habits, habitats, and biological variables.

There are 15 functional trophic groups in the CNP ecosystem (Table 5.1). This representation necessarily aggregates more species within lower-trophic-level groups, largely reflecting our ignorance as to the species compositions, biology, and ecology of these species. I specified a group as a teuthivore if squid constituted more than 40% of the published diet contents (by mass). The compositions of these groups are as follows:

SMALL ZOOPLANKTON: Copepods, salps, amphipods, etc.

LARGE ZOOPLANKTON: Euphasiids and crustacean decapods.

MESOPELAGIC FISHES: Fishes of the following families: Myctophidae, Alepisauridae, Gempylidae, Sternopychidae, Sudinae.

EPIPELAGIC FISHES: Fishes of the following families: Exocoetidae, Carangidae, Atherinidae, Molidae, Bramidae.

SMALL SQUID: Juveniles of all squid species, and all life stages of squids whose maximum mantle length is less than 30 cm. This includes members of families Brachioteuthidae, Bathyteuthidae, Cranchiidae, Enoploteuthidae, Cycloteuthidae, Pyroteuthidae, plus smaller members of the family Ommastrephidae.

LARGE SQUID: *Ommastrephes bartramii, Stenoteuthis oualaniensis, Onchyteuthis banksii, Thysanoteuthis rhombus,* and *Ancistrocheirus lesuerii.* The bulk of the information of food habits of this group is restricted to *Ommastrephes bartramii.*

SMALL EPIPELAGIC PISCIVORES: Skipjack tuna, dolphinfish *(Coryphaenus hippurus),* and juvenile stages of all tuna species (yellowfin, bigeye, bluefin, albacore).

LARGE EPIPELAGIC PISCIVORES: Marlins, sailfish, and large yellowfin tuna.

MESOPELAGIC TEUTHIVORES: swordfish *(Xiphias gladius).*

MESOPELAGIC PISCIVORES: Large bigeye tuna and large albacore.

TEUTHIVOROUS ELASMOBRANCHS: Blue sharks *(Prionace glauca).*

PISCIVOROUS ELASMOBRANCHS: Lamnids and carcharhinids other than blue sharks.

BALEEN WHALES: Mostly Bryde's whales *(Balaenoptera edeni),* plus limited numbers of sei whales *(Balaenoptera borealis).* Other baleen whales (e.g., humpback) do not apparently feed while residing in the CNP model region and, for that reason, are not considered in this model.

SPERM WHALES: Sperm whales are unique among the ondontocetes in the Central Pacific in that their population dynamics and stock sizes are reasonably well known. Given the paucity of information on the status and tends of odontocetes other than sperm whales, I considered only the relatively well-understood sperm whale dynamics in this model.

TABLE 5.1
Contemporary Ecopath Model, Showing Model Inputs (Normal Type) and Ecopath Estimates (Bold Type)

	Biomass (g m⁻²)	Production-to-biomass Ratio, PB (yr⁻¹)	Consumption-to-biomass Ratio, QB (yr⁻¹)	Proportion of Total Mortality from Fishing, P_Z	Fishery (g m⁻² yr⁻¹)	Rate of biomass Accumulation During 1990s, dB/dt (g m⁻² yr⁻¹)
Primary producers	7[a]	100[a]	0	**0.23**	0	0
Mesozooplankton	**2.85**	5[b]	50[c]	0.9	0	0
Microzooplankton	**3.57**	8[b]	80[c]	0.9	0	0
Mesopelagic fishes	2[d]	1.46[d]	7.3[c]	**0.67**	0	0
Epipelagic fishes	0.6[d]	2.3[d]	10[c]	**0.39**	0	0
Small squid	**0.45**	3.4[e]	10	0.65	0	0
Large squid	**0.20**	1.4[e]	6.67	0.65	0.0007[f]	0
Small epipelagic piscivores	0.0345[g]	1.68[g]	15.95[f]	**0.82**	0.01893[f]	0
Large epipelagic piscivores	0.0140[g]	0.94[g]	13.21[f]	**0.21**	0.00253[f]	0
Mesopelagic teuthivores	0.0009[g]	0.60[g]	5.00[f]	**0.62**	0.00033[f]	0
Mesopelagic piscivores	0.0159[g]	0.41[g]	7.26[f]	**0.29**	0.00177[f]	0
Piscivorous elasmobranchs	0.0122[f]	0.18[f]	2.79[f]	**0.49**	0.00107[f]	0
Teuthivorous elasmobranchs	0.0053[g]	0.30[g]	2.75[f]	**0.64**	0.00102[f]	0
Baleen whales	0.0708	0.078[h]	7.40[i]	**0.14**	0	0.00057
Sperm whales	0.0101	0.03	12.8[j]	**0.28**	0	7.8E-05

[a] Longhurst et al. (1995).

[b] Peterson and Wroblewski (1984).

[c] Assuming 10% gross conversion efficiency.

[d] Mann (1984).

[e] Hoenig (1983). Most squid have 1-year life cycles, and I adjusted large squid downward to account for reduced productivity of large-bodied members of each species.

[f] Cox et al. (2002a).

[g] Cox et al. (2002b), but adjusted downward based on recent stock assessments (Hampton and Kleiber 2003; Hampton et al. 2003; Langley et al. 2003).

[h] IWC assumes that mortality rate equals 0.07, so P:B equal to 0.078 gives a 0.8% increase per year.

[i] Kenney et al. (1997).

[j] Santos et al. (2001).

Contemporary Food Web Structure

I used a trophic mass-balance model (Pauly et al. 2000) to construct the contemporary food web structure. The model is essentially a trophic flow network depicting the biomasses of functional groups, the trophic linkages between them, and removals by fisheries and whaling. For each trophic group i, the model balances production against losses from predation, fishing, and other losses in the following way:

$$\frac{dB_i}{dt} = PB_iB_i - \sum_i C_{ji} - Y_i - M_{oi}B_i, \tag{5.1}$$

where B_i is the group biomass, PB_i is the mass-specific production rate (production-to-biomass ratio), C_{ji} is the consumption of group i by predator group j, Y_i is fishery yields from group i, and M_{oi} is mortality from other agents not explicitly represented in the model. To derive estimates of all model components, Ecopath requires the user to specify production-to-biomass ratios, consumption rates (consumption-to-biomass ratio, Q:B), diet compositions, fishery yields, and the rate of biomass accumulation (dB_i/dt; positive or negative). In addition, the model requires an estimate of either biomass or the proportion of each group's total mortality presumed to be explicitly explained by predation and fishing losses. Independent biomass estimates for all groups were available except for the squids and zooplankton biomass pools.

Sources of all parameter estimates are given in Table 5.1, but I describe the more detailed calculations used to generate cetacean biomass here. Biomass estimates for cetaceans were based on regional stock assessments, estimates of mean body size, and estimates of the proportion of unit stocks residing within the CNP area. I assumed that one-half of the western North Pacific Bryde's whale stock resides in the CNP (Nemoto and Kawamura 1977) and that average body size was 15 tonnes (Trites and Pauly 1998). I assumed that only 10% of the North Pacific sei whale population resides in the

CNP area and that its mean body size is 17 tonnes (Trites and Pauly 1998). I used the most recent estimates of stock abundance (Bryde's whale = 26,000 [IWC 1997]; sei whale = 10,000 [Tillman 1997]), as provided by the International Whaling Commission (IWC), and assumed that baleen biomass was increasing by 0.8% per year based on IWC assessments of Bryde's whales (IWC 1997).

Sperm whale biomass in the CNP was calculated from Whitehead (2002), who indicated that total Indo-Pacific sperm whale abundance decreased from 594,000 to 216,000 between 1930 and 1980. If 75% of the Western Pacific, 5% of the Eastern Temperate Pacific, and 100% of the Hawai'i stocks are within the CNP region, then 31% of the total Indo-Pacific population is located in the CNP region. I therefore multiplied this proportion by the historical time series of Indo-Pacific sperm whale abundances provided by Whitehead (personal communication). I assumed an average sperm whale body mass equal to 18 tonnes (Trites and Pauly 1998). Given this information and the Whitehead (2002) assessments, this analysis indicated an average rate of increase of 0.094 kg km^{-2} yr^{-1} during the 1990s as estimated by linear regression.

The food habits data were based largely on Cox et al. (2002a), but with additional information included to describe cetacean diets. Sperm whale diets were taken from Flinn et al. (2002), and baleen whale diets were taken from Nemoto and Kamamura (1977). Baleen whales, especially Bryde's whales, are much more piscivorous in the CNP than they are in more northerly locations, so diet matrix parameters reflect higher consumption of epipelagic and mesopelagic fish than is indicated in other reports (Pauly et al. 1998).

The last parameter to specify was the proportion of the total mortality believed to be explicitly represented by the model (P_z; this is also referred to as Ecotrophic Efficiency in Ecopath documentation, but this terminology potentially misleads a reader into believing that it represents an ecosystem property). P_z values are needed only for those groups that do not have an independent biomass estimate (zooplankton and squids). I specified P_z for both zooplankton groups to equal 0.9 (a standard default assumption), but for squid groups I considered a range of values of P_z from 0.65 (the smallest value possible that produced an energetically balanced model) to 0.9. Because the qualitative results did not depend on P_z, I used the lowest possible value (0.65) for squids, which is a conservative estimate of the amount of squid turnover explained by the specified predator-prey interactions. For all groups, the portion of total mortality that is not accounted for ($1 - P_z$) is treated as a time-invariant, density-independent mortality term in the dynamic simulations.

Historical Food Web Structure

I used this "snapshot" of the CNP food web to initialize a dynamic biomass model (Ecosim) that simulated changes in food web structure accompanying a cessation of fishing and recovery of whale populations. This simulation was then "tuned" to produce steady-state biomasses that resembled estimated pre-exploitation biomass for apex predators and cetaceans. Historical population biomasses for fish apex predators (all scombrids, billfish, and sharks) were taken from Cox et al. (2002a,b), but some biomass estimates were modified downward based on more recent assessments (see Table 5.1). Historical whale biomass estimates were taken from the estimated North Pacific carrying capacity for baleen cetaceans (Bryde's whales = 40,000 [IWC 1997]; sei whales = 42,000 [Tillman 1977]), and from Whitehead (2002) for sperm whales (50,150).

The dynamic food web was based on the Ecosim model (Walters et al. 1997). Essentially, Ecosim extends the base Ecopath model (Eq. 5.1) by making consumption rates dynamic according to a specified functional response model, so that each group's production rate and predation mortality rate are varied depending on food and predator abundance:

$$\frac{dB_i}{dt} = GCE \sum_j C_{ij} - \sum_j C_{ji} - M_{oi}B_i - F_iB_i, \qquad (5.2)$$

where GCE is the gross conversion efficiency ($GCE = PB/QB$), F_i is the fishing mortality rate, and all other parameters are the same as previously described. Details as to how parameters are calculated from Ecopath initialization state can be found in Cox et al. (2002a).

The consumption rates (C) follow from a predator-dependent functional response relationship, whereby the degree of predator dependence is specified by the user before simulations are run. The functional response equation is described in detail in Cox et al. (2002a) and Walters et al. (1997) and is derived based on notions regarding risk-sensitive foraging and the flow of prey between safe and vulnerable habitats. Assuming a first-order process dictating prey movement between habitat types with a rate parameter denoted v, and assuming that predators attack vulnerable prey at a per-biomass rate of a, then the per-biomass rate of consumption by predator i on prey type j (C_{ij}/B_i) equals

$$\frac{C_{ij}}{B_i} = \frac{a_{ij}v_{ij}B_j}{2v_{ij} + a_{ij}B_i} \qquad (5.3)$$

An important attribute of this functional response is that, depending on the attack rate and the vulnerability exchange parameters, the model exhibits either weak predator dependence (in which per-capita consumption does not depend on predator abundance) or strong predator dependence (per-capita consumption declines with predator abundance). The latter induces density dependence into Eq. 5.2 and makes predation mortality induced by predator i on prey j ($M_{p,ij}$) insensitive to changes in predator biomass. Functional response parameters a_{ij} and v_{ij} are calculated from the initial Ecopath estimates of B_j, B_i and C_{ij}, and from a user-specified value describing the degree of predator dependence in the functional response. This scaling parameter, V^*, describes the sensitivity of $M_{p,ij}$ to predator biomass at the Ecopath biomass estimate compared to when predator biomass is very low (i.e., when there would be little predator dependence). Thus, V^* scales between 0 and 1, where 0 indicates strong density dependence and 1 indicates no predator dependence in consumption rates. The tuning exercise (see "Results") was conducted by iteratively adjusting the magnitude of predator dependence until the unfished equilibrium matched estimates of historical population size.

Systems-level properties of the historical and present-day food web were compared by calculating the primary production required (PPR) to support each functional group in each food web configuration. This calculation essentially traces back through all of the trophic pathways linking each functional group to primary producers and accounts for energy lost via assimilation and respiration along these pathways. I used an algorithm that was modified from the one employed in the Ecopath software in the following ways: I included cannibalism as viable pathway of energy flow (which can be substantial for certain groups, such as squid), and I did not consider the unexplained mortality term (M_o) to be an energy loss term analogous to respiration and assimilation costs. PPR is a useful index of the changes in individual functional groups, and these numbers can be easily compared across food web configurations. However, it is difficult to interpret this metric as a systemwide property that provides meaningful measures of energy flow in the food web. This is simply because some portion of the PPR for prey groups is also "counted" in the PPR for predator groups that feed upon them. To avoid this "multiple-counting" problem, I also calculated the net PPR, which subtracts the fraction of each group's PPR that is ultimately consumed by a higher trophic level. In this way, it is possible to evaluate systemwide shifts in the distribution and fate of primary production in the food web.

Results

Sperm whales have the highest trophic level in the contemporary food web (trophic level = 4.7), followed by the two teuthivorous fish groups (mesopelagic teuthivores and teuthivorous elasmobranchs). The trophic level of piscivorous elasmobranchs closely follows those groups, followed by mesopelagic piscivores, large squid, and large epipelagic piscivores. If apex predators are defined as those groups whose trophic level exceeds 4, the total apex predator biomass presently in the system is 0.23 g m^{-2}. The trophic level of the high-seas fisheries catch is quite high (trophic level = 4.2), especially compared to continental shelf–based fisheries, for which trophic level of catches tend to be around 3 (Essington, unpublished analysis).

The best available estimates of historical and contemporary abundance of apex predators and whales indicate a pronounced decline in the abundance of exploited functional groups (Figure 5.2). Six of the eight functional groups have been reduced to less than 60% of their historical biomass. Elasmobranch groups showed the most pronounced declines,

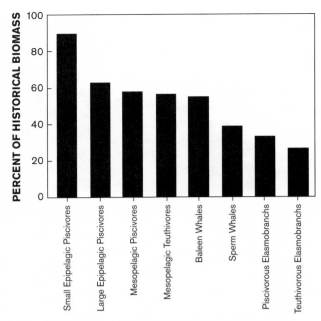

FIGURE 5.2. Ratio of contemporary to historical biomasses for functional groups exploited in commercial fishing or whaling operations.

with contemporary biomass roughly 30% of pre-exploitation levels. The small epipelagic piscivores showed a more modest decline, around 80% of historical biomass. Skipjack tuna make up the vast majority of the biomass of this functional group.

I simulated the equilibrium biomasses of all trophic groups in the absence of fishing by setting all fishing mortalities to zero and iteratively adjusting the functional response "predator dependence" parameters so that the model predicted the historical population biomasses of exploited fishes and cetaceans. I used a default value of $V^* = 0.3$ for most predator-prey linkages except for mesopelagic teuthivores ($V^* = 0.05$), large epipelagic piscivores ($V^* = 0.75$), baleen whales ($V^* = 0.85$), and sperm whales ($V^* = 0.7$). In other words, it was necessary to strengthen density dependence acting on feeding rates of mesopelagic teuthivores to prevent them from "overshooting" the historical biomass. Conversely, it was necessary to weaken this density dependence on large epipelagic piscivores and baleen and sperm whales to allow them to recover to historical abundances.

Visual representation of the changes in food web structure associated with the onset of commercial whaling and fishing is presented in Figure 5.3. Although the model analysis does not predict wholesale restructuring or radically altered energy flow associated with the onset of commercial fishing and whaling operations, it predicts some noteworthy changes in food web structure. The most noteworthy is a general depletion of high-trophic-level functional groups. This depletion induced a top-down release of predation on large squid and epipelagic fishes, resulting in slightly increased biomass of both groups (Figure 5.3). The shift in large squid biomass is largely due to the reduction in sperm whales, whereas the

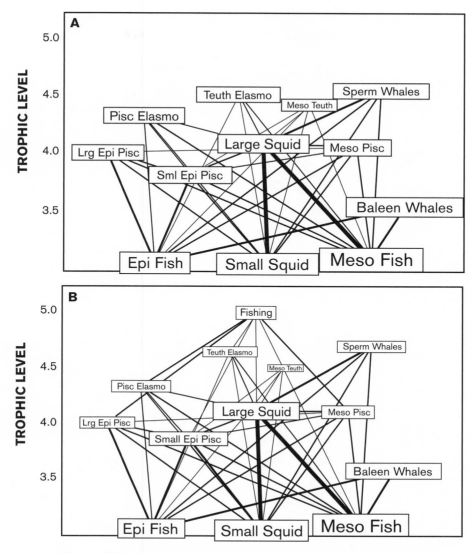

FIGURE 5.3. Historical (A) and contemporary (B) food web structure. Size of the font is proportional to ln(biomass). Only the upper food web (trophic level >3) is shown.

increase in epipelagic fishes is due to the reduced abundance of small epipelagic piscivores and baleen whales. The model, however, does not predict any substantial changes among lower-trophic-level groups. This lack of change is partly due to the small direct effects of commercially exploited stocks on these trophic levels. For example, a fully restored baleen whale functional group is predicted to account for only a small percentage (2%) of the total mortality rate on macrozooplankton and only 10% of the total mortality of mesopelagic fishes.

The total biomass of apex predators (trophic level >4) was only marginally greater in the historical state than in the contemporary state. However, the composition of the apex predator group was considerably different (Figure 5.4). Specifically, large squid constitute over 75% of the biomass of the contemporary apex predator guild but contributed less than 50% of the historical apex predator biomass. In other words,

the food web has shifted from an apex predator guild composed of relatively long-lived organisms with moderate productivities to one of short-lived organisms with high productivities.

This model indicates the potentially important role of large squid in the food web. Large squid account for 36% of the predation mortality on mesopelagic fishes and 28% of the mortality of small squid. Further, 10% of all the fishery landings are supported by predation on large squid. That is, 10% of all landed biomass at one point resided in the "large squid" biomass pool. The model predicts that the historical controls on large squid dynamics may have been different from those operating today. In the contemporary food web, predation mortality on large squid predominantly consists of cannibalism (38% of total) and sperm whale predation (36% of total). Historically, sperm whale predation was the dominant source of predation mortality (54% of total), with

TABLE 5.2
Contemporary Diet Composition (% by Mass)

Prey \ Predators	Macro-zooplankton	Micro-zooplankton	Mesopelagic Fishes	Epipelagic Fishes	Small Squid	Large Squid	Small Epipelagic Piscivores	Large Epipelagic Piscivores	Mesopelagic Teuthivores	Mesopelagic Piscivores	Piscivorous Elasmobranchs	Teuthivorous Elasmobranchs	Baleen Whales	Sperm Whales
Primary producers	0.9	0.995	–	–	–	–	–	–	–	–	–	–	–	–
Macro zooplankton	–	–	0.475	0.7	0.5	–	0.25	0.01	–	0.15	–	–	0.5	–
Micro zooplankton	0.1	0.005	0.475	0.3	0.5	–	–	–	–	–	–	–	–	–
Mesopelagic fishes	–	–	0.05	–	–	0.5	0.1	0.1	0.2	0.25	0.4	0.25	0.25	0.1
Epipelagic fishes	–	–	–	–	–	–	0.45	0.63	0.15	0.2	0.2	0.1	0.25	0.1
Small squid	–	–	–	–	–	0.45	0.15	0.11	0.2	0.1	0.1	0.1	–	0.2
Large squid	–	–	–	–	–	0.05	0.05	–	0.3	0.15	0.1	0.55	–	0.6
Small epipelagic piscivores	–	–	–	–	–	–	–	0.06	0.15	0.05	0.2	–	–	–
Large epipelagic piscivores	–	–	–	–	–	–	–	0.001	–	–	–	–	–	–
Mesopelagic teuthivores	–	–	–	–	–	–	–	–	–	–	–	–	–	–
Mesopelagic piscivores	–	–	–	–	–	–	–	–	–	–	–	–	–	–
Piscivorous elasmobranchs	–	–	–	–	–	–	–	–	–	–	–	–	–	–
Teuthivorous elasmobranchs	–	–	–	–	–	–	–	–	–	–	–	–	–	–

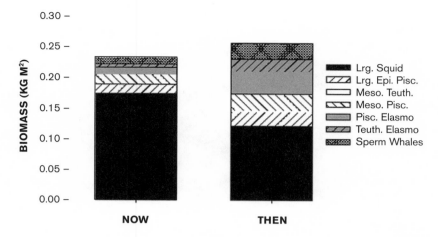

FIGURE 5.4. Biomass of functional groups with trophic level >4 in contemporary (now) and historical (then) food webs.

cannibalism only representing 22% of the total predation losses.

Analysis of the net primary production required to support each trophic group revealed compensatory food web responses that permitted the recovery of apex predator guild to historical abundances (Table 5.3). In the contemporary food web, 161 g m^{-2} yr^{-1} of primary production is required to support the depleted baleen whale, sperm whale, teuthivore, and piscivore functional groups. Of this, 105 g m^{-2} yr^{-1} is required to support apex (trophic level >4) predators (Table 5.3). This is only a small fraction of the estimated 700 g m^{-2} yr^{-1} of total primary production (Longhurst et al. 1995). In the restored food web, the exploited groups require 196 g m^{-2} yr^{-1}, and of this 140 g m^{-2} yr^{-1} is required to support the apex predator guild. These increases in net PPR are modest compared to the overall, ca. twofold increase in biomass associated with the recovery of these groups. The discordance between biomass and net PPR shifts partly reflects reductions in the mass-specific productivities of recovered functional groups and shifts in feeding that reduced the trophic level of these groups by as much as 0.13. Accompanying these shifts was enhanced trophic coupling between apex predators and their prey groups in the historic food web configuration, as evidenced by a reduction in the net PPR required to support squids and epi- and mesopelagic fishes (Table 5.3). This reduction indicates that a much higher proportion of energy reaching prey groups was ultimately sequestered by predator groups.

Discussion

Based on information on the contemporary food web structure, the intensity of fishing and whaling, and best estimates of historical abundances of exploited species, the model suggests that fishing and whaling have not had the wholesale restructuring influence in this blue-water ecosystem as seen in coastal ecosystems, estuaries, and semienclosed seas. This conclusion, however, is highly contingent on the quality and quantity of information that is presently available. Sadly, there is dearth of information on details of trophic interactions for any species other than those that are commercially exploited. Only a handful of studies have documented food habits of meso- and epipelagic fishes (Clarke 1978; Seki and Bigelow 1993; Moku et al. 2000), small squids (Rodhouse and Nigmatullin 1996) and large squids (Seki 1993). Given the limited information that is presently available, the model results need to be interpreted cautiously. In the following paragraphs, I describe some of the issues that need to be addressed in evaluating models of this sort.

The model necessarily aggregated numerous species into distinct functional groups. As an example, I aggregated over thirty different species of squid into two groups: small and large squid. By aggregating them in this way, there is an implicit assumption that functional redundancy or complimentarity is built into those functional groups, so that the buildup or decline of any one species is partially offset with compensatory changes in abundance of others. Punt and Butterworth (1995) demonstrated that model predictions can be highly sensitive to "aggregation errors," particularly when the aggregated species have unique functional roles in the food web. Yet most strong top-down effects are contingent on the particular interactions of species within and across functional groups. For example, the classic example of cascading trophic interactions in lakes (Carpenter et al. 2001) is contingent on size-selective feeding of zooplanktivorous fishes that remove those zooplankton that have the highest grazing rates. Given that whales (Whitehead et al. 2003) and fish selectively feed on particular squid and fish species, it is quite likely that certain species aggregated within the functional groups are more heavily preyed upon than others. Moreover, squid are highly teuthivorous themselves, often showing high rates of cannibalism, introducing even more complexity in the form of

TABLE 5.3

Net Primary Production Required (nPPR) to Support Functional Groups

Functional Group	Contemporary nPPR (g m^{-2} yr^{-1})	Historical nPPR (g m^{-2} yr^{-1})	Historical nPPR (g m^{-2} yr^{-1})
Macrozooplankton	33.9	34.3	34.3
Microzooplankton	36.9	36.6	36.6
Mesopelagic fishes	103.3	108.0	108.0
Epipelagic fishes	71.5	53.2	53.2
Small squid	36.6	43.7	43.7
Large squid	63.9	37.7	37.7
Small epipelagic piscivores	27.4	11.4	27.4
Large epipelagic piscivores	22.3	30.1	168.6
Mesopelagic teuthivores	1.4	1.2	4.5
Mesopelagic piscivores	20.6	21.7	120.3
Piscivorous elasmobranchs	9.8	14.4	47.0
Teuthivorous elasmobranchs	4.9	5.3	13.9
Baleen whales	28.5	45.0	45.0
Sperm whales	46.2	67.3	67.3
PPR all functional groups	507	510	791
PPR exploited groups	161	196	478

[1]Cox et al. 2002b.
[2]Myers and Worm 2003.

strong predator-prey interactions within the functional groups defined here.

Put in this light, the results of this modeling effort are not that surprising: Assuming functional group compensatory responses, we do not expect wholesale restructuring of the food web as a consequence of commercial whaling and fishing. That caveat might make the model results appear trivial, but that conclusion would not be correct either. That the model does not predict large-scale changes in patterns of energy and matter is an important conclusion that distinguishes this ecosystem from more heavily impacted coastal and estuarine ecosystems. For example, commercial exploitation of oysters in Chesapeake Bay led to profound changes in ecosystem processes because these species remove substantial volumes of organic matter from the water column (Jackson et al. 2001). That is, the effect of oyster harvest in Chesapeake Bay does not heavily depend on the species composition of the phytoplankton consumed by oysters, because oysters' main functional role is to transfer vast amounts of particulate organic matter from the water column to the benthos (Newell 1988). The modeling results presented here lead to a fairly robust conclusion that this type of restructuring of ecosystem processes probably has not occurred in the central North Pacific.

Another limitation to interpreting biomass-dynamic models such as the one used here is that that strong interaction strengths in food webs rarely follow from strong energetic links (Paine 1980). The lack of coherence between interaction and energetic links can be due to variable predator dependence in the functional response, behaviorally

induced effects, or nonconsumptive functions of animals (e.g. ecosystem engineers). Indeed, predicting interaction strengths from knowledge of food habitats and consumption rates is exceptionally difficult, and some have argued that they are essentially impossible to predict without a detailed understanding of the entire food web (Bax 1998; Yodzis 2001). In contrast, I have suggested elsewhere (Essington 2004) that models that are parameterized to be consistent with past food web behavior are much more robust to parameter and structural uncertainty. Regardless, modeling analyses of food web dynamics of the types performed here are best viewed as attempts to explore the potential for dramatic alterations in flux of energy and matter in the food web and to generate hypotheses regarding which food web linkages might be important in producing these alterations.

One emerging hypothesis from this modeling exercise is that the composition of the apex predator guild has shifted from large, widely dispersing, relatively unproductive, long-lived vertebrates (fishes and whales) to smaller, short-lived, and highly productive squids. This shift could potentially induce greater variability in top-down controls in this food web. Whereas sperm whale population dynamics show variability along decadal scales, squid population dynamics are much more likely to show striking interannual variability (Dawe and Warren 1990; Dawe et al. 2000; Hatfield 2000; Perez et al. 2002; Pierce and Boyle 2003). Appolonio (2002) has stressed that long-lived species in ecosystems perform functions that cannot be readily replaced by more dynamic species. By virtue of their long generation times, these species constrain the dynamics of ecosystem components at lower hierarchical levels and provide ecological "memory" to the system. Further, because these animals are capable of searching widely for patches of abundant prey, they can rapidly locate and exploit these patches and thereby reduce temporal and spatial variations in ecosystem structure. Thus, the simplistic perspective that looks only at the *quantity* of apex predators might overlook the complexities that arise through the *composition* of apex predators.

One obvious policy concern is whether a food web containing fully restored cetacean populations will be less productive for fishes currently exploited by commercial fishing fleets. Potential competition between marine mammals and fisheries is a concern in many fisheries, although DeMaster et al. (2001) concluded that the most immediate threat was the direct exploitation of food items for marine mammals. In the central Pacific, however, the conclusions of DeMaster et al. (2001) have less bearing, because fisheries predominantly exploit apex predator fishes, and these species are not important prey for whales. Rather, competitive interactions in the central Pacific may be more likely to reflect those described by Trites et al. (1997), who described this type of interaction as "food web competition" in which cetaceans could reduce the productivity of tunas and billfishes by shunting production away from those groups. Whether or not this is a concern is not yet clear, and the answer will depend on the degree to which these fishes and cetaceans share similar prey items. On a coarse scale, there is overlapping consumption on mesopelagic fishes and small and large squids, but on a finer, species-specific scale these diet overlaps could conceivably disappear. Clarke and Young (1998) found that sperm whale diets near Hawai'i consisted in order of importance: *Histioteuthis hoylei*, *Ommastrephes bartrami*, and *Architeuthis* sp. In comparison, fishes tend to feed primarily on *O. bartrami* but also on Onychoteuthidae and Enoploteuthidae (King and Ikehara 1956; Seki 1993).

Recently, Myers and Worm (2003) found that longline catch rates have declined by nearly 90% in the Pacific and other oceans, and they concluded from this that the apex predator guild has been reduced to 10% of its historic state. These results have been challenged on the basis of inappropriate analysis of spatially complex data (Walters 2003) and of the inconsistency of that conclusion with time series of total fisheries landings, which have increased severalfold since the initial rapid decline in longline catch rate (Figure 5.1; Hilborn 2004). Detailed stock assessments that use both fisheries-dependent and fisheries-independent (tagging) data suggest more modest declines in abundance (Table 5.1; Maunder and Watters 2001; Watters and Maunder 2002).

Can a food web model and simple principles of ecosystem energetics be useful in evaluating whether it is energetically plausible for apex predator biomass to be tenfold higher than it is now? The energetic constraint on the system is that there is a finite rate of primary production available to support these fish and marine mammal stocks. The best available estimate of primary production rate is 700 g m^{-2} yr^{-1} (Longhurst et al. 1995), yet some fraction of this is lost to sedimentation, detritus, and other exports. If there were no further changes in mass-specific productivities, trophic level, or trophic coupling between apex predators and their prey from that represented in the "historic" state in Table 5.3, then the additional pre-exploitation biomass of these groups posited by Myers and Worm (2003) would require 478 g m^{-2} yr^{-1}, and the entire food web would require 791 g m^{-2} yr^{-1}, a value exceeding available production by roughly 25%. This estimate is clearly within the range of variability and uncertainty in total primary production, but it suggests a very different food web from the one that we see presently, with the majority of primary production ultimately ending up in the apex predator guild.

The network analysis illustrates that there are three kinds of shifts that could permit a superabundant apex predator biomass as hypothesized by Myers and Worm: reduced mass-specific production rates, reduced trophic level of apex predators, and enhanced trophic coupling. The first is not only reasonable, but expected: The central tenet of fisheries population biology is that fish stocks undergo compensatory shifts in per-capita production at low stock sizes that permit exploitation. The unfished biomass, especially the high biomass suggested under the Myers and Worm hypothesis, must have had extremely low productivities compared to what is presently seen in these populations. The second mechanism, reduced trophic level, is difficult to evaluate. On one hand, a superabundant apex predator guild would certainly produce large depletions in large squid and other intermediate-trophic-level functional groups, and this should lead to shifts in feeding toward more abundant and productive low-trophic-level fishes. Indeed, the Ecosim model predicted just this kind of shift in reconstructing a more modest historical biomass of apex predators. However, the biomass dynamic model used here does not account for changes in the size-structure of the exploited population and therefore cannot depict changes in feeding and trophic position that would accompany an unfished population with many large-bodied fishes. These shifts can be considerable (Badalamenti et al. 2002) and could easily overwhelm any compensatory shifts in feeding that would accompany altered prey availability. The last mechanism, enhanced coupling between trophic levels, is eminently possible and suggests a highly coupled food web in pre-exploitation periods, wherein the vast majority of lower-trophic-level production is funneled into apex predator biomass.

In summary, this model analysis has suggested the possibility of compensatory shifts in food web structure in response to industrial fishing and whaling. This shift is characterized by a transition from an apex predator guild dominated by sperm whales, tunas, and billfishes to one dominated by smaller tuna species and large squid. Yet, overall, the extent of human influence on this ecosystem appears to be somewhat modest, at least compared to many nearshore marine ecosystems. Information on the details of species interactions in this food web is sparse, however, so the possibility that profound changes in food web structure accompanied fishing and whaling cannot be disregarded.

Acknowledgments

I thank the organizers of this symposium for inviting my participation. I also thank Hal Whitehead for kindly providing his estimates of sperm whale abundances, and I thank George Watters, Jim Estes, Terrie Williams, and an anonymous reviewer whose comments improved this manuscript. This work was funded by the National Science Foundation Award # 0220941 and New York Sea Grant.

Literature Cited

Apollonio, S. 2002. *Hierarchical perspectives on marine communities.* New York: Columbia University Press.

Badalamenti, F., G. D'Anna, J. K. Pinnegar, and N. V. C. Polunin. 2002. Size-related trophodynamic changes in three target fish species recovering from intensive trawling. *Marine Biology* 141: 561–570.

Bax, N. J. 1998. The significance and prediction of predation in marine fisheries. *ICES Journal of Marine Science* 55: 997–1030.

Caddy, J. F. 1993. Toward a comparative evaluation of human impacts on fishery ecosystems of enclosed and semi-enclosed seas. *Reviews in Fish Biology and Fisheries* 1: 57–95.

Carpenter, S. R., J. J. Cole, J. R. Hodgson, J. F. Kitchell, M. L. Pace, D. Bade, K. L. Cottingham, T. E. Essington, J. N. Houser, and D. E. Schindler. 2001. Trophic cascades, nutrients, and lake productivity: whole-lake experiments. *Ecological Monographs* 71: 163–186.

Clarke, M. and R. Young. 1998. Description and analysis of cephalopod beaks from stomachs of six species of odontocete cetaceans stranded on Hawaiian shores. *Journal of the Marine Biological Association of the United Kingdom* 78: 623–641.

Clarke, T. A. 1978. Diel feeding patterns of 16 species of mesopelagic fishes from Hawaiian waters. *Fishery Bulletin* 76: 495–513.

Cox, S. P., T. E. Essington, J. F. Kitchell, S. J. D. Martell, C. J. Walters, C. Boggs, and I. Kaplan. 2002a. Reconstructing ecosystem dynamics in the central Pacific Ocean, 1952–1998. II. The trophic impacts of fishing and effects on tuna dynamics. *Canadian Journal of Fisheries and Aquatic Sciences* 59: 1736–1747.

Cox, S. P., S. J. D. Martell, C. Walters, T. E. Essington, J. F. Kitchell, C. Boggs, and I. Kaplan. 2002b. Reconstructing ecosystem dynamics in the central Pacific Ocean, 1952–1998. I. Estimating population biomass and recruitment of tunas and billfishes. *Canadian Journal of Fisheries and Aquatic Sciences* 59: 1724–1735.

Dawe, E. G., E. B. Colbourne, and K. F. Drinkwater. 2000. Environmental effects on recruitment of short-finned squid (*Illex illecebrosus*). *ICES Journal of Marine Science* 57: 1002–1013.

Dawe, E. G. and W. G. Warren. 1990. Recruitment of short-finned squid in the northwest Atlantic Ocean and some environmental relationships. *Journal of Cephalopod Biology* 2: 1–21.

DeMaster, D. P., C. W. Fowler, S. L. Perry, and M. E. Richlen. 2001. Predation and competition: the impact of fisheries on marine-mammal populations over the next one hundred years. *Journal of Mammalogy* 82: 641–651.

Essington, T. E. 2004. Getting the right answer from the wrong model: evaluating the sensitivity of multispecies fisheries advice to uncertain species interactions. *Bulletin of Marine Science* 74: 563–581.

Flinn, R. D., A. W. Trites, E. J. Gregr, and R. I. Perry. 2002. Diets of fin, sei, and sperm whales in British Columbia: an analysis of commercial whaling records, 1963–1967. *Marine Mammal Science* 18: 663–679.

Fogarty, M. J. and S. A. Murawski. 1998. Large-scale disturbance and the structure of marine system: fishery impacts on Georges Bank. *Ecological Applications* 8 (Supplement): S6–S22.

Hampton, J. 2000. Natural mortality rates in tunas: Size really does matter. *Canadian Journal of Fisheries and Aquatic Sciences* 57: 1002–1010.

Hampton, J. and P. Kleiber. 2003. *Stock assessment of yellowfin tuna in the western and central Pacific Ocean.* Standing Committee on Tuna and Billfish, 16th Meeting, Working Paper YFT-1.

Hampton, J., P. Kleiber, Y. Takeuchi, H. Kurota, and M.N. Maunder. 2003. *Stock assessment of bigeye tuna in the western and central Pacific Ocean, with comparisons to the entire Pacific Ocean.* Standing Committee on Tuna and Billfish, 16th Meeting, Working Paper BET-1.

Hatfield, E. M. C. 2000. Do some like it hot? Temperature as a possible determinant of variability in the growth of the Patagonian squid, *Loligo gahi* (Cephalopoda: Loliginidae). *Fisheries Research* 47: 27–40.

Hilborn, R. 2004. Ecosystem-based management: the carrot or the stick? *Marine Ecology Progress Series* 274: 275–278.

Hoenig, J. M. 1983. Empirical use of longevity data to estimate mortality rates. *Fishery Bulletin* 81: 898–903.

IWC. 1997. Report of the Scientific Committee, Annex G. Report of the sub-committee on North Pacific Bryde's whales. *Report of the International Whaling Commission* 47: 163–168.

Jackson, J. B. C., M. X. Kirby, W. H. Berger, K. A. Bjorndal, L. W. Botsford, B. J. Bourque, R. H. Bradbury, R. Cooke, J. Erlandson, J. A. Estes, T. P. Hughes, S. Kidwell, C. B. Lange, H. S. Lenihan, J. M. Pondolfi, C. H. Peterson, R. S. Steneck, M. J. Tegner, and R. R. Warner. 2001. Historical overfishing and the recent collapse of coastal ecosystems. *Science* 293: 629–638.

Kenney, R. D., G. P. Scott, T. J. Thompson, and H. E. Winn. 1997. Estimates of prey consumption and trophic impacts of cetaceans in the USA Northeast continental shelf ecosystem. *Journal of Northwest Atlantic Fishery Science* 22: 155–171.

Kiddon, J. A., J. F. Paul, H. W. Buffum, C. S. Strobel, S. S. Hale, D. Cobb, and B. S. Brown. 2003. Ecological condition of US Mid-Atlantic estuaries, 1997–1998. *Marine Pollution Bulletin* 46: 1224–1244.

King, J. E. and I. I. Ikehara. 1956. Comparative study of food of bigeye and yellowfin tuna in the central Pacific. *Fishery Bulletin* 57: 61–85.

Langley, A., M. Ogura, and J. Hampton. 2003. *Stock assessment of skipjack tuna in the western and central Pacific Ocean.* Standing Committee on Tuna and Billfish, 16th Meeting, Working Paper SKJ-1.

Longhurst, A., S. Sathyendranath, T. Platt, and C. Caverhill. 1995. An estimate of global primary production in the ocean from satellite radiometer data. *Journal of Plankton Research* 17: 1245–1271.

Mann, M. H. 1984. Fish production in open ocean systems, in *Flows of energy and materials in marine ecosystems.* M. R. R. Fasham, ed. New York: Plenum Press pp. 435–458.

Maunder, M. N. and G. M. Watters. 2001. Status of yellowfin tuna in the eastern Pacific Ocean. *Inter-American Tropical Tuna Commission, Stock Assessment Report* 1: 5–86.

McCanahan, T. R., M. Nugues, and S. Mwachireya. 1994. Fish and sea-urchin herbivory and competition in Kenyan coral-reef lagoons: the role of reef management. *Journal of Experimental Marine Biology and Ecology* 184: 237–254.

Moku, M., K. Kawaguchi, H. Watanabe, and A. Ohno. 2000. Feeding habits of three dominant myctophid fishes, *Diaphus theta, Stenobrachius leucopsarus*, and *S. nannochir*, in the subarctic and transitional waters of the western North Pacific. *Marine Ecology Progress Series* 207: 129–140.

Myers, R. A. and B. Worm. 2003. Rapid worldwide depletion of predatory fish communities. *Nature* 423: 280–283.

Nemoto, T. and A. Kawamura. 1977. Characteristics and distribution of baleen whales with special reference to the abundance of North Pacific sei and Bryde's whales. *Report of the International Whaling Commission* Special Issue 1: 80–87.

Newell, R. I. E. 1988. Ecological changes in Chesapeake Bay: Are they the result of overharvesting the Eastern oyster (*Crassostrea virginica*)? in *Understanding the estuary: advances in Chesapeake Bay research*. E. C. Krome, ed. Gloucester Point, VA: Chesapeake Research Consortium, pp. 536–546.

Paine, R. T. 1980. Food webs: linkage, interaction strength and community infrastructure. *Journal of Animal Ecology* 49: 667–685.

Pauly, D., V. Christensen, and C. Walters. 2000. Ecopath, Ecosim, and Ecospace as tools for evaluating ecosystem impact of fisheries. *ICES Journal of Marine Science* 57: 697–706.

Pauly, D., A. W. Trites, E. Capuli, and V. Christensen. 1998. Diet composition and trophic levels of marine mammals. *ICES Journal of Marine Science* 55: 467–481.

Perez, J. A. A., D. C. de Aguiar, and U. C. Oliveira. 2002. Biology and population dynamics of the long-finned squid *Loligo plei* (Cephalopoda: Loliginidae) in southern Brazilian waters. *Fisheries Research* 58: 267–279.

Peterson, G. D. 2000. Political ecology and ecological resilience: an integration of human and ecological dynamics. *Ecological Economics* 35: 323–336.

Peterson, I. and J. S. Wroblewski. 1984. Mortality rate of fishes in the pelagic ecosystem. *Canadian Journal of Fisheries and Aquatic Sciences* 41: 1117–1120.

Pierce, G. J. and P. R. Boyle. 2003. Empirical modelling of inter-annual trends in abundance of squid (*Loligo forbesi*) in Scottish waters. *Fisheries Research* 59: 305–326.

Punt, A. E. and D. S. Butterworth. 1995. The effects of future consumption by the Cape fur seal on catches and catch rates of the Cape hakes. 4. Modelling the biological interaction between Cape fur seals *Arctocephalus pusillus pusillus* and the Cape hakes *Merluccius capensis* and *M. paradoxus. South African Journal of Marine Science* 16: 255–285.

Read, A. J. and Wade, P. R. 2000. Status of marine mammals in the United States. *Conservation Biology* 14: 929–940.

Rodhouse, P. G. and C. M. Nigmatullin. 1996. Role as consumers. *Philosophical Transactions of the Royal Society of London Series B–Biological Sciences* 351: 1003–1022.

Santos, M. B., M. R. Clarke, and G. J. Pierce. 2001. Assessing the importance of cephalopods in the diets of marine mammals and other top predators: problems and solutions. *Fisheries Research* 52: 121–139.

Scheffer, M., S. Carpenter, J. A. Foley, C. Folke, and B. Walker. 2001. Catastrophic shifts in ecosystems. *Nature* 413: 591–596.

Seki, M. P. 1993. The role of the neon flying squid, *Ommastrephes bartrami*, in the North Pacific pelagic food web. *Bulletin of the North Pacific Commission* 53: 207–215.

Seki, M. P. and K. A. Bigelow. 1993. Aspects of the life history and ecology of the Pacific pomfret *Brama japonica* during winter occupation of the subtropical frontal zone. *Bulletin of the North Pacific Commission* 53: 273–583.

Tillman, M. F. 1997. Estimates of population size for the North Pacific sei whale. *Report of the International Whaling Commission* (Special Issue) 1: 98–106.

Trites, A. W., V. Christensen, and D. Pauly. 1997. Competition between fisheries and marine mammals for prey and primary production in the Pacific Ocean. *Journal of Northwest Atlantic Fishery Science* 22: 173–187.

Trites, A. W. and D. Pauly. 1998. Estimating mean body masses of marine mammals from maximum body lengths. *Canadian Journal of Zoology* 76: 886–896.

Vitousek, P. M., H. A. Mooney, J. Lubchenco, and J. M. Melillo. 1997. Human domination of Earth's ecosystems. *Science* 277: 494–499.

Walters, C. 2003. Folly and fantasy in the analysis of spatial catch rate data. *Canadian Journal of Fisheries and Aquatic Sciences* 60: 1433–1436.

Walters, C. J., V. Christensen, and D. Pauly. 1997. Structuring dynamic models of exploited ecosystems from trophic mass-balance assessments. *Reviews in Fish Biology and Fisheries* 7: 39–172.

Watters, G. M. and M. N. Maunder. 2002. Status of bigeye tuna in the eastern Pacific Ocean. *Inter-American Tropical Tuna Commission, Stock Assessment Report* 2: 147–246.

Whitehead, H. 2002. Estimates of the current global population size and historical trajectory for sperm whales. *Marine Ecology-Progress Series* 242: 295–304.

Whitehead, H., J. Christal, and S. Dufault. 1997. Past and distant whaling and the rapid decline of sperm whales off the Galapagos Islands. *Conservation Biology* 11: 1387–1396.

Whitehead, H., C. D. MacLeod, and P. Rodhouse. 2003. Differences in niche breadth among some teuthivorous mesopelagic marine mammals. *Marine Mammal Science* 19: 400–406.

Yodzis, P. 2001. Must top predators be culled for the sake of fisheries? *Trends in Ecology and Evolution* 16: 78–84.

SIX

Evidence for Bottom-Up Control of Upper-Trophic-Level Marine Populations
Is It Scale-Dependent?

GEORGE L. HUNT, JR.

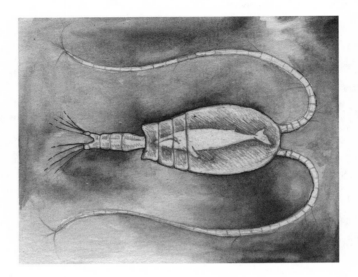

Understanding mechanisms of population control has been an area of concern since at least the time of Malthus, and early work in ecology emphasized the importance of energy flow for the trophic structure of ecosystems (Lindeman 1942; Odum and Odum 1955; Hutchinson 1959; Slobodkin 1960, 1962). Beginning in the 1960s, the focus was on assessing the importance of resource limitation (bottom-up mechanisms) and predation (top-down mechanisms) for population control and the structure of terrestrial (e.g., Hairston et al. 1960), lake (e.g., Brooks and Dodson 1965; Smith 1969; Zaret and Paine 1973), and rocky intertidal and subtidal ecosystems (e.g., Paine 1966, 1974; Estes et al. 1996). More recent work has emphasized that top-down and bottom-up mechanisms act in concert, together determining population size and ecosystem structure (Power 1992; Hunter and Price 1992; Menge 1992; Hunt and McKinnell 2006). These authors and others (Strong 1992; Hunt et al. 2002; Shurin et al. 2002; Sinclair et al. 2003) stress that a variety of factors affect the duration and strength of top-down and bottom-up effects.

Bottom-up control of a population occurs when the availability of food limits population growth. This definition does not imply that predation is not taking place; it is just not the source of population control. Species that have no predators must be limited by resources. Evidence of the importance of bottom-up processes on marine organisms includes large-scale patterns of distribution that reflect patterns of primary production or standing stocks of prey, and changes in the distribution, abundance, or condition of individuals in response to changes in primary or secondary production. For example, climate variability affects bottom-up processes by altering nutrient availability and the timing or amount of primary and secondary production. Responses to food limitation include density-dependent changes in growth, weight at age, age at maturity, productivity, stress-related physiological changes, and non-predation-related mortality. However, these density-dependent responses do not, in and of themselves, demonstrate bottom-up limitation, because predators may also be placing constraints on the population. Nevertheless, in the absence of significant predation or change in predator populations, these density-dependent changes are the best evidence that bottom-up mechanisms are limiting populations.

Top-down control occurs when the rate of predation is sufficient to limit the size of a population. In the open ocean, increases in prey populations upon the removal of a predator (e.g., by fisheries) have been taken as evidence of top-down limitation (Furness 2002; Worm and Myers 2003). Much of the progress in understanding the role of predators in controlling prey and in shaping ecosystems has resulted from studies in which the role of predation could be experimentally altered, either by removal of predators or by their introduction to the system under study. In the marine environment, such experiments have been largely confined to hard substrates in the intertidal or nearshore subtidal zone (e.g., the work of Paine). To my knowledge, no controlled experimental studies of large, pelagic marine systems have investigated the relative importance of top-down and bottom-up effects.

Two sources of perturbations of pelagic marine ecosystems—climate change (at time scales varying from the interannual to the decadal) and fisheries removals—have the potential of providing information on the responses of marine ecosystems to perturbation. However, in most cases, the marine ecosystems have been previously shorn of their upper-trophic-level components, such as large predatory fish (Pauly et al. 1998; Tegner and Dayton 1999; Jackson et al. 2001; Myers and Worm 2003) and whales (Merrick 1997; Hacquebord 1999; Moore et al. 1999; Roman and Palumbi 2003; Springer et al. 2003), thus making interpretation of change more difficult. Retrospective studies can, in some cases, provide an indication of the effects of fisheries removals (Furness 2002; Worm and Myers 2003; Jahncke et al. 2004), but for the most part we can only hypothesize what the effect may have been.

By diminishing the biomass of fish competing for prey, fisheries removals might be expected to reduce the strength of coupling between the fished species and its prey, thus diminishing the likelihood of documenting bottom-up effects of climate change. Therefore, the demonstration that climate change affects commercially fished species is taken as evidence that bottom-up processes play an important role in these populations.

For the purposes of this chapter, I use the term *top predator* to refer to those species that have no predators in the marine system, such as large piscivorous sharks, as well as killer whales *(Orcinus orca)*, skuas, falcons, eagles, and terrestrial predators (foxes, bears, hyenas, etc.) that attack marine species on land. I refer to those species that are near the top of their food web (such as large fish that have few predators as adults, marine birds, pinnipeds, and cetaceans other than killer whales) as upper-trophic-level species (or predators). Most likely all of the fish classified as upper-trophic-level predators are severely impacted by predation in their ichthyoplankton and juvenile stages. This predation on early life stages may affect the number of recruits to the population, but not the eventual biomass of the population.

The trophic role of a species may depend on its life stage. This is particularly true for marine organisms that begin life as tiny plankton that are prey for a wide variety of predators, but grow to be top predators in later life. A species may also cannibalize smaller individuals of its own species. Such cannibalism may or may not limit the population. Given the variety of trophic levels that an individual may occupy during its

lifespan, assessing the relative importance of top-down and bottom-up control can be difficult. Changes in trophic level are less an issue in marine birds, which complete most of their growth as nestlings. However, because most populations experience both predation pressure and some level of density-dependent impact from resource availability, the challenge is to assess the relative importance of these processes.

As discussed later in this contribution, the relative importance of top-down and bottom-up mechanisms may be dependent on the spatial scale at which the question is posed. Thus, it is more useful to seek evidence for situations in which top-down or bottom-up mechanisms may have their greatest impact, rather than trying to demonstrate that one or the other mechanism is controlling a population or trophic level. For example, along the coast of Namibia, seabirds nest almost exclusively on artificial islands where access of predators is prevented (top-down control) (Hunt, personal observation 2004). However, the number of birds nesting on the islands varies with the abundance of fish in nearby waters (bottom-up control). In this system, both predators and prey availability interact to determine the populations of seabirds.

The relative importance of resource availability and predation for the limitation of recruitment of commercially exploited populations is important in a management context (Hunt et al. 2002). If a population is primarily resource-limited, removal of individuals may be compensated for by more vigorous recruitment or growth of individuals. In contrast, if predation is the more important limit to the population, removal of individuals may result in intensified per capita predation and no compensatory increase in numbers of biomass.

Although disease is another form of "top-down" control, it is a separate issue from the role of predation. We know that disease can be important for both marine vertebrates and invertebrates. However, disease outbreaks often occur in conjunction with starvation or other stresses on the population, and it is difficult to untangle the effects of disease from other factors. Thus, its role will not be discussed here.

Many contributions in this volume emphasize the importance of predators, or top-down control of marine populations and the resulting impact of top predators on the composition of marine ecosystems. In assembling the volume, it was recognized that there should be some discussion of the importance of bottom-up processes in marine ecosystems. In this chapter, I examine evidence for bottom-up limitation of upper-trophic-level fish, marine birds, and marine mammals. I present evidence of the role of food limitation on population size and on individual body condition and growth, both of which can affect fecundity in fish. In marine birds and mammals, body condition similarly affects fecundity and the potential for survival to age of first breeding, as well as adult survival. I then examine briefly aspects of the life history characteristics that would argue for a weak role of predation in the regulation of populations of upper-trophic-level species in the marine environment, the spatial scales at which one might expect top-down or bottom-up control to play a more decisive role, and how the relative mobility of predator and prey affect the likelihood of top-down control.

Bottom-Up Effects on Fish Populations

During their early life stages, upper-trophic-level fish are subject to both predation and starvation. However, food availability plays a major role in their potential vulnerability to predators. One of their best defenses against predation is to grow fast and become too large for capture. Thus, at these early life stages, top-down and bottom-up influences on populations are closely linked. Later in life, age at maturity affects generation time, and size is a major determinant of fecundity. The growth rate of individuals in a population is a key factor in determining the biomass of the population, a feature of interest to fisheries. As juveniles and adults, upper-trophic-level fish are subject to predation by marine mammals and predatory fish and to a lesser extent by marine birds. Bottom-up limitation is determined by the availability of zooplankton and forage fish.

Climate and Fish Stocks

Climate variation on a variety of timescales is now widely recognized to influence fish production (Bakun and Broad 2003; Baumgartner et al. 1992). The responses of fish populations, including those whose populations have been depleted by fishing, to climate variability are most likely mediated through bottom-up processes. These processes include changes in the amount or timing of primary production, temperature-related changes in the growth and maturation of zooplankton, or the effects of temperature on the growth and survival of the fish themselves. Additionally, changes in water temperature may influence the likelihood of the co-occurrence of predators and prey. These top-down effects of climate change have been observed in the southeastern Bering Sea (Ohtani and Azumaya 1995; Wylie-Echeverria and Wooster 1998; Wespestad et al. 2000) and in the California Current system (Field et al. 2006). However, in most cases where climate shifts have been seen as influencing fish populations, the effect appears to be from the bottom up.

Scientists have long been aware of the potential for water temperature to influence growth rates and survival of juvenile fish (e.g., Javaid and Anderson 1967; Garside 1973; Dwyer and Piper 1987). However, it is only since the late 1980s that we have begun to focus on the potential for variations in climate to influence the dynamics of fish stocks and the species composition of the communities to which they belong (Lluch-Belda et al. 1989; Baumgartner et al. 1992; McFarlane and Beamish 1992; Kawasaki 1992a,b; Beamish and Bouillon 1993; Hollowed and Wooster 1992).

Evidence of past shifts in marine carrying capacity for fish is widespread (Baumgartner et al. 1992; Ishida et al. 2001; Finney et al. 2002). For instance, the 2,000-year record of

sediment scale depositions in the Santa Barbara basin shows that Pacific sardine (*Sardinops sagax*) populations have varied with an approximately 60-year cycle (Baumgartner et al. 1992), a periodicity similar to a 65–70-year global climate oscillation identified by Schlesinger and Ramankutty (1994). Sardine harvests throughout the northern and southern hemispheres fluctuate in synchrony, suggesting that they are linked by global climate events, as do harvests of anchovy (Lluch-Belda et al. 1989, 1993; Kawasaki 1992a). If the populations of these small pelagic fish were limited by predators, it seems unlikely that they would vary synchronously with climate cycles. Many commercially harvested fish species show coincident decadal-scale variation in both the Atlantic and Pacific oceans, though the fluctuations are not always in phase (Klyashtorin 1998).

In both the North Pacific and the North Atlantic, climate may force variation in zooplankton abundance, especially copepod abundance, either through production processes or through changes in mortality rates (e.g., Clark et al. 2003). Copepod abundance, in turn, influences the abundance and condition of fish. When salmon first enter the ocean, the availability of zooplankton prey influences their survival (Beamish and Mahnken 2001). Additionally, salmonids in the North Pacific have responded to climate-driven changes in zooplankton abundance in the sub-Arctic gyre (Brodeur and Ware 1992; Brodeur et al. 1996) with changes in population abundance and size and age at maturity (Beamish 1993; Hare and Francis 1995; Hinch et al. 1995; Helle and Hoffman 1995, 1998; Beamish et al. 1999). Salmon returns have varied on a decadal scale and have reflected changes in the atmospheric forcing of ocean circulation patterns and zooplankton abundance (Brodeur et al. 1996; Francis et al. 1998; Hollowed et al. 2001). It seems unlikely that changes in predation pressure played a significant role in these population-level responses of salmon.

Changes in water temperature affect the metabolic rates and food requirements of fish (e.g., Despatie et al. 2001). Sharp seasonal changes in the pelagic distribution of salmonids are likely caused by strong gradients in ocean temperature. The seasonal change in distribution indicates that salmonids (and likely other species of marine fish) must adjust their pelagic distributions to remain within regions with adequate prey densities to meet metabolic requirements (Welch et al. 1995, 1998a,b). These large spatial adjustments of foraging distributions suggest that prey concentrations are seasonally inadequate to support these fish over much of their foraging range. There is no evidence that these seasonal distribution shifts are a response to predator pressure.

Groundfish populations have also been affected by climate change (Hollowed and Wooster 1992; McFarlane and Beamish 1992; Hare and Mantua 2000; McFarlane et al. 2000). For example, in the North Pacific, environmental variability has been responsible for most of the variation in recruitment of Pacific halibut *(Hippoglossus stenolepis)*, whereas growth rates appear to be density-dependent responses to changes in stock size (Clark et al. 1999; Clark and Hare 2002). Other groundfish, including hake *(Merluccius productus)*, sablefish *(Anoplopoma fimbria)*, English sole *(Parophrys vetulus)*, Pacific cod *(Gadus macrocephalus)*, and walleye pollock *(Theragra chalcogramma)*, have shown strong shifts in population size or biomass in response to the 1976–1977 or 1989 regime shifts (Bailey, 2000; King et al. 2000; McFarlane et al. 2000). In at least some cases, these climate-forced changes in the abundance of stocks appear to result from changes in zooplankton abundance forced by shifts in water temperature (McFarlane and Beamish 1992; Coyle and Pinchuk 2002; Hunt et al. 2002; Baier and Napp 2003). Correlations between recruitment and environmental conditions may be strongest at the limits of a species' geographical range (Myers 1998), but likely occur throughout the range of a stock.

The North Atlantic Oscillation (NAO) atmospheric circulation pattern plays a pivotal role in forcing currents in the northern North Atlantic (reviewed in Parsons and Lear 2001). There, stocks of many fish species appear to be sensitive to changes in ecological conditions forced by changes in the NAO. For example, over much of this region, capelin *(Mallotus villosus)*, a small planktivorous fish, is a principal prey for Atlantic cod *(Gadus morhua)* and other predators (Carscadden and Vilhjálmsson 2002; Dolgov 2002; Ushakov and Prozorkevich 2002). The abundance of its zooplankton prey is sensitive to sea temperatures and, in the Barents Sea, to inflows of plankton-rich Atlantic water forced by the NAO. Variations in capelin growth rates are correlated with the abundance of their prey and vary inversely with the biomass of capelin present, indicating density-dependent limitation of capelin (Gjøsæter et al. 2002).

The availability of capelin also influences the condition of their predators (Nilssen et al. 1994). For example, in the Barents Sea, the condition of cod decreases rapidly when capelin stocks drop below a million tons (Yaragina and Marshall 2000); in Icelandic waters, the mean weights of age 5–8 year cod, but not other age classes, were correlated with capelin abundance (Vilhjálmsson 2002). Capelin have withdrawn from some of their traditional habitats along the coast of Labrador, and it has been suggested that cod will not recover from their recent decline there until the capelin return to this region (Rose and O'Driscoll 2002). Poor condition and low energy reserves have been suggested to have contributed to the collapse of this cod stock in the northern Gulf of St. Lawrence (Lambert and Dutil 1997). Although cod are subject to predation from seals and fisheries removals, their sensitivity to the abundance of capelin points to a strong bottom-up influence on individual condition and stock size.

Because of its commercial importance throughout the northern North Atlantic, there has been great interest in what controls recruitment (survival to age 3 or 4) and growth in cod stocks (e.g., Jakobsson et al. 1994). There is significant correlation of recruitment between stocks, and emphasis has been placed on factors, such as climate, that can act over wide areas, as important in controlling cod recruitment (Koslow et al. 1987; Cohen et al. 1991). However, the way in which climate affects recruitment seems to vary from region to region (Daan 1994). For example, cod stocks inhabiting regions of typically low temperatures respond positively to increasing temperature, whereas cod in typically warm regions have a negative response to increasing water temperature (Sundby 2000). Factors identified as important include currents that affect the drift of larvae and their retention in the stock or delivery to nursery grounds (Sinclair 1988); water temperature (Brander 1994; Nakken 1994; Ottersen et al. 1994); primary production and/or zooplankton abundance (Rothschild 1994); predation on eggs, larvae, and juveniles (Nakken 1994); cannibalism (Bogstad et al. 1994; Nakken 1994), and, of course, fishing.

Temperature and the availability of zooplankton have a strong influence on the growth and, possibly, survival of larval and juvenile cod (Brander 1994; Nakken 1994; Ottersen et al. 1994; Sundby 2000), although whether coupling between zooplankton abundance and larval survival is tight (the match/mismatch hypothesis) is still open to question (Brander 1994). Growth of larvae and small juveniles is closely related to vulnerability to predation, and the relative importance of these two factors probably cannot be separated at this time, particularly because weak or starving juveniles will be most vulnerable to predators (Nakken 1994). Nevertheless, at least some authors, while recognizing that predation and cannibalism occur, and at some times and places may be important, suggest that these sources of mortality are probably less important in limiting cod recruitment than factors affecting foraging and growth (Pálsson 1994). As mentioned previously, the availability of forage fish such as capelin or juvenile herring appears to be critical for the growth of juvenile cod and the condition of adult cod across much of the North Atlantic (Nilssen et al. 1994; Øiestad 1994) and there are indications that, if these forage fish are not available in sufficient quantity, the condition of the cod may be too poor for them to spawn ("Spawning cod without roe and liver . . . ," Øiestad 1994). The pulsing of the North Arctic cod stock may be a delayed reflection of the pulsing of the herring (or capelin) population (Øiestad 1994).

ENSO and Fish Stocks

At a time scale of 3 to 7 years, El Niño–Southern Oscillation (ENSO) events affect fish production and marine community structure throughout the eastern Pacific Ocean (Barber and Chavez 1983; Pearcy et al. 1985; Arcos et al. 2001; Hollowed et al. 2001). Primary production is slowed or shut down by a cap of warm nutrient-depleted water that prevents upwelling of cold nutrient-rich water from depth. As a result, during ENSO phytoplankton and zooplankton production is curtailed and small pelagic fishes in the eastern boundary current upwelling systems have drastically reduced production (Barber and Chavez 1983; Rothschild 1994). Predatory fish that are dependent on small pelagics shift their distributions, and suffer reduced productivity (Bakun 2001). The effects of ENSO events are felt over a wide area of the North Pacific (e.g., Sugimoto et al. 2001; Lehodey 2001).

In summary, there is evidence in species from forage fish to salmonids and groundfish of responses to climate events on timescales of years to at least decades. These climate-forced shifts in ocean ecology result in changes in the abundance of zooplankton prey. Upper-trophic-level species respond with predictable changes in recruitment, growth, or biomass. These population changes that occur in the apparent absence of changes in predation provide evidence that bottom-up processes are playing an important role in the population dynamics of fish and in the structuring of marine ecosystems.

Bottom-Up Control of Marine Birds

Although marine birds are exposed to predation at their nesting colonies (e.g., Parrish et al. 2001; Oro and Furness 2002) and occasionally at sea (Hunt, personal observation 2001), it has been commonly accepted that their populations are constrained by the availability of food (e.g., Ashmole 1963; Lack 1968; Birkhead and Furness 1985; Cairns 1992b). The life-history characteristics of marine birds, including delayed commencement of reproduction, low annual rates of productivity, and long-lived adults, are indications that marine birds have not evolved to cope with high rates of predation. Indeed, when chronic, elevated rates of adult mortality are imposed on seabird populations, such as by entanglement in gill nets, affected local populations decline.

Recent studies have shown that adult survival rates covary with climate indices, indicating that marine birds are sensitive to large-scale variation in ocean productivity (Jones et al. 2002). Evidence for food limitation of seabird populations comes from starvation events (e.g., Bailey and Davenport 1972; Hays 1986; Baduini et al. 2001), decreased production of young with decreases in prey abundance or increases in competition for prey (e.g., Anderson 1989; Hamer et al. 1993; Vandenbosch 2000; Lewis et al. 2001), and the observation that elevated levels of stress hormones can result in adult mortality (Kitaysky et al. 2001). Indirect evidence for food limitation comes from the examination of the geographic distribution of colonies of different sizes (Furness and Birkhead 1984; Cairns 1989) and the need of many seabird species to exploit foraging locales where prey are concentrated at levels well above background (i.e., threshold foraging, Piatt 1990; Veit et al. 1993; see also Hunt et al. 1999).

Marine Bird Responses to Food Limitation

Because the reproductive ecology of marine birds is sensitive to variation in prey availability, marine birds are useful indicators of prey abundance in coastal marine ecosystems (Cairns 1987, 1992a; Ainley et al. 1996; Gill et al. 2002; Davoren and Montevecchi 2003; Parrish and Zador 2003). Seabird response variables include time budgets (Uttley et al. 1994; Zador and Piatt 1999; Lewis et al. 2001; Jodice et al. 2002), diet shifts (Adams and Klages 1989; Montevecchi and Myers 1995; Sydeman et al. 2001; Suryan et al. 2002), adult body condition and levels of circulating stress hormones

(Oro and Furness 2002; Kitaysky et al. 1999a,b), and success in rearing young (Uttley et al. 1994; Rindorf 2000, Sydeman et al. 2001; Carscadden et al. 2002; Suryan et al. 2002). These same responses have also proved indicative of the effects of climate change on marine systems (Montevecchi and Myers 1997; Thompson and Ollason 2001; Gjerdrum et al. 2003).

Adult seabirds lose body condition while raising young (Coulson et al. 1983; Pugesek 1987, 1990), and this loss may result in increased postbreeding mortality (Golet et al. 1998; Golet and Irons 1999), particularly in years of low prey abundance (Oro and Furness 2002). Adult black-legged kittiwakes (*Rissa tridactyla*) at a colony with poor food resources had poorer body condition and higher levels of circulating corticosterones than did adults from a colony with high levels of prey availability (Kitaysky et al. 1999a,b). Adult birds with experimentally elevated corticosterone levels had lower over-winter survival than controls. Food-stressed chicks also have elevated levels of corticosterones, and this may lead to suppressed immune function and lower postfledging survival (Kitaysky et al. 1999a,b, 2001; Tella et al. 2001). Thus, food stress does not have to lead to starvation to have an effect on the survival of adults or newly fledged young. Over time, elevated adult mortality rates result in population decline.

Starvation events, in which a significant portion of the local population of a seabird species dies, are relatively rare, except in the case of El Niño–Southern Oscillation (ENSO) events (Piatt and van Pelt 1997). Die-offs tend to occur after major storms (e.g., Kazama 1968; Underwood and Stowe 1984), or near the terminus of long migrations (Douglas and Setton 1955; Oka and Maruyama 1986). Whether these have population-level consequences depends on whether the birds killed are juveniles, excess nonbreeding adults unable to find a colony site or adequate food to commence breeding, or breeding adults.

Marine Birds and ENSO

During ENSO events, marine birds in upwelling regions suffer direct mortality (Barber and Chavez 1983; Hodder and Graybill 1985; Valle and Coulter 1987; Bodkin and Jameson 1991) and low breeding success (Hodder and Graybill 1985; Anderson 1989; Ainley et al. 1995; Guinet et al. 1998; Ramos et al. 2002) because of a lack of prey. Starvation events associated with ENSO events probably have the greatest impact on marine bird populations because they can result in the deaths of breeding birds that would otherwise be expected to have high rates of survival. Declines on the order of 70 to 90% of the adults previously present have been recorded (Hays 1986; Schreiber and Schreiber 1989; Kalmbach et al. 2001). In some regions, the effects of ENSO events result in improved prey availability and enhanced reproductive success (Perriman et al. 2000). When ENSO events come in rapid succession or are coupled with fishery removals of prey organisms, marine bird population recovery may be stalled (Hays 1986; Jahncke et al. 2004).

Marine Bird Foraging and Prey Availability

The at-sea distribution of marine birds reflects ocean productivity at large geographic scales, and ecologically profitable concentrations of prey at smaller scales (Hunt and Schneider 1987; Karnovsky and Hunt 2002). Murphy (1914) was one of the first to describe latitudinal differences in marine avifaunas that were due to differences in oceanic conditions and productivity, and Jesperson (1930) demonstrated a strong quantitative relationship between the abundances of zooplankton and marine birds at the scale of the North Atlantic Ocean. In the Pacific Ocean, Murphy (1936) recognized the importance of onshore-offshore gradients in water properties for marine bird assemblages. In the equatorial regions of the Pacific, Ballance et al. (1997) have shown that not only the biomass of birds but also the types of birds change with productivity regime, such that species of birds with lower wing loading are found over regions with lower productivity than where species of birds with high wing loading are found (see also Ainley 1977). Changes in marine productivity at smaller geographic scales are also reflected in the distribution and abundance of seabirds at sea (e.g., Aebischer et al. 1990; Veit et al. 1996, 1997; Hyrenbach and Veit 2003). Because birds at sea rarely suffer from predation, patterns of distribution at sea, particularly in the nonbreeding season, are the result of the availability of prey.

In summary, marine birds have few predators except at the colonies. Seabird distributions at sea reflect the distribution and abundance of prey over spatial scales from the regional to the world ocean. Seabirds are sensitive to temporal variability in the availability of prey, and their populations are generally limited by prey availability, though they rarely limit the populations of their prey.

Bottom-Up Effects on Pinniped and Cetacean Populations

There is less information on the relative importance of bottom-up and top-down control of populations of pinnipeds and cetaceans than of marine birds. However, there is considerable circumstantial evidence that marine mammal populations, and in particular populations of cetaceans and pinnipeds, have in the past been food-limited. Many of the great whales were near the top of their respective food chains, and when their numbers were severely reduced by commercial whaling, there is evidence that the populations of other species that shared similar food habits expanded. For example, in the Southern Ocean, with the demise of the great whales, populations of minke whale *(Balaenoptera acutorostrata)*, crabeater seal *(Lobodon carcinophagus)*, and Antarctic fur seal *(Arctocephalus gazella)* expanded (Laws 1985), as did populations of krill-eating penguins (Croxall 1984; Croxall et al. 1988). In the North Atlantic, subsequent to the severe depletion of many populations of great whales, populations of several species of plankton-eating birds, fish, and mammals, including the harp seal *(Pagophilus groenlandica)*, increased in

abundance (Hacquebord 1999). These population increases have been taken as evidence that when great whales were more plentiful, food supplies were in short supply for these pinniped and seabird populations, if not also for the great whales that were depressing prey stocks. Indeed, it is hypothesized that the anatomical and life history characteristics (including slow growth, large body mass, delayed maturity, and extended longevity) of the great whales, and in particular the bowhead whale *(Balaena mysticetus)*, evolved to cope with an environment where food was not plentiful (George et al. 1999; see also Sinclair et al. 2003 for a view on the negative relationship between body mass and likelihood of top-down limitation). As in marine birds, delayed onset of reproduction and long lifespans are indicative of species with high adult survival rates, and not of species subject to high levels of predation.

Are Some Marine Mammals Close to Carrying Capacity?

There is limited evidence from studies on several populations of large whales that present-day populations may be close to carrying capacity and that density-dependent impacts are occurring. In the Barents Sea, the condition of minke whales was weakly correlated with annual changes in prey abundance (Haug et al. 2002). However, when the condition of the whales was compared in years of high and low prey abundance, minke whales in the southern Barents Sea, particularly females and immatures, were in significantly better condition in years when immature herring were abundant. Similarly, interannual variation in body fat has been correlated with food availability in minke whales in the Antarctic (Ichii et al. 1998) and in fin whales *(B. physalus)* in Icelandic waters (Lockyer 1986).

In the northern Bering Sea, the eastern North Pacific gray whale *(Eschrichtius robustus)* forages primarily on benthic infaunal invertebrates (Rice and Wolman 1971). Although gray whale numbers were reduced by commercial whaling in the late nineteenth and early twentieth centuries, recent restriction on the hunting of these whales has led to a population rebound, and now this population may have exceeded the carrying capacity of the infaunal amphipod community on which it depends (Highsmith and Coyle 1992; Moore et al. 2003). As a result, the whales are moving north of Bering Strait to obtain their food, and there are now signs of declines in gray whale reproductive success and of higher mortality (Perryman et al. 2002; Moore et al. 2001, 2003). This mortality is apparently the result of starvation; there is no evidence that the beached whales have died from attacks by natural predators.

Cetacean Foraging and Prey Distributions

For profitable foraging, baleen whales require high prey concentrations, and they respond to variation in prey densities with a threshold response (e.g., Piatt and Methven 1992). In the Bay of Fundy, Canada, North Atlantic right whales and fin whales foraged in regions where their prey was significantly concentrated (Woodley and Gaskin 1996). In the Great South Channel of Georges Bank, Northwest Atlantic, right whales annually congregate to forage on high concentrations of the copepod *Calanus finmarchicus* in surface waters (Kenney et al. 1995; Kann and Wishner 1995). In 1986, only by foraging in the very densest prey patches could the whales have experienced a net gain in energy (Kenny et al. 1986; Wishner et al. 1988), whereas in other years whales focused their foraging near oceanographic features that concentrated zooplankton (Kann and Wishner 1995; Beardsley et al. 1996). In the case of the right whales, it appears as though they could be prey-limited without at the same time putting a top-down limit on their prey.

Evidence of present-day bottom-up limitation of pinnipeds is available from the North Atlantic. For harp seals density-dependent food limitation is indicated by inverse correlations between body growth rates and age-at-maturity in populations from both the Northwest Atlantic and the Barents Sea (Bowen et al. 1981; Innes et al. 1981; Kjellqwist et al. 1995). Variability in mean age at maturity of seals has been explained as an effect of changes in food availability on growth rates (e.g., Laws 1956, 1959), and has been found in crabeater seals in the Antarctic (Bengtson and Siniff 1981). Food limitation of harp seals could be the result of either population change, or density-independent changes in prey abundance from competition from other consumers, including commercial fishing, or ecosystem change (Frie et al. 2003). Declines in the Barents Sea stocks of harp seals coincided with severe depletions of important prey species, reduced body condition, and mass invasions of the Norwegian coast by these seals (Nilssen et al. 1997; Frie et al. 2003). Depression of walrus *(Odobenus rosmarus)* populations near Svalbard by overexploitation may have allowed populations of bearded seals *(Erignathus barbatus)* in the area to expand into the niche previously occupied by the walrus (Hjelset et al. 1999). Similarly, it has been suggested that the continuing exponential increase in gray seal abundance at Sable Island, Nova Scotia, may be the result of prey made available by the collapse of the stocks of large predatory fish there (Bowen et al. 2003).

Because ENSO events cause severe reductions in the abundance of forage fishes, they have the potential to impact upper-trophic-level predators, such as many pinnipeds, that are dependent on these resources (Trillmich and Ono 1991). In the eastern Pacific Ocean, the impact of ENSO events on pinnipeds varies with the latitude of the rookery (Trillmich et al. 1991). ENSO events begin near the equator and propagate north and south; pinnipeds breeding near the equator not only lose pups and fail to bring off young; they also suffer high rates of adult mortality. In contrast, populations at high latitudes may only lose a few pups or may not be affected (Trillmich et al. 1991). Pups, and then juveniles, are generally the most vulnerable individuals in a population, whereas adults are more resilient, perhaps because they are more efficient foragers, and because they have greater body

mass and potentially more reserves. Females also show marked reductions in fertility immediately after an ENSO event (Trillmich et al. 1991). These reductions have at least temporary impacts on the size of affected pinniped populations. The evidence supports the view that the reductions in pinniped populations associated with El Niño events are driven by bottom-up processes.

In summary, there is evidence that both pinnipeds and great whales are limited by their prey. However, unlike the case with seabirds, there is also evidence that, in some cases, coupling between marine mammals and their prey is sufficiently strong that the marine mammals may be limiting their prey (e.g., walruses at Spitzbergen, gray whales in the northern Bering Sea, great whales in the Antarctic).

Overview

Upper-trophic-level organisms, such as fish, marine birds, and marine mammals, are all subject to bottom-up limitation. Under some circumstances, they may also be subject to top-down limitation by predators. For most adult large fish and marine birds, top-down limitation appears to be rare and localized (e.g., Birt et al. 1987; Parrish et al. 2001; Nordstrom 2002). For marine mammals, top-down control by killer whales has been proposed (Springer et al. 2003), but evidence is lacking (DeMaster et al. 2006), and the existence of top-down control as the major mechanism remains to be demonstrated for marine mammals except in a few cases (e.g., Estes et al. 1998). Many of the species of upper-trophic-level predators are large, and species that grow to large size tend to have lower rates of natural mortality than smaller species (Pauly 1998, Figure 3; Sinclair et al. 2003). Given their life history characteristics of long lifespan and delayed reproduction (at least in some species), these upper-trophic-level organisms would seem to be poor candidates for control by predation. Perhaps this is in part why we have such difficulty in developing sustainable harvesting rates for many of the large predatory fish (Myers et al. 1997; Jackson et al. 2001; Pauly et al. 1998).

Scale Dependence of Control Mechanisms

Because it is likely that the mechanism of population control varies through time or space, it is useful to ask what circumstances may favor top-down or bottom-up control. First, the relative importance of top-down and bottom-up mechanisms is scale-dependent. At global scales, the highest biomass of fish, marine birds, and marine mammals is, and was, found in areas of the world ocean with high levels of primary production, such as the sub-Arctic seas and upwelling regions on the continental shelves. These large-scale distribution patterns are driven by the availability of food, not the absence of top predators. In fact, top predators are abundant in the very areas where these other upper-level consumers are abundant because that is where their prey is. At regional scales, it is likely that the threat of predation, or past predation events, has shaped local distributions, at least in marine birds and pinnipeds. The locations of colony, rookery, and haul-out sites reflect not only the need for access to prey (Hunt et al. 1999), but also a need for refuge from predators. And the presence of many individuals in one place provides further reduction of risk of predation for the individuals within the colony (Wittenberger and Hunt 1985). The lack of safe refuges may therefore constrain population expansion in times of plentiful prey. Within colonies, rookeries, or haulouts, at least within those that are only occupied on a seasonal basis (see later), resource availability seems to play a major role in determining the number of individuals occupying sites. Predators can remove individuals from marginal sites, or as they come and go from the colony or rookery, but there is substantial evidence that fluctuations in the numbers of animals at colonies and rookeries are responses to fluctuations in resource availability (e.g., El Niño events, and fluctuations at sites where predators are essentially absent) and not to predation.

Second, the relative mobility of the predator and its prey will also affect the relative importance of top-down or bottom-up control (Schneider 1978, 1992). If the predator is more mobile than its prey, the predator can inflict severe local depletion on its prey and then move on and repeat the process at a new site (e.g., shorebirds foraging on benthic prey). In contrast, if the prey is more mobile, then the predator's impact is limited by the ability of the prey to escape. In this case, the increase of the predator population will be limited by the availability of prey during the season when prey is least available to the predator, and top-down limitation of the prey is unlikely. Marine birds and pinnipeds that breed in colonies or rookeries and then disperse widely during the period of nonbreeding severely curtail the ability of a predator to limit their populations. Birds of prey and top predators such as sharks or killer whales may forage extensively on locally abundant populations of marine birds or pinnipeds, but populations of these top predators will be limited by their ability to secure prey once the birds or pinnipeds have dispersed and are no longer available at densities that result in a net gain in energy from foraging. Species of pinnipeds that remain on a rookery or at a haulout year round are more vulnerable to local depletion by a mobile predator. We know relatively little about the potential impact of cetaceans on their prey because, in many cases, their populations are greatly reduced from preharvest levels and their roles in an intact ecosystem are unknown. However, in at least two cases, significant impacts on prey populations were demonstrated (e.g., Estes et al. 1998; Moore et al. 2001). Likewise, sea otters are able to control the populations of their sea urchin prey (Estes 1996). In each of these cases, the prey species was relatively sedentary when compared to the predator.

It is less clear how widely the relative mobility of predators and prey affects the control of fish populations. Anadromous fish are often subject to substantial predation when entering or leaving the ocean (e.g., salmon smolts in the Columbia River estuary, Collis et al. 2002; Roby et al. 2003), but while they are dispersed on the high seas, predation is less likely a factor in their survival. However, since the life history pattern of fish involves a considerable period of time at small size and concomitant vulnerability to predation, it may be that it is at this phase of the life cycle that relatively mobile predators can exert a top-down control of species that would otherwise grow up to be top predators.

Third, to exert top-down control, the predator must remove individuals or biomass from the prey population at a rate sufficient to affect the number of recruits entering the prey population or to limit adult survival. For example, there is evidence that upper-trophic-level fish impact their prey significantly (e.g., Worm and Myers 2003). In contrast, it is only under special circumstances, such as in the vicinity of large breeding colonies or where high numbers of migrants aggregate, that marine birds have been shown to consume more than a tiny fraction of the standing stock of a prey species (e.g., Wiens and Scott 1975; Furness 1978; Birt et al. 1987; Karnovsky and Hunt 2002).

The presence of top-down or bottom-up control of upper-trophic-level predators has implications for the eventual fate of carbon in marine ecosystems (see Paine, Chapter 2 of this volume). Depending on the structure of the food web, strong top-down control by an upper-trophic-level species may result in a decoupling between phytoplankton and herbivores and in increased benthic-pelagic coupling. In contrast, if predators do not control herbivores, there will be strong coupling between the herbivores and primary producers, with more of the carbon entering a pelagic food web (e.g., Walsh and McRoy 1986). This carbon is likely to be respired to the water column and recycled to the atmosphere.

Summary

Despite a general lack of research that evaluates the relative importance of bottom-up and top-down population limitation in the ocean, based on the literature available, several issues seem clear:

1. At the scale of the world ocean, populations and biomass of fish, whales, pinnipeds, and seabirds are limited by bottom-up processes. The populations and the biomass of these taxa are greatest where primary production is greatest. Fish, marine mammals, and seabirds are not concentrated in productive waters because they are constrained by predators. On the contrary, these productive waters are where the predators of each group are most abundant.

2. At a regional scale, it is difficult to separate the population effects of bottom-up and top-down mechanisms, particularly for fish. Most often, both play a role in regulating population size and biomass. The tightness of the coupling between trophic levels is critical, as is whether the coupling is uni- or bidirectional.

3. Although density-dependent responses relative to prey do not demonstrate bottom-up control, in the absence of evidence for coincident changes in top-down control, density-dependent responses related to prey may be the best indication available that bottom-up control is occurring.

4. Strong responses of fish stocks and pinniped and seabird populations to climate variability at scales of years to decades are likely the manifestation of bottom-up forcing. Responses to climate variability include (a) changes in stock size (salmon, groundfish, small pelagics, pinnipeds, and seabirds), (b) changes in distribution (small pelagics, groundfish such as cod and hake), and (c) mortality from starvation (pinnipeds, seabirds). There is no evidence that these responses are the result of top-down control.

5. At the regional scale, the distribution and abundance of seabirds and pinnipeds is constrained by the availability of safe places to breed or haul out. This constraint is primarily an issue of top-down control by predators, though, because seabirds and pinnipeds when using a colony or haulout are central-place foragers, they are also potentially constrained by the distance to profitable foraging areas.

6. At the scale of individual colonies or rookeries, annual variation in reproductive success is usually determined by the availability of prey. Over multiple years, the numbers of seabirds or pinnipeds attending a site can be an indication of prey availability or the level of predation pressure.

7. For taxa such as salmon, migratory whales, pinnipeds, and seabirds, the potential for top-down forcing will be constrained by the relative mobility of the predator and its prey. If the prey species congregate seasonally and at other times of the year are widely dispersed, the predator may be limited to cropping the prey profitably only during a brief period. Predator populations will be constrained from increasing to match seasonally abundant prey resources by the challenge of surviving when the prey are dispersed (e.g., salmon at the mouth of streams versus in the open ocean, fur seals and seabirds at colonies versus dispersed across vast expanses of ocean). If the predator is more mobile than its prey, then top-down constraint of prey is more likely (shorebirds and benthos, cod and epibenthic crustacea, killer whales and sea otters and possibly Steller sea lions).

Acknowledgments

I thank Jim Estes for inviting me to the Whale Workshop. I thank Dick Beamish, Bob Furness, Jack Helle, Sandy McFarlane, and Dave Schneider for stimulating discussions about this subject and for their willingness to share ideas and data.

Andrew Trites, Doug DeMaster, and two anonymous referees provided helpful comments on an earlier draft of the manuscript.

Literature Cited

Adams, N. J. and N. T. Klages. 1989. Temporal variation in the diet of the gentoo penguin *(Pygoscelis papua)* at sub-Antarctic Marion Island. *Colonial Waterbirds* 12: 30–36.

Aebischer, N. J., J. C. Coulson, and J. M. Colbrook. 1990. Parallel long-term trends across four marine trophic levels and weather. *Nature* 347: 753–755.

Ainley, D. G. 1977. Feeding methods of seabirds: a comparison of polar and tropical nesting communities of the South Pacific. *Studies in Avian Biology* 8: 2–23.

Ainley, D. G., L. B. Spear, and S. G. Allen. 1996. Variation in the diet of Cassin's auklet reveals spatial, seasonal and decadal occurrence patterns of euphausiids off California, USA. *Marine Ecology Progress Series* 137: 1–10.

Ainley, D. G., W. J. Sydeman, and J. Norton. 1995. Upper trophic level predators indicate interannual negative and positive anomalies in the California Current food web. *Marine Ecology Progress Series* 118: 69–79.

Anderson, D. J. 1989. Differential responses of boobies and other seabirds in the Galapagos North Pacific Ocean to the 1986–87 El Niño–Southern Oscillation event. *Marine Ecology Progress Series* 52: 209–216.

Arcos, D. F., L. A. Cubillos, and S. P. Núñez. 2001. The jack mackerel fishery and El Niño 1997–98 effects off Chile. *Progress in Oceanography* 49: 597–617.

Ashmole, N. P. 1963. The regulation of numbers of tropical oceanic birds. *Ibis* 103b: 458–473.

Baduini, C. L., K. D. Hyrenbach, K. O. Coyle, A. Pinchuk, V. Mendenhall, and G. L. Hunt, Jr. 2001. Mass mortality of short-tailed shearwaters in the southeastern Bering Sea during summer 1997. *Fisheries Oceanography* 10: 117–130.

Baier, C. T. and J. M. Napp. 2003. Climate-induced variability in *Calanus marshallae* populations. *Journal of Plankton Research* 25: 771–782.

Bailey, E. P. and G. H. Davenport. 1972. Die-off of murres on the Alaska Peninsula and Unimak Island. *Condor* 74: 215–219.

Bailey, K. M. 2000. Shifting control of recruitment of walleye pollock *Theragra chalcogramma* after a major climatic and ecosystem change. *Marine Ecology Progress Series* 198: 215–224.

Bakun, A. 2001. "School-mix feedback": a different way to think about low frequency variability in large mobile fish populations. *Progress in Oceanography* 49: 485–511.

Bakun, A. and K. Broad. 2003. Environmental 'loopholes' and fish population dynamics: comparative pattern recognition with focus on El Niño effects in the Pacific. *Fisheries Oceanography* 12: 458–473.

Ballance, L. T., R. L. Pitman, and S. B. Reilly. 1997. Seabird community structure along a productivity gradient: importance of competition and energetic constraint. *Ecology* 78: 1502–1518.

Barber, R. T. and F. P. Chavez. 1983. Biological consequences of El Niño. *Science* 222: 1203–1210.

Baumgartner, T. R., A. Soutar, and V. Ferreira-Bartrina. 1992. Reconstruction of the history of Pacific sardine and northern

Pacific anchovy populations over the past two millennia from the sediments of the Santa Barbara basin. *California Cooperative Oceanic Fisheries Investigations Report* 33: 24–40.

Beamish, R. J. 1993. Climate and exceptional fish production off the west coast of North America. *Canadian Journal of Fisheries and Aquatic Sciences* 50: 2270–2291.

Beamish, R. J. and D. R. Bouillon. 1993. Pacific salmon production trends in relation to climate. *Canadian Journal of Fisheries and Aquatic Sciences* 50: 1002–1026.

Beamish, R. J. and C. Mahnken. 2001. A critical size and period hypothesis to explain natural regulation of salmon abundance and the linkage to climate and climate change. *Progress in Oceanography* 49: 423–437.

Beamish, R. J., D. J. Noaks, G. A. McFarlane, L. Klyashtorin, V. V. Ivanov, and V. Kurashov. 1999. The regime concept and natural trends in the production of Pacific salmon. *Canadian Journal of Fisheries and Aquatic Sciences* 56: 516–526.

Beardsley, R. C., A. W. Epstein, C. Chen, K. F. Wishner, M. C. Macaulay, and R. D. Kenney. 1996. Spatial variability in zooplankton abundance near feeding right whales in the Great South Channel. *Deep-Sea Research II* 43: 1601–1625.

Bengtson, J. L. and D. B. Siniff. 1981. Reproductive aspects of female crabeater seals *(Lobodon carcinophagus)* along the Antarctic Peninsula. *Canadian Journal of Zoology* 59: 92–102.

Birkhead, T. R. and R. W. Furness. 1985. Regulation of seabird populations, in *Behavioural ecology: ecological consequences of adaptive behaviour.* R. M. Sibley and R. H. Smith, eds. Oxford: Blackwell Scientific Publications, pp. 145–167.

Birt, V. L., T. P. Birt, D. Goulet, D. K. Cairns, and W. A. Montevecchi. 1987. Ashmole's halo: direct evidence for prey depletion by a seabird. *Marine Ecology Progress Series* 40: 205–208.

Bodkin, J. L. and R. J. Jameson. 1991. Patterns of seabird and marine mammal carcass deposition along the central California USA coast 1980–1986. *Canadian Journal of Zoology* 69: 1149–1155.

Bogstad, B., G. R. Lilly, S. Mehl, Ó. K. Pálsson, and G. Stefánsson. 1994. Cannibalism and year-class strength in Atlantic cod *(Gadus morhua* L.) in arcto-boreal ecosystems (Barents Sea, Iceland, and eastern Newfoundland). *ICES Marine Science Symposia* 198: 576–599.

Bowen, W. D., C. K. Capstick, and D. E. Sergeant. 1981. Temporal changes in the reproductive potential of female harp seals *Pagophilus groenlandicus. Canadian Journal of Fisheries and Aquatic Sciences* 38: 495–503.

Bowen, W. D., J. McMillan, and R. Mohn. 2003. Sustained exponential growth of grey seals at Sable Island, Nova Scotia. *ICES Journal of Marine Science* 60: 1265–1274.

Brander, K. M. 1994. Patterns of distribution, spawning, and growth in North Atlantic cod: the utility of inter-regional comparisons. *ICES Marine Science Symposia* 198: 406–413.

Brodeur, R. D., B. W. Frost, S. R. Hare, R. C. Francis, and W. J. Ingraham, Jr. 1996. Interannual variations in zooplankton biomass in the Gulf of Alaska and covariation with California Current zooplankton. *California Cooperative Oceanic Fisheries Investigations Report* 37: 80–99.

Brodeur, R. D. and D. M. Ware. 1992. Long-term variability in zooplankton biomass in the subarctic Pacific Ocean. *Fisheries Oceanography* 1: 32–38.

Brooks, J. L. and S. I. Dodson. 1965. Predation, body size, and the composition of plankton. *Science* 150: 28–35.

Cairns, D. K. 1987. Seabirds as indicators of marine food supplies. *Biological Oceanography* 5: 261–271.

———. 1989. The regulation of seabird colony size: a hinterland model. *American Naturalist* 134: 141–146.

———. 1992a. Bridging the gap between ornithology and fisheries science: use of seabird data in stock assessment models. *Condor* 94: 811–824.

———. 1992b. Population regulation of seabird colonies. *Current Ornithology* 9: 37–61.

Carscadden, J. E., W. A. Montevecchi, G. K. Davoren, and B. S. Nakashima. 2002. Trophic relationships among capelin *(Mallotus villosus)* and seabirds in a changing ecosystem. *ICES Journal of Marine Science* 59: 1027–1033.

Carscadden, J. E. and H. Vilhjálmsson. 2002. Capelin: what are they good for? Introduction. *ICES Journal of Marine Science* 59: 863–869.

Clark, R. A., C. L. J. Frid, and K. R. Nicholas. 2003. Long-term, predation-based control of a central-west North Sea zooplankton community. *ICES Journal of Marine Science* 60: 187–197.

Clark, W. G. and S. R. Hare. 2002. Effects of climate and stock size on recruitment and growth of Pacific halibut. *North American Journal of Fisheries Management* 22: 852–862.

Clark, W. G., S. R. Hare, A. M. Parma, P. J. Sullivan, and R. J. Trumble. 1999. Decadal changes in growth and recruitment of Pacific halibut *(Hippoglosssus stenolepis). Canadian Journal of Fisheries and Aquatic Sciences* 56: 242–252.

Cohen, E. B., D. G. Mountain, and R. O'Boyle. 1991. Local scale versus large-scale factors affecting recruitment. *Canadian Journal Fisheries and Aquatic Sciences* 48: 1003–1006.

Collis, K., D. D. Roby, D. P. Craig, S. Adamany, J. Y. Adkins, and D. E. Lyons. 2002. Colony size and diet composition of piscivorous waterbirds on the lower Columbia River: implications for losses of juvenile salmonids to avian predation. *Transactions of the American Fisheries Society* 131: 537–550.

Coulson, J. C., P. Monaghan, J. Butterfield, N. Duncan, C. Thomas, and C. Shedden. 1983. Seasonal changes in the herring gull in Britain: weight, moult, and mortality. *Ardea* 71: 235–244.

Coyle, K. O. and A. I. Pinchuk. 2002. Climate-related differences in zooplankton density and growth on the inner shelf of the southeastern Bering Sea. *Progress in Oceanography* 55: 177–194.

Croxall, J. P. 1984. Seabirds, in *Antarctic ecology.* R. M. Laws, ed. London: Academic Press, pp. 533–616.

Croxall, J. P., T. S. McCann, P. A. Prince, and P. Rothery. 1988. Reproductive performance of seabirds and seals at South Georgia and Signy Island, South Orkney Islands, 1976–1987. Implications for Southern ocean monitoring studies, in *Antarctic Ocean and resources variability.* D. Sahrhage, ed. Berlin: Springer, pp. 261–285.

Daan, N. 1994. Trends in North Atlantic cod stocks: a critical summary. *ICES Marine Science Symposia* 198: 269–270.

Davoren, G. K. and W. A. Montevecchi. 2003. Signals from seabirds indicate changing biology of capelin stocks. *Marine Ecology Progress Series* 258: 253–261.

DeMaster, D. P., A. W. Trites, P. Clapham, S. Mizroch, P. Wade, and R. J. Small. 2006. The sequential megafaunal collapse hypothesis: testing with existing data. *Progress in Oceanography* 68: 329–342.

Despatie, S.-P., M. Castonguay, D. Chabot, and C. Audet. 2001. Final thermal preferendum of Atlantic cod: effect of food ration. *Transactions of the American Fisheries Society* 130: 263–275.

Dolgov, A. V. 2002. The role of capelin *(Mallotus villosus)* in the foodweb of the Barents Sea. *ICES Journal of Marine Science* 59: 1034–1045.

Douglas, G. and A. Setton. 1955. Mortality of shearwaters. *Emu* 55: 259–262.

Dwyer, W. P. and R. G. Piper. 1987. Atlantic salmon growth efficiency as affected by temperature. *Progressive Fish-Culturist* 49: 57–59.

Estes, J. A. 1996. The influence of large, mobile predators in aquatic food webs: examples from sea otters and kelp forests. In *Aquatic predators and their prey*. S. Greenstreet and M. Tasker, eds. London: Blackwell, pp. 65–72.

Estes, J. A., M. T. Tinker, T. M. Williams, and D. F. Doak. 1998. Killer whale predation on sea otter linking oceanic and nearshore ecosystems. *Science* 282: 473–476.

Field, J. D., R. C. Francis, and K. Aydin. 2006. Top-down modeling and bottom-up dynamics: linking a fisheries-based ecosystem model with climate hypotheses in the Northern California Current. *Progress in Oceanography* 68: 238–270.

Finney, B. P., I. Gregory-Eaves, M. S. V. Douglas, and J. P. Smol. 2002. Fisheries productivity in the northeastern Pacific Ocean over the past 2,200 years. *Nature* 416: 729–733.

Francis, R. C., S. R. Hare, A. B. Hollowed, and W. S. Wooster. 1998. Effects of interdecadal climate variability on the oceanic ecosystems of the northeast Pacific. *Fisheries Oceanography* 7: 1–20.

Frie, A. K., V. A. Potelov, M. C. S. Kingsley, and T. Haug. 2003. Trends in age-at-maturity and growth parameters of female Northeast Atlantic harp seals, *Pagophilus groenlandicus* (Erxleben, 1777). *ICES Journal of Marine Science* 60: 1018–1032.

Furness, R. W. 1978. Energy requirements of seabird communities: a bioenergetics model. *Journal of Animal Ecology* 47: 39–53.

———. 2002. Management implications of interactions between fisheries and sandeel-dependent seabirds and seals in the North Sea. *ICES Journal of Marine Science* 59: 261–269.

Furness, R. W. and T. R. Birkhead. 1984. Seabird colony distributions suggest competition for food supplies during the breeding season. *Nature (London)* 331: 655–656.

Garside, E. T. 1973. Ultimate upper lethal temperature of Atlantic salmon *Salmo salar* L. *Canadian Journal of Fisheries and Aquatic Sciences* 51: 898–900.

George, J. C., J. Bada, J. Zeh, L. Scott, S. E. Brown, T. O'Hara, and R. Suydam. 1999. Age and growth estimates of bowhead whales *(Balaena mysticetus)* via aspartic acid racemization. *Canadian Journal of Zoology* 77: 571–580.

Gill, V. A., S. A. Hatch, and P. H. Lanctot. 2002. Sensitivity of breeding parameters to food supply in black-legged kittiwakes *Rissa tridactyla*. *Ibis* 144: 268–283.

Gjerdrum, C., A. M. J. Vallée, C. C. St. Clair, D. F. Bertram, J. L. Ryder, and G. S. Blackburn. 2003. Tufted puffin reproduction reveals ocean climate variability. *Proceedings of the National Academy of Sciences of the United States of America* 100: 9377–9382.

Gjøsæter, H., P. Dalpadado, and A. Hassel. 2002. Growth of Barents Sea capelin *(Mallotus villosus)* in relation to zooplankton abundance. *ICES Journal of Marine Science* 59: 959–967.

Golet, G. H. and D. B. Irons. 1999. Raising young reduces body condition and fat stores in black-legged kittiwakes. *Oecologia* 120: 530–538.

Golet, G. H., D. B. Irons, and J. A. Estes. 1998. Survival costs of chick rearing in black-legged kittiwakes. *Journal of Animal Ecology* 67: 827–841.

Guinet, C., O. Chastel, M. Koudil, J. P. Durbec, and P. Jouventin. 1998. Effects of warm sea-surface temperature anomalies on the blue petrel at the Kerguelen Islands. *Proceedings of the Royal Society of London Series B: Biological Sciences* 265: 1001–1006.

Hacquebord, L. 1999. The hunting of the Greenland right whale in Svalbard, its interaction with climate and its impact on the marine ecosystem. *Polar Research* 18: 375–382.

Hairston, N. G., Sr., F. E. Smith, and L. B. Slobodkin. 1960. Community structure, population control and competition. *American Naturalist* 94: 421–425.

Hamer, K. C., P. Monaghan, J. D. Uttley, P. Walton, and M. D. Burns. 1993. The influence of food supply on the breeding ecology of kittiwakes *Rissa tridactyla* in Shetland. *Ibis* 135: 255–263.

Hare, S. R. and R. C. Francis. 1995. Climate change and salmon production in the northeast Pacific Ocean. *Canadian Special Publications in Fisheries and Aquatic Sciences* 121: 357–372.

Hare, S. R. and N. J. Mantua. 2000. Empirical evidence for North Pacific regime shifts in 1997 and 1989. *Progress in Oceanography* 47: 103–145.

Haug, T., U. Lindstrom, and K. T. Nilssen. 2002. Variations in minke whale *(Balaenoptera acutorostrata)* diet and body condition in response to ecosystem changes in the Barents Sea. *Sarsia* 87: 409–422.

Hays, C. 1986. Effects of the 1982–1983 El Niño on Humboldt penguin *Spheniscus humboldti* colonies in Peru. *Biological Conservation* 36: 169–180.

Helle, J. H. and M. S. Hoffmann. 1995. Size decline and older age at maturity of two chum salmon *(Oncorhynchus keta)* stocks in western North America. *Canadian Special Publications in Fisheries and Aquatic Sciences* 121: 245–260.

Helle, J. H. and M. S. Hoffmann. 1998. Changes in size and age of maturity of two North American stocks of chum salmon *(Oncorhynchus keta)* before and after a major regime shift in the North Pacific Ocean. *North Pacific Anadromous Fish Commission Bulletin* 1: 81–89.

Highsmith, R. C. and K. O. Coyle. 1992. Productivity of arctic amphipods relative to gray whale energy requirements. *Marine Ecology Progress Series* 83: 141–150.

Hinch, S. G., M. C. Healey, R. E. Diewart, K. A. Thompson, R. Hourston, M. A. Henderson, and F. Juanes. 1995. Potential effects of climate change on marine growth and survival of Fraser River sockeye salmon. *Canadian Journal of Fisheries and Aquatic Science* 52: 2651–2659.

Hjelset, A. M., M. Andersen, I. Gjertz, C. Lydersen, and B. Gulliksen. 1999. Feeding habits of bearded seals *(Erignathus barbatus)* from the Svalbard area, Norway. *Polar Biology* 21: 186–193.

Hodder, J. and M. R. Graybill. 1985. Reproduction and survival of seabirds in Oregon during the 1982–1983 El Niño. *Condor* 87: 535–541.

Hollowed, A. B., S. R. Hare, and W. S. Wooster. 2001. Pacific basin climate variability and patterns of Northeast Pacific marine fish production. *Progress in Oceanography* 49: 257–282.

Hollowed, A. B. and W. Wooster. 1992. Variability of winter ocean conditions and strong year classes of northeast Pacific groundfish. *ICES Marine Science Symposium* 195: 433–444.

Hunt, G. L., Jr., F. Mehlum, R. W. Russell, D. Irons, M. B. Decker, and P.H. Becker. 1999. Physical processes, prey abundance, and the foraging ecology of seabirds, in *Proceedings of the 22nd International Ornithological Congress, Durban*. N. J. Adams and R. Slotow, eds. Johannesburg: BirdLife South Africa, pp. 2040–2056.

Hunt, G. L., Jr. and S. McKinnell. 2006. Interplay between top-down, bottom-up and wasp-waist control in marine ecosystems. *Progress in Oceanography* 68: 115–124.

Hunt, G. L., Jr. and D. C. Schneider. 1987. Scale dependent processes in the physical and biological environment of marine birds, in *Seabirds: feeding biology and role in marine ecosystems*. J. Croxall, ed. Cambridge: Cambridge University Press, pp. 7–41.

Hunt, G. L., Jr., P. Stabeno, G. Walters, E. Sinclair, R. D. Brodeur, J. M. Napp, and N. A. Bond. 2002. Climate change and control of the southeastern Bering Sea pelagic ecosystem. *Deep Sea Research Part II* 49: 5821–5853.

Hunter, M. D. and P. W. Price. 1992. Playing chutes and ladders: heterogeneity and the relative roles of bottom-up and top-down forces in natural communities. *Ecology* 73: 724–732.

Hutchinson, G. E. 1959. Homage to Santa Rosalia or why are there so many kinds of animals? *American Naturalist* 93: 145–159.

Hyrenbach, K. D. and R. R. Veit. 2003. Ocean warming and seabird communities of the southern California Current system (1987–98): response at multiple temporal scales. *Deep-Sea Research II* 50: 2537–2565.

Ichii, T., N. Shinohara, Y. Fujise, S. Nisiwaki, and K. Matsuoka. 1998. Interannual changes in body fat condition index of minke whales in the Antarctic. *Marine Ecology Progress Series* 175: 1–12.

Innes, S., R. E. A. Stewart, and D. M. Lavigne. 1981. Growth in the Northwest Atlantic harp seals *Phoca groenlandica*. *Journal of Zoology, London* 194: 11–24.

Ishida, Y., T. Hariu, J. Yamashiro, S. McKinnell, T. Matsuda, and H. Kaneko. 2001. Archeological evidence of Pacific salmon distribution in northern Japan and implications for future global warming. *Progress in Oceanography* 49: 539–550.

Jackson, J. B. C., M. X. Kirby, W. H. Berger, K. A. Bjorndal, L. W. Botsford, B. J. Bourque, R. H. Bradbury, R. Cooke, J. Erlandson, J. A. Estes, T. P. Hughes, S. Kidwell, C. B. Lange, H. S. Lenihan, J. M. Pandolfi, C. H. Peterson, R. S. Steneck, M. J. Tegner, and R. R. Warner. 2001. Historical overfishing and the recent collapse of coastal ecosystems. *Science* 293: 629–638.

Jahncke, J., D. Checkley, and G. L. Hunt, Jr. 2004. Trends in carbon flux to seabirds in the Peruvian upwelling system: effects of wind and fisheries on population regulation. *Fisheries Oceanography* 13: 208–223.

Jakobsson, J., O. S. Astthorsson, R. J. H. Beaverton, B. Bjornsson, N. Dann, K. T. Frank, J. Meincke, B. J. Rothschild, S. Sundby, and S. Tilseth, eds. 1994. Cod and climate change. Proceedings of a symposium held in Reykjavik, 23–27 August 1993. *ICES Marine Science Symposia* 198.

Javaid, M. Y. and J. M. Anderson. 1967. Thermal acclimation and temperature selection in Atlantic salmon, *Salmo salar*, and rainbow trout, *S. gairdneri*. *Journal of the Fisheries Research Board of Canada* 24: 1507–1513.

Jesperson, P. 1930. Ornithological observations in the North Atlantic Ocean. *Oceanographic Reports of the Dana Expedition* 7: 1–35.

Jodice, P. G. R., D. D. Roby, V. A. Gill, S. A. Hatch, R. B. Lanctot, and G. H. Visser. 2002. Does food availability constrain energy expenditure of black-legged kittiwakes raising young? A supplemental feeding experiment. *Canadian Journal of Zoology* 80: 214–222.

Jones, I. L., F. M. Hunter, and G. J. Robertson. 2002. Annual adult survival of least auklets (Aves, Alcidae) varies with large-scale climatic conditions of the North Pacific Ocean. *Oecologia* 133: 38–44.

Kalmbach, E., S. C. Ramsay, H. Wendeln, and P. H. Becker. 2001. A study of neotropic cormorants in central Chile: possible effects of El Niño. *Waterbirds* 24: 345–351.

Kann, L. M. and K. Wishner. 1995. Spatial and temporal patterns of zooplankton on baleen whale feeding grounds in the southern Gulf of Maine. *Journal of Plankton Research* 17: 235–262.

Karnovsky, N. J. and G. L. Hunt, Jr. 2002. Estimation of carbon flux to dovekies *(Alle alle)* in the North Water. *Deep-Sea Research Part II* 49: 5117–5130.

Kawasaki, T. 1992a. Climate-dependent fluctuations in Far Eastern sardine population and their impacts on fisheries and society, in *Climate variability, climate change and fisheries*. M. Glantz, ed. Cambridge: Cambridge University Press, pp. 325–355.

———. 1992b. Mechanisms governing fluctuations in pelagic fish populations. *South African Journal of Marine Science* 12: 873–879.

Kazama, T. 1968. On the mass destruction of *Rissa tridactyla* and *Calonectris leucomelas* and their migration at Kashiwazaki Prefecture. *Tori* 18: 260–266.

Kenney, R. D., M. A. M. Hyman, R. E. Owen, G. P. Scott, and H. E. Winn. 1986. Estimation of prey densities required by western North Atlantic right whales. *Marine Mammal Science* 2: 1–13.

Kenney, R. D., H. E. Winn, and M. C. Macaulay. 1995. Cetacens in the Great South Channel, 1979–1989: right whale *(Eubalaena glacialis)*. *Continental Shelf Research* 15: 385–414.

King, J. R., G. A. McFarlane, and R. J. Beamish. 2000. Decadal scale patterns in the relative year class success of sablefish, *Anoplopoma fimbra*. *Fisheries Oceanography* 9: 62–70.

Kitaysky, A. S., E. V. Kitaiskaia, J. C. Wingfield, and J. Piatt. 2001. Dietary restriction causes chronic elevation of corticosterone and enhances stress response in red-legged kittiwake chicks. *Journal of Comparative Physiology B* 171: 701–709.

Kitaysky, A. S., J. C. Wingfield, and J. F. Piatt. 2001. Corticosterone facilitates begging and affects resource allocation in the black-legged Kittiwake. *Behavioral Ecology* 12: 619–625.

Kitaysky, A. S., J. F. Piatt, J. C. Wingfield, and M. Romano. 1999a. The adrenocortical stress-response of black-legged kittiwake chicks in relation to dietary restrictions. *Journal of Comparative Physiology B* 169: 303–310.

Kitaysky, A. S., J. C. Wingfield, and J. F. Piatt. 1999b. Dynamics of food availability, body condition and physiological stress response in breeding black-legged kittiwakes. *Functional Ecology* 13: 577–584.

Kjellqwist, S. A., T. Haug, and T. Øritsland. 1995. Trends in age-composition, growth and reproductive parameters of Barents Sea harp seals, *Phoca groenlandica*. *ICES Journal of Marine Science* 52: 197–208.

Klyashtorin, L. B. 1998. Long-term climate change and main commercial fish production in the Atlantic and Pacific. *Fisheries Research* 37: 115–125.

Koslow, J. A., K. R. Thompson, and W. Silvert. 1987. Recruitment to Northwest Atlantic cod *(Gadus morhua)* and haddock *(Melanogrammus aeglefinus)* stocks: influence of stock size and climate. *Canadian Journal of Fisheries and Aquatic Sciences* 44: 26–39.

Lack, D. 1968. *Ecological adaptations for breeding in birds*. London: Chapman and Hall.

Lambert, Y. and J.-D. Dutil. 1997. Condition and energy reserves of Atlantic cod *(Gadus morhua)* during the collapse of the northern Gulf of St. Lawrence stock. *Canadian Journal of Fisheries and Aquatic Sciences* 54: 2388–2400.

Laws, R. M. 1956. Growth and sexual maturity in aquatic mammals. *Nature* 178: 193–194.

———. 1959. Accelerated growth in seals with special reference to the Phocidae. *Norsk Hvalfangsttidende* 9: 425–452.

———. 1985. The ecology of the Southern Ocean. *American Scientist* 73: 26–40.

Lehodey, P. 2001. The pelagic ecosystem of the tropical Pacific Ocean: dynamic spatial modeling and biological consequences of ENSO. *Progress in Oceanography* 49: 439–468.

Lewis, S., T. N. Sherratt, K. C. Hamer, and S. Wanless. 2001. Evidence of intra-specific competition for food in a pelagic seabird. *Nature* 412: 816–819.

Lindeman, R. L. 1942. The trophic-dynamic aspect of ecology. *Ecology* 23: 399–417.

Lluch-Belda, D., R. Crawford, T. Kawasaki, A. MacCall, R. Parrish, R. Shwartzlose, and P. Smith. 1989. World-wide fluctuations of sardine and anchovy stocks. The regime problem. *South African Journal of Marine Science* 8: 195–205.

Lluch-Belda, D., R. Shwartzlose, R. Serra, R. Parrish, T. Kawasaki, D. Hedgecock, and R. Crawford 1993. Sardine and anchovy regime fluctuations of abundance in four regions of the World Ocean: a workshop report. *Fisheries Oceanography* 2: 339–343.

Lockyer, C. 1986. Body fat condition in northeast Atlantic fin whales *(Balaenoptera physalus)* and its relationship with reproduction and food resource. *Canadian Journal of Fisheries and Aquatic Sciences* 43: 142–147.

McFarlane, G. A. and R. J. Beamish. 1992. Climatic influence linking copepod production with strong year-classes in sablefish, *Anoplopoma fimbria*. *Canadian Journal of Fisheries and Aquatic Sciences* 49: 743–753.

McFarlane, G. A., J. R. King, and R. J. Beamish. 2000. Have there been recent changes in the climate? Ask the fish. *Progress in Oceanography* 47: 147–169.

Menge, B. A. 1992. Community regulation: under what conditions are bottom-up factors important on rocky shores? *Ecology* 73: 755–765.

Merrick, R. L. 1997. Current and historical roles of apex predators in the Bering Sea ecosystem. *Journal of Northwest Atlantic Fisheries Science* 22: 343–355.

Montevecchi, W. A. and R. A. Myers. 1995. Prey harvests of seabirds reflect pelagic fish and squid abundance on multiple spatial and temporal scales. *Marine Ecology Progress Series* 117: 1–9.

Montevecchi, W. A. and R. A. Myers. 1997. Centurial and decadal oceanographic influences on changes in northern gannet populations and diets in the northwest Atlantic: implication for climate change. *ICES Journal of Marine Science* 54: 608–614.

Moore, M. J., S. D. Berrow, B. A. Jensen, P. Carr, R. Sears, V. J. Rowntree, R. Payne, and P. K. Hamilton. 1999. Relative abundance of large whales around South Georgia (1979–1998). *Marine Mammal Science* 15: 1287–1302.

Moore, S. E., J. M. Grebmeier, and J. R. Davis. 2003. Gray whale distribution relative to forage habitat in the northern Bering Sea: current conditions and retrospective summary. *Canadian Journal of Zoology* 81: 734–742.

Moore, S. E., J. R. Urban, W. L. Perryman, R. Gulland, H. M. Perez-Cortes, P. R. Wade, L. Rojas-Bracho, and T. Rowles. 2001. Are gray whales hitting "K" hard. *Marine Mammal Science* 17: 954–958.

Murphy, R. C. 1914. Observations of birds of the South Atlantic. *Auk* 31: 439–455.

———. 1936. *Oceanic birds of South America*. New York: American Museum of Natural History.

Myers, R. A. 1998. When do environment-recruitment correlations work? *Reviews in Fish Biology and Fisheries* 8: 285–305.

Myers, R. A., J. A. Hutchings, and N. J. Barrowman. 1997. Why do fish stocks collapse? The example of cod in Atlantic Canada. *Ecological Applications* 7: 91–106.

Myers, R. A. and B. Worm. 2003. Rapid worldwide depletion of predatory fish communities. *Nature* 423: 280–283.

Nakken, O. 1994. Causes of trends and fluctuations in the Arcto-Norwegian cod stock. *ICES Marine Science Symposia* 198: 212–228.

Nilssen, E. M., T. Pedersen, C. C. E. Hopkins, K. Thyholt, and J. G. Pope. 1994. Recruitment variability and growth of Northeast Arctic cod: influence of physical environment, demography, and predator-prey energetics. *ICES Marine Science Symposia* 198: 449–470.

Nilssen, K. T., T. Haug, and V. Potelov. 1997. Seasonal variation in body condition of adult Barents Sea harp seals *(Phoca groenlandica)*. *Journal of Northwest Atlantic Fisheries Science* 22: 17–25.

Nordstrom, C. A. 2002. Haul-out selection by Pacific harbor seals *(Phoca vitulina richardii)*: isolation and perceived predation risk. *Marine Mammal Science* 18: 194–205.

Odum, H. T. and E. P. Odum. 1955. Trophic structure and productivity of a windward coral reef community on Eniwetok Atoll. *Ecological Monographs* 25: 291–320.

Ohtani, K. and T. Azumaya. 1995. Influence of interannual changes in ocean conditions on the abundance of walleye pollock *(Theragra chalcogramma)* in the eastern Bering Sea. *Canadian Special Publication in Fisheries and Aquatic Sciences* 121: 87–95.

Øiestad, V. 1994. Historic changes in cod stocks and cod fisheries: Northeast Arctic cod. *ICES Marine Science Symposia* 198: 17–30.

Oka, N. and N. Maruyama. 1986. Mass mortality of short-tailed shearwaters along the Japanese coast. *Tori* 34: 97–104.

Oro, D. and R. W. Furness. 2002. Influences of food availability and predation on survival of kittiwakes. *Ecology* 83: 2516–2528.

Ottersen, G., H. Loeng, and A. Raknes. 1994. Influence of temperature variability on recruitment of cod in the Barents Sea. *ICES Marine Science Symposia* 198: 471–481.

Paine, R. T. 1966. Food web complexity and species diversity. *American Naturalist* 100: 65–75.

———. 1974. Intertidal community structure: experimental studies on the relationship between a dominant competitor and its principal predator. *Oecologia* 15: 93–120.

Pálsson, Ó. K. 1994. A review of the trophic interactions of cod stocks in the North Atlantic. *ICES Marine Science Symposia* 198: 553–575.

Parrish, J. K., M. Marvier, and R. T. Paine. 2001. Direct and indirect effects: interactions between bald eagles and common murres. *Ecological Applications* 11: 1858–1869.

Parrish, J. K. and S. G. Zador. 2003. Seabirds as indicators: an exploratory analysis of physical forcing in the Pacific Northwest coastal environment. *Estuaries* 26: 1044–1057.

Parsons, L. S. and W. H. Lear. 2001. Climate variability and marine ecosystem impacts: a North Atlantic perspective. *Progress in Oceanography* 49: 167–188.

Pauly, D. 1998. Beyond our original horizons: the tropicalization of Beverton and Holt. *Reviews in Fish Biology and Fisheries* 8: 307–334.

Pauly, D., V. Christiansen, J. Dalsgaard, R. Froese, and F. Torres, Jr. 1998. Fishing down the marine food webs. *Science* 279: 860–863.

Pearcy, W., J. Fisher, R. Brodeur, and S. Johnson. 1985. Effects of the 1983 El Niño on coastal nekton off Oregon and Washington, in *El Niño north*. W. S. Wooster and D. L. Fluharty, eds. Seattle: Washington Sea Grant Program, University of Washington, pp. 188–204.

Perriman, L., D. Houston, H. Steen, and E. Johannesen. 2000. Climate fluctuation effects on breeding of blue penguins *(Eudyptula minor)*. *New Zealand Journal of Zoology* 27: 261–267.

Perryman, W. L., M. A. Donahue, P. C. Perkins, and S. B. Reilly. 2002. Gray whale calf production 1994–2000: are observed fluctuations related to changes in seasonal ice cover? *Marine Mammal Science* 18: 121–144.

Piatt, J. F. 1990. The aggregative response of common murres and Atlantic puffins to school of capelin. *Studies in Avian Biology* 14: 36–51.

Piatt, J. F. and D. A. Methven. 1992. Threshold foraging behavior of baleen whales. *Marine Ecology Progress Series* 84: 205–210.

Piatt, J. F. and T. van Pelt. 1997. Mass-mortality of guillemot *(Uria aalge)* in the Gulf of Alaska in 1993. *Marine Pollution Bulletin* 34: 656–662.

Power, M. E. 1992. Top-down and bottom-up forces in food webs: do plants have primacy? *Ecology* 73: 733–746.

Pugesek, B. H. 1987. Age-specific survivorship in relation to clutch size and fledging success in California gulls. *Behavioral Ecology and Sociobiology* 21: 217–221.

———. 1990. Parental effort in the California gull: tests of parent-offspring conflict theory. *Behavioral Ecology and Sociobiology* 27: 211–215.

Ramos, J. A., A. M. Maul, V. Ayrton, I. Bullock, J. Hunter, J. Bower, G. Castle, R. Mileto, and C. Pacheco. 2002. Influence of local and large-scale weather events and timing of breeding on tropical roseate tern reproductive parameters. *Marine Ecology Progress Series* 243: 271–279.

Rice, D. W. and Wolman, A. A. 1971. The life history and ecology of the gray whale *(Eschrichtius robustus)*. *American Society of Mammology Special Publications* 3: 1–142.

Rindorf, A., S. Wanless, and M. P. Harris. 2000. Effects of changes in sandeel availability on the reproductive output of seabirds. *Marine Ecology Progress Series* 202: 241–252.

Roby, D. D., D. E. Lyons, D. P. Craig, K. Collis, and G. H. Visser. 2003. Quantifying the effect of predators on endangered species using a bioenergetics approach: Caspian terns and juvenile salmonids in the Columbia River estuary. *Canadian Journal of Zoology* 81: 250–265.

Roman, J. and S. R. Palumbi. 2003. Whales before whaling in the North Atlantic. *Science* 301: 508–510.

Rose, G. A. and R. L. O'Driscoll. 2002. Capelin are good for cod: can the northern cod stock rebuild without them? *ICES Journal of Marine Science* 59: 1018–1026.

Rothschild, B. J. 1994. Decadal transients in biological productivity, with special reference to cod populations of the North Atlantic. *ICES Marine Science Symposia* 198: 333–345.

Schlesinger, M. E. and N. Ramankutty. 1994. An oscillation in the global climate system of period 65–70 years. *Nature* 367: 723–726.

Schneider, D. C. 1978. Equalization of prey numbers by migratory shorebirds. *Nature* 271: 353–354.

———. 1992. Thinning and clearing of prey by predators. *American Naturalist* 139: 148–160.

Schreiber, E. A. and R. W. Schreiber. 1989. Insights into seabird ecology from a global natural experiment. *National Geographic Research* 5: 64–81.

Shurin, J. B., E. T. Borer, E. W. Seabloom, K. Anderson, C. A. Blanchette, B. Broitman, S. D. Cooper, and B. S. Halpen. 2002. A cross-ecosystem comparison of the strength of trophic cascades. *Ecology Letters* 5: 785–791.

Sinclair, A. R. E., S. Mbuma, and J. S. Brashares. 2003. Patterns of predation in a diverse predator-prey system. *Nature* 425: 288–290.

Sinclair, M. 1988. Marine populations: an essay on population regulation and speciation. Washington Sea Grant Program. Distributed by the University of Washington Press, Seattle.

Slobodkin, L. B. 1960. Ecological energy relationships at the population level. *American Naturalist* 94: 213–236.

———. 1962. *Growth and regulation of animal populations*. New York: Holt, Rinehart and Winston.

Smith, F. E. 1969. Effects of enrichment in mathematical models, in *Eutrophication: causes, consequences, corrections*. Washington, DC: National Academy of Sciences, pp. 631–645.

Springer, A. M., J. A. Estes, G. B. van Vliet, T. M. Williams, D. F. Doak, E. M. Danner, K. A. Forney, and B. Pfister. 2003. Sequential megafaunal collapse in the North Pacific Ocean: an ongoing legacy of industrial whaling? *Proceedings of the National Academy of Sciences of the United States of America* 100: 12223–12228.

Strong, D. R. 1992. Are trophic cascades all wet? Differentiation and donor-control in speciose ecosystems. *Ecology* 73: 747–754.

Sugimoto, T., S. Kimura, and K. Tadokoro. 2001. Impact of El Niño events and climate regime shift on living resources in the western North Pacific. *Progress in Oceanography* 49: 113–127.

Sundby, S. 2000. Recruitment of Atlantic cod stocks in relation to temperature and advection of copepod populations. *Sarsia* 85: 277–298.

Suryan, R. M., D. B. Irons, M. Kaufman, J. Benson, P. G. R. Jodice, D. D. Roby, and E. D. Brown. 2002. Short-term fluctuations in forage fish availability and the effect on prey selection and brood-rearing in the black-legged kittiwake *Rissa tridactyla*. *Marine Ecology Progress Series* 236: 273–287.

Sydeman, W. J., M. M. Hester, J. A. Thayer, F. Gress, P. Marin, and J. Buffa. 2001. Climate change, reproductive performance and diet composition of marine birds in the southern California Current system, 1969–1997. *Progress in Oceanography* 49: 309–329.

Tegner, M. J. and P. K. Dayton. 1999. Ecosystem effects of fishing. *Trends in Ecology and Evolution* 14: 261–262.

Tella, J. L., M. G. Forero, M. Bertellotti, J. A. Donázar, G. Blanco, and O. Ceballos. 2001. Offspring body condition and immunocompetence are negatively affected by high breeding densities in a colonial seabird: a multiscale approach. *Proceedings of the Royal Society of London B (Biology)* 268: 1455–1461.

Thompson, P. M. and J. C. Ollason. 2001. Lagged effects of ocean climate change on fulmar population dynamics. *Nature (London)* 413: 417–420.

Trillmich, F. and K. A. Ono, eds. 1991. *Pinnipeds and El Niño: responses to environmental stress. Ecological Studies*, vol. 88. Berlin: Springer-Verlag.

Trillmich, F., K. A. Ono, D. P. Costa, R. L. DeLong, S. D. Feldkamp, J. M. Francis, R. L. Gentry, C. B. Heath, B. J. LeBoeuf, P. Majluf, and A. E. York. 1991. The effects of El Niño on pinniped populations in the eastern Pacific, in *Pinnipeds and El Niño: responses to environmental stress*. F. Trillmich and K. A. Ono (eds.). *Ecological Studies*, vol. 88. Berlin: Springer-Verlag, pp. 247–270.

Underwood, L. A. and T. J. Stowe. 1984. Massive wreck of seabirds in eastern Britain, 1983. *Bird Study* 31: 79–88.

Ushakov, N. G. and D. V. Prozorkevich. 2002. The Barents Sea capelin—a review of trophic interrelations and fisheries. *ICES Journal of Marine Science* 59: 1046–1052.

Uttley, J. D., P. Walton, P. Monaghan, and G. Austin. 1994. The effects of food abundance on breeding performance and adult time budgets of guillemots *Uria aalge*. *Ibis* 136: 205–213.

Valle, C. A. and M. C. Coulter. 1987. Present status of the flightless cormorant, Galapagos penguin and greater flamingo populations in the Galapagos Islands, Ecuador, after the 1982–83 El Niño. *Condor* 89: 276–281.

Vandenbosch, R. 2000. Effects of ENSO and PDO events on seabird populations as revealed by Christmas Bird Count data. *Waterbirds* 23: 416–422.

Veit, R. R., J. C. McGowan, D. G. Ainley, T. R. Wahl, and P. Pyle. 1997. Apex marine predator declines ninety percent in association with changing oceanic climate. *Global Change Biology* 3: 23–28.

Veit, R. R., P. Pyle, and J. A. McGowan. 1996. Ocean warming and long-term change in pelagic bird abundance within the California Current System. *Marine Ecology Progress Series* 139: 11–18.

Veit, R. R., E. D. Silverman, and I. Everson. 1993. Aggregation patterns of pelagic predators and their principal prey, Antarctic krill, near South Georgia. *Journal of Animal Ecology* 62: 551–564.

Vilhjálmsson, H. 2002. Capelin *(Mallotus villosus)* in the Iceland-Greenland-Jan Mayen ecosystem. *ICES Journal of Marine Science* 59: 870–883.

Walsh, J. J. and C. P. McRoy. 1986. Ecosystem analysis in the southeastern Bering Sea. *Continental Shelf Research* 5: 259–288.

Welch, D. W., A. I. Chigirinsky, and Y. Ishida. 1995. Upper thermal limits on the oceanic distribution of Pacific salmon *(Oncorhynchus* spp.) in the spring. *Canadian Journal of Fisheries and Aquatic Sciences* 52: 489–503.

Welch, D. W., Y. Ishida, and K. Nagasawa. 1998a. Thermal limits and ocean migrations of sockeye salmon *(Oncorhynchus nerka)*: long-term consequences of global warming. *Canadian Journal of Fisheries and Aquatic Sciences* 55: 937–948.

Welch, D. W., Y. Ishida, K. Nagasawa, and J. P. Everson. 1998b. Thermal limits on the ocean distribution of steelhead trout *(Oncorhynchus mykiss)*. *North Pacific Anadromous Fish Commission Bulletin* 1: 396–404.

Wespestad, V. G., L. W. Fritz, W. J. Ingraham, and B. A. Megrey. 2000. On relationships between cannibalism, climate variability, physical transport, and recruitment success of Bering sea walleye pollock *(Theragra chalcogramma)*. *ICES Journal of Marine Science* 57: 272–278.

Wiens, J. A. and J. M. Scott. 1975. Model estimation of energy flow in Oregon coastal seabird populations. *Condor* 77: 439–452.

Wishner, K., E. Durbin, A. Durbin, M. Macaulay, H. Winn, and R. Kenney. 1988. Copepod patches and right whales in the Great South Channel off New England. *Bulletin of Marine Science* 43: 825–844.

Wittenberger, J. F. and G. L. Hunt. 1985. The adaptive significance of coloniality in birds, in *Avian Biology VIII*. D. S. Farner, J. R. King, and K. Parkes (eds.). New York: Academic Press, pp. 1–78.

Woodley, T.H. and D.E. Gaskin. 1996. Environmental characteristics of North Atlantic right and fin whale habitat in the lower Bay of Fundy, Canada. *Canadian Journal of Zoology* 74: 75–84.

Worm, B. and R. A. Myers. 2003. Meta-analysis of cod-shrimp interactions reveals top-down control in oceanic food webs. *Ecology* 84: 162–173.

Wyllie-Echeverria, T. and W. S. Wooster. 1998. Year-to-year variations in Bering Sea ice cover and some consequences for fish distributions. *Fisheries Oceanography* 7: 159–170.

Yaragina, N. A. and C. T. Marshall. 2000. Trophic influences on interannual and seasonal variation in the liver condition index of Northeast Arctic cod *(Gadus morhua)*. *ICES Journal of Marine Science* 57: 42–55.

Zador, S. G. and J. F. Piatt. 1999. Time-budgets of common murres at a declining and increasing colony in Alaska. *Condor* 101: 149–152.

Zaret, T. M. and R. T. Paine. 1973. Species introduction in a tropical lake. *Science* 182: 449–455.

WHALES AND WHALING

SEVEN

Evolutionary Patterns in Cetacea
Fishing Up Prey Size through Deep Time

DAVID R. LINDBERG AND NICHOLAS D. PYENSON

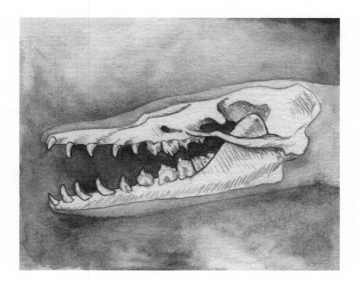

Food webs are dominant features of ecosystems, and predator-prey interactions are important linkages between the species that populate them. Community-level structuring forces are often invoked to explain food web dynamics, and these forces have been characterized as predominantly being either bottom-up or top-down processes (Hunter and Price 1992; Power 1992). In marine systems, the bottom-up perspective focuses on how resources and physical factors (e.g., light, nutrients, and temperature) influence food webs, beginning with phytoplankton and proceeding up through higher trophic levels. In contrast, a top-down view focuses on the influence of predators and follows their impact down to lower trophic levels (e.g., sea otters → sea urchins → kelp). Thus, in overtly simplified parlance, bottom-up systems are food- or resource-limited, whereas structure in top-down systems is driven primarily by predation.

Structuring in most marine ecosystems has overwhelmingly been viewed as resulting from bottom-up processes (Steele 1974; Lipps and Mitchell 1976; Baker and Clapham 2004). However, top-down effects are undeniably responsible for structuring communities in several nearshore marine ecosystems in the northern hemisphere (Paine 1966; Estes and Duggins 1995; Lindberg et al. 1998). Recently, Springer et al. (2003) have suggested that top-down effects may also have been historically important in structuring marine ecosystems through killer whale *(Orcinus orca)* predation on the great whales (primarily Mysticeti and Physeteridae). Although the data supporting the Springer et al. (2003) hypothesis are compelling, we contend that their hypothesis should be examined on an evolutionary timescale as well; at a scale that requires a deeper historical approach than is often used in contemporary ecological and oceanographic studies. From the perspective of a modern ecologist, historical approaches tend to lose resolution as temporal and spatial scales increase. However, a macroscopic view of time provides robust controls for other, larger-scale events in the history of a community or an ecosystem (Gould 1986; Cleland 2001). Moreover, when ecological interactions are especially strong, or when habitats (or niches) are exploited anew, the opportunity to see patterns and examine potential correlations becomes greater, with varying levels of mechanistic resolution (e.g., Barnosky 2001; Barnosky and Bell 2003).

Many ecological studies begin with the premise that modern-day communities and their interactions represent fully integrated settings that evolved synchronously. This approach circumvents the temporal and spatial histories of the constituent organisms, their probable interactions, and how a system may have changed with the addition or subtraction of key components (see also Fowler and MacMahon 1982; O'Hara 1988; Lipps and Culver 2002). When historical influences on taxa are considered, many researchers perceive the ecological consequences on the scale of human lifetimes (see Brown 1995). More recently, an expanded temporal view has entered some studies, as investigations have started to consider and acknowledge the presence of human intervention and alteration of ecosystems over scales several orders of magnitude greater than a human lifetime. In marine ecosystems, the ecological ramifications of anthropogenic influences—such as overhunting, habitat fragmentation, or the introduction of exotic species—have been demonstrated to culminate in the widespread extinction of multiple high-trophic-level taxa, significantly altering ecosystem structure (Estes et al. 1998; Castilla 1999; Jackson et al. 2001). Nonetheless, even the aforementioned temporal scales represent only a minuscule part of the geologic or "deep" time (see McPhee 1998) of a taxon's history and its niche.

Here we examine and discuss the evolution of Cetacea in the context of their role as pelagic and nearshore predators in modern ecosystems. Human historical accounts (Scammon 1874) and other popular anecdotes (Melville 1851; Mowat 1972; Scheffer 1969) describe cetaceans as familiar members of modern pelagic and nearshore systems, yet, from an evolutionary perspective on the scale of geologic time, cetaceans (whales, dolphins, and porpoises) have been present in marine systems for less than 10% of metazoan history. Furthermore, the presence of cetaceans in marine systems pales in comparison to that of other well-known community components such as mollusks, arthropods, and bony fish, which have been present for 100%, 100%, and 70% of marine metazoan history, respectively (data from Sepkoski 2002). Moreover, all of the latter taxa are organisms with which cetaceans have strong trophic interactions. Thus, the presence of cetaceans in marine systems represents a recent evolutionary event, compared with the over half-billion-year history of metazoan marine life. Within their comparatively short history in marine systems, cetaceans have made a dramatic transition from obscurity and nonexistence (terrestrial ancestry) to trophic dominance and marked ecological presence, relative to the other ecological players with which they interact.

Methods

We begin by examining cetacean diversification patterns through time using data derived from two recent compendia of genus-level originations and extinctions: the McKenna and Bell (1997) mammal database and the Sepkoski (2002) marine animal database. These data are placed in a chronological context, allowing comparisons with co-occurring patterns in morphology, size, ecology, and earth history, including tectonic, climatic, and oceanographic changes. We also discuss possible signals in co-occurring taxa as well as convergent events with earlier vertebrate reentrants into the sea. Data from the cetacean fossil record bear significantly on our understanding of their current structural and functional role in modern

marine ecosystems, because fossils elucidate the sequence of morphological and taxonomic diversity changes that provide the deep sequence of historical events, or chronicle (*sensu* Lindberg and Lipps 1996), that provides the much needed context for understanding the history of marine communities.

Cetacean Evolution

The mammalian heritage of cetaceans has been recognized since Aristotle. This evolutionary relationship was further resolved in the nineteenth century, as observations of skeletal and soft-tissue morphology prompted Flower (1885) to posit a close evolutionary relationship with ungulates. Van Valen (1966) first suggested that cetaceans were related to an extinct lineage of hoofed Paleogene mammals called mesonychids. Both molecular and morphological cladistic analyses now confirm that cetaceans are a monophyletic group nested within or sister to Artiodactyla (even-toed ungulates), although for many years molecular data alone identified the hippopotamids as cetaceans' extant sister taxon (Gatesy and O'Leary 2001) (Figure 7.1). Recent analyses of morphological (i.e., including fossil) data confirm that cetaceans are the sister taxon of hippopotamids (Boissièrie et al. 2005; Geisler and Uhen 2005), leaving the position of mesonychids unresolved, but likely not related to either cetaceans or hippopotamids (Geisler and Uhen 2005; see also Berta et al. 2006 for a review).

Extensive paleontological fieldwork in the past 30 years (Thewissen and Williams 2002; Gingerich et al. 2001; Fordyce and Muizon 2001) has clarified the picture of early cetacean evolution, documenting both the terrestrial ancestry of early cetaceans and the subsequent evolutionary transition of cetaceans to obligate marine lifestyles. The oldest known cetaceans have been recovered from coastal and fluvial deposits in the Indo-Pakistan region of South Asia, dating to approximately 55 Ma (million years ago). Collectively known as Archaeoceti, these early cetaceans had four limbs, heterodont (generalized mammalian) dentitions, and shared basicranial, facial, vertebral, and forelimb characters with modern cetaceans. Extant cetaceans, in contrast, exhibit much greater size range, retain only vestigial pelvic bones, and have a diversity of homodont tooth types, including independent losses of all dentition in several lineages. From the middle to late Eocene, the paraphyletic archaeocetes exhibited sequential changes from a quadrupedal terrestrial lifestyle to semiaquatic and aquatic to marine living, marked most obviously by the posterior telescoping of the facial bones and nares (Figure 7.2A–D), and the sequential decoupling of the sacrum from the vertebral column (Thewissen and Fish 1997, Thewissen and Williams 2002, Gingerich et al. 2001).

The chronology of Late Eocene archaeocete localities in the modern Indo-Pakistan region suggests a relatively rapid morphological radiation into distinct lineages (Thewissen and Williams 2002) in less than 10 million years. The five described families of archaeocetes (Pakicetidae, Ambulocetidae, Remingtonocetidae, Protocetidae, and Basilosauridae) range in size from pakicetids that were about the size of a fox

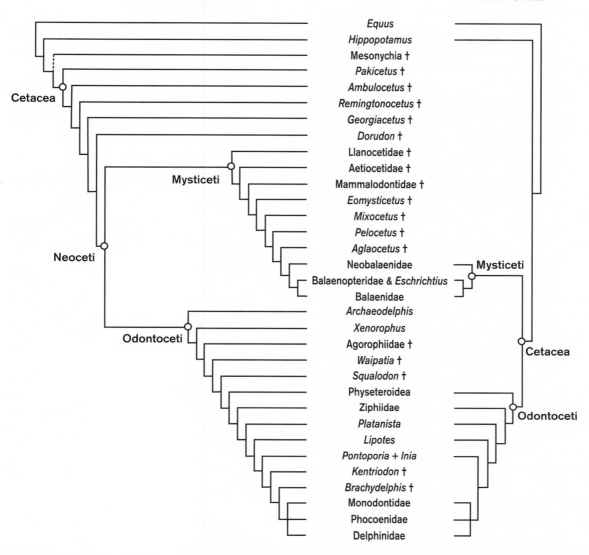

FIGURE 7.1. Composite phylogeny of the Cetacea based on morphological (Kimura and Ozawa 2002; Geisler and Sanders 2003; Uhen 2004) and molecular (Arnason et al. 2004) analyses. Outgroup relationships modified from Gatesy and O'Leary (2001).

(Thewissen and Bajpai 2001) to eel-shaped basilosaurids that exceeded 15 m in length (Gingerich et al. 1990). These taxa also display different feeding specializations, concomitant with different locomotor styles (Thewissen and Williams 2002). Later diverging protocetids and basilosaurids were also unique among archaeocetes for their dispersal throughout the Tethys seaway, producing significant fossil localities throughout present-day Jordan, Egypt, and the southeastern United States (Uhen 2004) (Figure 7.2A). Archaeocete fossils from the Eocene are also represented in higher temperate latitudes including Europe, North America, New Zealand, and Antarctica (Köhler and Fordyce 1997, Fordyce and Muizon 2001, Uhen 2004) (Figure 7.2A).

Cetacean diversity in the early Oligocene (~33 Ma) shows a marked reduction in archaeocetes, concurrent with the

first appearances of echolocating and baleen-bearing cetaceans (odontocetes and mysticetes, respectively, which together are also called Neoceti [Fordyce 2003]). Marine rock outcrop from the early Oligocene is generally poor (Uhen and Pyenson 2002); however, middle to late Oligocene localities in New Zealand, Japan, and North America indicate that early odontocetes and mysticetes were quite unlike modern forms. So-called archaic toothed mysticetes, such as *Aetiocetus*, exhibited heterodont dentition with denticulate postcanine cheek teeth (Fordyce 2003), whereas living mysticetes have only embryonic teeth, which are resorbed before birth. Early odontocetes, such as *Simocetus,* retained plesiomorphic characters (e.g., intertemporal constrictions) reminiscent of basilosaurids (Fordyce 2002a, Geisler and Sanders 2003). The mosaic of

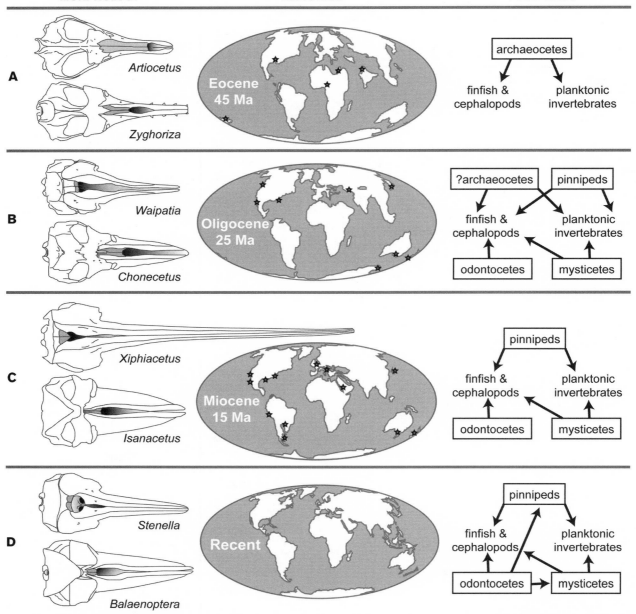

	MORPHOLOGY	TECTONICS	FOOD WEBS

A *Artiocetus* / *Zyghoriza* — Eocene 45 Ma

B *Waipatia* / *Chonecetus* — Oligocene 25 Ma

C *Xiphiacetus* / *Isanacetus* — Miocene 15 Ma

D *Stenella* / *Balaenoptera* — Recent

FIGURE 7.2. Exemplars of cetacean morphological and trophic evolution set against the tectonic context of the past 45 million years. **Morphology** (shaded areas = anterior position of nasal bones): A. *Artiocetus*, a protocetid, and *Zygorhiza*, a basilosaurid, representatives of the morphological diversity of archaeocetes in the Eocene. B. *Waipatia*, an early odontocete, and *Chonecetus*, a toothed mysticete reflecting the emergence of Neoceti in the Oligocene. C. *Xiphiacetus*, a long-beaked odontocete, and *Isanacetus*, a cetotheriid mysticete, reflecting morphological specializations during the Miocene. D. *Stenella* and *Balaenoptera* represent the morphological disparity between modern Odontoceti and Mysticeti. (*Artiocetus* modified from Gingerich et al. [2001]; *Zygorhiza*, *Waipatia*, and *Chonecetus* from Fordyce [2002b]; *Isanacetus* from Kimura and Ozawa [2002]; and *Balaenoptera* from Fordyce and Muizon [2001]). **Tectonics**: Paleogeographic reconstructions from 45 Ma to Recent (Holocene) (Smith et al. 1994). Stars indicate noteworthy fossil localities. **Food webs**: Likely feeding interactions of cetaceans over the past 45 million years based on fossil morphology and phylogenetic relationships of living taxa and their dietary preferences and specializations (Evans 1987 and references therein).

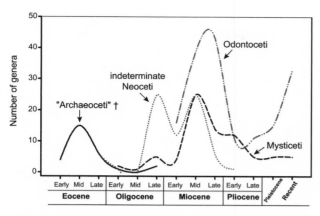

FIGURE 7.3. Changes in the generic diversity of cetaceans over time, including Archaeoceti, indeterminate taxa, Mysticeti, and Odontoceti. Data from Sepkoski (2002).

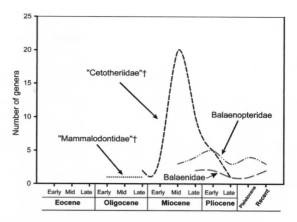

FIGURE 7.4. Changes in mysticete generic diversity over time, including the extinct "Mammalodontidae" and "Cetotheriidae" and the extant Balaenidae and Balaenopteridae (all *sensu* McKenna and Bell 1997). Data from Sepkoski (2002).

characters present in these early neocetes defy the boundaries of traditional taxonomic groupings into toothed whales and baleen-bearing whales and render many fossil neocetes indeterminate for higher nomenclatural ranks (Figure 7.3).

The largest diversification of cetaceans appears to have occurred during the Miocene (23.5–5 Ma) and especially the middle (14.3–11.1 Ma) and late Miocene (11.1–5.1 Ma) (Figure 7.3). Researchers have long recognized this spike in diversity (Barnes 1976), although this taxonomic peak may be the result of multiple historiographic, geologic, or taphonomic biases (Uhen and Pyenson 2002). By the middle Miocene, most representatives of extant families appear, including the mysticete families Balaenopteridae (rorquals) and Balaenidae (bowhead and right whales), as well as the odontocete families Delphinidae (true dolphins), Physeteridae (sperm whales), Monodontidae (belugas and narwhals), Phocoenidae (porpoises), and Ziphiidae (beaked whales) (Figure 7.3). Significant Miocene fossil localities are found throughout the world (Figure 7.2C).

The apex of mysticete radiation also occurs during the middle Miocene, coinciding with the diversification of Cetotheriidae, a paraphyletic group of basal mysticetes (Figure 7.2C). Cetotheriids, or "cetotheres," represent one of the largest taxonomic groups of named cetaceans, representing small- to medium-sized archaic mysticetes (Kimura and Ozawa 2002). Cetotheriids evolved during the late Oligocene and were extinct by the end of the late Pliocene, replaced taxonomically and possibly ecologically by members of the Balaenopteridae and Balaenidae beginning in the early Pliocene (Figure 7.4) (Bisconti 2003, Demere et al. 2005).

In general, the broad pattern of odontocete generic diversity parallels that of mysticete diversity during the Neogene, with the exception of a few specific taxa. Whereas mysticetes peaked in generic diversity in the middle Miocene, odontocete generic diversity reached a zenith in the late Miocene. By the Pliocene (5.1–2 Ma), odontocete generic diversity fell precipitously, also paralleling mysticete generic diversity

changes (Figure 7.5). Since the early Pliocene, mysticetes have maintained low generic diversity (20% of the early Miocene value), whereas odontocete generic diversity has rebounded and currently stands at 75% of its late Miocene high (Figure 7.5).

Within odontocetes this pattern of recovery of generic diversity is not universal. Physeteridae, Phocoenidae, Ziphiidae, and two extinct porpoiselike groups (the apparently paraphyletic Kentriodontidae and the grossly polyphyletic Acrodelphinidae) all reach maximum diversification in the middle and late Miocene and then show reduced levels of generic diversity for the remainder of the Neogene. In contrast, Delphinidae and Monodontidae obtain their greatest diversity in the Holocene (Figure 7.5). In the case of Monodontidae, generic diversity has always been low (1 to 3 taxa). However, the delphinid pattern is quite different from these other taxa.

A preponderance of undescribed material suggests that fossil record of delphinids (true dolphins) begins as early as 11–12 Ma, but, based on unequivocal, described taxa, delphinids originated in the early Pliocene, ~5 Ma (Barnes 1990) and underwent an exponential increase in generic diversity into the Holocene, or Recent (Figure 7.5). This initial delphinid increase tracks a rebound in physeterid generic diversity during the late Pliocene, although the physeterid increase plummets between the Pleistocene and the Holocene (Figure 7.5). Thus, Delphinidae appear to be the most recent odontocete diversification. They follow on the earlier diversifications (and subsequent reductions) of other porpoise and dolphin groups such as Phocoenidae, "Kentriodontidae" and "Acrodelphinidae" (see also Barnes et al. 1985). The current Ziphiidae diversification follows a reduction in physeterid diversity (Figure 5). This similar pattern in all three lineages (Delphinidae, Ziphiidae, and Physeteridae) may indicate a similar evolutionary response to increasing cephalopod resources during this time (Figure 7.5). We caution, however,

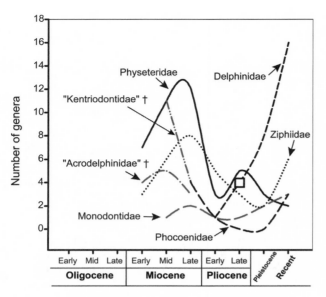

FIGURE 7.5. Changes in odontocete generic diversity over time, including the extinct taxa "Acrodelphidae" and "Kentriodontidae" and the living Monodontidae, Physeteridae, Ziphiidae, Phocoenidae, and Delphinidae (all *sensu* McKenna and Bell 1997). □ = first occurrence of *Orcinus* sp. Data from Sepkoski (2002).

that these patterns of changing diversity, especially as taxonomic groups approach recent geologic time (10,000 years to 1 Ma), might reflect the "pull of the Recent," a sampling problem in which more complete sampling of the Holocene biases diversity patterns toward that epoch (Raup 1979, Jablonski et al. 2003). However, given the adequacy of the fossil record (Benton et al. 2000), especially for cetaceans (Uhen and Pyenson 2002), the broad patterns of generic diversity changes in the cetacean fossil record clearly suggest an evolutionary signal that ought to be tested with additional paleontological data.

Physical Setting

Here we consider the multiple, complex physical environmental factors that occurred during the Cenozoic Era (65 Ma–present). This record of physical factors includes information about oceanographic, climate, and tectonic changes via proxies (e.g., isotopic signatures in foraminfera tests). The chronology of these factors over geological time provides a backdrop of strong environmental forces that affected the evolution of cetaceans. For most of the Paleocene (after the Cretaceous/Tertiary [K-T] extinction and before the earliest known fossil cetaceans), Earth's climate could be characterized as a global "greenhouse," with temperatures much higher than those of today. This regime persisted into the Eocene, which had a similarly warm and equable global climate and ice-free poles. Plate positions also allowed circumglobal seaways though equatorial regions, but they were restricted in high latitudes, especially in the Southern Hemisphere (Figure 7.2A). As the proto-Atlantic Ocean formed, marine transgressions occurred over Africa, Australia, and Siberia.

The Eocene/Oligocene boundary (~34 Ma) demarcates an unparalleled shift in global climate during the Cenozoic era, when a colder, drier "icehouse" world replaced a wet and warm "greenhouse" world (Prothero et al. 2002). This transition in global climate relates to the establishment of the circum-Antarctic current and a marked shift in the structure of the Southern Ocean in particular (Fordyce 2003). Many of these global changes have been attributed to separation of Antarctica from South America and the opening of the Drake Passage in the late Eocene (~41 Ma), which permitted circum-Antarctic flow and initiated global cooling (Scher and Martin 2006). The Southern Ocean cooled first, with expansive glaciation of the Antarctic by 43–42 Ma (Browning et al. 1996), although Ivany et al. (2006) have supplied evidence for much more extensive glaciation of Antarctica by the late Eocene. Paleogeographic reconstructions during the Oligocene (33.7–23.8 Ma) indicate that the previously insular subcontinent India docked with the Eurasian plate, and Gondwana ceased to exist as Australia and South America separated from Antarctica (Figure 7.2B).

The Miocene (23.5–14.3 Ma) marked a reversal of the Oligocene cooling trend, and global temperatures rose in the early Miocene, peaking ~17–14 Ma with the mid-Miocene Climatic Optimum (Zachos et al. 2001). By the late Miocene, cooling again prevailed, especially in the Northern Hemisphere, and an increase in high-latitude cooling in the North Pacific Ocean has been dated between 10.0 and 4.5 Ma (Barron 2003). This cooling in the Northern Hemisphere was accompanied by intensification of coastal upwelling (Jacobs et al. 2004), which further contributed to the cooling of coastal waters in the North Pacific (Barron 1998) (Figure 7.2C).

Global cooling of low and temperate latitudes culminated in Pliocene and Pleistocene glaciations, especially in the northern hemisphere. Tectonic and eustatic events were also prominent during this period, and several major tectonic events (such as the collision of Australia and Indonesia, the closure of the Panama portal, and the closure of the Bering Strait) limited cetacean dispersal and dispersion (Figure 7.2D). Cetacean habitats were also likely lost with the closure and diminution of the Mediterranean Sea (the relictual Tethys Sea) during the Messianian Salinity Crisis, when the entire Mediterranean evaporated and refilled repeatedly about 6 Ma (Hsu et al. 1973, Mackenzie 1999).

Eustatic events associated with Pleistocene glaciations would have substantially reduced nearshore feeding areas for cetaceans, especially in the North Pacific Ocean, where much of today's feeding areas for humpback (*Megaptera novaeangliae*) and, especially, gray whales (*Eschrichtius robustus*) are found in the relatively shallow waters of the Bering, Chukchi, and Okhotsk Seas. Repeated sea level drops transformed these feeding areas into exposed land (e.g., Beringia) and substantially altered migration routes by shrinking passages between today's islands. If population sizes were substantially diminished as a result of the loss of these northern feeding areas, it is likely that the effects of Plio-Pleistocene glaciations registered in the genetic history of these lineages (Medrano-Gonzalez et al. 2002, see also Berta and Sumich 1999: 309).

Evolution of Ecological Role

The evolution of cetaceans, like that of any other organism, did not take place in an ecological vacuum. Just as the physical setting played a role in cetacean evolution, countless trophic interactions, competitors, and predators have also undoubtedly had impacts on cetacean evolution over the last 50 million years. Moreover, because multiple feedback loops among physical environment, morphological evolution, and ecological interactions mediate the evolutionary history of any group, recognizing the significance of fundamental ecological constraints, as well as novel ecological opportunities, in historical reconstructions can often help delimit and falsify alternative evolutionary scenarios.

For example, cetacean gross morphology (streamlined body, forelimb flippers, tail fluke) is not unique—sirenians (manatees and dugongs) share all of these traits. Because sirenians and cetaceans evolved from distantly related terrestrial mammal ancestors, their convergence on an obligate marine existence hints at fundamental constraints for mammals living in a marine environment. This particular comparison carries additional significance because sirenians evolved within roughly the same temporal framework as the cetaceans, with the earliest sirenian fossils present in the Eocene of Jamaica, ~50 Ma (Domning 2001). Like early cetaceans, the earliest sirenians were fully quadrupedal and capable of terrestrial locomotion, with weight-bearing limbs and a multivertebral sacrum coupled to a pelvic girdle (Domning 2001). Over time, sirenians and cetaceans show parallel morphological adaptations to the aquatic environment, including the loss of hind limbs, enlargement of the tail to serve as the main propulsive structure, modification of the forelimbs into flippers, a hydrodynamic body shape, loss of hair, the use of blubber for insulation, and other physiological modifications. And even though both taxa started from very different mammalian lineages, there is no suggestion of morphological constraints limiting sirenian evolutionary potential. However, the comparative diversification between sirenians and cetaceans through time shows a substantial difference in their relative success (Figure 7.6). Whereas cetaceans show maximum generic rank diversity in the middle Miocene, when the oceans teemed with 90 genera, sirenian diversification hit an all-time high in the early Miocene with 8 genera. Total generic rank diversity in cetacean history includes over 220 taxa (39 extant; 17%), whereas sirenian history is represented by only 31 genera (2 extant; 6%) (data from Sepkoski 2002).

In an ecological context, we would argue that the convergent evolutionary trajectories of sirenians and cetaceans represent

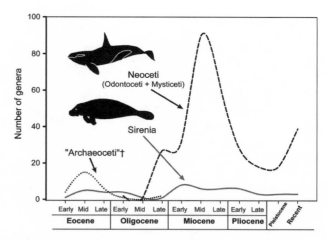

FIGURE 7.6. Comparison of Archaeoceti, Neoceti, and Sirenia generic diversity over time (data from McKenna and Bell 1997).

a *prima facie* case of the importance of trophic constraints. Sirenians are herbivorous, and they share common ancestry with other herbivorous mammals such as hyracoids and proboscideans (Gheerbrant et al. 2005). Cetaceans, on the other hand, are nested within artiodactyls—a group comprising living and extinct herbivorous, omnivorous, and carnivorous taxa. From an evolutionary standpoint, sirenians were trophically constrained by an herbivorous ancestry to shallow, photic-zone waters with abundant marine angiosperms or algae. Early cetaceans did not have similar feeding constraints, so they were free to feed literally anywhere across the ocean and to depths in excess of 1500 m.

The success of cetaceans as marine reentrants can be attributed, in part, to their trophic position and the lack of any obvious ecological constraints. The terrestrial-to-aquatic transition occurred rapidly, but archaeocetes could hear underwater early in their evolution (Luo and Gingerich 1999; Spoor et al. 2002), and fossil stomach contents from late Eocene basilosaurids reveal fish as prey items (Uhen 2004) (Figure 7.2A). By the early Oligocene (34 Ma), pelagic feeding had evolved (Fordyce 2003) and photic-zone cephalopods were likely food items for archaic toothed mysticetes, toothless baleen-bearing mysticetes, and deep-diving, echolocating odontocetes (Figure 7.2B). Ecological portioning among Oligocene neocetes likely differed greatly from that among living neocetes, as "archaic toothed" mysticetes likely used completely nonanalogous feeding strategies (Fordyce 2003) (Figure 7.2C).

Feeding

Traditionally, the classification of cetacean feeding ecology follows a sharp dichotomy: Odontocetes are active hunters pursuing mobile prey, whereas mysticetes are planktonic grazers, often ecologically equated with cows in a field. In reality, both clades are carnivorous, and the dichotomy of feeding styles reflects different feeding apparatus modes (raptorial verus filter-feeding) and different proportions of fish, cephalopod, and arthropod prey items (Gaskin 1982). Here we discuss the dietary components of different extant cetacean groups [summarized below from Evans (1987) and references therein], and the evolutionary implications thereof.

Among mysticetes, balaenids (bowhead and right whales) feed almost exclusively on arthropods (copepods and euphasiids), while about half of the members of Balaenopteridae include fish and cephalopods in their diets. Odontocetes such as physeterids and ziphiids are primarily cephalopod feeders, but in some regions fish can also be a substantial component of their diet. Similarly, phocoenids and delphinids feed primarily on fish and cephalopods. Monodontids feed on fish and cephalopods, but their diet incorporates crustaceans and other benthic invertebrates. This array of feeding compositions shows similar prey utilization patterns within both mysticetes and odontocetes: Each suborder has lineages that feed predominantly on either arthropods (Balaenidae) or cephalopods (Physeteridae and Ziphiidae). Notably, other lineages within odontocetes, such as Monodontidae and Delphinidae, also include arthropods in their diets, whereas several balaenopterid lineages also feed, sometimes predominantly, on fish and cephalopods.

When the phylogeny and the fossil record of cetaceans are combined with the feeding ecology of extant taxa, a polarity for these patterns emerges. Both morphological (Geisler and Sanders 2003) and molecular (Nikaido et al. 2001; Arnason et al. 2004; May-Collado and Agnarsson 2006) data sets (Figure 7.1) resolve the living clades of Balaenidae and Balaenopteridae as sister taxa. Balaenopterids appear in the middle Miocene (14.3–11.1 Ma), with a generic diversity almost twice as high as that of balaenids, which do not appear until the late Miocene (11.1–5.1 Ma) (Figure 7.4). Whereas balaenopterid diversity peaks in the early Pliocene, their generic maxima is only 25% of the middle Miocene cetotheriid diversity peak (Figure 7.4). Like balaenopterids, the size-diverse "cetotheres" most likely exploited a wide range of prey types—arthropods, fish, and cephalopods—all of which would have seen an increase of abundance during Miocene cooling and enhanced global upwelling. Balaenids, which appear later, appear to have narrowed their prey range to predominantly arthropods.

The odontocete pattern is also supported by both morphological (Geisler and Sanders 2003) and molecular (Nikaido et al. 2001; Arnason et al. 2004; May-Collado and Agnarsson 2006) data sets. The cephalopod specialists (Physeteridae and Ziphiidae) are basal odontocetes, and both achieve their greatest generic diversity during the late Miocene (Figure 7.5). All groups of marine odontocetes feed on fish and cephalopods, but some species of phocoenids, delphinids, and monodontids have added arthropods to this ancestral diet. We therefore hypothesize that the common diet at the time of divergence between odontocetes and mysticetes (i.e., the origin of Neoceti) most likely included cephalopods and fish, but probably not arthropods (Figure 7.2B).

The importance of cephalopod biodiversity in the evolution of cetaceans cannot be overstated. Clarke (1996) has documented that over 80% of all living odontocete species and two mysticete species regularly feed on cephalopods, and for 28 odontocete species cephalopods constitute their main food item.

This intensity of predation pressure on cephalopods had likely not occurred since another terrestrial vertebrate group (diapsid reptiles) reentered the ocean during the Mesozoic (245–65 Ma) (Vermeij and Dudley 2000). In contrast to the handful of mammalian reentries, the appearance and evolution of aquatic diapsid reptiles have occurred much more frequently, with more than 20 independent entries (Carroll 1988), most of which occurred during the Mesozoic Era (245–65 Ma). With the exception of some sea turtles, all were highly predaceous taxa. Of these re-entrants, ichthyosaurs, plesiosaurs, and mosasaurs were the most conspicuous denizens during the Mesozoic. Fossil stomach contents from these groups suggest that fish, other marine reptiles, and especially cephalopods (including the extinct belemnites and ammonites) were the primary food source of these marine reptiles (Sato and Tanabe 1998; Cicimurri and Everhart 2001; Kear et al. 2003). Marine reptiles ranged from salmon-sized to mysticete-sized (Carroll 1988)—remarkably similar to the size range seen in cetaceans today—and, based on the size of the eye sockets in some ichthyosaur species, it has been suggested that these animals were hunting in deep water for their prey (Motani et al. 1999).

Marine reptile genera declined sharply during the Cretaceous (146–65 Ma), and only turtles, crocodiles, sea snakes, and a single marine iguana remain in the oceans today (data from Sepkowski 2002; Vermeij and Dudley 2000). Between the extinction of plesiosaurs and mosasaurs at the K-T boundary (65 Ma) and the evolution of deep-diving odontocetes, there were very few, if any effective predators on mid- and deep-water cephalopods (see also Lipps and Culver 2002). With the exception of two shark species, surprisingly few living bony fish taxa feed exclusively on cephalopods (Smale 1996). Although other, extinct sharks may have exploited Paleogene cephalopod populations, cephalopods typically constitute less than 5–6% of any living sharks' diet (Australian Fisheries Management Authority 2001; McCord and Campana 2003). Consequently, it is likely that archaeocetes had little competition with other cephalopod predators during the middle to late Eocene radiation and that the subsequent evolution of large body size and deep diving in odontocetes provided access to unexploited cephalopod biodiversity that had been accumulating for nearly 35 million years.

Body Size

Although the word *whale* typically conjures up a thought of something enormous, living cetaceans show a wide range of sizes as well as some discrete bundling of size classes by taxon. The distribution of size classes among fossil

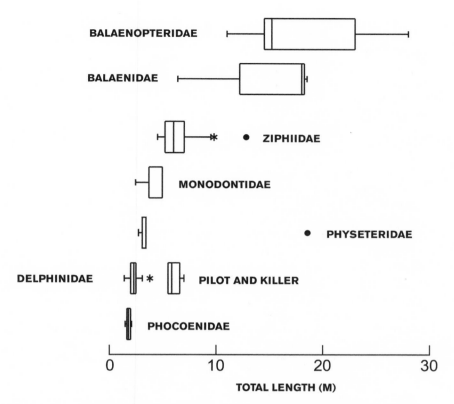

FIGURE 7.7. Boxplots of mean adult body lengths for extant odontocete and mysticete family rank taxa. Boxplots display sample mean, quartiles, and outliers (*, ●). A separate plot is shown for pilot and killer whale data in addition to their inclusion in the Delphinidae value. All outliers (*, ●) are predominantly cephalopod-feeding taxa. Data from Evans (1987) and references therein.

cetaceans remains an open question, and the absence of data from extinct taxa can be attributed to the lack of complete, articulated cetacean skeletons in the fossil record (Barnes et al. 1985). It may be possible to estimate size in fossil taxa using cranial proxies based on allometric relationships in extant species, but these methodologies are only now being investigated (Pyenson and Lindberg 2003; Pyenson, unpublished data).

Figure 7.7 shows adult size data for all living species of marine cetaceans, grouped at family-level rank (data from Evans 1987). Within odontocetes, family taxa are discretely grouped with little overlap in size ranges except for ziphiids and the largest delphinids where pilot and killer whales have converged on the size class structure seen in most ziphiids (Figure 7.7). Three clades feeding predominantly on larger cephalopods (Physeteridae, Ziphiidae, and Delphinidae) have members that are significant outliers from the rest of their taxon, resulting in the increased size variation within each clade. The two mysticete groups show more overlap in size class structure within each other than the odontocetes (Figure 7.7), as well as a wide range of internal size variation.

The increase in size variation is strongly correlated with cetacean length (Figure 7.8) ($r^2 = 0.9200$). In some cases this pattern appears to result because ancestral size ranges are not lost when a component of a clade diverges and becomes larger. This phenomenon can be observed in Figure 7.8, in which dual data points are presented for the Physeteridae, Ziphiidae, and Delphinidae. The solid circles represent values with the larger outlying species excluded, while the open circles represent values that include all members of the group. With the outliers excluded, cetacean taxa show a strong correlation between a taxon's mean size and the amount of size variation it contains. Furthermore, when killer and pilot whales are considered separately from their delphinid brethren, they also show a similar correlation (Figure 7.8). With the outliers included, all three taxa show the expected higher standard deviations and means. In all three cases the open circles (large taxa included) are offset at an angle of approximately 65° from the solid (large taxa excluded) data points. Although the vectors between excluded and included taxa for the ziphiid and delphinid groups are virtually identical in length, the physeterid vector is nearly seven times as long. These data may reflect deep phylogenetic divergence (i.e., long branch lengths), high extinction rate (Figure 7.5), or the outcome of developmental and macroevolutionary processes that would bypass intermediate sizes in physeterid taxa.

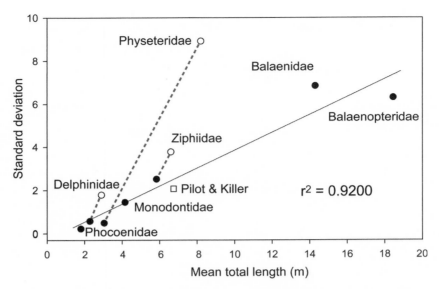

FIGURE 7.8. Disparity of variation in cetacean body size, with standard deviation plotted against mean length (meters). Dual data points are presented for the Physeteridae, Ziphiidae, and Delphinidae. Solid circles (●) represent values with the larger, predominantly cephalopod-feeding taxa excluded, whereas the open circles (○) include all members of the taxon. A separate data point for pilot and killer whales (□) is also shown. Data from Evans (1987) and references therein.

Discussion

Springer et al. (2003) hypothesize that the removal of the great whales by commercial whaling in the North Pacific Ocean set off a cascade of prey switches by killer whales that has ultimately led to a switch between alternate stable states in nearshore kelp forest communities (Estes 1996, Estes et al. 1998). Springer et al. (2003) propose an ecological mechanism to explain their hypothesis, similar to what Pauly et al. (1998) termed "fishing down the food web"—the removal of top consumers, trophic level by trophic level, successively extirpating each population of consumers. Springer et al. (2003), however, did not explicitly propose fishing down through trophic levels to explain the sequential collapse of marine mammal populations; rather, their pattern proceeded down marine mammal size classes. This use of size rather than trophic level is an important distinction, because it provides another perspective from which to examine the importance of top-down effects on trophic systems. More critically, body size can also be read directly and evaluated in the fossil record.

Fishing Up Prey Size in Deep Water and Deep Time

Our analysis of the evolutionary and ecological patterns of fossil and living cetacean lineages reveals two important points: (1) the role of cephalopod biodiversity in the cetacean evolution and (2) taxic size increases in different lineages specializing in cephalopod prey species (Physeteridae, Ziphiidae, large Delphinidae). The latter two observations have important implication in the context of Clarke's

(1996) observation that cetacean body size is positively correlated with the size of their cephalopod prey species. We propose that Clarke's correlation represents an evolutionary trajectory in cetacean evolution: As cetaceans independently evolved large body sizes (e.g., in Physeteridae, Ziphiidae, large Delphinidae), the ancestral cetaceans of each lineage successively "fished up" prey size through evolutionary time. The current, and perhaps upper, limit on this escalation (*sensu* Vermeij 1987) is the 18-m, 57,000-kg sperm whale feeding on an 18-m, 900-kg squid at depths in excess of 1000 m.

Fishing up prey size in the Cenozoic likely began with archaeocetes and their prey, although it may have occurred in the Mesozoic with marine reptiles that had similar size ranges, skull morphology, and tooth types to cetaceans (Ciampaglio et al. 2006; Massare 1987). Marine reptiles, however, apparently did not achieve engulfment filter feeding like that of balaenopterids. Collin and Janis (1997) have argued that diapsid oral and pharyngeal anatomy constrained this mode of foraging, whereas mammals evolved both facial musculature and coordinated swallowing reflexes early in their history. The recent discovery of mysticete-sized ichthyosaurs from the late Triassic of British Columbia that lack adult dentition (Nicholls and Manabe 2004) may necessitate reconsideration of Collin and Janis' hypothesis.

In addition to fishing up prey size through time, cetaceans also evolved to take relatively small but abundant arthropods in the photic zone, which likely selected for an increase in cetacean size not because their prey size increased, but rather because of scaling features associated with filter feeding, prey abundance, and metabolism (Webb 1984).

Did Killer Whales Evolve to Feed on Great Whales?

The answer to the question posed in the heading above is "Yes and no." It is unlikely that killer whales evolved large body size to feed on great whales. Rather, as we have hypothesized in the preceding paragraphs, the delphinid lineage that led to *Orcinus* fished up prey sizes, eventually producing a taxon with the *ability* to feed on great whales. Killer and pilot whales are clearly size outliers among the delphinids (Figure 7.7). With the exception of *Orcinus*, these larger delphinids are primarily deep-diving cephalopod feeders (Gaskin 1982; Clarke 1996), and even *Orcinus* feeds on both neritic and oceanic cephalopods (Clarke 1996). Because cephalopod feeding remains predominant in all delphinids (all but two species are known to feed on cephalopods; Clarke 1996) as well as in their marine odontocete outgroups (Physeteridae, Ziphiidae), we posit that killer whales did not become large to feed on great whales but rather that they feed on great whales because their ancestors were large. Great-whale feeding is thus exaptive in the *Orcinus* lineage (Gould and Vrba 1982; and see Diamond 1987 for similar arguments about the size of Komodo dragons).

The evolution of cooperative hunting also appears to be a crucial component in great-whale predation (Scammon 1874), but in the proposed evolutionary scenario for killer whales, cooperative hunting was likely present in the smaller, ancestral delphinids (Norris and Dohl 1980; Connor and Norris 1982), and therefore cooperative hunting was not an adaptation to become a whale killer. The only unique modifications of killer whales for feeding on large prey items such as great whales appears to be in tooth spacing, size, shape, and feeding musculature (Seagars 1983).

The fossil record of *Orcinus orca* is marked by the presence of dissociated, large delphinid teeth from the Pliocene of Italy (Heyning and Dahlheim 1988). A putative fossil relative, *O. citoniensis* (Capellini 1883), also from the Pliocene of Italy, is represented by partial cranial and postcranial material. Proportional estimates place *O. citoniensis* at about 4 m in length (Bianucci 1997, Pyenson, personal observation, 2005), a size range that falls below the maxima of living *Orcinus* but larger than most other delphinids—consistent with the evolutionary fishing-up hypothesis. The fossil record of *Orcinus*-like odontocetes requires much needed revision (Bianucci 2005), but more descriptions and discoveries may allow for rigorous testing of proposed delphinid evolutionary scenarios.

We explain the evolution of *Orcinus* as part of the Pliocene delphinid diversification (Figure 7.5), whereby successively larger delphinid ancestors fished up to larger prey items. This event was not unique in cetacean history but rather was a third track parallel to those already taken in the late Miocene by physeterids and ziphiids (Figure 7.5). In addition to the cetaceans, similar outcomes were likely realized for other predators on large marine vertebrates. Larger ichthyosaurs, plesiosaurs, and mosasaurs likely fished up prey size, perhaps culminating in the documented feeding on other smaller marine reptiles. *Carcharodon* (or *Carcharocles*) and its "giant-toothed" shark ancestors also likely fished up from larger fishes to prey on some of the earliest archaeocetes (Purdy 1996). Both the Miocene diversification and Pliocene attenuation of the mysticetes (Figure 7.4) coincide with the emergence of their likely predator, the titanic *Carcharodon megalodon* (see Purdy 1996). Interestingly, the extinction of *C. megalodon* is coincident with the appearance and diversification of some of the larger delphinids, including *Orcinus* (Figure 7.5). In both of these cases all available data suggest that the extinct predators on large marine vertebrates were always larger than their prey. Here the evolution scenario for *Orcinus* departs from the other examples, because the presence of cooperative hunting behavior in the delphinids did not require predator size to exceed prey size (Kliban 1976).

Top-down, Bottom-up, or Asteroid Impact?

Clearly, both top-down and bottom-up processes have been at work during the diversification of the Cetacea and in the evolution of *Orcinus* in particular. Sometimes these mechanisms appeared to have acted between trophic levels; at other times, between size classes; and both types of processes continue to act today. However, as with other events in the evolution of mammalian lineages, the stage for the evolution of *Orcinus* may have been set by events associated with the K-T impact event, over 65 Ma.

Extinction events associated with the K-T impact included the decimation of cephalopod diversity (Sepkoski and Hulver 1985), leaving few lineages at the beginning of the Tertiary *and no major cephalopod predators*. The survivors, which either were bathymetrically diverse at the outset or subsequently reinvaded depauperate depth ranges, rebounded to produce a diverse and little-exploited cephalopod fauna. Archaeocetes undoubtedly exploited the shallow-water components of this vast resource, but we emphasize that the increasing productivity during the Miocene, which increased biomass at all trophic and taxonomic levels, appears to be responsible for powering neocete diversification (Figure 7.3) (Lipps and Mitchell 1976). This same productivity also powered pinniped and sea bird diversifications (Warheit and Lindberg 1988), again across multiple size classes. As with cetaceans, we would expect that the primary prey items of Miocene pinnipeds and sea birds were the very same cephalopod, finfish, and arthropod stocks.

We hypothesized that the initial prey of Neogene mysticetes and odontocetes was most likely small fishes and cephalopods; the pleisomorphic diet shared with the archaeocetes. Within Odontoceti, larger members of the physeterid and ziphiid lineages dove deeper and fished up their cephalopod prey size; diversifying as they became larger and fed deeper. In marked contrast, the other odontocetes (phocoenids, monodontids, and the extinct "kentriodontids" and "acrodelphinids") all lost a large amount of their generic diversity during the late Miocene, perhaps due to their higher dependence on smaller-body-sized photic stocks of fish and cephalopods. In mysticetes, a similar scenario of reducing arthropod and fish stocks is also suggested by the extinction of cetotheriids by the late Pliocene. Generic diversity reductions of arthropod-feeding cetaceans lagged slightly behind those of the cetotheres, and the predominately cephalopod-feeding and deep-diving odontocetes lagged even further, but then they also showed a precipitous loss of generic diversity starting at the beginning of the late Miocene.

We interpret these patterns of decline to global loss of marine productivity in the Pliocene, especially in coastal upwelling systems, and the fishing down (*sensu* Pauly et al. 1998) of photic biomass exacerbated by decreasing productivity. These two factors would eventually affect offshore deep-water prey items as well, reducing physeterid and ziphiid generic diversity. Pleistocene glaciation and associated increases in marine productivity again powered generic diversification of surviving neocete taxa, with the exception of physeterids (Figure 7.5), which may have seen increasing competition with ziphiids and larger delphinids.

Orcinus's run up to feed on great whales appears to be linked with a shift from larger cephalopod and fish prey to pinnipeds and other cetaceans as prey items, concomitant with changes in skull morphology (Seagars 1983). Fishing up and down prey size does not necessarily require shifts between trophic levels, but the implications of such prey size shifts might cascade across trophic levels in both ecological and evolutionary time, with the actual temporal scales of fishing up versus fishing down prey size appearing to be highly asymmetric.

We have argued that fishing up prey size is an evolutionary phenomenon requiring diversification over hundreds of thousands and even millions of years. Body size changes, feeding and physiological modifications, and an array of other adaptations are all changes that, in the absence of plasticity, occur in evolutionary rather than ecological time. However, as demonstrated by Springer et al. (2003), fishing down prey size can occur in relatively short time periods (on the decadal timescale). From an evolutionary perspective, we explain this temporal asymmetry by *Orcinus's* ability to feed on smaller prey items as well—prey items that are likely to have been steps on the way up the evolutionary size gradient (e.g., fish → cephalopods → pinnipeds → other cetaceans). Similar abilities remain in other cetacean lineages as well (e.g., some mysticetes' ability to feed on fish and cephalopods in addition to arthropod stocks; Gaskin 1982, Clarke 1996).

Because body size represents a continuous variable that connects ecological interactions across these two timescales, we submit that resolution of the historical patterns of prey, predators, and the systems in which they swim need to be examined and compared on multiple temporal and spatial scales. These scales extend beyond the artificial constraints of the history of their human predators and provide several examples of the evolution of taxa that fished up prey size, as well as of the biotic consequences of short- and long-term changes in prey availability and productivity in oceanic systems. Historical data such as these, coupled with human history, oceanographic data, and contemporary ecological studies and observations, are all required to understand fully the role of *O. orca* and other great whales in today's oceans and to construct informed hypotheses regarding their past, present, and future.

Acknowledgments

We thank J. A. Estes for the opportunity to contribute to this volume. He, D.K. Jacobs, T.M. Williams, and two anonymous reviewers provided helpful criticism and comments on an earlier draft of the manuscript. Estes, Williams, A.D. Barnosky, D.P. Costa, P.D. Gingerich, J.H. Lipps, M.D. Uhen, G. Vermeij, and J. Voight provided us with important insights, comments, and encouragement; however, this in no way implies that any of these colleagues agree with our musings. This work was supported, in part, by the Remington Kellogg and Doris and Samuel P. Welles Funds of the University of California Museum of Paleontology (UCMP) and by a National Science Foundation Graduate Research Fellowship to NDP; this is UCMP Contribution No. 1880. Lastly, we are indebted to two departed colleagues: Steve Gould, for his formulation and advocacy of methodologies in historical science, and Jack Sepkoski, for compiling the database that facilitates the exploration of the patterns of life.

Literature Cited

Australian Fisheries Management Authority. 2001. Chair summary. Management Advisory Committee, Southern Squid Jig Fishery. Melbourne, Australia.

Arnason, U., A. Gullberg, and A. Janke. 2004. Mitogenomic analyses provide new insights into cetacean origin and evolution. *Gene* 333: 27–34.

Baker, C. S. and P. J. Clapham. 2004. Modelling the past and future of whales and whaling. *Trends in Ecology and Evolution* 19: 365–371.

Barnes, L. G. 1976. Outline of eastern Pacific fossil cetacean assemblages. *Systematic Zoology* 25: 321–343.

———. 1990. The fossil record and evolutionary relationships of the genus *Tursiops*, in *The bottlenose dolphin: biology and conservation*. J. E. Reynolds, R. S. Wells, and S. D. Eide, eds. San Diego, CA: Academic Press, pp. 3–26.

Barnes, L. G., D. P. Domning, and C. E. Ray. 1985. Status of studies on fossil marine mammals. *Marine Mammal Science* 1: 15–53.

Barnosky, A. D. 2001. Distinguishing the effects of the Red Queen and the Court Jester on Miocene mammal evolution in the northern Rocky Mountains. *Journal of Vertebrate Paleontology* 21: 172–185.

Barnosky, A. D. and Bell, C. J. 2003. Evolution, climatic change and species boundaries: perspectives from tracing *Lemmiscus curtatus* populations through time and space. *Proceedings of the Royal Society of London, Biological Sciences* 270: 2585–2590.

Barron, J. A. 1998. Late Neogene changes in diatom sedimentation in the North Pacific. *Journal of Asian Earth Sciences* 16: 85–95.

———. 2003. Planktonic marine diatom record of the past 18 My: appearances and extinctions in the Pacific and Southern Oceans. *Diatom Research* 18: 203–224.

Benton, M. J., M. A. Wills, and R. Hitchin. 2000. Quality of the fossil record through time. *Nature* 403: 534–537.

Berta, A. and J. L. Sumich. 1999. *Marine mammals: evolutionary biology*. San Diego, CA: Academic Press.

Berta, A., J. L. Sumich, and K. M. Kovacs. 2006. *Marine mammals: evolutionary biology*, 2nd edition. San Diego, CA: Academic Press.

Bianucci, G. 1997. *Hemisyntrachelus cortesii* (Cetacea, Delphinidae) from the Pliocene sediments of Campore Quarry (Salsomaggiore Terme, Italy). *Bollettino della Societa Paleontologica Italiana* 36: 75–83.

———. 2005. *Arimidelphis sorbinii*, a new small killer whale-like dolphin from the Pliocene of Marecchia River (central eastern Italy) and a phylogenetic analysis of the Orcininae (Cetacea: Odontoceti). *Rivista Italiana di Paleontologia e Stratigrafia* 111(2): 329–344.

Bisconti, M. 2003. Evolutionary history of Balaenidae. *Cranium* 20(1): 9–50.

Boissièrie, J.-R., F. Lihoreau, and M. Brunet. 2005. The position of Hippopotamidae within Cetartiodactyla. *Proceedings of the National Academy of Sciences* 102: 1537–1541.

Brown, J. A. 1995. *Macroecology*. Chicago: University of Chicago Press.

Browning, J. V., K. G. Miller, M. Van Fossen, C. Liu, M. -P. Aubry, D. K. Pak, and L. M. Bybell. 1996. Lower to middle Eocene sequences of the New Jersey coastal plain and their significance for global climate change. *Proceedings of the Ocean Drilling Program, Scientific Results* 150: 229–242.

Capellini, G. 1883. Di un' orca fossile scoperta a Cetona in Toscana. *Memorie dell'Accademia delle Scienze dell'Istituto di Bologna Series IV* 4: 665–687.

Carroll, R. L. 1988. Vertebrate paleontology and evolution. New York: W. H. Freeman and Company.

Castilla, J. C. 1999. Coastal marine communities: trends and perspectives from human-exclusion experiments. *Trends in Ecology and Evolution* 14: 280–283.

Ciampaglio, C. N., G. A. Wray, and B. H. Corliss. 2005. A toothy tale of evolution: convergence in tooth morphology among marine Mesozoic-Cenozoic sharks, reptiles, and mammals. *The Sedimentary Record* 3: 4–8.

Cicimurri, D. J. and M. J. Everhart, 2001. An elasmosaur with stomach contents and gastroliths from the Pierre Shale (late Cretaceous) of Kansas. *Transactions of the Kansas Academy of Sciences* 104: 129–143.

Clarke, M. R. 1996. Cephalopods as prey. III. Cetaceans. *Philosophical Transactions of the Royal Society of London, Biological Sciences* 351: 1053–1065.

Cleland, C. E. 2001. Historical science, experimental science, and the scientific method. *Geology* 29: 987–990.

Collin, R., and C. M. Janis. 1997. Morphological constraints on tetrapod feeding mechanisms: why were there no suspension-feeding marine reptiles? in *Ancient marine reptiles*. J. M. Callaway and E. L. Nicholls, eds. San Diego, CA: Academic Press, pp. 451–465.

Connor, R. C. and K. S. Norris. 1982. Are dolphins reciprocal altruists? *American Naturalist* 119: 358–374.

Demere, T. A., A. Berta, and M. R. Gowen. 2005. The taxonomic and evolutionary history of fossil and modern balaenopteroid mysticetes. *Journal of Mammalian Evolution* 12: 99–143.

Diamond, J. M. 1987. Did Komodo dragons evolve to eat pygmy elephants? *Nature* 326: 832.

Domning, D. P. 2001. The earliest known fully quadrupedal sirenian. *Nature* 413: 625–627.

Estes, J. A. 1996. The influence of large, mobile predators in aquatic food webs: examples from sea otters and kelp forests, in *Aquatic predators and their prey*. S. P. R. Greenstreet and M. L. Tasker, eds. Oxford: Fishing News Books, pp. 65–72.

Estes, J. A. and D. O. Duggins. 1995. Sea otters and kelp forests in Alaska—generality and variation in a community ecological paradigm. *Ecological Monographs* 65: 75–100.

Estes, J. A., M. T. Tinker, T. M. Williams, and D. F. Doak. 1998. Killer whale predation on sea otters linking oceanic and nearshore ecosystems. *Science* 282: 473–476.

Evans, P. G. H. 1987. *The natural history of whales and dolphins*. New York: Facts on File.

Flower, W. H. 1885. *An introduction to the osteology of the Mammalia*. London: Macmillan.

Fordyce, R. E. 2002a. *Simocetus rayi* (Odontoceti: Simocetidae) (new species, new genus, new family), a bizarre archaic Oligocene dolphin from the eastern North Pacific. *Smithsonian Contributions to Paleobiology* 93: 185–222.

———. 2002b. Neoceti. *Encyclopedia of Marine Mammals*. W. F. Perrin, B. Wursig, and J. G. M. Thewissen, eds. San Diego, CA: Academic Press pp. 789–791.

———. 2003. Cetacean evolution and Eocene-Oligocene oceans revisited, in *From greenhouse to icehouse: the marine Eocene-Oligocene transition*. D. R. Prothero, L. C. Ivany and E. A. Nesbitt, eds. New York: Columbia University Press, pp. 154–170.

Fordyce, R. E. and C. de Muizon. 2001. Evolutionary history of cetaceans: a review, in *Secondary adaptation of tetrapods to life in water*. J. -M. Mazin and V. de Buffrénil, eds. Munich, Germany: Friedrich Pfeil, pp. 169–233.

Fowler, C. W. and J. A. MacMahon. 1982. Selective extinction and speciation: their influence on the structure and functioning of communities and ecosystems. *American Naturalist* 119: 480–498.

Gaskin, D. E. 1982. *The ecology of whales and dolphins*. London: Heinemann.

Gatesy, J., and M. A. O'Leary. 2001. Deciphering whale origins with molecules and fossils. *Trends in Ecology and Evolution* 16: 562–570.

Geisler J. H. and A. E. Sanders. 2003. Morphological evidence for the phylogeny of Cetacea. *Journal of Mammalian Evolution* 10: 23–129.

Geisler, J. H. and M. D. Uhen. 2005. Morphological support for a close relationship between hippos and whales. *Journal of Vertebrate Paleontology* 23: 991–996.

Gheerbrant, E., D. P. Domning, and P. Tassy. 2005. Paenungulata (Sirenia, Proboscidea, Hyracoidea, and relatives), in *The rise of placental mammals: origin and relationships of the major extant clades*. K. D. Rose and J. D. Archibald, eds. Baltimore, Johns Hopkins University Press, pp. 84–105.

Gingerich, P. D., B. H. Smith, and E. L. Simons. 1990. Hind limbs of Eocene *Basilosaurus*: evidence of feet in whales. *Science* 249: 154–157.

Gingerich, P. D., M. ul Haq, I. S. Zalmout, I. H., Khan, and M. S. Malkani. 2001. Origin of whales from early artiodactyls: hands and feet of Eocene Protocetidae from Pakistan. *Science* 293: 2239–2242.

Gould, S. J. 1986. Evolution and the triumph of homology, or why history matters. *American Scientist* 74: 60–69.

Gould, S. J and E. S. Vrba. 1982. Exaptation: a missing term in the science of form. *Paleobiology* 8(1): 4–15.

Heyning, J. E. and M. E. Dahlheim. 1988. *Orcinus orca*. *Mammalian Species* 304: 1–9.

Hsu, K. J., W. B. F. Ryan, and M. B. Cita. 1973. Late Miocene desiccation of the Mediterranean. *Nature* 242: 240–244.

Hunter, M. D. and P. W. Price. 1992. Playing chutes and ladders: heterogeneity and the relative roles of bottom-up and top-down forces in natural communities. *Ecology* 73: 724–732.

Ivany, L. C., S. Van Simaeys, E. W. Domack, and S. D. Samson. 2006. Evidence for an earliest Oligocene ice sheet on the Antarctic Peninsula. *Geology* 34(5): 377–380.

Jablonski, D., K. Roy, J. W. Valentine, R. M. Price, and P. S. Anderson. 2003. The impact of the pull of the Recent on the history of marine diversity. *Science* 300: 1133–1135.

Jackson, J. B. C., M. X. Kirby, W. H. Berger, K. A. Bjorndal, L. W. Botsford, B. J. Bourque, R. Bradbury, R. Cooke, J. Erlandson, J. A. Estes, T. P. Hughes, S. Kidwell, C. B. Lange, H. S. Lenihan, J. M. Pandolfi, C. H. Peterson, R. S. Steneck, M. J. Tegner, and R. Warner. 2001. Historical overfishing and the recent collapse of coastal ecosystems. *Science* 293: 629–638.

Jacobs, D. K., T. A. Haney, and K. D. Louie. 2004. Genes, diversity, and geological process on the Pacific coast. *Annual Review of Earth and Planetary Science* 32: 601–652.

Kear, B. P., W. E. Boles, and E. T. Smith. 2003. Unusual gut contents in Cretaceous ichthyosaurs. *Proceedings of the Royal Society of London, Biological Sciences* (Supplement) 270: 206–208.

Kimura, T. and T. Ozawa. 2002. A new cetothere (Cetacea: Mysticeti) from the early Miocene of Japan. *Journal of Vertebrate Paleontology* 22: 684–702.

Kliban, B. 1976. *Never eat anything bigger than your head and other drawings*. New York: Workman Publishing Company.

Köhler, R. and R. E. Fordyce. 1997. An archaeocete whale (Cetacea: Archaeoceti) from the Eocene Waihao Greensand, New Zealand. *Journal of Vertebrate Paleontology* 17: 574–583.

Lindberg, D. R., J. A. Estes, and K. I. Warheit. 1998. Human influences on trophic cascades along rocky shores. *Ecological Applications* 8: 880–890.

Lindberg, D. R. and J. H. Lipps. 1996. Reading the chronicle of Quaternary temperate rocky shore faunas, in *Evolutionary Paleobiology*. D. Jablonski, D. H. Erwin, and J. H. Lipps, eds. Chicago: University of Chicago Press, pp. 161–182.

Lipps, J. H. and S. J. Culver. 2002. The trophic role of marine microorganisms through time. *Paleontological Society Papers* 8: 69–92.

Lipps, J. H. and E. D. Mitchell. 1976. Trophic model for the adaptive radiations and extinctions of pelagic marine mammals. *Paleobiology* 2: 147–155.

Luo, Z. -X. and P. D. Gingerich. 1999. Terrestrial Mesonychia to aquatic Cetacea; transformation of the basicranium and evolution of hearing in whales. *University of Michigan Papers on Paleontology* 31: 1–98.

MacKenzie, J. A. 1999. From desert to deluge in the Mediterranean. *Nature* 400: 613–614.

Massare, J. A. 1987. Tooth morphology and prey preference of Mesozoic marine reptiles. *Journal of Vertebrate Paleontology* 7(2): 121–137.

May-Collado, L. and Agnarsson, I. 2006. Cytochrome b and Bayesian inference of whale phylogeny. *Molecular Phylogenetics and Evolution* 38: 344–354.

McCord, M. E. and S. E. Campana. 2003. A quantitative assessment of the diet of the blue shark (*Prionace glauca*) off Nova Scotia, Canada. *Journal of Northwest Atlantic Fishery Science* 32: 57–63.

McKenna, M. C. and S. K. Bell (editors). 1997. *Classification of mammals above the species level*. New York: Columbia University Press.

McPhee, J. 1998. *Annals of the former world*. New York: Farrar, Straus and Giroux.

Medrano-Gonzalez, L., C. S. Baker, M. R. Robles-Saavedra, J. Murrell, M. J. Vazquez-Cuevas, B. C. Congdon, J. M. Straley, J. Calambokidis, J. Urban-Ramirez, L. Florez-Gonzalez, C. Olavarria-Barrera, A. Aguayo-Lobo, J. Nolasco-Soto, R. A. Juarez-Salas, and K. Villavicencio-Llamosas. 2002. Trans-oceanic population genetic structure of humpback whales in the North and South Pacific. *Memoirs of the Queensland Museum* 47: 465–479.

Melville, H. 1851. *Moby-Dick: or the whale*. Reprint edition. New York: Modern Library.

Motani, R., B. M. Rothschild, and W. Wahl Jr. 1999. Large eyeballs in diving ichthyosaurs. *Nature* 402: 747.

Mowat, F. 1972. *A whale for the killing*. New York: Bantam Books.

Norris, K. S. and T. P. Dohl. 1980. The structure and function of cetacean schools, in *Cetacean behavior*. L. M. Herman, ed. New York: Wiley Intersciences, pp. 211–262.

Nicholls, E. L. and M. Manabe. 2004. Giant ichthyosaurs of the Triassic—a new species of *Shonisaurus* from the Pardonet Formation (Norian: Late Triassic) of British Columbia. *Journal of Vertebrate Paleontology* 24: 838–849.

Nikaido, M., F. Matsuno, H. Hamilton, R. L. Brownell Jr., Y. Cao, W. Ding, Z. Zuoyan, A. M. Shedlock, R. E. Fordyce, M. Hasegawa, and N. Okada. 2001. Retroposon analysis of major cetacean lineages: the monophyly of toothed whales and the paraphyly of river dolphins. *Proceedings of the Academy of Natural Sciences* 98: 7384–7389.

O'Hara, R. J. 1988. Homage to Clio, or, toward an historical philosophy for evolutionary biology. *Systematic Zoology* 37: 142–155.

Paine, R. T. 1966. Food web complexity and species diversity. *American Naturalist* 100: 65–75.

Pauly, D., V. Christensen, J. Dalsgaard, R. Froese, and F. C. Torres, Jr. 1998. Fishing down marine food webs. *Science* 279: 860–863.

Power, M. E. 1992. Top-down and bottom-up forces in food webs: Do plants have primacy? *Ecology* 73: 733–746.

Prothero, D. R., L. C. Ivany, and E. Nesbitt (editors). 2002. *From greenhouse to icehouse*. New York: Columbia University Press.

Purdy, R. W. 1996. Paleoecology of fossil white sharks, in *Great white sharks: the biology of* Carcharodon carcharias. A. P. Klimley and D. G. Ainley, eds. New York: Academic Press, pp. 67–78.

Pyenson, N. D. and D. R. Lindberg. 2003. Phylogenetic analyses of body size in Neoceti: preliminary proxies for studying cetacean ecology in the fossil record. Abstracts. 15th Biennial Conference on the Biology of Marine Mammals, Greensboro, North Carolina.

Raup, D. M. 1979. Biases in the fossil record of species and genera. *Bulletin of the Carnegie Museum of Natural History* 13: 85–91.

Sato, T. and K. Tanabe. 1998. Cretaceous plesiosaurs ate ammonites. *Nature* 384: 629–630.

Scammon, C. M. 1874. *The marine mammals of the northwestern coast of North America.* San Francisco: John H. Carmony and Co. Reprint, New York: Dover Publications.

Scheffer, V. 1969. *The year of the whale.* New York: Charles Scribner's Sons.

Scher, H. and E. M. Martin. 2006. Timing and climatic consequences of the opening of Drake Passage. *Science* 312: 428–430.

Seagars, D. J. 1983. Jaw structure and functional mechanics in six delphinids (Cetacea: Odontoceti). Abstracts. 5th Biennial Conference on the Biology of Marine Mammals. Boston, Massachusetts.

Sepkoski, J. J., Jr. 2002. A compendium of fossil marine animal genera. *Bulletins of American Paleontology* 363: 1–560.

Sepkoski, J. J., Jr. and M. L. Hulver. 1985. An atlas of Phanerozoic clade diversity diagrams, in *Phanerozoic diversity patterns: profiles in macroevolution.* Valentine, J. W., ed. Princeton, NJ: Princeton University Press and American Association for the Advancement of Science, pp. 11–39.

Smale, M. J. 1996. Cephalopods as prey. IV. Fishes. *Philosophical Transactions of the Royal Society of London, Biological Sciences* 351: 1067–1081.

Smith, A. G., D. G. Smith, and B. M. Funnell. 1994. *Atlas of Mesozoic and Cenozoic coastlines.* Cambridge: Cambridge University Press.

Spoor, F., S. Bajpai, S. T. Hussain, K. Kumar, and J. G. M. Thewissen. 2002. Vestibular evidence for the evolution of aquatic behavior in early cetaceans. *Nature* 417: 163–166.

Springer, A. M., J. A. Estes, G. B. van Vliet, T. M. Williams, D. F. Doak, E. M. Danner, K. A. Forney, and B. Pfister. 2003. Sequential megafaunal collapse in the North Pacific Ocean: An ongoing legacy of industrial whaling? *Proceedings of the National Academy of Sciences* 100: 12223–12228.

Steele, J. H. 1974. *The structure of marine ecosystems.* Cambridge, MA: Harvard University Press.

Thewissen, J. G. M. and S. Bajpai. 2001. Whale origins as a poster child for macroevolution. *Bioscience* 51: 1037–1049.

Thewissen, J. G. M. and F. E. Fish. 1997. Locomotor evolution in the earliest cetaceans: functional model, modern analogues, and paleontological evidence. *Paleobiology* 23: 482–490.

Thewissen, J. G. M. and E. M. Williams. 2002. The early radiations of Cetacea (Mammalia): evolutionary pattern and developmental correlations. *Annual Review of Ecology and Systematics* 33: 73–90.

Uhen, M. D. 2004. Form, function, and anatomy of *Dorudon atrox* (Mammalia, Cetacea): an archaeocete from the middle to late Eocene of Egypt. *University of Michigan Papers on Paleontology* 34: 1–222.

Uhen, M. D. and N. D. Pyenson. 2002. Evolution of cetacean diversity. *Journal of Vertebrate Paleontology* 22 (Supplement): 116A.

Van Valen, L. M. 1966. Deltatheridia, a new order of mammals. *Bulletin of the American Museum of Natural History* 132: 1–126.

Vermeij, G. J. 1987. *Evolution and escalation: an ecological history of life.* Princeton, NJ: Princeton University Press.

Vermeij, G. J. and R. Dudley. 2000. Why are there so few evolutionary transitions between aquatic and terrestrial ecosystems? *Biological Journal of the Linnaean Society* 70: 541–554.

Warheit, K. I. and Lindberg, D. R. 1988. Interactions between seabirds and marine mammals through time: interference competition at breeding sites, in *Seabirds and other marine vertebrates: competition, predation, and other interactions.* Burger, J., ed. New York: Columbia University Press, pp. 292–328.

Zachos, J., M. Pagani, L. Sloan, E. Thomas, and K. Billups. 2001. Trends, rhythms, and aberrations in global climate 65 Ma to present. *Science* 292: 686–693.

Webb, P. W. 1984. Body form, locomotion and foraging in aquatic vertebrates. *American Zoologist* 24: 107–120.

EIGHT

A Taxonomy of World Whaling
Operations and Eras

RANDALL R. REEVES AND TIM D. SMITH

Whaling ranks along with some pelagic marine fishing as the world's most spatially extensive form of exploitation of wild living resources. An understanding of its history is therefore important for analyzing the role of humans in modifying marine ecosystems.

Whaling has involved most of the 14 mysticete (baleen) species, many of the 28 or so medium- to large-sized odontocetes (toothed whales), and numerous geographically distinct populations of these species (at least dozens). The scale of world whaling has been global, spanning bays and gulfs, continental and island shelves, and pelagic waters. Whaling began in antiquity (more than a thousand years ago) and continues to the present. Numerous maritime societies, from all inhabited continents and many oceanic islands, have been engaged in whaling at one time or another. The technologies employed to kill, secure, and process whales have ranged from primitive and nonmechanical to technically sophisticated and industrial. The economic complexities associated with whaling have been diverse, encompassing fluctuations in production rates, product prices, operation costs, labor characteristics, and other variables.

Whaling is defined here as the purposeful killing of large cetaceans to obtain economically useful products. It therefore embraces both "commercial" and "subsistence" whaling

(Reeves 2002). We define the large cetaceans to include all species of baleen whales (mysticetes) and the sperm whale *(Physeter macrocephalus)*. In addition, we include the hunting of some of the larger beaked whales (Baird's beaked and northern bottlenose whales—*Berardius bairdii* and *Hyperoodon ampullatus*, respectively) and other medium-sized toothed whales (e.g., killer and pilot whales, belugas or white whales, and narwhals—*Orcinus orca*, *Globicephala* spp., *Delphinapterus leucas*, and *Monodon monoceros*, respectively) because such hunting has often been ancillary to the hunting of large cetaceans and has involved similar technology and markets.

The literature on whaling is voluminous. Most of it follows disciplinary lines—biological, economic, technical, historical, anthropological/archaeological, political/regulatory, and even literary. Major works tend to be limited in scope, reflecting the author's interest in a particular nation, region, period, species, or fishery. Although some studies that focus on a single region or whale population are rigorous and data-rich (e.g., Henderson 1972; Ross 1975; Bockstoce and Botkin 1983; Mitchell and Reeves 1983), overviews of the entire history of whaling are rare and necessarily superficial (e.g., Jenkins 1921; Vaucaire 1941; Sanderson 1954; Spence 1980; Francis 1990; Ellis 1991). The monograph by Tønnessen and Johnsen (1982), and its more authoritative Norwegian-language antecedents (Johnsen

1959; Tønnessen 1967, 1969, 1970), are singularly comprehensive in their coverage of "modern" whaling.

In the present chapter we provide a systematic overview of world whaling, working toward a unified classification system—a taxonomy—that can help order and manage investigations of whaling history, particularly those requiring the allocation of whaling effort and whale removals to different geographical regions and time periods (e.g. Allison and Smith 2004). A preliminary attempt was described in a series of papers concerned solely with the exploitation of humpback whales (*Megaptera novaeangliae*) in the North Atlantic Ocean (Smith and Reeves 2002, 2003c; Reeves and Smith 2002). In that case, it was possible to fit all of the relevant whaling activities into a reasonably complete and coherent framework of named "fisheries" and "subfisheries." That framework had considerable heuristic value, but it dealt with only a single ocean basin and, indeed, a single species within that basin. In attempting to apply a similar approach on a broader scale, we found the fisheries/subfisheries concept inadequate. As an alternative, then, we have chosen in this chapter to group whaling activities or enterprises into *operations*, and operations into *eras*. We begin by explaining how operations are defined and by reviewing them briefly. We then describe the 11 eras that were defined after inspecting the characteristics of the operations.

With another objective in mind—to achieve a coherent understanding of whaling cultures through time—Klaus Barthelmess has developed a similar global taxonomy with different emphases (personal communciation, December 2003; and see Barthelmess 1992). We stress that both his taxonomy and ours are works in progress, by no means complete or definitive.

Operations and Eras

Two organizational schemes are used to describe whaling: operations and eras.

Operations

Mitchell and Reeves (1980) proposed what appeared to be a simple scheme for classifying various types of whaling: who, what, where, why, when, and how. Although each of these variables is multidimensional, and no single one can be used to define any particular category unambiguously, the basic scheme seems useful. As a first-order attempt to classify world whaling into manageable units, we propose the term *operation* as an alternative to *fishery* or *subfishery* (cf. Reeves and Smith 2002). An *operation* can be defined on the following basis:

WHO: The ethnic or national group involved. A whaling operation could be defined by ethnic or national origin of the people who kill the whales, by the nation on whose land a shore station is situated or under whose flag a ship operates, by who provides the capital, by who governs the waters where whaling is conducted, or by who profits from use or sale of the products. We indicate *Who* in terms of the geographic location, either the geographic region where the catches were taken or the nationality of those pursuing the operation.

WHAT: The targeted species. The capability of whaling for different species or groups of species varies with certain aspects of an operation, such as the technology required to catch, tow, and process the whales; the types and amounts of products that can be obtained; the importance of a species to the operation (i.e., principal or supplemental); and the seasonality and location, according to the animals' movement patterns. Often multiple species are targeted, and additional species are taken opportunistically or during periods when the principal targets are unavailable. Also, there has been a typical pattern of shifting from more to less valuable or accessible species over time as targeted species become increasingly scarce or elusive. A clear example in which the targeted species help to distinguish the character of the operations is Arctic whaling, where the three Arctic species—bowhead (*Balaena mysticetus*), narwhal, and beluga—have formed the basis for both "subsistence" and "commercial" whaling on a circumpolar scale (Vaughan 1984). Operations involving multiple target species are particularly complex, because the focus of the hunting process often varies across years as well as seasons. We list the species known to have been taken, distinguishing between the principal and supplemental targets for an operation. In some instances in which a species became a principal target only toward the end of an era when other target species had been depleted (e.g., the fin whale, *Balaenoptera physalus*, in the Gulf of Maine), it was considered supplemental.

WHERE: The geographical location of the whaling and its distance from the whalers' residences. This latter question has numerous implications, not least that it largely determines whether an operation is considered shore-based, coastal, or pelagic. The variable *Where* can be described at varying geographic scales, including ocean basin or region within an ocean basin, such as a whaling ground or an island area. The American open-boat pelagic whalers had the habit of naming grounds that were often visited on a predictable circuit, reflecting seasonal movements of the whales, wind and current patterns, ambient conditions for conducting whaling operations, and access to port facilities to obtain supplies (food, water, wood), recruit crew, and transship oil and baleen (Clark 1887; Townsend 1935). Such grounds have been used to delineate study areas for a number of analyses of catch history and trends in abundance (e.g., Bannister et al. 1981; Hope and Whitehead 1991). In the present context, a change in geographic location may imply the need to designate a new operation, especially for coastal whaling or a land station, but for a pelagic operation such a change may be interpreted as a mere adjustment in strategy within the same operation. As a result, the geographic scope of some operations has been vast, almost global. We summarize *Where* in terms of ocean basin (North and South Atlantic and Pacific, Indian, Arctic, and Antarctic) and identify regions within those basins where shore or coastal operations occurred.

WHY: The products of the hunt (e.g., oil, baleen, meat, ivory) and thus the incentive(s) driving it. Market factors are crucial in determining the intensity of whaling, species preferences, and, ultimately, the degree of stock depletion. In fact, in some population studies it has been difficult to determine whether reduced whale abundance (i.e., depletion) or market conditions (e.g., a change in product prices) were most responsible for the decline or cessation of a fishery (e.g., Best 1983; Davis et al. 1997). The traditional distinction between "commercial" and "subsistence" whaling hinges primarily on this variable *(Why)*, although the *How* element has also played a prominent role (Mitchell and Reeves 1980; Reeves 2002). The uses of whale products have shifted through time according to cultural preferences, availability of less expensive substitutes (e.g., spring steel replaced baleen as a garment stiffener in the early 1900s; Bockstoce 1986), and technical innovations (e.g., hydrogenation of oils to produce margarine dramatically enhanced the value of baleen whale oil in the 1920s and 1930s; Tønnessen and Johnsen 1982). Nevertheless, a change in this variable alone would not necessarily be the basis for defining a new whaling operation, and we did not systematically consider this characteristic.

WHEN: The years or decades in which the whaling took place; also, its seasonality. The latter might determine whether or not a given operation affects a migratory whale population. The temporal history of some whaling operations is imprecise or incomplete, and in a few cases it is lacking altogether. In many instances, especially involving shore-based operations, whaling has been episodic, punctuated by years or decades of closure. For example, modern whaling factories were opened, closed, and relocated periodically in Newfoundland and Labrador (Mitchell 1974), Iceland (Sigurjónsson 1988), northwestern North America (Pike and MacAskie 1969; Reeves et al. 1985; Webb 1988), and southern Africa (Best 1994). Although we describe the temporal span of an operation simply by noting the starting and ending years or decades, we recognize that doing so can mask differences through time in the intensity of whaling, catch composition (species or age/size class), catch level, and other features of interest.

HOW: The equipment, methods, and techniques for taking and processing the whales. The distinction between shore-based and pelagic operations is relevant here, as are questions of whether powered or sailing vessels are employed, which weapons are used to capture and kill the animals, and how those are delivered (e.g., manually or mechanically) (Mitchell et al. 1986). Specific aspects of this variable can be in an almost continuous state of flux as whalers experiment and innovate, as regional availability of the targeted whale species varies, and as product, labor, and capital markets fluctuate. Best (1983), for example, described eight major improvements in vessels and gear that emerged in the American pelagic sperm whale fishery between the 1760s and 1850s, quoting Scammon's (1874) observation that "there is hardly a fixture, or an implement, pertaining to the outfit that has not been improved upon." While it is generally assumed that such improvements would have made whaling more profitable, they also may have allowed vessel owners to "get by" with less proficient crews, contributing to lower voyage productivity (Davis et al. 1997). Technology and practices determine to a considerable degree how efficient an operation is—"efficiency" being defined in terms of hunting loss, humaneness, or profitability (e.g., O'Hara et al. 1999). Hunting loss (hidden mortality, as when animals are seriously injured or killed but not secured) can significantly affect the magnitude and rate of removals and therefore must be taken into account in population analyses that are premised on the availability of complete catch histories. We considered the mode of operation as shore, coastal, or pelagic and whether mechanical power was used. Finally, we considered the tool(s) or weapon(s) used to kill the whales and the method of delivery (e.g., thrown by hand or shot from a cannon).

Understanding changes in these six variables and their interrelationships is essential to defining whaling operations. The complexity of the relationships among the variables has meant that some of the decisions concerning which changes do and do not justify the designation of a "new" whaling operation are somewhat arbitrary, especially where our access to the source material has been limited.

Using these criteria, we identified more than 100 whaling operations worldwide. For each operation, we considered the catching method (e.g., harpoon, poison dart, net), delivery method (e.g., hand-thrown, deck-mounted cannon), mode of operation (shore, coastal, or pelagic), propulsion method (nonpowered or powered), likely cumulative secured catch level (to order of magnitude for all species, combined), start and end years, species (primary and supplemental), ocean basin, regions within ocean basins, and relevant data sources. The approximate location of each of the shore and coastal whaling operations and the ocean basins used by the pelagic operations are shown in Figure 8.1. The operations are indicated by the sequence numbers given in the Appendix at the end of this chapter.

As with any effort of this kind, there is an inherent tension between "lumping" and "splitting." Our bias has not been consistently in either direction. On one hand, too much lumping would gloss over a multitude of differences (in who, what, where, why, when, and how), some and perhaps many of which would be meaningful and important in certain types of analyses. On the other hand, too much splitting would defeat our purpose of organizing a highly fragmented, almost chaotic body of information on world whaling activities. The spread of Norwegian-style shore whaling in southern Africa between 1908 and 1930 might be seen as a single integrated operation involving primarily Norwegian capital and personnel (cf. Best 1994). The various stations used essentially identical technology, targeted roughly the same suite of species, and served similar product markets. We nevertheless "split" these activities into different operations according to the national jurisdictions (using present-day political geography) in which the shore stations were sited (e.g., Gabon, Angola, South Africa, Mozambique). Similarly, in the twentieth century Japanese personnel and capital became closely involved in whaling with "partners" in the waters of other countries (Kasuya 2002), and again, we ascribed these joint ventures to the host nation-states as separate operations rather than to Japan as extensions of its mechanized shore or pelagic operations.

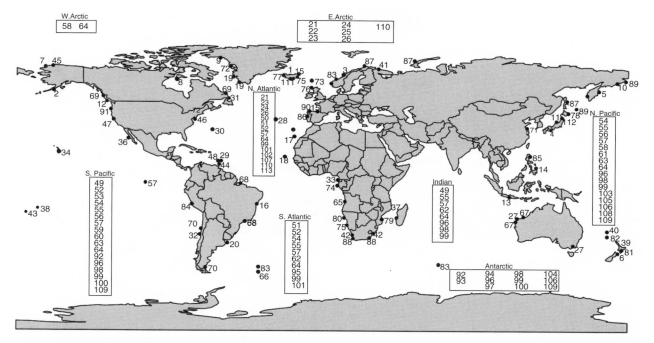

FIGURE 8.1. Locations of whaling operations (listed in the Appendix to this chapter) around the world.

Eras

Having established a rationale and procedure for naming *operations* (which can be viewed as parallel to *species* within a biological taxonomy), we needed to devise at least one higher level of organization to capture patterns or trends in our whaling taxonomy. The concept of *era* is familiar to geologists and historians, who frequently organize chronological data on such a basis. We used one of the *Oxford English Dictionary*'s definitions: "a portion of historical time marked by the continuance throughout it of particular influences, social conditions, etc." As demonstrated earlier in our definitions of operations, chronology alone is not an adequate basis for defining whaling eras. The history of whaling is marked most notably by changes in technology, such as transitions from spear-and-salvage to harpoon-line-float techniques (Lindquist 1993); from stripping whales at sea and carrying the blubber home for processing to using onboard tryworks to convert blubber to oil at sea (Ashley 1928; Whipple 1979: 54); from the hand lance to the shoulder- or darting-gun (Brown 1887); and from sail to steam power (Bockstoce 1986; Webb 2001). While the timing of such developments can usually be specified or reasonably approximated, their adoption has been highly variable in both space and time.

We defined 11 different eras based on multiple considerations (Table 8.1). The characteristic features of the eras included geography (arctic versus temperate versus tropical aboriginal), mode (shore versus pelagic), and methods/technology (e.g., poison, Basque-style, American-style). The temporal boundaries of most of the eras are reasonably clear (prehistoric and other "ancient" eras excepted). Figure 8.2

shows the approximate temporal limits of the eras, with those that began in antiquity denoted by the letter A on the left and those that include one or more continuing whaling operations with right-pointing arrows. The temporal overlap of several of the eras is a significant feature of world whaling.

The spatial boundaries of eras often overlap substantially. This spatial overlap is less amenable to graphic portrayal and is, therefore, only noted here with two (of many) examples. First, the term "bay whaling" in reference to nearshore whaling for southern right whales (*Eubalaena australis*) and humpback whales in Australia and New Zealand during the nineteenth century (Bannister 1986; Dawbin 1986) embraces the activities of both shore-based whalers (American-Style Shore era) and anchored pelagic whalers (American-Style Pelagic era). Secondly, Norwegian-style shore whaling and factory ship whaling took place alternately or contemporaneously in numerous areas of the Southern Hemisphere (Best 1994; Findlay 2000) and North Pacific Ocean (Ohsumi 1980; Webb 1988; Brownell et al. 2001).

A central objective in defining eras was to provide a framework for organizing the whaling operations such that a given operation could be attributed to only one era. Some of these assignments were not straightforward. As observed by Best and Ross (1986: 276), whaling activities often do not fall neatly within the parameters assigned to a particular category: "Towards the end of the open-boat whaling era, and before modern whaling proper began, some 'intermediate' technology was adopted, including the use of small, mounted harpoon guns and some powered craft such as launches." In fact, "experimental" whaling was particularly intensive in the North Atlantic from the late 1850s to early 1870s as American

TABLE 8.1
Eleven Whaling Eras

With Approximate Timing, Characteristic Features, and Spatial Extent

Era	Start[a]	End[b]	Characteristic Features	Spatial Extent
Poison	Antiquity	ca. 1900	Use of poison-tipped arrows, darts, or lances to kill, sometimes involving barrier nets as well	Norway/Iceland, Rim of North Pacific (Aleutians, Kodiak)
Net	Late 1600s	1910	Fiber, leather, or steel nets, sometimes used in conjunction with driving of animals (the many shut-in fisheries for belugas are not included here)	New Zealand, Japan, Kamchatka
Arctic Aboriginal	Antiquity	Ongoing	Skin boats, hand harpoons and lances grading into use of firearms and explosives in various forms, powered boats at least for towing whales to ice edge or shore for processing	Chukotka east to Greenland
Temperate Aboriginal	Antiquity	Early 1900s (or 1999 if Makah Indians are considered)	Dugout or skin boats, mainly hand-powered; hand harpoons and lances	Rim of North Pacific
Tropical Aboriginal	By 1600s	Ongoing	Open boats powered by hand or sail, hand-delivered weapons (harpoons, large hooks, blowhole plugs), shore processing	Indonesia, Philippines
Basque-Style	By 1050s	1911 (British Arctic); 1973 (Brazil)	Open boats, hand- and sail-propelled, deployed from shore or from ships along shore, in bays, or along ice edges; harpoon-line-float; hand lance; whales either towed to shore for processing, or stripped of blubber at sea, with blubber stowed on-board ships and delivered to processing sites on shore; in later years included steam power (Arctic)	Rim of North Atlantic including Arctic; a few sites in South Atlantic (and western Indian Ocean?)
American-Style Shore	Early 1600s	Ongoing (Bequia)	Whaleboats launched from shore, hand- or sail-powered, grading into powered boats at least for towing; hand harpoons and lances grading into use of firearms and explosives in various forms; in later years included steam power	Global except Antarctic
American-Style Pelagic	Mid 1700s	1920s	On-board tryworks; mother-ship operations with whaleboats, hand- and sail-powered; hand harpoons and lances grading into use of firearms and explosives in various forms; in later years included steam power	Global except Antarctic

TABLE 8.I (CONTINUED)

With Approximate Timing, Characteristic Features, and Spatial Extent

Era	Start[a]	End[b]	Characteristic Features	Spatial Extent
Norwegian-Style Shore	1860s	Ongoing	Powered catcher boats operating from shore stations; deck-mounted cannons; whales towed to shore processing plants; includes experimental whaling to some extent	Global
Factory Ship	Early 1900s	Ongoing (Japan in Antarctic)	Floating factories either moored near shore or pelagic; powered catcher boats with deck-mounted cannons; eventually stern slipways for on-board processing	Global
Small-Type	1870s	Ongoing	Powered catcher boats; deck-mounted harpoon guns and small cannons; whales either flensed at sea or towed to shore for processing; coastal or semipelagic	North Atlantic and Arctic (Great Britain, Norway), North Pacific (Japan)

[a] "Antiquity" means before AD 1000.
[b] "Ongoing" means as of 2006.

and European whalers competed to invent ways of killing and retrieving the fast-swimming rorquals (Schmitt et al. 1980; Tønnessen and Johnsen 1982). Our designations of American-style shore and American-style pelagic whaling were interpreted sufficiently broadly to accommodate activities involving "intermediate" technology and "experimental" whaling (e.g., coastal steam whaling in the Gulf of Maine, steam whaling for bowhead whales in the Arctic).

Implicit in some of the distinctions that we have made between eras are assumptions about technology invention or transfer. The validity of such assumptions is especially uncertain for areas and times in which the origins of whaling are poorly known. For example, it is uncertain whether, or to what degree, the Basque method of open-boat, hand-harpoon whaling influenced, or was influenced by, the Eskimos' use of skin boats and hand-thrown implements to capture large whales. Numerous uncertainties surround the origins of "aboriginal" whaling operations, but it seems likely that the ability to kill and secure whales developed independently in more than one place and time. Thus, Barnes (1996) pointed out that the people of Lamalera, Indonesia, were hunting whales long before the late eighteenth century, when they were first visited by American and European whalers. The unique design of local whaling boats and sails, and the islanders' way of leaping from a boat onto a whale's back to secure the harpoon, clearly distinguish this warm-water "aboriginal" whaling operation, which targets sperm whales, from the cold-water operation of the Eskimos, which targets bowhead whales. The Indonesian operation probably has a closer affinity to that of the people of Pamilacan, Philippines, who also leapt from their boats to embed a large hook in the back of their prey,

mainly Bryde's whales *(Balaenoptera edeni/brydei)* (Dolar et al. 1994). This method dates back more than a century and has no obvious link to the activities of visiting foreign whalers.

A common theme in efforts to establish when active whaling began is the problem of "drift" whales: whales that came ashore dead or dying ("stranded") or that were discovered as floating carcasses in coastal waters (Freeman 1979; Little and Andrews 1981; McCartney 1984). In some instances, the salvage of drift whales can be interpreted as a step preceding the invention of whaling *per se*. In other contexts, however, it seems clear that drift whales resulted from the active pursuit of and attempts to kill whales; that is, they were part of the struck-but-lost component of the fishery. The poison-dart whalers of the northern North Pacific (Crowell 1994) and the steam whalers of New England (Clark 1887; Webb 2001) made no attempt to "fasten" or attach floats to the whales that they killed. Rather, they took what we have described as a "shoot-and-salvage" approach (Reeves et al. 2002), in which the whale was simply struck and wounded or killed, the whalers hoping to recover it days later when the carcass floated to the surface or came ashore.

Systematic Summaries of Eras

The eleven eras are described in this section.

Poison

This era is defined entirely on the basis of *how* the catching or killing was accomplished. Poison whaling consisted of two main approaches: one using aconite (from the monkshood

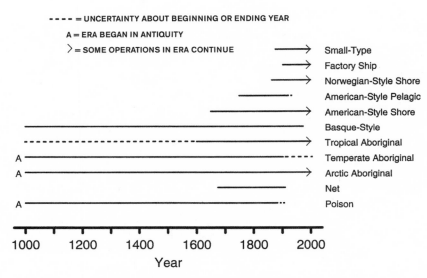

FIGURE 8.2. Approximate time periods for the eras defined in Table 8.1.

plant) and the other using bacteria (e.g., Øen 1997). There appear to have been multiple inventions of poison techniques at different times and in geographically distant places. Aconite poison whaling was used across the northern rim of the North Pacific (Heizer 1938; McCartney 1984; Crowell 1994), whereas bacterial poison whaling was practiced in Iceland and Norway (Heizer 1968). The latter involved impounding whales in bays using barrier nets, then darting them with a crossbow and waiting for septicemia to debilitate or kill the animal (Jonsgård 1955). A certain amount of experimentation, generally unsuccessful, was conducted with cyanide (Brown 1887: 248–249) and strychnine and curare (Thiercelin 1866) during the nineteenth century, and a variety of modern pharmaceutical drugs were tried during the twentieth century (Tønnessen and Johnsen 1982; references in Mitchell et al. 1986). The following are two key features of poison whaling:

1. Any species or size/age class that could be approached closely enough to dart could be targeted.

2. The killed:secured ratio could be very high.

Crowell (1994: 229) supposed that fin and humpback whales were targeted by North Pacific poison whalers, because these species were more likely to be available in bays during the late spring and summer months when whaling was practiced. Other authors have argued that gray whales *(Eschrichtius robustus)*, North Pacific right whales *(Eubalaena japonica)*, and possibly common minke whales *(Balaenoptera acutorostrata)* were targeted (Mitchell 1979). Norwegian bacterial whaling is said to have targeted minke whales primarily (Jonsgård 1955). Because no attempt was made to fasten to the whales, and it took several days for them to die (McCartney [1984: 85] referred to this as the "lance-and-wait" technique), hunting loss was very high in North Pacific poison whaling. The shut-in method practiced in Norway and Iceland probably involved much less hunting loss.

It is impossible to even guess reasonably at the numbers of whales killed by the poison whalers, and if it were possible, the allocation to species would be largely speculative.

Net

Like poison whaling, net whaling seems to have originated independently on several different occasions in different areas. Whaling with leather nets in Kamchatka is poorly documented but was being carried out in the 1730s and 1740s (Krasheninnikov 1972) and supposedly targeted gray whales (Rice and Wolman 1971). Japanese net whaling for humpback, right, gray, Bryde's, and probably other whale species is remarkably well documented, considering that it began in the seventeenth century and ended before 1900 (Omura 1984, 1986). Although some authors (see McCartney 1984: 86) have suggested that net whaling was introduced to Japan by the Dutch or Portuguese, it seems clear that the technique evolved locally as a way of improving the efficiency of traditional Japanese harpoon whaling (Kasuya 2002). Whaling for humpbacks in New Zealand with steel nets was conducted from 1890 to 1910 (Dawbin 1967). We have attributed the use of nets to block the escape of whales from bays in Norway and Iceland to the Poison era rather than the Net era. Also, we note that barrier, or shut-in, nets have been used extensively to catch belugas in the Arctic and Subarctic, but these fisheries are not included among the operations recognized here in view of our emphasis on large whales (see Introduction).

Arctic Aboriginal

Whaling on bowhead whales by Eskimos probably began at least 2,000 years ago (Stoker and Krupnik 1993). The basic approach involved hand-paddled skin boats launched from shore or ice, hand-thrown harpoons with rawhide lines attached to inflated sealskin floats, and hand lances. Unlike the Basque-style and American-style whalers (see below), the

Eskimos did not get "fast" to the whale; that is, keep the harpoon line attached to the boat as the whale attempted to escape. Rather, they depended on the float to tire the animal and to allow them to track it so that additional harpoon-line-float arrays could be deployed and they could approach the whale closely enough to kill it with the lance. A prominent feature of Eskimo whaling has been its selective incorporation of new technologies. Thus, the present-day hunt in Alaska incorporates "modern" (e.g., shoulder gun, bomb-lance, even the occasional use of aircraft for spotting) as well as traditional equipment and techniques (e.g., skin-covered boats for attacking and towing whales).

Although the original distribution of Eskimo bowhead whaling was "neither continuous nor homogeneous" (Stoker and Krupnik 1993: 591), the archaeological and ethnographic records of cultural exchange and transfers of technology imply circumpolar linkages. It is generally believed that the Thule whaling culture originated in the Bering Strait region and that there was "an unbroken technologic sequence which lasted for almost two millennia" across the Arctic from Siberia eastward to Greenland (McCartney 1984: 80). Separate operations have been provisionally defined on the basis of political geography. Thus, we have divided Siberian (Russian/Soviet), Alaskan (American), Canadian, and Greenlandic whaling into different operations within this era. In doing so, we recognize that these "operations" are not monolithic or homogeneous, nor are they altogether distinct from one another historically or culturally. In the Russian Far East, both bowhead and gray whales have been primary targets (Krupnik 1987), whereas in Greenland both bowhead and humpback whales were important historically (Kapel 1978; Caulfield 1997).

Few rigorous attempts have been made to generate long time series of removals for this era. Stoker and Krupnik (1993: 604) cite estimates of 790 bowheads secured in Alaska between 1910 and 1969, and 15 to 20 per year between 1914 and the 1980s. Circumpolar totals of removals for the entire era probably have been at least tens of thousands of large whales and hundreds of thousands of belugas and narwhals.

Temperate Aboriginal

We have generally followed McCartney (1984) in proposing the Temperate Aboriginal era to distinguish operations in southern Alaska, the Aleutian Islands (including Kodiak Island), and the "Northwest Coast" of North America from those in higher latitudes. The distinction is not clear, however, because some Eskimo societies (e.g., the Chugach Eskimos in the Gulf of Alaska) are included among the temperate-region aboriginal whalers, and there is considerable overlap, at least spatially if not also temporally, in the use of poison by kayak whalers (Poison era) and the use of standard harpoon-line-float arrays by whalers in open dugout or skin boats (this era). As McCartney (1984: 80) observed in reference to what he called "Arctic" and "Subarctic" aboriginal whaling, "to review the antiquity, spread, and patterns of whaling . . . is very much a matter of interpreting fragmentary evidence." Not

only is the evidence of aboriginal whaling in the North Pacific fragmentary, but we have probably overlooked a number of ancient, poorly documented operations in the North and South Atlantic, the South Pacific, and even the Indian Ocean that would belong in this era.

Principal target species appear to have been humpback and gray whales in the eastern Pacific, whereas the North Pacific right whale was likely a primary target of aboriginal whalers in the western Pacific.

Tropical Aboriginal

Relatively few operations have been identified that arose from local initiative and invention in tropical latitudes. We have provisionally included only two operations in the Indo-Pacific (Indonesia, Philippines). Operations in low latitudes that were (or in a few cases are) essentially extensions of the American nineteenth-century open-boat fishery were considered to belong in the American-Style Shore era (described subsequently) rather than this era. Thus, for example, the shore whaling for humpback and sperm whales in several parts of the West Indies and for humpback whales in Tonga and Equatorial Guinea are not considered part of the Tropical Aboriginal era, but those that arose independently in Indonesia and the Philippines are. (We recognize that this classification is not entirely consistent with the International Whaling Commission's terminology and management system, which classifies the humpback fishery in St. Vincent and the Grenadines as "aboriginal subsistence" whaling [Reeves 2002].) Humpback and sperm whales have been the principal species involved in Tropical Aboriginal whaling, except in Lamakera, Indonesia, where rorquals other than humpbacks are (or were) the primary targets (Barnes 1991).

Basque-Style

The Basques pursued whales in small open boats, attacking them with hand harpoons and lances. This basic technique was used for many centuries. Basques ventured to distant regions (southern Ireland, the English Channel, Iceland, Newfoundland, and Labrador), where they set up permanent or semipermanent shore stations for processing whales (Aguilar 1986). Many of the whaling activities by British, Dutch, Danish, German, and other European nationalities, in addition to the Basques themselves, have been assigned to this era. Often an operation mounted with foreign capital depended on a Basque crew with specialized expertise (Aguilar 1986).

Frederic Martens's description of British whaling in the Greenland Sea in 1671, frequently quoted (e.g., in Scammon 1874), defines the basic methods and techniques that characterized the overseas, non-Basque operations assigned to this era. A whale was sighted from the mother ship, oar-powered boats were launched in pursuit, and the whale was harpooned. The boats were thus made fast to, and often towed by, the whale, sometimes over considerable distances. After being killed with lances, the whale was towed to the mother ship and flensed (stripped of its blubber) alongside. Blubber was either packed as cargo aboard the ship and delivered to the home country ("ice whaling") or taken to nearby shores for trying out (cooking to oil) ("bay whaling" in, e.g., Spitsbergen and Labrador). Basque ships, alone among the Arctic fleet, supposedly sometimes cooked their blubber on board, despite the fire hazards, to avoid paying taxes for setting up their tryworks on shores claimed by European powers (Aguilar 1986; see Ellis 2002a for discussion of whether this really happened; also see Scoresby 1820, Vol. 2: 412). In areas such as Labrador, however, the normal procedure was for galleons to anchor in harbors while boats were sent away to search for and catch whales, which were then towed to shore for flensing and trying out (Aguilar 1986: 195).

For the most part, Basque-style whaling was confined to the North Atlantic and Arctic. We know of only one sustained operation in the South Atlantic, inaugurated in Brazil in 1603 (Peterson 1948). It presumably targeted southern right whales through the 1820s (Peterson 1948), after which humpback and sperm whales may have become more prominent (Tønnessen and Johnsen 1982: 205). We considered the continued catching of southern right whales into the 1970s (IWC 1986: 29) to represent an extension of this same Brazilian operation.

We provisionally assigned a brief attempt at shore whaling in the Canary Islands between 1778 and 1799 (Aguilar 1986) and the poorly documented shore whaling in the Cape Verde Islands from 1690 to 1912 (Hazevoet and Wenzel 2000) to this era, while recognizing that new information could lead us to reconsider and perhaps subdivide such operations, particularly the latter, which may have been influenced by American-style whaling (described in a subsequent section).

Balaenids, specifically the bowhead and the North Atlantic right whale *(Eubalaena glacialis)*, bore the brunt of this era's whaling. Sperm and humpback whales, together with narwhals, belugas, northern bottlenose whales, and other smaller sorts, are best regarded as supplemental targets. Aguilar (1986) estimated that 25,000–40,000 balaenids might have been taken off Newfoundland and Labrador by the Basques between 1530 and 1610. Other rough calculations indicate that perhaps 15,000 bowheads were taken around Svalbard between 1610 and 1669 and more than 85,000 between 1669 and 1800 (Hacquebord 1999); tens of thousands more were taken in Davis Strait and Baffin Bay between the 1720s and early 1900s (Mitchell and Reeves 1981). Catches of balaenids over the entire era (over five centuries) were in the hundreds of thousands, although it is important to recognize that some of the whaling operations assigned to the American-Style Pelagic era contributed to these catch estimates.

American-Style Shore

Whalemen of this era employed the basic Basque techniques of killing and processing whales. They sighted whales from lookouts on shore, pursued them in open boats, and attacked them with harpoons and lances, at least initially. However, the era was characterized by innovation, transition, and participant diversity, as described more fully subsequently for the American-Style Pelagic era. The essential elements of open-boat whaling, as practiced during this era, persisted at the Azores until 1984, at Tonga until 1981, and in the West Indies to the present day (see Reeves 2002). These and many of the other operations assigned to this era incorporated firearms, explosives, and mechanized vessels to some degree. With regard to mechanization, the initial appeal of steam propulsion was that it improved navigation of ice-infested waters in high latitudes, but it was also introduced to coastal operations, for example in the temperate waters of New England in 1880 (Clark 1887).

Open-boat shore whaling conducted in the United States from the seventeenth century to 1924 is assigned to this era (Edwards and Rattray 1932; Sayers 1984; Reeves and Mitchell 1986, 1988; Reeves et al. 1999), as are similar operations in Bermuda from the early seventeenth century to 1941 (Mitchell and Reeves 1983), in Australia and New Zealand from 1805 to 1932 (Baker 1983; Bannister 1986; Dawbin 1986) and in South Africa from 1789 to 1929 (Best and Ross 1986, 1989). Other operations, about which we know too little to comment further (e.g., Santa Maria, Chile, or Abrolhos Banks, Brazil; Klaus Barthelmess, in litt.), probably should also be included. As indicated above, some operations that have been classified recently in other contexts as "aboriginal" or artisanal (Bequia, Tonga, Azores) were assigned to this era in view of their direct historical links to the American open-boat fishery (Clarke 1954; Reeves 2002).

Right and humpback whales were the primary targets throughout most of this era. Off western North America, gray whales were also important, as were sperm whales in those areas with deep water close to shore (e.g., the Azores, Madeira, and the West Indies). Many hundreds of thousands of whales were taken.

Although this era eventually gave way during the second half of the nineteenth century to the Norwegian-Style Shore era, it did so incompletely. For example, the shore whalers on Long Island (New York) and the Outer Banks of North Carolina continued to launch their hand-powered open boats into the surf in pursuit of right whales, which they killed and processed in the old-fashioned manner, until the early twentieth century (Reeves and Mitchell 1986, 1988).

American-Style Pelagic

This era began in approximately the middle of the eighteenth century and ended with the last American voyages in the 1920s. Its most striking aspect was rapid expansion, in terms of both geography and the size and capacity of the whaling fleets. Basque-style whaling was confined almost entirely to the North Atlantic Ocean, including the Arctic Atlantic, whereas American-style pelagic whaling spanned the globe. The Basque-Style era was dominated by European nations, while the American-Style Pelagic era was dominated by the United States. The introduction of on-board tryworks (by 1762; Ashley 1928) facilitated the high-seas, long-distance voyages that typified the era (Ellis 2002b). Whaling equipment and practices were in an almost constant state of flux. Best (1983), for example, identified several important innovations in addition to the on-board tryworks, including the addition of sails to whaleboats in the 1820s, the toggle harpoon in 1848, and improvements in the bomb-lance in 1852. He also cited the demand for spermaceti oil in the manufacture of candles as a critical motivating force (also see Ellis 2002b).

There was substantial interpenetration and integration between the American-Style Pelagic and American-Style Shore whaling operations, so that in a sense these could be viewed as comprised in a single fishery (see Brown 1887; Clark 1887). New methods invented and adopted in one sector of the fishery, such as the shoulder gun and bomb-lance, soon found their way into the other.

The United States provided most of the capital, manpower, and expertise that defined this era. In 1846, near the chronological middle of the era, the world whaling fleet was estimated at approximately 1,000 vessels, of which 729 were U.S.-registered (Clark 1887: 192), and some of the vessels sailing under other nations' flags had American masters and were underwritten at least partly by American capital (Stackpole 1972; Du Pasquier 1982). By the 1880s, however, crews, even on American ships, were extremely diverse. As Brown (1887: 218) put it, "A more heterogeneous group of men has never assembled in so small a space than is always found in the forecastle of a New Bedford sperm whaler."

Several distinctions merit particular explanation and comment. Whaling historians recognize two distinct, partly concurrent British whale fisheries: the northern fishery, consisting of voyages to the Arctic Atlantic (the Spitsbergen and Davis Strait subfisheries) in pursuit of bowhead whales, and the southern (or South Seas) fishery, consisting of voyages to anywhere except the Arctic in pursuit of sperm and right whales (Jenkins 1921: 210). We regard the northern fisheries—not only that of Great Britain but also those of France, Germany, Denmark, the Netherlands, and other European states—as part of the Basque-Style era, whereas we have subsumed the southern fisheries (again including those of Great Britain, France, Germany, Denmark, the Netherlands, and other European states) within our American-Style Pelagic era. Steam whaling by American vessels in the Western Arctic in the final decades of the nineteenth century and the early decades of the twentieth is also attributed to this era, because it arose directly from the sailing-vessel fishery (Bockstoce 1986).

The sperm whale was the focal species of the American-Style Pelagic era as a whole, followed closely by the right whales. Bowhead, humpback, and gray whales were primary targets in particular areas and seasons. Importantly, the latter two species were hunted mainly on their winter calving/breeding grounds. Pilot whales were significant supplemental and "practice" targets. Several authors have attempted to estimate total catches for large portions of this era. For example, Best (1987) estimated that American-registered vessels secured about 30,000 bowhead whales, 70,000–75,000

right whales, 14,000–18,000 humpback whales, and 2,500–3,000 gray whales between 1804 and 1909, and Scarff (2001) adjusted Best's results for the North Pacific by incorporating information on non-American fleets and by applying a loss rate factor to account for hunting loss, producing an estimate of 26,500–37,000 North Pacific right whales killed between 1839 and 1909. Best (1983, 2004) estimated global sperm whale landings in the nineteenth century at 270,000, and detailed studies by Bannister et al. (1981) and Hope and Whitehead (1991) provide important empirical and methodological background for global or basinwide estimation of sperm whale catches (also see Whitehead 2002; Smith and Reeves 2003b).

Norwegian-Style Shore

By the 1860s and 1870s, when the Norwegian inventor Svend Foyn was developing the modern basis for mechanized whaling, using harpoons with explosive heads fired from cannons mounted in the bows of steam-powered catcher boats, the infrastructure and motivation were already in place for its global proliferation. Norwegian shore whaling spread worldwide during the late nineteenth and early twentieth centuries (e.g., to both coasts of North and South America, East Asia, the Caribbean, and East and West Africa). Many of the same sites where Basque-Style and American-Style Shore whaling had taken place previously became the sites of "modern" land stations (e.g., in Australia and New Zealand; Tønnessen and Johnsen 1982: 220–221). With innovative technology that allowed the exploitation of blue and fin whales, however, opportunities arose to establish whaling stations in new areas as well, most notably on the subantarctic islands, of which South Georgia was by far the most noteworthy. Norwegian skill and enterprise were as central to this era as Basque and American contributions had been to earlier eras. In the 1930s, some 70% of the aggregate world whaling crews were composed of Norwegians (K. Barthelmess, in litt., December 2003). We have defined operations largely on the basis of where the whaling took place (i.e., national jurisdictions), but we stress that Norway provided the capital and expertise for many of the shore stations, even as recently as the 1960s and early 1970s (e.g., Webb 1988). Typically, a Norwegian captain and gunner accompanied vessels that started shore-based operations in far-flung locations. After the Second World War, the export of whale meat and blubber to Japan became a central feature of many shore-based whaling operations, particularly in South and North America and in eastern Asia.

The numbers of whales killed in this era have totaled many hundreds of thousands, particularly if one includes the minke, killer, pilot, and other "small" whales that have been supplemental to many of the operations. All commercially valuable species have been hunted from shore stations. The numbers of right whales taken were relatively low because they had been depleted everywhere before this era began, and also because they were legally protected from the mid-1930s onward.

Factory Ship

The modern era of factory ship whaling began in the first decade of the twentieth century, when several vessels operated in the Southern Ocean and North Atlantic as "floating factories" (Tønnessen and Johnsen 1982). Many of the first factory ships operated essentially as floating shore stations, moored or anchored in bays while the catcher boats fanned out in search of prey. Best and Ross (1986) classified these as "shore-based establishments" (see Tønnessen and Johnsen [1982: 264, 503, 654] regarding the technical, legal, and biological implications). It was not until 1923 that truly pelagic factory ship whaling got under way in the Southern Ocean, and the stern slipway was not introduced until 1925, when the Norwegian ship *Lancing* operated off the Congo, in the Antarctic, and off Patagonia (Tønnessen and Johnsen 1982: 354–355). The capability of catching and processing with no links to shore stations gave the industry access to the final, and most profitable, whaling frontier: the Antarctic. Over the course of the twentieth century, factory ship operations from at least 15 countries (not counting so-called "pirate" operations) accounted for more than a million whales worldwide (see Clapham and Baker 2002). Relatively little factory ship whaling occurred in the North Atlantic (Jonsgård 1977), and most of such whaling in the South Atlantic was centered along the African coast or near the offshore subantarctic islands (Findlay 2000). The Antarctic and North Pacific were, by far, the most productive grounds during this whaling era, each yielding many hundreds of thousands of baleen and sperm whales. The last Soviet expedition to the Antarctic took place in 1981–1982. Only Japan has continued to engage in factory ship whaling since the 1982 International Whaling Commission (IWC) moratorium, taking roughly 400 minke whales annually.

Small-Type

For this era we relied largely on the IWC's formal definition of *small-type whaling*: "catching operations using powered vessels with mounted harpoon guns hunting exclusively for minke, [northern] bottlenose, [long-finned] pilot or killer whales" (IWC 1977: 34). We used a somewhat broader interpretation that encompasses not only the Norwegian hunt for those four species in the northern North Atlantic (clearly the intended subject of the IWC's schedule amendment just quoted; see Christensen 1975), but also the Japanese shore-based hunts for minke, short-finned pilot, killer, Baird's beaked, and other beaked whales in the western North Pacific (Ohsumi 1975; Kasuya 2002). Also, the British and Norwegian fisheries for northern bottlenose whales from the 1870s to early 1900s, using steam power and mounted harpoon guns, are included here, along with the recent and ongoing minke whale hunt in Iceland (Sigurjónsson 1982). In contrast, the taking of minke whales from powered vessels

in Greenland (Caulfield 1997) has been subsumed within a single Norwegian-style shore operation that has targeted fin and humpback as well as minke whales. The total numbers of whales taken by the operations included within this era are at least in the hundreds of thousands for the North Atlantic and the North Pacific.

Overlooked Operations

Determinations of the characteristics, or indeed the existence, of many ancient whaling operations are limited by the fragmentary nature of the historical record. Even in some instances where in-depth analyses have been conducted by ethnographers (Lindquist 1993) or archaeologists (McCartney 1984), a detailed understanding of ancient whaling activities is lacking. Moreover, in the present study, time and budget constraints have meant that our list of operations in the Appendix to this chapter falls well short of being comprehensive with respect to the existing literature. For example, although Heizer's (1968) bibliography provides tantalizing references to whaling in Zanzibar in 1295, in "Arabia" in the ninth century, in Tierra del Fuego in the 1600s and 1700s, and in Mozambique (possibly meaning Madagascar) at some time in the ancient past, we did not examine and evaluate his sources, and therefore those "operations" are missing from our list. Also, Barthelmess (in litt.) has called our attention to, *inter alia*, possible early whaling in northwestern Europe (Old Norse, Saami), the Mediterranean region, Sokotra (?), Namibia, and in Hainan (China) in the seventeenth century. Thus, we expect that other investigators will be able to expand considerably the number of operations recognized, including both recent and ancient ones. Our list of operations, like any species list in biological taxonomy, remains provisional and subject to updating and revision.

We expect our list of higher-order taxa (i.e., eras) to be relatively stable in comparison to our list of operations. In other words, we do not expect newly proposed operations to be so distinctive in character as to require the erection of additional eras to accommodate them.

Applying the Taxonomy to Estimate Historical Whale Catches

The taxonomy just outlined provides an overview of the development and extent of whaling worldwide. As such, it should assist anyone interested in, for example, the ecological effects of whaling, to understand the full complexity of that enterprise at any scale of analysis (local, regional, or global). Although similar global overviews have been developed previously for other purposes (e.g., Barthelmess 1992), this is the first attempt of which we are aware to provide a formal classification system in support of quantitative constructions of whaling "catch histories."

Given that the goal is to produce definitive summaries of removals from biological populations of whales, our lists of eras and operations can be used to guide efforts to identify, obtain, and organize the relevant "catch statistics." Sources of the needed information can be classified broadly into four categories: archaeological, ethnographic/historical, production-centered, and whale-centered. These categories have approximate time boundaries (with some overlap) and can be characterized by representative source types (Table 8.2).

To understand what is known and can be known about each operation requires an understanding of the characteristics of each source category and, more specifically, of which sources are relevant to a given task. The sources vary greatly in their level of resolution and reliability and therefore in their usefulness for estimating whaling removals. For example, the primary data embedded in archaeological and ethnographic/historical sources tend to be very limited for this purpose, and in some instances they may yield nothing beyond confirmation (or supposition) that whaling took place. In contrast, production-centered and individual whale-centered sources are generally much richer and more useful for constructing "catch histories." *Primary* sources (i.e., the sources of original or "raw" data) are generally viewed as more authoritative and reliable than *secondary* sources (i.e., those that are derived from primary sources). However, primary sources can be difficult to use for a variety of reasons (e.g., they are unavailable, their handling is prohibitively expensive and/or time-consuming), and analyses have therefore often relied on secondary sources, such as published lists (e.g., Starbuck 1878; Hegarty 1959; Lund 2001; *Whalemen's Shipping List*; compilations by the Bureau of International Whaling Statistics) and charts (e.g., Maury 1852; Townsend 1935; Scarff 1991). Although secondary sources are useful, especially when primary sources are no longer available, it is important to understand the relationship between the secondary (derivative) sources and the primary sources underlying them. For example, in the cases of the Maury (1852) and Townsend (1935) charts of whale distribution, it is necessary to ensure that those authors' extractions from primary materials (voyage logbooks and journals) were accurate and complete (see, e.g., Reeves et al. 2004). Similarly, the Starbuck (1878) and Hegarty (1959) lists of American whaling voyage returns (oil and baleen) need to be evaluated in relation to their primary sources (e.g., custom-house or company records, newspapers). The discipline of history has well-developed procedures for evaluating the nature, reliability, and completeness of any source—for example, by examining who created it and for what purpose, and why the artifact or document has survived to the present day.

The *absence* of sources for specific areas or time periods of an operation also needs to be considered. Gaps in the catch history cannot, for example, be treated as "no catch" in population analyses unless there is positive evidence for the suspension of whaling activities for the fishery, area, or time period in question. When sources of data are insufficient or lacking entirely, it is often necessary to fill gaps through interpolation or extrapolation (e.g., Smith and Reeves 2002; Reeves et al., in press). Failure to account for such gaps could help explain the failure of population models to "fit" observed or estimated values of current abundance or population growth

rates, such as in the case of eastern North Pacific gray whales (IWC 1993: 248–250; Punt and Butterworth 2002). However, even when a spatially coarse compilation of catch history seems sufficient to support modeling of hemisphere-wide trends, as in the case of southern right whales (IWC 2001), the need to estimate abundance and trends for individual "management units based on breeding stocks" (IWC 2001: 26) creates a requirement for finer-scaled catch histories.

Secondary sources are adequate, in some cases, for supplying the data needed to support analyses. However, in those cases where (a) there is reason to believe that the data derived from secondary sources are either incomplete or ambiguous (e.g., in regard to species taken or loss rates), (b) spatial resolution is critical, or (c) information on statistical precision is important, the need to consult primary sources may be inescapable. For example, Best (1983) provided an extremely useful summary of American-Style Pelagic era sperm whaling based on various secondary sources, one of which (Lyman in IWC 1969) gives decadal production-derived catch estimates for sperm whales from 1800 to 1910. However, not only were Lyman's estimates negatively biased because of failure to account fully for non-U.S. voyages (Best 1983: 43) but they also provided no way to disaggregate the data so that catches could be estimated at a less than global level. Further, Lyman's compilation is not amenable to quantification of bias or measurement of statistical precision. This defect may be considered typical of secondary sources.

In an important recent analysis of the effects of whaling on world stocks of sperm whales, Whitehead (2002: his Figure 1) appears to have derived his global catch series for the "open-boat hunt" between 1800 and the 1920s from either IWC (1969), Best (1983), or a combination of the two. As Whitehead acknowledges (2002: 302), certain of his modeling results conflict with the evidence in primary sources (logbooks) concerning rates of decline in regional sperm whale abundance (Tillman and Breiwick 1983; Whitehead 1995). It is difficult to see how understanding of the ecological effects of whaling at the population or ecosystem level can be greatly improved without more studies of primary sources (e.g., Bannister et al. 1981; Bockstoce and Botkin 1983; Hope and Whitehead 1991; Smith and Reeves 2003a, b).

Even when quantified or quantifiable data exist, the validity of the sources may be in doubt. For twentieth-century examples, the primary data submitted from some Soviet pelagic expeditions (Best 1988; Zemsky et al. 1995a, b; Mikhalev 1997) and some Japanese (Kasuya 1999) and possibly South African (Best 1989) shore stations are known to have been falsified in some way. As indicated by Best (1989), there are two broad categories of falsified data: one consisting of relatively minor issues such as sizes or sexes of whales taken, the other of large-scale illegal catches, such as those by Soviet pelagic and Japanese shore whaling. Although alternative "actual" catches have been reported for some of the falsified Soviet expeditions (e.g., Clapham and Baker 2002: their Table II), it remains unclear how the other unreliable or invalid data might be corrected (Allison and Smith 2004).

TABLE 8.2
Sources of Whaling Data

Divided into Four Broad Categories, with Basic Features, Time Period, and Representative Types

Category	Basic Features	Time Period	Representative Source Types
Archaeological	Prehistoric, artifact-based, with limited ability to make inferences from written materials (e.g., early travel narratives)	Antiquity to 18th century	Hunting tools (e.g., harpoons); whale bones in middens and shelter or ritual structures, on beaches, or incorporated into art objects; illustrations on cave walls or scenes depicted in carvings and other art/craft forms; historic toponymy
Ethnographic/ Historical	Written or printed materials, generally based on first-hand observations by the writer	1700 to early 1900s	Descriptions in nonwhaling trade newspapers, anthropological field studies, diaries or journals of whalemen, travelogues, personal account books
Production-centered	Records of oil, baleen (whalebone), and other whale products, usually compiled on an annual or voyage basis	1750s to present	Whaling-trade newspapers, whaling voyage logbooks and account books, customs-house records, British colonial Blue Books
Individual whale-centered	Records of numbers of whales caught and processed	1870 to present	Lists maintained by company or government officials, data sheets submitted to national or international agencies (Bureau of Whaling Statistics, International Whaling Commission)

The clear evidence of data manipulation and falsification in the twentieth-century has created what may be an unwarranted degree of skepticism toward earlier primary sources of whaling data. For example, Roman and Palumbi (2003), while acknowledging that whaling logbooks (in the context referring to pre-twentieth-century material) "provide clues" to how many whales were removed by whaling, claim that such sources "may be incomplete, intentionally underreported, or fail to consider whales that were struck and lost." Indeed, any responsible user of logbooks and other primary historical sources needs to be aware of those possibilities, weigh their importance, and incorporate measures in the analysis to account for them (see, e.g., Smith and Reeves 2002, 2003c). Sherman (1983) outlined in detail many of the limitations of logbooks as sources of catch and other data. However, until management of whaling began in the twentieth century, whalers had no incentive to "intentionally underreport" catches, and logbooks recorded struck and lost whales more or less consistently, depending on the keepers' abilities and predilections. The assumption that whalers were strategic liars about their catches long before any form of regulation was imposed on their activities is, in our judg-

ment, facile and unfounded, based as it is on perceptions that have been shaped by modern realities in regard to incentives to deceive. In fact, the catch data reported in Yankee logbooks have been shown to be consistent with amounts of oil reported and measured on shore at the end of the voyages (Mitchell and Reeves 1983; Smith and Reeves 2004).

Final Comments

In designing and developing our proposed taxonomy of world whaling, we conducted a relatively superficial review of whaling literature. A more thorough study would likely have led us to split apart or combine some of the operations provisionally defined here, prompted us to assign some of the operations to different eras, and allowed us to include at least a few (and possibly many) additional operations. Although a more thorough study might also have caused us to define eras somewhat differently, we believe that our definitions are fairly robust and that they reflect the key differences among whaling operations. The temporal, geographical, technological, and "platform" (i.e., shore versus pelagic) differences used to distinguish eras are sharper and less subject to interpretation (or misinterpretation) than many of the differences used to define operations.

Our main goal in this chapter has been to establish the conceptual utility of a taxonomy that can serve as a heuristic tool for studying the history of whaling on any scale of time or space and for analyzing the effects of whaling, whether at a species, population, or ecosystem level. Although it may not be possible to make reliable estimates of removals, or indeed to identify the species taken, in all whaling operations or eras, a first step must be to determine what is known, what can be known, and what is essentially unknowable. That step will depend upon a clear and complete understanding of the whaling activities in the area of interest or on the population of concern.

Acknowledgments

We acknowledge the financial support of the Alfred E. Sloan Foundation through the History of Marine Animal Populations project of the Census of Marine Life. In addition, Reeves received support from the National Marine Mammal Laboratory, the University of California Santa Cruz, and the U.S. Geological Survey. We are especially grateful to Klaus Barthelmess, who kindly reviewed an earlier draft in great detail and shared his expert knowledge of whaling history. We also thank Elizabeth Josephson for her assistance in preparing Figure 8.1, and Fred Serchuk and Bob Brownell for helpful reviews.

Literature Cited

Aguilar, A. 1986. A review of old Basque whaling and its effect on the right whales *(Eubalaena glacialis)* of the North Atlantic. *Report of the International Whaling Commission* (Special Issue) 10: 191–199.

Allison, C. and T. D. Smith. 2004. *Progress on the construction of a comprehensive database of twentieth century whaling catches.* Scientific Committee Document SC/56/O39. Cambridge, UK: International Whaling Commission.

Ashley, C. W. 1928. *The Yankee whalers.* London: George Routledge and Sons.

Baker, A. N. 1983. *Whales and dolphins of New Zealand and Australia: an identification guide.* Wellington, New Zealand: Victoria University Press.

Bannister, J. L. 1986. Notes on nineteenth century catches of southern right whales *(Eubalaena australis)* off the southern coasts of Western Australia. *Report of the International Whaling Commission* (Special Issue) 10: 255–259.

Bannister, J. L., S. Taylor, and H. Sutherland. 1981. Logbook records of 19th century American sperm whaling: a report on the 12 month project, 1978–79. *Report of the International Whaling Commission* 31: 821–833.

Barnes, R. H. 1991. Indigenous whaling and porpoise hunting in Indonesia, in *Cetaceans and cetacean research in the Indian Ocean Sanctuary.* S. Leatherwood and G. P. Donovan, eds. Marine Mammal Technical Report 3. Nairobi: United Nations Environment Programme, pp. 99–106.

———. 1996. *Sea hunters of Indonesia: fishers and weavers of Lamalera.* Oxford: Clarendon Press.

Barthelmess, K. 1992. Auf Walfang—Geschichte einer Ausbeutung [A-whaling—the story of an exploitation], in *Von Walen und Menschen [Of whales and humans].* Knuth Weidlich, ed. Hamburg: Historika Photoverlag, pp. 4–51, 157–159.

Best, P. B. 1983. Sperm whale stock assessments and the relevance of historical whaling records. *Report of the International Whaling Commission* (Special Issue) 5: 41–55.

———. 1987. Estimates of the landed catch of right (and other whalebone) whales in the American fishery, 1805–1909. *Fishery Bulletin* 85: 403–418.

———. 1988. Right whales at Tristan da Cunha: a clue to the "non-recovery" of depleted species? *Biological Conservation* 46: 23–51.

———. 1989. Some comments on the BIWS catch record data base. *Report of the International Whaling Commission* 39: 363–369.

———. 1994. A review of the catch statistics for modern whaling in southern Africa, 1908–1930. *Report of the International Whaling Commission* 44: 467–485.

———. 2004. *Estimating the landed catch of sperm whales in the nineteenth century.* Scientific Committee Document SC/56/IA5. Cambridge, UK: International Whaling Commission.

Best, P. B. and G. J. B. Ross. 1986. Catches of right whales from shore-based establishments in southern Africa, 1792–1975. *Report of the International Whaling Commission* (Special Issue) 10: 275–289.

———. 1989. Whales and whaling, in *Oceans of life off southern Africa.* A. I. L. Payne and R. J. M. Crawford, eds. Vlaeberg: Vlaeberg Publishers, pp. 315–338.

Bockstoce, J. R. 1986. *Whales, ice, and men: the history of whaling in the western Arctic.* Seattle: University of Washington Press.

Bockstoce, J. R. and D. B. Botkin. 1983. The historical status and reduction of the Western Arctic bowhead whale *(Balaena mysticetus)* population by the pelagic whaling industry, 1848–1914. *Report of the International Whaling Commission* (Special Issue) 5: 107–142.

Brown, J. T. 1887. The whalemen, vessels and boats, apparatus, and methods of the whale fishery, in *The fisheries and fishery industries of the United States, Section V. History and methods of the fisheries. Part XV. The whale fishery.* G. B. Goode, ed. Washington, DC: Government Printing Office, pp. 218–293.

Brownell, R. L., Jr., P. J. Clapham, T. Miyashita, and T. Kasuya. 2001. Conservation status of North Pacific right whales. *Journal of Cetacean Research and Management* (Special Issue) 2: 269–286.

Caulfield, R. A. 1997. *Greenlanders, whales, and whaling: sustainability and self-determination in the Arctic.* Hanover, NH: University Press of New England.

Christensen, I. 1975. Preliminary report on the Norwegian fishery for small whales: expansion of Norwegian whaling to Arctic and northwest Atlantic waters, and Norwegian investigations of the biology of small whales. *Journal of the Fisheries Research Board of Canada* 32: 1083–1094.

Clapham, P. J. and C. S. Baker. 2002. Whaling, modern, in *Encyclopedia of marine mammals.* W. F. Perrin, B. Würsig, and J. G. M. Thewissen, eds. San Diego: Academic Press, pp. 1328–1332.

Clark, A. H. 1887. History and present condition of the fishery, in *The fisheries and fishery industries of the United States, Section V. History and methods of the fisheries. Part XV. The whale fishery.* G. B. Goode, ed. Washington, DC: Government Printing Office, pp. 3–218.

Clarke, R. 1954. Open boat whaling in the Azores: the history and present methods of a relic industry. *Discovery Reports* 26: 281–354, pls. XIII–XVIII.

Crowell, A. 1994. Koniag Eskimo poisoned-dart whaling, in *Anthropology of the North Pacific Rim*. W. W. Fitzhugh and V. Chaussonnet, eds. Washington, DC: Smithsonian Institution Press, pp. 217–242.

Davis, L. E., R. E. Gallman, and K. Gleiter. 1997. *In pursuit of Leviathan: technology, institutions, productivity, and profits in American whaling, 1816–1906*. Chicago: University of Chicago Press.

Dawbin, W. H. 1967. Whaling in New Zealand waters 1791–1963, extract from *An Encyclopaedia of New Zealand*. Wellington: Government Printer.

———. 1986. Right whales caught in waters around south eastern Australia and New Zealand during the nineteenth and early twentieth centuries. *Report of the International Whaling Commission* (Special Issue) 10: 261–267.

Dolar, M. L. L., S. J. Leatherwood, C. J. Wood, M. N. R. Alava, C. L. Hill, and L. V. Aragones. 1994. Directed fisheries for cetaceans in the Philippines. *Report of the International Whaling Commission* 44: 439–449.

Du Pasquier, T. 1982. *Les baleiniers français au XIXème siècle 1814–1868*. Grenoble, France: Terre et Mer.

Edwards, E. J., and J. E. Rattray. 1932. *"Whale off!" The story of American shore whaling*. New York: Frederick A. Stokes.

Ellis, R. 1991. *Men and whales*. New York: Alfred A. Knopf.

———. 2002a. Whaling, early and aboriginal, in *Encyclopedia of marine mammals*. W. F. Perrin, B. Würsig, and J. G. M. Thewissen, eds. San Diego: Academic Press, pp. 1310–1316.

———. 2002b. Whaling, traditional, in *Encyclopedia of marine mammals*. W. F. Perrin, B. Würsig, and J. G. M. Thewissen, eds. San Diego: Academic Press, pp. 1316–1328.

Findlay, K. P. 2000. A review of humpback whale catches by modern whaling operations in the Southern Hemisphere. *Memoirs of the Queensland Museum* 47: 411–420.

Francis, D. 1990. *A history of world whaling*. Markham, Ontario: Viking Penguin Books Canada.

Freeman, M. M. R. 1979. A critical review of Thule Culture and ecological adaptation, in *Thule Eskimo culture: an anthropological retrospective*. A. P. McCartney, ed. Mercury Series, Paper No. 88. Ottawa: Archaeological Survey of Canada, pp. 278–285.

Hacquebord, L. 1999. The hunting of the Greenland right whale in Svalbard, its interaction with climate and its impact on the marine ecosystem. *Polar Research* 18: 375–382.

Hazevoet, C. J. and Wenzel, F. W. 2000. Whales and dolphins (Mammalia, Cetacea) of the Cape Verde Islands, with special reference to the humpback whale *Megaptera novaeangliae* (Borowski, 1781). *Contributions to Zoology* 69(3): 197–211.

Hegarty, R. B. 1959. *Returns of whaling vessels sailing from American ports: a continuation of Alexander Starbuck's "History of the American Whale Fishery" 1876–1928*. New Bedford, MA: Old Dartmouth Historical Society/New Bedford Whaling Museum.

Heizer, R. F. 1938. Aconite poison whaling in Asia and America: an Aleutian transfer to the New World. *Bulletin of the Bureau of American Ethnology* 133: 417–468.

———. 1968. A bibliography of aboriginal whaling. *Journal of the Society for the Bibliography of Natural History* 4(7): 344–362.

Henderson, D. A. 1972. *Men and whales in Scammon's Lagoon*. Los Angeles: Dawson's Book Shop.

Hope, P. L. and H. Whitehead. 1991. Sperm whales off the Galápagos Islands from 1830–50 and comparisons with modern studies. *Report of the International Whaling Commission* 41:273–286.

IWC. 1969. Report of the IWC-FAO Working Group on Sperm Whale Stock Assessment. *Report of the International Whaling Commission* 19: 39–83.

———. 1977. Chairman's report of the twenty-eighth meeting. *Report of the International Whaling Commission* 27: 22–35.

———. 1986. Report of the workshop on the status of right whales. *Report of the International Whaling Commission* (Special Issue) 10: 1–33.

———. 1993. Report of the Special Meeting of the Scientific Committee on the Assessment of Gray Whales. *Report of the International Whaling Commission* 43: 241–259.

———. 2001. Report of the Workshop on the Comprehensive Assessment of Right Whales: a worldwide comparison. *Journal of Cetacean Research and Management* (Special Issue) 2: 1–60.

Jenkins, J. T. 1921. *A history of the whale fisheries from the Basque fisheries of the tenth century to the hunting of the finner whale at the present date*. Port Washington, NY: Kennikat Press.

Johnsen, A. O. 1959. *Den moderne hvalfangst historie: opprinnelse og utvikling*. Vol. 1. Oslo: H. Aschehoug & Co.

Jonsgård, Å. 1955. Development of the modern Norwegian small whale industry. *Norsk Hvalfangst-tidende* 44(12): 697–718.

———. 1977. Tables showing the catch of small whales (including minke whales) caught by Norwegians in the period 1938–75, and large whales caught in different North Atlantic waters in the period 1868–1975. *Report of the International Whaling Commission* 27: 413–426.

Kapel, F. O. 1978. Exploitation of large whales in West Greenland in the twentieth century. *Report of the International Whaling Commission* 29: 197–214.

Kasuya, T. 1999. Examination of the reliability of catch statistics in the Japanese coastal sperm whale fishery. *Journal of Cetacean Research and Management* 1: 109–122.

———. 2002. Japanese whaling, in *Encyclopedia of marine mammals*. W. F. Perrin, B. Würsig, and J. G. M. Thewissen, eds. San Diego: Academic Press, pp. 655–662.

Krasheninnikov, S. P. 1972. *Explorations of Kamchatka: North Pacific Scimitar. Report of a journey made to explore eastern Siberia in 1735–1741, by order of the Russian Imperial Government*. Portland: Oregon Historical Society.

Krupnik, I. I. 1987. The bowhead vs. the gray whale in Chukotkan aboriginal whaling. *Arctic* 40: 16–32.

Lindquist, O. 1993. Whaling by peasant fishermen in Norway, Orkney, Shetland, the Faeroe Islands, Iceland and Norse Greenland: mediaeval and early modern whaling methods and inshore legal regimes, in *Whaling and history: perspectives on the evolution of the industry*. B. L. Basberg, J. E. Ringstad, and E. Wexelsen, eds. Publication No. 29. Sandefjord, Norway: Kommandoer Chr. Christensens Hvalfangstmuseum, pp. 17–54.

Little, E. A. and J. C. Andrews. 1981. Drift whales at Nantucket: the kindness of Moshup. *Man in the Northeast* 23: 17–38.

Lund, J. N. 2001. *Whaling masters and whaling voyages sailing from American ports: a compilation of sources*. New Bedford, MA: New Bedford Whaling Museum.

Maury, M. F. 1852 et seq. Whale Chart of the World. Series F (Wind and Current Charts) Sheet 1 (1852), sheets 2–4 (no date). Washington, DC: U. S. Naval Observatory.

McCartney, A. P. 1984. History of native whaling in the Arctic and Subarctic, in *Arctic whaling: proceedings of the international symposium Arctic Whaling, February 1983.* H. K. s'Jacob, K. Snoeijing, and R. Vaughan, eds. Works of the Arctic Centre, No. 8. Groningen, The Netherlands: University of Groningen, pp. 79–111.

Mikhalev, Y. A. 1997. Humpback whales *Megaptera novaeangliae* in the Arabian Sea. *Marine Ecology Progress Series* 149: 13–21.

Mitchell, E. 1974. Present status of Northwest Atlantic fin and other whale stocks, in *The whale problem: a status report.* W. E. Schevill, ed. Cambridge, MA: Harvard University Press, pp. 108–169.

———. 1979. Comments on the magnitude of early catch of East Pacific gray whale *(Eschrichtius robustus). Report of the International Whaling Commission* 29: 307–314.

Mitchell, E. and R. R. Reeves. 1980. The Alaska bowhead problem: A commentary. *Arctic* 33: 686–723.

———. 1981. Catch history and cumulative catch estimates of initial population size of cetaceans in the eastern Canadian Arctic. *Report of the International Whaling Commission* 31: 645–682.

———. 1983. Catch history, abundance, and present status of northwest Atlantic humpback whales. *Report of the International Whaling Commission* (Special Issue) 5: 153–212.

Mitchell, E. D., R. R. Reeves, and A. E. Evely. 1986. Bibliography of whale killing techniques. *Report of the International Whaling Commission* (Special Issue) 7.

Øen, E. O. 1997. Norsk hvalfangst i gamle dager—og en veterinær oppdagelse. *Norsk Veterinærtidsskrift* 109: 371–376.

O'Hara, T. M., T. F. Albert, E. O. Oen, L. M. Philo, J. C. George, and A. L. Ingling. 1999. The role of Eskimo hunters, veterinarians, and other biologists in improving the humane aspects of the subsistence harvest of bowhead whales. *Journal of the American Veterinary Medical Association* 214: 1193–1198.

Ohsumi, S. 1975. Review of Japanese small-type whaling. *Journal of the Fisheries Research Board of Canada* 32: 1111–1121.

———. 1980. Catches of sperm whales by modern whaling in the North Pacific. *Report of the International Whaling Commission* (Special Issue) 2: 11–18.

Omura, H. 1984. History of gray whales in Japan, in *The gray whale* Eschrichtius robustus. M. L. Jones, S. L. Swartz and S. Leatherwood, eds. Orlando, FL: Academic Press, pp. 57–77.

———. 1986. History of right whale catches in the waters around Japan. *Report of the International Whaling Commission* (Special Issue) 10: 35–41.

Peterson, B. W. 1948. South Atlantic whaling: 1603–1830. Ph.D. thesis, University of California.

Pike, G. C. and I. B. MacAskie. 1969. *Marine mammals of British Columbia.* Bulletin No. 171. Ottawa: Fisheries Research Board of Canada.

Punt, A. E. and D. S. Butterworth. 2002. An examination of certain of the assumptions made in the Bayesian approach used to assess the eastern North Pacific stock of gray whales *(Eschrichtius robustus). Journal of Cetacean Research and Management* 4: 99–110.

Reeves, R. R. 2002. The origins and character of "aboriginal subsistence" whaling: a global review. *Mammal Review* 32: 71–106.

Reeves, R. R., J. M. Breiwick, and E. D. Mitchell. 1999. History of whaling and estimated kill of right whales, *Balaena glacialis,* in the northeastern United States, 1620–1924. *Marine Fisheries Review* 61(3): 1–36.

Reeves, R. R., E. Josephson, and T. D. Smith. 2004. Putative historical occurrence of North Atlantic right whales in mid-latitude offshore waters: "Maury's Smear" is likely apocryphal. *Marine Ecology Progress Series* 282: 295–305.

Reeves, R. R., S. Leatherwood, S. A. Karl, and E. R. Yohe. 1985. Whaling results at Akutan (1912–39) and Port Hobron (1926–37), Alaska. *Report of the International Whaling Commission* 35: 441–457.

Reeves, R. R. and E. Mitchell. 1986. The Long Island, New York, right whale fishery: 1650–1924. *Report of the International Whaling Commission* (Special Issue) 10: 201–220.

———. 1988. *History of whaling in and near North Carolina.* NOAA Technical Report NMFS-65.

Reeves, R. R. and T. D. Smith. 2002. Historical catches of humpback whales in the North Atlantic Ocean: An overview of sources. *Journal of Cetacean Research and Management* 4: 219–234.

Reeves, R. R., T. D. Smith, and E. A. Josephson. In press. Near-annihilation of a species: right whaling in the North Atlantic, in *The urban whale: North Atlantic right whales at the crossroads.* S. D. Kraus and R. M. Rolland, eds. Cambridge, MA: Harvard University Press.

Reeves, R. R., T. D. Smith, R. L. Webb, J. Robbins, and P. J. Clapham. 2002. Humpback and fin whaling in the Gulf of Maine from 1800 to 1918. *Marine Fisheries Review* 64(1): 1–12.

Rice, D. W. and A. A. Wolman. 1971. *The life history and ecology of the gray whale* (Eschrichtius robustus). American Society of Mammalogists, Special Publication No. 3. Lawrence, KS: Allen Press.

Roman, J. and S. Palumbi. 2003. Whales before whaling in the North Atlantic. *Science* 301: 508–510.

Ross, W. G. 1975. *Whaling and Eskimos: Hudson Bay 1860–1915.* Publications in Ethnology No. 10. Ottawa: National Museum of Man.

Sanderson, I. T. 1954. *Follow the whale.* London: Bramhall House. [Reprinted 1993 as *A history of whaling.* New York: Barnes & Noble.]

Sayers, H. 1984. Shore whaling for gray whales along the coast of the Californias, in *The gray whale* Eschrichtius robustus. M. L. Jones, S. L. Swartz and S. Leatherwood, eds. Orlando, FL: Academic Press, pp. 121–157.

Scammon, C. M. 1874. *The marine mammals of the north-western coast of North America, described and illustrated with an account of the American whale-fishery.* New York: John H. Carmany and Co.

Scarff, J. E. 1991. Historic distribution and abundance of the right whale *(Eubalaena glacialis)* in the North Pacific, Bering Sea, Sea of Okhotsk and Sea of Japan from the Maury whale charts. *Report of the International Whaling Commission* 41: 467–89.

———. 2001. Preliminary estimates of whaling-induced mortality in the 19th century North Pacific right whale *(Eubalaena japonicus)* fishery, adjusting for struck-but-lost whales and non-American whaling. *Journal of Cetacean Research and Management* (Special Issue) 2: 261–268.

Schmitt, F. P., C. De Jong, and F. H. Winter. 1980. *Thomas Welcome Roys: America's pioneer of modern whaling.* Charlottesville: University Press of Virginia.

Scoresby, W., Jr. 1820. *An account of the Arctic regions with a history and description of the northern whale-fishery.* 2 volumes. Edinburgh: Archibald Constable. Reprint, New York: Augustus M. Kelley, 1969.

Sherman, S. C. 1983. The nature, possibilities and limitations of whaling logbook data. *Report of the International Whaling Commission* (Special Issue) 5: 35–39.

Sigurjónsson, J. 1982. Icelandic minke whaling 1914–1980. *Report of the International Whaling Commission* 32: 287–295.

———. 1988. Operational factors of the Icelandic large whale fishery. *Report of the International Whaling Commission* 38: 327–333.

Smith, T. D., and R. R. Reeves. 2002. Estimating historical humpback whale removals from the North Atlantic. *Journal of Cetacean Research and Management* 4 (Supplement): 242–255.

———. 2003a. Estimating American 19th century catches of humpback whales in the West Indies and Cape Verde Islands. *Caribbean Journal of Science* 39: 286–297.

Smith, T. D. and R. R. Reeves, eds. 2003b. *Design of a program of research on sperm whale catch history: results of a workshop.* Census of Marine Life, History of Marine Animal Populations. Available at http://www.cmrh.dk/Cachalot.htm. Accessed May 18, 2006.

Smith, T. D. and R. R. Reeves. 2003c. Estimating historical humpback whale removals from the North Atlantic: an update. *Journal of Cetacean Research and Management* 5 (Supplement): 301–311.

———. 2004. *Estimating whaling catch history.* Scientific Committee Document SC/56/O22. Cambridge, UK: International Whaling Commission.

Spence, B. 1980. *The story of whaling.* Greenwich, UK: Conway Maritime Press.

Stackpole, E. A. 1972. *Whales and destiny: the rivalry between America, France, and Britain for control of the southern whale fishery, 1785–1825.* Amherst: University of Massachusetts Press.

Starbuck, A. 1878. *History of the American whale fishery from its earliest inception to the year 1876.* Report of the U. S. Fish Commission 4, 1875–6, Appendix A.

Stoker, S. W., and I. I. Krupnik. 1993. Subsistence whaling, in *The bowhead whale.* J. J. Burns, J. J. Montague, and C. J. Cowles, eds. Society for Marine Mammalogy, Special Publication No. 2. Lawrence, KS: Allen Press, pp. 579–629.

Thiercelin, L. 1866. Action des sels solubles de strychnine, associés au curare, sur les gros cétacés. *Comptes Rendus des Séances de l'Académie des Sciences, Paris* 63: 924–927.

Tillman, M. F. and J. M. Breiwick. 1983. Estimates of abundance for the western North Pacific sperm whale based upon historical whaling records. *Report of the International Whaling Commission* (Special Issue) 5: 257–269.

Tønnessen, J. N. 1967. *Den moderne hvalfangsts historie: opprinnelse og utvikling.* Vol. 2. Sandefjord, Norway: Norges Hvalfangstforbund.

———. 1969. *Den moderne hvalfangsts historie: opprinnelse og utvikling.* Vol. 3. Sandefjord, Norway: Norges Hvalfangstforbund.

———. 1970. *Den moderne hvalfangsts historie: opprinnelse og utvikling.* Vol. 4. Sandefjord, Norway: Norges Hvalfangstforbund.

Tønnessen, J. N. and A. O. Johnsen. 1982. *The history of modern whaling.* Berkeley: University of California Press.

Townsend, C. H. 1935. The distribution of certain whales as shown by logbook records of American whaleships. *Zoologica* 19: 1–50, 4 charts.

Vaucaire, M. 1941. *Histoire de la pêche à la baleine.* Paris: Payot.

Vaughan, R. 1984. Historical survey of the European whaling industry, in *Arctic whaling: proceedings of the international symposium Arctic Whaling, February 1983.* H. K. s'Jacob, K. Snoeijing, and R. Vaughan, eds. Works of the Arctic Centre, No. 8. Groningen, The Netherlands: University of Groningen, pp. 121–134 .

Webb, R. L. 1988. *On the Northwest: commercial whaling in the Pacific Northwest 1790–1967.* Vancouver: University of British Columbia Press.

———. 2001. Menhaden whalemen: nineteenth-century origins of American steam whaling. *American Neptune* 60(3): 277–288.

Whipple, A. B. C. 1979. *The whalers.* Alexandria, VA: Time-Life Books.

Whitehead, H. 1995. Status of Pacific sperm whale stocks before modern whaling. *Report of the International Whaling Commission* 45: 407–412.

———. 2002. Estimates of the current global population size and historical trajectory for sperm whales. *Marine Ecology Progress Series* 242: 295–304.

Zemsky, V. A., A. A. Berzin, Yu. A. Mikhaliev, and D. D. Tormosov. 1995a. Soviet Antarctic pelagic whaling after WW II: review of actual catch data. *Report of the International Whaling Commission* 45: 131–135.

———. 1995b. *Soviet Antarctic whaling data (1947–1972).* Moscow: Center for Russian Environmental Policy.

Appendix: A Taxonomy of World Whaling Operations and Eras

What follow are details for 113 whaling operations grouped in the 11 whaling eras listed in Table 8.1. Within each era, operations are listed alphabetically by people involved and/or region of operation (Who/Where), then numbered sequentially. "Antiquity" means before AD 1000. "Ongoing" means an operation that was still ongoing as of May 2006.

Era	Operation	Who/Where	Year Start	Year End	Species Principal	Species Supplemental	Basins
Poison	1	Iceland	1600s	Unk			Na
	2	North Pacific	Antiquity	Unk	Fi,Hb,Gr	Mi,Ri	Np
	3	Norway	1600s	1900	Mi		Na
Net	4	Japan	1674	1901	Fi,Hb,Br,Gr,Ri	Mi	Np
	5	Kamchatka	1700s	Unk	Gr		Np
	6	New Zealand	1890	1910	Hb		Sp
Arctic Aboriginal	7	Alaskan Arctic	Antiquity	ongoing	Bh,Ot	Gr,Mi	Ar
	8	Canadian Arctic	Antiquity	ongoing	Bh,Ot		Ar
	9	Greenland Arctic	Antiquity	1920s	Hb,Bh,Ot		Ar
	10	Siberian Arctic	Antiquity	ongoing	Gr,Bh,Ot	Hb	Ar
Temperate Aboriginal	11	Japan	900	1880s	Fi,Hb,Gr,Ri,Ot		Np
	12	North Pacific	1500	1999	Hb,Gr	Bh,Ri	Np
Tropical Aboriginal	13	Indonesia	1600s	ongoing	Sp	Ot	Io
	14	Philippines	1800s	1996	Br		Sp
Basque-Style	15	Basque	1059	1688	Sp,Bh,Ri		Ar,Na
	16	Brazil	1603	1973	Sp,Hb,Ri		Sa
	17	Canary Islands (Spain)	1778	1799	Sp		Na
	18	Cape Verde Islands	1690	1912	Sp,Hb		Na
	19	Denmark/ Greenland	1740s	1923	Hb,Bh		Ar,Na
	20	Spain	1789	1797	Ri		Sa
	21	Basque	1350	1766	Bh,Ri		Ar,Na
	22	Denmark/ Norway	1620	1790	Bh		Ar
	23	Dutch	1610	1824	Bh	Sp,Ri	Ar,Na
	24	French	1610	1868	Bh	Ri,Ot	Ar,Na
	25	Germany	1640	1801	Bh		Ar
	26	Great Britain	1570	1911	Bh,Ri	Ot	Ar,Na
American-Style Shore	27	Australia	1805	1932	Sp,Hb,Ri		Sp,Io
	28	Azores	1851	1984	Sp		Na
	29	Barbados	1868	1913	Hb		Na
	30	Bermuda	1607	1941	Sp,Hb		Na
	31	Canada	1775	1850	Hb,Ri		Na
	32	Chile	1870s	1908	Sp,Hb,Ri		Sp
	33	Equatorial Guinea	1850	1975	Hb		Sa
	34	Hawaii	1840s	Unk	Sp,Hb		Np
	35	Madeira	1941	1981	Sp	Fi,Hb,Ri	Na
	36	Mexico	1850s	1885	Gr		Np
	37	Mozambique	1805	Unk	Hb		Io
	38	New Hebrides	1800s	Unk	Hb		Sp
	39	New Zealand	1825	1933	Hb,Ri	Sp	Sp

Era	Operation	Who/Where	Year Start	Year End	Species Principal	Species Supplemental	Basins
	40	Norfolk Island	1858	1910	Hb		Sp
	41	Russia	1850	1873	Bh,Ri	Gr	Ar,Np
	42	South Africa	1789	1929	Ri	Hb	Sa,Io
	43	Tonga	1890s	1981	Hb		Sp
	44	Trinidad	1826	1865	Hb		Na
	45	U.S. Arctic	1884	1914	Bh		Ar
	46	U.S. East Coast	1650	1924	Hb,Gr,Ri	Fi,Ot	Na
	47	U.S. West Coast	1854	Unk	Hb,Gr		Np
	48	West Indies	1876	ongoing	Sp,Hb	Ot	Na
American-Style Pelagic	49	Australia	1828	1896	Sp,Hb,Ri		Sp,Io
	50	Azores	1875	1900	Sp		Na
	51	Bermuda	1786	Unk	Sp,Ri		Na,Sa
	52	Canada	1804	1893	Sp,Hb,Ri	Bl,Fi	Na,Sa,Sp
	53	Denmark	1800s	Unk	Ri		Sp
	54	Dutch	1840	1860	Gr		Sa,Np,Sp
	55	French	1784	1868	Sp,Ri	Gr	Sa,Np,Sp,Io
	56	Germany	1800s	Unk	Gr,Ri		Np,Sp
	57	Great Britain	1775	1850s	Sp,Ri	Ot	Na,Sa,Np, Sp,Io
	58	Hawaii	1832	1878	Sp,Bh,Ri		Ar,Np
	59	New Zealand	1800s	Unk	Sp,Ri		Sp
	60	Portugal	1774	1890s	Sp,Ri		Sp
	61	Russia	1852	1860	Gr,Bh,Ri		Np
	62	South Africa	1816	1846	Ri		Sa,Io
	63	Tahiti	1800s	Unk	Sp		Np,Sp
	64	United States	1730	1925	Sp,Hb,Gr,Bh,Ri	Ot	Ar,Na,Sa,Np, Sp,Io
Norwegian-Style Shore	65	Angola	1910	1928	Bl,Fi,Sp,Hb,Se,Br	Ri	Sa
	66	Argentina	1904	1960	Bl,Fi,Sp,Hb,Se,Ri		Sa
	67	Australia	1912	1978	Sp,Hb	Bl	Sp,Io
	68	Brazil	1910	1986	Fi,Sp,Hb,Se,Br,Mi	Bl	Na,Sa
	69	Canada	1898	1971	Bl,Fi,Sp,Hb,Se	Ot	Na,Np
	70	Chile	1900	1981	Bl,Fi,Sp,Hb,Se,Br	Ri	Sa,Sp
	71	China	1953	1981	Fi,Hb,Br,Mi	Gr	Np
	72	Denmark/ Greenland	1924	ongoing	Fi,Hb,Mi		Na
	73	Faroes	1894	1987	Bl,Fi,Sp,Hb,Se	Mi,Ri,Ot	Na
	74	Gabon (Congo)	1922	1959	Hb		Sa
	75	Germany	1883	1914	Bl,Hb		Na,Sa
	76	Great Britain	1903	1951	Bl,Fi,Sp,Hb,Se,Ri	Ot	Na
	77	Iceland	1883	1989	Bl,Fi,Sp,Hb,Se	Ri	Na
	78	Japan	1896	1985	Bl,Fi,Sp,Hb,Se, Br,Gr		Np
	79	Mozambique	1912	1923	Hb	Bl,Fi,Sp,Br	Io
	80	Namibia (SW Africa)	1912	1930	Bl,Fi,Hb	Sp,Se	Sa
	81	New Zealand	1910	1964	Hb	Sp,Br	Sp

Era	Operation	Who/Where	Year Start	Year End	Species Principal	Species Supplemental	Basins
	82	Norfolk Island	1948	1962	Hb		Sp
	83	Norway	1864	1971	Bl,Fi,Sp,Hb,Se	Ri,Ot	Na,Sa,Io
	84	Peru	1952	1980s	Bl,Fi,Sp,Se,Br		Sp
	85	Philippines	1983	1985	Br		Np
	86	Portugal	1925	1951	Fi	Bl,Sp	Na
	87	Russia	1883	1912	Bl,Fi,Hb,Se,Gr		Ar,Na,Np
	88	South Africa	1908	1976	Bl,Fi,Sp,Hb,Se, Br,Ri	Mi	Sa,Io
	89	Soviet Union	1932	1992	Bl,Fi,Sp,Hb,Se,Gr, Bh,Ri		Ar,Np
	90	Spain	1921	1985	Fi,Sp,Se		Na
	91	United States	1911	1972	Bl,Fi,Sp,Hb,Ri	Se,Ot	Np
Factory Ship	92	Chile	1906	1914	Bl,Hb,Ri		Sp,An
	93	Denmark	1930	1932			An
	94	Dutch	1946	1964	Bl,Fi,Sp,Hb	Se	An
	95	French	1949	1959	Hb	Sp,Se,Br	Sa
	96	Germany	1936	1939	Bl,Fi,Sp,Hb	Se,Ri	Np,Sp,Io,An
	97	Great Britain	1919	1963	Bl,Fi,Sp,Hb,Se		An
	98	Japan	1934	ongoing	Bl,Fi,Sp,Hb,Se, Br,Mi		Np,Sp,Io,An
	99	Norway	1907	1968	Bl,Fi,Sp,Hb,Se, Gr,Ri		Na,Sa,Np, Sp,Io,An
	100	Onassis	1950	1956	Bl,Fi,Sp,Hb	Se,Ri	Sp,An
	101	Pirates	1968	1979	Bl,Fi,Sp	Hb,Ri	Na,Sa
	102	Portugal	1925	1925	Fi,Sp	Bl	Na
	103	Russia	1903	1904			Np
	104	South Africa	1909	1957	Bl,Fi,Sp		An
	105	South Korea	1946	1986	Fi,Mi	Se,Gr	Np
	106	Soviet Union	1932	1985	Bl,Fi,Sp,Hb,Se,Br, Gr,Mi,Ri		Np,An
	107	Spain	1924	1934	Fi,Sp		Na
	108	Taiwan	1955	1979	Hb,Br		Np
	109	United States	1921	1941	Bl,Fi,Sp,Hb,Gr		Np,Sp,An
Small-Type	110	Great Britain	1877	1896	Ot		Ar,Na
	111	Iceland	1914	ongoing	Mi		Na
	112	Japan	1948	ongoing	Mi,Ot		Np
	113	Norway	1920s	ongoing	Mi,Ot	Ot	Na

NOTE: Species names are abbreviated as follows: Bl = blue, Sp = Sperm, Hb = humpback, Se = sei, Br = Bryde's, Gr = gray, Mi = minke, Bh = bowhead, Ri = right, and Ot = others.

Ocean basins are abbreviated as follows: Ar = Arctic, Na = North Atlantic, Sa = South Atlantic, Np = North Pacific, Sp = South Pacific, Io = Indian, and An = Antarctic.

NINE

The History of Whales Read from DNA

STEPHEN R. PALUMBI AND JOE ROMAN

Of what value is knowledge about the history of a threatened species? Is it possible to chart the future of a species without information about its past? Or is knowledge about past population sizes and carrying capacities crucial to future management plans? These issues are particularly relevant to the management of populations of the great whales.

Populations of all the baleen whales were dramatically reduced by whaling, and unprecedented international cooperation has established a global moratorium on the hunting of whales for commercial purposes until stocks recover. But what does recovery entail? For gray whales that migrate along the western coastline of North America, removal from the United States Endangered Species list in 1994 (IUCN 1994) was heralded as the first time a whale population had recovered sufficiently so that it was no longer in danger of human-caused extinction. Removal from the list depended on population size increases that brought gray whales to about the same levels that were thought to have existed before whaling began (Scammon 1874). For this species, the estimated population sizes that existed before whaling have long served as a benchmark against which to evaluate recovery.

In addition, knowledge of the historical numbers of whales may be instrumental in determining how ocean ecosystems are constructed. Large numbers of whales may have been a key component of marine ecosystems before the oceans were disrupted by humans. Recent reports suggest many populations

of large consumers, including whales, turtles, and sharks and other pelagic fish, have plummeted since the advent of global commercial fisheries (Jackson 1997; Baum et al. 2003; Myers and Worm 2003; Roman and Palumbi 2003). As a result, our current view of the oceans is missing a crucial set of organisms. How did marine ecosystems function before the near-extirpation of large consumers? The answer depends on values for historical population size.

In this chapter we describe a new method that employs the analysis of genetic variation to measure the numbers of whales before whaling. We present published accounts of prior methods of whale population estimation using catch data and compare results from both approaches. Genetic estimates of whale populations are far higher than those previously obtained from catch records. It is important to note that the fundamental data for both methods have uncertainty; for example, records may be missing in the whaling data, and the mutation rate may be higher than phylogenetic analyses suggest. This uncertainty may have a strong impact on the conclusions. We end with a discussion of potential ways to bring divergent estimates of whaling history into accord.

The Current View of the Past

For many species of whales, historical population values serve as a backdrop to management. The International Whaling

Commission (IWC) set targets for whale population recovery based on the idea that a population below about 54% of its prewhaling level should be protected to enable it to quickly increase. For populations above this level, careful management should regulate any allowable hunt, largely based on the trajectory of current populations.

Conventional estimates of current and past numbers, however, seem at odds with current protection schemes. For example, it has often been quoted that fin whales (*Balaenoptera physalus*) in the North Atlantic Ocean had population sizes of 30,000–50,000 individuals before whaling (Sergeant 1977). Yet current population size for this species in the North Atlantic is estimated at about 56,000 (Bérubé et al. 1998). Because fin whales were extensively hunted in the nineteenth and twentieth centuries, and because they were fully protected by the IWC only in 1984, it seems unlikely that current populations could exceed historical values. Are fin whales fully recovered in the North Atlantic, and thus open to exploitation? Have they breached their carrying capacity, or will populations continue to increase? Estimates for historical population size can help answer these questions.

Similar questions can be raised for one of the best-studied whale species in the North Atlantic—the humpback whale, *Megaptera novaeangliae*. Extensive survey work suggests that there are now about 12,000 humpback whales in the North Atlantic (Stevick et al. 2003). Using readily available published and unpublished works, including scientific papers, nineteenth-century annual reviews of the whale fishery, whale charts, and a sample of whaling logbooks, Mitchell and Reeves (1983) estimated a minimum number of humpback whales in the western North Atlantic of about 4,700 individuals. More recent work shows a record of kills totaling more than 29,000 humpbacks from the 1600s through the early 1900s (Smith and Reeves 2002). Although the accuracy of these historic reconstructions has increased greatly in the past decade, placing these data into a framework in which they can be related to historical numbers, and understanding the limits to precision of these estimates, has been challenging.

Genetic Approaches to Whale Populations

Recent theoretical work in population genetics has provided a new source of data for the measurement of historical population numbers. Patterns of genetic diversity within populations are controlled by many factors, including mutation, selection and migration. But one of the primary determinants of genetic diversity is the long-term average population size of a species. Mutation is always the primary source of genetic variation, and it tends to increase levels of DNA variation at a steady, slow rate. This variation is weeded out by natural selection against deleterious mutants or culled by *genetic drift*: the process by which alleles are lost randomly from one generation to the next, which tends to be more common when populations are small. In general, a population loses about $1/(2N_e)$ of its genetic diversity per generation

through drift (Hartl and Clark 1997), where N_e is the genetically effective population size—roughly the number of breeding adults. This simple relationship shows that small populations are subject to higher levels of inbreeding and a faster loss of genetic diversity.

For populations that remain at the same size for long periods of time, an equilibrium is reached between the addition of variation generated by mutation and the loss of variation deleted by genetic drift. Assuming that only mutation and drift are acting to control variation, the relationship between genetic diversity (measured by the parameter θ), mutation rate per generation (measured by μ), and effective population size is expected to be $\theta = 4N_e\mu$. If the population does not stay the same size for long periods, the equation can still apply, but the average genetic diversity depends on the long term average effective population size (Hartl and Clark 1997). This relationship applies to nuclear genes. For mitochondrial genes, the relationship between diversity and mutation and population size is similar, with two modifications. The population size is for females only, because only females transmit mitochondrial DNA (mtDNA) to offspring, and the factor of 4 in the formula drops to 2 because mtDNA is haploid, not diploid. As a result, for mitochondrial genes, $\theta = 2N_f\mu$, where N_f is the effective population size for females.

Armed with a measure of genetic diversity for a species and knowledge of the mutation rate, it should be possible to calculate values for effective population size. These values will not reflect the size of current whale populations, but they will tend to reflect the accumulation of genetic diversity over long periods of time. As a result, this diversity provides a molecular record of past population size that is independent of historical whaling records.

The level of mtDNA diversity in a population reflects accumulation of mutations over about N_f generations. Thus, estimates of historic numbers based on genetic diversity give a population size typical for the species over the past 10,000–100,000 years (for species with $N_f = 500$–5,000 and a generation time of 20 years), not just the past few centuries. If population size cycles over time, then the effective size can be smaller than the typical observed size. Populations that have grown steadily over time—such as humans—may have a genetic diversity that is far lower than expected (Takahata 1995). By contrast, populations that have experienced a bottleneck in the past may have a higher diversity than current populations would allow. Recent methods in DNA analysis can sometimes allow these circumstances to be distinguished based on branching patterns in gene genealogies (see, for example, Shapiro et al. 2004).

Just as for any type of historical data, historical levels of genetic diversity are subject to uncertainty. Mutation rate, unsampled populations, and variation between genetic loci can affect population estimates. Other factors besides population size and mutation rate are also known to affect levels of genetic diversity. The most important of these other factors are natural selection and population structure.

For moderate to large populations, selection generally weeds out mutations, and so decreases levels of genetic variation, more quickly than drift does. So-called *genetic sweeps* may reduce genetic variation in a region of the genome to very low levels, while leaving levels of variation on other parts of the genome untouched. Because the mitochondrial genome is one long molecule, all genes in this genome are linked, and selection on one of them can reduce genetic variability in all of them. Selection can also *increase* genetic diversity when its direction varies over space or time, if it varies with the frequency of alleles in a population, or if individuals with heterozygous genotypes have an advantage.

Population structure can also increase overall genetic variation. When separate populations experience genetic drift independently, they may come to be dominated by different alleles. Diversity within any population may be low because of this domination, but genetic diversity in the species overall still remains high in these cases. For species with separate populations in separate ocean basins, such as many whales, it is critical to be able to correct measures of genetic diversity for potentially high levels of population substructure. If these separated populations are exchanging migrants at a low level, then genetic diversity in each local population may be enhanced by the immigration of novel genes from elsewhere. If this migration occurs often enough, then the diversity of a local population will be nearly as high as the diversity of the whole species. In such cases, local diversity would provide a measure of the population size of the entire species, not just the size of the local population. Because genetic studies of baleen whales, such as humpbacks and fins, have shown substantial divergence between oceans with occasional gene flow (Baker et al. 1998; Bérubé et al. 1998), this transfer of genetic variation could artificially elevate long-term population estimates in ocean basins such as the North Atlantic and must be taken into account.

In summary, genetic measures of past population sizes require (1) estimates of genetic diversity corrected for gene flow, (2) estimates of generation time, and (3) estimates of mutation rate. In the following sections we discuss the data currently available to estimate each of these values for several whale populations and show how these values are combined to provide estimates of historic population size.

Measuring Mutation Rate

Divergence Rate

Estimating the mutation rate for a genetic locus used to measure diversity requires data on the genetic difference at this locus between two species. It also requires information about their divergence time. Such information is often the bottleneck in establishing rates of molecular evolution, because divergence dates between species are often hard to estimate robustly. Instead, many calibrations of molecular rates rely on divergence of genera or families. In the case of cetaceans, well-established dates of species divergence are rare because species-level fossil data are scarce.

The divergence of odontocetes and mysticetes (see also Lindberg and Pyenson, Chapter 7 in this volume) is fairly well

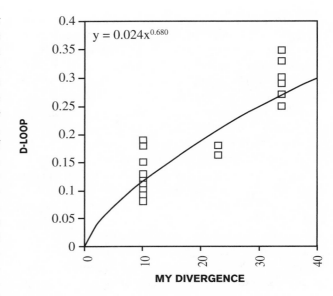

FIGURE 9.1. Genetic divergence in D-loop sequences among baleen whales. Genetic distances were calculated using a Tamura-Nei model with a gamma shape parameter of 0.2. Divergence dates are discussed in the text.

dated at about 35 million years ago, and the divergence of the mysticete families Balaenidae and Balaenopteridae is set at the base of the Miocene about 23 million years ago. Diversification of the species of the genus *Balaenoptera* is thought to have occurred by 6–20 million years ago, and the genera of Balaenids are thought to have diversified in the early Pliocene 3–5 million years ago (Rychel et al. 2004). Overall, these dates provide a few good temporal points on which to hang a measurement of substitution rates, but there are relatively few dates that are recent. The paucity of recent calibration points is important because rapidly evolving regions of mitochondrial DNA might have so many multiple substitutions that sequences will be *saturated* with changes—new changes will overwrite older ones making inferences about substitution rates difficult.

Fortunately, baleen whales tend to show some of the lowest rates of molecular evolution among mammals (Martin et al. 1990; Kimura and Ozawa 2002), reducing the problem of saturation. Yet it is crucial to understand the extent of multiple substitutions in whale DNA. One way to explore this issue is to examine patterns of molecular divergence at a series of timescales. If a measured rate of substitution is higher when the diverging taxa are more closely related and lower for older taxa, then saturation may be a problem. Typically, genetic divergence is plotted against time; if the curve asymptotes strongly, then saturation is suggested. In this situation, the estimated slope of the curve near the origin is the best estimate of short-term substitution rate.

For the mitochondrial control region, known as the *D-loop* in mammals, graphs of genetic difference versus temporal divergence in baleen whales show modest curvature, and the slope near the origin is about 1.5–2.0% change per million years (Figure 9.1). Because this value is the rate of divergence

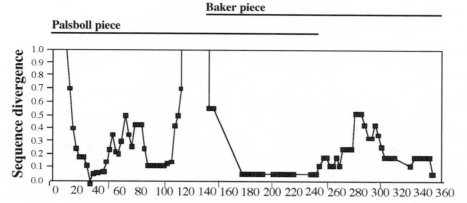

FIGURE 9.2. Sliding-window view of genetic differences between humpback and blue whales along the mitochondrial D-loop. Genetic differences (measured as corrected % sequence divergence) are calculated within 20 bp windows along the length of the D-loop region sequenced by Palsbøll et al. (1995) and Baker et al. (1993). Regions where divergence is greater than 1.0 represent areas where large numbers of insertions and deletions make comparisons across 20 base pairs difficult.

of two species evolving independently, the rate of evolution of each species is half this value, or about 0.75–1.0% per million years. The accumulated D-loop divergence within balaenopterids is about 10–15%, climbing to about 25–35% between odontocetes and baleen whales. This increase suggests that the D-loop values seen among balaenopterids are not as strongly saturated as they are in other mammalian lineages that have been separated for similar amounts of time.

Cautions about Rate Variation

Because different regions of the D-loop vary in their level of sequence conservation, this section of DNA can present special problems in molecular calibrations. Most models of DNA evolution assume that each base in a sequence has the same probability of change as every other base. If some bases have a higher chance of substitution (usually because selection is weaker at these positions), then a simple model of DNA evolution may underestimate the number of substitutions at these positions. One way to estimate the severity of this problem is to examine genetic distances between species in a sliding window that runs the length of the sequence. These graphic plots can reveal areas of particularly high substitution or show regions that are under such strong selection that no variation is allowed—in effect, they provide a topographic map of evolutionary rates along a stretch of DNA.

Such maps are important because the DNA sequences used to measure population structure in whales are from a small section of the control region. The D-loop has been shown to have alternating sections of conserved and highly variable regions (Franz et al. 1985; Andersen et al. 2003). If a population data set includes large portions of a highly conserved region, it may have lower genetic variation, reducing the estimated mutation rate. By contrast, if a data set includes

mostly variable regions, then both genetic diversity and mutation rate may increase. Because of this variation, it is critical that the rate calibration and the genetic diversity measurement be made from exactly the same section of DNA.

Population genetic comparisons have emphasized two partially overlapping pieces of the D-loop region. The first piece was a 264-bp region corresponding to positions 16043 to 16307 in the fin whale mtDNA sequence (Baker et al. 1993). Subsequent authors have used a section that is shifted about 130 bases upstream of this region (Palsbøll et al. 1995, Bérubé et al. 1998). Across the combined Palsbøll/Baker regions, there are some sections in which the amount of sequence change is so great that comparisons among species are difficult (Arnason et al. 1993). The most difficult section is in the middle of the Palsbøll piece, a region of the D-loop that is just before the beginning of the Baker piece. In this region (about position 120 of Figure 9.2), a comparison of humpback and blue whale sequences shows as many as 10 or 11 differences and 5 deleted bases within a given 20-bp window. Sequences with such a degree of variation are difficult to align, presenting challenges in the estimation of overall genetic divergence.

Within the region of DNA included in our analyses (approximately positions 130–360 in Figure 9.2), there are few insertions or deletions, and comparisons from one species to another are relatively straightforward. Most 20-bp windows show one to three substitutions. The largest divergence occurs at about position 240–340, where as many as 7 substitutions in 20 bases occur. Overall divergence of blue and humpback whales is between 10% and 16% across the entire region, corresponding to a divergence rate of about 1.5% per million years if humpback and blue whales diverged 10 million years ago. If we compare just the 100 bases from positions 240–340, however, then divergence increases to

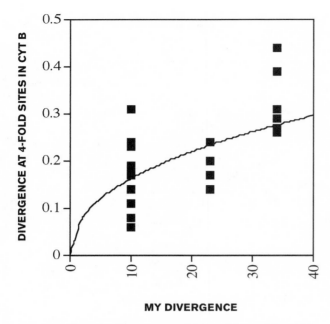

FIGURE 9.3. Genetic divergence at third positions of fourfold degenerate sites among baleen whales. Genetic distances were as the K4 statistic of Wu and Li (1985). Taxa and divergence dates are the same as in Figure 9.1.

13–25%, corresponding to a divergence rate of about 2% per million years.

If different sections of the D-loop show different levels of divergence, which section is best taken as a measure of neutral substitution rate? One way to address this is to examine a set of species for D-loop divergence and for divergence at *silent* positions of protein-coding genes—that is, positions at which a change does not affect the resulting amino acid sequence. Fourfold-degenerate codons are those for which any changes in the third position are silent. For these positions, the rate of change of bases is likely to be the closest measure of the neutral rate of substitution, although there are some cases in which codon bias may interfere (Li 1997). We measured the rate of change across species at third positions of fourfold codons in cytochrome b genes from the same species used in Figure 9.1 to estimate the rate of silent substitution in whale mitochondrial DNAs (Figure 9.3). The resulting estimate of slope near the origin is very close to the 1.5–2.0% measured from D-loop data, suggesting that divergence in silent positions across the mitochondrial genome occurs at about this rate.

Cautions about Hypermutation

Rate variation in the mitochondrial D-loop can occur on genetic spatial scales smaller than those seen in Figure 9.2. Some investigations of patterns of genetic divergence show that change occurs more commonly at some positions than at others. If the chance of substitution is the same among all bases in a DNA sequence, then in theory the number of times

a particular base shows a change should be approximately Poisson-distributed. In practice, different bases show different probabilities of substitution, and the number of times a particular position shows a substitution tends to follow a gamma distribution (Li 1997). The shape of the gamma distribution varies depending on the level of rate heterogeneity among nucleotide positions. For example, as the level of heterogeneity changes from strong to moderate to weak, the shape parameter a varies from below 0.5 to 1.0.

The shape parameter of the gamma distribution can be estimated from data on the number of polymorphisms in a collection of DNA sequences (Posada and Crandall 1998). Once estimated, this shape parameter can be included in estimates of genetic divergence, in essence correcting for rate heterogeneity in the sequence data. In the case of whale D-loop data, rate heterogeneity is strong ($a = 0.2$), and estimates of genetic divergence that take this heterogeneity into account are higher than those that do not. For example, divergence between humpback and blue whale D-loop sequences from the Baker segment range from 9% to 11% using a simple Kimura two-parameter model of DNA divergence that assumes no rate heterogeneity. This is 30–50% lower than the values we used, which are based on a model of DNA substitution that accounts for high levels of rate heterogeneity in the data. We conservatively used the higher rate estimate because it produces the lowest estimates of effective population size.

Extreme rate heterogeneity can occur if there are some nucleotide positions that accumulate changes very rapidly. If these positions exist, then comparisons among closely related sequences may show differences at these sites. More distant comparisons also show changes at these sites, but multiple changes overlay one another, and a later change can wipe out any evidence of the previous changes. Thus, a large number of hypermutable sites can increase rates of nucleotide diversity within species but decrease apparent rates of substitution between species. We estimated the number of these hypermutable sites by examining the number of times each base in our data set was estimated to change along the intraspecific phylogenetic tree (Roman and Palumbi 2003, supplement). For example, there were 14 nucleotide positions that were hypothesized to change more than three times in the phylogenetic history of humpback whales. One of these sites could have changed as many as eight times, although this number varies with the minimum-length phylogenetic tree used. Removing these 14 sites from the analysis reduced genetic diversity but also reduced the estimated mutation rate. This analysis suggests that D-loop hypermutation does not completely account for the high intraspecific variation we see in whale populations. However, the reliance on D-loop data for all analyses of whale genetic diversity is the single biggest problem with this approach and must be addressed by further data collection.

Currently, the best information we have suggests a whale D-loop substitution rate of about 0.75–1.0% per million years (note that substitution rate is half the divergence rates detailed above). In our analyses we conservatively doubled this best estimate to account for hypermutation, providing a substitution rate range of 1.5–2% per million years.

The equations relating genetic diversity and mutation rate require that we calculate the mutation rate per generation, not per million years. To do this, we used an estimate of generation time derived from the average age of sexually mature females. These data derive from whaling studies of age at capture, and we also used a long-term photo-identification study of humpback whales to estimate generation time for this species (Table 9.1).

Measuring Diversity in North Atlantic Whales

We used the D-loop region of DNA to measure genetic diversity in North Atlantic populations of three whale species. We initially chose these species and these populations based on the abundance of existing data and the growing need to understand the population status of these populations.

Humpback Whales

Found in all the oceans of the world, the humpback whale is a highly migratory species, feeding in temperate waters and wintering in tropical calving grounds. A coastal species, with strong fidelity to natal feeding and calving grounds, humpbacks were often the first whales to be hunted in a newly discovered area (Clapham 2002). It was also one of the first whale species to be analyzed genetically (Baker et al. 1993) and has the best global representation, with sequences from the mitochondrial control region currently available from the North Pacific, Southern Hemisphere, and North Atlantic. Using two overlapping DNA fragments (Palsbøll et al. 1995; Baker et al. 1998; Bérubé et al. 1998), we assembled a data set of 312 individuals: 188 whales from the North Atlantic, 31 from the North Pacific, and 93 from the Southern Hemisphere. Geographic structure within the North Atlantic has been found between feeding populations in Iceland and western Atlantic populations (Palsbøll et al. 1995). Migration is high between these two regions, and analyses separating Iceland from the west Atlantic produced similar results to runs incorporating the entire ocean basin as a single population.

For humpbacks, the North Atlantic is populated by two independent genetic lineages: the IJ clade, which includes a large number of differentiated North Atlantic individuals, and the CD clade, which is dominated by Pacific and Southern animals (Baker et al. 1993). Phylogenetic analysis suggests that the IJ clade migrated into the North Atlantic from the Southern Oceans long ago and diversified there. Thus, the genetic patterns of the IJ clade may provide a view of North Atlantic humpback history that is independent of the view derived from the rest of the population.

Fin Whales

The fin whale, *Balaenoptera physalus*, is a cosmopolitan rorqual, found both along the continental shelf and in deep pelagic regions. Fin whales are fast-swimming whales; they cruise at 5–8 knots and are capable of bursts of up to 15 knots (Aguilar 2002). Since the development of bomb-lance technology in the mid-nineteenth century, they have been a primary target of whalers because of their large size. We analyzed data from 475 samples of Northern Hemisphere fin whales collected by Bérubé et al. (1998, 2002) from the eastern North Pacific Ocean (Sea of Cortez and coastal California), North Atlantic Ocean, and Mediterranean Sea. Sampling areas in the North Atlantic ($n = 235$) included the Gulf of Maine, Gulf of St. Lawrence, West Greenland, Iceland, and Spain. Fifty-six samples from the Gulf of California, 10 from California, and 69 samples from the Mediterranean were also available for comparison to the Atlantic.

To date, Southern Hemisphere control region samples are not available for fin whales. However, Wada and Numachi (1991) found a significant difference in gene frequencies for two of three polymorphic allozyme loci, suggesting low historical gene flow between the North Atlantic and the Southern Hemisphere.

Minke Whales

Minke whales are the smallest of the rorquals. There are two recognized species: *Balaenoptera acutorostrata*, found in the Northern Hemisphere, and *B. bonaerensis* in the south. The Northern Hemisphere minke, *B. acutorostrata*, is currently divided into three subspecies: the North Atlantic, North Pacific, and dwarf minke whale of the Southern Hemisphere. Although the capture of minke whales in Norway dates back to the Middle Ages, minkes were not heavily exploited until the twentieth century (Horwood 1990). In the North Atlantic, they have been hunted by both shore-based and pelagic whalers. It is presumed that most minke whale stocks are in better shape than those of other large whales (Perrin and Brownell 2002), and both aboriginal and commercial hunting of minke whales continues.

Because of the species and subspecies divisions between ocean basins, we analyzed minkes from the North Atlantic as a distinct, monophyletic population. North Atlantic minke whale data ($n = 87$) are derived from Bakke et al. (1996). A recent publication, including a 500-bp fragment for 306 individuals from the North Atlantic (Andersen et al. 2003), showed higher diversity values, suggesting that our values for minke whales may be conservative.

Estimating Migration and Genetic Diversity of Whale Populations

Given a data set showing the degree of variability in a region of DNA, we next need an estimate of the amount of intraspecific diversity within and between populations. In practice, comparisons between sequences within and between populations are used to estimate the level of population structure and genetic variability. Recently, several approaches to measuring genetic variation in populations with genetic structure have been devised.

TABLE 9.1
Historical Population Estimates Based on Genetic Diversity and Generation Time of
Baleen Whales in the North Atlantic Ocean

Species	n	θ		Generation time (years)	$N_{e(f)}$ (thousands)		Genetic population estimates (thousands)		Current estimates (thousands)
		Mean	95% CI		Mean	95% CI	Mean	95% CI	
Humpback	188	0.0216	0.0179–0.0274	12–24	34	23–57	240	156–401	9.3–12.1
Fin	235	0.0430	0.0346–0.0526	25	51	38–65	360	249–481	56.0
Minke	87	0.0231	0.0161–0.0324	17	38	26–57	265	176–415	149.0
Total							865		214–217

Peter Beerli, Joe Felsenstein, Mary Kuhner, and colleagues have developed an approach using maximum-likelihood techniques to estimate effective population size and migration between populations using sequence or microsatellite data. The program MIGRATE (Beerli and Felsenstein 2001) employs a likelihood analysis to calculate migration events and long-term population size simultaneously. In this program, Markov chain Monte Carlo methods are used to estimate migration rates and θ, a measure of genetic diversity. Simulations have been used to test how well MIGRATE operates. Although the measurement of migration rate appears to be problematic with this approach, MIGRATE has been shown to provide fairly precise estimates of θ under a wide variety of population diversities and structures (Abdo et al. 2004). For example, when two populations connected by migration were simulated with $\theta = 0.025$, MIGRATE estimated $\theta = 0.0251$ to 0.0266 for a wide variety of different migration rates (Abdo et al. 2004, Table 9.1).

Migration

For all species, migration rates between oceanic populations were low. For humpback whales, results from MIGRATE indicate that the long-term average for transequatorial migration between the southern ocean and the North Atlantic is less than one female per generation. MIGRATE estimates of migration have high standard errors but are best under conditions that seem to pertain to whale populations: when θ is high and migration is low (Abdo et al. 2004). Indeed, similar migration values derive from different analyses. Baker et al. (1994) estimated the interoceanic divergence as $F_{ST} = 0.3$ and showed with phylogenetic analysis of gene flow that successful interoceanic migration was rare. For fin whales, the majority of genetic diversity is found in the North Atlantic, with low levels of gene flow between the North Pacific and North Atlantic. Based on MIGRATE, the average migration rate between these two basins was 0.19, or slightly less than one female per five generations. Bérubé et al. (1998) also

estimated high genetic divergence between ocean basins for fin whales ($F_{ST} = 0.42$–0.60), indicating low gene flow.

Migration rates between subpopulations within the North Atlantic are, unsurprisingly, much higher than transequatorial rates. For example, the long-term average of fin whale migration between the Mediterranean and the North Atlantic is approximately 7 females per generation. Migration rates between the West Atlantic and Iceland are also high for humpback whales, with approximately 11 females migrating per generation between these areas. These long-term averages, however, may be the result of recent colonization of Iceland by western Atlantic whales.

Genetic Diversity

Using sequence data from several populations, MIGRATE generates a likelihood profile for the parameter θ. This profile allows the calculation of 95% confidence intervals for each value of θ. Because the program uses a Monte Carlo simulator, it provides a slightly different result each time it analyzes a data set. To account for this variability from run to run, we used the program to generate 10 different values for θ and for the 95% confidence limits of θ. Independent runs gave very similar results, and we averaged the ten means and confidence limits to provide the most robust estimate of genetic diversity in each population.

These data sets provided estimates of genetic diversity for whale populations in the North Atlantic (Table 9.1). Values of θ were higher than would be expected if traditional historical values for fin and humpback whales were accurate. For North Atlantic humpback whales the mean value of θ is 0.022, with a 95% confidence interval (CI) of 0.018–0.027. For North Atlantic fin whales, the mean θ is 0.043, with 95% CI of 0.035–0.053. Estimates of θ for North Pacific fin whales were much lower than for those in the North Atlantic. (This difference may be the result of the relatively small sampling area for Pacific fins: Most whales were biopsied in the Sea of Cortez, with ten samples from coastal California.) The North Atlantic minke whale, which was not under heavy commercial exploitation until the twentieth century, has estimates of θ that are roughly consistent with hunting records (mean value 0.023, with 95% CI of 0.016–0.032).

Genetic diversity estimates for the minke whale are restricted to the North Atlantic, because this is a recognizable separate subspecies that shares no mitochondrial lineages with the subspecies in the Pacific. For humpback whales the diversity value in the Atlantic is estimated in reference to diversity in four other population regions, including extensive sampling in the Southern Hemisphere. For fin whales, data from the Southern Hemisphere are the most limited, and diversity values are estimated only in reference to populations in the Mediterranean—which are genetically distinct—and the Pacific.

From Genetic Diversity to Effective Size to Population Estimates

Once mutation rate (μ) is established and we have reliable estimates of genetic diversity (θ), we can use the relationship $\theta = 2N_f\mu$ to calculate estimates of the genetic effective population size for females in each population (Table 9.1). For example, for humpback whales a mean genetic diversity value of 0.0216, generation time of 18 years, and mutation rate of 1.75% per million years translates into a genetically effective size for females of 34,000. This is the mean estimate, sitting inside a fairly wide 95% confidence limit resulting from uncertainty about the value of θ (Table 9.1). Because all estimated parameter values have uncertainty, we employed a Monte Carlo method of estimating population size and its 95% confidence limits (see below).

However, this number is not an estimate of the ancestral census size. To determine census size from effective size, we relied on three conversion factors. First, we converted N_f to total effective population size, N_e, by multiplying by two, because sex ratio of these whale species is 1:1 (Lockyer 1984). Second, N_e was converted to N_T, the total number of breeding adults. The N_T:N_e ratio approaches 2.0 in perfect populations with a constant population size (Nunney 1993). Numerous genetic studies suggest that this is a very conservative estimate. For example, Nunney (2000) has shown that for fluctuating populations the ratio would be closer to 10. Similarly, the average N_T:N_e ratio for 33 mammal species reviewed by Nei and Graur (1984) was 10, identical to the value Frankham (1995) found in his review of wildlife species. Mace and Lande (1991) have proposed an N_T:N_e ratio of 5–10 for a wide array of species; the IUCN (now the World Conservation Union) employs a ratio of 5. To ensure conservative estimates for baleen whales, we used Nunney's theoretical estimate, employing a multiplier of 2 to derive N_T from N_e. Third, to take into account the number of juveniles, we multiplied N_T by an estimate of (number of adults + juveniles)/(number of adults) derived from catch data. This ratio is 1.6 to 2.0 for humpbacks, assuming that whales in year classes 1 through 5 are equally abundant in the population (Chittleborough 1965). The ratio is 1.5 in gray whales

(Weller et al. 2002) and 2.5 to 3.0 in bowheads (Angliss et al. 1995). Considering these ranges, we used a multiplier of 1.5 to 2.0.

Thus, we estimate total population size as 6–8 times the number of breeding females ($N_C = 2 \times 2 \times 1.5N_{e(f)}$ or $2 \times 2 \times 2N_{e(f)}$). This highly conservative rate ignores fluctuation in population size, variation in female fecundity, and polygyny, all of which tend to reduce the ratio between effective population and census size (Avise et al. 1988; Nunney 2000). In particular, Nunney (1993) has noted that effective population size depends strongly on the mating system. Although rorqual whale mating systems are largely unknown (Lockyer 1984), Clapham (1996) has proposed that humpbacks form floating leks, a type of dominance polygyny, with male songs helping to establish dominance on the breeding grounds. Leks also increase values of N_T:N (Nunney 1993). If we had used a more likely estimate of 5–10, our suggested historical population sizes would be several times higher than presented here. Empirically estimating N_T:N_e for whale populations should be an important part of future research efforts.

Because several parameters are estimated as ranges, we employed a Monte Carlo resampling scheme to estimate historical population size. We randomly sampled μ and θ from within ranges estimated from the data analyses described in previous sections (assuming a uniform distribution) and used these chosen values to estimate N_f. We then randomly chose a multiplier from the range 6–8 to convert N_f to N_T and used 1,000 replicates to estimate mean values and 95% confidence limits on the number of breeding females and total census size (Roman and Palumbi 2003).

The final results suggest that North Atlantic whales had a population far in excess of those typically assumed (Table 9.1). For humpback whales, the genetic estimate of 240,000 animals (including juveniles and subadults) is about ten times higher than prior estimates of about 20,000. Confidence limits for these estimates are large, from 156,000 to 401,000, but even the lowest value is much higher than previous estimates. For fin whales, genetic diversity patterns suggest populations of about 360,000. This is also several times higher than previous estimates (Sergeant 1977). Minke whales show genetic estimates of population sizes that are closer to expectations, but in this case, commercial hunting began only in the twentieth century.

Reliability of Genetic Data and Analysis

Because nucleotide variation can be maintained through population bottlenecks of brief duration, genetic data may provide practical population estimates, independent of catch data and logbooks, for species that have been heavily exploited. This can be especially useful for species that are at historically low population levels. Baleen whales, which are long-lived and were exploited to levels that brought some species to the brink of extinction, are ideal organisms for exploring the practicality of this type of approach.

One important question is whether we are looking at populations from deep in the evolutionary history of North Atlantic rorquals or at more recent populations, when commercial hunting began. The data and their analysis suggest that we are looking at both. The large amount of genetic diversity of current whale populations has required more than a million years to accumulate; whale populations have probably been large over at least that time span. If population size fluctuates, genetic diversity is particularly vulnerable to periods of reduced numbers of reproductively active individuals, because such bottlenecks can rapidly winnow genetic diversity. As a result, fluctuation in population size has been suggested as a cause for widespread underestimates of population size of terrestrial and marine species from genetic data (Nei and Graur 1984; Avise 1994).

Could it be that whale populations have been typically large for long periods of time but fell to low levels *before* the start of industrial whaling? Although this has not been suggested by whale biologists previously, it is a formal possibility that could explain the data. If true, this decline in whale numbers would have had to be global in scale. Global levels of genetic diversity for humpback whales suggest world populations of about 1–1.5 million for this species (based on global diversity estimates of approximately 0.10), far higher than the 115,000 individuals estimated for this species at the start of whaling.

In addition, current analyses suggest that if a recent global bottleneck were the explanation, this depleted population size could not have extended for very long, because such an event would have reduced the diversity found in rorquals and would have changed the shape of the phylogenetic tree relating different mitochondrial haplotypes to one another. For whales, the shape of the tree (monitored using Tajima's [1989] D statistic) shows no signs of population decline. However, the power of this test to detect population trends is low, and further investigation using more sensitive techniques is warranted. Further evaluation of the hypothesis that whale population plummeted during human prehistory awaits evidence of natural declines in whale numbers over the last 10,000 years.

If such a crash occurred and was due to a climatological or ecological shift, this could signal dire consequences for current whale populations. We may be poised for a severe climate shift in coming decades. How will reduced populations of baleen whales respond? In fact, we have very little idea of how whale populations weathered the series of Ice Ages over the past two million years. The depth of phylogenetic trees in whales shows that many species stretch back this far in time, and research focused on the impact of climate change on whale populations would be very valuable.

A second concern is that analyses based on one genetic locus might incur uncertainty, resulting in an inaccurate population estimate. Analysis of nuclear data and other mitochondrial loci are a key part of future approaches to this problem (Roman and Palumbi 2003). Palsbøll et al. (2004) have found that genetic diversity of microsatellite data in humpbacks is largely concordant with the diversity signature

in mtDNA. However, their data set did not address the effect of population subdivision on genetic diversity, and they could not yet answer the question of whether the nuclear genetic signature of large populations in North Atlantic humpbacks persisted until the start of industrial whaling. Thus the microsatellite data could not distinguish between a recent and an ancient population crash for humpbacks. Microsatellite data from all ocean basins will be needed to take the next step in this analysis.

A third concern is that absolute genetic measures of past population size require good estimates of sequence mutation rate. As detailed above, these rates are typically based on phylogenetic comparison of sequences from species with known fossil records. However, some studies suggest that rates estimated from known historical pedigrees may be higher than phylogenetic and fossil records suggest. A recent paper by Howell et al. (2003) suggests that the pedigree mutation rate is tenfold higher than rates obtained from phylogenetic analyses. If true, this discrepancy could reduce genetic estimates by an order of magnitude. However, there is considerable disagreement over whether pedigree rates are indeed higher than phylogenetic rates. In particular, use of the Howell rate to date the age of the diversification of humans and the spread of human lineages out of Africa gives a date of less than 10,000 years, whereas mtDNA and other loci agree that the date for these events is in excess of 150,000 years. Thus, the pedigree rate in this case is in conflict with rates derived from other loci and is not likely to provide a better view of the history of populations.

In our view, the most serious limitation of current data sets is that they are tied to a single piece of DNA whose patterns of molecular evolution are complex enough to engender caution. Providing data from additional loci is the key to overcoming this limitation, but it is challenging because the slow rate of mutation in whales produced few nuclear polymorphisms for analysis. Hare et al. (2003) analyzed intron polymorphism at four loci for delphinid dolphins and showed that intron lengths of 500–1,000 bp were needed in order to discover an adequate sampling of phylogenetically informative positions. Palumbi et al. (2001) and Palumbi and Baker (1994) showed a similarly low level of polymorphim in baleen whales. Nevertheless, such data sets for baleen whales will be required to provide data parallel to the mtDNA data discussed here.

As a consequence of the uncertainty of basing conclusions on a single locus, we have built a set of highly conservative assumptions into our analysis. Yet the final result remains that populations are typically high. For example, we have used a low value for the conversion of female effective size to total size (a multiplier of 6–8 instead of a multiplier of 15–30 or more, typical of other mammals). An empirically derived value for this multiplier—perhaps based on reproductive success in well-known populations—would be very valuable. It is possible that such empirical values will increase our estimate of population size.

Additionally, we would like to have a better understanding of the dynamics of whale population size in the past. Have whale populations been higher in the past than in historical times? Have whale populations been stable during the serial Ice Ages of the past? Is there any evidence that whale populations plummeted before whaling began? Analysis of DNA patterns can reveal population trajectories in the past (Shapiro et al. 2004), but these approaches are more powerful when more than a single locus is applied.

Though these caveats are serious, we are left with a set of empirical conclusions that challenge current thinking about the history of whale populations. It seems to us that the genetic analysis of whale populations is inconsistent with a view that prewhaling population numbers were as low as previously suggested. These conclusions open a debate on the management of current whale populations that is timely and important.

Though initially surprising, our results are consistent with findings for other large marine vertebrates, such as green turtles (Jackson et al. 2001), pelagic sharks (Baum et al. 2003), and other fish (Myers and Worm 2003). The removal of turtles may have had an impact on the Caribbean ecosystem, and emerging research suggests a similar impact for the removal of pelagic predatory fish (R. A. Myers, personal communication). In parallel, it is likely that the removal of great whales has also had important ecological effects. Such changes may include prey shifting in predators (Springer et al. 2003) or perhaps alteration of the nitrogen cycle in coastal regions.

Management Implications of High Genetic Diversity

How might these findings affect the management of whales? In its original New Management Procedure, the IWC established that catches should not be allowed on stocks below 54% of the estimated carrying capacity. Under this criterion, humpbacks and fin whales in the North Atlantic might be considered exploitable even though they have been protected for only a generation or two. The Revised Management Procedure retains full protection for populations below 54% of carrying capacity, while recognizing the need for additional population models to set carefully monitored take limits above this level (Cooke 1999, Holt 2004).

In the North Atlantic, current populations of approximately 12,000 humpback and 56,000 fin whales are far lower than long-term population sizes based on genetic estimates. Instead of being close to exploitable, our new analyses suggest that these populations will require decades of continued full protection.

Only the minke whale, little exploited until the twentieth century, approaches the population size recommended by IWC models. However, genetic estimates of population size are usually several orders of magnitude too low—especially for fluctuating populations (Soulé 1976; Avise et al. 1988; Frankham 1996; Nunney 2000). If minkes are more vulnerable to such perturbations, genetic techniques could seriously underestimate population size. We also note that our diversity estimates are from the eastern Atlantic stock of minkes only, and the IWC estimates of 120,000 to 182,000 for the North Atlantic do not include the Canadian East Coast. Clearly the population trends for this species—which is still hunted by Greenland, Iceland, and Norway—are in need of further analysis.

Regular hunting for minke whales in Greenland began in 1948, after fishing vessels were equipped with harpoon cannons (MacLean et al. 2002). Current catch limits are for 187 minke whales for both east and west Greenland (Reeves 2002). If greater genetic diversity were found in the western Atlantic (Andersen et al. 2003), pre-exploitation estimates would have to be corrected. A cautionary approach suggests that well above 150,000 minkes should be present in the North Atlantic before commercial exploitation is considered for other nations.

It is important to note that long-term reductions in some populations, such as right whales (Eubalaena glacialis), have winnowed genetic diversity to the point where it is difficult to use contemporary DNA to make historical population estimates (Rosenbaum et al. 2000, Waldick et al. 2002). Unlike rorquals, which were heavily exploited only in the nineteenth and twentieth centuries, right whales have been commercially hunted for more than five centuries and are now completely extirpated from important breeding grounds. Low frequency microsatellite alleles are rare in E. glacialis, suggesting that genetic diversity has been reduced in this species (Waldick et al. 2002). Fortunately, humpbacks, fins, and minkes do not appear to have sustained such extended reductions in population size (Baker et al. 1993).

There is an important distinction between the regulatory apparatus of the IWC, which emphasizes carrying capacity, and population genetics, which produces long-term typical population size. Carrying capacity is an intrinsically ecological concept and can change if the state of an ecosystem changes. The current capacity of the oceans to support whales may be lower than before humans began removing 80 million metric tons of seafood a year from the sea. Yet, long-term population size may be a very good estimate of historical ocean carrying capacity, because long-term population size is a reflection of the numbers of whales that were supported by the oceans in the past. Current-day managers may find it necessary to limit whale populations to less than this value for economic reasons, but such debates have not been fully played out. The reason for lack of debate is that the *possibility* that oceans may have supported more whales than previously thought has not been thoughtfully advanced before. With new genetic results on the table, this debate should have greater scope.

Is a Synthesis of Two Historical Views Possible?

Estimates of prewhaling population sizes from summaries of whaling catch data suggest an order of magnitude fewer whales than do genetic estimates from humpback and fin whales. The discrepancy for minke whales in the North Atlantic is about threefold. Discrepancies for other species are

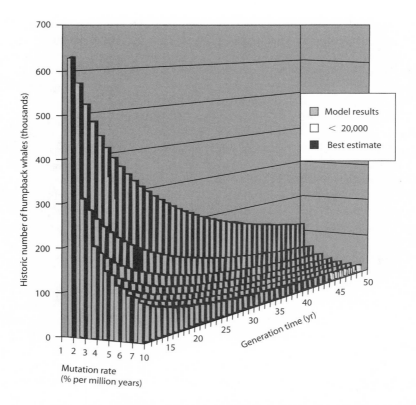

FIGURE 9.4. A sensitivity analysis of the impact of assumptions about mutation rate and generation time on conclusions about numbers of North Atlantic humpback whales before whaling. In this figure, we allowed mutation rate (μ) to vary between 1% per million years and 8% per million years and generation time to vary between 10 and 50 years. The product of these numbers is the amount of DNA sequence change per generation and is the value we need in the equation to calculate population size. The results of the analysis show that only extreme values of parameters will give population estimates consistent with previously published values for prewhaling population size (in white).

higher. Are these numbers reconcilable? Or do these severe differences signal a serious breach in one of these two approaches?

One way to approach this question is through *sensitivity analyses*—a process in which the same basic data are used in conjunction with a set of different assumptions to explore how the conclusions vary with the assumption sets used. For example, we estimated North Atlantic humpback genetic population sizes by measuring genetic diversity and mutation rate and applying a generation time estimate from the whaling literature. But what if either the mutation rate or the generation time were incorrect? To explore the impact of varying parameters on our conclusions, we recalculated population size for a variety of different mutation rates and generation times. The result (Figure 9.4) shows that an estimate of 10,000–20,000 humpback whales requires mutation rates of about 8% per million years and a generation time of 45 years. If generation time is held constant at 25 years, then the mutation rate must be about 16% per million years in order for our population estimate to be 10,000–20,000. These

values are outside reasonable ranges. For example, a mutation rate this high would imply that the *Balaenoptera* species, which have diverged by 10–20% in mtDNA at silent sites (Figures 9.1, 9.3) had diverged only 750,000–1.5 million years ago. For these reasons, we consider it unlikely that altering the assumptions of the genetic analysis will bring genetic and conventional numbers into full agreement.

Other analyses are not as clear-cut. For example, an estimate of 100,000 humpbacks in the North Atlantic requires mutation rates of 3% per million years and generation times of about 25 years. Furthermore, the best analysis—yielding an average of 240,000 humpback whales—has wide 95% confidence limits, ranging from 150,000 to 400,000. These explorations of the data analysis suggest that historical numbers of as few as 100,000 humpbacks probably could not be rejected by the mtDNA genetic analysis.

A minimum estimate of 100,000 animals from mtDNA or future nuclear analysis would remain far higher than suggested by current analyses of whaling records. Summaries of whale mortality from hunting by Smith and Reeves (2003) show a total of about 29,000 killed from 1650 to 1910, and these data are thought to be consistent with a population of 40,000 (T. D. Smith, personal communication). Yet this analysis, like the one for genetic data, is based on a series of assumptions as well as a model relating the basic data (recorded whale kills) to conclusion about ancient populations. Whitehead (2000) has shown how a variety of different assumptions about population growth rates, generation times, and mortality rates can dramatically change estimates of past population sizes, but a rigorous sensitivity analysis has never been applied to the humpback whale data.

Although we cannot provide a rigorous sensitivity analysis here, we suggest there may be room for changes that would yield higher numbers of whales in the past. Two critical assumptions serve as examples. First, current analyses appear to assume that the loss rate during industrial hunting was only 2%; that is, the analysis assumes that 98% of all humpback whales struck by harpoons were killed and taken by whaling ships. Loss rates in earlier eras were far higher—at least 50%, according to Mitchell and Reeves (1983)—and it is not clear whether calves, injured to lure a mother toward a whale ship, were counted as individuals killed but not landed. If loss rates were higher than 2%, then the current views of mortality from whaling may be underestimates (see also IWC 2003).

Second, catch records are never complete, and a comprehensive analysis must take into account the whales that were killed but never recorded. This is a prodigious task for historians and remains a serious challenge. Estimates of the completeness of historical data are an important part of reconstruction of past views of many aspects of human industry and culture. Understanding the true history of whaling must include understanding the best ways to correct current data compilations for missing records.

The impact of both of these simple sources of error on analyses is strong. If it were discovered that whale loss rates from hunting were 50% instead of 2%, and that only half of the whales killed in the oceans during the three centuries of whaling were known to us, then we would need to adjust the total number of whales killed by whalers from about 30,000 to about 120,000. We do not suggest here that this fourfold correction is warranted at this time. We merely point out that correcting catch records upward is possible and could potentially help bridge the genetic gap.

The final assumption set needed to convert hunting mortality to standing population size is a set of models relating mortality to population trajectory over time (e.g., Whitehead 2002). Such models do not function well for humpback whales, suggesting that estimates of whaling mortality might be low (IWC 2003). In the absence of such quantitative models, it has been suggested that the minimum number of animals can be estimated if a pulse of hunting is followed by a dramatic reduction in catch. In such cases, Mitchell and Reeves (1983) argue that the sum of the animals killed during the hunting pulse is a good starting value for a historical population estimate.

A pulse of whaling from about 1870 to 1900 took 20,000 humpback whales from the North Atlantic (Smith and Reeves 2003). However, this number jumps to 80,000 if we change the loss rate and recording rate assumptions as described above. This number is similar to the lower bound of the estimate from genetics and could represent a value that might reconcile these approaches.

These changes in how catch data could be handled are heuristic devices only. Both whaling and genetic data are subject to uncertainty and rely on critical assumptions to make estimates about past population size (Clapham et al. 2004). Future work should concentrate on the validity of assumption sets for both types of data and analytical approaches (Clapham et al. 2004) in a framework that might reveal common ground.

Conclusions

DNA data add a new tool to our ability to research the past. Written into DNA sequence variation is a record of the population history of a species, and new theoretical tools are allowing us to read that variation in more and more powerful ways. Genetic data suggest that humpback and fin whales were much more abundant before whaling than conventional estimates suggest. For example, instead of a global total of 115,000 humpback whales, genetics estimates a world-wide abundance of about 1.5 million. Future analyses and additional data may refine these genetic estimates, but the major conclusion is that the number of whales that the oceans can support has been substantially underestimated.

For either DNA data or written history, it is a mistake to assume we have a perfect record of the past. The challenges for the future are to understand how different kinds of data jointly illuminate the past and to use as many perspectives as possible to try to reconstruct the history of whales before whaling.

Literature Cited

Abdo, Z., K. A. Crandall, and P. Joyce. 2004. Evaluating the performance of likelihood methods for detecting population structure. *Molecular Ecology* 13: 837–851.

Andersen, L., E. Born, R. Dietz, T. Haug, N. Øien, and C. Bendixen. 2003. Genetic population structure of minke whales *Balaenoptera acutorostrata* from Greenland, the North East Atlantic and the North Sea probably reflects different ecological regions. *Marine Ecology Progress Series* 247: 263–280.

Angliss, R. P., D. J. Rugh, D. E. Withrow, and R. C. Hobbs. 1995. Evaluations of aerial photogrammetric length measurements of the Bering-Chukchi-Beaufort Seas stock of bowhead whales (*Balaena mysticetus*). *Report of the International Whaling Commission* 45: 313–324.

Arnason, U., A. Gullberg, and B. Widegren. 1993. Cetacean mitochondrial-DNA control region: sequences of all extant baleen whales and 2 sperm whale species. *Molecular Biology and Evolution* 10: 960–970.

Avise, J. 1994. *Molecular Markers, Natural History and Evolution.* NY: Chapman and Hall.

Avise, J., R. Ball, and J. Arnold. 1988. Current versus historical population sizes in vertebrate species with high gene flow: a comparison based on mitochondrial DNA lineages and inbreeding theory for neutral mutations. *Molecular Biology and Evolution* 5: 331–344.

Baker, C., L. Medrano-Gonzalez, J. Calambokidis, A. Perry, F. Pichler, H. Rosenbaum, J. Straley, J. Urban-Ramirez, M. Yamaguchi, and O. Von Ziegesar. 1998. Population structure of nuclear and mitochondrial DNA variation among humpback whales in the North Pacific. *Molecular Ecology* 7: 695–707.

Baker, C., A. Perry, J. Bannister, M. Weinrich, R. Abernethy, J. Calambokidis, J. Lien, R. Lambertsen, J. Ramirez, O. Vasquez, P. Clapham, A. Alling, S. O'Brien, and S. Palumbi. 1993. Abundant mitochondrial-DNA variation and world-wide population structure in humpback whales. *Proceedings of the National Academy of Sciences* 90: 8239–8243.

Baker, C., R. Slade, J. Bannister, R. Abernethy, M. Weinrich, J. Lien, J. Urban, P. Corkeron, J. Calmabokidis, O. Vasquez, and S. Palumbi. 1994. Hierarchical structure of mitochondrial-DNA gene flow among humpback whales *Megaptera novaeangliae*, worldwide. *Molecular Ecology* 3: 313–327.

Bakke, I., S. Johansen, O. Bakke, and M. ElGewely. 1996. Lack of population subdivision among the minke whales *(Balaenoptera acutorostrata)* from Icelandic and Norwegian waters based on mitochondrial DNA sequences. *Marine Biology* 125:1–9.

Baum, J. K., R. A. Myers, D. G. Kehler, B. Worm, J. Harley, and P. A. Doherty. 2003. Collapse and conservation of shark populations in the northwest Atlantic. *Science* 299: 389–392.

Beerli, P. and J. Felsenstein. 2001. Maximum likelihood estimation of a migration matrix and effective population sizes in n subpopulations by using a coalescent approach. *Proceedings of the National Academy of Sciences* 98: 4563–4568.

Bérubé, M., A. Aguilar, D. Dendanto, F. Larsen, G. Di Sciara, R. Sears, J. Sigurjónsson, J. Urban-Ramirez, and P. Palsbøll. 1998. Population genetic structure of North Atlantic, Mediterranean Sea and Sea of Cortez fin whales, *Balaenoptera physalus* (Linnaeus 1758): analysis of mitochondrial and nuclear loci. *Molecular Ecology* 7: 585–599.

Bérubé, M., J. Urban, A. Dizon, R. Brownell, and P. Palsbøll. 2002. Genetic identification of a small and highly isolated population of fin whales *(Balaenoptera physalus)* in the Sea of Cortez, Mexico. *Conservation Genetics* 3: 183–190.

Chittleborough, R. 1965. Dynamics of two populations of the humpback whale, *Megaptera novaeangliae* (Borowski). *Australian Journal of Marine and Freshwater Research* 16: 33–128.

Clapham, P. J. 1996. The social and reproductive biology of humpback whales: an ecological perspective. *Mammal Review* 26: 27–49.

———. 2002. Humpback whale, in *Encyclopedia of marine mammals*. W. F. Perrin, B. Wursig, and J. G. M. Thewissen, eds. San Diego: Academic Press, pp.

Clapham, P., P. Palsbøll, L. Pastene, T. Smith and L. Walløe. 2004. Estimating pre-whaling abundance. *Report of the Scientific Committee*. International Whaling Commission. Annex S.

Cooke. J. 1999. Improvement of fishery-management advice through simulation testing of harvest algorithms. *ICES Journal of Marine Science* 56: 797–810.

Frankham, R. 1995. Effective population size/adult population size ratios in wildlife: a review. *Genetical Research* 66: 95–107.

———. 1996. Relationship of genetic variation to population size in wildlife. *Conservation Biology* 10: 1500–1508.

Franz, G., D. Tautz, and G. Dover. 1985. Conservation of major nuclease S1-sensitive sites in the non-conserved spacer region of ribosomal DNA in *Drosophila* species. *Journal of Molecular Biology* 183: 519–527.

Hare, M., F. Cipriano, and S. R. Palumbi. 2003. Genetic evidence on the demography of specuation in ailopatric dolphin species. *Evolution* 56: 804–816.

Hartl, D. and A. Clark. 1997. *Principles of population genetics*. Sunderland, MA: Sinauer Associates.

Holt, S. 2004. Counting whales in the North Atlantic. *Science* 303: 39b–40.

Horwood, J. 1990. *Biology and exploitation of the minke whale*. Boca Raton, FL: CRC Press.

Howell, N., C. B. Smejkal, D. A. Mackey, P. F. Chinnery, D. M. Turnbull, and C. Herrnstadt. 2003. The pedigree rate of sequence divergence in the human mitochondrial genome: There is a difference between phylogenetic and pedigree rates. *American Journal of Human Genetics* 72: 659–670.

IUCN. 1994. *1994 IUCN Red List of threatened species*. Gland, Switzerland: World Conservation Union.

IWC. 2003. Report of the Subcommittee on the Comprehensive Assessment of Humpback Whales. *Journal of Cetacean Research and Management* 5 (Suppl.): 293–323.

Jackson, J. B. C. 1997. Reefs since Columbus. *Coral Reefs* 16 (Suppl.): S23–S32.

Jackson, J. B. C., M. X. Kirby, W. H. Berger, K. A. Bjorndal, L. W. Botsford, B. J. Bourque, R. Bradbury, R. Cooke, J. Erlandson, J. A. Estes, T. P. Hughes, S. Kidwell, C. B. Lange, H. S. Lenihan, J. M. Pandolfi, C. H. Peterson, R. S. Steneck, M. J. Tegner, and R. Warner. 2001. Historical overfishing and the recent collapse of coastal ecosystems. *Science* 293: 629–638.

Kimura, T. and T. Ozawa. 2002. A new cetothere (Cetacea: Mysticeti) from the early Miocene of Japan. *Journal of Vertebrate Paleontology* 22: 684–702.

Li, W. 1997. *Molecular evolution*. Sunderland, MA: Sinauer Associates.

Lockyer, C. 1984. Review of baleen whale (Mysticeti) reproduction and implications for management. *Report of the International Whaling Commission* (Special Issue) 6: 27–50.

Mace, G. and R. Lande. 1991. Assessing extinction threats: toward a reevaluation of IUCN threatened species categories. *Conservation Biology* 5: 148–157.

MacLean, S., G. Sheehan, and A. Jensen. 2002. Inuit and marine mammals, in *Encyclopedia of marine mammals*. W. F. Perrin, B. Wursig, and J. G. M. Thewissen, eds. San Diego: Academic Press, pp.

Martin, A., B. Kessing, and S. Palumbi. 1990. Accuracy of estimating genetic-distance between species from short sequences of mitochondrial-DNA. *Molecular Biology and Evolution* 7: 485–488.

Mitchell, E. and R. R. Reeves. 1983. Catch history, abundance, and present status of northwest Atlantic humpback whales. *Report of the International Whaling Commission* (Special Issue) 5: 153–212.

Myers, R. A. and B. Worm. 2003. Rapid worldwide depletion of predatory fish communities. *Nature* 423: 280–283.

Nei, M. and D. Graur. 1984. Extent of protein polymorphism and the neutral mutation theory. *Evolutionary Biology* 17: 73–118.

Nunney, L. 1993. The influence of mating system and overlapping generations on effective population-size. *Evolution* 47: 1329–1341.

———. 2000. The limits to knowledge in conservation genetics: the value of effective population size. *Evolutionary Biology* 32: 179–194.

Palsbøll, P., P. Clapham, D. Mattila, F. Larsen, R. Sears, H. Siegismund, J. Sigurjónsson, O. Vasquez, and P. Arctander. 1995. Distribution of mtDNA haplotypes in North-Atlantic humpback whales: the influence of behavior on population-structure. *Marine Ecology-Progress Series* 116: 1–10.

Palsbøll, P., M. Bérubé, J. Robbins, A. Nettel-Hernanz, P. J. Clapham, and D. Mattila. 2004. Whales before whaling in the North Atlantic: nuclear DNA markers. Paper SC/56/O25 presented to the International Whaling Commission Scientific Committee.

Palumbi, S. R. and C. S. Baker. 1994. Contrasting population structure from no clear intron sequences and mtDNA of humpback whales. *Molecular Biology and Evolution* 11: 426–435.

Palumbi, S. R., F. Cipriano and M. Hare. 2001. Predicting nuclear gene coalescence from mitochondrial data: the three-times rule. *Evolution* 55: 859–868.

Perrin W. and R. Brownell. 2002. Minke whales, in *Encyclopedia of marine mammals*. W. F. Perrin, B. Wursig, and J. G. M. Thewissen, eds. San Diego: Academic Press, pp.

Posada, D., and K. Crandall. 1998. Model test: testing the model of DNA substitution. *Bioinformatics* 14: 817–818.

Reeves, R. R. 2002. The origins and character of "aboriginal subsistence" whaling: a global review. *Mammal Review* 32: 71–106.

Roman, J. and S. R. Palumbi. 2003. Whales before whaling in the North Atlantic. *Science* 301: 508–510.

Rosenbaum, H., M. Egan, P. Clapham, R. Brownell, S. Malik, M. Brown, B. White, P. Walsh, and R. Desalle. 2000. Utility of North Atlantic right whale museum specimens for assessing changes in genetic diversity. *Conservation Biology* 14: 1837–1842.

Rychel, A. L., T. W. Reeder, and A. Berta. 2004. Phylogeny of mysticete whales based on mitochondrial and nuclear data. *Molecular Phylogenetics and Evolution* 32: 892–901.

Scammon, C. M. 1874. *The marine mammals of the north-western coast of North America, described and illustrated with an account of the American whale-fishery.* New York: John H. Carmany and Co. Reprint, Riverside, CA: Manessier, 1969.

Sergeant, D. E. 1977. Stocks of fin whales *Balenoptera physalus* L. in the North Atlantic Ocean. *Report of the International Whaling Commission* 27: 460–473.

Shapiro, B., A. J. Drummond, and 25 others. 2004. The rise and fall of the Beringian Steppe bison. *Science* 306: 1561–1565.

Smith, T. D., and R. R. Reeves. 2002. Estimating historical humpback whale removals from the North Atlantic. *Journal of Cetacean Research and Management* 4 (Supplement): 242–255.

———. 2003. Estimating historical humpback whale removals from the North Atlantic: an update. *Journal of Cetacean Research and Management* 5 (Supplement): 301–311.

Soulé, M. 1976. Allozyone variation: its determinants in space and time. in *Molecular evolution.* F. J. Ayala, ed. Sunderland, MA: Sinauer, pp. 60–77.

Springer, A. M., J. A. Estes, G. B. van Vliet, T. M. Williams, D. F. Doak, E. M. Danner, K. A. Forney, and B. Pfister. 2003. Sequential megafaunal collapse in the North Pacific Ocean: an ongoing legacy of industrial whaling? *Proceedings of the National Academy of Sciences* 100: 12223–12228.

Stevick, P. T., J. Allen, P. J. Clapham, N. Friday, S.K. Katona, F. Larsen, J. Lien, D. K. Mattila, P. J. Palsbøll, J. Sigurjónsson, T. D. Smith, N. Øien, and P. S. Hammond. 2003. North Atlantic humpback whale abundance and rate of increase four decades after protection from whaling. *Marine Ecology Progress Series* 258:263–273.

Tajima, F. 1989. The effect of change in population size on DNA polymorphism. *Genetics* 123: 597–601.

Takahata, N. 1995. A genetic perspective on the origin and history of humans. *Annual Review of Ecology and Systematics* 26: 343–372.

Wada, S. and K. Numachi. 1991. Allozyme analyses of genetic differentiation among the populations and species of the Balaenoptera. *Report of the International Whaling Commission* (Special Issue) 13: 125–154.

Waldick, R., S. Kraus, M. Brown, and B. White. 2002. Evaluating the effects of historic bottleneck events: an assessment of microsatellite variability in the endangered, North Atlantic right whale. *Molecular Ecology* 11: 2241–2250.

Weller, D., A. Burdin, B. Wursig, B. Taylor, and R. Brownell. 2002. The western gray whale: a review of past exploitation, current status and potential threats. *Journal of Cetacean Research and Management* 4: 7–12.

Whitehead, H. 2000. Density-dependent habitat selection and the modeling of sperm whale *(Physeter macrocephalus)* exploitation. *Canadian Journal of Fisheries and Aquatic Sciences* 57: 223–230.

———. 2002. Estimates of the current global population size and historical trajectory for sperm whales. *Marine Ecology Progress Series* 242: 295–304.

Wu, C.-I. and W.-H. Li. 1985. Evidence for higher rates of nucleotide substitution in rodents than in man. *Proc. Nat. Acad. Sci.* (USA) 82: 1741–1745.

TEN

Changes in Marine Mammal Biomass in the Bering Sea/Aleutian Islands Region before and after the Period of Commercial Whaling

BETE PFISTER AND DOUGLAS P. DEMASTER

The Bering Sea is one of the most productive and diverse ecosystems in the world, supporting approximately 266 species of phytoplankton, 300 species of zooplankton, 450 species of fish and invertebrates, and 38 species of seabirds (Loughlin et al. 1999). Currently, 26 species of marine mammals, including large whales, beaked whales, small cetaceans, pinnipeds, the polar bear *(Ursus maritimus)*, and the sea otter *(Enhydra lutris)*, inhabit the area (National Research Council 1996). Many of these species were severely impacted by large-scale commercial harvesting from the mid-1700s to the mid-1900s (Clapham and Baker 2002; Webb 1988). During this time period, the overall abundance of marine mammals changed markedly as the Steller sea cow *(Hydrodamalis gigas)* became extinct and populations of northern fur seal *(Callorhinus ursinus)*, walrus *(Odobenus rosmarus divergens)*, sea otter, and large whales declined. Aside from sea otters in coastal marine ecosystems (Estes and Palmisano 1974; Estes and Duggins 1995; Estes et al. 2004), the ecological effects of this large-scale removal of top predators in the Bering Sea/Aleutian Islands region are largely unknown, despite studies that have attempted to quantify changes in trophic interactions from pre– to post–commercial harvest periods (Trites et al. 1999; Sobolevsky and Mathisen 1996).

Marine mammals occupy three ecological roles as consumers, prey, and detritus (see Estes, Chapter 1 of this volume). If they are important drivers of ecological process in ocean ecosystems, their removal could affect various food web interactions. Springer et al. (2003) recently speculated that the removal of large whales in the North Pacific led directly to the observed declines in pinniped (i.e., harbor seal *[Phoca vitulina richardii]*, northern fur seal, Steller sea lion *[Eumetopius jubatus]*) and sea otter abundance. The proposed mechanism for this relationship was increased predation by killer whales *(Orcinus orca)* on pinnipeds and sea otters following the reduction in availability of the killer whale's primary prey in this region—large whale species (e.g., sperm whale *[Physeter macrocephalus]*, fin whale *[Balaenoptera physalus]*, humpback whale *[Megaptera novaeangliae]*, and sei whale *[B. borealis]*).

A basic understanding of pre–commercial harvest and current abundance of marine mammals is critical before trophic effects can be evaluated. Therefore, one of the primary objectives of this chapter is to provide such a foundation to facilitate further evaluation of the nature and magnitude of the impacts of commercial whaling on the North Pacific ecosystem. Two factors must be considered when addressing this question.

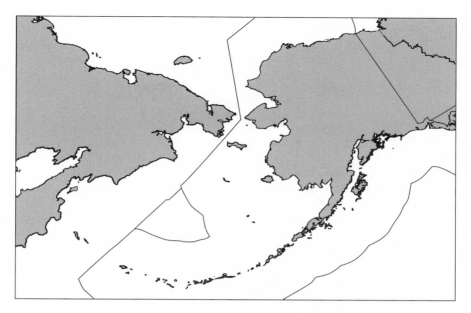

FIGURE 10.1. Map of the Bering Sea/Aleutian Islands region. The black line delineates the U.S. Exclusive Economic Zone and the boundary of the study area.

First, it is necessary to account for seasonal changes in biomass. All of the large whale species and several of the pinniped species (e.g., northern fur seals, ice seals, walrus) included in this report are considered migratory. Most of the large whale species migrate out of the Bering Sea during the late fall and early winter. In contrast, bowhead whales, walrus (except for the adult males), and ringed seals migrate into the Bering Sea during the winter and leave in April.

Second, some species of marine mammals have age- or sex-specific migratory patterns that must also be accounted for when evaluating the effects of commercial whaling on the marine mammal community in the North Pacific. For example, only adult male sperm whales are reported to occur in the Bering Sea, whereas the other age and sex classes are typically found south of the Aleutian Islands (Tomilin 1967; Berzin and Rovnin 1966; Gosho et al. 1984; Omura 1955, as cited in Gosho et al. 1984; Rice 1974). Another example of this life history phenomenon is the walrus. In the North Pacific, adult male walruses typically spend the summer in the Bering Sea at various male haulouts, while the other segments of the population follow the sea ice north into the Chukchi and Beaufort Seas (Sease and Chapman 1988).

Studies that document the temporal and spatial removal of large whales and quantify the available information on changes in biomass of marine mammal species in the ecosystem provide a foundation for more complex studies regarding the ecological effects of the removal of large whales from the world's oceans. Here, we document the overall change in marine mammal biomass of one ecosystem, the Bering Sea and Aleutian Islands region, from the onset of commercial harvesting of large whales and pinnipeds to the present. Population structure and age- and sex-specific seasonal movement patterns are taken into account.

Methods

Marine Mammal Biomass

Pre–commercial harvest and current biomass estimates of 19 species of marine mammals were calculated for the U.S. portion of the Bering Sea/Aleutian Islands region (Figure 10.1). Seven species of large whales, three species of small cetaceans, eight species of pinnipeds, and the sea otter were included in the analysis (Table 10.1).

Biomass of each marine mammal species was calculated as a product of the abundance, average body weight, and number of days spent in the Bering Sea/Aleutian Islands region divided by 365, representing the average biomass by species on an annual basis (see appendix for details). We attempted to account for population structure in two ways. For estimates of biomass reported in the literature at the species level, we used reported values to determine what proportion of the species distribution or which populations for a given species occurred in the area of interest (i.e., Bering Sea/Aleutian Islands). If estimates of biomass were reported at the population level, we used the published data directly for populations in the area of interest. When appropriate, the biomass calculation accounted for population structure (age and sex classes) and seasonal changes in biomass occurring in the area of interest.

Data were derived from a variety of published and unpublished documents and databases. When possible, pre–commercial harvest abundance estimates were taken from the time period before large-scale commercial exploitation of a species, and current abundance estimates were reported as the most recent available. When pre–commercial harvest abundance estimates of populations were not available for the time

TABLE 10.1

Marine Mammal Species Included in the Biomass Analyses

Common Name	Scientific Name	Stock/Population
Bowhead whale	*Balaena mysticetus*	Western Arctic
North Pacific right whale	*Eubalaena japonicus*	Eastern North Pacific
Fin whale	*Balaenoptera physalus*	Northeast Pacific
Humpback whale	*Megaptera novaeangliae*	Central North Pacific
Minke whale	*Balaenoptera acutorostrata*	Alaska
Gray whale	*Eschrictius robustus*	Eastern North Pacific
Sperm whale	*Physeter macrocephalus*	North Pacific
Harbor porpoise	*Phocoena phocoena*	Bering Sea
Dall's porpoise	*Phocoenoides dalli*	Bering Sea, Aleutian Islands
Beluga whale	*Delphinapterus leucas*	Beaufort, Eastern Chukchi, Eastern Bering Sea, Bristol Bay
Northern fur seal	*Callorhinus ursinus*	Eastern Pacific
Steller sea lion	*Eumetopius jubatus*	Western United States
Harbor seal	*Phoca vitulina richardii*	Bering Sea, Gulf of Alaska
Spotted seal	*Phoca largha*	Alaska
Bearded seal	*Erignathus barbatus*	Alaska
Ringed seal	*Phoca hispida*	Alaska
Ribbon seal	*Phoca fascista*	Alaska
Pacific walrus	*Odobenus rosmarus divergens*	Alaska
Sea otter	*Enhydra lutris*	Southwest Alaska

period prior to the onset of the commercial harvest, we used the estimate that was as close to that time period as possible.

All average body weight estimates, except for that for the sea otter (Kenyon 1969), were taken from Trites and Pauly (1998). In the latter study, the authors reported mean mass (kg) for males and females. To obtain the average body weight estimate for each species, we averaged the two weights and then converted to metric tons (t). This procedure likely introduced a positive bias into the estimate of average weight for some species. This is because for most species of marine mammals, the proportion of females in the population is greater than the proportion of males, and because adult males tend to be larger than adult females. However, for the following reasons we have assumed that this bias is negligible: (1) for many of the species considered in this paper the adult size of male and females is approximately equal (e.g., ice seals, baleen whales, and small cetaceans), and (2) the difference in average weight of males and females for most species is insignificant until after the age of maturation.

The biomass data were analyzed and broken down into three main categories: annual biomass, summer biomass (May to October), and winter biomass (November to April). Within each of these categories, pre–commercial harvest and current biomass estimates were calculated for all marine mammal species and for the three subgroups—large whales, small cetaceans, and pinnipeds. In addition, estimates for all marine mammal species and the large whale subgroup were also calculated with and without sperm whales because (1) the results of the biomass calculations are highly sensitive

to the inclusion of sperm whale data, and (2) estimates of pre–commercial harvest and current abundance of sperm whales in the study area are very poorly known (Croll et al., Chapter 16 of this volume). Clearly, assuming that there are no sperm whales in the region of interest biases the analysis. On the other hand, the existing estimates of pre–commercial harvest and current abundance are considered unreliable by scientists familiar with these data (i.e., the Scientific Committee of the International Whaling Commission [IWC]). Given the lack of adequate stock-specific data used in estimating pre–commercial harvest abundance, it is likely that these estimates are positively biased to some unknown degree. Further, because the estimate of current abundance is based on survey data from only a portion of the Bering Sea/Aleutian Islands region and is not corrected for probability of sighting, this estimate is likely negatively biased. Therefore, calculating relative composition of the marine mammal biomass in this region, as well as the percent change in biomass from the period prior to commercial whaling to the present both with and without sperm whale data, seems a robust approach to inferring the actual changes that likely occurred.

Caveats and Assumptions

The estimation of biomass required us to make several assumptions. Regarding the input parameters (see the appendix), we assumed that the information on seasonal presence of the population and average body weight was the same for both pre–commercial harvest and current populations of marine mammal species. Abundance estimates, percentage of the population present in the region, and the areas for which the estimates were made varied between the two time periods.

Most of the pre–commercial harvest abundance estimates used in this chapter are less certain than current estimates. In fact, the pre–commercial harvest and current abundance estimates used in the biomass calculation are only considered reliable for three species: bowhead whale, gray whale, and northern fur seal. Pfister (2004) presented a detailed account of the caveats and assumptions used to calculate the biomass of the fin whale, humpback whale, minke whale, sperm whale, harbor porpoise, Dall's porpoise, beluga whale, Steller sea lion, harbor seal, Pacific walrus, and sea otter. These caveats and assumptions can be classified into the following three broad categories.

OLD, UNAVAILABLE, OR UNRELIABLE ABUNDANCE ESTIMATES

Current abundance estimates for humpback whale, Aleutian Dall's porpoise, the Beaufort and Eastern Chukchi stocks of beluga whale, walrus, both harbor seal stocks, and the four ice seals were calculated from surveys that were completed between 1987 and 1996; some of these data are over a decade old. Pre–commercial harvest estimates were not available for eight species: minke whale, harbor porpoise, Dall's porpoise, beluga whale, and the four species of ice seals. Therefore, we substituted the current abundance estimates for pre–commercial harvest abundance in an effort to provide some information to the model. These current estimates of abundance are likely negatively biased because surveys are typically incomplete (i.e., the entire distribution is not surveyed) and correction factors for the probability of sighting are not available. For the minke whale and small cetacean species, the assumption that pre–commercial harvest and current abundance is approximately equal is supported by the available literature (Horwood 1990; Tomilin 1967; Ohsumi 1991; IWC 1991; Bjøre et al. 1994; Jones 1984; Braham 1984; Lensink 1961; Seaman and Burns 1981; Burns and Seaman 1985; Sherrod 1982; Fraker 1980, as cited in Hazard 1988). Specifically it is thought that populations of harbor porpoise, Dall's porpoise and beluga whale in the eastern North Pacific have not fluctuated greatly between the two time periods because they were subjected to low direct takes in the commercial fishery (IWC 1991; Bjøre et al. 1994; Gaskin 1984; Jones 1984; Braham 1984; Lensink 1961; Seaman and Burns 1981; Burns and Seaman 1985; Sherrod 1982; Stickney 1982; Wolfe 1982; and Fraker 1980). The population of Dall's porpoise in the western North Pacific did experience a high rate of incidental mortality in the high seas salmon driftnet fishery that operated from 1952 to 1987 (Jones 1984; Jones et al. 1985; Turnock et al. 1995; Hobbs and Jones 1993; IWC 1991). However, the magnitude of the decline cannot be reliably estimated. For some species, such as the sperm whale, estimates of pre-commercial harvest

and current abundance exist, however, they are considered unreliable because they are at most a "best guess" or are based on data that are no longer considered reliable by the International Whaling Commission (IWC).

Several species of Alaskan marine mammals were not included in the analysis. The blue whale (*Balaenoptera musculus*), sei whale (*B. borealis*), Baird's beaked whale (*Berardius bairdii*), Cuvier's beaked whale (*Ziphius cavirostris*), and Stejneger's beaked whale (*Mesoplodon stejnegeri*) are species whose presence has been documented in the Bering Sea, but either reported numbers were very low (Sobolevsky and Mathisen 1996; Berzin and Rovnin 1966; Moore et al. 2002; Mizroch et al. 1984a; Masaki 1976, as cited in Mizroch et al. 1984b; Nasu 1974) or there were no data available for abundance estimates (Balcomb 1989; Rice 1986; Kasuya and Ohsumi 1984; Tomilin 1967; Loughlin and Perez 1985).

PRE-COMMERCIAL HARVEST AND CURRENT ABUNDANCE ESTIMATES REPRESENTATIVE OF DIFFERENT AREAS

In some cases, the pre–commercial harvest range of certain species was more expansive than the current one (i.e., the northern right whale and the sea otter). In others, the survey area for both time periods may not be representative of the entire Bering Sea/Aleutian Islands region or may not be specific to the region. For example, current abundance estimates for the fin whale, humpback whale, minke whale, harbor porpoise, and Dall's porpoise are only applicable to the central eastern and southeastern portion of the Bering Sea and were not extrapolated to the entire study area. Those estimates are likely negatively biased, although that determination depends on information regarding the demographics, distribution, and movement patterns of the species that can only be known through more frequent and consistent surveys. On the other hand, pre–commercial harvest estimates for the fin whale, humpback whale, sperm whale, Steller sea lion, harbor seal, and sea otter were for larger areas such as the entire North Pacific or the entire Bering Sea, including Russian waters. In this case, an effort was made to adjust the pre–commercial harvest estimate to represent the same area as the current, using information on the temporal and spatial distribution of the population. However, for some species like sperm whales adequate information to prorate the historic abundance estimate to the Bering Sea region was not available (but see Wade et al. 2006, who did prorate this estimate). As noted in Pfister (2004), our preference was to estimate biomass with and without sperm whale data to provide an estimate of the range of the decline.

LACK OF INFORMATION ABOUT THE DEMOGRAPHICS OR THE DISTRIBUTION OF SOME POPULATIONS

Many of the estimates were calculated from surveys done at a particular time of year. Technically, those estimates and patterns of distribution are specific to the time of year the animals were observed. However, we assumed that information was representative of the population year round or for the

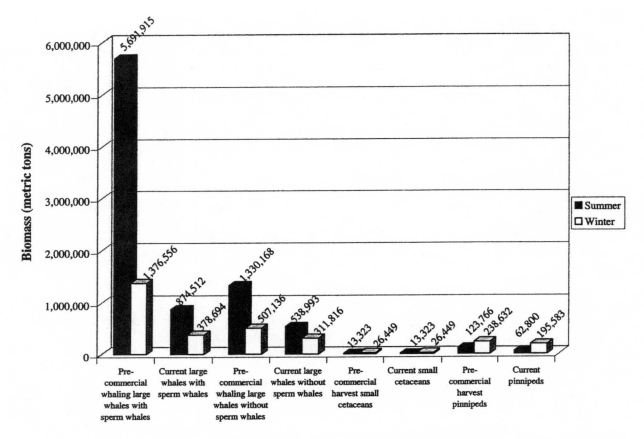

FIGURE 10.2. A comparison of the summer and winter biomass of the pre-whaling and current biomass of large whales (with and without sperm whales), small cetaceans, and pinnipeds, showing the annual and seasonal dominance of large whale biomass over all other marine mammal subgroups.

entire summer or winter period, as appropriate. There are major gaps in information regarding the seasonal distribution and movement patterns of some marine mammal populations in the North Pacific including, but not limited to, the sperm whale, Northern right whale, fin whale, minke whale, and Dall's porpoise.

We further assumed that all populations were at carrying capacity 150 years ago and that carrying capacity has been constant over time. This assumption implies that humans are the only variable in the system and excludes natural cycles of top-down and bottom-up forcing. Given the recent literature on the effects of El Niño–Southern Oscillation (ENSO) events and basin-wide regime changes (e.g., the Pacific Decadal Oscillation), this assumption is a clear oversimplification.

We urge the users of this information to carefully consider the caveats in the data and incorporate them, as possible, into future analyses.

Results

The data presented are the best available and provide a general picture of changes in pre–commercial harvest and current marine mammal biomass during the defined time period.

Percent Reduction in Biomass

Overall, marine mammal biomass in the Bering Sea/Aleutian Islands region has declined from the onset of modern industrial whaling to the present day, despite the apparent increase in a few populations of large whales and pinnipeds during this time period. This pattern was seen in all taxa on an annual and seasonal basis, regardless of whether sperm whale biomass was included in the analysis (Figure 10.2). The estimated percent reduction in the biomass of all marine mammals and of the large whale subgroup from the early 1800s to the present is shown in Table 10.2. There were no changes in the percent of small cetacean biomass because pre–commercial harvest and current abundance estimates were assumed to be equal. Pinniped biomass experienced the following declines: 30% annual, 49% summer, and 18% winter.

Analysis of Annual and Seasonal Biomass by Subgroup

In most cases, the pre–commercial whaling and current biomass of large whales dominated the biomass of all other species of marine mammals combined, regardless of whether sperm whales were included in the analysis (Figure 10.2). In

TABLE 10.2

Percent Reduction in the Biomass of All Marine Mammals and of the Large Whale Subgroup from the Early 1800s to the Present

Calculated with and without Sperm Whale Biomass

Biomass	With Sperm Whales	Without Sperm Whales
Annual		
Total	80%	49%
Large whale	82%	54%
Summer		
Total	84%	54%
Large whale	85%	59%
Winter		
Total	63%	31%
Large whale	72%	39%

TABLE 10.3

Percent Composition of Small Cetaceans on an Annual and Seasonal Basis

Species	Annual	Summer	Winter
Beaufort beluga	**30%**	<10%	**46%**
Norton Sound beluga	19%	**43%**	22%
Dall's porpoise	18%	40%	14%
Harbor porpoise	<10%	11%	<10%

NOTE: No distinction is made between pre–commercial harvest and current biomass percentages because pre–commercial harvest and current abundance estimates were assumed to be equal.

LARGE WHALES

On an annual basis, sperm whales dominated pre–commercial whaling large whale biomass (74%), yet composed only 32% of current large whale biomass. The other large whale species composed the remainder of the current whale biomass in the following percentages: bowhead (25%), gray (19%), fin (14%), and humpback (10%). When sperm whales were excluded from the analysis, pre–commercial whaling biomass was fairly evenly distributed between the bowhead (25%), fin (25%), gray (21%), humpback (15%), and right whale (13%), and the same species (minus the northern right whale) dominated the current biomass, although the percentage of each species' composition changed slightly: bowhead (36%), gray (29%), fin (20%), and humpback (14%).

In the summer, sperm whales composed 77% of the pre–commercial whaling biomass and 38% of the current biomass. Other large whale species that contributed to the current biomass were gray (28%), fin (19%), and humpback (14%) whales. When sperm whales were excluded, pre–commercial whaling and current biomass was distributed among the other large whale species in the following way: fin (34% and 31%), gray (29% and 45%), humpback (21% and 23%), and northern right (15%, pre–commercial harvest only). In the winter, sperm whales dominated pre–commercial whaling biomass (63%), followed by the bowhead whale (34%), but the contribution of each species was reversed in the current biomass composition: bowhead (82%) and sperm whale (18%). It follows that when sperm whales were excluded, bowhead whales dominated pre–commercial whaling (92%) and current (99%) biomass.

SMALL CETACEANS

The percentage of small cetacean biomass did not change from pre–commercial harvest to current times because pre–commercial harvest and current estimates of abundance were assumed to be equal. In general, four populations composed the majority of annual and seasonal small cetacean biomass in the Bering Sea/Aleutian Islands region, although the percent composition varied as shown in Table 10.3.

the following comparisons, large whales contributed greater than an order of magnitude more biomass to the system than small cetaceans and pinnipeds:

1. Pre–commercial harvest and current biomass of large whales (with and without sperm whales) versus pre–commercial harvest and current small cetacean biomass, respectively.

2. Pre–commercial harvest biomass of large whales (when sperm whale biomass was included) and the pre–commercial harvest biomass of pinnipeds.

3. Summer large whale biomass and summer small cetacean and pinniped biomass in every combination of biomass comparisons (pre–commercial harvest vs. current, with sperm whales vs. without) except one (current large whale biomass excluding sperm whales and current pinniped biomass).

4. Winter large whale biomass and winter small cetacean biomass in all combinations of biomass comparisons (pre–commercial harvest vs. current, with sperm whales vs. without).

The annual and seasonal biomass of small cetaceans ranged from 1 to 4% and represented a minute percentage of the overall biomass. Small cetacean biomass was larger in the winter, mainly due to the presence of the Beaufort stock of belugas.

Species Percent Composition of Annual and Seasonal Biomass

Species that are not listed in the following summaries of large whale, small cetacean, and pinniped annual and seasonal biomass composed less than 10% of the total biomass reported for each subgroup.

Pinniped biomass experienced the following changes from pre–commercial harvest to current times: 30% annual, 49% summer, and 18% winter. Pre–commercial harvest, the walrus (42%), Steller sea lion (15%), northern fur seal (14%), and bearded seal (10%) dominated annual pinniped biomass. The walrus (54%), bearded seal (13%), and ringed seal (10%) composed the majority of current pinniped biomass. In pre–commercial harvest and present times, four species, the northern fur seal (36% and 21%), walrus (27% and 47%), Steller sea lion (21% and 11%), and ribbon seal (12%, current only) have dominated summer pinniped biomass. Four species also composed the majority of pre–commercial harvest and current winter biomass: the walrus (56% and 61%), bearded seal (13% and 15%), Steller sea lion (11%, pre–commercial harvest only), and the ringed seal (10% and 12%).

Discussion

Major Changes in Composition of the Marine Mammal Community

Commercial harvesting of marine mammals over the last 150 years appears to have caused marked changes in their biomass in the Bering Sea/Aleutian Islands region, the most dramatic being a reduction in the percent composition of large whales. This change was likely due to the dominance of sperm, right, humpback, bowhead, fin, and gray whales in the pre-harvest community. Our results show an 80% decline in annual total marine mammal biomass from pre–commercial harvest to present times when sperm whales were included in the analysis and 49% decline when they were excluded. The relative change in annual great whale biomass from pre–commercial harvest to present compared with the relative change in biomass of the other marine mammals collectively over the same period was 82% (with sperm whales) and 54% (without sperm whales), compared to a 0% change in small cetacean biomass from pre–commercial harvest to present and a 30% change in overall pinniped biomass. Nonetheless, throughout this time period, large whales have dominated the annual marine mammal biomass, composing 95% (with sperm whales) and 83% (without sperm whales) of the pre–commercial harvest total marine mammal biomass and 82% (with sperm whales) and 75% (without sperm whales) of current marine mammal biomass. Small cetacean biomass ranged from 0.5 to 3.5% of the total pre–commercial harvest and current marine mammal biomass, and pinniped biomass ranged from 5 to 21%, taking into account calculations with and without sperm whales. This pattern of large whale dominance of pre–commercial harvest and current marine mammal biomass was also seen in the summer and winter seasonal estimates.

The Importance of Seasonality in Characterizing Marine Mammal Biomass in the Region

More total marine mammal biomass was present in the summer than in the winter, as illustrated in Figure 10.2. Also, the percent reduction of large whale biomass in the summer is similar to the percent reduction in total marine mammal biomass for the summer, and to the percent reduction of the annual total and large whale biomass (Table 10.2). These patterns can be explained by the influx of large whale migrants into the Bering Sea during this season.

The percent reduction of winter biomass, and therefore large whale biomass, is less than in the summer, mostly likely because the bowhead is the only species of large whale that is found in the Bering Sea/Aleutian Islands region during the winter. In addition, some species of winter pinnipeds that were subject to an intense harvest have recovered to a greater extent than many of the great whale species; therefore, changes in their percent biomass from pre–commercial harvest to current times did not vary as greatly, and they have a lesser influence on changes in total marine mammal biomass. In the winter, pinniped biomass was greater than in the summer, as many animals from several species enter the Bering Sea to breed on or forage from the extensive pack ice that forms during this season. The percent reduction of pinnipeds was greater in the summer (49%) than in the winter (18%), perhaps due to the fact that the biomass of non–ice inhabiting pinnipeds (Northern fur seal, Steller sea lion, and harbor seal) that have experienced population declines in the recent past was greater in the summer than in winter.

General Patterns in Great Whale Biomass

A majority of the decline in large whale biomass occurred 50–100 years ago with the depletion of the following populations: (1) eastern North Pacific right whale, (2) western Arctic population of bowhead whale, (3) eastern North Pacific gray whale, and (4) all three populations of North Pacific humpback whale. Unfortunately, considerable uncertainty exists in the estimated pre–commercial whaling and current abundance of several key species, including sperm, fin, and humpback whales, so it is difficult to make quantitative inferences regarding changes in each species' biomass during the last 50 years. On the other hand, we know that some populations of large whales have increased in abundance during this time period. Therefore, their current abundance estimates are likely higher than they were immediately at the end of the harvest. These populations include the western Arctic population of bowhead whale and the eastern North Pacific population of gray whale. We found that the contribution of bowhead whale and gray whale biomass to total marine mammal biomass increased during the period from pre–commercial whaling to the present. Conversely, sperm whale and northern right whale biomass relative to total marine mammal biomass decreased during the same time period.

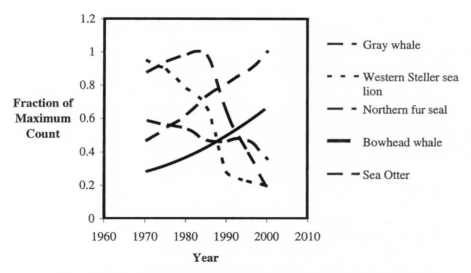

FIGURE 10.3. Post–commercial harvest changes in abundance of five species of marine mammals from 1970 to 2000, showing a decline in the western stock of Steller sea lion, the Northern fur seal, and the sea otter and an increase in the gray and bowhead whale populations in the Bering Sea/Aleutian Islands region.

The increase seen in the percent composition of bowhead and gray whale stocks was consistent with the observed recovery of these populations after they were protected from further commercial harvests. The annual rate of increase was estimated at 3.4% (George et al. 2003) for the western Arctic bowhead whale stock from 1978 to 2001, and at between 2.4 and 4.4% for the eastern North Pacific gray whale stock, which was removed from the List of Endangered and Threatened Wildlife in 1994 (Buckland et al. 1993; Wade and DeMaster 1996; IWC 1995; Breiwick 1999). The decrease of percent composition of the sperm whale and northern right whale was consistent with the documented removal of both species by modern commercial whalers (Scarff 1991; Doroshenko 2000; Clapham et al. 2001; Brownell et al. 2001; Brownell et al. 2000) and a lack of a documented increase in abundance over the last 30 years.

During the last 30 years, increases in the abundance of gray, bowhead, humpback, and fin whales likely instigated an overall increase in large whale biomass (and therefore total marine mammal biomass) in the late 1960s through the 1990s (Figures 10.3 and 10.4). However, this conclusion should be considered preliminary because of a lack of quantitative information on sperm and fin whale abundance and information on the age- and sex-specific distribution patterns of these two species.

Additional Caveats

An additional factor to consider when evaluating the impacts of commercial harvests on marine mammal community structure in the North Pacific has to do with the timing of the commercial harvest and the timing of the protection measures implemented to recover depleted populations. For example, the peak of commercial harvest for the northern right whale,

bowhead whale, gray whale, and sea otter occurred before 1950 (Scarff 2001; Doroshenko 2000; Clapham et al. 2001; Bockstoce and Burns 1993; IWC 2003; Rotterman and Simon-Jackson 1988), whereas peak harvests for the fin whale, sperm whale, humpback whale, harbor seal, and ice seals occurred after 1950 (IWC 2002; Pitcher 1984; Hoover 1988; Quakenbush 1988; Kelly 1988a,b,c). The post-1950 harvest for great whales varied in intensity as humpback whales were protected from commercial whaling in the mid-1960s (Johnson and Wolman 1984), while commercial hunting of fin and sperm whales occurred throughout the 1960s and 1970s (both legally and illegally) (Brownell et al. 2000; Yablokov and Zemsky 2000; Danner et al., Chapter 11 of this volume). Northern fur seal, Steller sea lion, and walrus populations were subjected to a period of heavy commercial harvesting during both time periods (York and Hartley 1981; Kenyon 1962; Fay et al. 1989).

Complicating the interpretation of these data is the wide range of recovery rates exhibited by different species of marine mammals. For example, as mentioned previously, the eastern North Pacific population of gray whale was reported to have increased at approximately 2.4–4.4% per year over the last 30 years, whereas some populations of sea otters have recovered at rates exceeding 20% per year (Estes 1990).

Therefore, to begin to assess the ecological impacts of changes in large whale abundance, information on the following is needed for each of the harvested species: (1) how severely a given population was depleted, (2) the distribution and timing of the harvest, (3) stock structure, (4) the onset of conservation measures, (5) the current abundance, (6) net production relative to status, and (6) the ability of the population to recover. This information is necessary to produce reliable estimates of pre–commercial whaling abundance and subsequent recovery (or in the case of North Pacific right whale populations, lack of

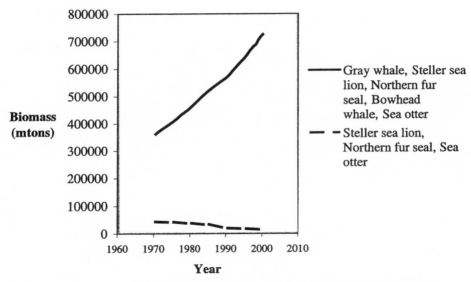

FIGURE 10.4. Yearly biomass trends by species groups depict an overall increase in marine mammal biomass post–commercial harvesting from 1970 to 2000 despite population declines in the Steller sea lion, Northern fur seal, and sea otter. These increases can be attributed to the recovery of gray and bowhead whale populations, illustrating the influence of large whale biomass on the total biomass of marine mammals in the study area.

recovery). An example of this approach is reported by Eberhardt and Breiwick (1992). They derived a reliable estimated trajectory of abundance over time for the western Arctic bowhead whale population. These estimates of abundance over time were generated using information on current abundance, an estimate of the maximum rate of net production, estimates of annual removals by year, and evidence to indicate that all harvested whales from this time series were from the same stock of whales. Similar data sets for other stocks of large whales are currently unavailable (e.g., all stocks of sperm whale, fin whale, and North Pacific right whale). Because of the dominance of large whale species in marine mammal biomass in the North Pacific, quantitative inferences regarding changes in total marine mammal biomass reported from the data herein for the last 30 years are not possible at this time.

An additional caveat that needs to be mentioned is the possibility that the estimates of pre–commercial whaling abundance based on the back-calculation method mentioned earlier (e.g., for the bowhead whale, Eberhardt and Breiwick 1992) may underestimate true pre–commercial whaling and historic abundance, as proposed by Roman and Palumbi (2003) and Palumbi and Roman (Chapter 9 of this volume). Given the time it takes to reach equilibrium and the relative quickness with which commercial whaling decimated large whale populations, they argue that an estimate of historic abundance (thousands of years ago) is a good proxy for pre-whaling abundance. They calculated historic abundance for three species of large whale in the North Atlantic using a technique based on the genetic variability in a population, and their estimates were approximately 4–10 times larger than the back-calculated estimates for pre-whaling abundance. The Scientific Committee of the IWC has reviewed the approach

used by Roman and Palumbi and has concluded that these estimates of historic abundance should be considered unreliable as estimates of pre–commercial whaling abundance at this time (IWC 2005). Further, the Committee recommended that a series of studies be undertaken prior to completing an evaluation of the use of genetic methods for estimating pre–commercial whaling abundance. Clearly, if the Roman and Palumbi hypothesis is correct and is appropriately applied to the North Pacific, the impact of commercial whaling on the marine mammal community in the Bering Sea/Aleutian Islands region would be significantly greater than shown by the findings reported herein. However, at this time, it would be premature to draw such a conclusion.

Possible Causes for the Declines

It is difficult to identify all of the likely causal mechanisms associated with the changes in relative biomass reported in this chapter. Much of the reported decline of marine mammal biomass is a consequence of direct human removal either by commercial harvest, incidental fisheries loss, or illegal killing. Clearly, commercial whaling was a major factor leading to the decline in relative biomass of several species of large whales (e.g., sperm whale, bowhead whale, North Pacific right whale). Similarly, harvesting of sea otters, walrus, and northern fur seals led to severe depletion of these species at various times. However, the declines in some cases have far more complex and poorly understood explanations. For example, considerable literature exists concerning the potential causes of the decline of the western stock of Steller sea lion (National Research Council 2003). These purported causes include (1) subsistence hunting, (2) mortality incidental to commercial fishing, (3) competition with commercial and subsistence fisheries, (4) nutritional stress caused by environmental regime changes, (5) disease agents, (6) contamination, and (7) predation (National Research Council 2003). The identification of such factors for this relatively well-studied species may serve as an example for understanding the dynamics of many of the less-studied species that occur in the North Pacific. Another case in point is the sea otter, which was historically subject to an intense commercial harvest that nearly decimated the North Pacific population (Doroff et al. 2003). Here, there is evidence indicating that killer whale predation and not direct human take is the likely cause of recent declines documented in the Aleutian Islands (Estes et al. 1998). This example, like that of the Steller sea lion decline, could possibly be explained as a complex food web interaction.

Recommendations

Based on our review of the available literature on pre–commercial whaling abundance of large whale populations, it is clear that additional research is needed. As noted by Roman and Palumbi (2003) and Palumbi and Roman (Chapter 9 of this volume), methods to estimate pre–commercial whaling abundance of large whales have concentrated on the evaluation of whaling records. Pre–commercial whaling abundance has been estimated using back-calculation models that require at a minimum (1) an estimate of current abundance, (2) a time series of total removals, and (3) the relationship between abundance and net production. Bias in any one of these data requirements will lead to bias in the estimation of pre-whaling abundance. Therefore, further research on these elements for species (or populations) for which pre-whaling abundance estimates are available is needed to improve the quality of the estimate, as is further research needed for species (or populations) for which such estimates do not exist. Given the difficulty in determining with any confidence the density-dependent relationship between abundance and net production, efforts to date have focused, and should continue to focus, on improving estimates of abundance, the time series of commercial catches, and the rate of struck-and-loss associated with commercial harvests. In addition, other methods to estimate pre–commercial whaling and historic abundance exist. One such method, described by Roman and Palumbi (2003), uses the patterns of DNA variation in current populations. However, an evaluation of the utility of this method for providing estimates of pre-whaling abundance must be undertaken before it is further applied to other large whale populations (IWC in press).

Acknowledgments

B. Pfister was supported by the National Marine Mammal Laboratory under contract numbers AB133F-03-SE-0662, 40ABNF100829, and AB133F-03-SE-1409, and by the University of California Santa Cruz Natural Science Department under contract number P0164377. We thank Robyn Angliss, Marilyn Dahlheim, Brian Fadley, Nancy Friday, Rod Hobbs, Kristin Laidre, Keri Lodge, Tom Loughlin, Sally Mizroch, Bruce Robson, Dave Rugh, Kim Shelden, Paul Wade, Janice Waite, Dave Withrow, and Anne York of the National Marine Mammal Laboratory and Joel Garlich-Miller and Morgan Robertson from the U.S. Fish and Wildlife Service for reviewing portions of this document and providing data. We also thank Sonja Kromann from the National Marine Mammal Laboratory for her help in acquiring resources.

Literature Cited

Alaska Department of Fish and Game. 1973. *Alaska's wildlife and habitat*. Anchorage, AK: Van Cleve Printing, 143p.

Angliss, R. P. and K. L. Lodge. 2002. *Alaska marine mammal stock assessments, 2002*. U.S. Department of Commerce, NOAA Technical Memorandum NMFS-AFSC-133. Seattle: Alaska Fisheries Science Center.

———. 2004. Alaska marine mammal stock assessments, 2003. U.S. Department of Commerce, NOAA Technical Memorandum NMFS-AFSC-144. Seattle: Alaska Fisheries Science Center.

Balcomb, K. C. 1989. Baird's beaked whale, *Berardius bairdii* Stejneger, 1883, and Arnoux's beaked whale *Iberardius arnouxii* Douvernoy, 1851, in *Handbook of marine mammals, Volume 4: River dolphins and the larger toothed whales*. S. H. Ridgway and R. Harrison, eds. New York: Academic Press, pp. 261–288.

Berzin, A. A. and A. A. Rovnin. 1966. The distribution and migrations of whales in the northeastern part of the Pacific, Chukchi, and Bering Seas. *Izvestia TINRO* 58: 179–207.

Bessonov, B., V. V. Mel'nikov, and V. A. Bobkov. 1990. Distribution and migration of cetaceans in the Soviet Chukchi Sea, in *Conference Proceedings*, Third Information Transfer Meeting. U.S. Department of the Interior, Minerals Management Service, Alaska OCS Region, Anchorage, AK, pp. 25–31.

Bigg, M. A. 1969. The harbour seal in British Columbia. *Fisheries Research Board of Canada Bulletin* 172: 33p.

———. 1981. Harbour seal, *Phoca vitulina*, (Linnaeus, 1758) and *Phoca largha*, (Pallas, 1811), in *Handbook of Marine Mammals, Vol. 2: Seals*. S. H. Ridgway and R. J. Harrison, eds. New York: Academic Press, pp. 1–27.

———. 1986. Arrival of northern fur seals, *Callorhinus ursinus*, on St. Paul Island, Alaska. *Fisheries Bulletin* 84: 383–394.

Bjøre, A., R. L. Brownell Jr., G. P. Donovan, and W. F. Perrin. 1994. Significant direct and incidental catches of small cetaceans. *Report of the International Whaling Commission* (Special Issue) 15: 76–130.

Bockstoce, J. R. and J. J. Burns. 1993. Commercial whaling in the North Pacific sector, in *The bowhead whale*. J. J. Burns, J. J. Montague, and C. J. Cowles, eds. [Society of Marine Mammals Special Publication vol. 2], Lawrence KS: Society for Marine Mammology, pp. 563–577.

Bogoslovskaya, L. S., L. M. Votrogov, and I. I. Krupnik. 1982. The bowhead whale off Chuckotka: migrations and aboriginal whaling. *Report of the International Whaling Commission* 32: 391–399.

Boveng, P. L., J. L. Bengtson, D. E. Withrow, J. C. Cesarone, M. A. Simpkins, K. J. Frost, and J. J. Burns. 2003. The abundance of harbor seals in the Gulf of Alaska. *Marine Mammal Science* 19(1): 111–127.

Braham, H. W. 1984. Review of reproduction in the white whale, *Delphinapterus leucas*, narwhal, *Monodon monoceros*, and

Irrawaddy dolphin, *Orcaella brevirostris*, with comments on stock assessment. *Report of the International Whaling Commission* (Special Issue) 6: 81–89.

Braham, H. W., J. J. Burns, G. A. Fedoseev, and B. D. Krogman. 1984. Habitat partitioning by ice-associated pinnipeds: distribution and density of seals and walruses in the Bering Sea, April 1976, in *Soviet-American cooperative research on marine mammals, Volume 1: Pinnipeds.* F. H. Fay and G. A. Fedossev, eds. U.S. Department of Commerce, NOAA Technical. Report NMFS-12.

Braham, H. W., M. A. Fraker, and B. D. Krogman. 1980. Spring migration of the western Arctic population of bowhead whales. *Marine Fisheries Review* 42(9–10): 36–46.

Braham, H. W., B. D. Krogman, and G. M. Carroll. 1984. *Bowhead and white whale migration, distribution, and abundance in the Bering, Chukchi, and Beaufort Seas, 1975–78.* U.S. Department of Commerce, NOAA Technical. Report NMFS SSRF-778.

Breiwick, J. W. 1999. Gray whale abundance estimates, 1967/68–1997/98: ROI, RY, and K, in *Status review of the Eastern North Pacific stock of gray whales.* D. J. Rugh, M. M. Muto, S. E. Moore, and D. P. DeMaster, eds. U.S. Department of Commerce, NOAA Technical Memo NMFS-AFSC-103, p. 62.

Brodie, P. F. 1971. A reconsideration of aspects of growth, reproduction, and behavior of the white whale with reference to the Cumberland Sound, Baffin Island, population. *Journal of the Fisheries Research. Board of Canada* 28: 1309–1318.

Brownell, R. L., Jr., P. J. Clapham, T. Miyashita, and T. Kasuya. 2001. Conservation status of North Pacific right whales. *Journal of Cetacean Research and Management* Special Issue 2: 269–286.

Brownell, R. L., Jr., A. V. Yablokov, and V. A. Zemsky. 2000. USSR pelagic catches of North Pacific sperm whales, 1949–1979: conservation implications, in *Soviet whaling data (1949–1979).* A. V. Yablokov and V. A. Zemsky, eds. Moscow: Center for Russian Environmental Policy, pp. 123–130.

Brueggeman, J. J. 1982. Early spring distribution of bowhead whales in the Bering Sea. *Journal of Wildlife Management* 46: 1036–1044.

Brueggeman, J. J., B. Webster, R. Grotefendt, and D. Chapman. 1987. Monitoring the winter presence of bowhead whales in the Navarin Basin through association with sea ice. U.S. Department of Commerce, National Technical Information Service Report No. PB88-101258, 179 pp. Available from NTIS, 5285 Port Royal Road, Springfield, VA 22167.

Buckland, S. T., J. M. Breiwick, K. L. Cattanach, and J. L. Laake. 1993. Estimated population size of the California gray whale. *Marine Mammal Science* 9(3): 235–249.

Buckland, S. T., K. L. Cattancach, and R. C. Hobbs. 1993. Abundance estimates of Pacific white-sided dolphin, northern right whale dolphin, Dall's porpoise and northern fur seal in the North Pacific, 1987–1990. *INPFC Bulletin* 53(III): 387–407.

Burns, J. J. 1981. Bearded seal *Erignathus barbatus* (Erxleben, 1777), in *Handbook of marine mammals, Volume 2: Seals.* S. H. Ridgway and R. J. Harrison, eds. London: Academic Press, pp. 145–170.

———. 1986. Ice seals, in Marine mammals of eastern North Pacific and arctic waters, 2nd edition, D. Haley, ed. Seattle, WA: Pacific Search Press, pp. 216–229.

Burns, J. J. and G. A. Seaman. 1985. Investigations of belukha whales in coastal waters of western and northern Alaska: II. Biology and ecology. Final report prepared for U. S. Department of Commerce, NOAA, National Ocean Service, Anchorage, Alaska, contract NA81RAC00049. Alaska Department of Fish and Game, Fairbanks.

Calambokidis, J. and J. Barlow. 1991. Chlorinated hydrocarbon concentrations and their use for describing population discreteness in harbor porpoises from Washington, Orgeon, and California, in *Marine mammal strandings in the United States: proceedings of the second Marine Mammal Stranding Workshop: 3–5 December 1987.* J. E. Reynolds III and D. K. Odell, eds. Miami, FL: NMFS, NOAA Technical Report NMFS 98: 101–110.

Calambokidis, J., G. H. Steiger, J. M. Straley, T. Quinn, L. M. Herman, S. Cerchio, D. R. Salden, M. Yamaguchi, F. Sato, J. R. Urban, J. Jacobson, O. VonZeigesar, K. C. Balcomb, C. M. Gabriele, M. E. Dahlheim, N. Higashi, S. Uchida, J. K. B. Ford, Y. Miyamura, P. Ladrón de Guevara, S. A. Mizroch, L. Schlender, and K. Rasmussen. 1997. Abundance and population structure of humpback whales in the North Pacific basin. *Final Contract Report 50ABNF500113.* Southwest Fisheries Science Center, P.O. Box 271, La Jolla, CA 92038.

Clapham, P. J. and C. S. Baker. 2002. Whaling, modern, in *Encyclopedia of marine mammals.* W.F. Perrin, B. Wursig, and J.G.M. Thewissen, eds. San Diego: Academic Press, pp. 1328–1332.

Clapham, P. J., C. Good, S. E. Wetmore, and R. L. Brownell Jr. 2001. Distribution of North Pacific right whales (*Eubalaena japonica*) as shown by 19th and 20th century whaling catch and sighting records. Paper SC/53/BRG15 presented to the International Whaling Commission Scientific Committee.

Clapham, P. J., S. B. Young, and R. L. Brownell, Jr. 1999. Baleen whales: conservation issues and the status of the most endangered populations. *Mammal Review* 29(1): 35–60.

Darling, J. D. 1984. Gray whales off Vancouver Island, British Columbia, in *The gray whale* Eschrichtius robustus. M. L. Jones, S. L. Swartz, and S. Leatherwood, eds. Orlando, FL: Academic Press, pp. 267–287.

DeMaster, D. P. 1996. Minutes from the 11-13 September 1996 meeting of the Alaska Scientific Review Group, Anchorage, Alaska. 20 pp. plus appendices. (Available upon request from D. P. DeMaster, Alaska Fisheries Science Center, 7600 Sand Point Way, NE, Seattle, WA 98115.)

Doroff, A. M., J. A. Estes, M. T. Tinker, D. M. Burn, and T. J. Evans. 2003. Sea otter population declines in the Aleutian Archipelago. *Journal of Mammalogy* 84(1): 55–64.

Doroshenko, N. V. 2000. Soviet whaling for blue, gray, bowhead and right whales in the North Pacific Ocean, 1961–1979, in *Soviet whaling data (1949–1979).* A. V. Yablokov and V. A. Zemsky, eds. Moscow: Center for Russian Environmental Policy, pp. 96–103.

Duval, W. S. 1993. Proceedings of a workshop on Beaufort Sea beluga: February 3–6, 1992. Vancouver, B.C. Environmental Studies Research Foundation Report No 123. Calgary.

Eberhardt, L. L. and J. M. Breiwick. 1992. Impact of recent population data on historical population levels of bowhead whales inferred from simulation studies. *Report of the International Whaling Commission* 42: 485–489.

Estes. J. A. 1990. Growth and equilibrium in sea otter populations. *Journal of Animal Ecology* 59: 385–401.

Estes, J. A., E. M. Danner, D. F. Doak, B. Konar, A. M. Springer, P. D. Steinberg, M. T. Tinker, and T. M. Williams. 2004. Complex trophic interactions in kelp forest ecosystems. *Bulletin of Marine Sciences* 74: 621–638.

Estes, J. A. and D. O. Duggins. 1995. Sea otters and kelp forests in Alaska: generality and variation in a community ecological paradigm. *Ecological Monographs* 65: 75–100.

Estes, J. A. and J. F. Palmisano. 1974. Sea otters: their role in structuring nearshore communities. *Science* 185: 1058–1060.

Estes, J. A., M. T. Tinker, T. M. Williams, and D. F. Doak. 1998. Killer whale predation on sea otters linking oceanic and nearshore ecosystems. *Science* 282: 473–476.

Fay, F. H. 1981. Walrus, *Odobenus rosmarus*, (Linnaeus, 1758), in *Handbook of Marine Mammals, Volume 1: The walrus, sea lions, fur seals and sea otter*. S. H. Ridgway and R. J. Harrison, eds. London: Academic Press, pp. 1–23.

———. 1982. Ecology and biology of the Pacific walrus, *Odobenus rosmarus divergens*. *North American Fauna* 74: 1–279.

Fay, F. H., G. Carleton Ray, and A. A. Kibal'Chich. 1985. Time and location of mating and associated behavior of the Pacific walrus, *Odobenus rosmarus divergens*, Illiger in *Soviet-American cooperative research on marine mammals, Volume 1: Pinnipeds*. F. H. Fay and G. A. Fedoseev, eds. U.S. Department of Commerce, NOAA Technical Report NMFS-12, pp. 89–99.

Fay F. H., L. L. Eberhardt, B. P. Kelly, J. J. Burns, and L. T. Quakenbush. 1997. Status of the Pacific walrus population, 1950–1989. *Marine Mammal Science* 13(4): 537–565

Fay, F. H., B. P. Kelly, and J. L. Sease. 1989. Managing the exploitation of Pacific walrus: a tragedy of delayed response and poor communication. *Marine Mammal Science* 5(1): 1–16.

Fedoseev, G. A. 1973. The morphological-ecological characteristics of ribbon seal populations and factors affecting the conservation of usable stocks. *Izv. TINRO* 86: 158–177. (Translated from Russian by Canadian Fisheries Marine Service, 1975, Transl. Ser.676795.)

Fiscus, C. H. 1980. Marine mammal-salmonid interactions: a review, in *Salmonid ecosystems of the North Pacific*. W. J. McNewil and D. C. Himsworth, eds. Corvallis, OR: Oregon State University Press, pp. 121–132.

Fisher, H. D. 1952. The status of the harbour seal in British Columbia, with particular reference to the Skeena River. *Fisheries Research Board of Canada Bulletin* 93.

Fraker, M. A. 1980. Status and harvest of the Mackenzie stock of white whales *(Delphinapterus leucas)*. *Report of the International Whaling Commission* 30: 451–458.

Frost, K. J. 1985. The ringed seal *(Phoca hispida)*. *Alaska Department of Fish and Game, Game Technical Bulletin* 7: 79–87.

Frost, K. J. and L. F. Lowry. 1995. *Radio tag based correction factors for use in beluga whale population estimates*. Working paper for Alaska Beluga Whale Committee Scientific Workshop, Anchorage, AK, 5–7 April 1995. (Available upon request from Alaska Dep. Fish and Game, 1300 College Rd., Fairbanks, AK 99701.)

Frost, K. J., L. F. Lowry, and G. Carroll. 1993. Beluga whale and spotted seal use of a coastal lagoon system in the northeastern Chukchi Sea. *Arctic* 46: 8–16.

Frost, K. J., L. F. Lowry, and D. P. DeMaster. 2002. Alaska beluga whale committee surveys of beluga whales in Bristol Bay, Alaska, 1999–2000. *Alaska Beluga Whale Committee Report 02–1*.

Gaskin, D. E. 1984. The harbour porpoise, *Phocoena phocoena* (L.): regional populations, status, and information on direct and indirect catches. *Report of the International Whaling Commission* 34: 569–586.

Gentry, R. L. 1981. Northern fur seal, *Callorhinus ursinus*, in *Handbook of Marine Mammals, Volume 1: The walrus, sea lions, fur seals and sea otter*. S. H. Ridgway and R. J. Harrison, eds. London: Academic Press, pp. 143–160.

George, J. C., J. Zeh, R. Suydam, and C. Clark. 2003. N_4/P_4 abundance estimate for the Bering-Chukchi-Beaufort Seas stock of bowhead whales, *Balaena mysticetus*, based on the 2001 census off Point Barrow, Alaska. Paper SC/55/BRG7 presented to the International Whaling Commission Scientific Committee.

Gilbert, J. R., G. A. Fedoseev, D. Seagars, E. Razlivalov, and A. LaChugin. 1992. Aerial census of Pacific walrus, 1990. *U.S. Fisheries and Wildlife Service Administrative Report R7/MMM 92–1*.

Gosho, M. E., P. J. Gearin, J. Calambokidis, K. M. Hughes, L. Cooke, and V. E. Cooke. 1999. *Gray whales in the waters of northwestern Washington in 1996 and 1997*. Unpublished document submitted to the International Whaling Commission (SC/51/AS9).

Gosho, M. E., D. W. Rice, and J. M. Breiwick. 1984. The sperm whale, *Physeter macrocephalus*. *Marine Fisheries Review* 46(4): 54–64.

Harrison, C. S. and J. D. Hall. 1978. Alaskan distribution of the beluga whale, *Delphinapterus leucas*. *Canadian Field-Nature* 92(3): 235–241.

Harwood, L. A., S. Innes, P. Norton, and M. C. S. Kingsley. 1996. Distribution and abundance of beluga whales in the Mackenzie Estuary, southeast Beaufort Sea and west Amundsen Gulf during late July 1992. *Canadian Journal of Fisheries and Aquatic Science* 53: 2262–2273.

Hazard, K. 1988. Beluga whale *(Delphinapterus leucas)*, in *Selected marine mammals of Alaska: species accounts with research and management recommendations*. J. W. Lentfer, ed. Washington, DC: Marine Mammal Commission, pp. 195–235.

Hobbs, R. C. and L. L. Jones. 1993. Impacts of high seas driftnet fisheries on marine mammal populations in the North Pacific. *North Pacific Commission Bulletin* 53(III): 409–434.

Hobbs, R. C. and J. A. Lerczak. 1993. Abundance of Pacific white-sided dolphin and Dall's porpoise in Alaska estimated from sightings in the North Pacific Ocean and the Bering Sea during 1987 through 1991 in *Marine Mammal Assessment Program: status of stocks and impacts of incidental take*. H. W. Braham and D. P. DeMaster, eds. U.S. Department of Commerce (National Marine Mammal Laboratory—MMPA studies of 1992). Seattle: Alaska Fisheries Science Center, National Marine Fisheries Service, NOAA.

Hobbs, R. C. and J. M. Waite. Forthcoming. Harbor porpoise abundance in Alaska, 1997–1999. *Marine Mammal Science*.

Hoover, A. A. 1988a. Steller sea lion *(Eumetopias jubatus)*, in *Selected marine mammals of Alaska: Species accounts with research and management recommendations*. J. W. Lentfer, ed. Washington, DC: Marine Mammal Commission, pp. 159–193.

———. 1988b. Harbor seal *(Phoca vitulina)*, in *Selected marine mammals of Alaska: species accounts with research and management recommendations*. J. W. Lentfer, ed. Washington, DC: Marine Mammal Commission, pp. 125–157.

Horwood, J. 1990. *Biology and exploitation of the minke whale*. Boca Raton, FL: CRC Press.

IWC. 1991. The conservation of small cetaceans: a review. Paper SC/43/SM3 presented to the International Whaling Commission Scientific Committee.

———. 1995. Report of the Scientific Committee. *Report of the International Whaling Commission* 45: 53–95.

———. 2002. International Whaling Commission catch database. Version 2 (December, 2002). Cambridge, UK: International Whaling Commission.

———. 2003. *Catch data for the North Pacific.* Cambridge, UK: International Whaling Commission.

———. 2005. Report of the Scientific Committee. *Journal of Cetacean Research and Management* 7: 393–400.

Johnson, A. M. 1982. Status of Alaska sea otter populations and developing conflicts with fisheries. *North American Wildlife Confederation* 47: 293–299.

Johnson, J. H. and A. A. Wolman. 1984. The humpback whale, *Megaptera novaeangliae. Marine Fisheries Review* 46(4): 1–100.

Jones, L. 1984. Incidental take of the Dall's porpoise and the harbor porpoise by Japanese salmon driftnet fisheries in the western North Pacific. *Report of the International Whaling Commission* 34: 531–538.

Jones, L., D. W. Rice, and A. A. Wolman. 1985. Biological studies of Dall's porpoise taken incidentally by the Japanese mothership fishery. Paper SC/35/SM9 presented to the International Whaling Commission Scientific Committee.

Kajimura, H. R., H. Lander, M. A. Perez, A. E. York, and M. A. Bigg. 1980. *Further analysis of pelagic fur seal data collected by the United States and Canada during 1958–74, Part 1.* Unpublished Report, 23rd Annual Meeting, Standing Science Committee, North Pacific Fur Seal Committee, 7–11 April 1980, Moscow, USSR.

Kasuya, T. and S. Ohsumi. 1984. Further analysis of the Baird's beaked whale stock in the western North Pacific. *Report of the International Whaling Commission* 34: 587–595.

Kawamura, A. 1975. Whale distribution in Bering Sea and northern North Pacific in the summer of 1974: results of a visual sighting study aboard the University of Hokkaido training vessel OSHORO MARU. *Bulletin of Japan Society of Fisheries and Oceanography* 26: 120–128. In Japanese. (Translated by R. Brownell, U.S. Fisheries and Wildlife Services, Washington, D.C., 9 pp.)

Kelly, B. P. 1988a. Bearded seal *(Erignathus barbatus),* in *Selected marine mammals of Alaska: species accounts with research and management recommendations.* J. W. Lentfer, ed. Washington, DC: Marine Mammal Commission, pp. 77–94.

———. 1988b. Ribbon seal *(Phoca fasciata),* in *Selected marine mammals of Alaska: species accounts with research and management recommendations.* J. W. Lentfer, ed. Washington, DC: Marine Mammal Commission, pp. 97–106.

———. 1988c. Ringed seal *(Phoca hispida), Selected marine mammals of Alaska: species accounts with research and management recommendations.* J. W. Lentfer, ed. Washington, DC: Marine Mammal Commission, pp. 59–75.

Kenyon, K. W. 1962. History of the Steller sea lion at the Pribilof Islands, Alaska. *Journal of Mammalogy* 43(1): 68–75.

———. 1969. The sea otter in the eastern Pacific Ocean. *North American Fauna* 68: 1–352.

———. 1981. Sea otter *Enhydra lutris,* (Linnaeus, 1758), in *Handbook of marine mammals, Volume 1: The walrus, sea lions, fur seals and sea otter.* S. H. Ridgway and R. J. Harrison, eds. London: Academic Press, pp. 209–223.

Lander, R. H. and H. Kajimura. 1982. *Status of northern fur seals, in Mammals in the seas, Volume 4: Small cetaceans, seals, sirenians and otters.* Advisory Committee of Marine Resources Research, FAO Fisheries Service 5. Food and Agriculture Organization, United Nations, Rome, pp. 319–345.

Leatherwood, S., A. E. Bowles, and R. R. Reeves. 1983. Endangered whales of the eastern Bering Sea and Shelikof Strait, Alaska; results of aerial surveys, April 1982 through April 1983

with notes on other marine mammals seen. U.S. Department of Commerce, NOAA, *OCSEAP Final Report* 42(1986): 147–490.

LeDuc, R. G., W. L. Perryman, J. W. Gilpatrick Jr., J. Hyde, C. Stinchcomb, J. V. Carretta, and R. L. Brownell, Jr. 2001. A note on recent surveys for right whales in the southeastern Bering Sea. *Journal of Cetacean Research Management* (Special Issue) 2: 287–289.

Lensink, C. J. 1961. *Status report: beluga studies.* Juneau: Alaska Department of Fish and Game.

Ljungblad, D. K., S. E. Moore, J. T. Clarke, and J. C. Bennett. 1988. Distribution, abundance, behavior and bioacoustics of endangered whales in the western Beaufort and northeastern Chukchi seas, 1979–1987. *Technical Report (1111).* San Diego, CA: Naval Oceans Systems Center.

Loughlin, T. R. and M. A. Perez. 1985. *Mesoplodon stejnergeri. Mammalian Species* 250, pp. 1–6.

Loughlin, T. R., D. J. Rugh, and C. H. Fiscus. 1984. Northern sea lion distribution and abundance: 1956–1980. *Journal of Wildlife Management* 48(3): 729–740.

Loughlin, T. R., I. N. Sukhanova, E. H. Sinclair, and R. C. Ferrero. 1999. Summary of biology and ecosystem dynamics in the Bering Sea, in *Dynamics of the Bering Sea.* T. R. Loughlin and K. Ohtani, eds. Fairbanks: University of Alaska Sea Grant, pp. 387–407.

Loughlin, T. R. I. N. Sukhanova, E. H. Sinclair and R. C. Ferrero. 1999. Summary of biology and ecosystem dynamics in the Bering Sea, p. 387-407. In T. R. Loughlin and K. Ohtani (editors), *Dynamics of the Bering Sea.* University of Alaska Sea Grant, Fairbanks, Alaska.

Lowry, L. F. 1985a. The spotted seal *(Phoca fasciata).* Alaska Department of Fish and Game, *Game Technical Bulletin* 7: 89–96.

———. 1985b. The ribbon seal *(Phoca fasciata).* Alaska Department of Fish and Game, *Game Technical Bulletin* 7: 71–78.

Lowry, L. F., V. N. Burkanov, K. J. Frost, M. A. Simpkins, R. Davis, D. P. DeMaster, R. Suydam, and A. Springer. 2000. Habitat use and habitat selection by spotted seals *(Phoca largha)* in the Bering Sea. *Canadian Journal of Zoology* 78: 1959–1971.

Lowry, L. F. and K. J. Frost. 1981. Feeding and trophic relationships of phocid seals and walruses in the eastern Bering Sea, in *The eastern Bering Sea shelf: Oceanography and resources, Volume 2.* D. W. Hood and J. A. Calder, eds. Seattle: University of Washington Press, pp. 387–407.

Lowry, L. F., K. J. Frost, D. G. Calkins, G. L. Swartzman, and S. Hills. 1982. *Feeding habits, food requirements, and status of Bering Sea marine mammals.* North Pacific Fishery Management Council Document No. 19. North Pacific Fisheries Management Council, P. O. Box 103136, Anchorage, AK 99510.

Lowry, L. F., K. J. Frost, R. Davis, D. P. DeMaster, and R. S. Suydam. 1998. Movements and behavior of satellite-tagged spotted seals *(Phoca largha)* in the Bering and Chukchi seas. *Polar Biology* 19: 221–230.

Masaki, Y. 1976. Biological studies on the North Pacific sei whale. *Bulletin of Far Seas Fisheries Research Laboratory (Shimizu)* 14: 1–104.

McAlister, W. B. 1981. *Estimates of fish consumption by marine mammals in the eastern Bering Sea and Aleutian Island area.* Unpublished Report.

Mizroch, S. A., D. W. Rice, and J. M. Breiwick. 1984a. The blue whale, *Balaenoptera musculus. Marine Fisheries Review* 46(4): 1–100.

———. 1984b. The sei whale, *Balaenoptera borealis. Marine Fisheries Review* 46(4): 1–100.

Moore, S. E. and R. R. Reeves. 1993. Distribution and movement, in *The bowhead whale*. J. J. Burns, J. J. Montague, and C. J. Cowles, eds. Society of Marine Mammalogy, Special Publication No. 2, pp. 313–386.

Moore, S. E., J. M. Waite, N. A. Friday, and T. Honkaleto. 2002. Cetacean distribution and relative abundance on the central-eastern and the southeastern Bering Sea shelf with reference to oceanographic domains. *Progress in Oceanography* 55: 249–261.

Nasu, K. 1974. Movement of baleen whales in relation to hydrographic conditions in the northern part of the North Pacific Ocean and the Bering Sea, in *Oceanography of the Bering Sea with emphasis on renewable resources*. D. W. Hood and E. J. Kelley, eds. Fairbanks: Institute of Marine Science, University of Alaska, pp. 345–361.

National Research Council. 1996. *The Bering Sea ecosystem*. Washington, DC: National Academies Press.

———. 2003. *The decline of the Steller sea lion in Alaskan waters: untangling food webs and fishing nets*. Washington, DC: National Academies Press.

Nelson, R. R., J. J. Burns, and K. J. Frost. 1985. The bearded seal (*Erignathus barbatus*). Alaska Department of Fish and Game, *Game Technical Bulletin* 7: 55–63.

Nikulin, P. G. 1946. Distribution of cetaceans in seas surrounding the Chukchi Peninsula. *Izv. TINRO* 22: 255–257. In Russian. (Translated by U.S. Naval Oceanography Office, Washington, D.C., Traslation No. 428, 1969.)

Ohsumi, S. 1966. Sexual segregation of the sperm whale in the North Pacific. *Scientific Report of the Whales Research Institute* 20: 1–16.

Ohsumi, S. 1991. A review on population studies of the North Pacific minke whale stocks. Paper SC/43/Mi26 presented to the International Whaling Commission Scientific Committee. Omura, H. 1955. Whales in the northern part of the North Pacific. *Norsk Hvalfangst-Tidende* 44(6): 323–345.

Ohsumi, S. and Y. Fukuda. 1974. *Revised sperm whale population model and its application to the North Pacific sperm whale*. Unpublished document. Submitted to the International Whaling Commission (SC/25/8).

Ohsumi, S. and Y. Masaki. 1977. Stocks and trends of abundance of the sperm whale in the North Pacific. *Report of the International Whaling Commission* 27: 167–175.

Ohsumi, S., Y. Masaki., and S. Wada. 1977. Seasonal distribution of sperm whales sighted by scouting boats in the North Pacific and Southern Hemisphere. *Report of the International Whaling Commission* 27: 308–323.

Ohsumi, S. and S. Wada. 1974. Status of whale stocks in the North Pacific, 1972. *Report of the International Whaling Commission* 24: 114–126.

Omura, H. 1955. Whales in the northern part of the North Pacific. *Nor. Hvalfangst-Tidende* 44(6):323–345.

Osmek, S., P. E. Rosel, A. E. Dizon, and R. L. DeLong. 1994. Harbor porpoise, Phocoena phocoena, population assessment studies for Oregon and Washington in 1993. Annual report to the MMPA Assessment Program, Office of Protected Resources, NMFS, NOAA, 1335 East-West Highway, Silver Spring, MD 20910.

Perez, M. A. 1990. Review of marine mammal and prey population information for Bering Sea ecosystem studies. U.S. Department of Commerce NOAA Technical Memorandum NMFS/NWC-186.

Perez, M. A., and W. B. McAlister. 1993. Estimates of food consumption by marine mammals in the Eastern Bering Sea. U.S. Department of Commerce, NOAA Technical. Memorandum NMFS-AFSC-14.

Pfister, B. 2004. Computations of historic and current biomass estimates of marine mammals in the Bering Sea. Alaska Fisheries Science Center Processed Report 2004-05. Seattle: National Marine Fisheries Service, NOAA. Available online at http://nmml.afsc.noaa.gov/library/publish.htm.

Pike, G. C. 1962. Migration and feeding of the gray whale (*Eschrichtius gibbosus*). *Journal of the Fisheries Research Board of Canada* 19(5): 815–38.

Pitcher, K. W. 1984. The harbor seal (*Phoca vitulina richardsi*), in *Marine mammal species accounts*. J. J. Burns, ed. Wildlife Technical Bulletin No. 7. Juneau: Alaska Department of Fish and Game, pp. 1–4.

Quakenbush, L. T. 1988. Spotted seal (*Phoca largha*), in *Selected marine mammals of Alaska: species accounts with research and management recommendations*. J. W. Lentfer, ed. Washington, DC: Marine Mammal Commission, pp. 107–124.

Rice, D. W. 1974. Whales and whale research in the eastern North Pacific, in *The whale problem: a status report*. W. E. Schevill, ed. Cambridge, MA: Harvard University Press, pp. 170–195.

———. 1986. Beaked whales, in *Marine mammals of the eastern North Pacific and Arctic waters*. D. Haley, ed. Seattle: Pacific Search Press, pp. 102–109.

Rice, D. W. and A. A. Wolman. 1971. *The life history and ecology of the gray whale (Eschrichtius robustus)*. American Society of Mammalogists, Special Publication No. 3. Lawrence, KS: Allen Press.

Roman, J. and S. R. Palumbi. 2003. Whales before whaling in the North Atlantic. *Science* 301: 508–510.

Rosel, P. E., A. E. Dizon, and M. G. Haygood. 1995. Variability of the mitochondrial control region in populations of the harbour porpoise, *Phocoena phocoena*, on inter-oceanic and regional scales. *Canadian Journal of Fisheries and Aquatic Science* 52: 1210–1219.

Rotterman, L. M. and T. Simon-Jackson. 1988. Sea otter (*Enhydra lutris*), in *Selected marine mammals of Alaska: species accounts with research and management recommendations*. J.W. Lentfer, ed. Washington, DC: Marine Mammal Commission, pp. 237–275.

Rugh, D. J., J. M. Breiwick, R. C. Hobbs, and J. A. Lerczak. 2002. *A preliminary estimate of abundance of the eastern North Pacific stock of gray whales in 2001 and 2002*. Unpublished document. Submitted to International Whaling Commission (SC/54/BRG6).

Rugh, D. J., K. E. W. Shelden, and D. E. Withrow. 1995. Spotted seals in Alaska 1992–93. *Annual report to the MMPA Assessment Program*, Office of Protected Resources, NMFS, NOAA.

Scarff, J. E. 1991. Historic distribution and abundance of the right whale, *Eubalaena glacialis*, in the North Pacific, Bering Sea, Sea of Okhotsk and Sea of Japan from the Maury Whale Charts. *Report of the International Whaling Commission* 41: 467–487.

———. 2001. Preliminary estimates of whaling-induced mortality in the 19th century North Pacific right whale (*Eubalaena japonicus*) fishery, adjusting for struck-but-lost whales and non-American whaling. *Journal of Cetacean Research and Management* Special Issue 2: 261–268.

Scheffer, V. B., and J. W. Slipp. 1944. The harbor seal in Washington state. *American Midlands Naturalist* 32: 373–416.

Seaman, G. A. and J. J. Burns. 1981. Preliminary results of recent studies of belukhas in Alaskan waters. *Report of the International Whaling Commission* 31: 567–574.

Seaman, G. A., K. J. Frost, and L. F. Lowry. 1985. Distribution, abundance, and movements of belukha whales in western and northern Alaska. Draft final report. Prepared for U.S. Department of Commerce, NOAA, National Ocean Service, Anchorage, Alaska. Fairbanks: Alaska Department of Fish and Game.

Sease, J. L. 1992. Status review: harbor seals (*Phoca vitulina*) in Alaska. NMFS-AFSC Processed Report 92-15, Alaska Fisheries Science Center, National Marine Fisheries Service, NOAA.

Sease, J. L. and D. G. Chapman. 1988. Pacific walrus *Odobenus rosmarus divergens*, in *Selected marine mammals of Alaska: species accounts with research and management recommendations*. J. W. Lentfer, ed. Washington, DC: Marine Mammal Commission, pp. 17–38.

Sease, J. L. and C. J. Gudmundson. 2002. Aerial and land-based surveys of Steller sea lions (*Eumetopias jubatus*) from the western stock in Alaska, June and July 2001 and 2002. U. S. Department of Commerce, NOAA Technical Memorandum NMFS/AFSC-131.

Shelden, K. E. W. and D. J. Rugh. 1996. The Bowhead Whale, *Balaena mysticetus*: its historic and current status. *Marine Fisheries Review* 57(3–4): 1–20.

Sherrod, G. 1982. *Eskimo Walrus Commission's 1981 research report: the harvest and use of marine mammals in fifteen Eskimo communities*. Report prepared for Eskimo Walrus Commission Subsistence Section. Juneau: Alaska Department of Fish and Game.

Shustov, A. P. 1969. Relative indices and possible causes of mortality of Bering Sea ribbon seals, in *Marine mammals (Morskikh mlekopitaiushchie)*. V. A. Arsenyev, V. A. Zenkovich, and K. K. Chapskii, eds. Moscow: Nauka (Izdatel'stvo "Nauka", Moskva 1969), pp. 83–92. (Translation by F. Fay.)

Sobolevsky, Y. I. and O. A. Mathisen. 1996. Distribution, abundance, and trophic relationships of Bering Sea cetaceans, in *Ecology of the Bering Sea: a review of Russian literature*. O. A. Mathisen and Kenneth O. Coyle, eds. Alaska Sea Grant College Program Report No. 96-01. Fairbanks: University of Alaska, pp. 265–275.

Springer, A. M., J. A. Estes, G. B. van Vliet, T. M. Williams, D. F. Doak, E. M. Danner, K. A. Forney, and B. Pfister 2003. Sequential megafaunal collapse in the North Pacific Ocean: an ongoing legacy of industrial whaling? *Proceedings of the National Academy of Sciences of the United States of America* 100(21): 12223–12228.

Stickney, A. 1982. Memorandum from A. Stickney to R. J. Wolfe, Alaska Department of Fish and Game, Division of Subsistence, Bethel, September 24, 1982.

Tomilin, A. G. 1967. *Mammals of the USSR and adjacent countries. Mammals of eastern Europe and northern Asia. Vol. IX. Cetacea*. A. G. Tomlinin, ed. Jerusalem, Israel Program for Scientific Translations.

Trites, A. W., P. A. Livingston, S. Mackinson, M. C. Vasconcellos, A. M. Springer, and D. Pauly. 1999. *Ecosystem change and the decline of marine mammals in the eastern Bering Sea: testing the ecosystem shift and commercial whaling hypothesis*. Fisheries Centre Research Report 1999, 7(1). Vancouver: University of British Columbia.

Trites, A. W. and D. Pauly. 1998. Estimating mean body masses of marine mammals from maximum body lengths. *Canadian Journal of Zoology* 76: 886–896.

Turnock, B. J., S. T. Buckland, and G. C. Boucher. 1995. Population abundance of Dall's porpoise *(Phocoenoides dalli)* in the western

North Pacific Ocean. *Report of the International Whaling Commission* (Special Issue) 16: 381–397.

Wada, S. 1980. Japanese whaling and whale sighting in the North Pacific 1978 season. *Report of the International Whaling Commission* 30: 415–424.

———. 1981. Japanese whaling and whale sighting in the North Pacific 1979 season. *Report of the International Whaling Commission* 31: 783–792.

Wade, P. R. 2002. A Bayesian stock assessment of the eastern Pacific gray whale using abundance and harvest data from 1967 to 1996. *Journal of Cetacean Research Management* 4(1): 85–98

Wade, P. R. and D. P. DeMaster. 1996. A Bayesian analysis of eastern Pacific gray whale population dynamics. Paper SC/48/AS3 presented to the International Whaling Commission Scientific Committee.

Wade, P. R., L. G. Barrett-Lennard, N. A. Black, V. N. Burkanov, A. M. Burdin, J. Calambokidis, S. Cerchio, M. E. Dalheim, J. K. B. Ford, N. A. Friday, L. W. Fritz, J. K. Jacobsen, T. R. Loughlin, C. O. Matkin, D. R. Matkin, S. M. McCluskey, A. V. Mehta, S. A. Mizroch, M. M. Muto, D. W. Rice, R. J. Small, J. M. Straley, G. R. VanBlaricom, and P. J. Clapham. In press. Killer whales and marine mammal trends in the North Pacific—a re-examination of evidence for sequential megafauna collapse and the prey switching hypothesis. *Marine Mammal Science*.

Wahl, T. R. 1979. Observation of Dall's porpoise in the northwestern Pacific Ocean and Bering Sea in June 1975. *Murrelet* 60: 108–110.

Webb, R. L. 1988. *On the Northwest: commercial whaling in the Pacific Northwest 1790–1967*. Vancouver: University of British Columbia Press.

Withrow, D. E., and T. R. Loughlin. 1996a. Haulout behavior and a correction factor estimate for the proportion of harbor seals missed during molt census surveys near Cordova, Alaska. Annual report to the MMPA Assessment Program, Office of Protected Resources, NMFS, NOAA.

———. 1996b. Abundance and distribution of harbor seals (*Phoca vitulina richardsi*) along the north side of the Alaska Peninsula and Bristol Bay during 1995. *Annual report to the MMPA Assessment Program*, Office of Protected Resources, NMFS, NOAA.

Wolfe, R. J. 1982. Letter from R. J. Wolfe, Alaska Dep. Fish and Game, Div. Subsistence, Bethel, to J. J. Burns, Alaska Dep. Fish and Game, Dive. Game, Fairbanks, 24 September 1982.

Wolman, A. A. 1985. Gray whale *Eschrichtius robustus* (Lilljeborg, 1861), in Handbook of marine mammals, Volume 3: The Sirenians and baleen whales. S. H. Ridgway and R. Harrison, eds. London: Academic Press, pp. 67–90.

Yablokov, A. V. and V. A. Zemsky. 2000. *Soviet whaling data (1949–1979)*. Moscow: Center for Russian Environmental Policy.

York, A. E. and J. R. Hartley. 1981. Pup production following harvest of female northern fur seals. *Canadian Journal of Fisheries and Aquatic Sciences* 38: 84–90.

Zenkovich, B. A. 1955. The migration of whales, whale fishing in the waters of the Soviet Far East, in *Kitoiony Promysel Sovetskogo Soyuza, VNIRO, Part I*. S. E. Kleinenberg and T. I. Makarova, eds., pp. 51–68. In Russian. (Translation by Israel Program Science Translation for U.S. Department of the Interior and National Science Foundation, 1968.)

Appendix: Parameters Used in the Biomass Calculations

Species[a]	Stock/Population	Historic Abundance	Current Abundance	Body Weight (metric tons)[b]	Percentage of Population in Study Area	# of Days Spent in Study Area	Sources
Bowhead whale (*Balaena mysticetus*)	Western Arctic	15,000 (range 12,000–18,000)	10,020 (95% CI = 7,800–12,900; SE = 1,290)	31.1*	100%	November to April; 181	Eberhardt and Breiwick 1992; George et al. 2003; Bogoslovskaya et al. 1982; Brueggeman 1982; Ljungblad et al. 1988; Moore and Reeves 1993; Braham et al. 1984a; Braham et al. 1980; Brueggeman et al. 1987; Bessonov et al. 1990; Shelden and Rugh 1996
Northern right whale (*Eubalaena japonicus*)	Eastern North Pacific	31,750	50	23.4*	33% historic[c]; 100% current	April to September; 183	Scarff 2001; Clapham et al. 1999; LeDuc et al. 2001; Clapham et al. 2001; Scarff 1991; Berzin and Rovnin 1966
Fin whale (*Balaenoptera physalus*)**	Northeast Pacific	43,500 (range 42,000–45,000)	4,051	55.6	25% historic; 100% current	Late June to October; 138	Ohsumi and Wada 1974; Moore et al. 2002; Nikulin 1946; Berzin and Rovnin 1966; Nasu 1974
Humpback whale (*Megaptera novaeangliae*)**	Central North Pacific	9,060	4,005 (CV = 0.095)	30.4	100%	May to October; 184	Rice 1974; Calambokidis et al. 1997; Tomilin 1967; Zenkovich 1955; Berzin and Rovnin 1966
Minke whale (*Balaenoptera acutorostrata*)**	Alaska	1,813	1,813	6.6	100%	May to July; 92	Moore et al. 2002; Tomilin 1957; Horwood 1990; Ohsumi 1991
Gray whale (*Eschrictius robustus*)	Eastern North Pacific	28,240 (range 24,640–31,840)	17,414 (95% CI = 14,322–21,174)	15.4*	90%	May to October; 184 days	Wade 2002; Rugh et al. 2002; Pike 1962; Rice and Wolman 1971; Darling 1984; Gosho et al. 1999; Wolman 1985
Sperm whale (*Physeter macrocephalus*)**	North Pacific	195,000	15,000	26.9[d]	100%	April to September; 183 days	Ohsumi and Fukuda 1974; Perez 1990[e]; Tomilin 1967; Berzin and Rovnin 1966; Ohsumi and Masaki 1977; Ohsumi et al. 1977; Ohsumi 1966; Gosho et al. 1984
Harbor porpoise (*Phocoena phocoena*)**	Bering Sea	50,127	50,127	0.03	100%	365 days	Moore et al. 2002; Hobbs and Waite, forthcoming data; Bjøre et al. 1994; Gaskin 1984; Jones 1984; IWC 1991; Osmek et al. 1994; Rosel et al. 1995; Calambokidis and Barlow 1991

Species[a]	Stock/Population	Historic Abundance	Current Abundance	Body Weight (metric tons)[b]	Percentage of Population in Study Area	# of Days Spent in Study Area	Sources
Dall's porpoise (Phocoenoides dalli)**	Bering Sea/Aleutian Islands	90,141	90,141	0.06*	100%	365 days	Moore et al. 2002; Hobbs and Lerczak 1993; Kawamura 1975; Wahl 1979; Fiscus 1980; Buckland et al. 1993a
Beluga whale (Delphinapterus leucas)**	Beaufort	39,258	39,258	0.31[f]	100%	November to April; 181 days	Duval 1993; Harwood et al. 1996; Hazard 1988; Harrison and Hall 1978; Leatherwood et al. 1983; Fay, personal communication, as cited in Lensink 1961; Seaman et al. 1985
Beluga whale (Delphinapterus leucas)**	Eastern Chukchi	3,710	3,710	0.31	100%	October to April, 212 days	Frost et al. 1993; Frost and Lowry 1995; Brodie 1971; Hazard 1988
Beluga whale** (Delphinapterus leucas)	Eastern Bering Sea	18,142 (CV = 0.24)	18,142 (CV = 0.24)	0.31	100%	365 days	R.C. Hobbs, unpublished data, as cited in Angliss and Lodge 2004; Hazard 1988
Beluga whale** (Delphinapterus leucas)	Bristol Bay	1,642	1,642	0.31	100%	365 days	Frost et al. 2002; Hazard 1988
Northern fur seal (Callorhinus ursinus)	Eastern Pacific	3,000,000	888,120	0.03	May to October; 50% November to April; 5%g	May to October; 184 days; November to April 181 days	Lander and Kajimura 1982; Angliss and Lodge 2004; Bigg 1986; Perez and McAlister 1993
Steller sea lion** (Eumetopius jubatus)	Western United States	126,900	35,194[h]	0.2	100%	365 days	Loughlin et al. 1984, as cited in Hoover 1988a; Sease and Gudmundson 2002; Lowry et al. 1982; Trites et al. 1999
Harbor seal (Phoca vitulina richardii)	Bering Sea	30,000[j]	13,312 (CV = 0.062)[i]	0.06[k]	100%	365 days	Sease 1992; Withrow and Loughlin 1996a,b; DeMaster 1996; ADF&G 1973; Scheffer and Slipp 1944; Fisher 1952; Bigg 1969, 1981; and Perez 1990
Harbor seal** (Phoca vitulina richardii)	Gulf of Alaska	140,000	22,376	0.06	100%	365 days	Sease 1992; Boveng et al. 2003; Hoover 1988b; Angliss and Lodge 2004; ADF&G 1973; Scheffer and Slipp 1944; Fisher 1952; Bigg 1969, 1981; and Perez 1990
Spotted seal (Phoca largha)	Alaska	140,000[m]	140,000	0.04*	100%	November to June; 212 days	Perez 1990; Lowry et al. 1998, 2000

APPENDIX (CONTINUED)

Species	Location						References
Bearded seal[n] (Erignathus barbatus)	Alaska	150,000[o]	150,000	0.20	100% winter <5% summer	November to April; 181 days May to October; 184 days	Burns 1981; Lowry and Frost 1981; Nelson et al. 1985; Braham et al. 1984b; Perez 1990
Ribbon seal[n] (Phoca fasciata)	Alaska	85,000	105,000	0.07*	100%	365 days	Shustov 1969, as cited in Fedoseev 1973; Lowry 1985b
Ringed seal[n] (Phoca hispida)	Alaska	600,000[p]	600,000	0.04	100%	November to April; 181 days	Perez 1990; Braham et al. 1984b
Pacific walrus** (Odobenus rosmarus divergens)	Alaska	225,000	201,039	0.59	100% winter 22% summer	December to April; 151 days May to November; 214 days	Fay 1982; Gilbert et al. 1992; Fay 1981; Sease and Chapman 1988; Fay et al. 1985; Fay et al. 1997
Sea otter** (Enhydra lutris)	Southwest Alaska	118,000	41,474	0.01*	100%	365 days	Johnson 1982, as cited in Rotterman and Simon-Jackson 1988; Doroff et al. 2003; Angliss and Lodge 2002; Kenyon 1969, 1981

a Caveats and assumptions for biomass calculations are elaborated upon in the text for species that are marked with a double asterisk (**).

b All body weight estimates were taken from Trites and Pauly 1998. Weights marked with an asterisk (*) were calculated without incorporating age structure of the population.

c It was assumed that the historic population of North Pacific right whales was evenly split between the Sea of Okhotsk, the Bering Sea, and the Gulf of Alaska (Scarff 1991).

d Body weight is for male sperm whales only.

e We adopted the estimate that Perez (1990) generated using the following sources: Ohsumi and Masaki 1977; Gosho et al. 1984; Ohsumi 1966; Ohsumi et al. 1977; Wada 1980, 1981.

f Trites and Pauly (1998) reported an average body weight of 0.31 tons for the beluga, which is used here for all four stocks.

g There is variability in the interannual arrival and departure dates of different age and sex classes of northern fur seals to the Pribilof Islands (Kajimura et al. 1980; McAlister 1981; Gentry 1981; Bigg 1986).

h This is a minimum abundance based on pup count data from 2001 and 2002 and non-pup count data from 2002.

i The historic abundance estimates for both stocks of harbor seals gives an indication of the overall magnitude of the population, not an exact estimate of population size (Pitcher 1984). They are likely negatively biased.

j The Alaska SRG considered the correction factor used to calculate this abundance estimate conservative (DeMaster 1996).

k Trites and Pauly (1998) reported an average body weight of 0.06 tons for the harbor seal, which is used here for both stocks.

l There is no reliable historic or current abundance estimate for the spotted seal (Rugh et al. 1995) in the Bering Sea/Aleutian Islands region. However, they are present in the Bering Sea ecosystem (Braham et al. 1984b) and should therefore be included in the analysis.

m Perez's method was used for calculating abundance (Lowry and Frost 1981; Lowry 1985a; Burns 1986; Braham et al. 1984b, as cited in Perez 1990).

n No reliable historic or current abundance estimates exist for the bearded, ribbon, or ringed seal (Angliss and Lodge 2004). However, they are present in the Bering Sea ecosystem and should therefore be included in the analysis (Braham et al. 1984b). For the purpose of this analysis, historic and current abundance estimates for the bearded seal and ringed seal were considered equal.

o We adopted the estimate that Perez (1990) generated using the following sources: Burns 1981; Lowry and Frost 1981; Nelson et al. 1985; Braham et al. 1984b.

p We adopted the estimate that Perez (1990) generated using the following sources: Lowry and Frost 1981; Lowry et al. 1982; Frost 1985; Brahm et al. 1984b.

ELEVEN

Industrial Whaling in the North Pacific Ocean 1952–1978

Spatial Patterns of Harvest and Decline

ERIC M. DANNER, MATTHEW J. KAUFFMAN, AND
ROBERT L. BROWNELL, JR.

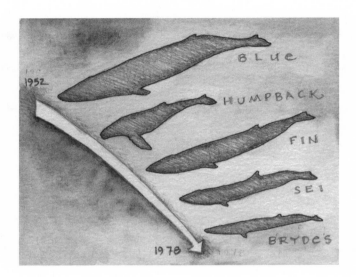

For many species of conservation concern, efficient management, conservation, and harvesting strategies require information about population spatial structure. Ecologists are increasingly aware that species rarely exist as large, well-mixed populations, but rather as multiple subpopulations separated by geography and connected by various levels of dispersal. Compared to more easily studied marine mammals, relatively little is known about the population spatial structure of commercially exploited whale species (Hoelzel 1991). Accurate characterization of large regional populations as sets of smaller subpopulations can improve the accuracy of harvesting quotas (Jonzen et al. 2001), population monitoring efforts (Shelden et al. 2001), spatial location of reserves (Hooker et al. 1999), and the maintenance of genetic diversity (Hoelzel 1998). More central to the topic of this volume is that our ability to detect the impacts on ocean food webs resulting from the removal of the great whales may be enhanced by accounting for the spatial dimension of whale exploitation. For example, short-term local harvesting could lead to dramatic local declines of an exploited whale species existing as a small subpopulation with high affinity for a particular (spatially constrained) summer feeding ground and low interchange of individuals with nearby subpopulations.

In such a case, local declines could occur long before regional trends are detected. Thus, the perturbation to the system in terms of whale removal might be more pronounced at local spatial scales when compared with regional impacts, and perhaps more important, community-level effects of local whale removal could be manifested years before regional stock assessments alerted managers to the impact on whale stocks.

Such spatial population structure seems to exist for humpback whales (*Megaptera novaeangliae*) in the North Atlantic. These whales winter in the West Indies, but are thought to segregate into four subpopulations (Gulf of Maine, Newfoundland, Greenland, and Iceland; Baker et al. 1990, 1993) during migration to summer feeding grounds. As with most whaling efforts, humpbacks are hunted on their feeding grounds. Thus, targeted exploitation at one of the summer feeding grounds could dramatically reduce a local population, even though such spatially constrained hunting may have much less noticeable effects on regional population numbers. Also consistent with this type of migratory pattern is the expectation that the whaling fleet will have to continually shift its focus geographically to new summer feeding grounds as local stocks are depleted. Thus, if strong spatial structure exists, regional catch limits might not adequately protect local whale stocks.

In his 1994 review, Hoelzel (reviewing Palsbøll 1990 and Bakke et al. 1996) suggested that minke whales *(Balaenoptera acutorostrata)* may have a contrasting type of population structure, in which separate breeding populations may mix in shared summer feeding grounds in the North Atlantic. If these types of migratory patterns exist, then whalers could be targeting multiple subpopulations that forage in the same local area during summer, but are spatially separated on their breeding grounds during the winter. In such a case, the impact of whale hunting should have a more diffuse effect on breeding subpopulations. In contrast to the mobile fleet behavior just described in which distinct feeding subpopulations exist, a whaling fleet hunting on a mixed feeding ground could successfully return to the same local area over many years because such stocks, derived from several breeding populations, should be more resilient to depletion.

In this chapter, we investigate population spatial structure of exploited whale species in the North Pacific Ocean during the period of industrial whaling, 1952–1979. Little detailed information, such as fidelity to feeding or breeding grounds, is known about the whale species that were hunted in this fishery. Our objective was to gain insights into the spatial dimensions of the North Pacific whale harvest and the whale populations that inhabited this area by analyzing the spatial pattern of local harvest and decline.

The North Pacific is also the regional context for one of the largest and most dramatic shifts in marine food webs ever suggested (Springer et al. 2003). Springer et al. (2003) hypothesize that the removal of great whales out of the North Pacific triggered a behavioral switch in killer whales *(Orcinus orca)*, whereby they shifted their diets to—and caused the decline of—four smaller marine mammal species (harbor seals, *Phoca vitulina*; fur seals, *Callorhinus ursinus*; Steller sea lions, *Eumetopias jubatus*; and sea otters, *Enhydra lutris*). In support of their theory, Springer et al. inferred depletion of local whales stocks from the movements of the whaling fleets out of the Bering Sea during the early 1960s. However, new regulations and a shift of whaling effort toward more southerly distributed species may also explain these geographic trends. Thus, a second objective of our work was to more formally assess the spatial and temporal patterns of whale catch records in the North Pacific and Bering Sea and to determine whether or not the timing of the whaling fleet movement out of these local areas are indicative of substantial whale stock depletion.

Methods

Our analyses were based entirely on the annual catch records in the International Whaling Commission (IWC) database for the North Pacific region (IWC 2002). These data include detailed information on each reported whale harvested, including species, length, position (latitude and longitude), date, stomach contents, sex, and expedition name (i.e., the

identity of the processor ship) and nationality. The database consists of 410,373 records, each corresponding to an individual whale taken during North Pacific harvests from 1924 to 2001. To simplify our analysis, we only used a subset of these data that met three criteria. First, we only included data from pelagic landings. Whaling expeditions in the North Pacific were generally either pelagic or shore-based operations. Because we were interested in the spatial patterns of the whale harvest, the limited geographic range of the shore-based operations could confound our results. Also, pelagic landings provided a better representation of how the fleet moved from year to year to most efficiently exploit the whale stocks. Pelagic whaling in the North Pacific was conducted almost exclusively by Japanese and Soviet fleets. Second, because it is known that some of the Soviet data has been falsified (Yablokov 1994), we further restricted our data set to include only Japanese pelagic landings. Third, we limited our analysis to baleen whales, excluding sperm whales *(Physeter macrocephalus)* from our analysis. Four species of baleen whales were harvested in appreciable numbers within the extent of our study: blue whales *(Balaenoptera musculus)*, humpback whales, fin whales *(B. physalus)*, sei whales *(B. borealis)*, and Bryde's whales *(B. brydei)*. Although significant numbers of sperm whales were harvested, pelagic whaling expeditions generally did not harvest baleen and sperm whales on the same days in order to avoid contaminating whale oil (R. Brownell, unpublished data). Therefore for days on which sperm whales comprised greater than 94% of the catch we excluded all whales from our analysis to prevent the sperm whale data from confounding our results (see calculation of effort later). These criteria produced a subset of 50,542 records, covering a period from 1952 to 1979, that was used for the remainder of the analysis.

Our analysis relies on three main components: *spatial grid coordinates* (subdivisions of the North Pacific region into smaller local areas), *catch* (the number of whales or amount of whale biomass harvested), and *effort* (the amount of time put toward the catch). The spatial grid provides a framework to compare the inherent differences in abundance between local areas, as well as how different local whale stocks responded to annual whale harvest within the North Pacific. The catch data provides information about the magnitude of whale harvest at both regional and local spatial scales. The quotient of the catch and effort provides the catch-per-unit-effort (CPUE), which we used as an index of whale stock abundance in both space and time. To facilitate this analysis, we built a Geographic Information System (GIS) database that mapped the spatial location of each whale harvested along with its associated attribute data (e.g., species identity, size, expedition).

Spatial Grid

To more accurately estimate the catch and effort values, and to establish a local spatial scale for analysis, we compartmentalized the data into distinct cells or geographic subsets. This was done by subdividing the region where harvesting occurred into 250 equally sized cells measuring 500 km on each side (250,000 km^2). Because the catch of particular species was highly variable across the region, measures of catch and effort were derived at the cell level to better reflect the dynamics of both the harvesting effort and stock abundance indices. In particular, this method allowed us to refine our measure of effort put toward a given species in a local area, providing a more accurate measure of the CPUE. Partitioning the region into cells also enabled us to make comparisons between regional and local trends in CPUE over time and provide insights into spatial population structure of the whale species exploited in the North Pacific.

Catch

Because of the variation in whale sizes, both within and between species, we present the catch data primarily in terms of biomass of whales landed, with a summary of the total numbers harvested. We believe that biomass is a more appropriate metric for comparison of the harvests of different species, and unless otherwise noted, all statistics such as catch, landings, and CPUE are presented in terms of tons of whale biomass harvested. Because biomass values were not included in the IWC database we estimated the biomass of each harvested whale using previously derived allometric relationships (Trites and Pauly 1998). We used the following equation: biomass = $c \times L^x$, where L is the length of the individual whale, and c and x are species-specific constants. The constant terms for each species were as follows: for blue, fin, and humpback whales, $c = 0.007996$ and $x = 2.9$; for sei whales, $c = 0.025763$ and $x = 2.43$; and for Bryde's whales, $c = 0.0126$ and $x = 2.76$. Allometric constants were obtained from McAlpine (1985) for fin, sperm, and sei whales and from Perrin et al. (2001) for Bryde's whales. Because independent data was not available for blue and humpback whales, the allometric equation for fin whales was also used for these two species. Biomass values were summed by species for each cell within each year.

Effort

We defined effort as the number of days each expedition vessel harvested whales during each year. Summing across each expedition (i.e., each processor ship, typically 2–3 in the Japanese fleet) produced the total number of expedition-days, which was used as the standard measure of effort in our analyses. CPUE was then calculated by dividing the catch by the number of expedition-days.

The partitioning of effort among species is problematic in a multispecies fishery because it cannot be assumed that all species are targeted equally. In the case of North Pacific whaling, there was a shifting of preference for individual species of baleen whales, with blue whales being the most preferred, followed by humpback, fin, sei, and Bryde's

whales. To a large extent, the catch data indicate that the Japanese whaling fleet adaptively expanded their harvest each year to include species of lower preference as stocks of high-preference species declined. This approach is consistent with optimal foraging theory, which predicts that consumers forage in a manner that maximizes their energy intake per unit time (Charnov 1976); this topic is further addressed with respect to whaling by Mangel and Wolf (Chapter 21 of this volume). Thus, before they became targeted, landings of less preferred species were part of an incidental catch, which was unrelated to annual levels of whaling effort. We attempted to minimize the effect of this error on our CPUE calculation by both spatially and temporally constraining our estimate of effort. The spatial constraint was accomplished in part by compartmentalizing the effort into cells within the spatial grid. This produced an effort value that was limited to (1) only the whale species that occurred within a cell and (2) the amount of effort (in expedition-days) that was apportioned within that cell. The temporal constraint was accomplished by determining the time period over which whalers were actively targeting each species.

For fin, sei, and Bryde's whales, most cells had an incidental level of catch (less than 20% of the annual harvest) for several years until these species were targeted by the whaling fleet. Once they became targeted, they often represented more than 50% of the annual catch in a given cell. To make our estimates of species-specific CPUE more reliable, we sought to include only the catch and effort put toward a species in each cell during the appropriate period of time when the effort was directed toward that species. This string of years, which we termed the Directed Effort Window (DEW), was unique for each species within each local (cell) area. The DEW for a particular species, within a particular cell, was determined by identifying the year in which the species first comprised more than 50% of the total annual catch. If this criterion was met, we assumed that whalers were actively targeting that species in that cell during that year. We further refined our measure of effort by including only cells where a minimum of 100 tons of biomass were harvested within that cell for that year. This criterion prevented small harvests from disproportionately influencing the results. Once the DEW began for a particular cell it remained in effect until whaling stopped for that species, most often due to a moratorium. For fin, sei, and Bryde's whales all effort estimates are calculated using the DEW. The DEW was not applied to estimates for blue and humpback whales; we assumed that these species were targeted consistently until the moratorium on their harvest.

After using these methods to estimate effort, we used regression analyses to investigate temporal and spatial trends in CPUE. Our analysis of CPUE trends was conducted at two spatial scales: (1) within each local cell area, and (2) over the entire region (i.e., the North Pacific and Bering Sea). At the local scale, we regressed CPUE on year

TABLE 11.1
Summary of Japanese Catch Statistics, 1952–1979

Species	Catch		Biomass		Area of Harvest (km²)	Regional CPUE Trend[a]		
	No. Harvested	%	Tons	%		First Year	Slope	P Value
Blue	977	2%	74,089	6%	5,500,000	1952	−2.59	<0.004
Humpback	469	0.9%	5,779	0.4%	4,000,000	1952	−0.46	<0.151
Fin	21,115	42%	830,558	62%	17,250,000	1955	−29.47	<0.001
Sei	25,279	50%	383,953	29%	16,500,000	1964	−17.27	<0.013
Bryde's	2,702	5%	38,779	3%	8,250,000	1974	−3.99	<0.393
Totals	50,542		1,333,159					

[a] Regional CPUE trend was calculated using linear regression of annual CPUE on year. The start year for fin, sei, and Bryde's whales was determined using the DEW method (see text).

for fin and sei whales using values of catch and effort from individual cells within our spatial grid. This provided a measure of the temporal trends in CPUE at the local level for these two species. At the regional scale, the catch and effort values for each individual species from all cells were summed across the entire region to generate a single regional CPUE value. This annual regional CPUE value was then regressed against year to test for temporal trends in this regional abundance index. Comparing CPUE trends at both spatial scales allowed us to compare estimates of mean CPUE and rates of decline from local cells with the regional values and draw inferences regarding population structure of these two species.

Results

Catch

Japanese pelagic expeditions reported harvesting 50,542 baleen whales (1,333,159 tons) from 1952 to 1979 (Table 11.1). The catch comprised blue, Bryde's, humpback, fin, and sei whales; the proportions of each species varied with time (Figure 11.1). Fin and sei whales made up the majority of the harvested biomass during most years, comprising as much as 91% (1961) and 73% (1968) of the annual catch, respectively. Although sei whales represented 50% of the total number of whales harvested over the entire period, they made up only 29% of the total biomass. In contrast, 42% of the total number of whales harvested were fin whales, which, due to their larger size, totaled 62% of the biomass harvested. Within the DEW for fin and sei whales, the harvested biomass of each species generally declined over time until moratoriums were declared on their take in 1975. Blue and humpback whales were less significant components of the catch, representing an average of less than 10% and less than 3% of the harvested biomass, respectively. Harvesting of Bryde's whales did not begin until 1971, when relatively few were taken. By 1976

Bryde's whales were the only baleen whale harvested until the end of pelagic commercial whaling in 1979.

The geographic distribution of the harvest varied considerably between species, covering as little as 4 million km² for humpback whales to as much as 17 million km² for fin whales (Table 11.1, Figure 11.2). The majority of the blue and humpback whales were harvested immediately south of the Aleutian Islands, whereas fin whales were harvested in the Bering Sea, around the Aleutian Islands, in the Gulf of Alaska, and south into the pelagic areas of the North Pacific (Figures 11.2a, 11.2b, and 11.2c). Sei whales were harvested almost entirely south of the Aleutians in the North Pacific and in the Gulf of Alaska (Figure 11.2d). Bryde's whales where harvested only in the pelagic waters of the central Pacific (Figure 11.2e). Substantial proportions of the total catch occurred in localized areas, particularly around the eastern Aleutian Islands (Figure 11.2f). In other areas, including large portions of the Bering Sea and North Pacific, there was little or no harvesting of baleen whales.

There was a general shift southward in the harvest over time as whalers switched from blue, humpback, and fin whales to sei whales, and later to Bryde's whales (Figure 11.1). During the first decade of the harvest the expeditions remained in the vicinity of the Aleutian Islands and the Bering Sea (Figures 11.3a and 11.3b). The expeditions then spread east to the Gulf of Alaska and south into the Pacific Ocean during the 1960s (Figures 11.3c and 11.3d). By the 1970s whalers had almost entirely abandoned the Bering Sea and Aleutian Islands region, and were harvesting in the pelagic waters of the north and central Pacific Ocean (Figures 11.3e and 11.3f). The number of whales harvested per km² had declined substantially by this time.

Catch Per Unit Effort

The regional CPUE declined over time for all species, although the trend was not statistically significant for

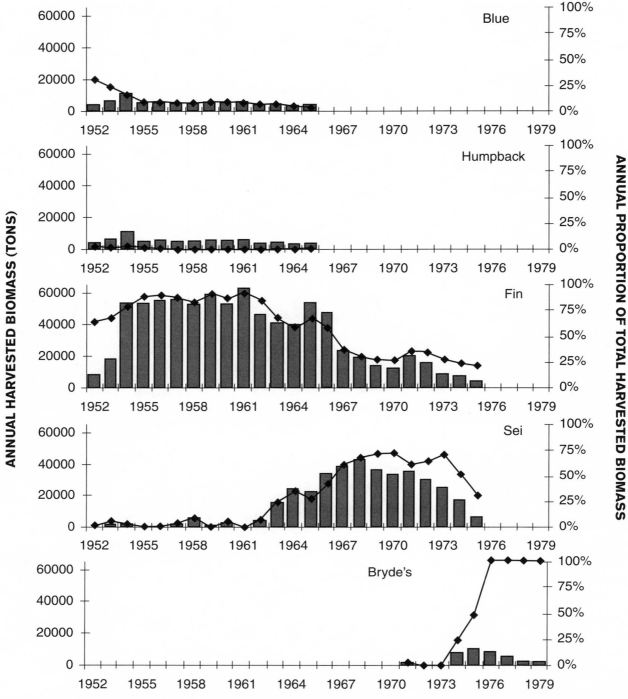

FIGURE 11.1. Biomass harvested (gray bars) and proportion of total harvest (black lines) of baleen whales for the North Pacific region. These data include days on which sperm whales were harvested.

humpback and Bryde's whales (Table 11.1). Interannual variation in CPUE declined over time for both fin and sei whales, perhaps as a consequence of the different habitats that the fleet exploited in the early years of these fisheries. On the local scale, CPUE values for fin and sei whales (the only species for which we evaluated local CPUE trends) varied considerably in space and time, and the majority of the cells did not have trends that were significantly different from zero (Figure 11.2, non-highlighted cells). The cells with significant declines in CPUE for fin whales all bordered the Aleutian Islands, with the exception of one cell in the western Gulf of Alaska (cell I3 in the western Aleutians had a significant *increase* in CPUE). The regression statistics (year range, slope, *P* value) for the cells with significant

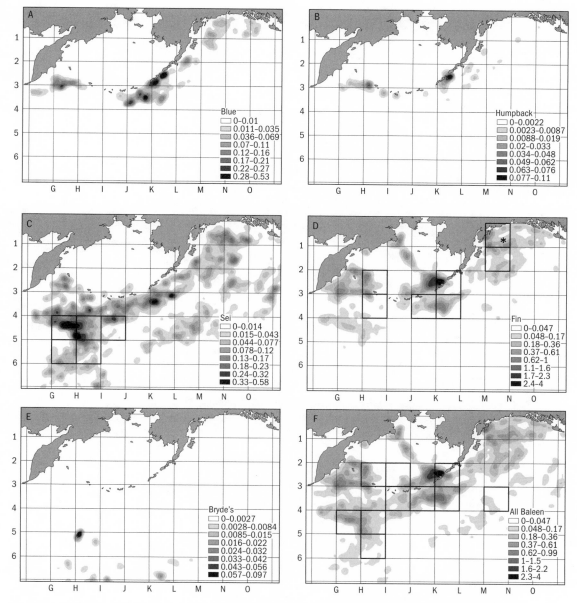

FIGURE 11.2. Density of biomass harvested (whales/km²) by species from 1952 to 1978. The spatial grid is denoted by light squares; cells with significant changes in CPUE over time are denoted by highlighted cell borders ($P < 0.05$, with the exception of cell N1 (D) [marked with an asterisk (*)], which was considered significant at $P = 0.053$). The map is centered on areas with highest density of whale harvest; some harvesting occurred in areas out of the map's range. Note that there is a different biomass scale for each species.

declines in fin whale CPUE are as follows: K4 (1957–1971, –21.33, 0.007), L3 (1954–1965, –48.44, 0.012), L4 (1957–1971, –47.44, 0.012), N1 (1961–1966, –114.35, 0.053), and N2 (1961–1966, –53.75, 0.038). All of the cells with significant declines in sei whale CPUE were located south of the central and western Aleutian Islands in the area of the highest concentration of the sei whale harvest. These local declines were observed in the following cells: H4 (1968–1974, –50.97, 0.033), I4 (1967–1972, –109.05, 0.012), J4 (1967–1971, –66.02, 0.025), H5 (1967–1974, –44.89, 0.001), and I5 (1967–1973, –35.53, 0.009). In most cases,

the rates of these local declines were faster than the rate of regional decline of fin and sei whale CPUE (Figure 11.4).

Discussion

Our analysis indicates that the interaction of multiple temporal and spatial trends produced the declines in raw harvest numbers and CPUE observed at local and regional levels. As CPUE is a measure of the time worked by a catcher vessel, it may be subject to various errors arising from changes in efficiency. This is particularly true in pelagic operations with a

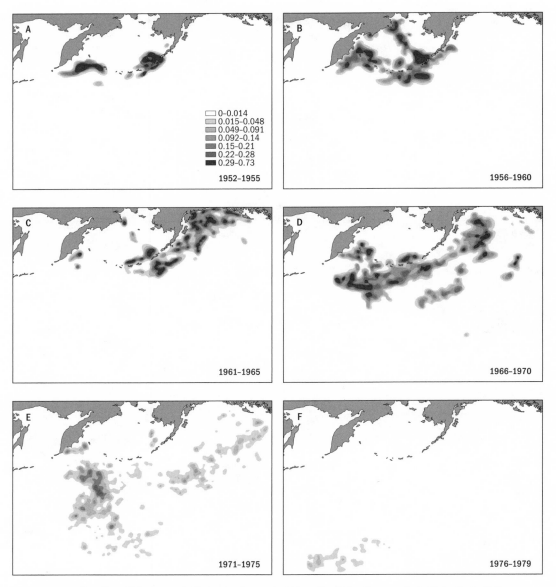

FIGURE 11.3. Density of biomass of all species harvested (tons/km²) by time period. Note that all time periods have the same density scale.

progressive increase in the size and power of the catcher (Allen 1980). Nonetheless, there has been a sequential exploitation of whale species of lesser and lesser value in the North Pacific (Figure 11.1), similar to patterns that have occurred in almost every ocean where commercial whaling has been conducted (Branch and Williams, Chapter 20 of this volume; Ballance et al., Chapter 17 of this volume). Historically, blue and humpback whales constitute a predominant share of the harvest when whaling is first initiated; however, in the case of post–World War II industrial whaling in the North Pacific, these two highly sought-after species had already been heavily exploited (Rice 1974). Regional whaling efforts from 1925 to World War II alone removed 1,016 blue and 1,799 humpback whales (104% and 386% of

their respective post–World War II catch), whereas fin, sei, and Bryde's whales were all harvested at levels below 10% of their post–World War II catch. Because of their reduced population sizes and consequent restrictions on their harvest, blue and humpback whale landings were less than 3% of the total landings over our study period (Table 11.1).

Due to these circumstances, whalers were beginning to switch to harvesting fin whales by the start of the pelagic fishery in 1952. By 1954, fin whales represented 78% of the total annual catch. Fin whales remained the dominant target species, averaging more than 85% percent of the annual landings, over the next 8 years. After reaching a peak in 1961, the fin whale fraction of the catch began to decline as whalers gradually turned their focus to sei whales. Sei whales

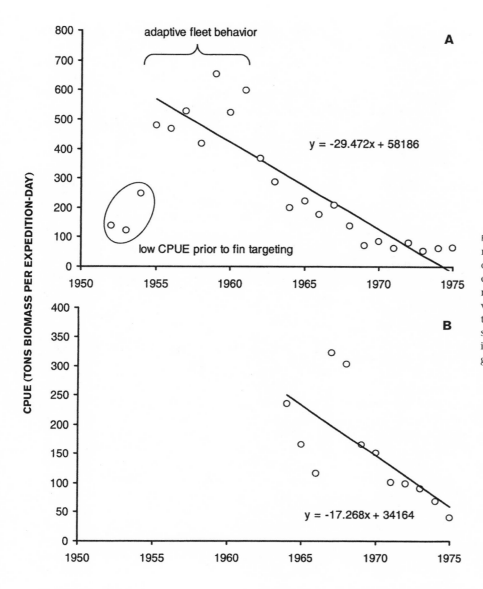

FIGURE 11.4. Regional CPUE estimates for fin (A) and sei (B) whales over time. CPUE values were generated from summed catch (biomass) and effort (expedition-days) values for each cell within the spatial grid for each year. Both regressions indicate a significant decline in CPUE over time; statistics are given in Table 11.1.

constituted an increasing proportion of the catch, beginning at only 8% in 1962 and growing steadily to include 62% of the catch in 1967. After 1967, sei whales represented a fairly constant proportion of the catch, fluctuating around 70% until 1973. The last species targeted by the Japanese fleet was the Bryde's whale. It wasn't until 1974 that this species represented a nontrivial fraction of the annual catch (23%). By 1975 it represented 48% of the catch, and in 1976—when the moratorium on fin and sei whales was enacted—the annual North Pacific Japanese pelagic harvest was made up entirely of Bryde's whales.

To a large extent, we believe this switching behavior is indicative of declining whale stocks. Comparing regional trends in harvest levels with CPUE estimates suggests that the fishery switched over to the next (relatively unexploited) species once the previously targeted species became less profitable, presumably due to regionwide stock depletion. However, there is a strong spatial component to the North Pacific whale harvest as well. Next we discuss these spatial components, first patterns of the fin and then those of the sei whale harvest and decline.

Fin Whales: Spatial Components of Harvest and Decline

One notable aspect of the regional trends for fin whales is that the catch levels and CPUE remained consistently high during the peak harvest years of 1955–1961 before beginning a sharp decline (Figure 11.4). A likely explanation for this pattern is that during this period the whaling fleet was able to maintain high CPUE levels by continually moving east across the Bering Sea and Aleutian Islands to more productive, unexploited stocks. (The low CPUE values for the years 1952–1954 are probably caused by the fleet's targeting of blue and humpback whales in those early years.)

The analyses of local CPUE trends suggest that stocks were beginning to decline in the southern Bering Sea and along the Aleutian Islands (cell K4) when the fleet shifted its geographical focus east and north to the western Gulf of Alaska. In this local area, our analysis indicates that local CPUE measures were initially high (844 tons/expedition-day in cell N1 in 1961) but declined sharply with the focused harvesting that occurred in this area, declining to 219 tons/expedition-day in 1966 (Figure 11.2). The two significant declines in this area (N1 and N2) occurred over a 6-year period ending in 1966, after which the fleet shifted its geographical focus back to the west and south. Thus, the stock depletion of the area around the Gulf of Alaska provides a fairly robust example of how focused exploitation in an area can lead to local declines that are much more dramatic and rapid than the rate of regional decline. A comparison of the slope of the regression lines between the local area represented by cells N1 and N2 and the regional values indicates that the rate of local decline in this area (–114.35 and –53.75 tons/expedition-day/year, respectively) was 2–3 times as great as the regional rate of decline (–29.47 tons/expedition-day/year).

What created this rapid local decline? A plausible explanation is that these whales have a high propensity to return to this local area year after year, which allowed their numbers to be substantially reduced by local hunting. Fin whales in adjacent waters must have had similar affinity for those areas, preventing spillover into the locally depleted area, which would have eroded the strong local decline that we found. Thus, we suspect that there is some degree of isolation between fin whales in the local area represented by cells N1 and N2 (i.e., the western Gulf of Alaska) and fin whales farther to the west.

This example of local depletion seems to contrast with the stock dynamics in cells H3 and I3 in the western North Pacific, which border the westernmost extent of the Aleutian Chain. These local areas received some of the highest exploitation, with estimates of 76,933 and 50,711 tons of fin whales removed out of H3 and I3, respectively (compared with a cell mean of 12,214 tons of fin whale biomass). Yet there is no evidence of a decline in CPUE in these local areas. Instead, CPUE remained relatively high and constant throughout the period 1952–1972, when these two cells were targeted by the whaling fleet. In fact, cell I3 indicates an *increase* in CPUE over this time period, a trend that is almost certainly due to sampling error. A possible explanation for the lack of local declines in cells H3 and I3 is that this area is a migration corridor for whales coming from multiple breeding grounds or going to multiple local feeding grounds farther north and east of this area. Thus, directed harvesting in this area did not deplete a local whale subpopulation because whaling in this area may have actually sampled a much larger regional population that used this corridor.

The last large local harvests of fin whales occurred in the areas just south of the islands of Umnak and Unalaska in the Aleutian Chain (cells K4, L3, and L4). These local areas had significant declines in CPUE with slopes of –21.33, –48.44, and –47.44, respectively, two of which were greater than the regional rate of decline of –29.47. Therefore we suspect that this area represents a second, larger subpopulation of fin

whales in the North Pacific. Aside from the local declines near the western Gulf of Alaska (cells N1 and N2), there was only one other cell (I4) that showed a significant decline in CPUE. Given the limited data points and low annual catch levels, we regard this as a mostly spurious result, and we do not feel that this isolated local decline provides any additional insights into fin whale population structure. Thus, we interpret our findings to support the existence of two subpopulations of fin whales: a large subpopulation that represents the majority of the individuals in the North Pacific (cells K4, L3, and L4), and a smaller subpopulation that resides near the western Gulf of Alaska. We suspect that the population structure created by these local subpopulations had a strong influence on the temporal trend in regional CPUE levels. Through the period 1955–1961 the fleet was able to maintain consistently high CPUE levels by modifying the harvest geographically to exploit new local areas (Figure 11.4). Only after these local areas had been exploited was a regional decline in CPUE evident in the trend data.

Sei Whales: Spatial Components of Harvest and Decline

The temporal trends in CPUE for sei whales at the regional level display some peculiar patterns. In particular, the CPUE estimates are relatively low in the mid-1960s (1964–1966), when sei whales were first being targeted, and then reached a peak in 1967. This type of CPUE trend would be rare in a typical single-species fishery, where the CPUE would be expected to be highest at the beginning of the fishery and then decline as the stock becomes depleted. In the multi-species whale fishery in the North Pacific, we believe that these early low CPUE values for sei whales were a result of the whaling fleet being in waters where sei whales were relatively less abundant. Fin whales, along with blue and humpback whales, were the majority of the catch during these years, and these landings were mostly to the north and east of the high concentration of sei whales near cells H4–5 and I4–5. Thus, we suspect that these early low CPUE values reflect inherent habitat differences for sei whales and were mainly the result of whaling in local areas of low sei whale abundance.

The peak in CPUE in 1967 fits well with this general interpretation. In this year, the geography of the fleet shifted dramatically south and west of their prior focal areas just south of the Aleutian Islands and the western Gulf of Alaska. These new whaling grounds (represented by cells H4–5 and I4–5) may be a distinct subpopulation and certainly represent the densest concentration of sei whales in the North Pacific. The area represented by these four local cells had (1) by far the highest sei whale total biomass harvest (103,723 tons), (2) initially high CPUE values (H4 = 396, H5 = 375, I4 = 602, I5 = 276 tons/expedition-day), and (3) some of the most dramatic local declines in CPUE. The rates of decline in these four cells were –50.97 (H4), –44.85 (H5), –109.05 (I4), and –35.53 (I5), with three of these declines being steeper than the regional rate of decline (calculated from 1964 to 1975 = –17.27; Figure 11.4). Thus this local area appears to have harbored a relatively high abundance of sei whales that experienced a fairly rapid depletion once the whaling fleet targeted this geographic area.

It seems likely that this local area (cells H4–5 and I4–5) represents a subpopulation of sei whales, or at the least these spatial patterns indicate that sei whales do not distribute themselves uniformly across their summer feeding grounds in the North Pacific. One interpretation of the high CPUE values experienced in 1967 when the Japanese fleet first began hunting in this local area is that these were largely unexploited stocks. Such reasoning suggests that earlier whaling prior to 1967, which was to the north and east of this local subpopulation, did little to diminish the abundance of whales at this local area. If whales in this local area did not have some affinity to these feeding grounds, but mixed freely with whales in waters to the north and east, then their numbers would have been diminished by earlier whaling efforts in the 1950s and early 1960s.

These results indicate that the population spatial structure of exploited species can interact critically with, and alter the analysis of, temporal trends in stock abundance. It is particularly worrisome that numerous rapid and severe local declines might be masked by analyzing data at long-term, regional scales, as was evident from our analysis. Aside from misrepresenting local population dynamics, ignoring spatial structure would also lead to false conclusions about the population health of whale stocks in the North Pacific region.

Conclusions

In this chapter, we have investigated the inherent spatial structure of whale stocks in the North Pacific Ocean by analyzing spatial and temporal patterns of harvesting together with declines in CPUE for individual species. Although our analysis makes several simplifying assumptions, we believe the results support the existence of several subpopulations of fin and sei whales that influenced temporal trends in indices of regional stock abundance. For fin whales, geographically separated stocks appear to exist around the central Aleutian Islands and the western Gulf of Alaska. This spatial structure allowed regional CPUE estimates to remain high at the beginning of this fishery as the Japanese whaling fleet adaptively exploited one stock and then shifted eastward to exploit the second stock. For sei whales, our results support the existence of a large subpopulation that is spatially constrained to pelagic waters southeast of Kamchatka. CPUE estimates for sei whales reached a peak in 1967, when the whaling fleet moved into this local area. Local declines in CPUE within this area occurred rapidly after 1967 and at a rate much faster than regional trends in CPUE. This study shows that understanding spatial structure is critical to harvesting, managing, and conserving exploited whale stocks. It also suggests that community-level effects of whaling may be manifested more rapidly at local scales when strong spatial structure exists.

Our analysis generally supports the inferences made by Springer et al. (2003) that fleet movement to areas south of the Bering Sea and Aleutian Islands were in response to local whale population declines. Further, this work indicates that in some local areas, declines (Figure 11.2) in whale populations may have been more dramatic than Springer et al. suggested. The magnitude of these declines suggests that they were a substantial perturbation to the larger oceanic food web, provided that strong interaction strengths exist between these whale species and their killer whale predators. The rapid local declines in fin whales moving roughly west to east across the Aleutians to the western Gulf of Alaska suggest that the timing and spatial distribution of the subsequent declines in marine mammal populations (harbor and fur seals, sea lions, and sea otters) may follow similar spatiotemporal patterns. This work also suggests that spatial structure may play an important, but largely unexplored, role in oceanic food webs.

Literature Cited

Allen, K. R. 1980. *Conservation and management of whales.* Seattle: University of Washington Press.

Baker, C. S., S. R. Palumbi, R. H. Lambertsen, M. T. Weinrich, J. Calambokidis, and S. J. O'Brien. 1990. Influence of seasonal migration on geographic distribution of mitochondrial DNA haplotypes in humpback whales. *Nature* 344: 238–240.

Baker, C. S., A. Perry, J. Bannister, M. Weinrich, R. Abernethy, J. Calambokidis, J. Lien, R. Lambertsen, J. Ramirez, O. Vasquez, P. Clapham, A. Alling, S. O'Brien, and S. Palumbi. 1993. Abundant mitochondrial-DNA variation and world-wide population structure in humpback whales. *Proceedings of the National Academy of Sciences* 90: 8239–8243.

Bakke, I., S. Johansen, O. Bakke, and M. R. Elgewely. 1996. Lack of subdivision among the minke whales *(Balaenoptera acutorostrata)* from Icelandic and Norwegian waters based on mitochondrial DNA sequences. *Marine Biology* 125: 1–9.

Charnov, E. L. 1976. Optimal foraging: the marginal value theorem. *Theoretical Population Biology* 9: 129–136.

Hoelzel, A. R. 1991. Whaling in the dark. *Nature* 352: 481.

———. 1994. Genetics and ecology of whales and dolphins. *Annual Review of Ecology and Systematics* 25: 377–399.

———. 1998. Genetic structure of cetacean populations in sympatry, parapatry, and mixed assemblages: implications for conservation policy. *Journal of Heredity* 89: 451–458.

Hooker, S. K., H. Whitehead, and S. Gowan. 1999. Marine protected area design and the spatial and temporal distribution of cetaceans in a submarine canyon. *Conservation Biology* 13: 592–602.

IWC. 2002. IWC Technical Note No. 15, Revision 4, 22 November 2002.

Jonzen, N., P. Lundberg, and A. Gardmark. 2001. Harvesting spatially distributed populations. *Wildlife Biology* 7: 197–203.

McAlpine, D. F. (1985). Size and growth of heart, liver, and kidneys in North Atlantic fin *(Balaenoptera physalus)*, sei *(B. borealis)*, and sperm *(Physeter macrocephalus)* whales. *Canadian Journal of Zoology* 63: 1402–1409.

Perrin, W. F., B. Würsig, and J. G. M. Thewissen. 2002. *Encyclopedia of marine mammals.* San Diego: Academic Press.

Palsbøll, P. J. 1990. Preliminary results of restriction fragment length analysis of mitochondrial DNA in minke whales from the Davis Strait, northwest and central Atlantic. Paper SC/42/NHMi35 presented to the International Whaling Commission Scientific Committee.

Rice, D. W. 1974. Whales and whale research in the eastern North Pacific, in *The whale problem: a status report*. W. E. Schevill, ed. Cambridge, MA: Harvard University Press, pp. 170–195.

Shelden, K. E., D. P. DeMaster, D. J. Rugh, and A. M. Olson. 2001. Developing classification criteria under the U. S. Endangered Species Act: bowhead whales as a case study. *Conservation Biology* 15: 1300–1307.

Springer, A. M., J. A. Estes, G. B. van Vliet, T. M. Williams, D. F. Doak, E. M. Danner, K. A. Forney, and B. Pfister. 2003. Sequential megafaunal collapse in the North Pacific Ocean: an ongoing legacy of industrial whaling? *Proceedings of the National Academy of Sciences* 100: 12223–12228.

Trites, A. W. and D. Pauly. 1998. Estimating mean body masses of marine mammals from maximum body lengths. *Canadian Journal of Zoology* 76: 886–896.

Yablokov, A. V. 1994. Validity of whaling data. *Nature* 369: 108.

TWELVE

Worldwide Distribution and Abundance of Killer Whales

KARIN A. FORNEY AND PAUL R. WADE

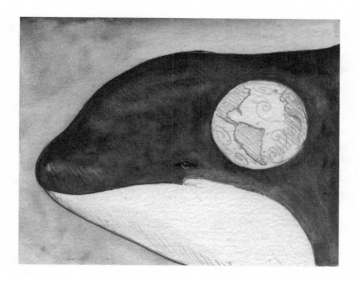

Killer whales have long been recognized as formidable predators in virtually all of the world's oceans (Leatherwood and Dahlheim 1978; Heyning and Dahlheim 1988). They are known to be common in many coastal areas, particularly at high latitudes, but they also occur in offshore and tropical waters (Leatherwood and Dahlheim 1978). Seasonal movement patterns, often associated with increased availability of prey species (Dahlheim 1981), have been documented or suggested in the eastern North Pacific (Braham and Dahlheim 1982; Baird and Dill 1995), the coast of Japan (Kasuya 1971), the North Atlantic and Norwegian waters (Sigurjónsson and Leatherwood 1988; Similä et al. 1996), the South Atlantic (Iñíguez 2001), and the Southern Hemisphere (Mikhalev et al. 1981; Kasamatsu and Joyce 1995). In a few coastal regions, most notably off British Columbia and the state of Washington, regularly occurring populations have been studied for decades, and much has been learned about their ecology, population structure, demography, social behavior, and genetics (see Bigg et al. 1990; Olesiuk et al. 1990; Ford et al. 1994, 1998; Baird and Dill 1995, 1996; Hoelzel et al. 1998, Barrett-Lennard 2000). In contrast, very little is known about killer whales in most other regions of the world. Several studies have suggested important functional roles for killer whale predation in the North Pacific and Southern Oceans, and

killer whale predation has been implicated in various recent marine mammal declines in the North Pacific and Southern Ocean (Barrat and Mougin 1978; Estes et al. 1998; Springer et al. 2003). These processes are, however, controversial and not well substantiated. A first step toward gaining a better understanding of relevant processes is to evaluate patterns of killer whale abundance, distribution, and foraging habits.

In this chapter, we provide a review of the published literature on worldwide killer whale population structure, abundance, distribution, and known prey. We discuss differences in the methods that have been used to identify and assess killer whale populations, including sighting and whaling records, photo-identification methods, mark-recapture estimates, and distance sampling methods from shipboard and aerial platforms. Each method has limitations and strengths and provides somewhat different information on population size, structure, and distribution. When combined within a single region, these sources can complement one another to provide a clearer picture of killer whale biology; however, taken individually they make comparisons between regions difficult. Nonetheless, in the following sections we have attempted to synthesize some of the diverse literature on killer whale populations worldwide into a general overview.

Populations

Killer whales are presently considered to form a single cosmopolitan species, *Orcinus orca*, although separate species status has been suggested for different groups found in the Southern Ocean (Mikhalev et al. 1981; Berzin and Vladimirov 1983; Pitman and Ensor 2003) and for different forms found in the North Pacific (Baird et al. 1992; Baird 1994), based on color pattern, diet, and morphological traits. Substantial differences in food habits, behavior, genetics, morphology, and movement patterns have led to the description of distinct populations throughout the world (see Baird 2000; Dahlheim and Heyning 1999), including several sympatric populations that do not apparently interact or interbreed. The well-studied killer whales in the eastern North Pacific are known to consist of at least two and maybe three distinct types, commonly termed *resident, transient,* and *offshore* killer whales (Ford et al. 1994). They differ in their food habits and in a variety of morphological, ecological, and behavioral characteristics (Baird 2000). Recent genetic investigations have revealed marked differences between the *resident* and *transient* types of killer whales (Stevens et al. 1989; Hoelzel and Dover 1991; Hoelzel et al. 1998; Barrett-Lennard 2000; Barrett-Lennard and Ellis 2001). In addition, smaller genetic differences were shown between different communities of *resident* killer whales (Hoelzel et al. 1998; Barrett-Lennard 2000; Barrett-Lennard and Ellis 2001).

Although the terminology just mentioned refers only to the coastal eastern North Pacific killer whale populations, similar differences in food habits and ecology have been observed in other regions of the world. Antarctic killer whales comprise three types, provisionally termed Types A, B, and C, with clear differences in coloration, food habits, and some morphological traits (Pitman and Ensor 2003). Killer whale populations that forage primarily on marine mammals have been observed in all ocean basins of the world. Populations that forage primarily on fish and other invertebrates can be divided further into killer whales that exploit one or more seasonally abundant, high-latitude coastal fish species and those that forage widely on pelagic fish and invertebrates. To illustrate some of the differences in more detail, we provide a brief summary of information for these three general types of killer whales, with emphasis on the well-studied eastern North Pacific killer whale types.

Mammal-eaters

Mammal-eating killer whales have been documented worldwide in both coastal and offshore waters (Jefferson et al. 1991), but mammal-eating specialists have been confirmed only in the North Pacific and Southern Ocean. Eastern North Pacific *transient*-type killer whales are among the best-studied mammal-eating killer whales and are known to prey on a wide variety of marine mammals, including whales, dolphins, porpoises, and pinnipeds (Baird and Dill 1996). In inshore waters of British Columbia (Ford et al. 1998) and Prince William Sound, Alaska (Saulitis et al. 2000), two regions where *transient* predation has been studied systematically, the most frequently observed prey have been pinnipeds and porpoises. They have also occasionally been reported feeding on sea turtles and seabirds (Caldwell and Caldwell 1969; Ford et al. 1998; Baird and Dill 1996). Groups are generally small (<10) with a dynamic composition (Baird and Whitehead 2000). Individual *transient* killer whales have been documented to move great distances, reflecting large home ranges as the killer whales follow mobile prey resources (Goley and Straley 1994; Black et al. 1997). Some eastern North Pacific *transients* have been photographed in locations ranging from southeastern Alaska to California, a distance of at least 2,600 km. In the eastern North Pacific and in the Southern Ocean, mammal-eating killer whales are known to be sympatric with other forms of killer whales, but they have not been seen intermixing in social groups with other forms. In Antarctic waters there are two known types of mammal-eating killer whales (Pitman and Ensor 2003). Type A killer whales are thought to follow the seasonal migration of their main prey (Antarctic minke whales, *Balaenoptera borderensis*) and are generally found in open-water habitat. Type B killer whales may be found sympatrically with Type A whales in the open water around Antarctica, but they are more commonly observed in loose pack ice, where they prey upon seals.

Coastal Fish-eaters

Resident killer whales in the North Pacific feed primarily on salmon and other coastal fish species and are seen regularly during summer in the inland waterways of Alaska, British Columbia, and Washington state. Groups of *resident* whales can be distinguished from groups of sympatric *transient* or *offshore* killer whales based on differences in dorsal fin shape and saddle patch coloration. They include several distinct and well-known populations, including the Southern, Northern, and Gulf of Alaska *residents*. *Resident*-type killer whales are also found in the Aleutian Islands and the Bering Sea and are also thought to be found along the Kamchatka Peninsula of Russia, but population structure in these areas has not been fully determined. *Resident* whales are usually found in large groups ranging from approximately 5 to 50 whales, but they are occasionally found in even larger aggregations of 100 whales or more. Where well studied, it has been found that *resident*-type whales occur in stable social groups that are matrilines because offspring do not disperse from their mother's social group (Bigg et al. 1990). Seasonal movement patterns are not fully understood (Bigg et al. 1990; Ford et al. 1998), but during winter the southern *residents* have been reported along the outer coasts of British Columbia, Washington, and California.

Coastal fish-eating killer whales have also been documented in waters off Iceland, the Faroe Islands, the British Isles, and Norway, where they feed primarily on seasonally abundant herring, *Clupea harengus,* and other coastal fish species (Sigurjónsson et al. 1988, Jonsgård and Lyshoel 1970, Similä et al. 1996). Based on photo-identification studies, it has been suggested that the social system of coastal Norwegian whales is similar to that of eastern North Pacific *resident* whales, with stable pods comprising males and females with their offspring, and with no dispersal of either sex from their natal group (Bisther 2002). Pod sizes for Norwegian killer whales are reported to range from 6 to 30 whales with a median of 15. During the course of a 4-year study in northern Norway, pods with known individuals were seen repeatedly and regularly in the study area when herring were present (Similä et al. 1996).

The Antarctic Type C killer whale, known primarily from deep within the Antarctic pack ice, may also broadly be considered a coastal fish-eater. It is found in large groups of 10 to at least 150 (mean 46) and is known to feed on Antarctic toothfish (*Dissostichus mawsoni*). Some reports suggest they may also take penguins and seals off the ice (Pitman and Ensor 2003), but no takes of these species have actually been observed. Stomach contents of 629 "yellow" killer whales taken in Soviet whaling operations during 1979–1980, which would have represented either Type B or C killer whales, contained 98.5% (by frequency of occurrence) fish, 1.1% squid, and 0.5% marine mammal prey (Berzin and Vladimirov 1983). Seasonal movement patterns are not well understood for the Type C killer whale (Pitman and Ensor 2003). Although some individuals have been reported to overwinter in Antarctic waters, there are also winter records of Type C whales at lower latitudes, including waters off New Zealand (Pitman and Ensor 2003).

Oceanic and Neritic Killer Whales

Killer whales have been reported from oceanic and neritic waters worldwide, but little is known about these whales. Killer whales are found at low densities throughout the eastern tropical Pacific (Wade and Gerrodette 1993), but their diet and population structure are little understood. One form, thought to be mostly neritic in distribution, occurs in the eastern North Pacific and is referred to as the *offshore* type killer whale. Morphologically they are similar to *resident* whales (i.e., their dorsal fins appear to be more rounded at the tip than those of the *transient* type, and they share saddle patch coloration patterns), and they are commonly seen in groups ranging from 10 to 70 whales. *Offshore* killer whales have been found from central coastal Mexico to Alaska and are thought to occur mostly on the continental shelf, but they occasionally come into coastal waterways and have also been seen in offshore waters extending at least 200 miles from the U.S. West Coast (Black et al. 1997). Their main foraging target is assumed to be fish, but observational data on feeding events are extremely limited. *Offshore* whales are not known to intermingle in social groups with *resident* or *transient* whales. Genetic analyses suggest that *offshores* may be reproductively isolated, but they appear to be more closely related to *resident*-type whales (Hoelzel et al. 1998; Barrett-

Lennard 2000). In other regions of the world, data on oceanic or neritic killer whales are scarce. Off Japan, killer whales are reported to feed on cod, flatfish, and cephalopods (Nishiwaki and Handa 1958), and these animals may represent a similar neritic form.

Methods of Population Estimation

Information on the abundance and distribution of killer whales ranges widely in age, reliability, detail, and quantitative nature. Population estimates are available for some areas of the eastern North Pacific, North Atlantic, and Southern Oceans and for scattered locations elsewhere in the world. However, methods used to assess populations of killer whales are not strictly comparable, because they estimate somewhat different population parameters and are subject to different biases. In this section we summarize the most common methods of assessing populations of killer whales and other cetaceans, along with an overview of strengths and weaknesses of each method (Table 12.1). After consideration of these differences, we have attempted to synthesize the available information to provide a worldwide summary of patterns of abundance and distribution of killer whales.

Photo-identification Catalogs (CAT)

Killer whales are amenable to long-term studies of local populations because individual whales can readily be identified on the basis of unique characteristics of their dorsal fins and saddle patches (Bigg et al. 1987), and whales often return to the same areas each year. Such studies provide a wealth of information on behavior, ecology, demography, genealogy, population structure, and movement patterns and can be used to establish a catalog of known individuals (see Bigg et al. 1990; Olesiuk et al. 1990; Ford et al. 1994). In cases where populations are small and individual whales are regularly seen in areas of photographic sampling, photo-identification catalogs can provide a minimum or actual population count, particularly if deaths can be documented through time. Such methods do not, however, work well in large, widely dispersed populations, because not all animals are likely to be photographed. For example, the catalog of eastern North Pacific *transient* killer whales contained 251 known individuals in 1999 (Ford and Ellis 1999; Matkin et al. 1999, summarized in Carretta et al. 2001), based on two decades of sampling off Washington state, British Columbia, and Alaska. Subsequently, photographs of killer whales seen off California between 1987 and 1996 were examined (Black et al. 1997), revealing ten matches to known *transient* whales and 95 previously unknown *transient* whales. The catalog count thus increased by 95 animals (38%) to 346 *transient* killer whales when photographs from a different part of this population's range were included. This suggests that not all individuals use the entire range, and catalogs developed within a subset of the entire population's range can underestimate total population size. Thus, photo-identification catalogs will provide

TABLE 12.1
Summary of Methods of Population Estimation for Killer Whales

Method	Information Type	Advantages	Disadvantages	Reference Examples
OBS: Anecdotal or descriptive information, such as whaling records, stranding data, incidental observations	Presence/absence; qualitative descriptions of occurrence, broad distribution patterns	Widely available in most inhabited/ exploited regions of the world; provides historical information	Non-quantitative and difficult to compare between records; limited information for many unexploited regions of the world	Diverse historical literature, whaling records, natural history notes; see for example summaries in Leatherwood et al. 1984; Sigurjónsson and Leatherwood 1988; Dahlheim et al. 1982
CAT: Photo-identification studies and catalogs	Individual-based information on behavior, birth/death rates, ecology and movement patterns; catalog of known individuals; long-term genealogical records	Provides detailed demographic data on population of interest	Photographic sampling may not capture entire population; takes many years and large effort to build comprehensive photo-identification catalog; if deaths are not documented, catalog can overestimate population size	Sigurjónsson et al. 1988; Leatherwood et al. 1990; Ford et al. 1994; Dahlheim et al. 1997; Dahlheim 1997; Heise et al. 1991; Black et al. 1997; Matkin et al. 1999; Iñíguez 2001; Guerrero-Ruiz et al. 1998; See also Hammond et al. 1990
MR: Mark-recapture methods	Estimate of the total number of whales that use a given study area during the study period; with sufficient data, demographic parameters can be estimated	Estimates proportion of unobserved whales to provide more complete estimate of population size; cost-effective method that can be carried out mostly from small boats; model selection and careful design of sampling sites can minimize biases	Requires at least two sampling periods to conduct analysis; if sampling not well-designed, can be negatively biased by individual heterogeneity in sighting probabilities	Poncelet et al. 2002; other cetacean mark-recapture studies: Hansen 1990; Wilson et al. 1999; Matthews et al. 2001; Read et al. 2003; Clapham et al. 2003; Calambokidis and Barlow 2004
SURV: Non-standardized surveys and observations with a measure of effort	Encounter rates, distribution information, instantaneous measure of relative abundance within surveyed region	Can be obtained incidentally to other surveys; semiquantitative data	Non-standardized effort; limited comparability between platforms, methods, regions, generally small sample sizes	Kasuya 1971; Eyre 1995; Miyashita et al. 1996; O'Sullivan and Mullin 1997; Aguayo et al. 1998; Gannier 2000; Garrigue and Greaves 2001; Kahn and Pet 2003; Ballance et al. 2001; Secchi et al. 2002
LTR: Line-transect surveys (Buckland et al. 1993)	Standardized estimate of instantaneous abundance and density	Comparable estimate across studies, as long as potential biases addressed	Costly, requires moderate number of sightings (>30) for analysis; does not provide movement data or information on animals outside of study area	Hammond 1984, Øien 1990; Wade and Gerrodette 1993; Barlow 1995; Forney et al. 1995; Branch and Butterworth 2001; Waite et al. 2002

accurate estimates of population size only if the animals are sampled throughout their range or if all members of the population are known to enter areas of photographic sampling during the study period.

Mark-recapture Methods (MR)

Mark-recapture methods (Seber 1982; Hammond 1986) can yield improved estimates of abundance based on photo-identification data, and they have been used successfully for a variety of individually recognizable cetacean species, including bottlenose dolphins (*Tursiops truncatus*; Hansen 1990; Wilson et al. 1999; Read et al. 2003), humpback whales (*Megaptera novaeangliae*; Clapham et al. 2003; Calambokidis and Barlow 2004), blue whales (*Balaenoptera musculus*; Calambokidis and Barlow 2004), and sperm whales (*Physeter macrocephalus*; Matthews et al. 2001). The method involves collecting a "mark" data set, in which a number of individuals are photographed and identified, and a "recapture" data set, from which a proportion of recaptured (previously identified) individuals is calculated. In the simplest form of the method (for a closed population) the population size is estimated as the number of individuals in the initial sample divided by the proportion of recaptured individuals. The result is an estimate of the total number of animals in the population during the study period. More sophisticated models are available for open populations (i.e., those with births, deaths, immigration, or emigration during the study period).

Mark-recapture methods are, however, sensitive to several assumptions. One is that all individuals in a population have equal capture probability within each sample. This assumption may be violated in various ways, such as if some individuals have indistinct identification characteristics or behave in a way that makes them less likely than others to be photographed, or if the population is not distributed homogeneously. These effects can be mitigated to a large extent by limiting the mark and recapture samples to the best photographs (to allow indistinct marks to be reliably identified, Friday et al. 2000), by making efforts to photograph all individuals in a sighted group, and by selecting broad geographic sampling locations throughout the population's range (Calambokidis and Barlow 2004). Other important assumptions are that marks are not lost through time and that marked animals can correctly be recognized upon recapture. If any of these assumptions are not met during sampling or addressed analytically, this can lead to substantial (usually downward) bias in the abundance estimates (see Hammond et al. 1990). Calambokidis and Barlow (2004) present a mark-recapture estimate for blue whales that was initially biased downward because of limited and uneven photographic sampling. This bias was largely eliminated once broader geographic sampling was achieved.

Despite the wealth of photographic data on killer whales in many regions of the world, there are no published mark-recapture estimates of abundance for this species except for that of Poncelet et al. (2002). Ongoing investigations in the eastern North Pacific are expected to include mark-recapture estimates in the future (J. Durban, personal communication).

Line-transect Surveys (LTR)

Dedicated line-transect surveys using established distance-sampling protocols (Buckland et al. 1993) have been used extensively to provide quantitative and reliable estimates of abundance for cetaceans at sea. Such surveys are generally conducted from medium-sized or large vessels or from aircraft, and the sighting efficiency of each platform is taken into account in estimating densities. The resulting abundance estimates represent a snapshot of the average number of animals within the study region during the study period. Many potential sources of bias can be accounted for during survey planning (e.g., spatial coverage and transect design) or during data analysis (e.g., correction factors for animals missed because of poor weather conditions or diving behavior). Line-transect methodology has been widely used to estimate the abundance of diverse cetacean species worldwide (Hammond 1984; Wade and Gerrodette 1993; Barlow 1995; Forney et al. 1995; Branch and Butterworth 2001) and is generally considered a reliable and effective technique when surveys are designed properly (Buckland et al. 1993). In a number of cases where both line-transect and mark-recapture estimates of abundance are available for a single population, as for blue whales and humpback whales off the U.S. West Coast (Calambokidis and Barlow 2004), abundance estimates from the two methods have been remarkably similar. In general, line-transect surveys yield the most directly comparable abundance estimates across studies. They do not, however, provide information on population identity or movement of individuals unless photo-identification or biopsy efforts are included in the survey.

Non-standardized Surveys (SURV)

In many cases marine mammal surveys are conducted incidentally during other marine investigations, such as by fishery observers, during seismic or oceanographic cruises, on ferry routes, or on other platforms of opportunity. As with dedicated line-transect surveys, opportunistic surveys provide a snapshot of the distribution and abundance of animals within the study area. Methodology has varied widely, however, and it is difficult or impossible to compare results directly from one study to another. Nonetheless, such surveys often provide new information on the distribution, relative abundance, and species composition of cetaceans, particularly in poorly studied regions (Ballance et al. 2001; Gannier 2000; Eyre 1995). Broad patterns of cetacean occurrence can be inferred if one takes into account characteristics of each platform, such as sighting efficiency and potential sources of bias. For example, the effective area surveyed will be greater for surveys conducted from large vessels, at greater height above the water, using multiple observers, and with high-powered binoculars than during surveys conducted from small boats, from platforms closer to the water surface, with a single observer, and without high-powered binoculars. Bias is introduced if animals avoid the platform from which observations are made (e.g., during seismic surveys; Stone 1997), or if they are attracted to the vessel while observations are underway (e.g., killer whales are known to depredate fish from longline vessels and may be attracted to longline survey vessels, particularly during set operations; Secchi and Vaske 1998, Pinedo et al. 2002).

Observations and Anecdotal Information (OBS)

The final category of information comprises a wide variety of descriptive, anecdotal, or incidental observations collected by whalers, fishermen, naturalists, oceanographers, biologists, and others. It may include, for example, presence/absence data, whaling or fishery catch records, or details of observations made (with or without measures of effort). Such observational information generally represents the least quantifiable source of data and is difficult to interpret in a broader context; however, in many regions of the world the only information available on killer whales is of this qualitative nature. In this study we have obtained descriptive information from published reviews (e.g., Sigurjónsson and Leatherwood 1988; Dahlheim et al. 1982), published notes and articles, and gray literature that was accessible in academic libraries or on the Internet. The quality and relevance of information from each source was evaluated individually before inclusion in the overall synthesis presented in the following section.

Patterns of Killer Whale Distribution and Abundance

Using information from all of the types of studies just presented, we have compiled a worldwide summary of patterns of abundance and distribution of killer whales (Table 12.2). Information fell into five general categories: observational information (OBS); photo-identification catalogs (CAT); mark-recapture estimates (MR); non-standardized surveys (SURV); and line-transect surveys (LTR). Because of the differences in accuracy and reliability of these methods, data were standardized into four broad categories of abundance (Table 12.2), defined as follows, based on the range of available line-transect estimates of killer whale densities:

- Rare (0–0.10 whales per 100 km²)
- Uncommon (0.10–0.20 whales per 100 km²)
- Common (0.20–0.40 whales per 100 km²)
- Abundant (>0.40 whales per 100 km²)

Encounter rates (sightings per linear distance) from non-standardized surveys were compared to encounter rates during line-transect surveys, bearing in mind differences in platform and methodologies, to allow classification of each survey result into one of the four density categories. For mark-recapture abundance estimates and photo-identification catalogs, the number of known or estimated killer whales was divided by the size of the area thought to be inhabited by the whales. This provided a crude density estimate, which was then used as a guideline for assigning an abundance category to each study region, taking into consideration uncertainties in whale movements or population size. Finally, anecdotal information was subjectively categorized on the basis of descriptions of killer whale occurrence (e.g., "seasonally common," "rare visitors"), reported frequency and regularity of sightings, and estimated numbers of killer whales seen within each study region. The assigned categories should be considered provisional for all regions that do not have quantitative estimates of killer whale density. Nonetheless, they provide a basis for examining patterns of distribution and abundance worldwide.

Killer whales have previously been described to be very widely distributed throughout the world's oceans (Leatherwood and Dahlheim 1978; Heyning and Dahlheim 1988). They are found in tropical, temperate, and high-latitude waters and in pelagic and coastal habitats (Dahlheim et al. 1982; Dahlheim and Heyning 1999). Whales inhabiting coastal areas often enter shallow bays, estuaries, and river mouths (Leatherwood et al. 1976). The global picture of killer whale distribution obtained in this study is largely consistent with these past observations but provides further detail and quantitative information. Killer whale densities increase by 1–2 orders of magnitude from the tropics to the highest sampled latitudes in the Arctic and Antarctic (Figure 12.1). The available data also support previous observations that killer whales are more common along continental margins (Dahlheim et al. 1982); however, there is some variation in this general pattern that appears linked to ocean productivity (Figure 12.2). Killer whales appear to be less common in western boundary currents, such as the Gulf Stream or the Kuroshio, than in more productive eastern boundary currents, such as the California Current (Figure 12.2). Known areas of locally higher density often coincide with greater productivity (e.g., off Argentina). Specific patterns are discussed in the following paragraphs for each oceanic region, based on published accounts. It is important to bear in mind, however, that whaling catches (e.g., Nishiwaki and Handa 1958; Rice 1974; Ohsumi 1975; Dahlheim 1981; Mikhalev et al. 1981; Reeves and Mitchell 1988b) may have reduced the size of some local killer whale populations prior to abundance assessments.

North Pacific Ocean and Adjacent Arctic Waters

In the North Pacific Ocean, killer whales have been documented as far north as the Chukchi and Beaufort Seas in the Arctic Ocean (Lowry et al. 1987). Killer whales occur commonly in the eastern Bering Sea (Braham and Dahlheim 1982; Waite et al. 2002) and are frequently observed near the Aleutian Islands (Scammon 1874; Murie 1959; Waite et al.

2002). They occur year-round in the waters of southeastern Alaska (Scheffer 1967) and the intracoastal waterways of British Columbia and Washington State (Balcomb and Goebel 1976; Bigg et al. 1987; Osborne et al. 1988). Killer whales are commonly found along the coast of Washington and Oregon and are regularly found but less abundant off California (Norris and Prescott 1961; Fiscus and Niggol 1965; Rice 1968; Black et al. 1997; Barlow 1995; Forney et al. 1995). Killer whales are distributed sparsely but continuously throughout the eastern tropical Pacific (Dahlheim et al. 1982; Wade and Gerrodette 1993), including both coasts of Baja California (Guerrero-Ruiz et al. 1998; Gerrodette and Palacios 1996). Data provided by Gerrodette and Palacios (1996), combined with detection function estimates from Wade and Gerrodette (1993), allow estimation of killer whale densities by the Exclusive Economic Zones (EEZs) of eastern tropical Pacific countries, ranging from Mexico to Ecuador. The overall pattern appears to be one of decreasing density as one approaches the equator from either hemisphere, with a rare occurrence overall.

In the western North Pacific, killer whales are common along the Russian coast in the Bering Sea, the Sea of Okhotsk, and the Sea of Japan and along the eastern side of Sakhalin and the Kurile Islands (Tomilin 1957; Berzin and Vladimirov 1989). There are accounts of their occurrence off China (Wang 1985) and Japan (Nishiwaki and Handa 1958; Kasuya 1971, Ohsumi 1975), but killer whale encounters are rare or uncommon. Whaling catches of killer whales in Japanese waters may have lowered their current abundance relative to pristine levels. Central Pacific records are quite rare, but sightings have been documented around the Hawaiian Islands (Mobley et al. 2001; Barlow 2006) and between Hawaii and California (NMFS, 1997 Sperm Whale Abundance and Population Structure Cruise, unpublished).

South Pacific

In contrast to the North Pacific, records of killer whales are considerably scarcer in the South Pacific, mostly as a reflection of more limited sampling effort. The most systematically sampled areas south of the equator are those covered during eastern tropical Pacific dolphin cruises between 1986 and 1993 (Wade and Gerrodette 1993, Gerrodette and Palacios 1996), extending southward to about 15°S. Patterns of occurrence are similar to those north of the equator, indicating that killer whales are widespread but rare. Waters off Ecuador (excluding the Galápagos Islands), were estimated to contain about 75 killer whales (derived from Gerrodette and Palacios [1996] using the detection function estimates from Wade and Gerrodette [1993]). Sighting records off Galápagos indicate that killer whales are regular but rare in waters surrounding the archipelago (Merlen 1999). The only other quantitative estimate of killer whale abundance within the South Pacific Ocean was derived from photo-identification studies around New Zealand (Visser 2000), where an estimated 119 whales were found.

In other South Pacific regions, records indicate that killer whales are rare or uncommon throughout most of their range. Killer whales were sighted on three occasions during nonstandard sighting surveys in 1993–1995 between waters of the Peru Current off Chile and oceanic waters surrounding Easter Island (Aguayo et al. 1998). Although the total distance searched is not provided, the number of days of search effort suggests that killer whales are rare in oceanic waters of the southeastern Pacific (Table 12.2), although they appear to be more common in the more coastal waters of the Peru Current (Aguayo et al. 1998; Miyashita et al. 1995). In the Solomon Islands a 1993 sighting survey aboard two large research vessels resulted in one killer whale sighting during 3,700 km of search effort (Shimada and Pastene 1995). Taking platform and observer efficiency into account, this indicates that killer whales are rare in waters around the Solomon Islands. Surveys in French Polynesia during 1996–2000 (Gannier 2000; Laran and Gannier 2001) resulted in only one sighting of killer whales during more than 11,000 km. Ling (1991) compiled records from 1982 to 1990 of killer whales in South Australia and reported that they appeared to be present in most South Australian waters. Seasonal concentrations have been reported off Tasmania and Macquarie Island and in association with pinniped breeding activities. The available descriptions suggest that killer whales are uncommon in most South Australian waters, with localized areas in which they may be considered common.

Indian Ocean

Records of killer whales in the Indian Ocean and adjacent seas are generally rare. Most information on their distribution and relative abundance has been obtained during nonstandard sighting surveys and opportunistic studies. They have been reported taking catches from longlines targeting tuna in the Indian Ocean (Sivasubramanian 1965). A 1993 east-west survey transect through the Indian Ocean, at about 12°S, yielded one sighting of killer whales roughly midway between the Seychelles and the east coast of Africa (Eyre 1995), indicating that killer whales are rare in these waters. No killer whales were seen during a 1998 survey covering about 1,700 km in waters of the Maldives (Ballance et al. 2001). Visual and acoustic surveys within Komodo National Park, Indonesia, indicate that killer whales are present but rare, with two sightings in over 8,700 km surveyed. Photo-identification studies conducted during 1964–2000 in waters of the Crozet Archipelago, Southern Indian Ocean, indicate a small, regularly occurring population (Poncelet et al. 2002). Killer whales are also regular visitors at Marion Island in the Southern Indian Ocean during the southern elephant seal (*Mirounga leonina*) breeding season, with about 30 individually identified whales (Keith et al. 2001; Pistorius et al. 2002). In a synthesis of 1965–1988 data from Japanese whale scouting and whale sighting surveys, Miyashita et al. (1995) suggest that killer whales are more abundant, perhaps uncommon to common, in the Southern Indian Ocean.

TABLE 12.2
Regional Population Estimates for Killer Whales

Region	Source of abundance information	Data Type	Years	Area size (km^2)
Atlantic Ocean and Adjacent Seas				
Norwegian Sea	Øien 1990	LTR	1988	477,727
Northern Norway	Similä et al. 1996	CAT	1990–1993	67,000
Iceland & Faroe Islands	Gunnlaugsson and Sigurjónsson 1990; Sigurjónsson et al. 1989	LTR	1987	2,281,630
NW Scotland	MacLeod et al. 2003	LTR	1998	104,000
North Sea/West of Great Britain	Stone 1997	SURV	1996	
Newfoundland/Labrador	Lien et al. 1988	SURV	1979–1986	
Southeastern U.S. shelf/slope	Garrison et al. 2003	LTR	2002	263,564
Western North Atlantic (SE U.S.)	Mullin and Fulling 2003	LTR	1998	573,000
Gulf of Mexico (Oceanic Northern Gulf)	Derived from Mullin and Hoggard 2000	LTR	1991–1997	398,960
Gulf of Mexico (Oceanic Northern Gulf)	Mullin and Fulling 2004	LTR	1996–2001	380,432
Gulf of Mexico (GulfCet I area)	Derived from Mullin and Hoggard 2000	LTR	1991–1997	154,621
Gulf of Mexico (aerial surveys)	Mullin and Hoggard 2000	LTR	1996–1998	82,796
Gulf of Mexico (shelf waters)	Fulling et al. 2003	LTR	1998–2001	245,800
Northern Spain	López et al. 2004	SURV	1998–1999	9,842
Mediterranean	Notarbartolo-di-Sciara 1987	OBS	1985	
Southern Brazil	Pinedo et al. 2002	SURV		
Patagonia, Argentina	López and López 1985; Iñíguez 2001	CAT	1985–1997	20,002
Southern Ocean				
Area II	Hammond 1984	LTR	1981–1982	1,830,660
Area III	Hammond 1984	LTR	1979–1980	1,795,778
Area IV	Hammond 1984	LTR	1978–1979	1,431,045
Area V	Hammond 1984	LTR	1980–1981	1,868,464
S of 60°S	Branch and Butterworth 2001	LTR	1978–1983	9,935,989
S of 60°S	Branch and Butterworth 2001	LTR	1985–1990	11,655,723
S of 60°S	Branch and Butterworth 2001	LTR	1991–1997	10,922,924
Southern Ocean	Kasamatsu and Joyce 1995	LTR	1976–1988	28,765,576
Marion Island, Southern Ocean	Keith et al. 2001; Pistorius et al. 2002	OBS	1973–2000	
Pacific Ocean and Adjacent Arctic Waters				
Central Bering Sea	derived from Waite et al. 2002	LTR	1999	196,885
SE Bering Sea	Waite et al. 2002	LTR	2000	158,561
Aleutian Islands	Forney (unpublished data)	LTR	1994	634,042
Aleutian Islands, west of Unimak	Zerbini et al. 2006	LTR	2001–2003	109,933
Gulf of Alaska, east of Unimak	Zerbini et al. 2006	LTR	2001–2003	107,680
Gulf of Alaska *transients*	Matkin et al. 1999	CAT		214,307
Western Alaska (excluding Kodiak)	Dahlheim 1997; Waite, personal communication	CAT		*94,998*
BC/Washington *residents*, summer	Ford et al. 2000; Carretta et al. 2002	CAT	1970–2003	129,889
Alaska Southeast to Kodiak *residents*	Matkin et al. 1999; Dahlheim, personal communication; Angliss and Lodge 2002	CAT		150,515
U.S. West Coast *transients*	Matkin et al. 1999; Matkin, personal communication; Angliss and Lodge 2002	CAT		400,904

Effort (km)	No. Si	No. Ani	Si/1000 km	Abundance Estimate	CV	Density (Animals per 100 km²)	Catalog or Minimum count	Abundance Category
Atlantic Ocean and Adjacent Seas								
5,742	7	67	1.22	3,100	0.63	0.65		Abundant
						0.61	408	Abundant
21,827	25	199	1.15	6,618	0.32	0.29		Common
2,157	0	0	0.00	0		0.00		Rare
7100 hrs	17	120						Uncommon
85,273	58		0.68					
3,744	0	0	0.00	0		0.00		Rare
4,163	0	0	0.00	0		0.00		Rare
				246	0.39	0.06		Rare
12,162	5	12	0.41	133	0.49	0.03		Rare
				92	0.48	0.06		Rare
4,101	0	0	0.00	0		0.00		Rare
2,196	0	0	0.00	0		0.00		Rare
8,128	0	0	0.00	0		0.00		Rare
								Rare
3,324	5	15	1.50					Common
	408						30	Uncommon
Southern Ocean								
11,810	13	301	1.10	12,367	0.69	0.68		Abundant
12,812	22	1,608	1.72	38,278	0.63	2.13		Abundant
12,792	24	946	1.88	16,399	0.55	1.15		Abundant
10,014	24	6,014	2.40	136,500	0.69	7.31		Abundant
65,979	117	2,002	1.77	91,310	0.34	0.92		Abundant
67,550	114	817	1.69	27,168	0.26	0.23		Common
52,334	68	836	1.30	24,790	0.23	0.23		Common
130,036	129	1,135	0.99	80,400	0.15	0.28		Common
							25–30	
Pacific Ocean and Adjacent Arctic Waters								
1,761	2	10	1.14	121		0.06		Rare
2,194	11	50	5.01	391	0.43	0.25		Common
2,780	14	75	5.04	2,594	0.44	0.41		Abundant
3,560	16	113	4.49	584	0.51	0.54		Abundant
5,494	14	192	2.55	655	0.54	0.61		Abundant
				32		0.01	32	Rare
				180		0.19	180	Uncommon
				295		0.23	295	Common
				440		0.29	440	Common
				344		0.09	344	Rare

TABLE 12.2 (CONTINUED)

Region	Source of abundance information	Data Type	Years	Area size (km²)
Pacific Ocean and Adjacent Arctic Waters				
Oregon/Washington	Barlow 2003	LTR	1996–2001	325,018
California	Barlow 2003	LTR	1991–2001	817,549
California (aerial surveys)	Forney et al. 1995	LTR	1991–1992	264,270
Mexico (Gulf of California)	Guerrero-Ruiz et al. 1998	CAT	1972–1997	210,000
Eastern Tropical Pacific	Wade and Gerrodette 1993	LTR	1986–1990	19,148,000
Mexico (Gulf of California)	Gerrodette and Palacios 1996	LTR	1986–1993	262,125
Mexico (Pacific Coast EEZ)	Gerrodette and Palacios 1996	LTR	1986–1993	2,054,192
Central America EEZ	Gerrodette and Palacios 1996	LTR	1986–1993	323,013
Costa Rica EEZ	Gerrodette and Palacios 1996	LTR	1986–1993	475,737
Panama EEZ	Gerrodette and Palacios 1996	LTR	1986–1993	188,045
Columbia EEZ	Gerrodette and Palacios 1996	LTR	1986–1993	329,492
Ecuador EEZ	Gerrodette and Palacios 1996	LTR	1986–1993	229,863
Galapagos	Merlen 1999	OBS	1948–1997	
Eastern Temperate Pacific	NMFS, 1997 Sperm Whale Abundance and Population Structure Cruise, unpublished	LTR	1997	7,786,000
Hawaii	Barlow 2006	LTR	2002	2,452,916
Sea of Okhotsk	Berzin and Vladimirov 1989	OBS	1979–1984	
Kamchatka and Commander Islands	Miranova et al. 2002	OBS	1992–2000	
Japan (aerial surveys)	Kasuya 1971	SURV	1959–1970	
Western North Pacific	Miyashita et al. 1996	SURV	1993–1995	
Marquesas Islands/French Polynesia	Laran and Gannier 2001	SURV	1998–2000	
Society Islands/French Polynesia	Gannier 2000	SURV	1996–1999	
New Zealand	Visser 2000	CAT	1992–1997	
Solomon Islands	Shimada and Pastene 1995	SURV	1993	
Eastern South Pacific (Chile to Easter Island)	Aguayo et al. 1998	OBS	1993–1995	
Indian Ocean and Adjacent Seas				
Indian Ocean	Eyre 1995	SURV	1993	
Maldives	Ballance et al. 2001	SURV	1998	
South Australia	Ling 1991	OBS	1982–1990	
Komodos Island, Indonesia	Kahn and Pet 2003	SURV	1999–2001	
Crozet Archipelago	Poncelet et al. 2002	M-R	1977–2000	

NOTE: Worldwide estimates of abundance and density of killer whales. (Si = Sightings, Ani = Animals, CV = Coefficient of Variation).

North Atlantic and Adjacent Seas

The first comprehensive summary of killer whale occurrence in the North Atlantic (Sigurjónsson and Leatherwood 1988) was based largely on whaling records, incidental sightings, anecdotal information, strandings, and some photo-identification efforts in coastal waters. Overall, killer whales are rare to uncommon in lower latitudes of the North Atlantic, becoming increasingly common at higher latitudes. Quantitative survey-based estimates of killer whale abundance have been made in the eastern North Atlantic and adjacent seas, including waters off Norway, Iceland, and the Faroe Islands (Sigurjónsson et al. 1989;

Gunnlaugsson and Sigurjónsson 1990; Øien 1990). Killer whales appear common to very abundant in most of these waters and are seasonally well known as they follow herring in coastal areas off Norway (Sigurjónsson et al. 1988; Similä et al. 1996). Whaling catches of killer whales in Norwegian waters may have lowered their current abundance compared to pristine abundance levels. They have been documented infrequently in the North Sea north of Great Britain during seismic surveys (Stone 1997).

Information for the western North Atlantic is primarily qualitative, but killer whales appear to be common in areas of the Canadian Arctic (Reeves and Mitchell 1988a,

Effort (km)	No. Si	No. Ani	Si/1000 km	Abundance. Estimate	CV	Density (Animals per 100 km²)	Catalog or Minimum count	Abundance Category
Pacific Ocean and Adjacent Arctic Waters								
7,482	7	52	0.94	898	0.35	0.28		Common
33,327	11	64	0.33	511	0.35	0.06		Rare
13,042	2	2	0.15	65	0.69	0.02		Rare
	156	843				0.04	86	Rare
135,300	57	308	0.42	8,500	0.37	0.04		Rare
4,377	3	13	0.12	146		0.06		Rare
25,356	15	56	0.59	852		0.04		Rare
5,251	3	12	0.57	143		0.04		Rare
8,465	4	14	0.47	153		0.03		Rare
3,692	1	1	0.27	10		0.01		Rare
5,856	1	6	0.17	64		0.02		Rare
1,164	1	2	0.86	75		0.03		Rare
								Uncommon
8,100	4	31	0.49					Rare
17,050	2	13	0.12	349	0.98	0.01		Rare
								Common
	274	1,619					700–800	Common
318,190	26	151	0.08					Rare
20,179	7	31	0.35					Rare
4,896	1	–	0.20	0		0.00		Rare
6,452	0	–	0.00	0		0.00		Rare
				119	0.20		115	Uncommon
3,704	1	5	0.27					Rare
581 hrs	3	8						Rare
Indian Ocean and Adjacent Seas								
23,030	1	2	0.04					Rare
1,700	0	–	0.00	0		0.00		Rare
	26							Common
8,716	2	–	0.23					Rare
							43–93	

Mitchell and Reeves 1988), uncommon but seasonally regular in Labrador and Newfoundland (Lien et al. 1988), and rare in the Bay of Fundy and in U.S. Atlantic waters (Katona et al. 1988; Reeves and Mitchell 1988b; CeTAP 1982). During recent surveys conducted along mid-latitudes of the U.S. East Coast, there were no killer whale sightings (Mullin and Fulling 2003). Killer whales are considered rare in the Gulf of Mexico (O'Sullivan and Mullin 1997; Mullin and Hoggard 2000, Fulling et al. 2003), the Caribbean (Katona et al. 1988; Mitchell and Reeves 1988), and the Mediterranean Sea (Notarbartolo-di-Sciara 1987). López et al. (2004) report that killer whales are absent or rare in waters of the eastern North Atlantic, based on 1998–1999 sighting records from fishery observers onboard vessels off northwestern Spain. In the broader eastern North Atlantic, killer whales are considered widespread in coastal and offshore areas, but nowhere common (Hammond and Lockyer 1988).

South Atlantic

Although the presence of killer whales has been documented in the South Atlantic for decades (Budylenko 1981), quantitative estimates of abundance for this region are not

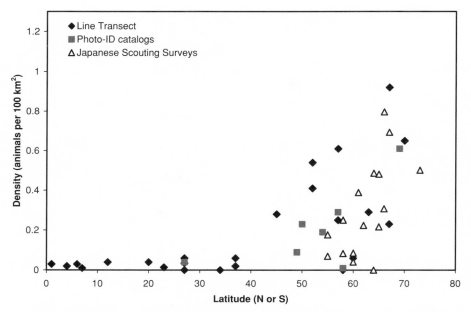

FIGURE 12.1. Worldwide killer whale density by latitude. Estimates are plotted separately for line-transect studies (diamonds), photo-identification catalog estimates of killer whale densities (squares), and Japanese whale scouting survey data (triangles; Kasamatsu and Joyce 1995), because the different types of estimates are not directly comparable.

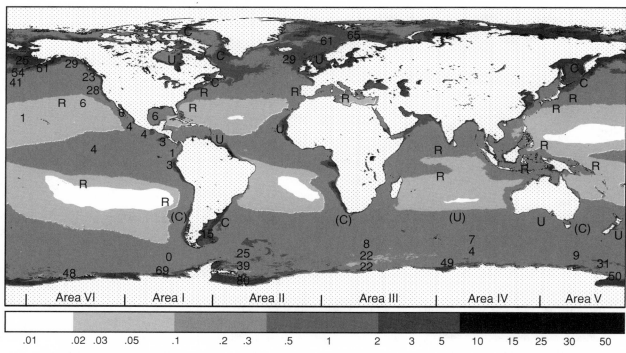

FIGURE 12.2. Worldwide killer whale densities, relative to ocean productivity as measured by the average chlorophyll-a concentration (mg/m³) from SeaWiFS images, 1997–2002. Killer whale density categories are defined as (R) rare, 0–0.10 per 100 km²; (U) uncommon, 0.10–0.20 per 100 km²; (C) common, 0.20–0.40 per 100 km²; and (A) abundant, >0.40 per 100 km². Labels in parenthesis represent greater uncertainty. Antarctic areas correspond to those used for IWC/IDCR surveys (Branch and Butterworth 2001). [Background image courtesy of NASA/Goddard Space Flight Center and ORBIMAGE, Inc.]

available. The best-studied population in the South Atlantic is regularly found off Patagonia, coincident with pinniped breeding activities, and is estimated to contain about 30 animals based on photo-identification efforts (López and López 1985; Iñíguez 2001). Killer whales are also known from pelagic waters off southern Brazil, where they have a history of depredation on tuna and swordfish longline catches (Secchi and Vaske 1998). During vessel of opportunity surveys conducted from fishing vessels off Southern Brazil in 1996–1999, killer whales were the second most common cetacean seen (Pinedo et al. 2002); however, it is possible that sighting rates were inflated because killer whales may have been attracted to the fishing vessels from which observations took place. When one takes into account only sightings that were made during vessel transits, not in association with longline operations, the overall encounter rate suggests that killer whales are nonetheless common off southern Brazil (Table 12.2). Sighting records and whaling catches indicate that killer whales are also found seasonally in the southeastern Atlantic (Budylenko 1981; Mikhalev et al. 1981), but insufficient effort data are available to categorize their abundance in this region. A map of data from Japanese whale scouting surveys (1964–1988) and whale sighting surveys (1972–1990) indicates that killer whales are common off the tip of Southern Africa (Miyashita et al. 1995).

Southern Ocean

Killer whales have long been known to be abundant in the Southern Ocean. Between 1961 and 1980, Soviet whaling operations are reported to have taken over 1,600 killer whales, including 906 killer whales during the 1979–1980 season (Mikhalev et al. 1981; Berzin and Vladimirov 1983). These operations may have reduced recent abundance in these areas. Several analyses of line-transect surveys have yielded abundance estimates for killer whales around Antarctica (Ohsumi 1981, Hammond 1984, Kasamatsu and Joyce 1995); however, some of the estimates were considered biased by methodology and survey coverage (Branch and Butterworth 2001). More recent estimates that account for some of these biases indicate that about 25,000 killer whales inhabit waters south of 60°S (Branch and Butterworth 2001); however, there are still uncertainties related to coverage of areas in the pack ice. Densities are known to vary locally within Antarctic waters, ranging from very abundant to uncommon (Secchi et al. 2002; Pitman and Ensor 2003), and it has been recognized that killer whale densities are higher closer to the ice edge, where Type B and C killer whales can occur in large aggregations of tens to hundreds of animals (Berzin and Vladimirov 1983, Pitman and Ensor 2003). Kasamatsu and Joyce (1995) calculated killer whale abundance separately for six east-west Antarctic areas and by distance from the ice edge. Although their absolute density values may be biased by the analytical methods used, their study provides a unique overview of relative patterns of Antarctic killer whale density by latitude; it

is therefore included separately in the overall assessment of worldwide latitudinal patterns (Figure 12.1).

Conclusions

The abundance categories in this study should be considered provisional, particularly for areas with limited sampling. Overall, the observed patterns of worldwide killer whale distribution are in general agreement with previous descriptions, which indicate that killer whales are more common at higher latitudes and in coastal areas. Killer whale occurrence also appears broadly tied to regions of higher ocean productivity, as indicated by remotely sensed chlorophyll levels (Figure 12.2), and the latitudinal and inshore/offshore patterns of abundance may simply be a reflection of the higher productivity in coastal and high-latitude areas. Regions of similar latitude exhibit differences in abundance that may be tied to patterns of productivity and prey availability; for example, killer whales are rare along the U. S. northeast coast although they are common along the coast of Oregon and Washington, at a similar latitude.

Many portions of the world's ocean have not been adequately studied, but it is noteworthy that fish-eating specialists appear to occur only at high latitudes in the Pacific, Atlantic, and Southern Oceans. In both the Pacific and Southern Oceans, fish-eating specialists occur in sympatry or parapatry with other types of killer whales that appear to specialize on marine mammal prey, suggesting that in high-productivity areas killer whales have undergone a form of niche separation to take advantage of relatively abundant prey, and this may have occurred independently several times. Although relatively fewer studies have occurred at lower latitudes and in less productive areas, there is some suggestion that killer whales in these areas may be more generalist predators (Baird 2002). For example, killer whales that are thought to represent a single population in the Gulf of California have been observed to take both marine mammal and fish prey (Silber et al. 1990, Guerrero-Ruiz et al. 2002).

Although the available data are far from complete, abundance estimates for the areas that have been sampled provide a *minimum* worldwide abundance estimate of about 50,000 killer whales. It is likely that the total abundance is considerably higher, because estimates are not available for many high-latitude areas of the northern hemisphere and for large areas of the South Pacific, South Atlantic, and Indian Oceans. The impact of killer whales on the oceanic ecosystem is difficult to evaluate without further targeted studies, particularly given their broad range of prey species. Ultimately, their impact depends fundamentally on four things: how many there are, where they are, what they eat, and how much they eat. This study has provided a worldwide overview of the first two of these issues and provided limited information on the third. Combined with more detailed studies on diet, energetics, and foraging ecology of individual populations, this represents a step towards gaining a better understanding of the role of killer whales in the ecosystem.

Acknowledgments

This paper was enhanced through thoughtful discussions with many people. We are grateful to Janice Waite, who reviewed recent information on eastern North Pacific killer whales and estimated the areas inhabited by these populations. Joan Parker at Moss Landing Marine Laboratories provided valuable assistance in tracking down references. We thank Robin Baird, Jay Barlow, Dan Doak, Jim Estes, and Richard Neal for their helpful comments on previous drafts of this manuscript.

Literature Cited

Aguayo, A., R. Bernal, C. Olavarría, V. Vallejos, and R. Hucke-Gaete. 1998. Cetofauna off central Chile, including Easter Island. *Revista de Biología de Marina y Oceanografía* 33(1): 101–123.

Angliss, R. P. and K. L. Lodge. 2002. *Alaska marine mammal stock assessments, 2002.* U. S. Department of Commerce, NOAA Technical Memorandum NMFS-AFSC-133. Seattle: Alaska Fisheries Science Center.

Baird, R. W. 1994. Foraging behavior and ecology of transient killer whales. Ph.D. thesis, Simon Fraser University, Burnaby, British Columbia.

———. 2000. The killer whale: foraging specializations and group hunting, in *Cetacean societies: field studies of dolphins and whales.* J. Mann, R. C. Connor, P. L. Tyack, and H. Whitehead, eds. Chicago: University of Chicago Press, pp. 127–153.

———. 2002. *Killer whales of the world: natural history and conservation.* Stillwater, MN: Voyageur Press.

Baird, R.W., P.A. Abrams, and L.M. Dill. 1992. Possible indirect interactions between transient and resident killer whales: implications for the evolution of foraging specializations in the genus *Orcinus. Oecologia* 89: 125–132.

Baird, R. W. and L. M. Dill. 1995. Occurrence and behaviour of transient killer whales: seasonal and pod-specific variability, foraging behavior, and prey handling. *Canadian Journal of Zoology* 73: 1300–1311.

———. 1996. Ecological and social determinants of group size in transient killer whales. *Behavioral Ecology* 7(4): 408–416.

Baird, R. W. and H. Whitehead. 2000. Social organization of mammal-eating killer whales: group stability and dispersal patterns. *Canadian Journal of Zoology* 78: 2096–2105.

Balcomb, K. C. and C. A. Goebel. 1976. *A killer whale study in Puget Sound.* Final Report, Contract No. NASO-6-35330. Seattle: National Marine Fisheries Service.

Ballance, L. T, C. R. Anderson, R. L. Pitman, K. Stafford, A. Shaan, Z. Waheed, and R. L. Brownell, Jr. 2001. Cetacean sightings around the Republic of the Maldives, April 1998. *Journal of Cetacean Research and Management* 3: 213–218.

Barlow, J. 1995. The abundance of cetaceans in California waters. Part II: Ship surveys in summer and fall of 1991. *Fishery Bulletin* 93: 1–14.

———. 2003a. *Preliminary estimates of the abundance of cetaceans along the U.S. west coast: 1991–2001.* Administrative Report LJ-03-03. La Jolla, CA: Southwest Fisheries Science Center.

———. 2006. Cetacean abundance in Hawaiian waters during summer/fall 2002. *Marine Mammal Science* 22: 446–464.

Barrat, A. and J. L. Mougin. 1978. The Southern elephant seal, *Mirounga leonina*, of Possession Island, Crozet Archipelago, 46°25′ south, 51°45′ east. *Mammalia* 42: 143–174 (in French).

Barrett-Lennard, L. G. 2000. Population structure and mating patterns of killer whales as revealed by DNA analysis. Ph.D. thesis, University of British Columbia, Vancouver.

Barrett-Lennard, L. G. and G. M. Ellis. 2001. *Population structure and genetic variability in northeastern Pacific killer whales: towards an assessment of population viability.* Research document 2001/065. Nanaimo, BC: Fisheries and Oceans Canada.

Berzin, A. A. and V. L. Vladimirov. 1983. A new species of killer whale (Cetacea, Delphinidae) from Antarctic waters. *Zoologicheskii Zhurnal* 62: 287–295 (in Russian).

———. 1989. Recent distribution and abundance of cetaceans in the Sea of Okhotsk. *Biologiya morya (Vladivostok)* 15(2): 84–90 (in Russian).

Bigg, M. A., G. M. Ellis, J. K. B. Ford, and K. C. Balcomb. 1987. *Killer whales: A study of their identification, genealogy, and natural history in British Columbia and Washington State.* Nanaimo, BC: Phantom Press and Publishers, Inc.

Bigg, M. A., P. F. Olesiuk, G. M. Ellis, J. K. B. Ford, and K. C. Balcomb III. 1990. Social organization and genealogy of resident killer whales *(Orcinus orca)* in the coastal waters of British Columbia and Washington State. *Report of the International Whaling Commission* (Special Issue) 12: 386–406.

Bisther, A. 2002. Intergroup interactions among killer whales in Norwegian coastal waters; tolerance vs. aggression at feeding areas. *Aquatic Mammals* 28:14–23.

Black, N.A., A. Schulman-Janiger, R.L. Ternullo, and M. Guerrero-Ruiz. 1997. *Killer whales of California and Western Mexico: a catalog of photo-identified individuals.* NOAA Technical Memorandum NMFS-SWFSC-247. La Jolla, CA: Southwest Fisheries Science Center.

Braham, H. W. and M. E. Dahlheim. 1982. Killer whales in Alaska documented in the Platforms of Opportunity Program. *Report of the International Whaling Commission* 32: 643–646.

Branch, T. A. and D. S. Butterworth. 2001. Estimates of abundance south of 60 degrees S for cetacean species sighted frequently on the 1978/79 to 1997/98 IWC/IDCR-SOWER sighting surveys. *Journal of Cetacean Research and Management* 3: 251–270.

Buckland, S. T., D. R. Anderson, K. P. Burnham, and J. L. Laake. 1993. *Distance sampling: estimating abundance of biological populations.* London: Chapman and Hall.

Budylenko, G. A. 1981. Distribution and some aspects of the biology of killer whales in the South Atlantic. *Report of the International Whaling Commission* 31: 523–525.

Calambokidis, J. and J. Barlow. 2004. Abundance of blue and humpback whales in the eastern North Pacific estimated by capture-recapture and line-transect methods. *Marine Mammal Science* 20: 63–85.

Caldwell, D. K., and M. C. Caldwell. 1969. Addition of the leatherback sea turtle to the known prey of the killer whale, *Orcinus orca. Journal of Mammalogy* 50(3): 636.

Carretta, J. V., J. Barlow, K. A. Forney, M. M. Muto, and J. Baker. 2001. *U.S. Pacific marine mammal stock assessments: 2001.* NOAA Technical Memorandum NMFS-SWFSC-317. La Jolla, CA: Southwest Fisheries Science Center.

Carretta, J. V., M. M. Muto, J. Barlow, J. Baker, K. A. Forney, and M. Lowry. 2002. *U. S. Pacific marine mammal stock assessments: 2002.* NOAA Technical Memorandum NMFS-SWFSC-346. La Jolla, CA: Southwest Fisheries Science Center.

CeTAP. 1982. *A characterization of marine mammals and turtles in the mid- and north Atlantic areas of the U. S. outer continental shelf.* Cetacean and Turtle Assessment Program, University of Rhode Island, Final Report AA551-CT8-48. Washington, DC: Bureau of Land Management.

Clapham, P., J. Barlow, M. Bessinger, T. Cole, D. Mattila, R. Pace, D. Palka, J. Robbins, and R. Seton. 2003. Abundance and demographic parameters of humpback whales from the Gulf of Maine, and stock definition relative to the Scotian Shelf. *Journal of Cetacean Research and Management* 5: 13–22.

Dahlheim, M. E. 1981. A review of the biology and exploitation of the killer whale, *Orcinus orca*, with comments on recent sightings from Antarctica. *Report of the International Whaling Commission* 31: 541–546.

———. 1997. *A photographic catalog of killer whales,* Orcinus orca, *from the Central Gulf of Alaska to the Southeastern Bering Sea.* NOAA Technical Report NMFS 131. Seattle: National Marine Fisheries Service.

Dahlheim, M. E., D. K. Ellifrit, and J. D. Swenson. 1997. *Killer whales of Southeast Alaska: a catalogue of photo-identified individuals.* Seattle: National Marine Fisheries Service.

Dahlheim, M. E. and J. E. Heyning. 1999. Killer whale, in *Handbook of marine mammals, Vol. 6.* S. Ridgway and R. Harrison, eds. San Diego: Academic Press, pp. 281–322.

Dahlheim, M. E., J. S. Leatherwood, and W. E. Perrin. 1982. Distribution of killer whales in the warm temperate and tropical eastern Pacific. *Report of the International Whaling Commission* 32: 647–653.

Estes, J. A., M. T. Tinker, T. M. Williams, and D. F. Doak. 1998. Killer whale predation on sea otters linking oceanic and nearshore ecosystems. *Science* 282: 473–476.

Eyre, E. J. 1995. Observations of cetaceans in the Indian Ocean Whale Sanctuary, May-July 1993. *Report of the International Whaling Commission* 45: 419–426.

Fiscus, C.H. and K. Niggol. 1965. *Observations of cetaceans off California, Oregon and Washington.* Special Scientific Report, Fisheries No. 498. Washington, DC: U.S. Fish and Wildlife Service.

Ford, J. K. B. and G. M. Ellis. 1999. *Transients: mammal-hunting killer whales of British Columbia, Washington, and southeastern Alaska.* Vancouver: University of British Columbia Press.

Ford, J. K. B., G. M. Ellis, and K. C. Balcomb. 1994. *Killer whales: the natural history and genealogy of* Orcinus orca *in British Columbia and Washington State.* Vancouver: University of British Columbia Press.

———. 2000. *Killer whales: the natural history and genealogy of* Orcinus orca *in British Columbia and Washington.* 2nd edition. Seattle: University of Washington Press.

Ford, J. K. B., G. M. Ellis, L. G. Barrett-Lennard, A. B. Morton, R. S. Palm, and K. C. Balcomb. 1998. Dietary specialization in two sympatric populations of killer whales *(Orcinus orca)* in coastal British Columbia and adjacent waters. *Canadian Journal of Zoology* 76(8): 1456–1471.

Forney, K.A., J. Barlow, and J.V. Carretta. 1995. The abundance of cetaceans in California waters. Part II: Aerial surveys in winter and spring of 1991 and 1992. *Fishery Bulletin* 93: 15–26.

Friday, N., T. D. Smith, P. T. Stevick, J. Allen. 2000. Measurement of photographic quality and individual distinctiveness for the photographic identification of humpback whales, *Megaptera novaeangliae*. *Marine Mammal Science* 16: 355–374.

Fulling, G. L., K. D. Mullin, and C. W. Hubard. 2003. Abundance and distribution of cetaceans in outer continental shelf waters of the U. S. Gulf of Mexico. *Fishery Bulletin* 101: 923–932.

Gannier, A. 2000. Distribution of cetaceans off the Society Islands (French Polynesia) as obtained from dedicated surveys. *Aquatic Mammals* 26.2: 111–126.

Garrigue, C. and J. Greaves. 2001. Cetacean records for the New Caledonian area (southwest Pacific Ocean). *Micronesica* 34: 27–33.

Garrison, L. P., S. L. Swartz, A. Martinez, C. Burks, and K. Stamates. 2003. *A marine mammal assessment survey of the southeast U.S. continental shelf: February–April 2002.* NOAA Technical Memorandum NMFS-SEFSC-492. Miami: Southeast Fisheries Science Center.

Gerrodette, T. and D. M. Palacios. 1996. *Estimates of cetacean abundance in EEZ waters of the eastern tropical Pacific.* SWFSC Administrative Report LJ-96-10. La Jolla, CA: Southwest Fisheries Science Center.

Goley, P.D. and J.M. Straley. 1994. Attack on gray whales in Monterey Bay, California, by killer whales previously identified in Glacier Bay, Alaska. *Canadian Journal of Zoology* 72: 1528–1530.

Guerrero-Ruiz, M., D. Gendron, and J. Urbán Ramirez. 1998. Distribution, movements and communities of killer whales *(Orcinus orca)* in the Gulf of California, Mexico. *Report of the International Whaling Commission* 48: 537–543.

Guerrero-Ruiz, M., J. Urban-Ramirez, D. Gendron, and V. Flores de Sahagun. 2002. Current knowledge of killer whales in the Mexican Pacific, in *Proceedings of the Fourth International Orca Symposium and Workshops, September 23–28, 2002.* Villiers en Bois, France: CEBC-CNRS, p. 76.

Gunnlaugsson, T. and J. Sigurjónsson. 1990. NASS-87: estimation of whale abundance based on observations made onboard Icelandic and Faroese survey vessels. *Report of the International Whaling Commission* 40: 571–580.

Hammond, P.S. 1984. Abundance of killer whales in Antarctic Areas II, III, IV, and V. *Report of the International Whaling Commission* 34: 543–548.

———. 1986. Estimating the size of naturally marked whale populations using capture-recapture techniques. *Report of the International Whaling Commission* (Special Issue) 8: 253–282.

Hammond, P.S. and C. Lockyer. 1988. Distribution of killer whales in the eastern North Atlantic. *Rit Fiskideildar* 11: 24–41.

Hammond, P.S., S.A. Mizroch, and G.P. Donovan (editors). 1990. Individual recognition of cetaceans: use of photo-identification and other techniques to estimate population parameters. *Report of the International Whaling Commission* (Special Issue) 12.

Hansen, L.J. 1990. California coastal bottlenose dolphins, in *The bottlenose dolphin.* S. Leatherwood and R.R. Reeves, eds. San Diego: Academic Press, pp. 403–420.

Heise, K., G. Ellis, and C. Matkin. 1991. *A catalogue of Prince William Sound killer whales.* Homer, AK: North Gulf Oceanic Society.

Heyning, J. E. and M. E. Dahlheim. 1988. *Orcinus orca.* *Mammalian Species* 304: 1–9.

Hoelzel, A. R., M. Dahlheim, and S. J. Stern. 1998. Low genetic variation among killer whales *(Orcinus orca)* in the eastern north Pacific and genetic differentiation between foraging specialists. *Journal of Heredity* 89(2): 121–128.

Hoelzel, A.R. and G.A. Dover. 1991. Genetic differentiation between sympatric killer whale populations. *Heredity* 66: 191–195.

Iñíguez, M. A. 2001. Seasonal distribution of killer whales *(Orcinus orca)* in Northern Patagonia, Argentina. *Aquatic Mammals* 27.2: 154–161.

Jefferson, T. A., P. J. Stacey, and R. W. Baird. 1991. A review of killer whale interactions with other marine mammals: predation to co-existence. *Mammal Review* 21: 151–180.

Jonsgård, Å. and Lyshoel, P. B. 1970. A contribution to the biology of the killer whale, *Orcinus orca. Nytt Magasin for Zoologi* 18, 41–48.

Kahn, B. and J. Pet. 2003. Long-term visual and acoustic cetacean surveys in Komodo National Park, Indonesia, 1999–2001: management implications for large migratory marine life, in *Proceedings and Publications of the World Congress on Aquatic Protected Areas 2002.* Lyneham, Australia: Australian Society for Fish Biology.

Kasamatsu, F. and G. G. Joyce. 1995. Current status of odontocetes in the Antarctic. *Antarctic Science* 7: 365–379.

Kasuya, T. 1971. Consideration of distribution and migration of toothed whales off the Pacific coast of Japan based upon aerial sighting record. *Scientific Reports of the Whales Research Institute* 23: 37–60.

Katona, S. K., J. A. Beard, P. E. Girton, and F. Wenzel. 1988. Killer whales *(Orcinus orca)* from the Bay of Fundy to the Equator, including the Gulf of Mexico. *Rit Fiskideildar* 11: 205–224.

Keith, M., M. N. Bester, P. A. Bartlett, and D. Baker. 2001. Killer whales *(Orcinus orca)* at Marion Island, Southern Ocean. *African Zoology* 36(2): 163–175.

Laran, S. and A. Gannier. 2001. Distribution of cetaceans in the Marquesas Islands (French Polynesia) and comparison with a precedent survey. Abstract presented at the 15th Conference of the European Cetacean Society, Rome, 2001.

Leatherwood, S., A. E. Bowles, E. Krygier, J. D. Hall, S. Ignell. 1984. Killer whales *(Orcinus orca)* in Southeast Alaska, Prince William Sound, and Shelikof Strait: a review of available information. *Report of the International Whaling Commission* 34: 521–530.

Leatherwood, S., D. K. Caldwell, and H. E. Winn. 1976. *Whales, dolphins and porpoises of the western North Atlantic: a guide to their identification.* NOAA Technical Report No. 396. Seattle: National Marine Fisheries Service.

Leatherwood, J. S. and M. E. Dahlheim. 1978. *Worldwide distribution of pilot whales and killer whales.* Report No. 443. San Diego: Naval Ocean Systems Center.

Leatherwood, S., C.O. Matkin, J.D. Hall, and G.M. Ellis. 1990. Killer whales, *Orcinus orca,* photo-identified in Prince William Sound, Alaska 1976 to 1987. *Canadian Field-Naturalist* 104: 362–371.

Lien, J., G. B. Stenson, and P. W. Jones. 1988. Killer whales *Orcinus orca* in waters off Newfoundland and Labrador, Canada, 1978–1986. *Rit Fiskideildar* 11: 194–201.

Ling, J. K. 1991. Recent sightings of killer whales, *Orcinus orca* (Cetacea: Delphinidae), in South Australia. *Transactions of the Royal Society of South Australia* 115: 95–98.

López, A., G.J. Pierce, X. Valeiras, M.B. Santos, and A. Guerra. 2004. Distribution patterns of small cetaceans in Galician waters. *Journal of the Marine Biological Association of the UK* 84: 283–294.

López, J.C. and D. López. 1985. Killer whales *(Orcinus orca)* of Patagonia and their behavior of intentional stranding while hunting nearshore. *Journal of Mammalogy* 66: 181–183.

Lowry, L. F., R. R. Nelson, and K. J. Frost. 1987. Observations of killer whales, *Orcinus orca,* in Western Alaska: sightings, strandings, and predation on other marine mammals. *Canadian Field-Naturalist* 101: 6–12.

MacLeod, K., M. P. Simmonds, and E. Murray. 2003. Summer distribution and relative abundance of cetacean populations off north-west Scotland. *Journal of the Marine Biological Association of the UK* 83: 1187–1192.

Matkin, C., G. Ellis, E. Saulitis, L. Barrett-Lennard, and D. Matkin. 1999. *Killer whales of southern Alaska.* Homer, AK: North Gulf Oceanic Society.

Matthews, J. N., L. Steiner, and J. Gordon. 2001. Mark-recapture analysis of sperm whale *(Physeter macrocephalus)* photo-ID data from the Azores (1987–1995). *Journal of Cetacean Research and Management* 3: 219–226.

Merlen, G. 1999. The orca in Galápagos: 135 sightings. *Noticias de Galápagos* 60: 2–8.

Mikhalev, Y. A., M. V. Ivashin, V. P. Savusin, and F. E. Zelenaya. 1981. The distribution and biology of killer whales in the Southern Hemisphere. *Report of the International Whaling Commission* 31: 551–566.

Mironova, A. M., A. M. Burdin, E. Hoyt, E. L. Jikiya, V. S. Nikulin, N. N. Pavlov, H. Sato, K. K. Tarasyan, and O. A. Filatova. 2002. *O. orca* abundance, distribution, seasonal presence, predation and strandings in the waters around Kamchatka and the Kommander Islands: an assessment based on reported sightings 1992–2000, in *Proceedings of the Fourth International Orca Symposium and Workshops, September 23–28, 2002.* Villiers en Bois, France: CEBC-CNRS, pp. 106–108.

Mitchell, E. and R. R. Reeves. 1988. Records of killer whales in the western North Atlantic, with emphasis on eastern Canadian waters. *Rit Fiskideildar* 11: 161–193.

Miyashita, T., H. Kato, and T. Kasuya (editors). 1995. *Worldwide map of cetacean distribution based on Japanese sighting data,* Vol. 1. Shimizu, Japan: National Research Institute of Far Seas Fisheries.

Miyashita, T., T. Kishiro, N. Higashi, F. Sata, K. Mori, and H. Kato. 1996. Winter distribution of cetaceans in the western North Pacific inferred from sighting cruises 1993–1995. *Report of the International Whaling Commission* 46: 437–441.

Mobley, J. R., L. Mazzuca, A. S. Craig, M. W. Newcomer, and S. S. Spitz. 2001. Killer whales *(Orcinus orca)* sighted west of Ni`ihau, Hawai`i. *Pacific Science* 55: 301–303.

Mullin, K. D. and G. L. Fulling. 2003. Abundance of cetaceans in the southern U. S. North Atlantic Ocean during summer 1998. *Fishery Bulletin* 101: 603–613.

———. 2004. Abundance of cetaceans in the oceanic northern Gulf of Mexico, 1996–2001. *Marine Mammal Science* 20: 787–807.

Mullin, K. D. and W. Hoggard. 2000. Visual surveys of cetaceans and sea turtles from aircraft and ships, in *Cetaceans, sea turtles and seabirds in the northern Gulf of Mexico: distribution, abundance and habitat associations. Volume 2: Technical report.* R. W. Davis, W. E. Evans, and B. Würsig, eds. New Orleans: Minerals Management Service, pp. 111–171.

Murie, O. J. 1959. *Fauna of the Aleutian Islands and Alaska Peninsula. North American fauna.* Report No. 61. Washington, DC: U. S. Fish and Wildlife Service.

Nishiwaki, M. and C. Handa. 1958. Killer whales caught in coastal waters off Japan. *Scientific Reports of the Whales Research Institute* 13: 85–96.

Norris, K. S. and J. H. Prescott. 1961. Observations of Pacific cetaceans of Californian and Mexican waters. *University of California Publications in Zoology* 63: 291–402.

Notarbartolo-di-Sciara, G. 1987. Killer whale *Orcinus-orca* in the Mediterranean Sea. *Marine Mammal Science* 3: 356–360.

Ohsumi, S. 1975. Review of Japanese small-type whaling. *Journal of the Fisheries Research Board of Canada* 32(7): 1111–1121.

———. 1981. Distribution and abundance of killer whales in the Southern Hemisphere. Paper SC/Jn81/KW10 submitted to the Scientific Committee of the International Whaling Commission.

Øien, N. 1990. Sightings surveys in the Northeast Atlantic in July 1988: distribution and abundance of cetaceans. *Report of the International Whaling Commission* 40: 499–511.

Olesiuk, P. F., M. A. Bigg, and G. M. Ellis. 1990. Life history and population dynamics of resident killer whales (*Orcinus orca*) in the coastal waters of British Columbia and Washington State. *Report of the International Whaling Commission* (Special Issue) 12: 209–244.

Osborne, R., J. Calambokidis, and E. M. Dorsey. 1988. *A guide to marine mammals of greater Puget Sound.* Friday Harbor, WA: Island Publishers.

O'Sullivan, S. and K. D. Mullin. 1997. Killer whales (*Orcinus orca*) in the northern Gulf of Mexico. *Marine Mammal Science* 13: 141–147.

Pinedo, M. C., T. Polacheck, A. S. Barreto, and M. P. Lammardo. 2002. A note on vessel of opportunity sighting surveys for cetaceans in the shelf edge region off the southern coast of Brazil. *Journal of Cetacean Research and Management* 4: 323–329.

Pistorius, P. A., F. E. Taylor, C. Louw, B. Hanise, M. N. Bester, C. De Wet, A. du Plooy, N. Green, S. Klasen, P. Podile, and J. Schoeman. 2002. Distribution, movement and estimated populations of killer whales at Marion Island, December 2000. *South African Journal of Wildlife Research* 32: 86–92.

Pitman, R. L. and P. Ensor. 2003. Three forms of killer whales (*Orcinus orca*) in Antarctic waters. *Journal of Cetacean Research and Management* 5(2): 131–139.

Poncelet, E., C. Guinet, S. Mangin, and C. Barbraud. 2002. Life history and decline of killer whales in the Crozet Archipelago, Southern Indian Ocean. Poster presentation. 4th International Orca Symposium and Workshops, September 23–28, 2002. Available at http://www.circe-asso.org/files/poster_orque_crozet_en.PDF. Accessed May 25, 2006.

Read, A. J., K. W. Urian, B. Wilson, and D. M. Waples. 2003. Abundance of bottlenose dolphins in the bays, sounds, and estuaries of North Carolina. *Marine Mammal Science* 19: 59–73.

Reeves, R. R. and E. Mitchell. 1988a. Distribution and seasonality of killer whales in the eastern Canadian Arctic. *Rit Fiskideildar* 11: 136–160.

———. 1988b. Killer whale sightings and takes by American pelagic whalers in the North Atlantic. *Rit Fiskideildar* 11: 7–23.

Rice, D. W. 1968. Stomach contents and feeding behaviour of killer whales in the eastern North Pacific. *Norsk Hvalfangsttidende* 57: 35–38.

———. 1974. Whales and whale research in the eastern North Pacific, in *The whale problem: a status report.* W.E. Schevill, ed. Cambridge, MA: Harvard University Press, pp. 170–195.

Saulitis, E., C. Matkin, L. Barrett-Lennard, K. Heise, and G. Ellis. 2000. Foraging strategies of sympatric killer whale (*Orcinus orca*) populations in Prince William Sound, Alaska. *Marine Mammal Science* 16: 94–109.

Scammon, C. M. 1874. *The marine mammals of the northwestern coast of North America.* San Francisco: John H. Carmany and Co. Reprint, New York: Dover Publications.

Scheffer, V. B. 1967. The killer whale. *Pacific Search* 1(7): 2.

Seber, G. A. F. 1982. *The estimation of animal abundance and related parameters.* 2nd edition. New York: Oxford University Press.

Secchi, E. R., L. Dalla Rosa, P. G. Kinas, M. C. O. Santos, A. N. Zerbini, M. Bassoi, and I. B. Moreno. 2002. Encounter rates of whales around the Antarctic Peninsula with special reference to humpback whales, *Megaptera novaeangliae*, in the Gerlache Strait: 1997/98 to 1999/2000. *Memoirs of the Queensland Museum* 47(2): 571–578.

Secchi, E. R. and T. Vaske, Jr. 1998. Killer whale (*Orcinus orca*) sightings and depredation on tuna and swordfish longline catches in southern Brazil. *Aquatic Mammals* 24.2: 117–122.

Shimada, H. and L.A. Pastene. 1995. Report of a sighting survey off the Solomon Islands with comments on Bryde's whale distribution. *Report of the International Whaling Commission* 45: 413–418.

Sigurjónsson, J., T. Gunnlaugsson, and M. Payne. 1989. NASS-87: Shipboard sighting surveys in Icelandic and adjacent waters June-July 1987. *Report of the International Whaling Commission* 39: 395–409.

Sigurjónsson, J. and S. Leatherwood, eds. 1988. North Atlantic killer whales. *Rit Fiskideildar* 11: 1–317.

Sigurjónsson, J., T. Lyrholm, S. Leatherwood, E. Jonsson, and G. Vikingsson. 1988. Photo-identification of killer whales, *Orcinus orca*, off Iceland 1981 through 1986. *Rit Fiskideildar* 11: 99–114.

Silber, G.K., M.W. Newcomer, and M.H. Perez-Cortes. 1990. Killer whales (*Orcinus orca*) attack and eat a Bryde's whale (*Balaenoptera edeni*). *Canadian Journal of Zoology* 68: 1603–1606.

Similä, T., J. C. Holst, and I. Christensen. 1996. Occurrence and diet of killer whales in northern Norway: seasonal patterns relative to distribution and abundance of Norwegian spring-spawning herring. *Canadian Journal of Fisheries and Aquatic Sciences* 53: 769–779.

Sivasubramanian, K. 1965. Predation of tuna longline catches in the Indian Ocean, by killer whales and sharks. *Bulletin of the Fisheries Research Station, Ceylon* 17: 221–236.

Springer, A. M., J. A. Estes, G. B. van Vliet, T. M. Williams, D. F. Doak, E. M. Danner, K. A. Forney, and B. Pfister. 2003. Sequential megafaunal collapse in the North Pacific Ocean: an ongoing legacy of industrial whaling? *Proceedings of the National Academy of Sciences* 100: 12223–12228.

Stevens, T. A., D. A. Duffield, E. D. Asper, K. G. Hewlett, A. Bolz, L. J. Gage, and G. D. Bossart. 1989. Preliminary findings of restriction fragment differences in mitochondrial DNA among killer whales, *Orcinus orca. Canadian Journal of Zoology* 67. 2592–2595.

Stone, C. J. 1997. *Cetacean observations during seismic surveys in 1996.* JNCC Report No. 228. Totnes, UK: Joint Nature Conservation Committee. Available at http://www.jncc.gov.uk/pdf/cetaceous%20observations%201996.pdf.

Tomilin, A. G. 1967. *Mammals of the USSR and adjacent countries. Mammals of eastern Europe and northern Asia.* Vol. IX. *Cetacea.* Israel Program for Scientific Translations, trans. Jerusalem: Israel Program for Scientific Translations.

Visser, I.N. 2000. Orca *(Orcinus orca)* in New Zealand waters. Doctoral dissertation, University of Auckland, Auckland, New Zealand.

Wade, P.R. and T. Gerrodette. 1993. Estimates of cetacean abundance and distribution in the eastern tropical Pacific. *Report of the International Whaling Commission* 43: 477–493.

Waite, J.M., N.A. Friday, and S.E. Moore. 2002. Killer whale *(Orcinus orca)* distribution and abundance in the central and southeastern Bering Sea, July 1999 and June 2000. *Marine Mammal Science* 18: 779–786.

Wang, P. 1985. *Distribution of cetaceans in Chinese waters.* Administrative Report No. LJ-85-24. La Jolla, CA: Southwest Fisheries Center.

Wilson, B., P.S. Hammond, and P.M. Thompson. 1999. Estimating size and assessing trends in a coastal bottlenose dolphin population. *Ecological Applications* 9: 288–300.

Zerbini, A.N., J.M. Waite, J.W. Durban, R. LeDuc, M.E. Dahlheim, and P.R. Wade. 2006. Estimating abundance of killer whales in the nearshore waters of the Gulf of Alaska and Aleutian Islands using line transect sampling. *Marine Biology* (in press).

THIRTEEN

The Natural History and Ecology of Killer Whales

LANCE G. BARRETT-LENNARD AND KATHY A. HEISE

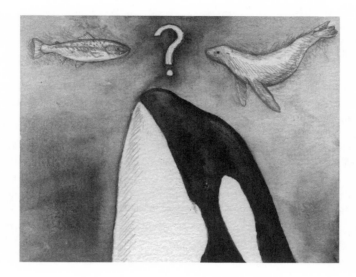

In his landmark book on marine mammals of the northeastern Pacific, whaling captain Charles Scammon remarked that killer whales "may be regarded as marine beasts, that roam over every ocean; entering bays and lagoons where they spread terror and death among mammoth balaenas and the smaller species of dolphins, as well as pursuing the seal and walrus, devouring, in their marauding expeditions up swift rivers, numberless salmon or other large fishes that may come in their way" (Scammon 1874:). Scammon's observations reflect the tendency, until very recent times, of most observers to describe killer whales—and other predators—from the perspective of their prey. It was not until aquariums began displaying killer whales in the late 1960s that they began to be seen as interesting creatures in their own right. Belying their bloodthirsty reputation in the wild, captive killer whales bonded quickly with their trainers and with other marine mammals, including much smaller species such as dolphins. They are now universally recognized as both highly social and by any measure highly intelligent. These features, along with their size and striking appearance, have earned them iconic status in popular culture, alongside such species as gorillas and elephants. Great advances in the scientific understanding of the cognitive and sensory abilities, social organization, population structure, and behavior of the killer whale were made throughout the 1970s, 1980s, and 1990s. Much of this research is continuing, but, in a trend that Scammon would appreciate,

there is now growing scientific interest in the impact of such a versatile group-foraging predator on the population dynamics, distribution, behavior, and evolution of its prey.

Population Structure

Killer whales (*Orcinus orca*) are the largest members of the family Delphinidae (marine dolphins). Only a single species is recognized at present. However, extensive interpopulational differences in morphology and ecology have been described, and distinct, reproductively isolated ecotypes persist in sympatry. The existence of fish-eating *resident* and mammal-hunting *transient* killer whales that co-occur, yet are socially and genetically isolated, is well documented in the northeastern Pacific Ocean (Bigg et al. 1987; Baird et al. 1992; Baird and Dill 1995, 1996; Ford et al. 1998, 2000; Hoelzel 1998; Hoelzel et al. 1998; Matkin et al. 1999; Barrett-Lennard 2000). A somewhat similar situation appears to exist in the Antarctic, where at least two and possibly three morphological and ecological variants persist in close proximity (Mikhalev et al. 1981; Berzin and Vladimirov 1983; Pitman and Ensor 2003). It is possible that taxonomic revisions dividing *Orcinus orca* into two or more species will eventually be accepted based on reproductive isolation and ecological specialization. However, genetic studies have so far failed to find genetic linkages between ecologically similar populations from the northern and southern hemispheres (Hoelzel et al. 1998), and no consensus has yet emerged on where species lines should be drawn. The known patterns are consistent both with a species that recently began an extensive adaptive radiation into a variety of new niches and with a species that has a propensity to live in xenophobic populations of at most a few hundred individuals. Although the two views are not mutually exclusive, they do support different conclusions regarding whether to split the species or leave it lumped. If an adaptive radiation is occurring, the divergent forms we see represent incipient species, and there is utility in naming them as such. On the other hand, if current patterns are to a great extent the result of social factors that cause populations that reach a certain size to divide, then the separations we see at present are likely ephemeral, and little is to be gained by according them species status. Until we better understand the processes leading to the diversity of closely related but well-differentiated forms of *Orcinus orca*, we believe that the best strategy is to resist the temptation to revise its taxonomy formally, and to simply describe it as a species complex.

Orcinus orca is distributed throughout all of the world's major ocean basins, and reaches its maximum abundance in productive coastal waters in temperate and polar regions. As such, it is probably the most widely distributed mammal on the planet after humans and their domesticated animals and pests. As would be expected of an apex predator, killer whales are relatively uncommon despite their extensive distribution. No reliable worldwide estimates of their numbers exist; however, they are believed to number in the low thousands in the northeastern Pacific (Ford et al. 2000; Barrett-Lennard and Ellis 2001), fewer than 1,000 in the Bering Sea (Waite et al. 2002), from 500 to 1,500 in the northeast Atlantic near Norway (Christensen 1988), from 200 to 300 in the northern Gulf of Mexico (Blaylock et al. 2000), and on the order of 25,000 in the southern ocean (Branch and Butterworth 2001). If these numbers typify the abundance of the species in similarly productive waters elsewhere, a rough filling-in-the-blanks exercise suggests that the worldwide population is probably in the range of 40,000–60,000 individuals.

Killer whales do not migrate to specific breeding or calving areas distant from their feeding grounds, unlike some other cetacean species. In the northeastern Pacific Ocean, where systematic research has focused on nearshore populations of the species for more than 30 years, killer whales shift their distributions and movement patterns seasonally to take advantage of prey aggregations (Nichol and Shackleton 1996), but they rarely if ever disperse permanently from their natal home ranges (Ford et al. 2000). Some authors (e.g., Budylenko 1981; Mikhalev et al. 1981) have described annual north-south movements in killer whale populations in the southern ocean that reflect seasonal changes in prey distribution.

Sympatric Killer Whale Ecotypes in the Northeastern Pacific Ocean

In 1970 Michael Bigg, a research scientist at the Pacific Biological Station in Nanaimo, British Columbia, began a sys-

tematic study of the killer whales in the waters of southern British Columbia and northern Washington State. By the late 1970s, he had determined that two socially isolated groups of killer whales shared the area (Bigg 1982). These groups, known as residents and transients, have been referred to variously as types, forms, races, populations, and ecotypes. Bigg's findings stimulated great interest in the species and led to an ongoing series of field studies by government, university, and independent researchers. This research has revealed that residents and transients specialize on fundamentally different prey—fish and marine mammals, respectively (Bigg et al. 1987; Ford et al. 1998). A third killer whale group, known as offshores, was identified in British Columbia in the late 1980s (Ford et al. 2000). Offshores are most often seen on the outer part of the continental shelf (Figure 13.1). They occasionally move into inshore waters commonly used by residents and transients, but have not been seen intermingling with either ecotype. Offshores are usually seen in relatively large groups (20 or more), and they resemble residents in their frequent and conspicuous use of echolocation and social calls.

Resident and transient killer whales differ subtly in pigmentation and dorsal fin shape (Bigg et al. 1987; Baird and Stacey 1998; Matkin et al. 1999; Ford et al. 2000). It is the subjective impression of the authors that transients attain greater average sizes than residents. Offshore killer whales have more rounded fins than residents and transients, and they appear smaller on average (Ford et al. 2000). In both residents and transients the dorsal fins of males elongate after puberty, making it easy to distinguish adult males from all others. This process is not nearly as marked in offshores, with the result that adult males and females are relatively similar in appearance. Although it seems most parsimonious to assume that the observed difference in male fin elongation has an underlying genetic basis, the possibility that it results from differences in the nutritional status of the ecotypes cannot be ruled out.

Transient and resident killer whales have been seen in relatively close proximity many times but have never been seen intermingling or socializing. On most occasions, members of the two ecotypes passed without any obvious interaction. However, avoidance behavior, such as grouping more tightly and changing the direction of travel, has been observed a number of times (Morton 1990; Barrett-Lennard 1992; Baird and Dill 1995), and on one occasion a large group of residents was seen to chase and ram a much smaller group of transients (Ford et al. 1998). Agonistic interactions between groups of killer whales appear to be more common in other areas, such as Norway, where "local" whales and "visitors" compete for herring, and New Zealand, where killer whales with extensive scarring from conspecifics have been reported (Visser 1998; Bisther 2002).

Subdivision of Ecotypes into Regional Populations

At least two of the three distinct killer whale ecotypes are further subdivided into well-defined, possibly discrete regional

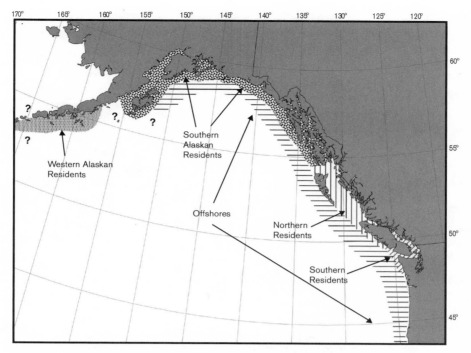

FIGURE 13.1. Approximate ranges of known resident and offshore killer whale populations.

populations. Three populations of resident killer whales have been described (Figure 13.1). The southern resident population inhabits northern Washington State and southern British Columbia in the summer and has about 90 members (Ford et al. 2000). Part of the population typically leaves the area for several months each winter. The extent of the winter range of the population is not known, but southern resident groups have been sighted twice in Monterey Bay and once off the northern Queen Charlotte Islands—1,200 km south and 800 km north of their summer ranges, respectively (G. Ellis, Fisheries and Oceans Canada, personal communication). The northern resident population extends from southern British Columbia to southeast Alaska and has about 200 members (Ford et al. 2000). The southern Alaskan resident subpopulation is not completely censused but contains at least 360 individuals (Matkin et al. 1999) and ranges from southeast Alaska to Kodiak Island. The northern residents and southern Alaskan residents are seen infrequently in winter, but the limited data that do exist suggest that their summer and winter ranges do not differ greatly.

The three resident populations occupy discrete adjacent ranges most of the time, but occasionally groups from one population pass through the range occupied by another. Groups from different populations have not been seen intermingling on these occasions, although groups from the northern residents and southern Alaskan residents have been seen in close proximity (Dahlheim et al. 1997). In contrast, groups belonging to the same resident population frequently intermingle and swim together for hours or days at a time. Aggregations of resident killer whales, referred to as superpods, which contain 50 or more individuals and persist for periods up to several days, form frequently in the late

summer. Based on a gestation period of 17 months and a peak in calving frequency in the late winter, superpods may play an important role in courtship and mating. What is significant for this discussion, however, is that members of different populations have not been observed in the same superpod aggregation, consistent with DNA fingerprinting studies indicating that intermating between populations is at most extremely rare (Barrett-Lennard 2000).

Transient killer whales also live in regional populations (Figure 13.2). The so-called West Coast transient population extends from the central part of the California coast to around 56°N latitude in southern Alaska and contains at least 220 individuals (Ford and Ellis 1999). The Gulf of Alaska transient subpopulation inhabits the waters from southern Alaska to at least as far as Kodiak Island. The size of this population is unknown, but 43 individuals have been photo-identified (Matkin et al. 1999), and the population is probably much larger. It contains at least two mitochondrial haplotypes (Barrett-Lennard 2000). A population of less than 11 transient killer whales, referred to as the *AT1s*, inhabits the waters of Prince William Sound and Kenai Fjords in the northernmost part of the Gulf of Alaska (Matkin et al. 1999). This population has not produced surviving calves since the mid-1980s, and extirpation appears inevitable. Members of the AT1 and Gulf of Alaska populations have been observed passing each other on rare occasions without intermingling (see Matkin and Saulitis 1994), in contrast to the affiliative behaviors usually noted during encounters between members of the same subpopulation. The social separation of the Gulf of Alaska and West Coast transients is less well documented. What may be most telling here is that the three transient populations each have different mitochondrial DNA haplotypes and differ

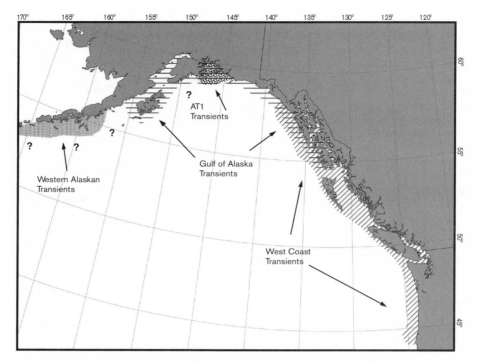

FIGURE 13.2. Approximate ranges of known transient killer whale populations.

significantly in nuclear allele frequencies (Barrett-Lennard 2000), suggesting that female migration and successful intermating are rare. Research on the acoustic behavior of each transient population indicates that they also have discrete social traditions (Saulitis 1993; Ford and Ellis 1999).

Social Structure of Resident and Transient Killer Whales

Resident killer whales live in *matrilines:* groups containing up to three generations of maternal descendents of a single living or recently deceased female. The matriline is the fundamental social unit for resident killer whales and is remarkably stable, with both male and female offspring remaining in their natal matrilines for life (Bigg et al. 1990). Members of matrilines travel together and rarely venture out of acoustic contact with each other. When a matriarch dies, her daughters' matrilines often diverge slowly into functionally independent social units. Matrilines are effectively closed to immigration, although in rare circumstances orphaned calves or juveniles may join a closely related matriline. Ford (1991) showed that the matrilines of the northern resident subpopulation fall into three distinct acoustic groups, or *clans*, based on the discrete, stereotyped calls that they use. The relatively small southern resident population has a single clan (Ford 1991), and there are two clans in the southern Alaskan resident population (Yurk et al. 2002). Early papers (e.g., Bigg et al. 1990) identified a social unit between the level of the matriline and clan, referred to as a *pod*, containing one or more matrilines that traveled together at least 50% of the time. Ford and Ellis (2002) challenged the utility of defining a social unit based on association,

showed that matriline associations are more dynamic than Bigg et al. (1990) predicted, and argued that resident social organization is best described by matrilines and clans alone. Barrett-Lennard (2000) showed that in the best-studied resident population (the northern residents), mating does not occur within matrilines. Most successful matings occur between individuals from matrilines with well-diverged call repertoires, usually belonging to different clans. Barrett-Lennard also showed that call similarity and genetic relatedness are positively correlated and that the mating preference just described minimizes inbreeding extremely effectively.

Transient killer whale social systems are less well understood than those of residents. Most transients live in groups of one to six individuals. Although rare, dispersal does occur (e.g., Bigg et al. 1987; Baird and Dill 1996). The dispersers may be males or females, juveniles or adults, and in some instances they have rejoined a group after years of apparent social separation (Ford and Ellis 1999). Transient groups sometimes contain a core of two or three maternally related members that stay together for many years, with other members that stay for shorter periods. Because dispersal is usually not compensated for entirely (or at all) by immigration, transient groups are smaller than those of residents, on average. This smaller size may be a result of group size constraints related to foraging (see Baird and Dill 1996). The fact that transient killer whales rely on stealth and minimize both communication and echolocation sounds when hunting their acoustically sensitive prey (Barrett-Lennard et al. 1996) means that the risk of alerting prey prior to an attack increases with larger group sizes. On the other hand, there are many accounts of large numbers of killer whales coalescing to kill or feed on baleen whales. In these cases, the joining of groups usually occurs after the attack has started and the element of surprise is lost. This raises the interesting possibility that transients live in a social milieu with well-established relationships that extend far beyond their immediate social group.

Diet and Foraging Behavior

The diets and foraging behaviors of killer whales present an interesting paradox: globally, they are predators with an extremely eclectic diet, but locally their diet and feeding behaviors can be extremely specialized, as described previously. In some, and possibly in many, areas of the world, killer whales feed on either fish or marine mammals, but not both. In some areas, killer whales display what appear to be relatively high-risk behaviors, from intentionally stranding themselves on beaches to capture pinniped prey (Lopez and Lopez 1985; Guinet 1991; Guinet and Bouvier 1995) to attacking prey many times larger than themselves (e.g., Arnbom et al. 1987; Silber et al. 1990; Goley and Straley 1994; Pitman et al. 2001). Such behaviors may reflect a low abundance of prey. In other areas, such as the coast of British Columbia, killer whales appear to be more cautious and are rarely, if ever, observed beaching themselves or taking large whales (Baird 2000; Ford et al. 1998, 2000).

At first glance, the morphology and physiology of killer whales do not appear to reflect their well-documented ability to tackle dangerous prey. They have no armor, and their skin is soft and sensitive. Their conical, slightly recurved teeth are adapted for puncturing and holding, rather than shearing, and are not as large or robust as might be expected of a predator of whales (Slijper 1979). Their jaws are not massive, and the mandible is positively fragile in the area of the pan bone, behind the back teeth. Dive depths and durations during foraging rarely exceed 300 m and 15 min (Baird 2000; Ford et al. 2000)—substantially less than many other cetaceans and some pinnipeds. Killer whales are smaller and slower than some of their prey and less agile than others. Many of the species they feed on seek refuge in the confusion of schooling aggregations. It is therefore apparent that their hunting success rarely depends on brute force. Rather, specialized foraging behaviors and group coordination are paramount. In this respect they resemble pack-hunting terrestrial species such as wolves, hunting dogs, and lions, and have relatively few marine counterparts.

Fish Predation

Killer whales feed on a wide variety of fish species around the world and use a correspondingly diverse set of foraging methods. In general, fish-eating killer whales are acoustically active when foraging, unlike mammal-eating killer whales (Ford 1989, 1991; Strager 1995; Barrett-Lennard et al. 1996). In British Columbia and Alaska, resident killer whales travel in groups but do not appear to catch prey cooperatively, although individual fish are sometimes shared. Their principal prey is salmon (Oncorhynchus spp.) (Ford et al. 1998; Saulitis et al. 2000). Based on stomach content analyses, residents in British Columbia are known to feed on 22 species of fish and 1 species of squid (Ford et al. 1998). Interestingly, they do not appear to take advantage of a superabundance of herring (Clupea pallasi) during the spring spawning period, unlike other marine mammals such as seals, sea lions, and gray whales. When foraging in the spring through fall, several matrilines often forage in loose groups that spread over areas up to several square kilometers while remaining in acoustic contact (Barrett-Lennard et al. 1996; Ford 1989). The feeding behavior of residents during winter is less well known.

The diet and foraging behavior of offshore killer whales is not well known. Like resident killer whales, they travel in large and acoustically active groups and therefore probably do not feed to a significant extent on marine mammals. The few carcasses of offshore killer whales that have been examined have had extremely worn teeth (G. M. Ellis and J. K. B. Ford, Fisheries and Ocean Canada, personal communication), suggesting that their diet includes prey with abrasive properties, such as sharks. Predation of sharks by killer whales has not been observed in the northeastern Pacific but has been reported in a number of other locations, including California (Pyle et al. 1999), Costa Rica (Fertl et al. 1996), and New Zealand (Visser et al. 2000). The stomachs of two offshore whales that died after becoming trapped in a lagoon in Alaska contained salmon bones and crab, but it is unknown whether these were typical items in their diet or unusual and in some way an artifact of the stranding event (Heise et al. 2003).

In Norway, whales use a coordinated strategy called carousel feeding to capture herring (Clupea harengus) (Similä and Ugarte 1993; Similä et al. 1996). Groups of up to 10 whales dive to depths in excess of 150 m to herd a school to the surface, and then slowly circle the school to maintain its coherence. Periodically, one of the whales slaps the tightly balled school with its tail, and members of the group eat debilitated fish that fall out of the school (Nottestad and Similä 2001). Fish feeding behaviors in other areas of the world have not been as well described as those in the northeastern Pacific and Norway. However, in the Antarctic the two and possibly three different forms of killer whales (types A, B, and C) appear to feed in distinctly different ways, a situation that may parallel that in the northeastern Pacific (Berzin and Vladimirov 1983; Pitman and Ensor 2003).

In many areas of the world, including Alaska and many of the sub-Antarctic islands, killer whales have learned to remove fish from longline gear as it is being raised to the surface. This has been documented for the Indian Ocean (Sivasubramanian 1965); Prince William Sound, Alaska (Matkin and Saulitis 1994); the Bering Sea (Yano and Dahlheim 1995); southern Brazil (Secchi and Vaske 1998); New Zealand, Prince Edward Island, and Marion Island (Visser 2000); and the Falklands (Nolan and Liddle 2000).

Many of the fish species that are being caught using this fishing gear are extremely rich in oil and would normally be unobtainable to killer whales because of the depths at which they are found.

Marine Mammal Predation

The foraging strategies of marine mammal–eating transient killer whales in the northeastern Pacific differ markedly from those of fish-eating residents in the same area. Transients travel silently much of the time, hunting by stealth to avoid alerting their acoustically sensitive prey to their presence (Barrett-Lennard et al. 1996). Their dive times are long relative to those of residents, and they often swim erratically (Morton 1990; Saulitis et al. 2000). When feeding on smaller prey such as seals, sea lions, and small cetaceans, transients are typically seen in groups of three to six (Baird and Dill 1996; Ford et al. 1998; Saulitis et al. 2000). They have also been observed on rare occasions attacking and feeding on river otters and swimming deer and moose (Pike and MacAskie 1969; Morton 1990; Ford and Ellis 1999; Matkin et al. 1999).

In both British Columbia and Prince William Sound, Alaska, the most common prey of transients are harbor seals, which they usually find while foraging within 20 m of shore (Baird and Dill 1995; Saulitis et al. 2000). Seals are generally caught and killed very quickly (less than 5 min) (Baird and Dill 1995; Ford et al. 1998), and often the only evidence at the surface is a sheen of oil or small fragments of blubber. Most other prey are taken in open water farther from shore. Attacks on harbor porpoises are usually successful, but Dall's porpoises escape up to 50% of the time (Ford et al. 1998; Saulitis et al. 2000). If successful capturing prey, transients often break their acoustic silence and vocalize, sometimes quite intensely, while sharing their meal (Ford 1984; Deecke et al. 2005). Transients also hunt Pacific white-sided dolphins, although such attacks are frequently unsuccessful (Ford et al. 1998). The killer whales typically drive a group of dolphins into a shallow or enclosed area or attack individuals that have been separated from the rest of the group (Ford et al. 1998; Dahlheim and Towell 1994).

Both California and Steller sea lions are preyed on by transient killer whales. Attacks on adult sea lions typically last between 1 and 2.5 hours as the whales repeatedly ram, jump on, or strike the sea lion with their tails (Ford et al. 1998; Saulitis et al. 2000; Heise et al. 2003). As the sea lion becomes increasingly debilitated, the whales drown it, but at least 50% of the time the sea lion escapes prior to this occurring. When a sea lion is killed, it is usually torn apart and shared by members of the group. Because of their large size and their ability to defend themselves during an attack, adult sea lions may present a fairly formidable prey item for killer whales. Juvenile sea lions may be taken much more quickly because of their smaller size, and such attacks are much less visible at the surface. Rice (1968) describes finding three intact yearling Steller sea lion skins in the stomach of one killer whale, and Heise et al. (2003) found flipper tags from 14 different Steller sea lions, all less than 4 years old, in the stomach of a single transient killer whale. Researchers based on sea lion rookeries in Alaska reported rarely seeing killer whale attacks on sea lions, and the majority of observed sea lion kills in Alaska and British Columbia involved subadults and adults (Heise et al. 2003). We suspect, however, that pups are targeted more frequently by transients than adult and subadult sea lions, but that such attacks are inconspicuous and therefore rarely observed. At a location in Argentina where attacks on southern sea lions by killer whales could be clearly seen, Hoelzel (1991) found that over 99% of observed kills of southern sea lions (*Otaria flavascens*) were of pups. Given the dramatic decline of the Steller sea lion population in western Alaska (more than 80% since the late 1970s, Trites and Larkin 1996, further research on the frequency of attacks and kills of sea lions is warranted, in order to better understand the possible impact of killer whale predation on Steller sea lion population dynamics.

The most dramatic killer whale hunting behavior is intentional beaching to capture pinniped prey. Although fascinating, such behaviors are quite rare globally, and there are to date only two areas of the world where such behaviors are known to occur on a regular basis: Patagonia, Argentina, where killer whales feed on southern elephant seals and sea lions (Lopez and Lopez 1985; Hoelzel 1991), and the Crozet Islands, where killer whales feed on southern elephant seals (Guinet 1991).

In Prince William Sound and western Alaska, killer whales have occasionally been observed attacking and apparently feeding on sea otters. Hatfield et al. (1998) describe nine attacks on otters, of which six appeared to be fatal, although the whales did not consume the carcass in all cases. Attacks typically involved one to four whales, and lasted from a few seconds to less than 5 minutes. Estes et al. (1998) argue that killer whale predation on sea otters has increased in recent years, likely as a result of declines in local pinniped populations, and has resulted in a decline in the sea otter population at a rate of approximately 25% per year during the 1990s. In British Columbia and southern Alaska, where pinnipeds are relatively abundant, killer whales have not been observed feeding on sea otters.

Predation on larger whales is rarely observed in the inshore waters of British Columbia and Alaska (Ford and Ellis 1999) but is well documented elsewhere. Killer whales have reportedly taken gray whales (Ljungblad and Moore 1983; Lowry et al. 1987; Goley and Straley 1994); narwhals (Steltner et al. 1984; Campbell et al. 1988); Bryde's whales (Silber et al. 1990); and sperm whales (Arnbom et al. 1987; Pitman et al. 2001). Raking scars from killer whale teeth are commonly seen on humpback whales in certain areas (Katona et al. 1988), indicating that some attacks do not result in kills, and most attacks probably involve some degree of risk for the killer whales. For example, Eschricht (1866) provides an account of a killer whale that was "most probably mortally wounded" by a blow to the head from a bowhead whale.

Transient killer whales, like terrestrial pack animals such as lions and wolves, hunt in small, stable groups (Packer 1994; Packard 2003; Ford and Ellis 1999). However, when a group of killer whales initiates an attack on a large cetacean, it may be joined by other groups, even hours after the attack has begun (e.g., Silber et al. 1990; Guinet et al. 2000; Pitman et al. 2001). These groups are not antagonistic toward each other and appear to cooperate fully. We are not aware of any counterpart in the terrestrial world where this type of behavior occurs among social carnivores. A flexible grouping strategy for mammal-eating killer whales has obvious benefits. Hunting in small groups helps to increase the encounter rate with prey that are widely dispersed and avoids competition between groups. However, a large cetacean may be difficult for a small group to hunt successfully, and when groups of killer whales join together, they are likely to be more successful in the kill. If the attack occurs in deep water, the killer whales can work together to prevent the carcass from sinking, thus extending the time it is available for consumption (Guinet et al. 2000). The fact that carcasses often sink beyond the reach of killer whales may explain why often only a few tissues of large baleen whales, such as the tongue, lips, and belly flaps, are eaten (e.g., Silber et al. 1990; Jefferson et al. 1991).

The role of male killer whales in hunting mammalian prey is not clearly understood. When the prey is relatively small, both male and female killer whales of all age classes participate fully during the hunt (Budylenko 1981; Jefferson et al. 1991; Baird and Dill 1995). However, when large whales are taken, often it is the females and subadult males that appear to be most aggressive initially, taking turns lunging at their prey and at times preventing it from surfacing. Adult males often remain at a distance until the prey is disabled (e.g., Silber et al. 1990; Dalheim and Towell 1994; Pitman et al. 2001). It is possible that males, which are significantly larger than females, lack the maneuverability to safely approach such powerful prey.

Surplus Killing

There are several accounts of killer whales engaging in surplus killing, particularly with seabirds (e.g., Stranek et al. 1983; Stacey et al. 1990; Williams et al. 1990). The whales often "play" with the target, batting it with their tails, jumping on it, or capturing it, and then releasing it dead or dying. It is possible that several, if not all, of the attacks on sea otters described by Hatfield et al. (1998) were examples of surplus killing. Over a 4-day period, Williams et al. (1990) reported that two whales killed, but did not eat, more than 290 young cormorants. Had such killing occurred on a regular basis, the population consequences for cormorants could have been quite severe, but subsequent visits to the colony revealed no evidence of significant killer whale–induced mortality. Surplus killing may be a part of the learning process of prey handling and capture for young killer whales, as seabirds are often attacked by subadult

whales, using techniques that are similar to those used to hunt pinnipeds and small cetaceans. Seabirds do not appear to be an important component of the killer whale diet in British Columbia or Alaska (Matkin and Saulitis 1994; Ford et al. 1998; Saulitis et al. 2000).

Dietary Flexibility and Prey Switching

Residents and transients in the northeastern Pacific show extraordinarily strong preferences for fish and marine mammal prey, respectively, and stomach content data do not provide any evidence that members of one group ever take prey of the other (Ford et al. 1998; Saulitis et al. 2000; Heise et al. 2003). The capture of three transient killer whales in British Columbia in 1970 illustrates just how strong these foraging traditions are. After refusing fish for 75 days in captivity, one transient killer whale died. On the 79th day, the other two whales began to eat salmon. When the two surviving whales were later returned to the wild they resumed their diet of marine mammals (Ford and Ellis 1999). The behavior of harbor seals also supports the suggestion that killer whales do not alternate between mammalian and pisciverous prey: Harbor seals hide at the surface when presented with the underwater vocalizations of transients but do not change their behavior in response to resident calls (Deecke et al. 2002). It seems unlikely that harbor seals would become habituated to the calls of residents if they were occasionally attacked by them.

Dietary preferences for certain species appear to be strong for both residents and transients. Resident killer whales favor chinook and coho salmon and feed on other species infrequently, even when they are more abundant (Ford et al. 1998; Saulitis et al. 2000). Likewise, West Coast and AT1 transient killer whales appear to prefer harbor seals and porpoises to larger pinniped prey, possibly because they take less energy to subdue (Saulitis et al. 2000; Heise et al. 2003).

We believe that if there are changes in the dietary preferences and hunting behaviors of killer whales, such changes occur slowly and in two phases. The first phase is innovation, in which a new feeding behavior is developed by one or a few individuals. This process generally takes place very slowly. The second phase, the cultural transmission of the newly acquired behaviors to the larger group occurs much more quickly. The raiding of longline gear for oil-rich prey illustrates these behavioral changes. Longline fisheries have occurred in areas for years before killer whales began to exploit them. In Prince William Sound, a single resident pod of killer whales began taking fish from lines, and this behavior spread to other pods. When the pattern became well established, the killer whales were remarkably difficult to deter and continued to take hooked fish despite frequent shootings and the use of underwater explosives (Matkin et al. 1986). Similarly, for many years sea otters in Alaska were observed in close proximity to killer whales on many occasions with no evidence of predation (Matkin and Saulitis 1994). However, in western Alaska, where there have been significant declines in fur seal, sea lion, and harbor seal populations, sea otters appear to have become extremely vulnerable to transient killer whales except in certain refugia, and this predation is believed to have rapidly depleted the population (Estes et al. 1998). Of greatest significance here is that this shift did not occur until sea otters were relatively abundant and other pinniped prey were relatively rare. This is consistent with the conservative behavior of killer whales as predators: Innovation occurs slowly, but new behaviors spread rapidly and whales maintain their preferences, possibly to the point of seriously depleting prey populations.

Anti-predation Strategies by Killer Whale Prey

What are the cumulative effects of a pack-hunting predator such as the killer whale on the oceans of the world? In stable ecosystems, one would expect a balance between the number of predators and their prey, provided the predator is not in turn regulated by higher-trophic-level predators. In systems that are experiencing significant shifts in species abundance, however, as in western Alaska, this balance may not occur. As previously discussed, predation by killer whales has been implicated in the population decline of sea otters (Estes et al. 1998), and it may have significant consequences for depleted populations of Steller sea lions (Barrett-Lennard et al. 1995) and Cook Inlet beluga whales (Shelden et al. 2003). In these species, prey populations have reached such low numbers that even occasional predation by killer whales is now sufficient to trap the prey population in a "predator pit" (Walters 1986) and prevent the population from recovering. Carrying this theme further, Springer et al. (2003) hypothesized that the depletion of large baleen whales in the north Pacific by commercial whalers by the 1960s deprived killer whales of a primary food source, and initiated a sequential cascade of population declines in Steller sea lions, harbor seals, and sea otters. Although certain details of the hypothesis are controversial (for example, it has not been established that large baleen whales ever constituted an important food source for killer whales in the north Pacific), the notion that killer whales have serially depressed populations of a number of prey species is entirely plausible, in our opinion.

Avoidance of predation by killer whales has been suggested to be the driving evolutionary force shaping the migration patterns of baleen and sperm whales (Corkeron and Connor 1999; Pitman et al. 2001). Corkeron and Connor (1999) hypothesize that humpback and gray whales leave rich feeding grounds in temperate areas to calve in warm but unproductive waters to avoid the risk of predation by killer whales, which are not common in tropical areas. Such migratory behavior may have evolved when feeding and calving areas were closer together, and the migration extended as the continents drifted. Pitman et al. (2001) suggest that killer whale predation may have been an important evolutionary force in the development of the complex social behavior and distribution patterns of different sex and age classes of sperm whales. Juveniles tend to remain in tropical and subtropical areas in matrilineal groups where killer whales are relatively rare, whereas adult males travel to higher latitudes, where killer whales are more abundant but food is too (Best 1979; Whitehead 2003).

Conclusions

Evidence of the behavioral versatility of killer whales is provided by (1) the wide variety of foraging methods they employ throughout their cosmopolitan range, (2) the correspondingly diverse set of prey species they are known to feed on, (3) the different types of social systems that characterize different populations, (4) interpopulational and in some cases intrapopulational variation in their use of acoustic signals for communication and echolocation, and (5) their well-known ability to learn complex and novel behaviors in captivity. Yet, paradoxically, a predominant feature of killer whales is their conservative nature—they are less innovative than one might expect of a large-brained, socially and behaviorally sophisticated animal. Indeed, innovation appears to play a relatively minor role in the lives of killer whales, which instead rely to a great extent on traditions (culturally transmitted suites of behaviors). To understand this point, consider that killer whales can be live-captured and held to the point of starvation using nets that they could easily jump over or break through, yet these whales demonstrate detailed knowledge of topography, tides, and currents by safely navigating surge channels and drying reefs in search of harbor seals. Or consider the strong foraging and dietary traditions demonstrated by the killer whale ecotypes in the northeastern Pacific Ocean and the extraordinary reluctance of captive killer whales to feed on unfamiliar food. Or consider the observation that killer whales may take years to begin depredating a new fishery, yet, when one of them eventually learns how to do so, the behavior quickly spreads throughout the population.

The role of tradition in the lives of killer whales should be taken into account when considering the impact of the species on prey populations. Their diet is unlikely to be predicted accurately by simple foraging models. For example, killer whales may subsist for years, possibly generations, on a limited suite of prey species, despite the availability of more abundant and apparently accessible alternate prey. Furthermore, it is plausible that prey switching or the expansion of dietary breadth may be largely dependent on chance events. For example, a switch from dietary reliance on pinnipeds to sea otters might begin with the killing of otters as play behavior in areas where otters are particularly abundant.

In light of the difficulty of understanding killer whale foraging traditions and of forecasting their cultural evolution, we predict that scientific progress on the impact of killer whales on prey species will probably be driven by (1) direct observations of the foraging behavior of killer whales in the wild, possibly supplemented by biochemical methods to estimate their diet composition, and (2) inferential evidence based on observations of (a) vigilance behavior by prey, (b) scarring of prey caused by killer whale attacks, and (c) comparison of the demographics and population trajectories of prey populations exposed to killer whales and those living in areas where killer whales are rare. At the least, it seems safe to assert that assessing the impact of the species on prey populations will remain an empirical science for some time to come.

Literature Cited

Arnbom, T., V. Papastavrou, L.S. Weilgart, and H. Whitehead. 1987. Sperm whales react to an attack by killer whales. *Journal of Mammalogy* 68: 450–453.

Baird, R.W. 2000. The killer whale: foraging specializations and group hunting, in *Cetacean societies*. J. Mann, R.C. Connor, P.L. Tyack, and H. Whitehead, eds. Chicago: University of Chicago Press, pp. 127–153.

Baird, R.W., P.A. Abrams, and L.M. Dill. 1992. Possible indirect interactions between transient and resident killer whales: implications for the evolution of foraging specializations in the genus *Orcinus*. *Oecologia* 89: 125–132.

Baird, R.W. and L.M. Dill. 1995. Occurrence and behaviour of transient killer whales: seasonal and pod-specific variability, foraging behaviour, and prey handling. *Canadian Journal of Zoology* 73: 1300–1311.

———. 1996. Ecological and social determinants of group size in transient killer whales. *Behavioral Ecology* 7: 408–416.

Baird, R.W. and P.J. Stacey. 1988. Variation in saddle patch pigmentation in populations of killer whales *(Orcinus orca)* from British Columbia, Alaska, and Washington State. *Canadian Journal of Zoology* 66: 2582–2585.

Barrett-Lennard, L.G. 1992. Echolocation in wild killer whales *(Orcinus orca)*. M.S. thesis, University of British Columbia.

———. 2000. Population structure and mating patterns of killer whales *(Orcinus orca)* as revealed by DNA analysis. Ph.D. thesis, University of British Columbia.

Barrett-Lennard, L.G. and G.M. Ellis. 2001. *Population structure and genetic variability in northeastern Pacific killer whales: towards an assessment of population viability.* Research Document 2001/065. Nanaimo, BC: Fisheries and Oceans Canada. Available at http:// www.dfo-mpo.gc.ca/csas/Csas/publications/ ResDocs-DocRech/2001/2001_065_e.htm. Accessed June 1, 2006.

Barrett-Lennard, L.G., J.K.B. Ford, and K.A. Heise. 1996. The mixed blessing of echolocation: differences in sonar use by fish-eating and mammal-eating killer whales. *Animal Behaviour* 51: 553–565.

Barrett-Lennard, L.G., K.A. Heise, E.L. Saulitis, G.M. Ellis, and C.O. Matkin. 1995. The impact of killer whale predation on Steller sea lion populations in British Columbia and Alaska. Report prepared for the Marine Mammal Research Unit, University of British Columbia.

Berzin, A.A. and V.L. Vladimirov. 1983. A new species of killer whale (Cetacea, Delphinidae) from Antarctic waters. *Zoologicheskii Zhurnal* 62: 287–295 (in Russian). Best, P.B. 1979. Social organization in sperm whales, *Physeter macrocephalus*, in *Behavior of marine animals*. H.E. Winn and B.L. Olla, eds. New York: Plenum Press, pp. 227–289.

Bigg, M.A. 1982. An assessment of killer whale *(Orcinus orca)* stocks off Vancouver Island, British Columbia. *Report of the International Whaling Commission* 32: 625–666.

Bigg, M.A., G.M. Ellis, J.K.B. Ford, and K.C. Balcomb III. 1987. Killer whales: a study of their identification, genealogy and natural history in British Columbia and Washington State. Nanaimo, Canada: Phantom Press.

Bigg, M.A., P.F. Olesiuk, G.M. Ellis, J.K.B. Ford, and K.C. Balcomb III. 1990. Social organization and genealogy of resident killer whales *(Orcinus orca)* in the coastal waters of British Columbia and Washington State. *Report of the International Whaling Commission* (Special Issue) 12: 383–405.

Bisther, A. 2002. Intergroup interactions among killer whales in Norwegian coastal waters; tolerance vs. aggression at feeding grounds. *Aquatic Mammals* 28: 14–23.

Blaylock, R.A., J.W. Hain, L.J. Hansen, D.L. Palka, and G.T. Waring. 2000. *U.S. Atlantic and Gulf of Mexico marine mammal stock assessments*. NOAA Technical Memorandum NMFS-SEFSC-363, National Marine Fisheries Service, Miami, Florida.

Branch, T.A. and D.S. Butterworth. 2001. Estimates of abundance south of 60°S for cetacean species sighted frequently on the 1978/79 to 1997/98 IWC/IDCR-SOWER sighting surveys. *Journal of Cetacean Research and Management* 3: 251–270.

Budylenko, G.A. 1981. Distribution and some aspect of the biology of killer whale in the South Atlantic. *Report of the International Whaling Commission* 31: 523–525.

Campbell, R.R., D.B. Yurick, and N.B. Snow. 1988. Predation on narwhals *Monodon-monoceros*, by killer whales *Orcinus orca* in the eastern Canadian Arctic. *Canadian Field Naturalist* 102: 689–696.

Christensen, I. 1988. Distribution, movements and abundance of killer whales *(Orcinus orca)* in Norwegian coastal waters, 1982–1987, based on questionnaire surveys. *Rit Fiskideildar* 11: 79–88.

Corkeron, P.J. and R.C. Connor. 1999. Why do baleen whales migrate? *Marine Mammal Science* 15: 1228–1245.

Dahlheim, M.E., D.K. Ellifrit, and J.D. Swenson. 1997. *Killer whales* (Orcinus orca) *of Southeast Alaska: a catalogue of photo-identified individuals.* Seattle: National Marine Mammal Laboratory, National Marine Fisheries Service.

Dahlheim, M.E. and R.G. Towell. 1994. Occurrence and distribution of Pacific white-sided dolphins *(Lagenorhynchus obliquidens)* in southeastern Alaska, with notes on an attack by killer whales *(Orcinus orca)*. *Marine Mammal Science* 10: 458–464.

Deecke, V.B., J.K.B. Ford, and P.J.B. Slater. 2005. The vocal behaviour of mammal-eating killer whales: communicating with costly calls. *Animal Behaviour* 69: 395–405.

Deecke, V.B., J.K.B. Ford, and P. Spong. 2000. Dialect change in resident killer whales: implications for vocal learning and cultural transmission. *Animal Behaviour* 60: 629–638.

Deecke, V.B., P.J.B. Slater, and J.K.B. Ford. 2002. Selective habituation shapes acoustic predator recognition in harbor seals. *Nature* 420: 171–173.

Eschricht, D.F. 1866. On the species of the genus *Orca* inhabiting the northern seas, in *Recent memoirs of the cetacea*. W.H. Flower, ed. London: Ray Society, pp. 153–188.

Estes, J.A., M.T. Tinker, T.M. Williams, and D.F. Doak. 1998. Killer whale predation on sea otters linking oceanic and nearshore ecosystems. *Science* 282: 473–476.

Fertl, D., A. Acevedo-Gutierrez, and F.L. Darby. 1996. A report on killer whales *(Orcinus orca)* feeding on a carcharhinid shark in Costa Rica. *Marine Mammal Science* 12: 606–611.

Ford, J.K.B. 1984. Call traditions and dialects of killer whales *(Orcinus orca)* in British Columbia. Ph.D. thesis, University of British Columbia.

———. 1989. Acoustic behaviour of resident killer whales *(Orcinus orca)* off Vancouver Island, British Columbia. *Canadian Journal of Zoology* 67: 727–745.

———. 1991. Vocal traditions among resident killer whales *(Orcinus orca)* in coastal waters of British Columbia. *Canadian Journal of Zoology* 69: 1454–1483.

Ford, J.K.B. and G.M. Ellis. 1999. *Transients: mammal-hunting killer whales of British Columbia, Washington, and southeastern Alaska.* Vancouver: University of British Columbia Press.

———. 2002. Reassessing the social organization of resident killer whales in British Columbia, in *Proceedings of the 4th International Orca Symposium and Workshops.* L.G. Barrett-Lennard, J.K.B. Ford, C. Guinet, T. Simila, and F. Ugarte, eds. Chize, France: Centre National de la Recherche Scientifique, pp. 72–74.

Ford, J.K.B., G.M. Ellis, and K.C. Balcomb. 2000. *Killer whales,* 2nd edition. Vancouver: University of British Columbia Press.

Ford, J.K.B., G.M. Ellis, L.G. Barrett-Lennard, A.B. Morton, and K.C. Balcomb III. 1998. Dietary specialization in two sympatric populations of killer whales *(Orcinus orca)* in coastal British Columbia and adjacent waters. *Canadian Journal of Zoology* 76: 1456–1471.

Goley, P.D. and J.M. Straley. 1994. Attack on gray whales *(Eschrichtius robustus)* in Monterey Bay, California, by killer whales *(Orcinus orca)* previously identified in Glacier Bay, Alaska. *Canadian Journal of Zoology* 72: 1528–1530.

Guinet, C. 1991. Intentional stranding apprenticeship and social play in killer whales *(Orcinus orca)*. *Canadian Journal of Zoology* 69: 2712–2716.

Guinet, C., L.G. Barrett-Lennard, and B. Loyer. 2000. Co-ordinated attack behavior and prey sharing by killer whales at Crozet Archipelago: strategies for feeding on negatively buoyant prey. *Marine Mammal Science* 16: 829–834.

Guinet, C. and J. Bouvier. 1995. Development of intentional stranding hunting techniques in killer whale *(Orcinus orca)* calves at Crozet Archipelago. *Canadian Journal of Zoology* 73: 27.

Hatfield, B.B., D. Marks, M.T. Tinker, K. Nolan, and J. Peirce. 1998. Attacks on sea otters by killer whales. *Marine Mammal Science* 14: 888–894.

Heise, K.H., L.G. Barrett-Lennard, E. Saulitis, C.O. Matkin, and D. Bain. 2003. Examining the evidence for killer whale predation on Steller sea lions in British Columbia and Alaska. *Aquatic Mammals* 29: 325–334.

Hoelzel, A.R. 1991. Killer whale predation on marine mammals at Punta Norte, Argentina: food sharing, provisioning and foraging strategy. *Behavioral Ecology and Sociobiology* 29: 197–204.

———. 1998. Genetic structure of cetacean populations in sympatry, parapatry, and mixed assemblages: implications for conservation policy. *Journal of Heredity* 89: 451–458.

Hoelzel, A.R., M. Dahlheim, and S.J. Stern. 1998. Low genetic variation among killer whales *(Orcinus orca)* in the Eastern North Pacific, and genetic differentiation between foraging specialists. *Journal of Heredity* 89: 121–128.

Jefferson, T.A., P.J. Stacey, and R.W. Baird. 1991. A review of killer whale interactions with other marine mammals: predation to co-existence. *Mammal Review* 21: 151–180.

Katona, S.J., J.A. Beard, P.E. Girton, and F. Wenzel. 1988. Killer whales *(Orcinus orca)* from the Bay of Fundy to the equator, including the Gulf of Mexico. *Rit Fiskideildar* 11: 205–224.

Ljungblad, D.K. and S.E. Moore. 1983. Killer whales *Orcinus orca* chasing gray whales *Eschrichtius robustus* in the northern Bering Sea. *Arctic* 36: 361–364.

López, J.C. and D. López. 1985. Killer whales *(Orcinus orca)* of Patagonia and their behavior of intentional stranding while hunting nearshore. *Journal of Mammalogy* 66: 181–183.

Lowry, L.T., R.R. Nelson, and K.J. Frost. 1987. Observations of killer whales, *Orcinus orca*, in western Alaska: sightings, strandings, and predation on other marine mammals. *Canadian Field Naturalist* 101: 6–12.

Matkin, C.O., G.M. Ellis, E.L. Saulitis, L.G. Barrett-Lennard, and D.R. Matkin. 1999. *Killer whales of southern Alaska.* Homer, AK: North Gulf Oceanic Society.

Matkin, C.O., G. Ellis, O. von Ziegesar, and R. Steiner. 1986. *Killer whales and longline fisheries in Prince William Sound Alaska 1986.* Seattle: National Marine Mammal Laboratory.

Matkin, C.O. and E.L. Saulitis. 1994. Killer whale *(Orcinus orca)* biology and management in Alaska. Report for the Marine Mammal Commission, Washington, D.C.

Mikhalev, Y.A., M.V. Ivashin, V.P. Savusin, and F.E. Zelenaya. 1981. The distribution and biology of killer whales in the southern hemisphere. *Report of the International Whaling Commission* 31: 551–566.

Morton, A.B. 1990. A quantitative comparison of the behaviour of resident and transient forms of killer whales off the central BC coast. *Report of the International Whaling Commission* (Special Issue) 12: 245–248.

Nichol, L. and D.M. Shackleton. 1996. Seasonal movements and foraging behavior of northern resident killer whales *(Orcinus orca)* in relation to the inshore distribution of salmon *(Oncorhynchus spp.)* in British Columbia. *Canadian Journal of Zoology* 74: 983–991.

Nolan, C.P. and G.M. Liddle. 2000. Interactions between killer whales *(Orcinus orca)* and sperm whales *(Physeter macrocephalus)* with a longline fishing vessel. *Marine Mammal Science* 16: 658–664.

Nottestad, L. and T. Similä. 2001. Killer whales attacking schooling fish: why force herring from deep water to the surface? *Marine Mammal Science* 17: 343–352.

Packard, J.M. 2003. Wolf behavior: reproductive, social, and intelligent, in *Wolves: behavior, ecology, and conservation.* Mech, D.L., and L. Boitani, eds. Chicago: University of Chicago Press, pp.

Packer, C. 1994. *Into Africa.* Chicago: University of Chicago Press.

Pike, G.C. and I.B. MacAskie. 1969. Marine mammals of British Columbia. *Journal of the Fisheries Research Board of Canada* 171: 1–54.

Pitman, R.L., L.T. Ballance, S.L. Mesnick, and S.J. Chivers. 2001. Killer whale predation on sperm whales: observations and implications. *Marine Mammal Science* 17: 494–507.

Pitman, R.L. and P. Ensor. 2003. Three forms of killer whales *(Orcinus orca)* in Antarctic waters. *Journal of Cetacean Research and Management* 5: 131–139.

Pyle, P., M.J. Schramm, C. Keiper, and S.D. Anderson. 1999. Predation on a white shark *(Carcharodon carcharias)* by a killer whale *(Orcinus orca)* and a possible case of competitive displacement. *Marine Mammal Science* 15: 563–568.

Rice, D.W. 1968. Stomach contents and feeding behavior of killer whales in the eastern North Pacific. *Norsk Hvalfangst-Tidende* 57: 36–38.

Saulitis, E.L. 1993. The behavior and vocalizations of the "AT" group of killer whales *(Orcinus orca)* in Prince William Sound, Alaska. M.S. thesis, University of Alaska, Fairbanks.

Saulitis, E., C. Matkin, L. Barrett-Lennard, K. Heise, and G. Ellis. 2000. Foraging strategies of sympatric killer whale *(Orcinus orca)* populations in Prince William Sound. *Marine Mammal Science* 16: 94–109.

Scammon, C.M. 1874. *The marine mammals of the northwestern coast of North America.* Reprint, New York: Dover Publications, 1968.

Secchi, E.R. and T.J. Vaske. 1998. Killer whale *(Orcinus orca)* depredation on tuna and swordfish longline catches in southern Brazil. *Aquatic Mammals* 24: 117–122.

Shelden, K.E.W., D.J. Rugh, B.A. Mahoney, and M.E. Dahlheim. 2003. Killer whale predation on belugas in Cook Inlet, Alaska: implications for a depleted population. *Marine Mammal Science* 19: 529–544.

Silber, G.K., M.W. Newcomer, and M.H. Perez-Cortes. 1990. Killer whales *(Orcinus orca)* attack and eat a Bryde's whale *(Balaenoptera edeni). Canadian Journal of Zoology* 68: 1603–1606.

Similä, T., J.C. Holst, and I. Christensen. 1996. Occurrence and diet of killer whales in northern Norway: seasonal patterns relative to the distribution and abundance of Norwegian spring-spawning herring. *Canadian Journal of Fisheries and Aquatic Sciences* 53: 769–779.

Similä, T. and F. Ugarte. 1993. Surface and underwater observations of cooperatively feeding killer whales in northern Norway. *Canadian Journal of Zoology* 71: 1494–1499.

Sivasubramanian, K. 1965. Predation by tuna longline catches in the Indian Ocean, by killer whales and sharks. *Bulletin of the Fisheries Research Station, Ceylon* 17: 221–236.

Slijper, E.J. 1979. *Whales,* 2nd edition. Ithaca, NY: Cornell University Press.

Springer, A.M., J.A. Estes, G.B. van Vliet, T.M. Williams, D.F. Doak, E.M. Danner, K.A. Forney, B. Pfister. 2003. Sequential megafaunal collapse in the North Pacific Ocean: an ongoing legacy of industrial whaling? *Proceedings of the National Academy of Sciences of the United States of America* 100: 12223–12228.

Stacey, P.J., R.W. Baird, and A.B. Hubbard-Morton. 1990. Transient killer whale *(Orcinus orca)* harassment, predation and "surplus killing" of marine birds in British Columbia. *Pacific Seabird Group Bulletin* 17: 38.

Steltner, H., S. Steltner, and D.E. Sergeant. 1984. Killer whales *(Orcinus orca)* prey on narwhals *(Monodon monoceros):* an eyewitness account. *Canadian Field Naturalist* 98: 458–462.

Strager, H. 1995. Pod-specific call repertoires and compound calls of killer whales *(Orcinus orca)* in the waters of northern Norway. *Canadian Journal of Zoology* 73: 1037–1047.

Stranek, R., B.C. Livezey, and P.S. Humphrey. 1983. Predation on steamer duck by killer whale. *Condor* 8: 255–256.

Trites, A.W. and P.A. Larkin. 1996. Changes in the abundance of Steller sea lions *(Eumetopias jubatus)* from 1956–1992: how many were there? *Aquatic Mammals* 22: 153–166.

Visser, I.N. 1998. Prolific body scars and collapsing dorsal fins on killer whales *(Orcinus orca)* in New Zealand waters. *Aquatic Mammals* 24: 71–81.

———. 2000. Killer whale *(Orcinus orca)* interactions with longline fisheries in New Zealand waters. *Aquatic Mammals* 26: 241–252.

Visser, I.N., J. Berghan, R. van Meurs, and D. Fertl. 2000. Killer whale *(Orcinus orca)* predation on a shortfin mako shark *(Isurus oxyrinchus)* in New Zealand waters. *Aquatic Mammals* 26: 229–231.

Waite, J.M., N.A. Friday, and S.E. Moore. 2002. Killer whale *(Orcinus orca)* distribution and abundance in the central and southeastern Bering Sea, July 1999 and June 2000. *Marine Mammal Science* 18: 779–786.

Walters, C. 1986. *Adaptive management of renewable resources.* New York: Macmillan.

Whitehead, H. 2003. *Sperm whales: social evolution in the ocean.* Chicago: University of Chicago Press.

Williams, A.J., B.M. Dyer, R.M. Randall, and J. Komen. 1990. Killer whales *Orcinus orca* and seabirds: "play", predation and association. *Marine Ornithology* 18: 37–41.

Yano, K. and M.E. Dahlheim. 1995. Killer whale *(Orcinus orca)* depredation on longline catches of bottomfish in the southeastern Bering Sea and adjacent waters. *Fishery Bulletin* 93: 355.

Yurk, H., L.G. Barrett-Lennard, J.K.B. Ford, and C.O. Matkin. 2002. Cultural transmission within maternal lineages: vocal clans in resident killer whales in southern Alaska. *Animal Behaviour* 63: 1103–1119.

FOURTEEN

Killer Whales as Predators of Large Baleen Whales and Sperm Whales

RANDALL R. REEVES, JOEL BERGER, AND PHILLIP J. CLAPHAM

Three or four of these voracious animals do not hesitate to grapple with the largest baleen whales; and it is surprising to see those leviathans of the deep so completely paralyzed by the presence of their natural, although diminutive enemies. Frequently the terrified animal—comparatively of enormous size and superior strength—evinces no effort to escape, but lies in a helpless condition, or makes but little resistance to the assaults of its merciless destroyers.

CHARLES M. SCAMMON (1874), an American whaleman

The position of the killer whale *(Orcinus orca)* at the top of the marine trophic pyramid is unquestioned. It consumes a remarkable variety of organisms, ranging in size from small schooling fish to blue whales *(Balaenoptera musculus)* and as taxonomically diverse as seabirds (ducks and alcids; Bloch and Lockyer 1988), marine reptiles (leatherback turtles, *Dermochelys coriacea;* Caldwell and Caldwell 1969), elasmobranchs (Fertl et al. 1996; Visser 1999a), and terrestrial mammals as they swim across coastal channels (e.g., cervids; Dahlheim and Heyning 1999). If killer whales have any natural predators, these would be other killer whales, as there is some evidence to suggest "cannibalism" (Shevchenko 1975; but see Pitman and Ensor 2003).

The interactions of killer whales with other marine mammals range from predation to coexistence (Jefferson et al. 1991; Weller 2002). Foraging specialization is a typical feature:

The *transient* ecotype preys regularly, if not exclusively, on marine mammals, whereas the *resident* type preys mainly on fish (Ford and Ellis 1999; Baird 2000). In recent years, field observations of predatory behavior have led to several hypotheses concerning relationships between killer whales and the great whales—the large baleen whales and the sperm whale *(Physeter macrocephalus)*. Corkeron and Connor (1999) argued that the low-latitude migrations of some baleen whales represent a strategy for avoiding killer whale predation on calves. Although Clapham (2001) forcefully contested that idea, citing the humpback whale *(Megaptera novaeangliae)* as an example, Pitman et al. (2001) subsequently proposed that "killer whales, through their predatory habits, represent a much more important selective force in shaping life history traits of individual marine mammal species, and in structuring their communities, than has generally been acknowledged." After witnessing several attacks on sperm whales off Mexico and California, Pitman et al. suggested that both the social organization and the sexually segregated distribution of sperm whales could be at least partly the result of predation pressure from killer whales. Earlier papers by Reeves and Mitchell (1988b) and Finley (1990) had argued that killer whale predation was likely an important factor in shaping the behavior, migration strategies, and life history of bowhead whales *(Balaena mysticetus)*. There are many more first-hand descriptions of killer whale attacks on gray whales *(Eschrichtius robustus)* than on any other species of large whale (cf. Jefferson et al. 1991; Melnikov and Zagrebin 2005; and subsequently in this chapter), and young gray whale calves are considered "particularly vulnerable, even while under the watchful eye of their mothers" (Weller 2002; also see Black et al. 2003; Ternullo and Black 2003).

Our goal in this chapter is to add to the discussion of killer whales as primary predators of large baleen and sperm whales. If attacks by killer whales on these large cetaceans were common, a set of outcomes might be expected. These would include a relatively high incidence of witnessed attacks, detection of maimed bodies of victims and escapees, and clear evidence of specific antipredator behavior. Assessment is complicated, however, by the logistics of making field observations in oceanic systems, where much of the relevant behavior occurs underwater—in contrast to open terrestrial landscapes, where the interactions between large carnivorous mammals and their prey can be more readily observed. Hence, we are forced to rely on two less direct approaches.

First, given existing knowledge of large or medium-sized land carnivores, we consider whether inferences about *their* current roles in shaping ecological and prey dynamics would remain valid if based exclusively on historical observations. In other words, would historical anecdotes stretching back some 200 years, without the benefit of marked individuals and current study methods, have been sufficient to appraise a terrestrial predator's ecological role? This question seems instructive, because we have little information apart from historical anecdotes with which to evaluate the importance of ecological interactions between killer whales and large whales in an earlier context when populations of the latter (if not also the former) were much greater than they are today.

Second, given the infrequency of witnessed predation by killer whales on large whales, we use indirect assessments of predation events and the antipredator behavior of prey to infer how killer whales might influence the spatial and population responses of other cetaceans. The following specific questions are addressed here:

1. Does predation occur sufficiently often to make it a major factor in regulating the populations of large whales?

2. What is the sex and age distribution of large whale prey taken by killer whales?

3. What can be inferred about the consumed:unconsumed ratio of the prey body mass when killer whales kill a large whale?

The observations and comparisons reported here span several centuries, stem from different parts of the world, and derive from accounts of scientists and hunters, whose astuteness and training have varied. Cumulatively, however, these diverse lines of evidence suggest that killer whales once played a role in determining the behavior and distribution of at least some populations of large cetaceans.

Historical Anecdotes about Terrestrial Carnivores: Were Past Insights Accurate?

While terrestrial and oceanic systems differ in important ways, the fundamental argument here is that predation as a selective force shapes the behavior of prey in similar fashion across systems. The primary issue, however, is not whether behavioral adaptations parallel species in respective systems, but whether sporadic, anecdotal observations of secretive or difficult-to-study species are likely to lead to accurate depictions of ecological dynamics. Before such tools as radio telemetry and biochemical analyses became available, speculations about prey choice on the part of hunting carnivores and about defense strategies of their prey were necessarily based on witnessed observations. Our understanding of the ecological dynamics of predators and prey, whether terrestrial or marine, large or small, will continue to be revised as new methods are applied and larger samples of observations become available. As a way of anticipating how views of killer whales in relation to their large whale prey may change in the future, we look back at what was once inferred about the ecological roles of three terrestrial carnivores and how understanding of them has been modified with studies involving improved techniques.

THE GRIZZLY BEAR: *Ursus arctos*, also called the brown bear in Europe and Asia, has been characterized ecologically in different ways through time. Early reports implied an ecological niche associated with herbivory and omnivory. During the 1830s and 1840s, Osborne Russell and John C. Fremont noted that bears in the Rocky Mountains fed in areas with willows and cherry trees, and William Kelly indicated that they dug for bulbs and roots in California (Bass 1996). Reports from as early as the 1600s also noted that grizzly bears fed on marine mammal carcasses along the California coast (Storer and Tevis 1955).

While grizzly bears are clearly omnivorous and subsist primarily on vegetation (Mattson et al. 1991), it is now clear that meat and fish are also important components of the diet (Hildebrand et al. 1999). Grizzlies have been shown to be capable predators of both juvenile and adult ungulates as large as elk *(Cervus elaphus)* and moose *(Alces alces)* (Gunther and Renkin 1990; Mattson 1997; Berger et al. 2001a). In fact, recent work with radio-collared neonatal moose in Alaska

TABLE 14.1
Chronologically Changing Insights or Speculations about Terrestrial Carnivore Ecology and Behavior

Species	Period	Insight	References
Grizzly bear	1830s–1880s	Primarily an omnivore and herbivore	Osborne and Fremont journals (Haines 1965; Schullery 1988; Clark and Casey 1996)
	Early 1990s–today	Effective (population-regulating?) predator; large effects on survival of neonatal caribou, moose, and elk	Gasaway et al. 1992; National Research Council 1997; Mech et al. 1998; P. J. White, personal communication
Wolverine	?–1980	Vicious scavenger and predator	Seton 1953
	1980–today	Uncertain	P. J. White, personal communication; Mech et al. 1998
Spotted hyena	1790–early 1960s	Scavenger	Kruuk 1972
	Mid 1960s–today	Effective predator	Kruuk 1972; Mills 1990

suggests that bear predation can limit population growth (Gasaway et al. 1992). In addition, grizzly bears were found to be more effective predators than wolves *(Canis lupus)* on 155 young radio-collared caribou *(Rangifer tarandus),* respectively accounting for 41% and 35% of the total mortality (Mech et al. 1998). Among 50 radio-collared Yellowstone elk fawns, bear (both grizzly and black, *Ursus americanus*) predation was responsible for 60% of the total mortality, with grizzly bears accounting for twice as much mortality as was caused by wolves (32% vs. 16%, respectively; P. J. White, personal communication).

Those findings (summarized in Table 14.1) illustrate how perceptions of bear ecology have changed over time and suggest that grizzly bears have played apex roles in ecological systems. Storer and Tevis (1955: 17) believed that grizzly bears shaped the "original native biota of California," their food sources including bulbous plants, clovers, and grasses as well as berries, seasonally (e.g., elderberry, manzanita, and blackberry). Those same authors speculated that Native Americans avoided areas used by grizzly bears, just as some hunters of ungulates in the Yellowstone ecosystem today avoid areas with high densities of grizzlies (Berger, unpublished data). The loss of bears in at least parts of this ecosystem has led to a cascade of events that ultimately decreased avian diversity via the release of herbivores, followed by intensified herbivory in riparian plant communities (Berger et al. 2001b). The sometimes-subtle influences of bears at multiple trophic levels would not have been detected or recognized without detailed studies made possible, in part, by modern research tools and techniques (Hildebrand et al. 1999).

THE WOLVERINE: *Gulo gulo* remains a relatively understudied forest and sub-Arctic carnivore across North America, Europe, and Asia. An early account described it as exceeding the weasel *(Mustela* sp.*)* by 50 times in "courage . . . slaughter, sleeplessness . . . and demonic fury" (Seton 1953). Initially, wolverines were considered major predators of some native ungulates, but that view has been modified, at least to some extent, as more information on food habits has become available (Magoun 1985; Landa 1997; Persson 2003). For instance, wolverines have a far

smaller impact on neonate, let alone adult, mortality of ungulates than bears or wolves have: <1% for juvenile caribou (Mech et al. 1998); <1% for juvenile elk (Smith and Anderson 1996; P. J. White, personal communication); <1% for juvenile moose (Ballard et al. 1991; Testa et al. 2000). Although predation on adult ungulates may occur occasionally, it is very infrequent, judging by the number of radio-collared ungulates that die from other causes. Clearly, our view of the wolverine's ecological role has changed during the past 50 years with the benefit of intensive research using modern tools and methods.

THE SPOTTED HYENA: *Crocuta crocuta* has been among the most misunderstood mammals, with misconceptions dating back to 1790 (Kruuk 1972). Early reports implied surprise when hyenas were observed to kill healthy prey. It was not until 1964 that Eloff (1964) inferred from tracks in the sand that hyenas hunted and killed more than just sickly prey. Subsequent study has confirmed that spotted hyenas in eastern and southern Africa are effective predators on large ungulates (Kruuk 1972; Mills 1990).

The three foregoing examples (see Table 14.1) show that reliance on anecdotal observations can cause us to misunderstand the relationships between carnivores and their prey. To be fair, anecdotes can also be highly informative. Indeed, ranchers had long insisted that coyotes *(Canis latrans)* were often responsible for low fawn recruitment in pronghorns *(Antilocapra americana),* a belief recently verified regionally by field work and explored by modeling (Byers 1997; Phillips and White 2003). Given our present knowledge about the ways that mammalian carnivores affect the dynamics of terrestrial and nearshore marine ecosystems (Estes et al. 1998; Soulé et al. 2003), however, it seems prudent to proceed cautiously in interpreting qualitative evidence dominated by historical anecdotes.

Observations by Whalemen

Jefferson et al. (1991) cautioned that reports from "whalers and other untrained observers" needed to be interpreted skeptically, yet at times biologists have been too eager to dismiss such observations. For example, the description by whalers of a killer whale attack on a group of sperm whales off southern Africa (Best et al. 1984) was considered inconclusive by Jefferson et al. (1991), who neglected the statement by Best et al (1984: their Fig. 7) that "killer whales do occasionally prey on sperm whale calves," a statement supported by photographs of a stranded animal. With the recent well-documented observations by biologists of successful attacks on sperm whales (Pitman et al. 2001; see later), historical reports by "untrained" whalemen seem more credible.

Descriptions of killer whales attacking large whales are nevertheless remarkably rare in the first-hand narratives of whalers. One of us (Reeves) has examined many hundreds of whaling logbooks from the eighteenth and nineteenth centuries. Although sightings of killer whales were frequently recorded, often on whaling grounds (Reeves and Mitchell 1988a), descriptions of attacks or evidence of attacks (e.g., finding dead or wounded whales) were uncommon. Other whaling historians who have handled large numbers of logbooks and journals as well as published memoirs of whalers and explorers report similar experience (Michael Dyer, New Bedford Whaling Museum Library, New Bedford, MA, personal communication, October 2003; Klaus Barthelmess, Cologne, Germany, personal communication, October 2003). We emphasize, however, that lack of observations of whale behavior and natural history is the norm in whaling logbooks and first-hand narratives; few whalemen had an interest in, or a knack for relating, such things.

While on a sperm whaling voyage to the Japan Ground, the ship *Phoenix* (1847–51MS) passed two right whales (*Eubalaena japonica*) and several humpbacks on February 28, 1849. The next day, at 35°52'N, 130°29'E, the ship's logbook recorded: "Saw a dead humpback surrounded by a school of killers." No further mention of this event appears in the logbook, so it is impossible to be certain that the humpback was a victim of predation.

Another account is much more detailed. While at anchor in Magdalena Bay on the Pacific coast of Baja California (February 14, 1858), the crew of the whaleship *Saratoga* observed an attack on gray whales (*Saratoga* 1857–58MS):

> saw 8 or 10 killers attacking a cow whale with her calf close in shore, the whale fighting bravely to protect her offspring, using . . . her flukes and fins, which were in constant motion, striking furiously . . . at her aggressors. . . . The battle had lasted about half an hour when we saw the water highly discolored with blood . . . when [a boat] arrived . . . both the whale and the killers had disappeared, they had no doubt killed the calf and taken it down.

The *Saratoga*'s log further describes killer whales and their typical way of attacking large whales: "Their favorite morsel of food is the tongue, they attack the whale in concert, worrying him until they force his mouth open, when they seize upon the tongue and soon dispatch him." Also, this logbook claims that whalers often took advantage of killer whales to

aid them in catching whales: "Whales are frequently taken by whalemen, when attacked by killers, being nearly exhausted, a boat pulls up to him, fastens and he becomes an easy victim to the lance, the killers not daring to attack the boat." This description, and others like it (e.g., Brown 1887: 284), anticipate the later accounts of "symbiosis-like" interactions between killer whales and shore whalers hunting right and humpback whales in Australia (Dakin 1934; Wellings 1964; Gaskin 1972; Mitchell and Baker 1980).

Killer whales often scavenged whales killed by whalers (see Whitehead and Reeves 2005). Charles Wilkes, commander of the U.S. Exploring Expedition, 1838–1842 (see Pond 1939), was told by whalemen that killer whales sometimes dragged whales away from boats as they were being towed to the ship (also see Scammon 1874: 90). In modern industrial whaling, carcasses flagged for retrieval or fastened to the catcher boat were subject to scavenging (McLaughlin 1962: 130; Ash 1964: 56; Gaskin 1972; Mitchell and Reeves 1988). Heptner et al. (1996: 689–690) mentioned that killer whales "regularly tore out the tongues of dead and air-filled whales" in the Antarctic; these authors included a photograph, as did Gaskin (1972: 120), and noted the economic implications of the oil lost as a result of this scavenging. Interestingly, though, Heptner et al. had received no reports of killer whales tearing chunks of blubber from, or biting the fins and flukes of, whale carcasses. Gaskin, in contrast, claimed that in some areas near the South Shetland Islands, flagged carcasses of rorquals that were not recovered within 24 hours sometimes had their blubber stripped away by killers. He agreed with Heptner et al., however, that "usually" only the tongues were consumed. It was reportedly very difficult to deter the killers, "even by opening fire" on them (Heptner et al. 1996).

Gray Whales

Rice and Wolman (1971: 98–99) downplayed the significance of killer whale predation on gray whales despite the evidence adduced by Andrews (1914, 1916b) indicating that they were frequently preyed upon or scavenged by killer whales in Korean waters. Since the early 1970s, the evidence of predation along the North American coast has mounted. Jefferson et al. (1991) listed nine successful attacks, five apparently unsuccessful, and at least seven in which the outcome was uncertain or the evidence circumstantial. The lack of confirmation of a successful outcome (i.e., a dead prey), however, was sometimes due to interruption of observations. For example, an observation of killer whales closing on a pod of gray whales in the northern Bering Sea in May 1981 ended when the observers "had to leave the area because the aircraft was low on fuel" (Ljungblad and Moore 1983). Jefferson et al. (1991) classified this report as a "No" kill, but in fact the observation team had no opportunity to confirm the outcome one way or the other.

The relatively numerous accounts of attacks on gray whales establish several things. First, they show that killer whale predation on gray whales is not new, dating at least as far back as the mid-1840s, when commercial whaling on gray whales began along the coast of Baja California (see the *Saratoga* event described in the preceding section; Scammon 1874: 90). Second, there is no basis for believing that such predation occurred only, or with greater frequency, when the gray whale population was especially large (i.e., before it was decimated by whaling). Attacks observed in about 1907 (Pond 1939), the 1940s (Walker 1949), and the early 1950s (Cummings and Thompson 1971) would have taken place when the eastern Pacific population of gray whales was well below its carrying capacity level. Third, predation is not limited to a particular segment or segments of the gray whale's annual distributional range; attacks have been observed in the calving/mating grounds off Baja California, at many points along the migration route between Mexico and Alaska, and on the feeding grounds in the Bering and Chukchi seas (Melnikov and Zagrebin 2005). Finally, even though calves and yearlings appear to be especially vulnerable, gray whales of any size are subject to attack.

Raymond M. Gilmore (1961) was an experienced gray whale watcher, having spent five field seasons in the lagoons of Baja California and many years observing the migration off San Diego, California. He had never seen killer whales in the lagoons, and although he saw them occasionally in California coastal waters during the gray whale migration, they usually "passed the gray whales without attack" or remained "at some distance without show of interest, or vice-versa" (Gilmore reported one unsuccessful attack by six killers on two grays off La Jolla in 1950). Theodore J. Walker (1975), Gilmore's contemporary with similarly long experience watching gray whales in Mexico and California, was aware of "isolated reports" of attacks that "occasionally" resulted in kills of gray whales, but he was, like Gilmore (and Rice and Wolman 1971), doubtful that killer whale predation was a significant feature in the lives of gray whales. Having spent six field seasons observing gray whales in Laguna San Ignacio (1977–1982), Jones and Swartz (1984: 342) reported that only one of the 32 stranded carcasses that they examined (an "immature") was the result of a killer whale attack. Swartz (1986) did not even mention killer whales in his review and analysis of the gray whale's migratory, social, and breeding behavior. Jorge Urbán-Ramírez (personal communication, December 2003), who had been studying gray whales in the lagoons and along the outer coast of Baja California annually for a decade (see Urbán Ramírez et al. 2003), had not seen killer whales in the lagoons, nor had his teams of researchers seen them close to gray whales during aerial and shipboard surveys outside the lagoons.

Jones and Swartz (2002) nevertheless referred to "frequent" reports of killers feeding on the tongues of gray whales and claimed that killer whale tooth rakes were "often" seen on the bodies of living gray whales. Those authors also speculated that avoidance of killer whale predation "might be a primary benefit to females leaving polar waters to give birth in the subtropics." Conversely, at the other end of the migration, Moore et al. (1986) suggested that the movement of

females and calves-of-the-year into the Chukchi Sea during the summer feeding season could represent a strategy to avoid killer whales, which were considered less common there than in the Bering Sea (cf. Forney and Wade, Chapter 12 of this volume). In fact, Moore et al. likened this situation to that on Isle Royale, Michigan, where moose cows with calves occupied suboptimal foraging areas that were "wolf-free," whereas solitary adult and yearling moose opted for prime foraging habitat and coexisted with wolves. Killer whale attacks on gray whales were said to be "fairly common" during the summer months in the Bering Strait region (George et al. 1994: 252; also see Melnikov and Zagrebin 2005). If gray whales migrate in part to reduce the risks of predation on calves, the long-range movements of killer whales can be seen as a counterstrategy. *Transient* killer whales from southeastern Alaska regularly appear in spring (April and May) in Monterey Bay (California), where they prey on northward-migrating gray whale calves, sometimes at least injuring, if not killing, mothers (Goley and Straley 1994; Black et al. 2003; Ternullo and Black 2003). Attacks on gray whales in Monterey Bay have occurred "on an almost predictable basis" since the late 1980s (Weller 2002: 993).

Although Jones and Swartz (2002: 533) concluded that "predation pressure does not appear to be a significant determinant in the gray whale's social organization," aspects of gray whale behavior have almost certainly been shaped by killer whale predation, as shown experimentally in playback experiments using recorded killer whale sounds (Cummings and Thompson 1971). Along the California coast, gray whales respond to the presence of killer whales by moving quickly and quietly into nearshore kelp beds and becoming quiescent; similar responsive behavior has been observed along the Korean coast (Andrews 1914, 1916b). When trying to avoid detection, gray whales typically "snorkel," meaning that they exhale underwater and barely expose their blowholes at the surface to inhale (Dahlheim and Heyning 1999). "Spyhopping," in which a whale stands vertically with the entire head exposed above the surface, is an oft-described feature of gray whale behavior, especially in the Mexican calving lagoons. Although gray whales spyhop under many circumstances when the threat of killer whale predation is not evident, one of the functions attributed to this behavior is visual scouting (Eberhardt and Evans 1962), including "looking for killer whales" (Cummings and Thompson 1971; Baldridge 1972). Cetaceans generally have good vision both in air and under water (Mass and Supin 2002). We are nevertheless skeptical of whether the eyes of a spyhopping gray whale would be far enough above the water surface to allow scanning at any but very close range, even in calm to moderate sea states.

Bowhead Whales

Most reports of killer whale attacks on bowhead whales are secondhand, having been described to biologists by Inuit hunters. Jefferson et al. (1991) listed 12 separate events, only two of which were witnessed by a biologist, both with an uncertain outcome (Finley 1990). Relatively few reports have emanated from the western Arctic, where most modern research and monitoring of bowhead whales have taken place (only one of the 12 events listed by Jefferson et al. was from outside the eastern Arctic, and that was from the Sea of Okhotsk; also see George et al. 1994; Melnikov and Zagrebin 2005). One explanation could lie in the same reasoning as mentioned in the previous section in relation to gray whales: Killer whales are much less common in the Chukchi and Beaufort seas (where most western Arctic bowhead whales spend the open-water season) than in the Bering Sea (where bowheads occur mainly during the dark, ice-affected season). George et al. (1994) doubted that the bowhead migration out of the "rich feeding grounds" of the Bering Sea and into the Beaufort Sea during the spring and summer would represent a predator avoidance strategy, yet they offered no clear basis for such skepticism. In those authors' view, the near absence of scarring on small bowheads taken by hunters was explained most plausibly by the fact that older bowheads have longer times to acquire scars ("greater exposure to killer whale attack") and/or that few small bowheads survive attacks.

In contrast to the western Arctic, predation in the eastern Arctic is thought to be relatively common. Bowhead whales and other Arctic marine mammals are well known for behavior that Finley (2001) called "killer whale phobia." Inuit of the eastern Canadian Arctic apply a specific word, *ardlungaijuq* or *aarlungajut*, to this fear of killer whales exhibited by whales and seals. Finley cited the belief of British whalemen that differences in the summer distribution of bowheads of various sex and age classes reflected differences in their vulnerability to killer whale predation. Adult males were believed to remain in the "most exposed and open situations," while females and juveniles penetrated farther into the ice. Finley (1990) interpreted the "coast-hugging" tendencies of autumn-migrating bowheads, "in the absence of protective ice cover," as an adaptive response to killer whale predation (this may also be true of gray whales, especially during their spring migration, although alternative hypotheses are that they follow the coastline to facilitate navigation or to take advantage of sublittoral food resources; Braham 1984). In several instances summarized by Mitchell and Reeves (1982), bowheads came close to shore or wedged themselves into narrow ice cracks to avoid killers, making them easy targets of hunters.

Finley (1990: 151) considered predation to be significant for depleted bowhead whale populations, "particularly when killer whales have an abundant alternate food source, as they do in the Eastern Arctic (i.e., the killer whale population was not food limited when the bowhead population declined)." The alternative prey base that Finley had in mind would include phocid seals, narwhals *(Monodon monoceros)*, belugas *(Delphinapterus leucas),* and possibly walruses *(Odobenus rosmarus)* (see also Reeves and Mitchell 1988b). His basic suggestion that bowhead whales in the eastern Arctic may be in a predator pit (Southwood 1975, as cited by Bergerud and Elliot 1986) is consistent with the views expressed by Moshenko et al. (2003: 22), citing a published review of hunter knowledge (NWMB 2000) and the opinions of Canadian Inuit who regard killer whales as "possibly the greatest threat to bowhead recovery." The idea that heavy ice confers protection from killer whale predation is implicitly supported by claims that in recent years with relatively light ice conditions in Foxe Basin (e.g., 1999), more killers were observed and unusually high numbers of bowheads were found dead (Moshenko et al. 2003: 24). It was reported that in western Greenland, during a four-day period in April 2002, "about 10 bowhead whales out of a pod of about 30 were killed by a large, uncounted pod of killer whales" (IWC 2003: 46). It has not been possible to verify that this event occurred, and we are very skeptical of its authenticity.

Sperm Whales

The infrequency of first-hand descriptions of attacks on sperm whales, which were the subjects of large-scale, worldwide whaling for approximately 200 years, is especially puzzling. Berzin (1972: 273) reported that remains of this species had not been found in the stomachs of killer whales and that Soviet scientists and whalers had reported "few cases of attack of killer whales on newborn sperms." He concluded that killer whales should not be considered "serious enemies of the sperm whale" and that adult male sperm whales were "too strong for killer whales and thus immune to attack while a herd [of females and young] possesses a sufficiency of strong instinct of mutual aid to give them protection"—a view shared by Heptner et al. (1996: 833). Jefferson et al. (1991) mentioned only five specific records of attacks and one generalized account. The aforementioned observation off southern Africa of a group of sperm whales in a typical defensive formation ("rosette"; see later) being encircled by killer whales was designated a "No" kill by Jefferson et al., yet their source (Best et al. 1984) provided no indication of the outcome (i.e., it should have been classified as "[?]"). Furthermore, Best et al. (1984) illustrated "killer whale damage" on a sperm whale calf that stranded near Cape Town (not mentioned by Jefferson et al.). Jefferson et al. (1991) also dismissed as inconclusive and "second-hand" the reference by Yukhov et al. (1975) to a filmed attack on a group of sperm whale females and calves.

Several lethal attacks have been observed or otherwise documented in recent years. Visser (1999b) summarized interactions between killer whales and sperm whales off New Zealand, including the finding of a fresh carcass of a 9.8-m female sperm whale that bore clear evidence of having been attacked and mauled by killer whales. Visser and Bonoccorso (2003) reported an observation off Papua New Guinea in which at least 20 killer whales "appeared to be hunting" a group of about 12 sperms, including calves, but they considered this secondhand record to have been nonpredatory (citing terminology of Jefferson et al. 1991). In Indonesian waters, Kahn (2003) witnessed the serious (probably lethal) wounding of a sperm whale calf and a subadult when five killers encountered a loosely associated pod of sperms. The attack did not appear to have been sustained to a point where either of the victims was killed outright, but the sperm whales were bleeding profusely and their mobility was seriously impaired before the killers moved slowly away from the area and beyond visual range of the observation vessel.

Pitman et al. (2001) provided the first well-documented record of a sustained, lethal attack. The first of three encounters (all in the eastern North Pacific) involved a group of eight sperm whales that formed a rosette—all animals huddled at the surface with their heads together and tails pointing outward—possibly in response to the presence of a group of about six killer whales. This event was confounded by the simultaneous presence of a school of short-finned pilot whales *(Globicephala macrorhynchus)*, and there was no direct evidence of an attack by the killer whales before observations had to be terminated because of deteriorating weather. In the second encounter, observations began when an attack was already under way (with a large slick of blood and oil in the water) on a group of about nine female and subadult sperm whales in rosette formation. Up to about 35 killers were in the area, although only female and immature individuals were seen attacking the sperms, following what the authors called a "wound and withdraw" strategy (16 separate "attacks" took place during about four hours of observation). At least one of the sperm whales was killed and at least partially eaten and others were seriously injured. Pitman et al. concluded that at least three or four of the "survivors" were mortally wounded, "and it is quite possible that the entire herd died as a result of injuries from the attack."

The third encounter described by Pitman et al. (2001) occurred only five days after the second one. During the course of a much less energetic and sustained attack, a series of small groups of sperm whales coalesced into an aggregation of about 50 individuals, apparently in response to the presence of five killer whales. The sperms formed a "spindle," consisting of one to five animals abreast, 12 to 15 animals long, all facing the same direction. An adult female killer made several forays into the midst of the sperms and caused an oily slick to appear, "suggesting that one or more had been bitten, although no blood was visible." This encounter ended with no apparently serious damage having been inflicted on the sperm whales. Whitehead (Chapter 25 of this volume) estimates that an average female sperm whale in the tropical Pacific might experience as many as 150 killer whale attacks in her lifetime.

Behavior of Killer Whales during Attacks

Some features of the specialized hunting behavior of killer whales were summarized by Jefferson et al. (1991): cooperation and coordination; biting the flukes and flippers to impair the prey's swimming ability; and swimming or "leaping" onto the whale's back to impede its progress or prevent it from respiring normally. One feature that they did not mention explicitly, but implied by reference to internal injuries, is body ramming. Typically, when hunting gray whales in Monterey Bay, the killers chase and tire the female and calf, separate the calf from its mother, and then ram and "body slam" the calf repeatedly (Ternullo and Black 2003; also see Goley and Straley 1994). Body ramming has also been observed when killers are hunting gray whales on the northern feeding grounds (Melnikov and Zagrebin 2005).

Like other large pack-hunting predators, killer whales combine stealth with probing and testing of the prey. Judging by the fact that so many observed attacks end in no kill, or at least in an uncertain outcome (Jefferson et al. 1991: their Appendix I), together with the relatively high incidence of killer whale-inflicted scars and injuries in populations of large whales (see subsequent paragraphs), many attacks are probably unsuccessful. In fact, some attacks may be initiated primarily to provide learning opportunities for young killer whales.

Some specialization of roles has been noted. Adult females often appear to be the most active and effective individuals during the actual killing phase of an attack on a large whale (Silber et al. 1990; Finley 1990) although pairs of adult males have also been seen working together to kill gray whale calves (Ternullo and Black 2003). An adult male participated in the successful attack on sperm whales just described only to "finish off" a whale that had first been critically wounded by the females and immature killers. Pitman et al. (2001) considered this feature of the attack reminiscent of lions *(Panthera leo)*, where adult males often use their larger size to move in and appropriate prey (Schaller 1972; Bertram 1979). Killer whales have been seen to continue feeding on a dead gray whale calf for up to 20 hours, with many more individuals participating than just those involved in the attack. Again, these features are in some respects reminiscent of lions.

Killer whales that prey on marine mammals usually stop vocalizing, or vocalize relatively little, while on the hunt (e.g., Cummings et al. 1974; Ljungblad and Moore 1983; Kahn 2003). Presumably, passive listening trumps active echolocation as a means of locating and tracking large prey that, like the killers themselves, must cross the air-water interface to breathe and that also depend on sound to communicate and echolocate. During a lethal attack on a female gray whale and calf in Monterey Bay, the killer whales reportedly "vocalized frequently during the entire episode" (Goley and Straley 1994). However, the attack had already been under way for about 40 minutes when Goley arrived on the scene and began audio recording. While silence may contribute to the stealth and prey-location phase of an attack, active sound processing presumably becomes allowable, and perhaps functionally important, once contact with the prey has been established. Although the sounds associated with an attack may attract distant killer whales to the site, allowing them to benefit should a kill occur, this possibility has been judged unlikely (Deecke et al. 2005).

Size/Age Preferences

The question of whether killer whales hunt selectively for whales of particular age or sex classes is critical for evaluating the ecological significance of predation. Scammon (1874: 90) was equivocal about the intensity and significance of killer whale predation on whales, concluding on one hand that "the larger Cetaceans" were attacked "rarely," but also that killers "chiefly prey with great rapacity" on the young of baleen whales. This latter suggestion appears largely consistent with the evidence amassed since his time (Dolphin 1987; Jefferson et al. 1991; Black et al. 2003; Melnikov and Zagrebin 2005). Also, however popular the notion may be that predators such as killer whales have a "purifying" effect on prey populations by selectively removing diseased, defective, and postreproductive individuals (cf. Bertram 1979: 231), the evidence of predation by killer whales provides little or no support for it.

The relatively frequent examples in the literature where killer whales attack and attempt to intimidate or debilitate large prey, but then abandon the hunt without making a kill, suggest that they are often wary of pressing beyond a certain threshold of energy investment, and perhaps of risk to their own health. We would expect there to be a gradation of some kind, from prey that can be attacked with little regard for the killer whales' own safety and need to conserve energy to prey that represent both a safety risk and a questionable cost:benefit tradeoff in energy. On one side of this threshold could be animals that are smaller or only slightly larger than an adult killer whale (9 m), e.g. minke whales (Balaenoptera acutorostrata and B. bonaerensis) or young Bryde's whales (Balaenoptera brydei and B. edeni). Attacks on such prey should represent little or no risk to the killers, and once contact has been made it is difficult to envision how these prey could avoid being killed (e.g., Hancock 1965; Wenzel and Sears 1988; Silber et al. 1990). On the opposite side of the size threshold, the outcome of an attack would be less certain, and the killers may be more tentative and cautious as they test their larger victims for speed, strength, and stamina (much as wolves do). Circumstances would, of course, play a large role. Groups of prey animals with the potential for mutual aid in defense presumably would be less vulnerable (e.g., unsuccessful attacks on right whales observed by Cummings et al. [1971, 1972; see the following section], on humpback whales by Whitehead and Glass [1985], and on sperm whales by Arnbom et al. [1987]). A chance encounter with a solitary large prey animal (e.g., the blue whale attack observed by Tarpy 1979) might give the killers an opportunity to obtain chunks of skin and meat while testing for weakness and vulnerability, thus deriving some energetic benefit even though the injured prey ultimately manages to escape. Calves of all large whale species would fall below the threshold of vulnerability. Although the presence of mothers and other adults could tip the balance in the calf's favor, it could also put the adults at risk of severe injury or death (e.g., Scammon 1874: 90; Goley and Straley 1994).

Resistance to Attack

Large whales (unlike minke whales) are clearly able to resist killer whale attacks. Weller (2002) singled out humpback whales in this regard, claiming that they were more aggressive than some other species in thrashing at the attackers with their flukes and flippers. This is consistent with the observations in Monterey Bay by Ternullo and Black (2003), who noted that although humpbacks there frequently have killer whale tooth rakes on their flukes, no lethal attacks have been observed. Dolphin's (1987) observations of interactions between humpback whales and killer whales in Alaska, discussed subsequently, present a different picture, implying that attempts at predation are exceptional there.

Cummings et al. (1971, 1972) described a "full-fledged attack" by approximately five killer whales (four females and a calf) on two courting southern right whales (Eubalaena australis) in nearshore waters off Patagonia. At times during the attack, the right whales were entirely surrounded, but they kept close together and vigorously slashed at the attackers with their flukes and flippers. "Most of the time the killer whales were in a swimming frenzy that took them over, between, and under the two big right whales" (Cummings et al. 1971: 266). After 25 minutes the killers abandoned the effort and moved away, and the right whales relocated into shallower water and became quiescent. The observers saw no signs of blood or other evidence of physical harm to the right whales. Cummings et al. (1972) were told by local fishermen of a previous similar, but apparently lethal, attack. Miguel Iñíguez (personal communication, July 2003) and colleagues also have observed successful (i.e., lethal) attacks on right whales off Patagonia, but in all such instances adult male killers were involved. In their experience, when only females, juveniles, and calves took part, the attack invariably ended with the right whales (mainly mothers and dependent calves) escaping alive.

The gray whale—in spite of its reputation as a "devilfish," dangerous to approach and difficult to kill (Scammon 1874)—has been described as exhibiting docile or submissive behavior on some occasions when under attack. A particularly graphic example was provided by Andrews (1916b: 200–201), in which a gray whale was delivered to a Korean whaling station after having been "shot in the breast between the fins." The whaling captain reported that he had encountered the animal surrounded by killer whales, "lying at the surface, belly up, with the fins outspread . . . absolutely paralyzed by fright." He was able to approach and fire the harpoon into its exposed chest region. This description is reminiscent of one by Norris and Prescott (1961: 358–359) of a female gray whale whose behavior changed abruptly after being chased by their capture vessel over "a considerable distance." The whale stopped trying to escape and rolled onto her back, exposing the ventral body surface with the flippers and lower jaw extended into the air. A possible explanation of this behavior is capture myopathy or capture stress, a syndrome often observed in ungulates (Chalmers and Barrett 1982).

There is also considerable evidence of nonresponsiveness, or at least lack of flight or avoidance responses, on the part of large whales to the presence of killer whales (Jefferson et al. 1991), as there is for savanna ungulates, which exhibit varying degrees of responsiveness to the presence of large predators. One explanation would be that the potential prey is somehow able to tell that the would-be predators are satiated or searching for easier targets (Dolphin 1987). Another explanation could be that the potential prey animals are aware of the food-preference differences between *resident* (fish-eating) and *transient* (mammal-eating) killer whales and that, therefore, the muted response or absence of response simply signifies an ability to distinguish between the two ecotypes, whether by visual, acoustic, or some other kind of cue.

Scarring and Mutilation as Evidence of Attack Frequency

Among the difficulties of interpreting scarring rates as indices of predation rates is that most predators fail in predation attempts much more often than they succeed (e.g., Mech 1970: 237). "Success rates" are difficult to measure accurately (Bertram 1979: 226). Hunter observations of attacks by killer whales on gray whales off Chukotka suggest a relatively high success rate, although it is difficult to judge whether the estimate of 80% success by Melnikov and Zagrebin (2005) is unbiased. Apart from the logistical challenges, high cost, and time requirements to obtain a representative sample of observations, predation by killer whales on whales is complicated by the fact that so much of the activity and evidence remains hidden below the sea surface and out of sight. Analyses of scarring and mutilation (e.g., bitten-off, frayed, or otherwise damaged appendages) are essentially studies of survivors. The timing of acquisition of scars, and the frequency of their occurrence on animals of different age classes, can nevertheless be instructive. Terrestrial mammals provide useful models in which scarring or maiming serves as an index for attacks by predators. Large herbivores such as elephants *(Loxodonta* sp.*)* and rhinoceroses were generally considered invulnerable to predation, but that conventional wisdom has changed. When tailless and earless black rhinoceros *(Diceros bicornis)* calves and adults were initially reported throughout sub-Saharan Africa, it was suggested that such anomalies reflected congenital problems (Goddard 1969). Later, the missing body parts (i.e., maiming) were tied to predation attempts by spotted hyenas (Hitchins 1986). Subsequent analyses of populations throughout Africa revealed an association between hyena density and degree of taillessness or earlessness in rhinos (Berger 1994). This idea was bolstered by the lack of maiming in regions where hyenas were absent (Berger and Cunningham 1994).

In populations of large whales, the percentages of individuals bearing rake-type scars (presumed to originate from killer whale attacks) vary by area, from a few percent to more than a third of the population (Mehta 2004; Mehta et al. in preparation). For example, 76 of 990 individually identified humpback whales from the Gulf of Maine population (about 7.5%) bear such scars. Seventy-four (97.4%) of the 76 had already acquired the scars prior to their first sighting on this feeding ground (Mehta 2004). Only two individuals that did not have such scars at first sighting were observed to have acquired them in a subsequent year. We recognize that Gulf of Maine humpback whales may not be representative of large whales in general, but the timing of scar acquisition in this population strongly suggests that the great majority of killer whale attacks target calves during the calves' first migration to high latitudes from the species' breeding grounds in tropical waters. Dolphin (1987) and Naessig and Lanyon (2004) reached similar conclusions from observations of humpback whales elsewhere, and analyses of scar acquisition in other mysticete populations indicate a similar pattern (Mehta 2004; Mehta et al. in preparation). As discussed in an earlier section of the present chapter, gray whale calves and yearlings appear to be more susceptible to predation than other cohorts in the eastern North Pacific population of that species.

The question of whether the rake-type scars on the bodies of large whales originate from killer whales has been debated. Other small or medium-sized odontocetes, particularly false killer whales *(Pseudorca crassidens)*, short-finned pilot whales, and pygmy killer whales *(Feresa attenuata)*, are known to associate with, and at least occasionally harass and bite, large whales (e.g., Earle and Giddings 1979; Palacios and Mate 1996; Weller et al. 1996). The "predatory advances" upon large whales by these smaller odontocetes, in combination with the "pronounced fear responses" exhibited by the larger animals (Weller 2002), imply that the potential for some kind of predator-prey interaction is at least perceived on both sides. However, comparison of rake marks on living whales with the dentition patterns of various potential predators (derived from examinations of museum specimens) strongly supports the belief that most rake-type scars do indeed originate with killer whales rather than other species (Mehta 2004; Mehta et al. in preparation). North Atlantic right whales *(Eubalaena glacialis)* appear to be an exception. They exhibit smaller and more narrowly spaced scars that are not consistent with killer whale dentition.

Consumption Patterns

Many descriptions of killer whale predation on large baleen whales stress that the killers were most interested in consuming the whale's tongue (Jefferson et al. 1991; Weller 2002). Often only the tongue, lips, and portions of the ventrum were consumed before the prey was abandoned. In some instances, such incomplete consumption of the carcass might be explained by the fact that the killers were disturbed or deliberately harassed by the observers. However, the frequent finding of beach-cast carcasses that have been bitten principally in the mouth and belly regions, with much of the rest of the body unmolested, implies that the killer whales routinely consume only certain portions of the large whales that they kill.

Silber et al. (1990) had the rare opportunity of watching an attack develop and conclude and then, two days later, locating the carcass of the Bryde's whale that had been killed. They noted that sections of the lower jaw had been removed, there was a large open wound on the abdomen, and parts of the viscera were extruded. Lowry et al. (1987) made similar observations on a 7-m gray whale killed and partially consumed by killer whales in Alaskan waters, and the observations of scavenging on whaler-killed carcasses (previously discussed) reinforce the finding that tongues are the killer whale's foremost interest.

However, in the observations by Pitman et al. (2001), the killer whales apparently fed heavily on the skin and blubber of the sperm whales, with no special attention given to the tongue or mouth area. A pod of approximately 30 killer whales observed attacking a young blue whale tore strips of skin, blubber, and muscle from its rostrum, sides, and back before abandoning their victim (Tarpy 1979). There was no opportunity to relocate the blue whale to see whether it survived or succumbed to its injuries, or to determine whether the killer whales obtained any additional nutrition for their efforts.

Discussion and Conclusions

Many biologists have expressed skepticism about the significance of killer whale predation on large whales. Heptner et al. (1996: 689), for example, claimed that even though killer whales were seen by Antarctic whalers in the vicinity of fin (*Balaenoptera physalus*), humpback, and other baleen whales, "no attempts by the killer whales to attack the others were ever recorded." According to those authors (p. 687), reports of attacks on baleen whales "have been copied from book to book" since the mid-nineteenth century, "without critical review." Dolphin (1987) claimed that reports of predation in the literature were "disproportionate to their frequency of occurrence." Andrews (1916a: 334), in contrast, was impressed by the frequency with which killer whale bite marks were found on the flippers and flukes of blue and fin whales (but not sei whales, *Balaenoptera borealis*) on the flensing decks of whaling stations in Asia. He considered killer whales "a menace" to the blue and fin whales around Japan, where killers were "very abundant." Andrews notwithstanding, Weller (2002) concluded that observed attacks on baleen whales were "not common." Until the 1990s, there was a fairly broad consensus among cetacean biologists that killer whale predation was an exceptional, rather than regular, feature in the lives of large whales, perhaps with two exceptions: gray whales on both sides of the Pacific and bowhead whales in the eastern Arctic.

Evidence of the last few decades has reinforced the impression that gray whales are common prey of killer whales and that calves, accompanied by their mothers, are particularly vulnerable as they move close along the west coast of North America from the winter breeding grounds in Mexico to the northern feeding areas. Recent observations on the feeding grounds near the Chukotka Peninsula suggest that young, perhaps recently weaned gray whales are similarly at risk of predation (Melnikov and Zagrebin 2005). Bowhead whales in the eastern Arctic, severely depleted by commercial whaling, are attacked at least occasionally even though killer whales do not appear to be particularly abundant there (suppressed in part by control hunting; Heide-Jørgensen 1988; Reeves and Mitchell 1988b). In contrast, there is little evidence of predation on bowhead whales in the western Arctic, where they were less severely reduced by whaling and where the ongoing hunt provides opportunities to examine carcasses for wounds and scars (George et al. 1994). With respect to sperm whales, remarkably little evidence of killer whale predation exists in whaling records, whereas recent field observations indicate that it occurs more frequently than was formerly assumed.

No definitive conclusions are possible concerning the scale and ecological significance of killer whale predation on large whales, whether viewed from a top-down or a bottom-up perspective. We can only speculate on the extent to which predation by killer whales helped regulate populations of large whales or, alternatively, killer whale populations were regulated by the availability of large whales as sources of nutrition. Killer whales definitely attack large whales, including adults, juveniles, and calves. For some species, such as the gray whale, attacks on calves are more likely to prove lethal, but adults are sometimes killed or seriously injured as well. The fact that killer whales have developed strategies for attacking, killing, and (often only partially) consuming large whales implies that these skills confer a selective advantage of some kind. Similarly, the fact that large whales have developed strategies for avoiding detection by, escaping from, and fighting off killer whales suggests that they have undergone selection pressure for these traits. However, Clapham (1993), noting that defense against predators is a major determinant of group size in a wide variety of taxa, suggested that the small, unstable groups typical of humpback whales could reflect (in part) a lack of need for large groups for communal defense or predator detection. In other words, predation events may not occur sufficiently often in the lives of humpback whales to warrant selection for large groups. Alternatively, other selective determinants of group size (e.g., foraging efficiency) may simply outweigh the risks of predation.

In assessing the importance of large whales in the diet of killer whales, it is important to consider two things in particular. First, calves and juveniles are probably substantially overrepresented in the class of animals removed from a population by killer whale predation. Second, consumption of large whale prey is often far from complete, with the tongues and ventral grooves of baleen whales clearly preferred over all other body parts. Incomplete consumption is evident, both from direct observations of killer whales abandoning their prey prematurely and from the occurrence of stranded whales that have been mauled and mutilated but only partially eaten by killer whales.

In his comparative analysis of killer whale–baleen whale relations and the predator-prey relations between African predators and their ungulate prey, Dolphin (1987) implicitly assumed that large savanna ungulates (specifically, wildebeests, *Connochaetes* spp.; Cape buffalos, *Syncerus caffer*, and elephants, *Loxodonta africana*) lived in a state of "nonbelligerent, if uneasy, coexistence" with their large, cooperatively hunting predators (specifically, lions, hyenas, and Cape hunting dogs, *Lycaon pictus*). He posited a similar state of coexistence between baleen whales and killer whales, assuming that successful attacks on such large prey were "rare" (citing Leuthold 1977 to support this assumption for attacks by lions on buffalos and elephants). Carrying Dolphin's reasoning further, we offer the hypothesis that before industrial hunting (particularly of elephants and whales), buffalos, elephants, and large whales were more filler fodder than staple fare for the big social predators. While some individual prides of lions, packs of canids, or pods of killer whales may have specialized in taking large prey and thus may have been, in some sense, obligate large-prey predators (and some prides, packs, and pods may still be), the essential condition of these species was to rely on small to moderate-sized prey as nutritional staples (e.g., pinnipeds, dolphins, porpoises, and minke whales in the case of mammal-eating killer whales) and to take large prey only opportunistically (facultatively) or in exceptional circumstances when their staple prey failed or the large prey were particularly susceptible (e.g., during migratory pulses). The concentrations of gray whale calves in the nursery lagoons and on their coast-hugging northbound migrations provide an example of the latter. Ice-edge aggregations of bowhead whales seeking access to their high-latitude summer feeding grounds might provide another example.

If the foregoing hypothesis were true, it would turn the argument by Springer et al. (2003) on its head. Thus, rather than the whaling-induced decline of large whales having a cascade of effects leading to the declines of pinnipeds, it would be the other way around. That is, declines of these formerly abundant and relatively catchable small to moderate-sized prey (e.g., harbor seals, *Phoca vitulina*, and sea lions, Otariidae) might help explain the recent surge in observations of attacks on large whales, including sperm whales. A number of considerations caution against that interpretation, however. The increase in human numbers, enhanced communication (reporting) opportunities, and growing interest in cetaceans, in combination, could explain the surge in reports of attacks on large whales. Also, California coastal waters, where many of the recently reported attacks occurred, host relatively large populations of the smaller prey species (at least California sea lions, *Zalophus californianus*, harbor seals, and elephant seals, *Mirounga angustirostris*), some of which may be more numerous now than they have been in decades, if not centuries. The

fact that killer whale predation on gray whales is not a recent innovation (as previously discussed) also could be interpreted as being inconsistent with our hypothesis. Finally, energetic/demographic analyses (Doak et al., Chapter 18 of this volume; Williams et al., in press) suggest that the smaller marine mammals, even at their estimated peaks of abundance, would not have been capable of sustaining the 300 or so *transient* killer whales estimated to inhabit the North Pacific today (Forney and Wade, Chapter 12 of this volume).

Acknowledgments

We are grateful to Michael Dyer, Librarian of the New Bedford Whaling Museum, for sharing his knowledge of old whaling manuscripts and providing the data from the nineteenth-century voyages of the *Phoenix* and *Saratoga*. We also thank Amee Mehta and Jooke Robbins for sharing unpublished data on the frequency of scarring on large whales and on patterns of marks derived from examination of museum specimens. Tom Jefferson, Jorge Urbán-Ramírez, Miguel Iñiguez, and Klaus Barthelmess kindly shared various types of information relevant to this study. Jim Estes provided an insightful review. Partial financial support for this work was provided by the Pew Fellows Program in Marine Conservation, the National Marine Fisheries Service, and the U.S. Geological Survey.

Literature Cited

Andrews, R.C. 1914. Monographs of the Pacific Cetacea, 1. The California gray whale (*Rhachianectes glaucus* Cope). *Memoirs of the American Museum of Natural History* New Series 1: 227–287.

———. 1916a. Monographs of the Pacific Cetacea, II. The sei whale (*Balaenoptera borealis* Lesson). *Memoirs of the American Museum of Natural History* New Series, 1: 289–388.

———. 1916b. *Whale hunting with gun and camera*. New York: D. Appleton and Co.

Arnbom, T., V. Papastavrou, L.S. Weilgart, and H. Whitehead. 1987. Sperm whales react to an attack by killer whales. *Journal of Mammalogy* 68: 450–453.

Ash, C. 1964. *Whaler's eye*. London: George Allen and Unwin.

Baird, R.W. 2000. The killer whale: foraging specializations and group hunting, in *Cetacean societies: field studies of dolphins and whales*. J. Mann, R.C. Connor, P.L. Tyack, and H. Whitehead, eds. Chicago: University of Chicago Press, pp. 127–153.

Baldridge, A. 1972. Killer whales attack and eat a gray whale. *Journal of Mammalogy* 53: 898–900.

Ballard, W.B., J.S. Whitman, and D.J. Reed. 1991. Population dynamics of moose in south-central Alaska. *Wildlife Monographs* 114: 1–49.

Bass, R. 1996. Foreword, in *California grizzly*. T.I. Storer and L.P. Tevis, eds. Berkeley: University of California Press, pp xii–xix.

Berger, J. 1994. Science, conservation, and black rhinos. *Journal of Mammalogy* 75: 298–308.

Berger, J. and C. Cunningham. 1994. Horns, hyenas, and black rhinos. *Research and Exploration* 10: 241–244.

Berger, J., J.E. Swenson, and I. Persson. 2001a. Recolonizing carnivores and naïve prey: conservation lessons from Pleistocene extinctions. *Science* 291: 1036–1039.

Berger, J., P.B. Stacey, L. Bellis, and M.P. Johnson. 2001b. A mammalian predator-prey imbalance: grizzly bears and wolf extinction affect avian neotropical migrants. *Ecological Applications* 11: 947–960.

Bergerud, A.T., and J.P. Elliot. 1986. Dynamics of caribou and wolves in northern British Columbia. *Canadian Journal of Zoology* 64: 1515–1529.

Bertram, B.C.R. 1979. Serengeti predators and their social systems, in *Serengeti: dynamics of an ecosystem*. A.R.E. Sinclair and M. Norton-Griffiths, eds. Chicago: University of Chicago Press, pp. 221–248.

Berzin, A.A. 1972. *The sperm whale*. Jerusalem: Israel Program for Scientific Translations.

Best, P.B., P.A.S. Canham, and N. MacLeod. 1984. Patterns of reproduction in sperm whales, *Physeter macrocephalus*. *Report of the International Whaling Commission* (Special Issue) 6: 51–79.

Black, N., T. Ternullo, A. Schulman-Janiger, G. Ellis, M. Dahlheim, and P. Stap. 2003. Behavior and ecology of killer whales in Monterey Bay, California. Poster presentation, 15th Biennial Conference on the Biology of Marine Mammals, Greensboro, North Carolina, December 14–19.

Bloch, D. and C. Lockyer. 1988. Killer whales *(Orcinus orca)* in Faroese waters. *Rit Fiskideildar* 11: 55–64.

Braham, H.W. 1984. Distribution and migration of gray whales in Alaska, in *The gray whale* Eschrichtius robustus. M.L. Jones, S.L. Swartz, and S. Leatherwood, eds. Orlando, FL: Academic Press, pp. 249–266.

Brown, J.T. 1887. Whalemen, vessels and boats, apparatus, and methods of the whale fishery, in *The fisheries and fishery industries of the United States, Section V. History and methods of the fisheries. Part XV. The whale fishery*. G.B. Goode, ed. Washington, DC: Government Printing Office, pp. 218–293.

Byers, J.A. 1997. *American pronghorn: social adaptations and the ghosts of predators past*. Chicago: University of Chicago Press.

Caldwell, D.K. and M.C. Caldwell. 1969. Addition of the leatherback sea turtle to the known prey of the killer whale, *Orcinus orca. Journal of Mammalogy* 50: 636.

Chalmers, G.A., and M.W. Barrett. 1982. Capture myopathy, in *Non-infectious diseases in wildlife*. J.W. Davis and G. Hieff, eds. Ames: Iowa State University Press, pp.

Clapham, P.J. 1993. Social organization of humpback whales on a North Atlantic feeding ground. *Symposia of the Zoological Society of London* 66: 131–145.

———. 2001. Why do baleen whales migrate? A response to Corkeron and Connor. *Marine Mammal Science* 17: 432–436.

Clark T.W. and D. Casey. 1996. *Tales of the grizzly*. Moose, WY: Homestead Publishing.

Corkeron, P.J. and R.C. Connor. 1999. Why do baleen whales migrate? *Marine Mammal Science* 15: 1228–1245.

Cummings, W.C. and P.O. Thompson. 1971. Gray whales, *Eschrichtius robustus*, avoid the underwater sounds of killer whales, *Orcinus orca. Fishery Bulletin* 69: 525–530.

Cummings, W.C., J.F. Fish, and P.O. Thompson. 1971. Bioacoustics of marine mammals off Argentina: R/V *Hero* cruise 71-3. *Antarctic Journal of the United States* 6: 266–268.

———. 1972. Sound production and other behavior of southern right whales, *Eubalaena glacialis. Transactions of the San Diego Society of Natural History* 17(1): 1–14.

Cummings, W.C., P.O. Thompson, and J.F. Fish. 1974. Behavior of southern right whales: R/V *Hero* cruise 72-3. *Antarctic Journal of the United States* 9: 33–38.

Dahlheim, M.E. and J.E. Heyning. 1999. Killer whale *Orcinus orca* (Linnaeus, 1758) , in *Handbook of marine mammals, Vol. 6*. S.H. Ridgway and R. Harrison, eds. San Diego: Academic Press, pp. 281–322.

Dakin, W.J. 1934. *Whalemen adventurers: the story of whaling in Australian waters and other southern seas related thereto, from the days of sails to modern times*. Sydney: Angus and Robertson.

Deecke, V.B., J.K.B. Ford, and P.J.B. Slater. 2005. The vocal behaviour of mammal-eating killer whales: communicating with costly calls. *Animal Behaviour* 69: 395–405.

Dolphin, W.F. 1987. Observations of humpback whale, *Megaptera novaeangliae*–killer whale, *Orcinus orca*, interactions in Alaska: comparison with terrestrial predator-prey relationships. *Canadian Field-Naturalist* 101: 70–75.

Earle, S.A. and A. Giddings. 1979. Humpbacks: the gentle giants. *National Geographic* 155(1): 2–17.

Eberhardt, R.L. and W.E. Evans. 1962. Sound activity of the California gray whale, *Eschrichtius robustus. Journal of the Audio Engineering Society* 10: 324–328.

Eloff, F.C. 1964. On the predatory habits of lions and hyaenas. *Koedoe* 7: 105–112.

Estes, J.A., M.T. Tinker, T.M. Williams, and D.F. Doak. 1998. Killer whale predation on sea otters linking oceanic and nearshore ecosystems. *Science* 282: 473–476.

Fertl, D., A. Acevedo-Gutierrez, and F.L. Darby. 1996. A report of killer whales *(Orcinus orca)* feeding on a carcharhinid shark in Costa Rica. *Marine Mammal Science* 12: 606–611.

Finley, K.J. 1990. Isabella Bay, Baffin Island: an important historical and present-day concentration area for the endangered bowhead whale *(Balaena mysticetus)* of the eastern Canadian Arctic. *Arctic* 43: 137–152.

———. 2001. Natural history and conservation of the Greenland whale, or bowhead, in the northwest Atlantic. *Arctic* 54: 55–76.

Ford, J.K.B. and G.M. Ellis. 1999. *Transients: mammal-hunting killer whales of British Columbia, Washington, and southeastern Alaska*. Vancouver: University of British Columbia Press.

Gasaway, W.C., R.D. Boertje, D.V. Grangaard, D.G. Kellyhouse, R.O. Stephenson, and D.G. Larsen. 1992. The role of predation in limiting moose at low densities in Alaska and Yukon and implications for conservation. *Wildlife Monographs* 120: 1–59.

Gaskin, D.E. 1972. *Whales, dolphins and seals, with special reference to the New Zealand region*. London: Heinemann.

George, J.C., L.M. Philo, K. Hazard, D. Withrow, G.M. Carroll, and R. Suydam. 1994. Frequency of killer whale *(Orcinus orca)* attacks and ship collisions based on scarring on bowhead whales *(Balaena mysticetus)* of the Bering-Chukchi-Beaufort seas stock. *Arctic* 47: 247–255.

Gilmore, R.M. 1961. *The story of the gray whale*. 2nd edition. San Diego: American Cetacean Society.

Goddard, J. 1969. A note on the absence of pinnae in the black rhinoceros. *East African Wildlife Journal* 7: 179–180.

Goley, P.D. and J.M. Straley. 1994. Attack on gray whales *(Eschrichtius probustus)* in Monterey Bay, California, by killer whales *(Orcinus orca)* previously identified in Glacier Bay, Alaska. *Canadian Journal of Zoology* 72: 1528–1530.

Gunther, K.A. and R.A. Renkin. 1990. Grizzly bear predation on elk calves and other fauna of Yellowstone National Park. *International Conference on Bear Research and Management* 8: 329–334.

Haines, A.L. (editor). 1965. *Journal of a trapper 1834–1843*. Lincoln: University of Nebraska Press.

Hancock, D. 1965. Killer whales kill and eat a minke whale. *Journal of Mammalogy* 46: 341–342.

Heide-Jørgensen, M.P. 1988. Occurrence and hunting of killer whales in Greenland. *Rit Fiskideildar* 11: 115–135.

Heptner, V.G., K.K. Chapskii, V.A. Arsen'ev, and V.E. Sokolov. 1996. *Mammals of the Soviet Union, Vol. 2, Part 3. Pinnipeds and toothed whales Pinnipedia and Odontoceti.* English-language edition. J.G. Mead, ed. Washington, DC: Smithsonian Institution Libraries and National Science Foundation (originally published in Russian by Vysshaya Shkola Publishers, Moscow, 1976).

Hildebrand, G.V., C.C. Schwartz, C.T. Robbins, M.E. Jacoby, T.A. Hanley, S.M. Arthur, and C. Servheen. 1999. The importance of meat, particularly salmon, to body size, population productivity, and conservation of North American brown bears. *Canadian Journal of Zoology* 77: 132–138.

Hitchins, P.A. 1986. Earlessness in the black rhinoceros—a warning. *Pachyderm* 7: 8–10.

IWC. 2003. Report of the Scientific Committee. *Journal of Cetacean Research and Management* 5 (Suppl.): 1–488.

Jefferson, T.A., P.J. Stacey, and R.W. Baird. 1991. A review of killer whale interactions with other marine mammals: predation to co-existence. *Mammal Review* 21: 151–180.

Jones, M.L. and S.L. Swartz. 1984. Demography and phenology of gray whales and evaluation of whale-watching activities in Laguna San Ignacio, Baja California Sur, Mexico, in *The gray whale* Eschrichtius robustus. M.L. Jones, S.L. Swartz, and S. Leatherwood, eds. Orlando, FL: Academic Press, pp. 309–374.

———. 2002. Gray whale *Eschrichtius robustus*, in *Encyclopedia of marine mammals*. W.F. Perrin, B. Würsig, and J.G.M. Thewissen, eds. San Diego: Academic Press, pp. 524–436.

Kahn, B. 2003. *Solor-Alor visual and acoustic cetacean surveys: interim report April–May 2003 survey period.* Sanur (Bali), Indonesia: The Nature Conservancy, Coral Triangle Center. Available at http://www.coraltrianglecenter.org/downloads/Alor3_Interim_Survey_Report_Final_150pdi.pdf. Accessed June 1, 2006.

Kruuk, H. 1972. *The spotted hyena: a study of predation and social behavior.* Chicago: University of Chicago Press.

Landa, A. 1997. *Wolverines in Scandinavia: ecology, sheep depredation, and conservation.* Trondheim: Norwegian University of Science and Technology.

Leuthold, W. 1977. *African ungulates: a comparative review of their ethology and behavioral ecology.* New York: Springer Verlag.

Ljungblad, D.K. and S.E. Moore. 1983. Killer whales *(Orcinus orca)* chasing gray whales *(Eschrichtius robustus)* in the northern Bering Sea. *Arctic* 36: 361–364.

Lowry, L.F., R.R. Nelson, and K.J. Frost. 1987. Observations of killer whales, *Orcinus orca,* in Western Alaska: sightings, strandings, and predation on other marine mammals. *Canadian Field-Naturalist* 101: 6–12.

Magoun, A.J. 1985. Population characteristics, ecology, and management of wolverines in north-western Alaska. Ph.D. Thesis, University of Alaska, Fairbanks.

Mass, A.M. and A. Ya. Supin. 2002. Vision, in *Encyclopedia of marine mammals.* W.F. Perrin, B. Würsig, and J.G.M. Thewissen, eds. San Diego: Academic Press, pp. 1280–1293.

Mattson, D.J. 1997. Use of ungulates by Yellowstone grizzly bears, *Ursus arctos. Biological Conservation* 81: 161–177.

Mattson, D.J., B.M. Blanchard, and R.R. Knight. 1991. Food habits of Yellowstone grizzly bears, 1977–1987. *Canadian Journal of Zoology* 69: 1619–1629.

McLaughlin, W.R.D. 1962. *Call to the south: a story of British whaling in Antarctica.* London: George G. Harrap and Co.

Mech, L.D. 1970. *The wolf: the ecology and behavior of an endangered species.* Garden City, NY: Natural History Press.

Mech, L.D., L.G. Adams, T.J. Meier, J.W. Burch, and B.W. Dale. 1998. *The wolves of Denali.* Minneapolis: University of Minnesota Press.

Mehta, A.V. 2004. How important are baleen whales as prey for killer whales *(Orcinus orca)* in high latitudes? M.S. Thesis, Boston University.

Melnikov, V.V., and I.A. Zagrebin. 2005. Killer whale predation in coastal waters of the Chukotka Peninsula. *Marine Mammal Science* 21: 550–556.

Mills, M.G.L. 1990. *Kalahari hyaenas.* London: Unwin Hyman.

Mitchell, E. and A.N. Baker. 1980. Age of reputedly old killer whale, *Orcinus orca,* "Old Tom" from Eden, Twofold Bay, Australia. *Report of the International Whaling Commission* (Special Issue) 3: 143–154.

Mitchell, E. and R.R. Reeves. 1982. Factors affecting abundance of bowhead whales *Balaena mysticetus* in the eastern Arctic of North America, 1915–1980. *Biological Conservation* 22: 59–78.

———. 1988. Records of killer whales in the western North Atlantic, with emphasis on eastern Canadian waters. *Rit Fiskideildar* 11: 161–193.

Moore, S.E., D.K. Ljungblad, and D.R. Van Schoik. 1986. Annual patterns of gray whale *(Eschrichtius robustus)* distribution, abundance, and behavior in the northern Bering and eastern Chukchi seas, July 1980–83. *Report of the International Whaling Commission* (Special Issue) 8: 231–242.

Moshenko, R.W., S.E. Cosens, and T.A. Thomas. 2003. *Conservation strategy for bowhead whales* (Balaena mysticetus) *in the eastern Canadian Arctic.* National Recovery Plan No. 24. Ottawa: Recovery of Nationally Endangered Wildlife (RENEW).

Naessig, P.J. and J.M. Lanyon. 2004. Levels and probable origin of predatory scarring on humpback whales *(Megaptera novaeangliae)* in east Australian waters. *Wildlife Research* 31: 163–170.

National Research Council. 1997. *Wolves, bears, and their prey in Alaska.* Washington, DC: National Academies Press.

Norris, K.S. and J.H. Prescott. 1961. Observations of Pacific cetaceans of Californian and Mexican waters. *University of California Publications in Zoology* 63: 291–402.

NWMB. 2000. *Final report of the Inuit bowhead knowledge study.* Iqaluit: Nunavut Wildlife Management Board.

Palacios, D.M. and B.R. Mate. 1996. Attack by false killer whales *(Pseudorca crassidens)* on sperm whales *(Physeter macrocephalus)* in the Galápagos Islands. *Marine Mammal Science* 12: 582–587.

Persson, J. 2003. *Population ecology of Scandinavian wolverines.* Umea: Swedish Agricultural University.

Phillips, G.E. and G.C. White. 2003. Pronghorn population response to coyote control: modeling and management. *Wildlife Society Bulletin* 31: 1162–1175.

Phoenix. 1847–51MS. *Logbook of the ship Phoenix of New Bedford, Joseph McCleave, master. August 3, 1847–May 28, 1851.* New Bedford, MA: Kendall Institute, New Bedford Whaling Museum, log number KWM 399.

Pitman, R.L., L.T. Ballance, S.L. Mesnick, and S.J. Chivers. 2001. Killer whale predation on sperm whales: observations and implications. *Marine Mammal Science* 17: 494–507.

Pitman, R.L. and P. Ensor. 2003. Three forms of killer whales (*Orcinus orca*) in Antarctic waters. *Journal of Cetacean Research and Management* 5: 131–139.

Pond, J.E. 1939. Truth stranger than fiction. *U.S. Naval Institute Proceedings* 65: 1309–1332.

Reeves, R.R. and E. Mitchell. 1988a. Killer whale sightings and takes by American pelagic whalers in the North Atlantic. *Rit Fiskideildar* 11: 7–23.

———. 1988b. Distribution and seasonality of killer whales in the eastern Canadian Arctic. *Rit Fiskideildar* 11: 136–160.

Rice, D.W. and A.A. Wolman. 1971. *The life history and ecology of the gray whale* (Eschrichtius robustus). American Society of Mammalogists, Special Publication No. 3. Lawrence, KS: American Society of Mammalogists.

Saratoga. 1857–58MS. *Logbook of the ship Saratoga of New Bedford, Frederick Slocum, master. April 23, 1857–December 12, 1858 (partial or incomplete voyage).* New Bedford, MA: Kendall Institute, New Bedford Whaling Museum, log number KWM 180.

Scammon, C.M. 1874. *The marine mammals of the north-western coast of North America, described and illustrated together with an account of the American whale-fishery.* San Francisco: John H. Carmany and Co.

Schaller, G.B. 1972. *The Serengeti lion: a study of predator-prey relations.* Chicago: University of Chicago Press.

Schullery, P. 1988. *The bear hunter's century: profiles from the golden age of bear hunting.* New York: Dodd, Mead, and Company.

Seton, E.T. 1953. *Lives of game animals.* Vol. 2. Boston: Charles T. Branford.

Shevchenko, V.I. 1975. Nature of correlations between killer whales and other cetaceans. *Morsk Mlekopitayushchie Chast'* 2: 173–174 (in Russian).

Silber, G.K., M.W. Newcomer, and M. Hector Perez-Cortes. 1990. Killer whales (*Orcinus orca*) attack and kill a Bryde's whale (*Balaenoptera edeni*). *Canadian Journal of Zoology* 68: 1603–1606.

Smith, B.L. and S.H. Anderson. 1996. Patterns of neonatal mortality of elk in northwestern Wyoming. *Canadian Journal of Zoology* 74: 1229–1237.

Soulé, M.E., J.A. Estes, J. Berger, and C. Martinez del Rio. 2003. Ecological effectiveness: conservation goals for interactive species. *Conservation Biology* 17: 1238–1250.

Southwood, T.R.E. 1975. The dynamics of insect population, in *Insects, science, and society.* D. Pimentel, ed. New York: Academic Press, pp. 151–199.

Springer, A.M., J.A. Estes, G.B. van Vliet, T.M. Williams, D.F. Doak, E.M. Danner, K.A. Forney, and B. Pfister. 2003. Sequential megafaunal collapse in the North Pacific Ocean: an ongoing legacy of industrial whaling? *Proceedings of the National Academy of Sciences* 100: 12223–12228.

Storer, T.I. and L.P. Tevis. 1955. *California grizzly.* Berkeley: University of California Press.

Swartz, S.L. 1986. Gray whale migratory, social and breeding behavior. *Report of the International Whaling Commission* (Special Issue) 8: 207–229.

Tarpy, C. 1979. Killer whale attack. *National Geographic* 155: 542–545.

Ternullo, R. and N. Black. 2003. Predation behavior of *transient* killer whales in Monterey Bay, California. Poster presentation, 15th Biennial Conference on the Biology of Marine Mammals, Greensboro, North Carolina, December 14–19.

Testa, J.W., E.F. Becker, and G.R. Lee. 2000. Temporal patterns in the survival of twin and single moose calves (*Alces alces*) in southcentral Alaska. *Journal of Mammalogy* 81: 162–168.

Urbán Ramirez, J., L. Rojas-Bracho, H. Pérez-Cortés, A. Gómez-Gallardo, S.L. Swartz, S. Ludwig, and R.L. Brownell, Jr. 2003. A review of gray whales (*Eschrichtius robustus*) on their wintering grounds in Mexican waters. *Journal of Cetacean Research and Management* 5: 281–295.

Visser, I.N. 1999a. Benthic foraging on stingrays by killer whales (*Orcinus orca*) in New Zealand waters. *Marine Mammal Science* 15: 220–227.

———. 1999b. A summary of interactions between orca (*Orcinus orca*) and other cetaceans in New Zealand. *New Zealand Natural Sciences* 24: 101–112.

Visser, I.N. and F.J. Bonoccorso. 2003. New observations and a review of killer whale (*Orcinus orca*) sightings in Papua New Guinea waters. *Aquatic Mammals* 29: 150–172.

Walker, L.W. 1949. Nursery of the gray whales. *Natural History* 58(6): 248–256.

Walker, T.J. 1975. *Whale primer, with special attention to the California gray whale.* San Diego: Cabrillo Historical Association.

Weller, D.W. 2002. Predation on marine mammals, in *Encyclopedia of marine mammals.* W.F. Perrin, B. Würsig, and J.G.M. Thewissen, eds. San Diego: Academic Press, pp. 985–994.

Weller, D.W., B. Würsig, H. Whitehead, J.C. Norris, S.K. Lynn, R.W. Davis, N. Clauss, and P. Brown. 1996. Observations of an interaction between sperm whales and short-finned pilot whales in the Gulf of Mexico. *Marine Mammal Science* 12: 588–594.

Wellings, H.P. 1964. *Shore whaling at Twofold Bay assisted by the renowned killer whales: an important industry now defunct.* Privately published, printed at The Magnet-Voice, Eden, Australia.

Wenzel, F. and R. Sears. 1988. A note on killer whales in the Gulf of St. Lawrence, including an account of an attack on a minke whale. *Rit Fiskideildar* 11: 202–204.

Whitehead, H. and C. Glass. 1985. The significance of the Southeast Shoal of the Grand Bank to humpback whales and other cetacean species. *Canadian Journal of Zoology* 63: 2617–2625.

Whitehead, H. and R.R. Reeves. 2005. Killer whales and whaling: the scavenging hypothesis. *Biology Letters* 1(4): 415–418.

Williams, T.M., J.A. Estes, D.F. Doak, and A.M. Springer. 2004. Killer appetites: assessing the role of predators in ecological communities. *Ecology* 85: 3373–3384.

Yukhov, V.L., Ye. K. Vinogradova, and L.P. Medvedev. 1975. Diet of killer whales (*Orcinus orca* L.) in Antarctic and adjacent waters. *Morsk Mlekopitayushchie Chast'* 2: 183–185 (in Russian).

PROCESS AND THEORY

FIFTEEN

Physiological and Ecological Consequences of Extreme Body Size in Whales

TERRIE M. WILLIAMS

The single most defining physical characteristic of whales in general and mysticete whales in particular is their great size. Among the odontocetes and mysticetes there is a 2000-fold range in body mass from the 50-kg vaquita *(Phocoena sinus)* to the 100,000-kg blue whale *(Balaenoptera musculus)*. The smallest odontocete is 22,000 times larger than the smallest terrestrial mammal, the Etruscan shrew *(Suncus etruscus)*, and of the 13 recognized species of mysticete (baleen) whales the smallest species, including the dwarf minke whale *(Balaenoptera acutorostrata subspecies)* and pygmy right whale *(Caperea marginata)*, are nearly 10 times larger than the biggest flesh-eating terrestrial mammals (Reeves et al. 2002). The largest of the cetaceans, the blue whale, is considered the biggest animal to have ever lived, theoretically outweighing even the largest of the dinosaurs (Alexander 1997).

This distinction in size for whales raises several interesting questions regarding the how and why of maintaining a large body mass, as well as limiting factors that dictate the size of mammalian form. In an evolutionary context, the constraints on size may have been driven internally (physiological systems must be capable of supporting a large body) or externally (adequate quantity and quality of food must be available in the environment to allow growth to take place). Differentiating between these evolutionary factors is difficult

at best, because of the absence of data concerning physiological function and foraging behavior in extinct animals. However, we can gain insight by examining the physiological and ecological consequences of large body size in modern-day species. Do all organ systems simply increase in size and function in concert with overall body mass? Assuming that larger bodies require more prey to meet metabolic demands, how do ecosystems support and respond to the presence or absence of large consumers?

The combination of great abundance, large body size, and endothermic metabolic demands make odontocete and mysticete whales some of the most voracious consumers in the oceans. As such, they have the potential to place extraordinary pressures on marine resources. Furthermore, the results of calculations of the prey biomass needed to support a population of whales, and the detritus generated in the form of digestive waste and carcasses, appear so overwhelming (see Croll et al., Chapter 16 in this volume; Smith, Chapter 22 in this volume) that it often seems impossible that large numbers of great whales could have ever existed.

In this chapter I examine some of the key physiological and physical limitations that dictate maximum body size in mammals. Rather than focus on the selective pressures that may have driven the evolution of extreme body size (see

Blanckenhorn 2000; Lindberg and Pyenson, Chapter 7 in this volume), I address the capacity of physiological processes to support large mammalian forms. With physiological requirements and capacities established, the ecological consequences of large predators within an ecosystem can then be assessed. Such an example is presented for the killer whale (*Orcinus orca*), which is one of the largest hunting odontocetes besides the sperm whale (*Physeter macrocephalus*) and has been implicated in the decline of marine mammal populations in the Aleutian archipelago (Estes et al. 1998; Springer et al. 2003) and the southern oceans (Branch and Williams, Chapter 20 in this volume). A comparative analysis for large grazing mysticetes can be found in Croll et al. (Chapter 16 in this volume).

The Extreme Size of Whales: Hunters versus Grazers

In relative terms, large body size provides numerous physiological and ecological advantages to an animal. Relative energetic demands of large mammals are lower and supported by comparatively high gastrointestinal capacity in both carnivores (Williams et al. 2001) and herbivores (Clauss et al. 2003). Competition and predation by other species may be reduced, and the capacity for large home ranges is increased. For aquatic mammals the latter increase provides access to enormous foraging areas both geographically and by depth. Thus, individual species of great whales may forage across entire ocean basins and throughout the water column, depending on the location of their prey (see Clapham et al. 1999 for a review; Reeves et al. 2002). Compared with smaller swimming mammals, the energetic cost of these endeavors per unit of biomass will be less for great whales (Williams 1999). This is especially apparent when relative distances traveled are compared for large and small swimmers; consider that a 100-m dive represents a travel distance of over 30 body lengths for a bottlenose dolphin (*Tursiops truncatus*) but only three body lengths for an adult blue whale. But what determines how big an animal can be?

As mentioned above, body mass in great whales far exceeds that of terrestrial mammals, a difference that undoubtedly reflects the physical forces encountered when trying to move on land versus through water (Dejours 1987). In accounting for the different physical environments, biologists and mechanical engineers commonly argue that marine mammals, including odontocetes and mysticetes, were able to attain enormous proportions because their bodies were supported by buoyant forces when immersed (Alexander 1997; Kardong 2002). On land, a similarly proportioned skeleton would collapse under its own weight, and body size is quickly limited by the requirement for supporting limbs so massive that locomotor movements are hindered.

Mechanical limitations notwithstanding, another important factor dictating body size is simply the ability for large animals to find, capture and assimilate enough food to maintain metabolic functions (Schmidt-Nielsen 1984). The size of a carnivorous hunter will influence its ability to cover large distances to locate prey as well as to handle and consume large quantities of prey at any one time. The size of a grazing herbivore will likewise determine foraging range and the body's capacity for processing low-quality food. In examining body size and food requirements of terrestrial herbivores and carnivores, Burness et al. (2001) suggested that maximum body mass in top species is determined by the land area available for acquiring food. For a given area of land the body size of grazing herbivorous mammals was larger than that of hunting carnivorous mammals. These authors also note that the largest cetaceans, including blue, killer, and sperm whales, have worldwide oceanic distributions that would be conducive to attaining large body size if these mammals followed the same patterns as found for terrestrial animals. For mammals in general, the quality of the diet (Case 1979) and accompanying digestive physiology (Clauss et al. 2003) also influence optimal body size. As a result, the maximum size of carnivores is often less than that of herbivores (Burness et al. 2001). Optimum body size in these top species decreases with increasing metabolic demands (Case 1979), indicating the influence of energetics in dictating both optimum and maximum body sizes.

Despite an advantage in relative energetic demands, large animals must contend with the simultaneous metabolic needs of the entire body—in simple terms, big animals must eat more than small animals. A survey of the largest species within different taxonomic groups also suggests that the method of feeding and digestion plays a role in determining or limiting body mass. In general, we find that, within mammalian groups, hunters gorging on relatively large prey are smaller than grazers that consume smaller prey items over multiple feeding bouts. Consequently, some of the largest mammals feed on the smallest prey or plants.

Examples can be found for several mammalian groups. Among terrestrial mammals, the largest hunter is the male Siberian tiger (*Panthera tigris altaica*, body mass approximately 300 kg) or polar bear (*Ursus maritimus*, 475 kg average, 800 kg maximum), depending on whether the latter is classified as a marine or land mammal (Stirling 2002). A common food item for the polar bear is the ringed seal (*Pusa hispida*), which can weigh up to 110 kg, or one-fourth of the average mass of the bear. The largest extant terrestrial mammal is the African elephant (*Loxodonta africana*, 3,900 kg average, 6,000 kg maximum), a grazing herbivore that is over seven times larger than the biggest terrestrial hunter (Estes 1991). Similar patterns are seen for the whales. Among odontocetes, the killer whale (4,900 kg average, 6,000 kg maximum) is the largest predator on other mammals and is capable of feeding on great whales equal to their size and larger (Dahlheim and Heyning 1999). The largest toothed whale that forages like a grazer is the sperm whale (> 45,000 kg), which exceeds the body mass of killer whales by a factor of 7 to 9 (Whitehead 2002; 2003) and feeds primarily on large and medium-sized squid that maximally are less than 1/400 the size of the

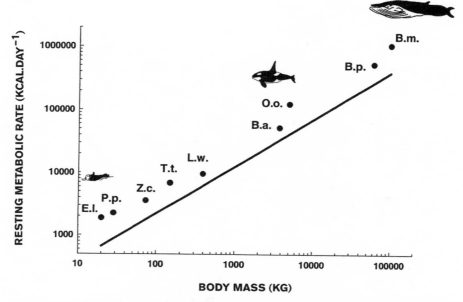

FIGURE 15.1. Metabolic rate in relation to body mass for marine mammals resting on the water surface. Each point represents the mean values for sea otters (E.l.: Williams 1989), harbor porpoise (P.p.: Kanwisher and Sundnes 1965), California sea lions (Z.c.: Liao 1990), bottlenose dolphins (T.t.: Williams et al. 2001), Weddell seals (L.w.: Williams et al. 2001), killer whales (O.o.: Kreite 1995), minke whale (B.a.: Lockyer 1981), fin whale (B.p.: Lockyer 1981), and blue whale (B.m.: Lockyer 1981). The solid line is the predicted regression for basal metabolic rate for domestic terrestrial mammals according to Kleiber (1975).

whale. Mysticete whales are the largest of all animals, attaining a body mass of nearly 100,000 kg in the blue whale; all are grazing animals feeding on small schooling fish and smaller zooplankton. By comparison, the largest hunting dinosaur was *Tyrannosaurus rex* at 6,000–8,000 kg (Alexander 1997). The body mass of this extinct hunter was approximately one tenth that of *Brachiosaurus*, a grazing herbivorous dinosaur estimated at 47,000–78,000 kg.

Physiological and Physical Constraints on Body Mass

To survive, animals must be capable of balancing caloric demand with caloric acquisition (Stephens and Krebs 1986). Although simple in concept, meeting the demand for food to support a massive body represents a physiological and physical challenge for both large hunters and large grazers. In physiological terms, two major pathways are involved in this process. The first defines the caloric or metabolic demands of the animal and involves the movement of oxygen along the aerobic pathway from lungs to the cardiovascular system and finally to the working tissues (see Weibel 1984 for a complete description of this pathway). The second pathway is associated with caloric acquisition, from the movement of food through the mouth and esophagus to the assimilation of nutrients within the gastrointestinal tract and eventual

processing by the liver. Complex coordination among organ systems must occur for either of these pathways to work efficiently.

The basic caloric demands of an animal can be reasonably estimated from its metabolic rate (Kleiber 1975). For both terrestrial and marine mammals, maintenance metabolism—that is, the number of calories needed to maintain basal biological functions—increases predictably with body mass (Figure 15.1). When measured under equivalent experimental conditions, the basal metabolic demands for marine mammals are 1.5–2.5 times higher than those predicted for terrestrial carnivores of similar body mass (Williams et al. 2001). The magnitude and reasons for these elevated costs have long been debated (Lavigne et al. 1986; Andrews 2002) and have been attributed to the extraordinary metabolic cost of endothermy in water (Irving 1973), support of the gastrointestinal tract (Williams et al. 2001), and a carnivorous lifestyle (McNab 1986) among others. Regardless, from sea otters *(Enhydra lutris)* to blue whales, marine mammals often display higher metabolic rates and therefore caloric demands than predicted for terrestrial mammals.

The elevated caloric demands of marine mammals might be expected to influence the size of the supporting organ systems. However, this is not the case for one of the two physiological pathways identified above. Despite differences in metabolic demands for marine and terrestrial mammals,

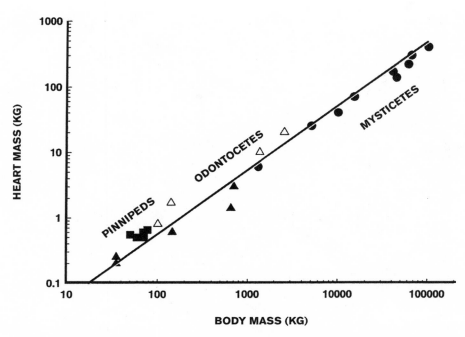

FIGURE 15.2. Heart mass in relation to total body mass for pinnipeds (squares), odontocetes (triangles), and mysticetes (circles). Each point represents the average value for males and females of each species. Open triangles denote the reportedly more athletic odontocetes, as discussed in the text. The solid line is the predicted relationship for terrestrial mammals from Brody (1945) presented in the text. Data are from Lockyer (1981, odontocetes and mysticetes), McAlpine (1985, mysticetes), Dahlheim and Heyning (1999, killer whales) and C. Calkins (unpublished data, pinnipeds).

the size of two primary organs of the aerobic metabolic pathway (the lungs and the heart) follow remarkably similar trends for both groups. From bats to whales, lung volume increases predictably with body mass (Tenney and Remmers 1963; Kooyman 1973) according to the relationship

$$\text{Lung volume} = 56.7\text{Mass}^{1.02}, \quad (15.1)$$

where lung volume is in ml and body mass is in kg. Lung volumes for marine mammals demonstrate considerable variation around this regression; such variation is often attributed to specific buoyancy requirements of the animal. Sperm whales, bottlenose whales (*Hyperoodon* and *Berardius*), and fin whales (*Balaenoptera physalus*) have lung volumes that are approximately half those expected for terrestrial mammals (Slijper 1976). In comparison, sea otters, dugongs (*Dugong dugon*), and several species of otariids (eared seals) have larger than expected lung volumes, which may enable the lungs to serve as a float (Williams and Worthy 2002).

Heart mass for marine mammals also follows the allometric pattern for terrestrial mammals. In terrestrial species, heart mass scales predictably with body mass according to the regression from Brody (1945),

$$\text{Heart mass} = 0.0059\text{Mass}^{0.984}, \quad (15.2)$$

where heart mass and body mass are in kg. Cetacean heart masses range from 0.2 kg in a 35-kg immature harbor porpoise *(Phocoena phocoena)* to 400 kg in a 100,000-kg blue whale and are generally within 20%–30% of the size predicted for terrestrial mammals (Figure 15.2). Exceptions do occur, however. In accordance with Brody's distinction between "athletic" and "nonathletic" species, the reportedly faster and more active odontocetes have relatively larger hearts compared to more sluggish species. Thus, Pacific white-sided dolphins *(Lagenorhynchus obliquidens)*, Dall's porpoise *(Phocoenoides dalli)*, and killer whales have heart sizes that are 1.3 to 2.2 times those predicted for terrestrial mammals (Figure 15.2). Interestingly, the size of the heart of purportedly one of the fastest mysticetes, the fin whale (known as the "greyhound of the oceans"; Reeves et al. 2002), is as predicted for terrestrial mammals. In part, this may reflect the difficulty of accurately measuring organ and body masses in the great whales as well as determining their swimming speeds in the wild. Overall, given these patterns for the heart and lungs, if the aerobic metabolic pathway defines the upper extreme of mammalian body size, then the limitations likely reside in elements other than the size of the primary organs.

In contrast to the aerobic pathway, the digestive pathway closely reflects the differences in metabolic demands between terrestrial and marine carnivores. This is particularly evident

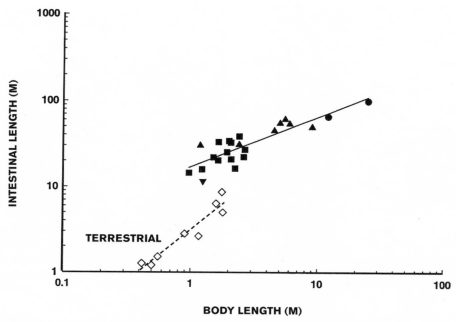

FIGURE 15.3. Intestinal length in relation to body length for marine and terrestrial mammals. In the upper regression (described in the text), data for sea otters (downward-pointing triangle), pinnipeds (squares), mysticete whales (circles), and odontocetes (upward-pointing triangles) including killer whales are compared. Terrestrial carnivores shown in the lower regression by open diamonds include mink, humans, cats, dogs, leopards, and African lions, described by intestinal length = 3.1Body length$^{1.22}$. Data are from Williams et al. (2001).

for the gastrointestinal tract. The extraordinary length of the small intestines of sea otters, pinnipeds (Eastman and Coalson 1974; King 1983), and cetaceans (Slijper 1976; Stevens and Hume 1995) has been noted by numerous investigators. For example, the small intestine of a 25-kg sea otter is 11.4 m long, over 30% longer than the small intestine of a 250-kg adult male African lion *(Panthera leo)* (Figure 15.3). Bottlenose dolphins, of similar body length to African lions, maintain a small intestine that is 3.6 times as long as that of the terrestrial carnivore (Williams et al. 2001). In the largest mammal, the blue whale, ingested material must pass through a remarkable 100 m of small intestine. For 24 species of marine mammal, small-intestine length is described by

$$\text{Intestinal length} = 17.02\text{Body length}^{0.762} \quad (15.3)$$

where intestinal length and body length are in m. Regardless of body size, the length of the small intestine for all marine mammals measured to date exceeds that of even the largest terrestrial carnivores and follows a different allometry (Figure 15.3).

The movement of food is only the first step in the digestive pathway; eventually, ingested nutrients must be processed and assimilated by organs such as the liver. Here again we find differences between marine and terrestrial mammals. Liver mass of odontocetes, mysticetes, and

pinnipeds increases with body mass according to the relationship

$$\text{Liver mass} = 0.063\text{Mass}^{0.851} \quad (15.4)$$

where liver mass and body mass are in kg (Figure 15.4). In general, liver mass follows the trend for metabolic rate; that is, liver mass is approximately 1.5 to 2.5 times as large in marine mammals as that predicted for terrestrial mammals of similar mass.

The size of individual organs in flesh-eating marine mammals is dictated primarily by the size of the body that must be supported rather than feeding style per se. The masses of the heart (Figure 15.2), lungs, intestines (Figure 15.3), and liver (Figure 15.4) follow the same scaling relationships for hunters (pinnipeds and odontocetes) as for grazers (mysticetes). The latter taxonomic group merely represents the upper extreme in body size. A distinction based on feeding style does occur when examining the relationship between the metabolic rates that set the caloric demand and the gastrointestinal tract that must deliver those calories to the tissues. For a given length of intestines, higher metabolic rates and body sizes are supported if a marine mammal feeds by grazing than by hunting (Figure 15.3; Williams et al. 2001). As observed for herbivorous vertebrates (Stevens and Hume 1995; Clauss et al. 2003), large body size and a

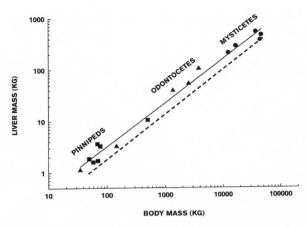

FIGURE 15.4. Liver mass in relation to total body mass for pinnipeds (squares), odontocetes (triangles), and mysticetes (circles). Each point represents the average value for males and females of each species. The solid line is the least-squares linear regression for marine mammals as described in the text. The predicted relationship for terrestrial mammals (dashed line) is from Brody (1945). Data are from Bryden (1972, odontocetes and elephant seals), McAlpine (1985, mysticetes), Dahlheim and Heyning (1999, killer whales) and C. Calkins (unpublished data, pinnipeds).

concomitant longer gastrointestinal tract enable large marine mammals to feed on prey of lower caloric value. In the case of mysticete whales, this characteristic also allows the animal to feed on prey that is spatially and temporally dispersed.

A disadvantage associated with a comparatively large gastrointestinal tract is the high energetic cost of maintaining the tissues. Because the gastrointestinal tract accounts for a disproportionately high fraction of energy utilization by animals (Stevens and Hume 1995), large mammals must balance the metabolic cost of supporting an exceptionally large intestinal tract and liver with the increased capability to assimilate nutrients to meet the daily energetic requirements of the entire animal (Williams et al. 2001). When costs exceed benefits, further increases in body mass may not be possible.

Fueling Large Bodies in the Wild: Consequences of Large Predators in Ecosystems

In addition to physiological limitations for processing food, available foraging area—or, more precisely, the availability of prey within a foraging area—represents a constraint on maximum attainable body size in vertebrates. In terrestrial hunters and grazers, the link between foraging area and body size follows from the large animals' need to consume more food, which in turn requires larger home ranges and, ultimately, larger available land masses than needed by smaller species (Burness et al. 2001). Consequently, body size scales positively with land area in terrestrial carnivores, but scales negatively with the number of competing carnivores per unit of available prey biomass (Carbone and Gittleman 2002).

Whether such relationships exist for marine carnivores, however, is not known—in part because of the lack of information concerning foraging biology, movement patterns, and ranges for many marine species. Because marine mammals often hunt at depth, direct assessment of foraging behavior and predator-prey interactions are impossible for most species (Davis et al. 1999). Only recently has satellite tag technology permitted large numbers of marine mammals to be tracked for prolonged periods while at sea. Some species move extraordinary distances (LeBoeuf 1994), but it is difficult to discern when and where the animals feed, or what constitutes a home range or a foraging range. Theoretically, the foraging range of large whales can span entire oceans. One such example has been reported for Alaskan killer whales that traveled over 2,660 km to California, where they fed on gray whales, *Eschrichtius robustus* (Goley and Straley 1994). High daily metabolic demands of killer whales (Figure 15.1) would suggest that the animals fed while traveling between Alaska to California. However, the activities of the killer whales were not observed during this period.

In the absence of direct observation, bioenergetic analyses combined with demographic models have been used to assess the caloric needs and potential impacts of whales on prey populations (Barrett-Lennard et al. 1995; Williams et al. 2004; Croll et al., Chapter 16 in this volume). The approach requires four types of basic energetic information about the predator and its prey: (1) caloric demands of the whale, (2) caloric value of the prey, (3) prey preferences of the whales, and (4) an estimate of the digestive efficiency of the whales. These elements provide a basic energetic profile for the individual whale, which can be subsequently refined to include factors affecting caloric balance including sex, seasonal changes in the availability or caloric value of prey, and increased metabolic demands associated with pregnancy, lactation, growth, locomotion and migratory movements, and environmental temperature (Kleiber 1975). Additional information about population size and social composition for the whales allows predation pressure on specific prey resources to be estimated.

Croll et al. (Chapter 16 in this volume) provide such an analysis for the impact of sperm whales and large, grazing baleen whales on primary productivity in the North Pacific. From their calculations the current population of large whales in this area consume an average of 55,200 metric tons of prey biomass per day; this compares with 183,100 metric tons per day for the prewhaling population. Based on these results, the authors suggested that the disappearance of large whales through industrialized whaling likely altered trophic interactions within North Pacific waters. Likewise, several studies have hypothesized that industrialized whaling may have instigated major ecological changes associated with dietary shifts in the largest hunting (mammal-eating) odontocete, the killer whale. Because killer whale predation has been proposed as being a primary factor in causing the population declines of sea otters (Estes et al. 1998), Steller sea

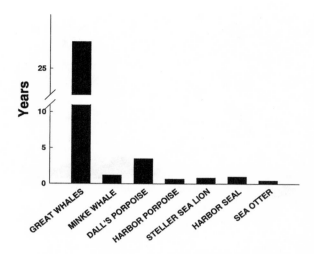

FIGURE 15.5. Time to consume the standing biomass of different prey species by mammal-eating killer whales in the Aleutian Islands. Seven common species or species complexes (i.e. great whales) of marine mammal taken by killer whales in this area are compared. Bar height represents the number of years required for 170 adult killer whales to eliminate each species at current population levels (see text). Note that seasonal availability of each species (Pfister and DeMaster, Chapter 10 in this volume) is taken into account in these calculations.

lions (*Eumetopias jubatus*; Williams et al. 2004), and other marine mammals (Springer et al. 2003) in the vicinity of the Aleutian Islands, the species provides an interesting case study for evaluating the effects of large predators on an ecosystem using the bioenergetic-demographic approach.

Daily caloric demands of killer whales have been estimated from pinniped ingestion rates (Baird 1994), activity budgets of wild killer whales (Kriete 1995), fish consumption rates of captive whales (Barrett-Lennard et al. 1995), and extrapolations from basal metabolic rate (Figure 15.1) and the field metabolic rate of smaller marine mammals (Williams et al. 2004). In the last study, the daily caloric needs of adult, nonpregnant, nonlactating free-ranging killer whales were estimated as 164,000 kcal day^{-1} (7,900 W) for a 2,800-kg female and 243,500 kcal day^{-1} (11,800 W) for a 4,733-kg male. Accounting for the assimilation efficiency associated with a carnivorous diet raised the required daily caloric ingestion rate to 193,000 kcal day^{-1} (9,400 W) in prey items for an adult female and 287,000 kcal day^{-1} (13,900 W) for an adult male killer whale.

The number of prey that would be needed to satisfy these caloric demands depends on the species eaten. Size, body composition, and therefore, total caloric value of the primary prey of killer whales varies widely. Caloric content determined for seven potential prey species in the Aleutian Island area (Figure 15.5) ranged from 41,600 to 61,500 kcal for adult sea otters to nearly 543,000 kcal for a Dall's porpoise and higher for larger whales. For the largest prey (i.e., adult Steller sea lions and mysticete whales), available food exceeds

stomach capacity of individual killer whales, with the result that large numbers of orca may feed on a single carcass (Jefferson et al. 1991).

Based on these caloric values, an adult killer whale requires 3 to 7 sea otters per day (1,100 to 2,500 per year) to meet basic nutritional demands depending on the size of the predator and prey (Williams et al. 2004). These numbers decrease as the size and caloric value of the prey increases. For example, an adult male killer whale requires nearly two harbor porpoises or half a Dall's porpoise per day. Alternatively, a killer whale could satisfy its caloric requirements by gorging on a small fraction of the total blubber and meat of a mysticete whale every other day. Here the feeding schedule is dictated by the capacity of the killer whale to ingest and process an overabundance of food provided by a carcass, rather than the number of calories in the prey item.

As impressive as these numbers appear, the potential impact of killer whales on an ecosystem becomes most obvious when the annual needs of a population are calculated. The minimum current estimate for mammal-eating killer whales in the Aleutian archipelago is 170 adults (Williams et al. 2004). To gauge the importance of these predators for marine mammal prey—and of the different prey to them—we can determine how long it would take this number of killer whales to consume the total number of representative populations, each comprising known or likely prey species in this area. It is important to point out that these analyses merely demonstrate the potential magnitude of interaction strengths between killer whales and their prey and do not imply responses by the entire marine mammal community. This approach is admittedly simplistic, because it does not account for the population dynamics of prey and the simultaneous use of multiple prey populations by the predator. However, it does provide a clear comparative measure of the strength of interactions between killer whales and each of several possible food sources. In particular, the measure of time to consume specific marine mammal populations (Figure 15.5) illustrates the potential importance of top-down forcing processes that heretofore have been either ignored or summarily regarded as inconsequential. Furthermore, it provides insight into the underlying physiology and allometric relationships that dictate the strength of these processes. More complex analyses involving prey dynamics and multiple species, while more realistic, will require considerably more data about killer whale foraging behavior, movement patterns, and their prey community than is currently available.

For seven common prey species or species complexes available in the Aleutian Island area (great whales, Dall's porpoises, harbor porpoises, Steller sea lions, harbor seals, and sea otters), only two—the standing crop of Dall's porpoises and large mysticete whales—provide enough calories to sustain the killer whale population for more than one year (Figure 15.5). At current population levels—which, admittedly, may be negatively biased (see Pfister and DeMaster, Chapter 10 in this volume)—none of these marine mammal

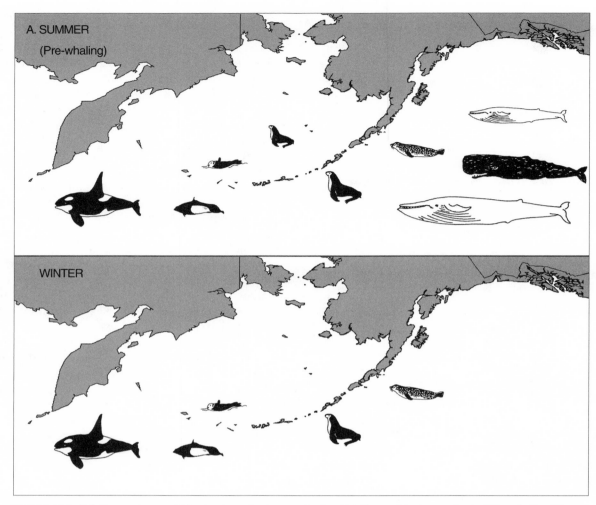

FIGURE 15.6. Relative presence of pinnipeds and cetaceans in the Aleutian archipelago prior to (A) and after (B) industrial whaling. Summer and winter months are compared for year-round resident (small cetaceans, Steller sea lion, harbor seal, sea otter) and transient (great whales, northern fur seal) prey species. Small mysticetes (minke whale), large mysticetes (blue whale, fin whale, gray whale, right whale, bowhead whale, and humpback whale), small odontocetes (Dall's porpoise, harbor porpoise, and northern right whale dolphin) and large odontocetes (killer whale, sperm whale) are represented generically by the animal symbols. Note, in particular, the "winterizing" of the killer whale's diet based on the remaining available prey species following removal of a large proportion of the biomass represented by great whales. In B the remaining great whale biomass is indicated by the minke whale (see Pfister and DeMaster, Chapter 10 in this volume, for species and biomass estimates).

species alone represents a sustainable resource except for the mysticetes, assuming reasonable rates of reproduction to replace these theoretical losses from predation. If the available estimates of abundance are correct, and if the population of mammal-eating killer whales fed exclusively and sequentially on these species, then minke whales, harbor seals *(Phoca vitulina)*, harbor porpoises, and Steller sea lions would be exterminated within 6 to 12 months; sea otters would be eliminated in only 2 days. Together these prey species would sustain the Aleutian killer whale population for less than three years.

The Dall's porpoise provides an example of the impact of focused predation by killer whales. Because an adult killer whale requires one Dall's porpoise every other day to meet basic caloric demands, a population of 170 killer whales

would need to take over 30,000 porpoises a year. With the current population of Dall's porpoises in the vicinity of the Aleutian Islands estimated at approximately 90,000 individuals (Pfister and DeMaster, Chapter 10 in this volume), this level of predation would remove the species in three years; a rate too fast to permit reproductive replacement.

One obvious corollary from these calculations is that the great whales, despite their reduced numbers, remain an important prey source for killer whales in this area (Figure 15.5). Otherwise, either the remaining marine mammal populations would be quickly decimated or killer whales would have to leave the area to feed.

It is difficult to predict the sequence of prey selection or nonspecific diet choices of killer whales given the lack of information concerning their long-term feeding preferences

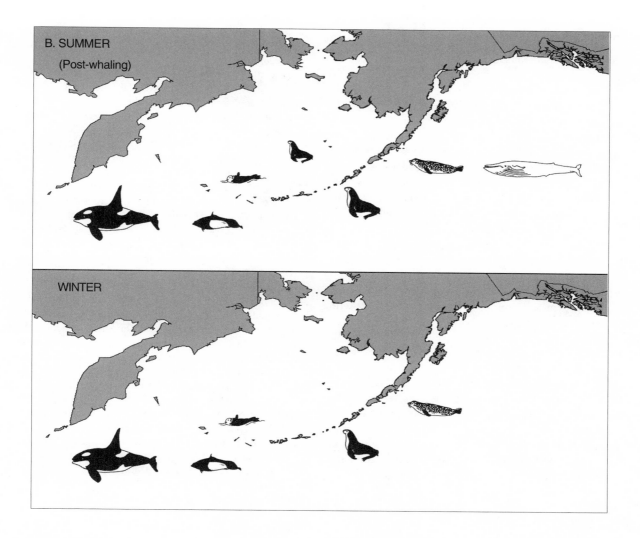

(Barrett-Lennard and Heise, Chapter 13 in this volume). Stomach contents of killer whales (Barrett-Lennard et al. 1995) indicate a gradual change from high- to low-caloric prey, although so few data are currently available that definitive conclusions cannot be made. In addition, changes in the caloric reserve provided by marine mammal prey (Williams et al. 2004), cultural transfer of hunting techniques for specific prey types (N. Black, personal communication), and evidence from large terrestrial predators (Seidensticker and McDougal 1993) suggest that mammal-eating killer whales in Alaska may switch prey and "eat down the caloric food chain" (see Mangel and Wolf, Chapter 21 in this volume). A potential complicating factor in this scenario is the effect of prey density on prey selection by killer whales, which is currently unknown. Alternatively, killer whales may demonstrate pod-specific eating habits that vary by region. Individual preferences for specific prey, as found for other mammals (Tinker 2004), would allow for simultaneous predation on many species of marine mammal by different segments of the killer whale population.

Another important factor influencing prey selection that was taken into account in the foregoing calculations is

seasonal changes in the availability of marine mammal prey in the Aleutian archipelago. Resident species were considered year-round prey for the killer whales, with the result that populations were quickly exhausted. In comparison, transient prey species could satisfy the caloric demands of killer whales for only 4 to 6 months of the year depending on migration patterns, and would experience attenuated population declines because predation pressure would be high for only part of the year. Thus, with the absence of most great whales and fur seals during the winter (Pfister and DeMaster, Chapter 10 in this volume), only small porpoises, harbor seals, Steller sea lions, and sea otters remained as common prey items for killer whales for several months (Figure 15.6a). A similar effect is observed with the reduction in mysticete and sperm whale populations following industrial whaling (Figure 15.6b). Elevated predation pressure created by this "winterizing" of the killer whale's diet could have easily contributed to the precipitous decline of harbor seal, Steller sea lion, and sea otter populations in the Aleutian archipelago during the past 20 years as suggested by Springer et al. (2003) and Williams et al. (2004).

Conclusions

In summary, present-day killer whales demonstrate the physiological and ecological consequences of high metabolic needs coupled with large body size in marine predators. As found for other marine mammals, the aquatic lifestyle of killer whales is associated with a high intake of food and a comparatively large gastrointestinal tract (Figures 15.3 and 15.4) to support elevated metabolic demands (Figure 15.1). Why killer whales never grew to the size of sperm whales or mysticete whales may be a function of how the animals meet these demands. By hunting and gorging on large prey items, killer whales may be subject to the same physiological constraints in body size observed for hunting terrestrial mammals that are absent in mammals that graze on organisms much smaller than themselves. Although marine mammals must take in nearly twice the daily calories of a terrestrial mammal (Figure 15.1), direct observations of predatory events, or monitoring programs that allow long-term individual prey preferences to be discerned, rarely occur even for the largest of whales. We have solved this problem by defining the physiological limitations of the predators based on fundamental biological laws, and taking into account ecologically effective population sizes of their prey. This procedure in turn enables us to assess the basic requirements critical for the survival of large whales. Such information can also be used to predict the impact of large predators within ecosystems and to evaluate the consequences of historical events such as whaling on the current status of both the predators and their prey.

Acknowledgments

I thank D. Calkins and S. Atkinson for support of the calorimeter studies through a grant to T. M. Williams from the Alaska SeaLife Center (Seward, Alaska). Tissue samples were generously supplied by National Marine Fisheries Service, U.S. Fish and Wildlife Service, USGS, the Alaska SeaLife Center, and Mystic Aquarium (Mystic, Connecticut).

Literature Cited

Alexander, R. McNeill. 1997. Engineering a dinosaur, in *The complete dinosaur*. J.O. Farlow and M.K. Brett-Surman, eds. Bloomington: Indiana University Press, pp. 414–425.

Andrews, R.D. 2002. The energetics of diving and the question of metabolic depression. *The Physiologist* 45(4): 325.

Baird, R.W. 1994. Foraging behavior and ecology of transient killer whales *(Orcinus orca)*. Ph.D. dissertation, Simon Fraser University, Burnaby, British Columbia.

Barrett-Lennard, L.G., K.A. Heise, E.L. Saulitis, G.M. Ellis, and C.O. Matkin. 1995. *The impact of killer whale predation on Steller sea lion populations in British Columbia and Alaska*. Vancouver: Marine Mammal Research Unit, University of British Columbia.

Blanckenhorn, W.U. 2000. The evolution of body size: What keeps organisms small? *Quarterly Review of Biology* 75(4): 385–407.

Bryden, M.M. 1972. Growth and development of marine mammals, in *Functional anatomy of marine mammals*. R.J. Harrison, ed. New York: Academic Press, pp. 1–79.

Brody, S. 1945. *Bioenergetics and growth*. New York: Reinhold.

Burness, G.P., J. Diamond, and T. Flannery. 2001. Dinosaurs, dragons, and dwarfs: the evolution of maximum body size. *Proceedings of the National Academy of Sciences* 98: 14518–14523.

Carbone, C. and J.L. Gittleman. 2002. A common rule for the scaling of carnivore density. *Science* 295: 2273–2276.

Case, T.J. 1979. Optimum body size and an animal's diet. *Acta Biotheoretica* 28: 54–69.

Clapham, P.J., S.B. Young, and R.L. Brownell, Jr. 1999. Baleen whales: conservation issues and the status of the most endangered populations. *Mammal Review* 29(1): 35–60.

Clauss, M., R. Frey, B. Kiefer, M. Lechner-Doll, W. Loehlein, C. Polster, G.E. Rossner, and W.J. Streich. 2003. The maximum attainable body size of herbivorous mammals: morphophysiological constraints on foregut, and adaptations in hindgut fermenters. *Oecologia* 136: 14–27.

Dahlheim, M.E. and J.E. Heyning. 1999. Killer whale *Orcinus orca* (Linnaeus, 1758), in *Handbook of marine mammals, Vol. 6*. S. Ridgway and R. Harrison, eds. San Diego: Academic Press, pp. 281–322.

Davis, R.W., L.A. Fuiman, T.M. Williams, S.O. Collier, W.P. Hagey, S.B. Kanatous, S. Kohin, and M. Horning. 1999. Hunting behavior of a marine mammal beneath the Antarctic sea ice. *Science* 283: 993–996.

Dejours, P. 1987. Water and air physical characteristics and their physiological consequences, in *Comparative physiology: life in water and on land*. P. Dejours, L. Bolis, C.R. Taylor, and E.R. Weibel, eds. Fidia Research Series 9. Berlin: Springer-Verlag, pp. 3–11.

Eastman, J.T. and R.E. Coalson. 1974. The digestive system of the Weddell seals *Leptonychotes weddellii*—a review, in *Functional anatomy of marine mammals*. R.J. Harrison, ed. New York: Academic Press, pp. 253–320.

Estes, J.A., M.T. Tinker, T.M. Williams, and D.F. Doak. 1998. Killer whale predation on sea otters linking oceanic and nearshore ecosystems. *Science* 282: 473–476.

Estes, R.D. 1991. *The behavior guide to African mammals*. Berkeley: University of California Press.

Goley, P.D. and J.M. Straley. 1994. Attack on gray whales in Monterey Bay, California, by killer whales previously identified in Glacier Bay, Alaska. *Canadian Journal of Zoology* 72: 1528–1530.

Irving, L. 1973. Aquatic mammals, in *Comparative physiology of thermoregulation*, Vol. 3. G.C. Whittow, ed. New York. Academic Press, pp. 47–96.

Jefferson, T.A., P.J. Stacey, and R.W. Baird. 1991. A review of killer whale interactions with other marine mammals: predation to co-existence. *Mammal Review* 21: 151–180.

Kanwisher, J., and G. Sundnes. 1965. Physiology of a small cetacean. *Hvalradets Skrifter* 48: 45–53.

Kardong, K.V. 2002. *Vertebrates: comparative anatomy, function, evolution*. Boston: McGraw-Hill.

King, J.E. 1983. *Seals of the world*. British Museum of Natural History, Ithaca, NY: Comstock Publishing Associates.

Kleiber, M. 1975. *The fire of life: an introduction to animal energetics*. Huntington, NY: R.E. Kreiger Publishing Co.

Kooyman, G.L. 1973. Respiratory adaptations of marine mammals. *American Zoologist* 13: 457–468.

Kreite, B. 1995. Bioenergetics of the killer whale, *Orcinus orca*. Ph.D. dissertation, University of British Columbia, Vancouver.

Lavigne, D.M., S. Innes, G.A.J. Worthy, K.M. Kovacs, O.J. Schmitz, and J.P. Hickie. 1986. Metabolic rates of seals and whales. *Canadian Journal of Zoology* 64: 279–284.

LeBoeuf, B.J. 1994. Variation in the diving pattern of northern elephant seals with age, mass, sex and reproductive condition, in *Elephant seals: population ecology, behavior, and physiology.* B.J. LeBoeuf and R.M. Laws, eds. Berkeley: University of California Press, pp. 237–252.

Liao, J.A. 1990. An investigation of the effect of water temperature on the metabolic rate of the California sea lion *(Zalophus californianus).* Master's thesis, University of California, Santa Cruz.

Lockyer, C. 1981. Growth and energy budgets of large baleen whales from the Southern Hemisphere, in *Mammals in the seas.* FAO Fisheries Series No. 5, Vol. 3. Rome: UN Food and Agricultural Organization, Fisheries Department, pp. 379–487.

McAlpine, D.F. 1985. Size and growth of heart, liver and kidneys in North Atlantic fin *(Balaenoptera physalus),* sei *(B. borealis),* and sperm *(Physeter macrocephalus)* whales. *Canadian Journal of Zoology* 63: 1402–1409.

McNab, B.K. 1986. The influence of food habits on the energetics of eutherian mammals. *Ecological Monographs* 56(1): 1–19.

Reeves, R.R., B.S. Stewart, P.J. Clapham, and J.A. Powell. 2002. *National Audubon Society guide to marine mammals of the world.* New York: Alfred A. Knopf.

Schmidt-Nielsen, K. (1984) *Scaling: Why is animal size so important?* New York: Cambridge University Press.

Seidensticker, J. and C. McDougal. 1993. Tiger predatory behaviour, ecology and conservation. *Symposia of the Zoological Society of London* 65: 105–125.

Slijper, E.J. 1976. *Whales and dolphins.* Ann Arbor: University of Michigan Press.

Springer, A.M., J.A. Estes, G.B. van Vliet, T.M. Williams, D.F. Doak, E.M. Danner, K.A. Forney, and B. Pfister. 2003. Sequential megafaunal collapse in the North Pacific Ocean: an ongoing legacy of industrial whaling? *Proceedings of the National Academy of Sciences* 100: 12223–12228.

Stephens, D.W. and J.R. Krebs. 1986. *Foraging theory.* Princeton, NJ: Princeton University Press.

Stevens, C.E. and I.D. Hume. 1995. *Comparative physiology of the vertebrate digestive system.* Cambridge, UK: Cambridge University Press.

Stirling, I. 2002. Polar bear, in *Encyclopedia of marine mammals.* W.F. Perrin, B. Wursig, and J.G.M. Thewissen, eds. San Diego: Academic Press, pp. 945–948.

Tenney, S.M. and J.E. Remmers. 1963. Comparative quantitative morphology of the mammalian lung diffusing area. *Nature* 197: 54–56.

Tinker, T. 2004. Demographics and behavior of the California sea otter population. Ph.D. thesis, University of California, Santa Cruz.

Weibel, E.R. 1984. *The pathway for oxygen: structure and function in the mammalian respiratory system.* Cambridge, MA: Harvard University Press.

Whitehead, H. 2002. Sperm whale, in *Encyclopedia of marine mammals.* W.F. Perrin, B. Wursig, and J.G.M. Thewissen, eds. San Diego: Academic Press, pp. 1165–1172.

———. 2003. *Sperm whales: social evolution in the ocean.* Chicago: University of Chicago Press.

Williams, T.M. 1989. Swimming by sea otters: adaptations for low energetic cost locomotion. *Journal of Comparative Physiology* 164: 815–824.

———. 1999. The evolution of cost efficient swimming in marine mammals: limits to energetic optimization. *Philosophical Transactions of the Royal Society London B* 354: 193–201.

Williams, T.M., J.A. Estes, D.F. Doak, and A.M. Springer. 2004. Killer appetites: assessing the role of predators in ecological communities. *Ecology* 85: 3373–3384.

Williams, T.M., J. Haun, R.W. Davis, L.A. Fuiman, and S. Kohin. 2001. A killer appetite: Metabolic consequences of carnivory in marine mammals. *Comparative Biochemistry and Physiology A* 129: 785–796.

Williams, T.M. and G.A.J. Worthy. 2002. Anatomy and physiology: the challenge of aquatic living, in *Marine mammal biology.* A.R. Hoelzel, ed. Oxford: Blackwell Science, pp. 73–97.

SIXTEEN

Ecosystem Impact of the Decline of Large Whales in the North Pacific

DONALD A. CROLL, RAPHAEL KUDELA, AND BERNIE R. TERSHY

Biodiversity loss can significantly alter ecosystem processes (Chapin et al. 2000), and ecological extinction can have similar effects (Jackson et al. 2001). For marine vertebrates, overharvesting is the main driver of ecological extinction, and the expansion of fishing fleets into the open ocean has precipitated rapid declines in pelagic apex predators such as whales (Baker and Clapham 2002), sharks (Baum et al. 2003), tuna, and billfishes (Cox et al. 2002; Christensen et al. 2003), leading to a trend in global fisheries toward exploitation of lower trophic levels (Pauly et al. 1998a). Globally, many fish stocks are overexploited (Steneck 1998), and the resulting ecological extinctions have been implicated in the collapse of numerous nearshore coastal ecosystems (Jackson et al. 2001). Nevertheless, although the declines of apex predators in pelagic ecosystems are well documented, the ecosystem impacts of such declines are unclear (Steneck 1998, Jackson et al. 2001).

The ecological role of large whales (baleen and sperm whales) in pelagic ecosystems and the consequences of the decline of their populations from whaling has been the focus of debate for pelagic ecologists, conservation biologists, fisheries managers, and the general public. The severe depletion of stocks of large whales is one of the best documented examples of the overexploitation of apex predators. In the North Pacific, at a minimum estimate, 62,858 whales, representing 1.8 million tons of whale biomass, were removed over a 150-year period (Springer et al. 2003). Although large whales are significant consumers of pelagic prey, such as schooling fish and euphausiids (krill), the trophic impacts of their removal is not clear (Trites et al. 1999). Indeed, it is possible that the biomass of prey consumed by large whales prior to exploitation exceeded that currently taken by commercial fisheries (Baker and Clapham 2002), but estimates of prey consumption by large whales before and after the period of intense human exploitation are lacking. Given the large biomass of pre-exploitation whale populations (see, e.g., Whitehead 1995; Roman and Palumbi 2003), their high mammalian metabolic rate, and their relatively high trophic position (Trites 2001), it is likely that the removal of large whales over a 150-year period by human harvest had cascading effects, leading to changes in energy flow and species composition at other trophic levels (Bowen 1997).

In Antarctica, Laws (1977, 1984, 1985) speculated that the removal of baleen whales increased krill availability to fur seals and penguins by as much as 150 million tons annually, and May et al. (1979) modeled how this could lead to significant increases in seal populations. Several studies have provided estimates of the prey requirements of current populations of cetaceans across multiple geographic areas using a variety of approaches (Sergeant 1969; Brodie 1975; Lockyer 1981; Lavigne et al. 1986; Innes et al. 1987; Vikingsson 1990; Armstrong and Siegfried 1991; Ichii and Kato 1991; Kawamura

1994; Vikingsson 1995; Sigurjónsson and Vikingsson 1997; Trites et al. 1997; Vikingsson 1997). However, with the exception of Laws' (1977) estimates for Antarctica and Trites et al.'s (1999) estimates for the Bering Sea, few studies have attempted to assess the impacts of commercial whaling on energy flow in pelagic ecosystems by comparing the percent of marine primary productivity consumed by whales before and after the period of intensive commercial whaling.

In the absence of empirical observations, one method to examine the trophic impact of consumers on ecosystems is to assess changes in the amount of net primary production required (PPR) to sustain them (Vitousek et al. 1986; Pauly and Christensen 1995; Kenney et al. 1997; Trites et al. 1997). Using this approach, Vitousek et al. (1986) estimated that humans consume 35%–40% of terrestrial primary production. In marine systems, Pauly and Christensen (1995) estimated that commercial fisheries required 8% of global aquatic primary production to sustain them, and Brooke (2004) estimated that seabirds consume 7% of global primary production. In this chapter we use a similar approach to assess the potential trophic impact of the historical removal of large whales from the North Pacific pelagic ecosystem.

Approach

An assessment of the impact of the removal of large whales on energy flow requires (1) estimates of prey biomass consumption rates for each whale species, (2) estimates of pre- and postexploitation whale populations, (3) estimates of the trophic level and trophic transfer efficiencies of individual whale species, and (4) estimates of the net primary production of the ecosystem.

Whale Prey Biomass Consumption

We used two approaches to estimate whale prey biomass consumption per individual whale of each large whale species: allometric estimates of whale metabolic rates and allometric estimates of whale ingestion rates. For the metabolic rate approach, we combined allometric estimates of resting/basal metabolic rates of whales with known diet, prey energy estimates, and assimilation efficiencies to calculate the mass of prey required daily by individuals of each large whale species. Three independent allometric metabolic rate models were used. The first model (Hemmingsen 1960) relied upon empirically derived measures of mammalian resting metabolic rates,

$$\text{Active Metabolic Rate (W)} = 12.3(\text{Body Mass (kg)})^{0.75}, \quad (16.1)$$

as did the second model (Kleiber 1961),

$$\text{Active Metabolic Rate (W)} = 9.84(\text{Body Mass (kg)})^{0.756}. \quad (16.2)$$

In both cases, we assumed that daily metabolic rates in free-living animals were 3 times resting/basal rates (Costa and Williams 1999). The third approach (Nagy et al. 1999) relied upon an allometric model of empirically measured field metabolic rates using doubly labeled water:

$$\text{Field Metabolic Rate (W)} = 8.88(\text{Body Mass (kg)})^{0.734}. \quad (16.3)$$

Prey energy required was converted to prey biomass required (PBR) using published data on diet composition (Pauly et al. 1998b), prey energy density (Clarke 1980; Boyd 2002), and consumer assimilation efficiency (84%; Lockyer 1981).

For the ingestion rate approach we used two allometric models of prey ingestion rates. The first model (Innes et al. 1987) was based upon empirically measured biomass ingestion rates in marine mammals:

$$\text{Prey Biomass Required (kg)} = 0.42(\text{Body Mass (kg)})^{0.67}. \quad (16.4)$$

The second model (Nagy 2001) was based upon empirically measured biomass ingestion rates in terrestrial and aquatic mammals using doubly labeled water:

$$\text{Prey Biomass Required (kg)} = 0.17(\text{Body Mass (kg)})^{0.773}. \quad (16.5)$$

An important caveat is that in all approaches, allometric estimates were calculated from extrapolations well beyond the range of empirically measured subjects.

Whale Population Estimates

We restricted our analysis to the North Pacific/Bering Sea region. Population numbers of this region prior to exploitation and at current levels were taken from published estimates (Table 16.1). Such estimates are difficult to derive, can be speculative, and can lead to inaccuracies in our estimates of trophic impact. Ranges of population estimates prior to exploitation are not published in the literature (only point estimates are available). Measures of dispersion of estimates of current populations in the study region have been published as confidence intervals, ranges, or standard deviations. To account for uncertainty in current population estimates, we estimated population prey consumption rates based on the high, low (i.e., dispersion) and published best estimates of the current population. Generally, dispersion measures varied by less than 25% of the best population estimate. For example, Perry et al. (1999) reviewed population estimates of large whales, and the largest maximum-minimum range difference in North Pacific large-whale populations they examined (fin whales) differed by only 22%. Differences of this magnitude have minimal impacts on biomass consumption estimates.

TABLE 16.1

Body Mass, Population Estimates, and Population Biomass of North Pacific Large Whales Used to Estimate Prey Consumption Rates

	Blue	Fin	Sei	Bryde's	Humpback	Minke	N. Right	Gray	Sperm
Body mass (kg)[a]	69,235[b]	55,600[c]	16,811[d]	16,143[d]	30,400[c]	6,600[c]	23,400[c]	15,400[c]	18,500[c]
Population									
Current Low	2,475[e]	14,620[e]	7,260[e]	22,500[f]	6,000[e]	22,500[g]	404[e]	24,477[h]	75,648[i]
Current High	4,125	18,630	12,620	37,500	8,000	37,500	2,108	28,793	83,958
Current Best	3,300	16,625	9,110	30,000	7,000	30,000	1,256	26,635	79,803
Pre-exploitation	4,900	43,500	42,000	39,000	15,000	30,000	31,750	26,635	204,454
Biomass (metric tons)									
Current Best	228,476	924,350	153,157	484,290	212,800	198,000	29,390	410,179	1,476,356
Pre-exploitation	339,252	2,418,600	706,062	629,577	456,000	198,000	742,950	410,179	3,782,399

[a] Body masses are averages of male and female values.

[b] Lockyer (1976).

[c] Pfister and DeMaster, Chapter 10 of this volume.

[d] Trites and Pauly (1998).

[e] Population estimates from Perry et al. (1999).

[f] Because of complexities of population structure and reporting uncertainty (i.e., confusion between whale stocks and other species such as sei whales), it is difficult to estimate the Northern Pacific population of Bryde's whales confidently. We used pre- and postexploitation estimates of the western North Pacific from the International Whaling Commission and combined these with an estimate of postexploitation numbers for the entire eastern tropical Pacific provided by Wade and Gerrodete (Carretta et al. 2002). Thus, both our pre- and postexploitation estimates are likely overestimates for Bryde's whales.

[g] Population estimates from Ohsumi (1991).

[h] Population estimates from Carretta et al. (2002).

[i] Sperm whale population estimates are derived from areal extrapolations of estimates published by Whitehead (2002).

TABLE 16.2

Allometric Estimates of Daily Metabolic Rates of Large Whales of the North Pacific

	Estimated Daily Metabolic Rate (W)					
	Eq. (16.1), 3 × Basal	Eq. (16.2), 3 × Basal	Eq. (16.3), Average Daily	Mean	SD	CV
Blue	52,499	44,904	31,720	43,041	10,514	0.24
Fin	44,536	38,043	27,003	36,527	8,864	0.24
Sei	18,159	15,401	11,223	14,928	3,492	0.23
Bryde's	17,615	14,936	10,894	14,482	3,384	0.23
Humpback	28,318	24,102	17,337	23,252	5,540	0.24
Minke	9,007	7,596	5,650	7,418	1,685	0.23
Northern right	23,271	19,775	14,307	19,118	4,518	0.24
Gray	17,004	14,413	10,524	13,980	3,262	0.23
Sperm	19,511	16,557	12,040	16,036	3,763	0.24

TABLE 16.3
Prey Biomass Requirements for North Pacific Large Whales Derived from Five Different Models

Whale	Prey Biomass Requirements (kg individual^{-1} day^{-1})							
	Eq. (16.1)	Eq. (16.2)	Eq. (16.3)	Eq. (16.4)	Eq. (16.5)	Mean	SD	CV
Blue	1,607	1,375	971	735	913	1,120	359	0.32
Fin	1,259	1,076	763	635	771	901	258	0.29
Sei	513	435	317	285	306	371	99	0.27
Bryde's	417	354	258	277	296	320	65	0.20
Humpback	711	605	435	423	483	532	123	0.23
Minke	235	199	148	152	148	176	39	0.22
Northern right	712	605	438	355	395	501	152	0.30
Gray	521	441	322	268	286	368	109	0.30
Sperm	476	404	294	304	329	361	77	0.21

When only ranges were available, we used median values as the best estimate of the population.

Trophic Level and Trophic Transfer Efficiencies

To estimate the PPR to sustain pre- and postwhaling populations, we converted estimated prey biomass requirements of whales (PBR) to PPR using estimates of whale trophic levels and trophic transfer efficiencies. We estimated whale trophic levels by combining whale diet composition with prey trophic level estimates (Pauly et al. 1998b). Using these estimates, we converted prey biomass requirements of whales to PPR using whale trophic level (TL) and an estimate of trophic transfer efficiency published by Pauly and Christensen (1995):

$$PPR = (PBR/9) \times 10^{(TL-1)} \qquad (16.6)$$

Net Primary Production

Net primary production (NPP) for the North Pacific was estimated using the vertically generalized production model (VGPM) derived from global samples of ^{14}C measures of primary production (Behrenfeld and Falkowski 1997). The VGPM provides a method to estimate primary production using satellite-derived measures of chlorophyll concentration. We used estimates of chlorophyll concentration from SeaWiFS satellite images of the North Pacific/Bering Sea, averaged 1998–2001, to estimate annual net primary production using the VGPM.

Model Results

Large-whale populations in the North Pacific declined to 47% of pre-exploitation levels—from approximately 437,000 pre-exploitation to the current level of approximately 204,000 (Table 16.1). This resulted in a 58% decline in large-whale biomass from 9.7 million metric tons to 4.1 million metric tons.

Individual Whale Prey Biomass Consumption

Based upon estimates from allometric models of respiration rate, the metabolic rates of large whales range from 43 kW for blue whales to 7.4 kW for minke whales (Table 16.2). Using these values, we estimated the mean PBR to sustain individual whales ranged from 1,120 kg day^{-1} (blue whales) to 176 kg day^{-1} (minke whales) (Table 16.3).

Whale Population Prey Biomass Consumption

Combining individual whale prey biomass consumption values with population estimates, we calculated the average daily prey biomass required to sustain the North Pacific populations of large whales before and after their declines from exploitation (Table 16.4). These numbers have uncertainties related to extrapolations of allometric equations beyond empirical data and errors in published estimates of whale populations. Given these caveats, we estimated daily population prey biomass requirements ranged from 629 metric tons day^{-1} to sustain the current population of northern right whales to approximately 74,000 metric tons day^{-1} to sustain the pre-exploitation population of sperm whales. Because of the large pre-exploitation population size of sperm whales, this species has the highest gross prey biomass requirements of North Pacific large whales. We estimate that whaling reduced the total daily prey biomass consumption for all North Pacific large-whale populations by 57%: Pre-exploitation daily prey biomass consumption totaled some 185,000 metric tons day^{-1}, whereas postexploitation consumption totals some 80,000 metric tons day^{-1}.

Primary Production Required to Sustain Whale Populations

We estimated average daily net primary production for the North Pacific to be 2.54×10^{10} kg C day^{-1}. Published information on the diet of North Pacific whales (Table 16.5) was combined with published trophic levels for whale diet items: large zooplankton, 2.2; fish, 2.7; small squid, 3.2; large squid, 3.7,

TABLE 16.4

Estimated Daily Prey Biomass Requirements (thousands of metric tons day^{-1}) for North Pacific Large-Whale Populations

Estimated Daily Prey Biomass Requirements (thousands of metric tons per day)

| | Respiration Models | | | | | | Ingestion Models | | | | Mean | |
| | Eq. (16.1) | | Eq. (16.2) | | Eq. (16.3) | | Eq. (16.4) | | Eq. (16.5) | | | |
Whale	Current	Pre-exploitation	Current	Pre-exploitation	Current	Pre-exploitation	Current	Pre-exploitation	Current	Pre-exploitation	Current	Pre-exploitation
Blue	5.3	7.9	4.5	6.7	3.2	4.8	2.4	3.6	3.0	4.5	3.7	5.5
Fin	20.9	54.8	17.9	46.8	12.7	33.2	10.5	27.6	12.8	33.5	15.0	39.2
Sei	4.7	21.6	4.0	18.3	2.9	13.3	2.6	12.0	2.8	12.8	3.4	15.6
Bryde's	12.5	16.3	10.6	13.8	7.7	10.1	8.3	10.8	8.9	11.6	9.6	12.5
Humpback	5.0	10.7	4.2	9.1	3.0	6.5	3.0	6.4	3.4	7.3	3.7	8.0
Minke	7.1	7.1	6.0	6.0	4.4	4.4	4.6	4.6	4.5	4.5	5.3	5.3
Northern right	0.9	22.6	0.8	19.2	0.6	13.9	0.4	11.3	0.5	12.5	0.6	15.9
Gray	13.9	13.9	11.8	11.8	8.6	8.6	7.2	7.2	7.6	7.6	9.8	9.8
Sperm	38.0	97.3	32.2	82.5	23.4	60.0	24.2	62.1	26.3	67.3	28.8	73.8
Total	108.2	252.0	91.9	214.2	66.6	154.8	63.2	145.4	69.7	161.6	79.9	185.6

TABLE 16.5
Trophic Position of Prey, Whale Diet, and Estimated Trophic Level of North Pacific Large Whales

Prey Species	Prey Trophic Position	Diet Composition (Proportion)[a]								
		Blue	Fin	Sei	Bryde's	Humpback	Minke	N. Right	Gray	Sperm
Large zooplankton	2.2	1	0.8	0.8	0.4	0.55	0.65	0	1	0
Fish	2.7	0	0.15	0.15	0.6	0.45	0.35	0	0	0.25
Small squid	3.2	0	0.05	0.05	0	0	0	0	0	0.1
Large squid	3.7	0	0	0	0	0	0	0	0	0.6
Small zooplankton	2.0	0	0	0	0	0	0	1	0	0.05
Trophic Level[b]		3.20	3.33	3.33	3.50	3.43	3.38	3.00	3.20	4.22

[a] Based on Pauly et al. (1998b).

[b] Based on weighted mean trophic position of diet items.

small zooplankton, 2.0 (Trites et al. 1999). This yielded estimates of the trophic position of North Pacific large whales (Table 16.5). We assumed a trophic transfer efficiency of 0.1 (Pauly and Christensen 1995) and used our estimates of trophic position and population prey biomass requirements to calculate biomass of PPR to sustain each whale species prior to and after exploitation (Table 16.6). For all large-whale populations in the North Pacific, we estimate that preexploitation populations required a total of 16 million metric tons of fixed carbon per day, whereas current populations require a total of 6.5 million metric tons of fixed carbon per day. Combining our estimates of primary production required to sustain large-whale populations with our estimate of average daily net primary production for the North Pacific, we estimated the percentage of average daily net primary production (NPP) required to sustain North Pacific large-whale populations at current numbers and at pre-exploitation numbers (Table 16.7). For all North Pacific whale populations, preexploitation populations required approximately 64.5% of NPP, whereas current populations require 26.3% of North Pacific NPP.

Discussion

Rapid Removal of Large Whales

Because of the large range, long-distance movement patterns, and logistical difficulties in conducting population surveys for large cetaceans, it is difficult to assess their current population numbers accurately (Clapham et al. 1999; Perry et al. 1999; Clapham et al. 2003). It is even more difficult to assess pre-exploitation population numbers accurately (Gerber et al. 2000). We used available estimates for current and postexploitation population numbers from the literature, which likely introduced error into our calculations. However, the magnitude of error introduced by

inaccuracies in population estimates is relatively small compared to the effect of whaling. Indeed, the range of estimated percent daily net primary production required (Table 16.7) varied by less than 2% using high, low, and best estimates of current population sizes from the literature. Regardless, there is a clear consensus that commercially exploited North Pacific whale populations experienced severe declines to a fraction of their original sizes during the past two centuries (Clapham et al. 1999). Thus, even substantial adjustments in pre-exploitation and current estimates of large-whale populations do not seriously affect conclusions drawn from the consumption estimates on which they are based.

Using best available population estimates, large-whale populations in the North Pacific were reduced by approximately 53% (from approximately 437,000 to 204,000 individuals) (Table 16.1). This reduction occurred in less than 150 years, with most of it taking place during approximately 100 years (from the mid-nineteenth to the mid-twentieth century; Clapham et al. 1999). Because larger species were preferentially harvested, total whale biomass in the North Pacific was reduced 58% from approximately 9.7 million metric tons to 4.1 million metric tons. A comparable decline took place in Antarctica, where commercial whaling was estimated to have removed 65% (Laws 1977) of whale stocks, reducing whale biomass by 85%.

Commercial whaling not only reduced the population and biomass of large whales in the North Pacific; it also changed the large-whale community composition. Pre-exploitation whale biomass of sperm whales represented 47% of total whale biomass, whereas sperm whales represent 39% of the present-day biomass in large whales. This may have important trophic implications, because the large-whale community has shifted from one dominated primarily by squid predators (sperm whales) to one dominated primarily by fish and zooplankton predators (rorquals). Sperm whales are almost a full trophic level above rorquals (Table 16.5) in marine food webs. Thus,

TABLE 16.6

Estimated Primary Production Required to Sustain Large-Whale Populations in the North Pacific

Primary Production Required (thousands of metric tons per day)

	Respiration Models						Ingestion Models					
	Eq. (16.1)		Eq. (16.2)		Eq. (16.3)		Eq. (16.4)		Eq. (16.5)		Mean	
Whale	Current	Pre-exploitation	Current	Pre-exploitation	Current	Pre-exploitation	Current	Pre-exploitation	Current	Pre-exploitation	Current	Pre-exploitation
Blue	93	139	80	119	56	84	43	63	53	79	65	97
Fin	492	1,286	420	1,099	298	780	248	648	301	787	352	920
Sei	110	506	93	429	68	313	61	281	65	302	79	366
Bryde's	440	572	373	485	272	353	292	380	312	406	338	439
Humpback	147	315	125	268	90	193	88	188	100	214	110	236
Minke	186	186	157	157	117	117	120	120	117	117	140	140
Northern right	10	251	8	214	6	155	5	125	6	139	7	177
Gray	244	244	207	207	151	151	126	126	134	134	172	172
Sperm	6,921	17,732	5,873	15,047	4,271	10,943	4,416	11,314	4,790	12,272	5,254	13,462
Total	8,643	21,232	7,337	18,024	5,329	13,088	5,398	13,245	5,879	14,451	6,517	16,008

TABLE 16.7
Percentage of Average Daily Net Primary Production of the North Pacific Required to Sustain
North Pacific Large-Whale Populations

| | Model Results | | | | | | |
| | Respiration Models | | | Ingestion Models | | Summary | |
	Eq. (16.1)	Eq. (16.2)	Eq. (16.3)	Eq. (16.4)	Eq. (16.5)	Mean	SD
Mysticetes and Sperm Whales							
Current N. Pacific Low	32.1	27.3	19.8	20.1	21.9	24.2	5.3
Current N. Pacific High	37.5	31.9	23.1	23.4	25.5	28.3	6.2
Current N. Pacific Best	34.8	29.6	21.5	21.7	23.7	26.3	5.8
Pre-exploitation N. Pacific	85.5	72.6	52.7	53.4	58.2	64.5	14.2
Mysticetes Only							
Current N. Pacific Low	5.7	4.8	3.5	3.2	3.6	4.2	1.1
Current N. Pacific High	8.2	7.0	5.0	4.7	5.2	6.0	1.5
Current N. Pacific Best	6.9	5.9	4.3	4.0	4.4	5.1	1.3
Pre-exploitation N. Pacific	14.1	12.0	8.6	7.8	8.8	10.3	2.7
Rorquals Only							
Current N. Pacific Low	4.8	4.0	2.9	2.7	3.1	3.5	0.9
Current N. Pacific High	7.1	6.0	4.3	4.1	4.6	5.2	1.3
Current N. Pacific Best	5.9	5.0	3.6	3.4	3.8	4.4	1.1
Pre-exploitation N. Pacific	12.1	10.3	7.4	6.8	7.7	8.9	2.3

their removal may have had more profound implications on marine ecosystems than the removal of rorquals.

Whale Metabolic Rates and Prey Consumption Rates

Our calculations of prey and primary production requirements are influenced by population estimates, estimates of trophic positions, and assumptions regarding metabolic rates. Boyd (2002) points out that error in such estimates is both additive (e.g., metabolic rates, assimilation efficiency) and multiplicative (e.g., populations, time). We used 3 independent allometric models to estimate whale metabolic rates. All models provided similar estimates: Across all species, the coefficient of variation was consistently around 24% (Table 16.2). Based upon various methods, Lockyer (1981) estimated the basal metabolic rate for a 70,790-kg blue whale at 12,269 to 24,539 W. Assuming an active metabolic rate of 3 times basal and using Lockyer's estimates yields a range of approximately 36.8 to 73.6 kW, within which our average estimate of 43.0 kW for a 69,200-kg blue whale falls. Lockyer also used a muscle equivalent method to estimate active metabolic rate. Her estimate of 232,778 W was considerably greater than our average estimate. However, it is unlikely that blue whales sustain an active metabolic rate 9.5 to 18 times basal. Our estimate for the daily metabolic rate

for minke whales (7.4 kW) lies within Markussen et al.'s (1992) estimates of the average daily energy requirements for minke whales of 9,214 W for females and 6,789 W for males.

Sigurjónsson and Vikingsson (1997) used two allometric methods to estimate the metabolic rates of blue, fin, sei, minke, humpback, and sperm whales. For each of these species, our estimates of metabolic rate were 3% (blue whale) to 52% (sperm whale) lower than their estimates of daily energy consumption made using two models. Costa and Williams (1999) concluded that the basal metabolic rate of marine mammals is actually 1.2 to 2 times Kleiber's (1961) allometric estimates. If that conclusion is true, then our estimates would be 20 to 100% lower than a calculation based upon Costa and Williams's values for marine mammal basal metabolic rates. Nonetheless, our values can be considered within the range of values estimated by most studies and tend to be somewhat conservative (i.e., lower).

Our estimates of individual consumption of prey biomass (Table 16.3) are also comparable to others'. Armstrong and Siegfried (1991) estimated Antarctic minke whale daily prey consumption using metabolic rate during the feeding season as 212 kg day[-1] for males and 252 kg day[-1] for females. These can be compared to our estimate of 176 kg day[-1]. Markussen et al. (1992) estimated minke whale prey consumption at 204 kg day[-1] for males and 277 kg day[-1] in females. Vikingsson

(1997) estimated fin whale prey consumption from stomach volume and passage rates at 677–1,356 kg day^{-1}, comparable to our estimate of 901 kg day^{-1}.

Trophic Importance of Current Populations of Large Whales

We estimate that large whales in the North Pacific currently consume some 80,000 metric tons of prey per day (Table 16.4). Several studies have estimated the prey biomass requirements for individual species as well as communities of marine mammals for a variety of marine areas, and these provide useful comparisons for our estimates. Based on our estimates, large whales in the North Pacific currently consume approximately 19% of the 418,688 metric tons day^{-1} Trites et al. (1997) estimated is consumed by marine mammals (including mysticetes, odontocetes, and pinnipeds) in the entire Pacific Ocean. Armstrong and Siegfried (1991) estimated the minke whale population prey biomass requirement in the Antarctic at 97,260 metric tons day^{-1}, whereas Markussen et al. (1992) estimated that northeast Atlantic minke whales consumed 14,667 metric tons day^{-1}.

The maximum biomass extracted by commercial fisheries in the North Pacific occurred in 1998 and averaged 75,468 metric tons of fish day^{-1} (FAO 2002). This is comparable to our estimate of prey biomass consumed by current populations of large whales in the North Pacific. It has been argued that most commercial fisheries, including the North Pacific fishery, are overexploited (Pauly et al. 1998a, Steneck 1998). Because commercially targeted species are often top predators, their declines can have cascading trophic impacts (Dayton et al. 1998, Worm and Myers 2003). The mean trophic level of commercial fisheries in the North Pacific was estimated by Pauly et al. (1998b) to have declined from a maximum of 3.4 in the early 1970s to 3.2 by 1994. Combining trophic level values (Table 16.5) with prey biomass consumption estimates (Table 16.4); we estimate the weighted mean trophic level of North Pacific large whales to be 3.4. Thus, in terms of both prey biomass consumption and trophic level, it can be argued that current populations of large whales are of similar importance in the North Pacific marine ecosystems to commercial fisheries. The impact of pre-exploitation levels of great whales in this area would have been much greater.

Several studies have estimated the primary production required to sustain consumer populations in other regions: Kenney et al. (1997) estimated that cetaceans in some coastal regions of the northwestern Atlantic consume 11.7%–20.4% of net primary production, while Trites et al. (1997) estimated that marine mammals in the Pacific consume 12%–17% of net primary production. We estimate that current populations of large whales (smaller odontocetes and pinnipeds excluded) in the North Pacific consume approximately 26% of net primary production. This is considerably greater than the weighted mean primary production required to sustain world fisheries (8% of net primary production; Pauly and Christensen 1995). Thus, assuming that commercial fishing has important trophic impacts in marine ecosystems, current populations of large whales also appear to be important trophic interactors.

Community Implications of the Decline of Large Whales

Virtually every ecosystem is characterized by both bottom-up (resource limitation) and top-down (consumer control) interactions (Hunter and Price 1992). The bottom-up view holds that populations of organisms on each trophic level are resource limited by nutrients or food, while the top-down view focuses on predators' control of their prey leading to the situation where, at successive trophic levels, populations are alternately resource or predator limited (Fretwell 1977). Traditionally, oceanographers have characterized pelagic marine ecosystems as being bottom-up regulated (Verity and Smetacek 1996), whereas the top-down view has dominated our understanding of freshwater and coastal ecosystems (Paine 1966; Estes and Palmisano 1974; Carpenter et al. 1987; Power 1992). Indeed, evidence of trophic cascades in coastal ecosystems has been documented from the removal of large vertebrates such as cod (Worm and Myers 2003), predatory reef fish (Hughes 1994), and sea otters (Estes and Palmisano 1974).

A key element of most trophic cascades is the dependence upon strong interactions by particular species (Pace et al. 1998). Because of their large body size and relatively high metabolic rate, marine mammals have the potential to be strong trophic interactors (Bowen 1997). However, although strong predator-induced trophic cascades have long been recognized in pelagic lake systems (Carpenter and Kitchell 1993), evidence of trophic cascades resulting from predator removals in pelagic marine systems is rare (Micheli 1999; Pace et al. 1999; Estes et al. 2001). Lack of evidence in pelagic systems may be a result of the logistical difficulty of documenting cascades in the open ocean (Estes et al. 2001), nonexistent baselines (Dayton et al. 1998; Jackson et al. 2001), weak coupling between phytoplankton and herbivores (Micheli 1999), or more reticulate pelagic food webs with higher degrees of omnivory that dampen classic trophic cascades (McCann et al. 1998).

The removal of large whales from pelagic food webs had several nonexclusive possible outcomes: (1) replacement of the trophic role of large whales by other predators, (2) changes in carbon turnover rates as longer-lived cetaceans are replaced by smaller species with shorter life spans, and (3) the initiation of trophic cascades. Laws (1977) proposed that the removal of large whales from the Southern Ocean ecosystem resulted in increases in their prey and their replacement by other krill predators such as penguins, fur seals, and seals. Essington (Chapter 5 in this volume) used a food web model to show that large whales—primarily sperm whales—were replaced by large squid in the tropical North Pacific. Unfortunately, information on trends in abundance, diet, and trophic interactions of pelagic species is lacking for most systems (DeMaster et al. 2001). This lack, coupled with the problems of assessing change in pelagic ecosystems outlined above, makes an evaluation of the relative importance of trophic replacement and trophic cascades initiated by the removal of large whales problematic.

It is certain, however, that whaling severely reduced the prey requirements of large whales. We estimate that North Pacific large whales currently consume approximately 43% of what they consumed prior to whaling (80,000 metric tons per day of prey biomass compared to 186,000 metric tons per day at prewhaling numbers). Laws et al. (1977) estimated that large-whale populations in Antarctica consume 17% of the levels they consumed prior to whaling (30,300 metric tons per day compared to 178,000 metric tons per day prewhaling) and that this decline in krill consumption increased the availability of krill to other consumers in the relatively simple Antarctic ecosystem (see also Ballance et al., Chapter 17 in this volume).

Trites et al. (1999) developed a mass balance model to examine complex trophic interactions resulting from marine mammal declines in the Bering Sea. They estimated that the commercial harvest of large whales in the Bering Sea accounted for 43.4% of Bering Sea net primary production during the 1950s. This does not account for the amount required to sustain the entire population of large whales (that is, it does not include the portion not harvested). Thus, our estimate of 64.5% of primary production requirement for the entire large-whale population of the North Pacific prior to large-scale population declines appears relatively congruent with theirs. In spite of these high consumption rates, Trites et al. (1999) concluded that the overall impact of marine mammal declines had little effect on other species in the Bering Sea ecosystem. They concluded that the removal of large whales may have had a positive effect on pollock (*Theragra chalcogramma*) abundance by reducing competition for food, but this could not explain the dramatic increase in pollock that occurred between the 1950s and 1980s. Instead, they attributed large-scale changes observed in the Bering Sea ecosystem to shifts in primary production due to oceanographic regime shifts.

However, recent studies are providing growing evidence that top-down interactions do play an important role in pelagic marine food webs (Pace et al. 1999). For example, fisheries exploitation has been shown to result in significant changes in the plankton community structure in the North Sea (Reid et al. 2000); microzooplankton have been hypothesized to control plankton community structure in the Bering Sea (Olson and Strom 2002); the relative biomass of phytoplankton and zooplankton in the North Pacific has been attributed to top-down control by pink salmon (Shiomoto and Hashimoto 2000); the profound changes witnessed recently in the Black Sea have been attributed to a trophic cascade initiated by the severe depletion of pelagic predator stocks (Daskalov 2002); and experimental studies have demonstrated that the top-down effect of copepod predation and grazing can profoundly affect pelagic community structure and primary production (Stibor et al. 2004).

While the food web models of Trites et al. (1999) and Essington (Chapter 5 in this volume) predict trophic replacement rather than trophic cascades from the removal of large whales, one assumption of their models is that no radical changes in ecosystem organization occurs. Scheffer and van Nes (2004) proposed that dramatic state shifts in marine systems can result from diffuse interactions among many species and feedbacks between organisms and the abiotic environment—scenarios not captured in typical food web models. Thus, it is possible that large changes in food web structure initiated by whaling coupled with large-scale climate changes (e.g., regime shifts) could lead to dramatic state shifts in pelagic ecosystems, including trophic cascades. Scheffer and van Nes (2004) speculated that open ocean state shifts might not arise easily but would tend to be impressive in magnitude and scale when they did occur. Both our estimates and those of Trites et al. (1999) demonstrate that a significant shift in energy flow occurred as a result of whale harvest. Large whales were (and continue to be) important consumers of North Pacific prey biomass; the decline of their populations certainly altered energy flow and trophic interactions; and such perturbations, coupled with climatic changes, have the potential to alter pelagic food web structure dramatically.

Sperm Whale Demands on Marine Ecosystems

Sperm whales may be one of the most important predators in pelagic ecosystems. At pre-exploitation levels, we estimated that sperm whales accounted for 84.1% of the 64.5% of average daily North Pacific NPP consumed by large whales. At current population sizes, we estimate that sperm whales require 80.6% of the total 26.3% of average daily North Pacific NPP consumed by large whales (see Table 16.7). We estimate that whaling has reduced sperm whale consumption of average daily North Pacific NPP to 39% of pre-exploitation levels. The impact of sperm whale reduction may be more significant than that of reductions in other whale species because of sperm whales' large body size, large population size, high trophic level, and potentially important role in top-down control of cephalopods.

Based upon their diet and NPP requirements, sperm whales in the North Pacific are significant cephalopod predators. Cephalopods (1) are active, fast-moving predators; (2) feed on a wide range of prey including crustaceans, fish, and other cephalopods; (3) generally live only one year and die after a single spawning event; (4) have rapid growth rates and high metabolic rates; (5) undergo rapid ontogenetic shifts in prey species within a year; and (6) have been estimated to consume 2.09 to 4.03 Gt of prey each year globally (Rodhouse and Nigmatullin 1996). Rodhouse and Nigmatullin (1996) argued that these traits indicate that predation by squid significantly affects pelagic ecosystem structure. For example, in the Gulf of California, Ehrhardt (1991) found that a strong migration of *Dosidicus gigas* led to high predation rates in the Gulf and a subsequent decline in sardine landings.

High rates of cephalopod predation by sperm whales, combined with the trophic importance of cephalopods in pelagic ecosystems, could lead to cascading trophic impacts as squid are released from top-down predation by sperm whales through whaling. Essington (Chapter 5 in this volume) predicted that declines of sperm whales in the tropical and subtropical North Pacific would lead to dramatic increases in the abundance of large squid (as they replace the trophic role vacated by sperm whales). Even in the absence of trophic cascades, the replacement of extremely large, long-lived predators (sperm whales) by smaller, short-lived predators (large squid) should dramatically increase interannual variability in large-predator abundance and thus cause both the resistance and resiliency of the ecosystem to decrease. In the presence of a strong trophic cascade, the removal of sperm whales would have an even greater capacity to alter food web structure significantly. This may be particularly important for less productive open-ocean systems, where sperm whale abundance is relatively high.

In summary, our estimates of NPP requirements demonstrate that large whales are important trophic interactors and that their declines from commercial harvest have significantly altered energy flow in marine food webs. However, a lack of historical data on pelagic ecosystems precludes assessing the relative importance of trophic replacement, trophic cascades, or other major ecosystem changes from this large alteration of energy flow. Nonetheless, there is recent support for the importance of top-down processes in pelagic ecosystems, and the high consumption rates of high-trophic-level species by large whales, particularly sperm whales, suggests that trophic cascades are possible. At the least, the reduction of large whales has probably led to increases in trophic competitors (Worm et al., Chapter 26 in this volume). Some of these competitors (e.g., squid) have short generation times and undergo dramatic interannual variability in abundance. Increased variability may alter the resilience and resistance of pelagic ecosystems to change and ultimately increase susceptibility to dramatic state changes.

Literature Cited

Armstrong, A.J. and W.R. Siegfried. 1991. Consumption of Antarctic krill by minke whales. *Antarctic Science* 3: 13–18.

Baker, C.S. and P.J. Clapham. 2002. Marine mammal exploitation: whales and whaling, in *Causes and consequences of global environmental change*. I. Douglas, ed. Chichester: John Wiley and Sons, pp. 446–450.

Baum, J.K., R.A. Myers, D.G. Kehler, B. Worm, J. Harley, and P.A. Doherty. 2003. Collapse and conservation of shark populations in the northwest Atlantic. *Science* 299: 389–392.

Behrenfeld, M.J. and P.G. Falkowski. 1997. Photosynthetic rates derived from satellite-based chlorophyll concentration. *Limnology and Oceanography* 42: 1–20.

Bowen, W.D. 1997. Role of marine mammals in aquatic ecosystems. *Marine Ecology Progress Series* 158: 267–274.

Boyd, I.L. 2002. Estimating food consumption of marine predators: Antarctic fur seals and macaroni penguins. *Journal of Applied Ecology* 39: 103–119.

Brodie, P.F. 1975. Cetacean energetics, an overview of intraspecific size variation. *Ecology* 56 :152–161.

Brooke, M.D. 2004. The food consumption of the world's seabirds. *Proceedings of the Royal Society of London Series B: Biological Sciences* 271: S246–S248.

Carpenter, S.R. and J.F. Kitchell. 1993. The trophic cascade in lakes: synthesis of ecosystem experiments. *Bulletin of the Ecological Society of America* 74: 186–187.

Carpenter, S.R., J.F. Kitchell, J.R. Hodgson, P.A. Cochran, J.J. Elser, M.M. Elser, D.M. Lodge, D. Kretchmer, X. He, and C.N. Vonende. 1987. Regulation of lake primary productivity by food web structure. *Ecology* 68: 1863–1876.

Carretta, J.V., M.M. Muto, J. Barlow, J. Baker, K.A. Forney, and M. Lowry. 2002. *U.S. Pacific marine mammal stock assessments: 2002*. NOAA Technical Memorandum NMFS-SWFSC-346. La Jolla, CA: Southwest Fisheries Science Center.

Chapin, F.S., III, E.S. Zavaleta, V.T. Eviner, R.L. Naylor, P.M. Vitousek, H.L. Reynolds, D.U. Hooper, S. Lavorel, O.E. Sala, S.E. Hobbie, M.C. Mack, and S. Diaz. 2000. Consequences of changing biodiversity. *Nature* 405: 234–242.

Christensen, V., S. Guenette, J.J. Heymans, C.J. Walters, R. Watson, D. Zeller, and D. Pauly. 2003. Hundred-year decline of North Atlantic predatory fishes. *Fish and Fisheries* 4: 1–24.

Clapham, P.J., P. Berggren, S. Childerhouse, N.A. Friday, T. Kasuya, L. Kell, K.-H. Kock, S. Manzanilla-Naim, G.N. Di Sciara, W.F. Perrin, A.J. Read, R.R. Reeves, E. Rogan, L. Rojas-Bracho, T.D. Smith, M. Stachowitsch, B.L. Taylor, D. Thiele, P.R. Wade, and R.L. Brownell, Jr. 2003. Whaling as science. *Bioscience* 53: 210–212.

Clapham, P.J., S.B. Young, and R.L. Brownell, Jr. 1999. Baleen whales: conservation issues and the status of the most endangered populations. *Mammal Review* 29(1): 35–60.

Clarke, A. 1980. The biochemical composition of krill (*Euphausia superba* Dana) from South Georgia. *Journal of Experimental Marine Biology and Ecology* 43: 221–236.

Costa, D.P. and T.M. Williams. 1999. Marine mammal energetics, in *Biology of marine mammals*. J.E. Reynolds and S.A. Rommel, eds. Washington, DC: Smithsonian Institution Press, pp. 176–217.

Cox, S.P., T.E. Essington, J.F. Kitchell, S.J.D. Martell, C.J. Walters, C. Boggs, and I. Kaplan. 2002. Reconstructing ecosystem dynamics in the central Pacific Ocean, 1952–1998. II. A preliminary assessment of the trophic impacts of fishing and effects on tuna dynamics. *Canadian Journal of Fisheries and Aquatic Sciences* 59: 1736–1747.

Daskalov, G.M. 2002. Overfishing drives a trophic cascade in the Black Sea. *Marine Ecology Progress Series* 225: 53–63.

Dayton, P.K., M.J. Tegner, P.B. Edwards, and K.L. Riser. 1998. Sliding baselines, ghosts, and reduced expectations in kelp forest communities. *Ecological Applications* 8: 309–322.

DeMaster, D.P., C.W. Fowler, S.L. Perry, and M.E. Richlen. 2001. Predation and competition: the impact of fisheries on marine-mammal populations over the next one hundred years. *Journal of Mammalogy* 82: 641–651.

Ehrhardt, N.M. 1991. Potential impact of a seasonal migratory jumbo squid (*Dosidicus gigas*) stock on a Gulf of California sardine (*Sardinops sagax caerulea*) population. *Bulletin of Marine Science* 49: 325–332.

Estes, J.A., K. Crooks, and R. Holt. 2001. Predators, ecological role of, in *The encyclopedia of biodiversity*. S. Levin, G.C. Daily, R.K. Colwell, J. Lubchenco, H.A. Mooney, E.-D. Schulze, and D. Tilman, eds. San Diego: Academic Press, pp. 857–878.

Estes, J.A. and J.F. Palmisano. 1974. Sea otters: their role in structuring nearshore communities. *Science* 185: 1058–1060.

FAO. 2002. *The state of the world fisheries and aquaculture 2002.* Rome: UN Food and Agriculture Organization.

Fretwell, S.D. 1977. Regulation of plant communities by food-chains exploiting them. *Perspectives in Biology and Medicine* 20: 169–185.

Gerber, L.R., D.P. DeMaster, and S.L. Perry. 2000. The conservation of endangered whale populations: Have management efforts succeeded? *American Scientist* 88: 316–324.

Hemmingsen, A.M. 1960. Energy metabolism as related to body size and respiratory surfaces, and its evolution. *Reports of the Steno Memorial Hospital and Nordinsk Insulin Laboratorium* 9: 6–110.

Hughes, T.P. 1994. Catastrophes, phase shifts, and large-scale degradation of a Caribbean coral reef. *Science* 265: 1547–1551.

Hunter, M.D. and P.W. Price. 1992. Playing chutes and ladders: heterogeneity and the relative roles of bottom-up and top-down forces in natural communities. *Ecology* 73: 724–732.

Ichii, T. and H. Kato. 1991. Food and daily food consumption of southern minke whales in the Antarctic. *Polar Biology* 11: 479–488.

Innes, S., D.M. Lavigne, W.M. Earle, and K.M. Kovacs. 1987. Feeding rates of seals and whales. *Journal of Animal Ecology* 56: 115–130.

Jackson, J.B.C., M.X. Kirby, W.H. Berger, K.A. Bjorndal, L.W. Botsford, B.J. Bourque, R. Bradbury, R. Cooke, J. Erlandson, J.A. Estes, T.P. Hughes, S. Kidwell, C.B. Lange, H.S. Lenihan, J.M. Pandolfi, C.H. Peterson, R.S. Steneck, M.J. Tegner, and R. Warner. 2001. Historical overfishing and the recent collapse of coastal ecosystems. *Science* 293: 629–638.

Kawamura, A. 1994. A review of baleen whale feeding in the Southern Ocean. *Report of the International Whaling Commission* 44: 261–271.

Kenney, R.D., G.P. Scott, T.J. Thompson, and H.E. Winn. 1997. Estimates of prey consumption and trophic impacts of cetaceans in the USA Northeast continental shelf ecosystem. *Journal of Northwest Atlantic Fishery Science* 22: 155–171.

Kleiber, M. 1975. *The fire of life: an introduction to animal energetics.* Huntington, NY: R.E. Kreiger Publishing Co.

Lavigne, D.M., S. Innes, G.A.J. Worthy, and K.M. Kovacs. 1986. Metabolic rate–body size relations in marine mammals. *Journal of Theoretical Biology* 122: 122–124.

Laws, R.M. 1977. Seals and whales of the Southern Ocean. *Philosophical Transactions of the Royal Society of London Series B: Biological Sciences* 279: 81–96.

———. 1984. Seals, in *Antarctic Ecology.* R.M. Laws, ed. London: Academic Press, pp. 621–715.

———. 1985. The ecology of the Southern Ocean. *American Scientist* 73: 26–40.

Lockyer, C. 1976. Body weights of some species of large whales. *Journal du Conseil International pour L'Exploration de la Mer* 36: 259–273.

———. 1981. Growth and energy budgets of large baleen whales from the Southern Hemisphere, in *Mammals in the seas.* FAO Fisheries Series No. 5, Vol. 3. Rome: UN Food and Agricultural Organization, Fisheries Department, pp. 379–487.

Markussen, N.H., M. Ryg, and C. Lydersen. 1992. Food consumption of the NE Atlantic minke whale (*Balaenoptera acutorostrata*) population estimated with a simulation model. *ICES Journal of Marine Science* 49: 317–323.

May, R.M., J.R. Beddington, C.W. Clark, S.J. Holt, and R.M. Laws. 1979. Management of multispecies fisheries. *Science* 205: 267–277.

McCann, K.S., A. Hastings, and D.R. Strong. 1998. Trophic cascades and trophic trickles in pelagic food webs. *Proceedings of the Royal Society of London Series B: Biological Sciences* 265: 205–209.

Micheli, F. 1999. Eutrophication, fisheries, and consumer-resource dynamics in marine pelagic ecosystems. *Science* 285: 1396–1398.

Nagy, K.A. 2001. Food requirements of wild animals: predictive equations for free-living mammals, reptiles, and birds. *Nutrition Abstracts and Reviews* 71: 21–33.

Nagy, K.A., I.A. Girard, and T.K. Brown. 1999. Energetics of free-ranging mammals, reptiles, and birds. *Annual Review of Nutrition* 19: 247–277.

Ohsumi, S. 1991. A review on population studies of the North Pacific minke whale stocks. Paper SC/43/Mi26 presented to the International Whaling Commission Scientific Committee.

Olson, M.B. and S.L. Strom. 2002. Phytoplankton growth, microzooplankton herbivory and community structure in the southeast Bering Sea: insight into the formation and temporal persistence of an *Emiliania huxleyi* bloom. *Deep-Sea Research Part II: Topical Studies in Oceanography* 49: 5969–5990.

Pace, M.L., J.J. Cole, and S.R. Carpenter. 1998. Trophic cascades and compensation: differential responses of microzooplankton in whole-lake experiments. *Ecology* 79: 138–152.

Pace, M.L., J.J. Cole, S.R. Carpenter, and J.F. Kitchell. 1999. Trophic cascades revealed in diverse ecosystems. *Trends in Ecology and Evolution* 14: 483–488.

Paine, R.T. 1966. Food web complexity and species diversity. *American Naturalist* 100: 65–75.

Pauly, D. and V. Christensen. 1995. Primary production required to sustain global fisheries. *Nature* 374: 255–257.

Pauly, D., V. Christensen, J. Dalsgaard, R. Froese, and F.C. Torres, Jr. 1998a. Fishing down marine food webs. *Science* 279: 860–863.

Pauly, A.W. Trites, E. Capuli, and V. Christensen. 1998b. Diet composition and trophic levels of marine mammals. *ICES Journal of Marine Science* 55: 467–481.

Perry, S.L., D.P. DeMaster, and G.K. Silber. 1999. The great whales: history and status of six species listed as endangered under the U.S. Endangered Species Act of 1973. *Marine Fisheries Review* 61: 1–74.

Power, M.E. 1992. Top-down and bottom-up forces in food webs: Do plants have primacy? *Ecology* 73: 733–746.

Reid, P.C., E.J.V. Battle, S.D. Batten, and K.M. Brander. 2000. Impacts of fisheries on plankton community structure. *ICES Journal of Marine Science* 57: 495–502.

Rodhouse, P.G. and C.M. Nigmatullin. 1996. Role as consumers. *Philosophical Transactions of the Royal Society of London Series B: Biological Sciences* 351: 1003–1022.

Roman, R. and S.R. Palumbi. 2003. Whales before whaling in the North Atlantic. *Science* 301: 508–510.

Scheffer, M., and E.H. van Nes. 2004. Mechanisms for marine regime shifts: Can we use lakes as microcosms for oceans? *Progress in Oceanography* 60: 303–319.

Sergeant, D.E. 1969. Feeding rates of cetaceans. *Fiskeridirektoratets Skrifter; Serie Havundersøkelser* 15: 246–258.

Shiomoto, A. and S. Hashimoto. 2000. Comparison of east and west chlorophyll a standing stock and oceanic habitat along the Transition Domain of the North Pacific. *Journal of Plankton Research* 22: 1–14.

Sigurjónsson, J. and G.A. Vikingsson. 1997. Seasonal abundance of and estimated food consumption by cetaceans in Icelandic and adjacent waters. *Journal of Northwest Atlantic Fishery Science* 22: 271–287.

Springer, A.M., J.A. Estes, G.B. van Vliet, T.M. Williams, D.F. Doak, E.M. Danner, K.A. Forney, and B. Pfister. 2003. Sequential megafaunal collapse in the North Pacific Ocean: an ongoing legacy of industrial whaling? *Proceedings of the National Academy of Sciences* 100: 12223–12228.

Steneck, R.S. 1998. Human influences on coastal ecosystems: Does overfishing create trophic cascades? *Trends in Ecology and Evolution* 13: 429–430.

Stibor, H., O. Vadstein, S. Diehl, A. Gelzleichter, T. Hansen, F. Hantzsche, A. Katechakis, B. Lippert, K. Loseth, C. Peters, W. Roederer, M. Sandow, L. Sundt-Hansen, and Y. Olsen. 2004. Copepods act as a switch between alternative trophic cascades in marine pelagic food webs. *Ecology Letters* 7: 321–328.

Trites, A.W. 2001. Marine mammal trophic levels and interactions, in *Encyclopedia of Ocean Sciences*. J.H. Steele, K.K. Turekian, and S.A. Thorpe, eds. London: Academic Press, pp. 1628–1633.

Trites, A.W., V. Christensen, and D. Pauly. 1997. Competition between fisheries and marine mammals for prey and primary production in the Pacific Ocean. *Journal of Northwest Atlantic Fishery Science* 22: 173–187.

Trites, A.W., P.A. Livingston, S. Mackinson, M.C. Vasconcellos, A.M. Springer, and D. Pauly. 1999. *Ecosystem change and the decline of marine mammals in the eastern Bering Sea: testing the ecosystem shift and commercial whaling hypothesis.* Fisheries Centre Research Report 1999, 7(1). Vancouver: University of British Columbia.

Trites, A.W. and D. Pauly. 1998. Estimating mean body masses of marine mammals from maximum body lengths. *Canadian Journal of Zoology* 76: 886–896.

Verity, P.G. and V. Smetacek. 1996. Organism life cycles, predation, and the structure of marine pelagic ecosystems. *Marine Ecology-Progress Series* 130: 277–293.

Vikingsson, G.A. 1990. Energetic studies on fin and sei whales caught off Iceland. *Report of the International Whaling Commission* 40: 365–373.

———. 1995. Body condition of fin whales during summer off Iceland, in *Developments in marine biology: whales, seals, fish and man.* A.S. Blix, L. Walloe, and O. Ultang, eds. Developments in Marine Biology, Vol. 4. Amsterdam and New York: Elsevier Science, pp. 361–369.

———. 1997. Feeding of fin whales *(Balaenoptera physalus)* off Iceland: diurnal and seasonal variation and possible rates. *Journal of Northwest Atlantic Fishery Science* 22: 77–89.

Vitousek, P.M., P.R. Ehrlich, A.H. Ehrlich, and P.A. Matson. 1986. Human appropriation of the products of photosynthesis. *Bioscience* 36: 368–373.

Whitehead, H. 1995. Status of Pacific sperm whale stocks before modern whaling. *Report of the International Whaling Commission* 45: 407–412.

———. 2002. Estimates of the current global population size and historical trajectory for sperm whales. *Marine Ecology Progress Series* 242: 295–304.

Worm, B. and R.A. Myers. 2003. Meta-analysis of cod-shrimp interactions reveals top-down control in oceanic food webs. *Ecology* 84: 162–173.

The Removal of Large Whales from the Southern Ocean

Evidence for Long-Term Ecosystem Effects?

LISA T. BALLANCE, ROBERT L. PITMAN, ROGER P. HEWITT,
DONALD B. SINIFF, WAYNE Z. TRIVELPIECE,
PHILLIP J. CLAPHAM, AND ROBERT L. BROWNELL, JR.

The Southern Ocean is the site of a vast uncontrolled experiment that began when commercial sealing and whaling activities in the nineteenth and twentieth centuries brought some seal and whale species near extinction.

R.M. LAWS (1977)

The Southern Ocean can broadly be defined as the oceanic region surrounding the continent of Antarctica and south of the Antarctic Convergence, or Polar Front (Figure 17.1A). Because of its latitude, it is an ecosystem of extreme seasonal variability. This variability is ultimately related to sunlight—its presence in the southern summer, and absence in the winter, results in dramatic changes in the extent of sea ice around the continent (Figure 17.1B, C), in the degree of primary productivity of the ocean, and in the distribution, abundance, and life history strategies of all organisms that live there.

Among these organisms are the baleen whales, including approximately 13 species that comprise the suborder *Mysticeti*. Each year, whales in the southern hemisphere migrate from low latitudes in the winter to the Southern Ocean in the summer to feed on seasonally abundant prey, primarily euphausiids, and particularly *Euphausia superba*, also known as krill.

Commercial whalers in the nineteenth and twentieth centuries exploited these whales, driving populations to extremely low levels. To this day, some populations are showing very limited signs of recovery despite the long period since the cessation of whaling on them, 65 and 85 years for blue (*Balaenoptera musculus*) and humpback whales (*Megaptera novaeangliae*), respectively, around South Georgia (Clapham and Hatch 2000).

What were the effects of this massive removal of large whales? In a seminal paper, Laws (1977) postulated that their removal resulted in a surplus of prey in the Southern Ocean ecosystem, with consequent increases in the numbers of other krill consumers via competitive release. Here, we provide a review of the literature published since that paper, and further investigate the effects of this massive removal of large whales and evaluate the potential ecosystem effects of this uncontrolled experiment.

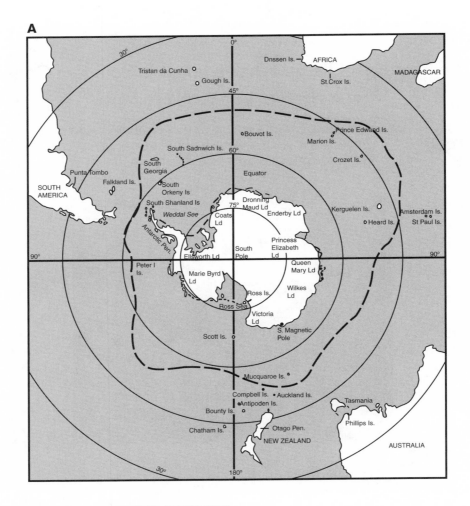

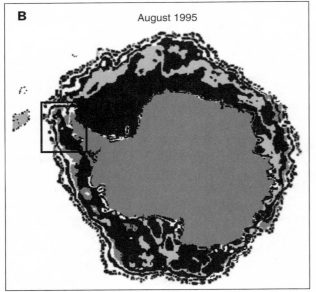

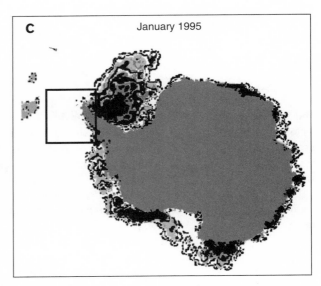

FIGURE 17.1. The Southern Ocean. (A) The Antarctic continent surrounded by the Southern Ocean, defined as that area south of the Antarctic Convergence, or Polar Front (indicated by the dashed line; from Williams 1995). The extent of sea ice varies dramatically by season, as shown for the southern winter (B) and the southern summer (C; from Hewitt 1997); rectangle indicates area used to construct index of seasonal sea ice extent.

Whaling in the Southern Ocean

The commercial hunting of whales in the 20th century represents what was arguably—in terms of sheer biomass—the greatest wildlife exploitation episode in human history. (Clapham and Hatch 2000)

The story of commercial whaling is a familiar one. From its start in the late 1800s, to the moratorium on commercial whaling imposed upon all member nations of the International Whaling Commission in 1986, some 2.7 million whales were reported to have been taken from the world's oceans (Gambell 1999). The kill peaked in the 1960s with some 70,000 whales recorded as killed annually over a period of several years. Commercial take was not distributed equally among all species through time. Instead, whalers focused on species with the largest body size first, then moved to progressively smaller species, one after another, so that whaling records indicate serial depletion of blue whales, followed by fin whales (*B. physalus*), then sei whales (*B. borealis*; Mackintosh and Brown 1974; Gambell 1999; Boyd 2002). Humpback, sperm (*Physeter macrocephalus*), and minke (*B. edeni*) whales were also heavily targeted (Clapham and Baker 2002).

What is less apparent is that the official records of number and species of whales taken were falsified, at times to a great degree, and illegal whaling occurred in a number of ocean basins (Yablokov 1994; Zemsky et al. 1995; Tormosov et al. 1998; Baker et al. 2000; Brownell and Yablokov 2002; Clapham and Baker 2002). For example, Soviet whalers took some 9,000 more pygmy blue whales (*B. musculus brevicauda*), 23,000 more sei whales, and 43,000 more humpback whales from the Southern Hemisphere than reported in official whaling records (Brownell and Yablokov 2002). The falsification of records, along with illegal whaling has complicated efforts to determine the extent of depletion of whale populations, the degree of recovery, and the current status, on a species- and population-specific basis. This also means that any ecosystem effects arising from the removal of these large-bodied predators as a group are potentially much greater than previously believed.

In fact, recent information indicates that some 2 million whales have been taken from the Southern Hemisphere alone (Table 17.1; IWC 1995; Clapham and Baker 2002). Of this 2 million, 400,000 were sperm whales, of which only males are known to migrate to the Southern Ocean in summer (Whitehead 2002). Assuming then, that half of the sperm whales were female (200,000 whales that did not migrate to Antarctic waters), some 1.8 million whales that were once a part of the Southern Ocean ecosystem were removed by commercial whaling.

The "Krill Surplus Hypothesis"

[I]t seems reasonable to conclude that a consequence of the dramatically reduced density of baleen whales would be a greater standing stock of krill, . . . that the remaining baleen whales are eating more, and growing and breeding faster . . .

TABLE 17.1

Total Number of Whales Taken Commercially in the Southern Hemisphere[a]

Species	Number Killed
Blue	360,644
Fin	725,116
Sei	203,538
Humpback	208,359
Bryde's	7,757
Minke	116,568
Right	4,338
Sperm	401,670
Other	11,631
Total	2,039,621

[a] Reprinted from Clapham and Baker 2002 with permission from Elsevier.

other groups have begun to consume more, with effects on their growth and reproductive rates . . . and their population sizes. (Laws 1977)

What were the consequences of the removal of some 1.8 million whales from the Southern Ocean? This question has intrigued marine ecologists for decades. Ideas from a series of papers dating back to at least 1964 (Sladen 1964) have resulted in a general hypothesis that has been referred to as the "Krill Surplus Hypothesis."

Laws (1977) calculated that the baleen whales taken by commercial whaling operations in the Southern Ocean may have consumed as much as 150 million tons of krill annually. To put this into perspective, it is estimated that the current populations of whales, birds, pinnipeds, fish, and squid in the Southern Ocean consume 250 million tons of krill annually (Miller and Hampton 1989), of which 60–90 million tons are believed to be consumed by 800,000 to 1.1 million whales (Hewitt and Lipsky 2002). Clearly, the baleen whales killed through commercial whaling consumed a significant proportion of the prey taken by krill predators in the Southern Ocean. Laws postulated that their removal resulted in a huge surplus of krill, with a number of potential consequences to the ecosystem.

One potential consequence was that this krill surplus became available to other krill predators, so biologists began to look for effects of this on other animals. For example, 150 million tons of krill was thought to be able to support the addition of 200–300 million penguins to the system *per year* (Sladen 1964; Emison 1968); commensurate with this idea, researchers reported increased growth in penguin populations of a number of species (Sladen 1964; Emison 1968; Conroy 1975; Croxall and Kirkwood 1979; Croxall et al. 1984; Rootes 1988). Increases in populations of Antarctic fur seals (*Arctocephalus gazella*) were also noted, and believed to be related to this surplus of prey (Payne 1977). Earlier migration to the Southern Ocean and a redistribution to more southerly latitudes by sei whales, reported increases in pregnancy rates of sei, blue, and fin whales, and decreased age of sexual maturity for sei and fin whales were all documented and attributed to this surplus of prey (Laws 1962; Gambell 1968; Lockyer 1972; Gambell 1973; Lockyer 1974).

This clear conclusion of competitive release for a great many species has since given way to a much more complex picture. Monitoring studies have continued and data sets now span decades. Technology has greatly improved the ability to measure physical habitat variables, and analyses of biological patterns now incorporate environmental variation. It has become clear that there are multiple hypotheses to explain many of the abundance, behavior, and life history trends identified in the studies initially supporting the Krill Surplus Hypothesis. We expand on these in the following sections.

Post-Whaling Trends in Antarctic Populations: How the Southern Ocean Ecosystem Has Changed

Twenty years have passed since the moratorium on commercial whaling and, although some whaling continues to this day, much more time has passed since the huge takes by commercial whalers in the Southern Ocean. What changes have occurred?

We address post-whaling trends by focusing on three trophic levels, in turn: large whale prey (krill), large whale competitors (pinnipeds and seabirds), and large whale predators (killer whales, *Orcinus orca*). In the following sections we will summarize what is known about temporal trends, and how these trends have been interpreted.

The Southern Ocean is not a single ecosystem. Although the Antarctic Circumpolar Current circulates completely around the Antarctic continent with no landmass to block its flow, the geographic relief of the continental coast is irregular, and the extent of sea ice varies from season to season, and year to year. The world's major seas influence the Southern Ocean in different longitudinal sections. It would be overly simplistic to assume that ecosystem trends in one area reflect those in another. Therefore, in our review of the literature, we will specify the geographical region to which particular trends apply. The implicit assumption here is that those trends may not apply elsewhere.

Large Whale Prey—Krill: Top-Down or Bottom-Up Regulation?

The numerically dominant euphausiid in deep waters of the Southern Ocean (i.e., off the shelf and slope) is *Euphausia superba*, which is consumed by all baleen whales in the region. Although other euphausiid and fish species have been observed in their diets, baleen whales foraging in the Southern Ocean do not exhibit strong species or size selectivity among their prey. Instead, the prevalence of *E. superba* in the diet appears to be related to its high abundance and widespread availability (Hewitt and Lipsky 2002).

It seems logical to assume that krill biomass would have increased over time as the large whales were removed. That is, overall natural mortality would have been reduced and the ultimate effect would have been higher biomass on average. However, krill reproduction and larval survival are characterized by extreme variability, both in space and time, and in at least some geographic areas, data support the idea that this variability is currently governed by bottom-up forces (see the following).

Large interannual variations in krill recruitment and density have been observed in the Antarctic Peninsula region (Loeb et al. 1997; Siegel et al. 1998; Fraser and Hofmann 2003). Although krill live to an average of five to seven years of age, during any particular year the age structure of the population is dominated by one or two age classes. It is also apparent that these strong year classes are autocorrelated in time; that is, several poor years of reproductive success are followed by one to two good years, describing a repeating cycle of a four- to five-year period (Figure 17.2A). Krill abundance, viewed as the sum of all age classes in the population, is cyclic as well, declining with successive decreases in reproductive success and increasing dramatically with the recruitment of strong year classes (Hewitt et al. 2003).

These abundance cycles have been associated with multi-year changes in the physical environment and consequent effects on primary and secondary production (Loeb et al. 1997; Nicol et al. 2000; White and Petersen 1996; Naganobu et al. 1999; Brierley et al. 1999). One manifestation of these changes is the extent of wintertime sea ice in the Antarctic Peninsula region (Figure 17.2B). During periods of equatorward excursions of the southern boundary of the Antarctic Circumpolar Current (SBACC), the seasonal development of sea ice (the underside of which is postulated to provide access to ice algae and refuge from predators for over-wintering adult and larval krill) is more extensive, populations of salps (*Salpa thompsonii*, a pelagic tunicate thought to be a competitor with krill for access to the springtime phytoplankton bloom) are displaced offshore, and both krill reproductive output and the survival of their larvae are enhanced. During periods of poleward excursions of the SBACC, the development of wintertime sea ice is less extensive, salps are more abundant closer to shore, and krill reproductive success is depressed. These cycles appear to also be coherent across the Scotia Sea (Brierley et al. 1999; Reid et al. 2002; Fraser and Hofmann 2003) and it may be concluded that large changes in the availability of krill to predators occur independently of predation pressure.

The interactions between sea ice, krill recruitment, and availability of krill to predators may be confounded by a warming trend observed in the Antarctic Peninsula region over the last 50 years (Vaughan and Doake 1996). While years of extensive sea ice continue to occur (Hewitt 1997), the frequency and duration of these "ice events" have decreased (Fraser et al. 1992; Fraser and Hofmann 2003). Fraser and Hofmann (2003) further suggest that as the time period between good krill year classes approaches the longevity of krill, cohort senescence (that is, the loss of a dominant cohort and its reproductive potential) will become an additional ecosystem stressor. These processes imply long-term changes

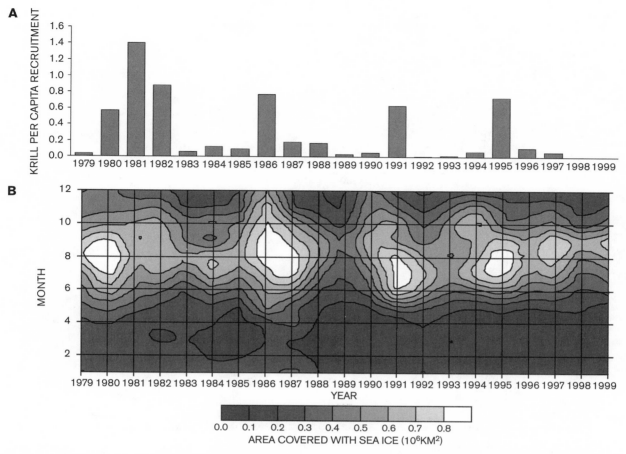

FIGURE 17.2. Temporal trends in krill recruitment and extent of sea ice cover in the Antarctic Peninsula region (from Hewitt and Linen Low 2000). (A) Krill recruitment (the number of recruits divided by the number of spawners the previous year) over time in the Elephant Island area, Scotia Sea. (B) Sea ice cover in the Antarctic Peninsula region, contoured by month and year.

in the pelagic food web (e.g., lower median krill population size, higher median salp population size) and a consequent change in energy transfer to vertebrate krill predators (Loeb et al. 1997; Hewitt and Linen Low 2000).

In summary, the current consensus is that krill abundance is driven by recruitment, which in turn appears to be regulated by aspects of the physical and biological environment (i.e., bottom-up regulation). Does this represent the ecosystem prior to commercial whaling? Three points are important. First, if there was strong control of krill by their predators and those predators were removed, one would expect to see an increase in krill abundance until bottom-up forces came into effect. So, the fact that bottom-up forces are now considered to be significant regulators for krill does not necessarily contradict the idea of top-down regulation in the past. Second, the research supporting these conclusions was conducted decades after the peak periods of commercial whaling and largely in geographic areas where whales remain depleted. Therefore, these conclusions apply to a system without the influence of abundant large whales. Third, in localized areas where predator demand is near the carrying capacity of the system, there is evidence that krill may be regulated by top-down effects. We discuss this aspect in detail in the next section.

Large Whale Competitors and Competitive Release for Pinnipeds and Seabirds

In the following sections, we consider temporal trends in abundance for large whale competitors: pinnipeds and seabirds, in various regions of the Southern Ocean.

ICE SEALS: CYCLIC PATTERNS, NO LONG-TERM TRENDS

Ice seals are notoriously difficult to monitor, simply because they live in a habitat that is constantly changing and logistically difficult and expensive to access. Nevertheless, a few long-term data sets do exist. For example, Boveng and Bengtson (1997) compiled a 44-year time series of cohort strength for crabeater seals (*Lobodon carcinophaga*) in the Antarctic Peninsula region (Figure 17.3A). There is obvious year-to-year variation in the data abundance, with patterns of coherent fluctuation that suggest longer-term cycles, but no apparent long-term trends. Boveng and Bengtson (1997) concluded that these fluctuations were responses to genuine demographic events (as opposed to statistical noise), and speculated that environmental events affecting production and survival of young animals may be the cause, at least in part. Similar

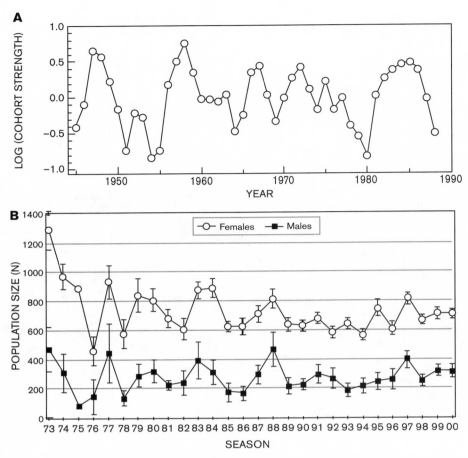

FIGURE 17.3. Temporal trends in abundance indices for ice seals in the Southern Ocean. (A) Cohort strength for crabeater seals in the Antarctic Peninsula region (from Boveng and Bengtson 1997). (B) Population estimates for Weddell seals in the Ross Sea (from Cameron and Siniff, 2004).

patterns seem to hold true for ice seals in other regions. A 27-year time series of population estimates for Weddell seals *(Leptonychotes weddellii)* in the Ross Sea also shows year-to-year variation (Cameron and Siniff 2004; Figure 17.3B), again suggestive of longer-term cycles but with no long-term trends. In this particular study, the cause of the cycles was suggested to be due to changes in immigration, and ultimately related to sea ice extent. The general consensus for data such as these is that these cyclic patterns are likely related to environmental cycles affecting sea ice extent, krill production, and, ultimately, prey abundance and availability. Again, it is worth noting that most of these studies of ice seals began well after the peak periods of commercial whaling, therefore their relevance to the ecosystem when it contained large numbers of whales is unknown.

PINNIPEDS AND SEABIRDS IN THE SOUTHERN INDIAN OCEAN AND EAST ANTARCTICA: POPULATION TRENDS CORRELATED WITH ENVIRONMENTAL CHANGES

The population sizes of nine species of seabirds and seals have been monitored at seven locations (island and continental) in the Southern Indian Ocean, with the time series

beginning in the 1950s in some cases. In a synthesis paper, Weimerskirch et al. (2003) found statistically significant correlations between population trends and environmental changes.

Of these nine species, seven showed marked declines beginning in the late 1960s and continuing through the 1970s (Figure 17.4A). These included Emperor, Adélie, and Rockhopper penguins (*Aptenodytes forsteri*, *Pygoscelis adeliae*, and *Eudyptes chrysochome*, respectively), Wandering and Black-browed albatross (*Diomedea exulans* and *Thalassarche melanophris*, respectively), and Southern Elephant seals *(Mirounga leonina)*. After the 1970s, some populations showed signs of recovery, some remained stable, and some continued to decline. At all seven locations, the mean annual air temperature increased from a minimum value during the 1960s until the mid-1980s, after which it stabilized at some locations and decreased at others (Figure 17.4B). The temperature increase was coincident with a decline in surface chlorophyll and zooplankton concentrations after the late 1970s.

Weimerskirch et al. (2003) interpreted these trends as evidence of environmental warming causing decreases in

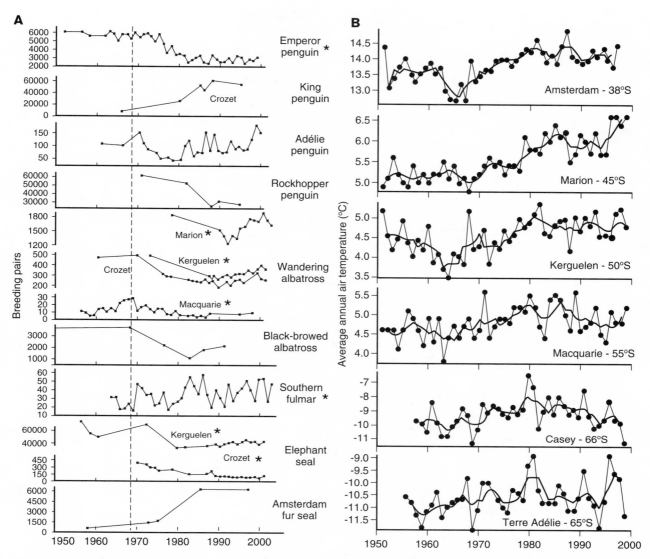

FIGURE 17.4. Temporal trends in population size and annual air temperature in the Southern Indian Ocean (from Weimerskirch et al. 2003; reprinted with the permission of Cambridge University Press). (A) Population size for nine species of seabirds and seals. The dashed vertical line indicates the approximate time when population changes began to occur. Asterisks indicate those populations for which a statistically significant correlation between abundance and temperature was found. (B) Changes in average annual air temperature at seven sites. Points and thin line represent average annual values; heavy line represents moving average over three years.

prey availability, with consequent decreases in predator populations. The existence of a statistically significant time lag between temperature increases and population declines for four species in five of the monitored sites supported the idea that these environmental changes affected fecundity, or recruitment, rather than adult survival.

Why two other populations, King penguin (*Aptenodytes patagonicus*) and Amsterdam fur seal (*Arctocephalus tropicalis*), showed increases in abundance during this warming trend was not clear. Weimerskirch et al. (2003) speculated that the warming trend may have favored the prey of these species (primarily myctophids, as opposed to squids, crustaceans, and other fish prey preferred by the seven declining predator species), or that both were exhibiting a recovery to pre-

exploitation levels following commercial harvest; both factors may be operating in combination.

PENGUINS IN THE ANTARCTIC PENINSULA REGION: COMPETITIVE RELEASE AND SUBSEQUENT EFFECTS OF ENVIRONMENTAL CHANGE

Increases in abundance of Chinstrap penguins (*Pygoscelis antarctica*) at sites on the Antarctic Peninsula, particularly between the 1930s and 1970s, have been interpreted as evidence in support of the krill surplus hypothesis (Sladen 1964; Emison 1968; Conroy 1975; Croxall and Kirkwood 1979; Croxall and Prince 1979; Croxall et al. 1984). More recent analyses suggest additional (but not necessarily mutually exclusive) interpretations.

A

PACK ICE EXTENT NORTH

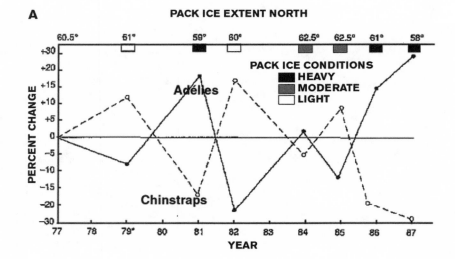

B

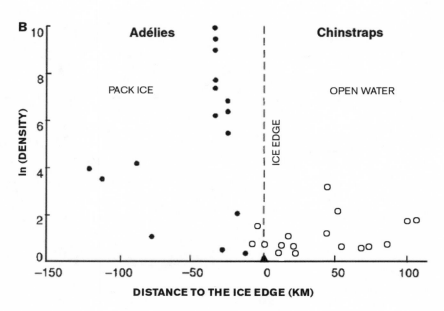

C

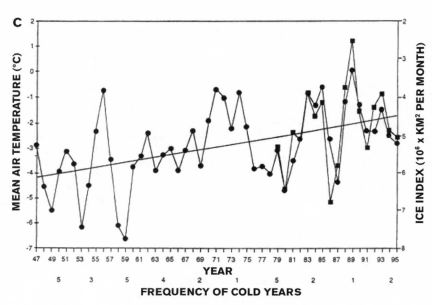

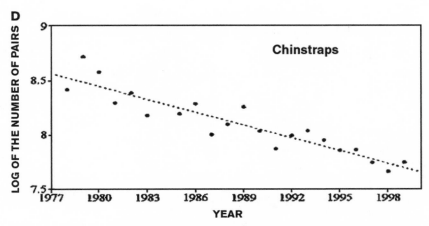

FIGURE 17.5. Temporal trends in population size of Adélie and Chinstrap penguins, and mean annual air temperature for the Antarctic Peninsula region. (A) Percent change in the number of Adélie (solid line) and Chinstrap (dashed line) breeding pairs on King George Island, Antarctica, as related to pack ice conditions during peak egg laying, and annual maximum pack ice extent (from Fraser et al. 1992). (B) Density and distribution of Adelie (closed circles) and Chinstrap (open circles) penguins relative to the ice edge in the Weddell Sea, Antarctica (from Fraser et al. 1992). (C) Long term trends in mean annual surface air temperature at Palmer Station, Antarctic Peninsula (circles), and regional sea-ice index (squares; modified from Loeb et al. 1997, with copyright permission from *Nature*). (D) Population trends for Chinstrap Penguins at Copacabana Colony, King George Island, Antarctic Peninsula (Trivelpiece and Trivelpiece, unpublished data).)

Trivelpiece et al. (1990) documented a 10-year time series where large annual increases in Adélie penguin populations at King George Island were coincident with large annual decreases in Chinstrap penguin numbers, and vice versa. These changes were also related to the extent of winter and spring sea ice cover, with Chinstraps increasing in years when sea ice extent diminished, and Adélies increasing when sea ice extent increased (Figure 17.5A). Although differences in prey type or abundance between the pack ice and pelagic wintering areas frequented by Adélie and Chinstrap penguins, respectively, are likely important factors (Trivelpiece et al. 1990), this converse relationship might ultimately be explained by habitat preferences, with Adélies being obligate pack ice penguins while Chinstraps prefer open water (Fraser et al. 1992; Ainley 2002a; Figure 17.5B).

The Antarctic Peninsula region has undergone a warming trend during the past 50 years (Fraser et al. 1992; King 1994; Loeb et al. 1997; Figure 17.5C). This warming trend has been coincident with a decreased frequency of cold years with extensive winter pack ice development (Fraser et al. 1992). These authors suggested that the increased abundance of Chinstrap penguins was due to an increase in their preferred habitat type, open water, rather than due to a surplus of krill.

In at least one study, Chinstrap penguin abundance has *decreased* during the past 20 years at several colonies on King George Island (Trivelpiece and Trivelpiece, unpublished data; Figure 17.5D). This decrease may be attributable to a warming trend, which would result in decreased krill recruitment and abundance, thereby negatively affecting populations of predators that rely on krill.

For penguins in the Antarctic Peninsula region then, the increase in abundance of two species, Chinstraps from the 1930s through the 1970s (Sladen 1964; Emison 1968; Conroy 1975; Croxall and Kirkwood 1979; Croxall and Prince 1979; Croxall et al. 1984), and Adélies in the 1950s and 1960s (Croxall et al. 2002), the fact that each has different foraging habitat requirements, and the timing of the increase, especially for Chinstraps, makes the Krill Surplus Hypothesis particularly compelling. Subsequently, these population patterns seem to have mainly been related to a warming trend and its effects on the krill prey base, largely mediated by changes in sea ice.

FUR SEALS AND PENGUINS IN THE SCOTIA SEA: COMPETITIVE RELEASE AS A RESULT OF THE LOSS OF LARGE WHALES

Since the 1950s, Antarctic fur seals have exhibited remarkable population growth. At South Georgia, the annual population growth rate was 16.6% during the 1950s and 1960s, 11.5% during the 1970s, and 9.8% during the 1980s; the breeding range of fur seals has also expanded to sites not previously occupied (Payne 1977; Croxall and Prince 1979; Boyd 1993; Figure 17.6A). Antarctic fur seals were hunted close to extinction during the first half of the nineteenth century and some 1.2 million were estimated to have been killed at South Georgia (Weddell 1825). The remarkable

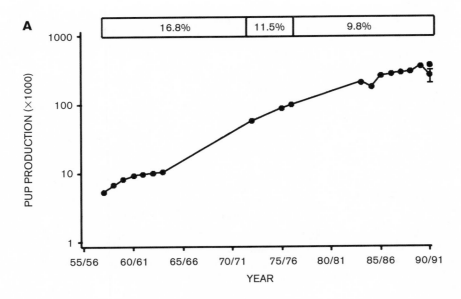

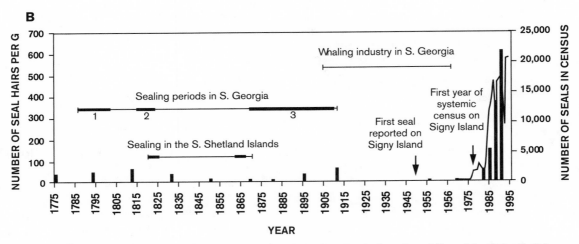

FIGURE 17.6. Temporal trends in abundance of Antarctic fur seals at South Georgia and Signy Island, South Orkney Islands. (A) Pup production of Antarctic fur seals at South Georgia, and annual population growth rate during three portions of the time series: 1950s and 1960s; 1970s; 1980s and 1990s (from Boyd 1993). (B) Number of Antarctic fur seal hairs in the upper 10 cm of a sediment core from Signy Island (vertical bars), seal census data (solid line), and sealing and whaling activities (horizontal bars; from Hodgson and Johnston 1997; with copyright permission from *Nature*).

growth of the population during the past 50 years could be continued recovery from this exploitation, competitive release due to the loss of baleen whales from the system, or a combination of both.

Hodgson and Johnston (1997) provide compelling evidence to suggest that competitive release has been a more significant factor for this species than post-sealing recovery. Sediment cores on Signy Island in the South Orkney Islands indicate that the number of visiting seals (measured by the number of fur seal hairs deposited in the sediments and closely correlated with seal census data) was historically low. In 1975, that number increased dramatically; by 1995 the number was estimated to be 78–94% greater than at any time during the past 6,570 radiocarbon years, and exceeding the range of natural variability (Figure 17.6B). The increase was not recorded until some 60 years after the end of sealing in

South Georgia, but did coincide with the end of whaling in that same area (Figure 17.6B).

At northwest South Georgia, including Bird Island, Reid and Croxall (2001) provide compelling evidence that, at least in this localized area and during the breeding season when foraging range is most restricted, a number of krill predators have reached the system's carrying capacity and have since affected the prey community. Populations of four krill predators, Antarctic fur seal, black-browed albatross, and Gentoo and Macaroni penguins (*Pygoscelis papua* and *Eudyptes chrysolophus*, respectively), were stable or increasing during the 1980s, but all four species declined during the 1990s (Figure 17.7A). Krill remained a key element in the diet of all species throughout the time series, but modal size taken by the fur seal and penguins declined, from 54–56 mm in the 1980s, to 42–44 mm in the 1990s (Figure 17.7B). The decline

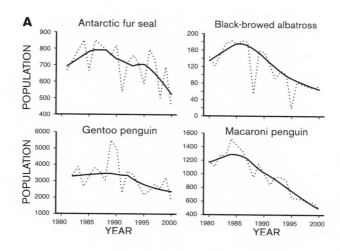

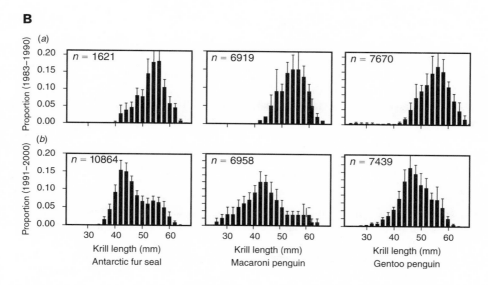

FIGURE 17.7. Temporal trends in population size and size of krill in the diet of krill predators at Bird Island, South Georgia (from Reid and Croxall 2001). (A) Population changes in four krill predators. Yearly values are represented by the dotted line; smoothed data by the solid line. (B) Length-frequency distribution of krill in the diet of Antarctic fur seals, and Gentoo and Macaroni penguins from (a) 1983 to 1990, and from (b) 1991 to 2000.

in the Black-browed albatross population was due, at least in part, to incidental mortality of birds from long-line fishing operations, but Reid and Croxall interpreted data for the other three predators to mean that the biomass of krill within the largest size class was sufficient to support predator demand in the 1980s, but not in the 1990s, with consequent effects on krill abundance and population structure. Their conclusion: "For krill-dependent predators at South Georgia the time of 'krill-surplus' . . . may now be at an end." (Reid and Croxall 2001). The significance of this, in light of the krill surplus hypothesis, is that, at least at South Georgia, there is evidence that krill predators can regulate krill abundance and size; the implication is that if those krill predators were removed, krill populations would respond with an increase in abundance and modal size.

Finally, there is evidence that direct competition may be occurring between at least two of these krill predators. Barlow et al. (2002) noted that, at Bird Island, Antarctic fur seals and Macaroni penguins both fed on krill of similar sizes, dove to similar depths, and were restricted in their foraging range while provisioning young (see references in Barlow et al. 2002), suggesting that they were exploiting the same prey population. On a year-to-year basis, there was variation in the proportion of the diet composed of krill, but the variation was greater in the diet of Macaroni penguins than it was for Antarctic fur seals, and the penguin diet always contained less krill (Figure 17.8A). Over a 12-year period, both predator populations declined, but the decline was considerably greater and significant for Macaroni penguins, while there was no statistically significant trend for fur seals (Figure 17.8B). They

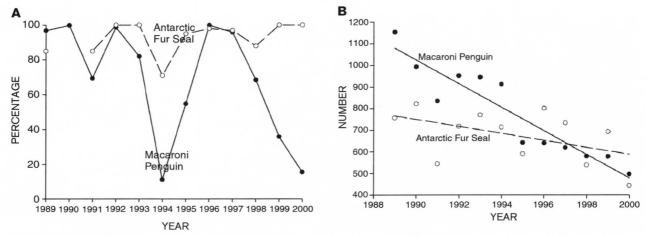

FIGURE 17.8. Temporal trends in population size and proportion of krill in the diet for two krill predators at Bird Island, South Georgia (from Barlow et al. 2002). (A) Percent krill in the diet of Macaroni penguins (filled circles and solid line) and Antarctic fur seals (open circles and dashed line). (B) Population changes for Macaroni penguins (filled circles and solid line) and Antarctic fur seals (open circles and dashed line).

concluded that these predator populations had reached the maximum that the prey resources could support, that the interaction between penguins, fur seals, and their mutual prey population was currently favoring the fur seals, and that considerable circumstantial evidence supported the idea of competition occurring between these two krill predators.

Large Whale Predators: A Loss of Prey for Killer Whales

The removal of large whales from the Southern Ocean should have resulted in a loss of prey for those species that consume large whales—killer whales. Is there any evidence that killer whale distribution or abundance in the Southern Ocean has changed, or that these predators have switched to other prey?

As is the case for large whale prey and their competitors, the fundamental problem is that there is little information on killer whales from pre-whaling times. Addressing this question is further complicated by the fact that historically, killer whales have been considered a single species with a worldwide distribution, whereas recent evidence confirms previous suggestions from Russian scientists that there are in fact three ecotypes of killer whales in the Southern Ocean which may comprise three distinct species (Pitman and Ensor 2003). Of these, only one ecotype, "Type A" (Pitman and Ensor 2003) is known to take baleen whales. The other two ecotypes are thought to specialize on pinnipeds ("Type B"), and Antarctic Toothfish, *Dissostichus mawsoni* ("Type C"; Pitman and Ensor 2003).

Given this, we can speculate as to possible cascade effects due to the removal of large whales. For example, currently no killer whales are known to prey mainly on large rorquals in Antarctica, but Type A killer whales appear to specialize on Antarctic minke whales. If the removal of large whales resulted in a krill surplus, then minke abundance may have increased as a result of competitive release (Antarctic minke whales prey primarily on euphausiids [Perrin and Brownell 2002]), and there may, therefore, be more minke whales available now for killer whales to eat. On the other hand, if Type

A killer whales primarily preyed on calves of larger whales, the removal of these large whales may have forced Type A killer whales to switch to minke whales, presumably less desirable prey. For the other types of killer whales, a krill surplus could have meant more prey for ice seals (and consequently, more prey for Type B killer whales), and possibly more toothfish (resulting in more prey for Type C killer whales).

Though the three killer whale ecotypes are easily distinguishable in the field based on body coloration, eyepatch shape and orientation, and, to a lesser extent, body size, group size, and habitat (Pitman and Ensor 2003), none of the published literature prior to 2003 distinguishes between the forms when reporting distribution, abundance, and prey preferences of Southern Ocean killer whales. Therefore, any trends in killer whale abundance and diet are confounded by the fact that they represent an unknown mix of the three ecotypes of killer whales.

Branch and Williams (Chapter 20, this volume) suggest that Southern Ocean killer whales have recently switched prey to southern elephant seals, as evidenced by declines in populations of this species in a number of locations throughout the Southern Ocean. Yet, other populations of southern elephant seals have remained stable over at least the past 45 years (Boyd et al. 1996). If killer whales are responsible for southern elephant seal declines, there remains the question of which ecotype(s) are taking them. Resolution of this, and other questions, must await further analysis of future trends, and, perhaps, efforts to resolve past datasets.

The Ross Sea: Where the Answer Lies

Ainley (2002b) makes a compelling case that the Ross Sea Shelf and Slope areas may be the last large marine ecosystems in the Southern Ocean where investigations of top-down forcing without the confounding effects of anthropogenic exploitation may be undertaken. The Ross Sea Shelf and Slope areas are unique amongst Southern Ocean ecosystems in that they have largely escaped anthropogenic alteration. Other than a recent and small-scale experimental fishery for Antarctic Toothfish on the shelf, and a scientific take of Antarctic minke whales mostly along the slope, there has been no direct human influence. Of most significance, with respect to the question posed by this paper, is that the great whales have never been a part of these ecosystems. Thus, their almost complete removal from the Southern Ocean had negligible effects here.

The Ross Sea hosts roughly 38% and 24% of the world population of Adélie and Emperor penguins, respectively; several million Antarctic Petrels (*Thalassoica antarctica*); 45%, 11%, and 12% of the Pacific sector populations of Weddell, Leopard (*Hydrurga leptonyx*) and Crabeater (*Lobodon carcinophagus*) seals, respectively; and on the order of 14,000 minke and 3500 killer whales (Ainley 2002b). Like deeper waters of the Southern Ocean, *Euphausia superba* is the principal prey species for top trophic-level predators in the Ross Sea Slope Ecosystem, but the most important middle-trophic level species in the Ross Sea Shelf Ecosystem are Crystal Krill (*Euphausia crystallorophias*) and Antarctic Silverfish (*Pleuragramma antarcticum*; Ainley 2002b).

As is the case in most other Southern Ocean ecosystems, a major gap in our understanding concerns the coupling between lower- and upper-trophic components. A number of lines of evidence, however, indicate that top-down regulation may be occurring here. As the breeding season progresses and chicks, and their demands for food, grow, Adélie penguins switch prey from krill to silverfish, travel farther from the colony to feed, forage at greater depths, take longer trips, and bring back smaller food loads, while penguins from larger colonies appear to exclude those from smaller colonies through increased density (Ainley et al. 1998; Ballard et al. 2001, 2002; Ainley et al. 2004). Weddell seals may deplete Antarctic Toothfish within breath-hold distance of their breeding areas and switch to smaller prey, such as silverfish (Ainley 2002b). Finally, minke whales killed through scientific whaling have enigmatically low body mass (Ichii et al. 1998), and Ainley (2002b) cites this as further evidence that top-down forcing may occur here.

During the 1970s, and especially from the 1980s to the present, Adélie penguin populations on Ross Island have been growing (Wilson et al. 2001). Part of the variability in annual population size is related to sea ice extent during winter (Wilson et al. 2001) and changes in the formation and decay of sea ice and polynyas are likely to be involved as well (Ainley 2002b). However, the possibility that the removal of several hundred minke whales annually from the Ross Sea Slope Ecosystem may have significantly increased penguin foraging success through competitive release cannot be discounted (Ainley 2002b).

The Effect of the Removal of Large Whales on the Southern Ocean Ecosystem

Antarctic marine systems have suffered major perturbation through hunting and fishing over the last two centuries, particularly amongst top predators. . . . The consequential changes in these ecosystems remain unknown or inferred. . . . (Croxall et al. 2002)

A great deal remains unknown. Ice seals are major krill predators, yet their past and present status and much of their ecology remains enigmatic. Squids are major components of the mesopelagic ecosystem and may consume significant quantities of krill; little is known about their distribution, abundance, and ecology, past or present. Nutrient cycles in the Southern Ocean ecosystem were likely altered by the loss of large whales, but little is known regarding the significance of fecal input from whales, or of the fact that krill, once removed from the system through whale predation, now remain. The effect of the krill fishery, a krill predator that was not present during pre-whaling times, is also a confounding factor in interpreting post-whaling trends. (In the Scotia Sea alone, some 80,000 tons of krill are harvested annually [Nicol and Endo 1999]). Finally, data used in ecological studies rarely extend back to pre-whaling, or even to commercial whaling times; in other words, baseline data are extremely difficult to come by. This is perhaps the most fundamental of problems, making it difficult to establish a cause and effect relationship simply because we know little about the system when it included an additional 1.8 million whales.

Despite this uncertainty, we do know some things. We know that krill seem to be largely regulated by bottom-up processes at the present time, and that this does not preclude the existence of top-down regulation in the past. Indeed, there is evidence that top-down regulation is occurring in localized areas. Specifically, when krill predators are near the carrying capacity of the ecosystem, they seem to change the abundance and population structure of their prey by increasing adult mortality, with a consequential decrease in modal krill age, and therefore, modal krill size.

We know that data support the idea of competitive release for some krill predators as a result of the loss of great whales from the system. In particular, Antarctic fur seals in the Scotia Sea have exhibited remarkable population growth, coincident with cessation of commercial whaling. This has seemingly negatively affected the success of other krill predators, especially Macaroni penguins, implying that competition for krill may currently be occurring, and therefore, may have occurred in the past.

We know that a great many krill predators are experiencing population changes due to long-term warming trends in the Southern Indian Ocean, East Antarctic, and Antarctic Peninsula regions, and that this environmental change is correlated with recruitment and, ultimately, abundance of krill. Other krill predators are recovering from commercial exploitation. These factors have complicated, perhaps overshadowed, the effects of the removal of large whales from the Southern Ocean in many cases (reviewed by Croxall et al. 2002).

Finally, and perhaps most significantly, we know that we have yet to understand the Southern Ocean ecosystem during pre-whaling times. Environmental and ecological changes documented in the studies have occurred over time scales at which we have been able to observe and measure them, whereas influences from whaling are less visible from the same perspective because the whales were essentially gone before data could be collected.

The ecosystem effect of the removal of large whales from the Southern Ocean continues to be an intriguing question, and one that, with future research, has the potential to be better resolved. Additional research on Antarctic food web dynamics and continued population monitoring has the possibility of allowing us, at some time in the future, to tease apart the tangled effects of post-whaling ecosystem alteration and climate change. Modeling approaches and attempts to uncover historical data will no doubt provide insights. Finally, a crucial key in understanding the significance of top-down forcing is to study a system where the top trophic predators are still relatively unexploited.

Acknowledgments

We wish to thank David Ainley for sharing his wealth of Antarctic experience through ideas and reviews of a number of versions of this paper. The paper benefited greatly from the comments of Jim Estes, Dan Doak, and one anonymous reviewer.

Literature Cited

Ainley, D.G. 2002a. *The Adélie penguin: bellwether of climate change*. New York: Columbia University Press.

———. 2002b. The Ross Sea, Antarctica, where all ecosystem processes still remain for study, but maybe not for long. *Marine Ornithology* 30(2): 55–62.

Ainley, D.G., C.A. Ribic, G. Ballard, S. Heath, I. Gaffney, B.J. Karl, K.R. Barton, P.R. Wilson, and S. Webb. 2004. Geographic structure of Adélie penguin populations: overlap in colony-specific foraging areas. *Ecological Monographs* 74(1): 159–178.

Ainley, D.G., P.R. Wilson, K.J. Barton, G. Ballard, N. Nur, and B.J. Karl. 1998. Diet and foraging effort of Adélie penguins in relation to pack-ice conditions in the southern Ross Sea. *Polar Biology* 20: 311–319.

Baker, C.S., G.M. Lento, F. Cipriano, M.L. Dalebout, and S.R. Palumbi. 2000. Scientific whaling: source of illegal products for market? *Science* 290: 1695.

Ballard, G., D.G. Ainley, J. Adams, K. Barton, S. Heath, M.C. Hester, B.J. Karl, H. Nevins, and S. Webb. 2002. Adélie penguin foraging behavior: variation depending on breeding season, colony, and individual. *Pacific Seabirds* 29: 30.

Ballard, G., D.G. Ainley, C.A. Ribic, and K.J. Barton. 2001. Effect of instrument attachment and other factors on foraging trip duration and nesting success of Adélie penguins. *Condor* 103: 481–490.

Barlow, K.E., I.L. Boyd, J.P. Croxall, K. Reid, I.J. Staniland, and A.S. Brierley. 2002. Are penguins and seals in competition for Antarctic krill at South Georgia? *Marine Biology* 140: 205–213.

Boveng, P.L. and J.L. Bengtson. 1997. Crabeater seal cohort variation: demographic signal or statistical noise? in *Antarctic communities: species, structure, and survival*. B. Battaglia, J. Valencia,

and D.W.H. Walton, eds. Cambridge, UK: Cambridge University Press, pp. 241–247.

Boyd, I.L. 1993. Pup production and distribution of breeding Antarctic fur seals *(Arctocephalus gazella)* at South Georgia. *Antarctic Science* 5(1): 17–24.

———. 2002. Antarctic marine mammals, in *Encyclopedia of marine mammals*. W.F. Perrin, B. Würsig, and J.G.M. Thewissen eds. San Diego: Academic Press, pp. 30–36.

Boyd, I.L., T.R. Walker, and J. Poncet. 1996. Status of southern elephant seals at South Georgia. *Antarctic Science* 8(3): 237–244.

Brierley, A.S., D.A. Demer, J.L. Watkins, and R.P. Hewitt. 1999. Concordance of interannual fluctuations in acoustically estimated densities of Antarctic krill around South Georgia and Elephant Islands: biological evidence of same-year teleconnections across the Scotia Sea. *Marine Biology* 134(4): 675–681.

Brownell, R.L., Jr. and A.V. Yablokov. 2002. Illegal and pirate whaling, in *Encyclopedia of marine mammals*. W.F. Perrin, B. Würsig, and J.G.M. Thewissen, eds. San Diego: Academic Press, pp. 608–612.

Cameron, M.F. and D.B. Siniff. 2004. Age-specific survival, abundance, and immigration rates of a Weddell seal *(Leptonychotes weddellii)* population in McMurdo Sound, Antarctica. *Canadian Journal of Zoology* 82: 601–615.

Clapham, P.J. and C.S. Baker. 2002. Whaling, modern, in *Encyclopedia of marine mammals*. W.F. Perrin, B. Würsig, and J.G.M. Thewissen, eds. San Diego: Academic Press, pp. 1328–1332.

Clapham, P.J. and L.T. Hatch. 2000. Determining spatial and temporal scales for population management units: lessons from whaling. Paper SC/52/SD2 submitted to the International Whaling Commission Scientific Committee.

Conroy, W.H. 1975. Recent increases in penguin populations in Antarctica and the subantarctic, in *The biology of penguins*. B. Stonehouse, ed. London: Macmillan, pp. 321–335.

Croxall, J.P. and E.D. Kirkwood. 1979. *The breeding distribution of penguins on the Antarctic Peninsula and islands of the Scotia Sea*. Cambridge, UK: British Antarctic Survey.

Croxall, J.P. and P.A. Prince. 1979. Antarctic seabird and seal monitoring studies. *Polar Record* 19: 573–595.

Croxall, J.P., P.A. Prince, I. Hunter, S.J. McInnes, and P.G. Copestake. 1984. The seabirds of the Antarctic Peninsula, islands of the Scotia Sea and Antarctic Continent between 80°W and 20°W: their status and conservation, in *Status and conservation of the world's seabirds*. J.P. Croxall, P.G.H. Evans, and R.W. Scheiber, eds. Cambridge: International Council for Bird Preservation, pp. 635–664.

Croxall, J.P., P.N. Trathan, and E.J. Murphy. 2002. Environmental change and Antarctic seabird populations. *Science* 297: 1510–1514.

Emison, W.B. 1968. Feeding preferences of the Adélie penguin at Cape Crozier, Antarctica in *Antarctic bird studies*. O.L. Austin, ed. Antarctic Research Series 12. Washington, DC: American Geophysical Union, pp. 191–212.

Fraser, W.R. and E.E. Hofmann. 2003. A predator's perspective on causal links between climate change, physical forcing and ecosystem response. *Marine Ecology Progress Series* 265: 1–15.

Fraser, W.R., W.Z. Trivelpiece, D.G. Ainley, and S.G. Trivelpiece. 1992. Increases in Antarctic penguin populations: reduced competition with whales or a loss of sea ice due to environmental warming? *Polar Biology* 11: 525–531.

Gambell, R. 1968. Seasonal cycles and reproduction in sei whales of the Southern Ocean. *Discovery Reports* 35: 31–134.

———. 1973. Some effects of exploitation on reproduction in whales. *Journal of Reproductive Fertility (Supplement)* 19: 533–553.

———. 1999. The International Whaling Commission and the contemporary whaling debate, in *Conservation and management of marine mammals*. J.R. Twiss, Jr. and R.R. Reeves, eds. Washington, DC: Smithsonian Institution Press, pp. 179–198.

Hewitt, R.P. 1997. Areal and seasonal extent of sea ice cover off the northwestern side of the Antarctic Peninsula: 1979 through 1995. *CCAMLR Science* 4: 65–73.

Hewitt, R.P., D.A. Demer, and J.H. Emery. 2003. An eight-year cycle in krill biomass density inferred from acoustic surveys conducted in the vicinity of the South Shetland Island during the austral summers of 1991/92 through 2001/02. *Aquatic Living Resources* 16(3): 205–213.

Hewitt, R.P. and E.H. Linen Low. 2000. The fishery on Antarctic krill: defining an ecosystem approach to management. *Reviews in Fisheries Science* 8(3): 235–298.

Hewitt, R.P. and J.D. Lipsky. 2002. Krill, in *Encyclopedia of marine mammals*. W.F. Perrin, B. Würsig, and J.G.M. Thewissen, eds. San Diego: Academic Press, pp. 676–684.

Hodgson, D.A. and N.M. Johnston. 1997. Inferring seal populations from lake sediments. *Nature* 387: 30–31.

Ichii, T., N. Shinohara, Y. Fujise, S. Nisiwaki, and K. Matsuoka. 1998. Interannual changes in body fat condition index of minke whales in the Antarctic. *Marine Ecology Progress Series* 175: 1–12.

IWC. 1995. Southern hemisphere catch data coding: position at 1 July 1994. *Reports of the International Whaling Commission* 45: 129–130.

King, J.C. 1994. Recent climate variability in the vicinity of the Antarctic Peninsula. *International Journal of Climatology* 14: 357–369.

Laws, R.M. 1962. Some effects of whaling on the southern stocks of baleen whales, in *The exploitation of natural animal populations*. E.D. Le Cren and M.W. Holdgate, eds. Oxford: Blackwell Press, pp. 137–158.

———. 1977. Seals and whales of the Southern Ocean. *Philosophical Transactions of the Royal Society of London Series B: Biological Sciences* 279: 81–96.

Lockyer, C.H. 1972. The age of sexual maturity of the southern fin whale *(Balaenoptera physalus)* using annual layer counts in the ear plug. *Journal du Conseil International pour L'Exploration de la Mer* 34: 276–294.

———. 1974. Investigation of the ear plug of the southern sei whale *Balaenoptera borealis* as a valid means of determining age. *Journal du Conseil International pour L'Exploration de la Mer* 36: 71–81.

Loeb, V., V. Siegel, O. Holm-Hansen, R. Hewitt, W. Fraser, W. Trivelpiece, and S. Trivelpiece. 1997. Effects of sea-ice extent and krill or salp dominance on the Antarctic food web. *Nature* 387: 897–900.

Mackintosh, N.A. and S.G. Brown. 1974. Whales and whaling, in *Antarctic Mammals, Folio 18, Antarctic Map Folio Series*. V.C. Bushnell, ed. New York: American Geographical Society, pp. 2–4.

Miller, D.G.M. and I. Hampton. 1989. Biology and ecology of the Antarctic krill. *Biomass Science Series* 9: 1–166.

Naganobu, M., K. Katsuwada, Y. Sasai, S. Taguchi, and V. Siegel. 1999. Relationships between Antarctic krill *(Euphausia superba)* variability and westerly fluctuations and ozone depletion in the Antarctic Peninsula area. *Journal of Geophysical Research* 104(C9): 20651–20665.

Nicol, S.T. and Y. Endo. 1999. Krill fisheries development, management and ecosystem implications. *Aquatic Living Resources* 12: 105–120.

Nicol, S.T., T. Pauly, N.L. Bindoff, S. Wright, D. Thiele, G.W. Hosie, P.G. Strutton, and E. Woehler. 2000. Ocean circulation off east Antarctica affects ecosystem structure and sea-ice extent. *Nature* 406: 504–507.

Payne, M.R. 1977. Growth of a fur seal population. *Philosophical Transactions of the Royal Society of London Series B: Biological Sciences* 279: 67–79.

Perrin, W.F. and R.L. Brownell, Jr. 2002. Minke whales, in *Encyclopedia of marine mammals*. W.F. Perrin, B. Würsig, and J.G.M. Thewissen, eds. San Diego: Academic Press, pp. 750–754.

Pitman, R.L. and P. Ensor. 2003. Three forms of killer whales *(Orcinus orca)* in Antarctic waters. *Journal of Cetacean Research and Management* 5(2): 131–139.

Reid, K. and J.P. Croxall. 2001. Environmental response of upper trophic-level predators reveals a system change in an Antarctic marine ecosystem. *Proceedings of the Royal Society of London Series B: Biological Sciences* 268: 377–384.

Reid, K., E.J. Murphy, V. Loeb, and R.P. Hewitt. 2002. Krill population dynamics in the Scotia Sea: variability in growth and mortality within a single population. *Journal of Marine Systems* 36: 1–10.

Rootes, D.M. 1988. The status of birds at Signey Island, South Orkney Islands. *British Antarctic Survey Bulletin* 80: 87–119.

Siegel, V., V. Loeb, and J. Gröger. 1998. Krill *(Euphausia superba)* density, proportional and absolute recruitment and biomass in the Elephant Island region (Antarctic Peninsula) during the period 1977 to 1997. *Polar Biology* 19: 393–398.

Sladen, W.J.L. 1964. The distribution of the Adélie and Chinstrap penguins, in *Biologie antarctique*. R. Carrick, M.W. Holdgate, and J. Prevost, eds. Paris: Hermann, pp. 359–365.

Tormosov, D.D., Y.Z. Mikhalev, P.B. Best, V.A. Zemsky, K. Sekiguchi, and R.L. Brownell, Jr. 1998. Soviet catches of southern right whales, *Eubalaena australis*, 1951–1971: biological data and conservation implications. *Biological Conservation* 86: 185–197.

Trivelpiece, W.Z., S.G. Trivelpiece, G. Gevpel, J. Kjelmyr, and N.S. Volkman. 1990. Adélie and Chinstrap penguins: their potential as monitors of the Southern Ocean marine ecosystem, in *Antarctic ecosystems: ecological change and conservation*. K.R. Kerry and G. Hempel, eds. Berlin: Springer-Verlag, pp. 191–202.

Vaughan, D.G. and S.M. Doake. 1996. Recent atmospheric warming and retreat of ice shelves on the Antarctic Peninsula. *Nature* 379: 328–331.

Weddell, J. 1825. *A voyage towards the South Pole, performed in the years 1822–24*. London: Longman and Hurst.

Weimerskirch, H., P. Inchausti, C. Guinet, and C. Barbraud. 2003. Trends in bird and seal populations as indicators of a system shift in the Southern Ocean. *Antarctic Science* 15(2): 249–256.

White, W.B. and R.G. Peterson. 1996. An Antarctic circumpolar wave in surface pressure, wind, temperature, and sea-ice extent. *Nature* 380: 699–702.

Whitehead, H. 2002. Sperm whale, *Physeter macrocephalus,* in *Encyclopedia of marine mammals.* W.F. Perrin, B. Würsig, and J.G.M. Thewissen, eds. San Diego: Academic Press, pp. 1165–1172.

Williams, T.D. 1995. *The penguins.* Oxford: Oxford University Press.

Wilson, P.R., D.G. Ainley, N. Nur, S.S. Jacobs, K.J. Barton, G. Ballard, and J.C. Comiso. 2001. Adélie penguin population change in the Pacific sector of Antarctica: relation to sea-ice extent and the Antarctic Circumpolar Current. *Marine Ecology Progress Series* 213: 301–309.

Yablokov, A.V. 1994. Validity of whaling data. *Nature* 367: 108.

Yablokov, A.V., V.A. Zemsky, Y.A. Mikhalev, V.V. Tormosov, and A.A. Berzin. 1998. Data on Soviet whaling in the Antarctic in 1947–1972 (population aspects). *Russian Journal of Ecology* 29: 38–42.

Zemsky, V.A., A.A. Berzin, Y.A. Mikhalev, and D.D. Tormosov. 1995. Soviet Antarctic pelagic whaling after WWII: review of actual catch data. *Reports of the International Whaling Commission* 45: 131–135.

EIGHTEEN

Great Whales as Prey

Using Demography and Bioenergetics to Infer Interactions in Marine Mammal Communities

DANIEL F. DOAK, TERRIE M. WILLIAMS, AND JAMES A. ESTES

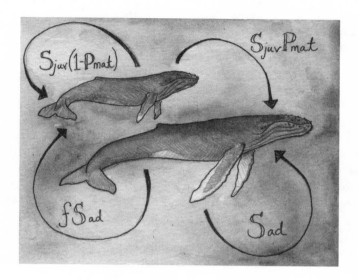

An increasing difficulty in many areas of conservation is the need to formulate and test theories about biological processes and management outcomes that occur over vast areas. Unfortunately, these large-scale problems are poorly matched with most research done in the name of conservation. Academic, and even nonacademic, conservation research typically focuses on one or a few species, often with detailed data, and usually at small scales, even as the scale of decision making and management undertaken by government agencies and nongovernmental organizations (NGOs) becomes ever larger. While many have decried this divorce between real-world conservation problems and conservation research, the best way to solve the problem is much less clear. Perhaps in no field of conservation ecology is this mismatch of scale more problematic than in the study and management of marine mammals, many of which have wide geographic ranges and long migrations and all of which are exceedingly difficult to observe in nature.

One way to address big-scale problems and processes is to concentrate on the collection of broad-scale data, including aerial surveys and remotely sensed oceanic productivity. While the advent of radio and satellite tagging of individual animals has helped to bring this type of data collection closer to the animals that are actually of management interest, the problem with most large-scale data is the difficulty of inferring the processes that influence the fates of individuals from data about factors many steps removed from their lives. Another way to bridge this gap is to use small-scale data to construct and test mechanistic models of events at higher levels of organization and at larger spatial scales. Even though this general approach to understanding ocean ecosystems underpins much of biological oceanography, it has not been so commonly used in efforts to understand the population and community dynamics of marine mammals, perhaps because of a reluctance to extrapolate or even hypothesize about events that are difficult to observe directly.

Here, we take the second of these approaches, employing demographic and bioenergetic tools to explore difficult-to-observe predator-prey interactions in marine mammal communities. Our specific goal is to understand better the interactions of mammal-feeding killer whales (*Orcinus orca*) with their prey species in the North Pacific region, particularly in the vicinity of the Aleutian archipelago. In the past, we and others have used this same tactic to test whether killer whales could plausibly have driven declines of sea otters (*Enhydra lutris*) (Estes et al. 1998) and Steller sea lions (SSLs, *Eumetopias jubatus*) (Springer et al. 2003; Williams et al. 2004) in western Alaska. Although the theory that killer whales may be a major driver of these declines remains controversial (Reeves et al., Chapter 14 in this volume; Clapham and Link, Chapter 24 in this volume; Branch and Williams, Chapter 20 in this volume), the least analyzed aspect of the scenario we have proposed is that human depredations of the great whales could have started this chain of events by removing a substantial source of killer whale food. After reviewing the bioenergetic-demographic approach that we have taken to understand the sea otter and sea lion declines, we develop demographic models to ask how many dead great whales would be produced by historic and current populations of great whales. These estimates are valuable not only for questions of predator-prey interactions but also in understanding the outputs of great whale populations (in terms of corpses) for other oceanic processes (Smith, Chapter 22 in this volume). However, we concentrate here on what these results imply about the number of killer whales that could have been supported by past and current great whale populations in the North Pacific, bringing in bioenergetic methods to make this link. We end by discussing the implications of our work for future research into the structure of marine mammal food webs.

The Predation Explanation for Sequential Marine Mammal Declines in the North Pacific

The idea that some common cause could link declines of several nearshore marine mammals in the Aleutian Islands of western Alaska arises from a simple plot of estimated catches of large whales and the numbers of pinnipeds and sea otters through time for the populations in this region (Springer et al. 2003, Figure 1). Although data are more limited for some of these species than for others (e.g., harbor seals [*Phoca vitulina*] vs. SSLs), the sequential timing of population collapses nonetheless suggests a predator that has steadily broadened its host range to include progressively less desirable prey, as would be predicted by general optimal foraging models (Mangel and Wolf, Chapter 21 in this volume). In contrast, this pattern of declines is not easily generated by the effects of a changing abiotic environment, which would be unlikely to strike different mammal species at different times and over such a protracted period, and with no evident recovery of populations after their initial collapse.

If the patterns of declines suggest that the actions of a predator might have caused these changes, the only remotely plausible candidate for this role is the killer whale. Killer whales are known predators of Steller sea lions, harbor seals, and sea otters (Jefferson et al. 1991; Weller 2002; Reeves et al., Chapter 14 in this volume), and they are relatively abundant in the vicinity of the Aleutians (Springer et al. 2003; Forney and Wade, Chapter 12 in this volume). Furthermore, by accounting for the difficulty of observing such events and the hours of observation made before and during the decline period, Estes et al. (1998) found statistically significant increases in observed attacks on sea otters in the Aleutians

during this population's collapse. Comparable data that could more definitively and directly link increased killer whale depredation to SSL and harbor seal declines do not exist, although killer whale predation on both of these species is not an uncommonly observed event (National Research Council 2003).

Although predation by killer whales on the two species with the best documented declines (SSLs and sea otters) is unquestionable based on simple observational data, the plausibility of predation as the *driver* of these declines must be evaluated more carefully. Note that we use the determination of *plausibility* here, rather than proof, as our goal. Very rarely do ecological data—especially on difficult-to-see events such as predation—allow unambiguous *proof* or *refutation* of a theory. Rather, we are almost always left assessing the plausibility of different explanations, sometimes on the basis of statistical significance, Akaike information criterion (AIC) values, or other formal measures of the strength of an explanation, and sometimes based only on expert opinion. Ideally, one should simultaneously estimate and compare the relative plausibilities (or likelihoods) of multiple explanations (Hilborn and Mangel 1997; Burnham and Anderson 1998) in order to gauge the relative support for each. We address only the plausibility of predation as an explanation of prey declines here, but we return to this theme in the discussion.

To assess whether killer whales could plausibly have driven the declines in abundance of other marine mammal populations seen in the vicinity of the Aleutians, we first built pre-decline demographic models for sea otters and SSLs, based on published age-based birth and death rate estimates for these species (see Williams et al. 2004). We then asked how much higher mortality rates must have been to generate the speed and extent of declines seen for each species. For the sea otter, we had only enough data to estimate an average annual increase in mortality, whereas for SSLs enough data exist during the declines to estimate changes in this increased mortality rate through the decline and also the possible age dependence of increased mortalities (Williams et al. 2004; see also Holmes and York's [2003] similar work on SSL declines). Finally, these estimates were turned into the numbers of excess deaths needed to generate the observed population declines. While we used these estimated numbers of deaths to evaluate the possibility of killer whale predation, they are also the basic problem that any hypothesis for these declines must explain: What caused the excess deaths that led to the observed population declines?

To ask whether killer whale predation could plausibly have created these excess deaths, we used bioenergetic models and data to estimate the average daily caloric needs of male and female killer whales. Along with the caloric yield of a SSL or sea otter, we could then assess whether a plausible number of killer whales could have caused the observed declines. It is easiest to express the results as the number of killer whales that, eating nothing but either otters or SSLs, would be needed to generate the declines. These turn out to be roughly 5–6 whales per year for the sea otters. For SSLs, the maximum

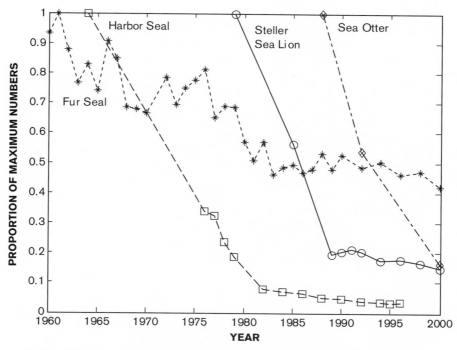

FIGURE 18.1. Sequential declines of pinnipeds and sea otters in Aleutians and Pribilof regions (modified from Springer et al. 2003). Data are shown as percentages of the maximum annual estimated population. See Springer et al. (2003) for sources and spatial extents of population estimates.

number of killer whales needed was 49 in the first year of the decline, an average of 29 over the first 5 years, and many fewer in subsequent years (Williams et al. 2004). However, given the expected foraging behavior of killer whales or any other predator, it is more likely that otters and SSLs did not become the sole food of any whales but, rather, were added to the diet of whales still feeding on other species. In the case of the SSLs, the dietary shift needed in an estimated 119 to 620 mammal-eating killer whales in the area (based on 95% confidence limits for the number of killer whales in this area and a guess as to the percentage of mammal-eaters [P. Wade, personal communication]; see Springer et al. 2003 and Forney and Wade, Chapter 12 of this volume, for this and other estimates of mammal-eating whale numbers) is modest: only a mean of 11% (range of 5%–26%) of the diet of this many whales would have to consist of SSLs to consume the average number of SSLs lost annually over the 5 steepest years of the decline.

Our merging of demographic estimates (to know how much mortality was needed to cause the declines) and bioenergetic methods (to know how many prey animals a killer whale needs to eat) suggests that predation rates required to generate the observed sea otter and SSL declines are not at all implausible. That is, the estimated numbers of killer whales—the only predator known to feed commonly on these two species in this area—are easily capable of killing and consuming the number of missing animals and thus generating the observed populationwide declines. In contrast, other possible explanations, evaluated with different types of data, do

not, in our view, appear likely to be the primary drivers of these declines (Estes et al 1998; Williams et al. 2004; National Research Council 2003; but see Trites and Donnelly 2003; Holmes and York 2003).

But if killer whales could cause declines in SSLs and sea otters (and other pinnipeds as well; see Figure 18.1), why did they only start when they did to feed on these species at high enough rates to cause declines? This brings up the most controversial part of our theory: that depletion of great whales by mechanized, industrial whaling in the mid-twentieth century removed a large and important food source for killer whales in the North Pacific, and that this in turn forced killer whales to broaden their prey choices, taking pinnipeds and sea otters at unsustainable rates. There are many aspects to the evaluation of this theory, some of which are taken up in detail in other chapters of this volume (Reeves et al., Chapter 14; Forney and Wade, Chapter 12). Here, we focus on the same demographic-energetic approach we have taken before, asking how many dead animals the great whale populations in this region produced before and after twentieth-century whaling, and how many killer whales this production might have supported then and now. In generating these estimates, we use a variety of information of varying certainty. However, one thing we do not assume (and have not previously assumed) is that great-whale populations were *controlled* by killer whale predation (see also Whitehead, Chapter 25 in this volume). That is, although we do suggest that great whales may have provided a substantial food source for killer whales, it is not necessary to assume that killer whales controlled great-whale numbers.

TABLE 18.1

Current and Pre-Exploitation Estimates of Great-Whale Numbers in the North Pacific

Species	Current Estimates[a]		Historical Estimates[a]	Stocks
	Low	High		
Blue	3,000	4,000	4,900	N. Pacific
Bryde's	30,000	30,000	39,000	N. Pacific
Fin	4,051	4,051	43,500	NE Pacific
Gray	14,322	21,174	28,240	Eastern N. Pacific
Minke	1,813	1,813	1,813	Alaska
Sei	9,110	9,110	42,000	N. Pacific
Humpback	4,005	4,005	9,060	Central N. Pacific
Bowhead	7,800	12,900	15,000	W. Arctic
Northern right	50	50	31,750	Eastern N. Pacific
Sperm	15,000	15,000	195,000	N. Pacific

[a] Estimates for blue and sei whales from Perry et al. (1999); for Bryde's estimates, see Croll (Chapter 16 in this volume); all other estimates are from Pfister and DeMaster (Chapter 10 in this volume).

Indeed, in forming our more conservative estimates of great-whale production, we explicitly assume that predation could not have been the force that controlled whale numbers historically.

Numbers and Life Histories of Common Whales of the North Pacific

Inferences about the natural production of deaths by populations of great whales must rely on two sets of estimates: population sizes (especially the historically stable population sizes, or carrying capacities) and the birth and death rates that govern the fates of individuals. Used together, these can give estimates of the number of animals dying each year—the production of the populations. We made production calculations using estimates of both current and pre-exploitation numbers of sperm and baleen whales for the North Pacific region and, where available, the region of the Aleutians in particular (Table 18.1). The ten species for which we have estimated numbers represent the majority of great-whale numbers in the North Pacific now and in the past.

Although the population estimates we have used are the best available of which we are aware, several caveats should be noted about the quality and interpretation of these data. First, the regions, or stocks, to which these estimates pertain are highly variable, ranging from the Bering Sea to the entire North Pacific Ocean (Table 18.1); we have used the most restrictive estimates available to tie our analyses to the Western Alaska/Aleutians region, but for the most part the population estimates we use are for much larger regions. We have used Pfister and DeMaster's (Chapter 10 in this volume) estimates whenever possible, but since they do not include data for blue, sei, or Bryde's whales, all of which were targets of twentieth-century whaling, we have sought other sources for these species. Second, most of these population estimates are

of unknown accuracy. This possible inaccuracy is especially important for sperm whales (Pfister and DeMaster, Chapter 10 in this volume), whose numbers have a large influence on our results. Third, the timing of whaling and the extent of subsequent population recovery differ among species. Some whale stocks were decimated long before the middle of the twentieth century (e.g., right whales), and others have recovered in numbers substantially since their nadir (e.g., gray whales). We do not believe that these effects are likely to change our overall results in any substantial way, and, to avoid presenting a horrible mire of different results, we have simply used prewhaling estimates and the most recent current population estimates for all species. However, the reader should be aware that our contrasts between prey resources available pre- and postwhaling factor in the whole history of whale removal; we do not try to parse out only the whaling conducted in the last years of these activities.

Full demographic schedules (birth and death rates by age or age class) have not been accurately estimated for most, or perhaps any, large-whale population. Instead, we relied on three commonly used summaries of life history patterns that have been estimated for most great whales: age at maturity, "natural" (nonanthropogenic) adult mortality rate, and interbirth interval. We were able to cobble together these estimates of each of our eight focal species, although we were forced to use rather uncertain allometric relationships to estimate most mortality rates (see footnotes to Table 18.2). Nonetheless, by using both high and low estimates for each value in our analyses, we hope to bracket the actual production rates of these populations.

Because whales invariably produce only single offspring, interbirth interval and age at maturity are sufficient to characterize reproductive rates. Thus, the only critical vital rates for which we could not find direct estimates were prereproductive mortality rates. In nearly all vertebrates, most age

Species	Age at Maturity[a] (years)		Annual Adult Mortality[b] (%)		Interbirth Interval[a] (years)
	Min	Max	Min	Max	
Blue	5	5	3.1	3.1	2
Bryde's	5	6	6.3	6.3	2
Fin	8	12	4.2	4.2	2
Gray	8	8	6.4	6.4	2
Minke	7.3	7.3	4.0	9.2	1
Sei	6	10	4.8	8.0	2
Humpback	11	11	5.7	5.7	2
Bowhead	4	4	5.4	5.4	3–4
Northern Right	4[c]	11[c]	1	3	3–4
Sperm	9	10	2	6.1	3–5

[a] From compiled data in Boyd et al. 1999.

[b] Multiple mortality estimates were made for each species. Empirical mortality estimates were available for sei (Horwood 1987) and minke (Butterworth et al. 1999) whales, though not specifically for the North Pacific stocks. Mortality rates for the southern right whale, *Eubalaena australis* (Best and Kishino 1998), were considered reasonable to use for northern right whales. We also used Ohsumi's (1979) allometric approximations to infer mortality from adult length. Minimum and maximum values shown are taken from all estimates obtained for each species. The lower estimate of sperm whale mortality comes from Whitehead (Chapter 25 in this volume).

[c] Boyd et al. (1999) provide no estimate for the right whale, so we use values spanning the range of bowhead and humpback values.

classes of young and juvenile animals have higher mortality probabilities than do presenescent adults. Killer whales show this same pattern, with age-averaged annual mortality rates 20 times higher for subadults than adults (Olesiuk et al. 1990; Brault and Caswell 1993). However, model fitting to estimate age-specific mortality of Southern hemisphere minke whales results in survival rates for young animals that range from 0.75 to 1.9 times that of adults (from Table 6 in Butterworth et al. 1999). Given this uncertainty, we ran our models with annual juvenile mortality rates (that is, rates for all prereproductive animals) equal to 0.5, 1.0, 1.5, or 2 times annual adult mortality rates for each species.

Demographic Models for North Pacific Whales

Ideally, demographic models are constructed to reflect the true complexity of birth and death rates and the many factors influencing them. For instance, the survival and birth rates of whales are certain to vary continuously with an animal's age, and they are also certain to vary over time and between geographic locations. However, the data available to parameterize the demography of most species is very poorly matched with this complexity, and most demographic models are constructed without spatial and temporal effects, and often with highly simplified age or stage structure (Caswell 2001; Morris and Doak 2002). Although this simplification is often alarming, well-understood principles of model construction and hypothesis testing show that more "realistic" models give much less reliable answers if they are supplied with only poorly estimated parameters (Ludwig and Walters 1985; Burnham and Anderson

1998). Conversely, many studies in conservation biology show that extraordinarily simple demographic models can give important insights into population processes (e.g., Crouse et al. 1985; Crowder et al. 1994, Harding et al. 2001). Here, we are interested in making predictions of average rates of population production, meaning that many more complicated aspects of demography can be safely ignored.

Given the data available for great whales (Table 18.2), we constructed a demographic model that uses only two age classes: young, or juvenile, animals (including all prereproductives) and adults (Figure 18.2). We base this model on a postbirth census, so the youngest animals seen would be newborns, and, as is typical, we base the calculations just on the female segment of the population. In its simplest form, this corresponds to the following 2 × 2 matrix model for annual population growth.

$$\begin{pmatrix} n_{juv} \\ n_{ad} \end{pmatrix}_{t+1} = \begin{pmatrix} S_{juv}\left(1 - P_{mat}\right) & fS_{ad} \\ S_{juv}P_{mat} & S_{ad} \end{pmatrix} \begin{pmatrix} n_{juv} \\ n_{ad} \end{pmatrix}_t \qquad (18.1)$$

Here, n_{juv} and n_{ad} are the number of juvenile and adult animals, respectively, while the matrix elements are composed of only four rates: S_{juv} is the annual juvenile survival rate, S_{ad} is the annual adult survival rate, f is the annual per capita birth rate of females (that is, average number of daughters produced per adult female per year), and P_{mat} is the probability of a juvenile transitioning to become an adult, given that she survives the year.

S_{juv} and S_{ad} are directly obtained from the adult mortality estimates. The annual fecundity rate, f, is one-half (to

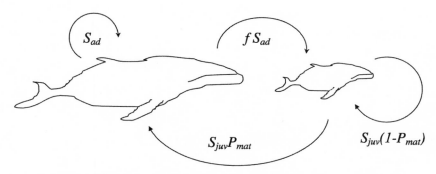

FIGURE 18.2. Schematic life history used for demographic models of great-whale populations. Individuals are classified as belonging to either an adult or juvenile stage (all ages prior to breeding). Rates associated with each arrow are the annual probabilities of each transition for females. See text for description of the meaning and estimation of these parameters.

account for female births alone) the reciprocal of interbirth interval. P_{mat}, the probability of maturation, is more complicated, relying on the age of maturity, T_M, and the juvenile survival rate (Caswell 2001):

$$P_{Mat} = \frac{S_{juv}^{T_M} - S_{juv}^{T_M - 1}}{S_{juv}^{T_M} - 1} \tag{18.2}$$

While it would be possible to include each year of the pre-reproductive period as a separate stage in the model, we do not use results that rely on transient dynamics, so this more compact method of modeling the prereproductive period will give identical results to a model that explicitly includes the multiple years of preadult life.

Note that this model does not include senescence—perhaps its most egregious violation of biological common sense. Although senescence is likely to occur in large whale species, there is no legitimate way to include it in the model without an estimate of the timing or degree of senescence (the extent of decreased survival or fertility). However, it has long been known that population growth is not especially sensitive to variation in longevity (Cole 1954); thus, for many long-lived vertebrates, addition of senescence also has little effect on demographic predictions (Doak, personal observation). Furthermore, if we use an overall estimate of adult survival (which will usually include data on older, senescent animals as well as those in their prime) for nonsenescent adult animals, inclusion of an extra penalty for senescence will actually yield a poorer prediction of dynamics than a model that leaves out senescence altogether (Lande 1988).

Modeling Dynamics and Deaths of North Pacific Whales

With the demographic rates and model structure just described, we can predict population growth and structure for each great-whale population and, as a byproduct of these values, the numbers of deaths occurring each year. For each

species we began by creating up to 32 different *Base* models for current populations: one for each combination of the numbers and rates for which we had high and low estimates (current population size, age at maturity, adult mortality, and juvenile mortality relative to adult mortality; for some species and rates there is only one estimate).

Before using these Base models to predict death rates, we asked whether these descriptions of whale dynamics were realistic, by examining their predicted rates of population increase (Figure 18.3). Predicted increase rates range from stable to nearly 20% increases per year. This range of values is in good accord with the 3 to 14% annual increases directly estimated for ten heavily depleted great-whale stocks (Best 1993; Reilly 1984), and thus we use the full range of models generated by our range of parameter estimates. However, the median predictions for all our models were relatively high (>5% per year), causing us to run other models for current predictions as well, which we describe shortly.

To predict current production of deaths from the Base models, we first assumed that the current population estimates are of total numbers of animals and are proportioned according to the stable stage distribution for each model. We also assume that male and female mortality patterns are identical, so that stage structure of the entire population matches that predicted by the female-only model. In this case, the number of deaths between years t and $t + 1$ is predicted to be $N_{juv,t}(1 - S_{juv})$, and the number of adult deaths, $N_{ad,t}(1 - S_{ad})$. Here $N_{juv,t}$ and $N_{ad,t}$ are the total (male and female) numbers of animals in the juvenile and adult stages at the start of a year, respectively.

Predicting historical production rates requires an additional set of assumptions. Since whale stocks were presumed to be more or less stable at these numbers, demographic rates must have been different from those currently estimated. However, which rates differed and by how much is unknown. At one extreme, it is possible that fecundities were unaffected by reductions in numbers (see Mizroch and York 1984), but mortality rates were considerably higher than currently estimated. This pattern of density-dependent effects would yield

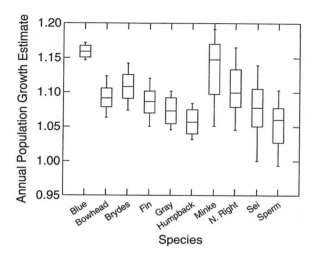

FIGURE 18.3. Boxplots of predicted current annual population multiplication rates for North Pacific whale stocks. All predictions are from the Base population models.

the maximum production of dead whales for a stable population of a given size. Conversely, as numbers increased, survival might be unaffected, but fecundities lowered to the point that populations are stable. This would result in the minimum number of deaths for a given stable population size. Since both extremes and many mixed responses to density are known in mammals, we modified each of the Basic models in two ways, resulting in up to 32 different *Reduced Fecundity* and 32 different *Reduced Survival* models for each species. In the case of changing survival rates, we assumed that the same proportional change in survival of juveniles and adults would occur to achieve a stable population. We then used these two types of models in conjunction with historical population estimates to arrive at estimated numbers of deaths.

The logic we used in formulating the two sets of historical models also led us to make two additional sets of predicted deaths under current conditions. Although the Base models are not unreasonable, not all whale stocks are currently increasing (e.g., the eastern North Pacific right whale and the gray whale; Highsmith, Chapter 23 in this volume; Springer et al., Chapter 19 in this volume), and therefore the Base models, which with one exception predict population growth, may be fairly inaccurate. We therefore used both the Reduced Fecundity or Reduced Survival models with estimated current numbers to make predictions about current death rates that would bracket the extremes of likely numbers of deaths.

Because we have a multitude of models, we present the median and the full range of the estimated numbers of deaths of adults and juveniles of each species historically and currently (Table 18.3). Median predictions indicate that the total numbers of juvenile and adult deaths are now only 21% and 24% of historic values, respectively. However, this overall figure hides a great deal of variation. Some species generate roughly similar numbers of deaths now and in the past (e.g., minke

and bowhead whales), whereas others are predicted to produce drastically fewer deaths now than in the past (fin, sei, and sperm whales, among others). As noted above, all these estimates must be viewed with some caution, given the unquantified uncertainty in the underlying population estimates upon which they are based.

Although these overall numbers give some comparison between current and past death rates, it is best to match carefully the predictions of models based on the same assumptions. We matched current and historical models by the values of vital rates used in each (i.e., minimum or maximum values from Table 18.2) and plotted the differences in predicted deaths of these model pairs (Figure 18.4). We separately compared historical Reduced Fecundity models with current Base and Reduced Fecundity models, and similarly we compared historical Reduced Survival models with either current Base or Reduced Survival results. We did not compare historical Reduced Survival and current Reduced Fecundity model results (or vice versa), since these models embodied opposing views of the most likely errors and density-dependent responses of whale populations. Note, too, that some historical estimates were used twice, to pair with current models that used either the low or high current population estimates.

The biggest factor influencing the estimated decrease in deaths is whether the Reduced Survival or Reduced Fecundity model is used for historical estimates (A and B vs. C and D in Figure 18.4). The choice to use the Base or corresponding Reduced model for current predictions has a less dramatic effect on the results, reflecting the lower current population sizes for most species. Within each of these sets of results, most sets of models for a species show fairly tight bounds— that is, there is not so much uncertainty as to make the exercise pointless. Reduced Fecundity models predict substantially smaller differences in death rates and, in particular, much lower differences in the numbers of juvenile deaths, whereas results that assume that population stability in the past was generated by increased mortality (the Reduced Survival model) predict sharp drops in the number of juvenile and adult deaths following industrial whaling.

Although the form of density dependence alters the magnitude of annual deaths, all pairings of current and historical models predict substantial declines in deaths. At a minimum, the historical Reduced Fecundity model with the current Basic model gives median predictions that sum to a reduction of 9,102 adult deaths and 2,572 juvenile deaths per year; using the current Reduced Fecundity model decreases the difference in adult deaths to 7,687 but increases juvenile deaths to 4,344. At the other extreme, the historical Reduced Survival model and current Basic model predict reductions of 26,995 juvenile and 14,949 adult deaths annually. As has been pointed out elsewhere in this volume (Croll et al., Chapter 16; Whitehead, Chapter 25), sperm whales tend to have disproportionate influence on most results about large whale effects, which we see here as well. The greatest estimated change in deaths is for sperm whales in all sets of models considered (Figure 18.4).

TABLE 18.3
Estimated Annual Numbers of Deaths of Great Whales, Current and Past

Species	Current		Historical	
	Juveniles	Adults	Juveniles	Adults
Blue	79	92	260	219
	(7–434)	(37–277)	(11–531)	(128–339)
Bryde's	1,569	1,302	2,686	1,840
	(244–3,537)	(683–2148)	(317–4,598)	(1,515–2,792)
Fin	169	101	2,569	1,234
	(23–425)	(41–182)	(247–4,565)	(944–1,954)
Gray	1,074	544	2,097	1,074
	(171–2,374)	(281–982)	(337–3,167)	(905–1,309)
Minke	135	57	223	70
	(9–311)	(16–92)	(9–311)	(51–92)
Sei	586	277	3,312	1,505
	(52–1,110)	(118–564)	(240–5,119)	(998–2,602)
Humpback	258	101	653	257
	(49–411)	(58–133)	(111–929)	(215–300)
Bowhead	313	458	532	814
	(40–1089)	(220–1,064)	(76–1,267)	(628–1,237)
Northern Right	1	1	1,094	864
	(0–4)	(0–4)	(6–2475)	(282–2,560)
Sperm	573	414	8,828	6,302
	(24–1,298)	(102–644)	(307–16,877)	(3,110–8,374)
Total	4,757	3,347	22,254	14,179

NOTE: Each cell of the table shows the median prediction over all current or all historical models, with the minimum and maximum estimate shown in parentheses. Totals are sums of median predictions for each species.

How Many Killer Whales Would Great-Whale Deaths Support?

The second issue we address is how much usable food for killer whales great-whale deaths might represent. Answering this question requires a number of further estimates and brings in the bioenergetic part of our work. In this section we present the data that are available to make these estimates, where the greatest uncertainties lie, and what our calculations reveal.

Metabolic Needs of Killer Whales

Active male killer whales need approximately 287,331 kcal/day, while females require about 193,211 kcal/day (these figures account for assimilation efficiency: Williams et al. 2004). How much whale tissue these needs correspond to depends on the type of meat or blubber primarily consumed. Although detailed data on the caloric content of whale tissues are not available, we use information on both whale tissue and other mammals to estimate that whale tongues provide 2.07 kcal/g; a mixture of whale meat and blubber has approximately 2.5 kcal/g; and blubber alone has about 4.0 kcal/g (Williams et al. 2004).

Attack Rates

Because we are basing our estimates of killer whale food on the output of dying whales from prey populations, we quantify attack rates indirectly, as the proportion of dying whales that die as a result of predation. Attacks by killer whales on many large-whale species have been observed, and a large fraction of living baleen and sperm whales show evidence of killer whale attacks. While attacks on calves have been more commonly observed, adult whales are also attacked and killed (Weller 2002; Reeves et al., Chapter 14 in this volume). Although whale biologists do not agree about the interpretation of these scarring rates, high scarring rates in other prey species are almost always taken to indicate high attack rates. These attacks may be only on young animals, leaving scars that are then carried for the rest of the survivors' lives, or the scars may indicate risks of attack on all age groups (Reeves et al., Chapter 14 in this volume; Clapham and Link, Chapter 24 in this volume; see also Siniff and Bengtson 1977). In either case, few predators—let alone highly intelligent predators—repeatedly attack prey that they have no chance of killing, and large, dangerous prey species that predators can only rarely take successfully show extremely low scarring rates (e.g., elephant, rhi-

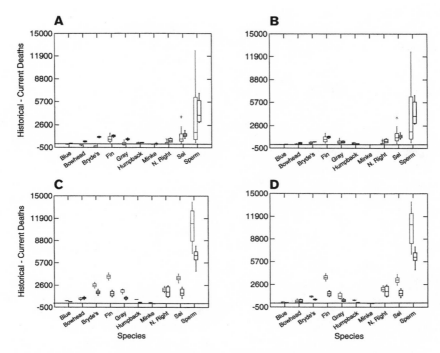

FIGURE 18.4. Estimated changes in annual numbers of juvenile and adult deaths of great whales from historical population sizes to current numbers. Each graph shows boxplots of the difference between estimated annual deaths prior to whaling and currently. These differences are between models matched for vital rate estimates, and how these may have changed to generate population stability. For each species the first boxplot, in gray, is of juvenile deaths, and the second, black plot is of adult deaths. (A) Historical Reduced Fecundity and current Base models; (B) historical and current Reduced Fecundity models; (C) historical Reduced Survival and current Base models; (D) historical and current Reduced Survival models. Horizontal lines indicate zero differences.

noceros, giraffe; personal observation). It is also worth noting that in many prey species that are not thought to be controlled by predators, predation is nonetheless a common way to die. Reeves et al. (Chapter 14 in this volume) mention Dolphin's (1987) idea that many large terrestrial prey live in "non-belligerent, if uneasy, coexistence" with large, cooperatively hunting predators. Whereas this may be true for elephants—one of the three prey species mentioned—predation is one of the most common ways for the other two species to die in intact savannah ecosystems: 23% to 28% of Cape buffalo deaths and 20% to 32% of wildebeest deaths in the Serengeti are due to predation (Schaller 1972). These species also make up a large fraction of all prey consumed by spotted hyenas and lions, even though neither population is believed to be controlled by these predators (Schaller 1972).

Nonetheless, current data do not allow quantitative estimates of the attack rates on great whales by killer whales. To reflect this uncertainty, we ran calculations with the percentage of deaths of juveniles, adults, or both due to killer whale predation ranging from 5% to 50%. Although different whale species are certain to vary in their risk of predation, we have no data with which to estimate these differences and

so used the same rates for all prey species when we ran each scenario.

Consumption Patterns and Limitations

Just as problematic as estimating the number of whales that may end up as prey for killer whales is the translation of this number into usable food for the predators. Although even a minke whale is a vast food resource for carnivores the size of killer whales, carcasses can sink rapidly, while killer whales usually attack and feed in groups, show strong preferences for certain tissues, and can eat only a limited amount of food at one time. We therefore considered three patterns, which reflect common ideas about the preferences of killer whales: (1) Only juvenile whales and adult minkes are predated, and since a mixture of blubber and meat would be consumed, the average energy gain is 2.5 kcal/g; (2) only the tongues of adult baleen whales are consumed, with 2.07 kcal/g gained from the tissue; and (3) juvenile and adult minkes are consumed (2.5 kcal/g), as are tongues (2.07 kcal/g) and blubber (4.0 kcal/g) of adults, but only a limited amount of blubber is used (up to the mass of the tongues for all baleen whales and up to 1,000 kg for sperm whales). Note that the first two

TABLE 18.4
Prey Mass Used for Estimation of Killer Whales Supported by Great-Whale Deaths

Species	Adult Total[a] (kg)	Adult Tongue[b] (kg)	Juvenile Total[c] (kg)
Blue	69,235	1,889	6,558
Bryde's	16,143	811	1,673
Fin	55,600	1,612	5,633
Gray	15,400	796	5,288
Minke	6,600	617	1,109
Sei	16,811	825	3,156
Humpback	30,400	1,071	4,330
Bowhead	31,100	1,129	2,328
Northern Right	23,400	487	2,374
Sperm	18,500		2,375

[a] Average adult masses provided by D. Croll (assembled from various IWC reports) and from articles in Perrin et al. (2002).

[b] Tongue masses estimated from separate regressions of tongue mass on total mass for humpback and right whales ($r^2 = 0.96$) and fin, blue, and gray whales ($r^2 = 0.98$), using data from Tomilin (1967) and Omura et al. (1969).

[c] Average of estimated weaning and birth masses for each species. Calculations were based on lengths at birth and weaning from Boyd et al. (1999) and species-specific allometric constants for mass-length relations from Lockyer (1976). Lockyer provides no values for gray whales, so for this species we used constants for humpbacks, which have similar adult mass and mass/length values. Similarly, some estimates for northern right whales and bowheads were used to fill some missing values for the other in these calculations.

scenarios were chosen to represent the most extreme views of how limited predation and feeding preferences may be. The approximate adult and juvenile masses, and the tongue masses of adult baleen and sperm whales we used, are shown in Tables 18.4 and 18.5. We conservatively estimated mean juvenile masses as the means of weaning and birth masses (values that are both estimable from the literature), although the average size of true juveniles will include many post-weaning individuals. We also estimate that only 66% of the total mass of juveniles or minke whales can be consumed, corresponding to the average percentage of total mass that is tongue, blubber, and meat for baleen whales as sampled by Tomilin (1967) and Omura et al. (1969).

There are further restrictions on the ability of killer whales to gain calories from great whale prey. Because of sinkage, we conservatively assume that each killer whale can feed only once on a great whale kill. The only estimate we know for the maximum intake rate of a killer whale is for a female, which could consume about 125 kg at one time (J. McBain, Seaworld, personal communication); we assume that the larger males can consume up to 186 kg at one time (estimated from the proportional metabolic needs of males and females). However, even this maximum may not be reached for each animal feeding on a whale kill, since killer whales are group hunters. Documented group sizes of hunting killer whales

TABLE 18.5
Predation and Consumption Parameters Used for Estimation of Killer Whales Supported by Great-Whale Deaths

Parameters	Values
Consumption pattern	1. Only tongues of adult whales
	2. Adult minkes and juveniles of all species
	3. Juveniles of all species, and tongue and blubber of adults
Attack rates	5% to 50% of deaths[a]
Killer whale group sizes	10, 20, or 30

[a] Depending on consumption pattern, this percentage applies to all minkes but only juveniles of other species; adults alone; or all individuals.

range from 5 up to 35, with many reports of larger, but unquantified packs (Reeves et al., Chapter 14 in this volume). We therefore ran calculations with pack sizes of 10, 20, or 30 animals feeding at once on each prey individual.

Great-Whale Deaths

Given the other uncertainties in making these estimates, we use the median numbers of juvenile and adult deaths across all other variables for each of the two historical (Reduced Fecundity and Reduced Survival) and three current population models.

Results

With these different estimates of attack patterns and rates, and the resulting calories that could be gained by killer whales, we generated multiple estimates of the supportable population of killer whales. We condense these into a reasonable form by showing boxplots of the current and future estimates for each consumption pattern and percentage of dying whales predated (Figure 18.5). Neither of the two basic trends in these plots is surprising: As the percentage of dying whales that are predated increases, more killer whales can be supported, and the difference between the number of killer whales that could have been supported by great-whale prey in the past (i.e., the last 150 years) and the number that can be so supported now also increases.

More interesting than these trends are the actual values predicted, especially those for the least intensive predation rates. If only adult tongues are consumed, and only 5% of whale deaths are attributable to predation, only a tiny number of killer whales could be supported (Figure 18.5A: medians of 3 currently and 8 historically). However, if only young animals are taken, again with only 5% of deaths caused by killer whales, median predictions are 12 killer whales currently and 50 historically—a difference in prey large enough to account for all the predation needed to drive the SSL

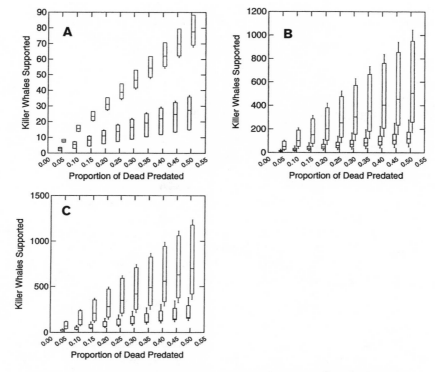

FIGURE 18.5. Numbers of killer whales that could be supported by historical and current great-whale populations. Boxplots show distributions of estimated killer whale numbers currently (gray) or historically (black) as functions of the fraction of dying whales that are predated and the types of individuals and tissues consumed: (A) only adult tongues consumed, (B) only young animals and minke whales consumed; and (C) young, minkes, and tongues and blubber of adults consumed. Note the different scales on each graph.

decline. With the scenario of feeding on young and selected adult tissues—which, we would argue, is the most reasonable guess at real killer whale feeding behaviors—these predictions rise to 16 and 70 animals for current and historical prey bases. Even though these numbers are not large, they begin to represent a substantial fraction of the likely number of mammal-specialist killer whales thought to exist near the Aleutian Islands (approximately 119–620, as described earlier in this chapter). Furthermore, it is quite plausible that much higher fractions of whales die by predation. If both adults and young are consumed and 25% of deaths are due to predation, a median of 349 killer whales could have been supported by these prey in the past, and 81 now. In presenting these results we primarily seek to show that even very selective and low rates of predation would still allow a substantial population of killer whales to subsist on great-whale stocks, especially before their depletion.

We also asked more systematically whether the differences in the numbers of killer whales that could be supported by great whales currently and historically match those needed to account for sea otter and SSL declines in the Aleutians. Although the timing of whaling and the different spatial extents of the stock estimates we use make these numbers difficult to compare directly, we can still ask whether wholesale slaughter of great whales by killer whales must be postulated in order for differences in supportable killer whales to match roughly the numbers needed to account for otter and SSL declines, or whether more plausible rates of attack would yield sufficient numbers. To present these differences, we again stratify our results by the types of population models, as well as by consumption pattern (Table 18.6). These differences are highly variable, but most are easily large enough to match the small numbers of killer whales (or small changes in diet of many killer whales) needed to account for SSL and sea otter declines. Only consumption solely of adult tongues will generally yield differences in supportable killer whales below the 20–50 needed to generate the SSL declines. Perhaps most notably, even low attack rates just on young whales are sufficient to generate substantial differences between historical and current support for killer whales; it is not necessary to postulate high rates of attack on calves or *any* attack on adults (even though we know that these do occur at some rate) for whaling to have substantially reduced the food resources for killer whales. If adults are also consumed, and if a higher percentage of deaths is attributable to predation, these changes between past and current food supplies increase even more. These results suggest that the change in abundance of the great whales has been large enough that,

TABLE 18.6
Differences in Killer Whales that could be Supported by Historical and Current Numbers of Great-Whale Prey

% of Dead Predated	Type of Prey and Tissues Consumed											
	Only Adult Tongues				Only Young and Minkes				Both Young and Adults			
	Population Models[a]				Population Models				Population Models			
	S-B	S-S	F-B	F-F	S-B	S-S	F-B	F-F	S-B	S-S	F-B	F-F
5	7	5	5	4	83	70	14	18	102	82	26	27
10	15	10	11	8	166	140	28	37	203	163	51	53
15	22	15	16	13	249	210	42	55	305	245	77	80
20	29	21	22	17	333	280	56	73	406	327	103	106
25	37	26	27	21	416	350	70	92	508	409	128	133
30	44	31	33	25	499	419	84	110	610	490	154	159
35	51	36	38	29	582	489	99	128	711	572	180	186
40	59	41	43	33	665	559	113	147	813	654	205	212
45	66	46	49	38	748	629	127	165	914	735	231	239
50	73	51	54	42	831	699	141	184	1016	817	257	266

NOTE: Data are median values across the three killer whale group sizes modeled. These potential changes in supportable killer whales are broken down by the fraction of dying whales that are predated, by the tissues consumed, and by the population model used for current and historical populations.

[a] Codes for population models indicate the Historical and Current models (S = reduced survival, F = reduced fecundity, and B = Base). For example, S-B columns show results for historical Reduced Survival models matched with current Base models.

even if they were and are only modestly used prey resources, industrial whaling could have accounted for a substantial drop in the prey resources available for killer whales.

Conclusions

As we said at the onset of this chapter, our goal has been to test the plausibility of a hypothesis, not to prove it. In doing so, we have deliberately chosen broad ranges for the parameters governing the behavior of our models, both to test the limits of the results and also to avoid accusations of bias. As the results show, one can find combinations of parameters that will show virtually no effect of whaling on killer whale food, or even on the numbers of annual deaths of great whales. However, the majority of our results show substantial declines in the annual supply of dying whales following industrial whaling. The question, then, is how much this possible source of food is used by killer whales—that is, how many of these deaths do they cause, and what do they consume when they do kill great whales? The uncertainties about the food habits of killer whales make answering this question with certainty impossible, but it is striking that, even if great whales are only minimally used by killer whales, they represented an extremely large food resource and still, even though to a much lesser extent, represent a large food resource at present. Although the data do not yet exist to make any conclusions about the frequency with which great whales fall prey to killer whales, these results suggest that the argument that industrial whaling substantially affected the

food resources of killer whales need not invoke rampant rates of attack nor complete consumption of these prey when killed. Indeed, the basic feeding biology of killer whales makes their seemingly finicky habits, such as favoring tongues, quite reasonable, even if great whales are in fact a large part of their diet.

Critics of the idea that killer whales obtain substantial amounts of food resources from great whales have pointed out that relatively frequent attacks on calves are better documented than attacks on adults. However, as we show in Table 18.6, attacks only on young whales would still generate substantial changes in the food available for killer whales, even if a minority of deaths of these young animals were caused by killer whales. The reduction in the great whales from whaling was so large that even if killer whales only use a small fraction of these populations, whaling is still likely to have reduced the resources available to support dozens to hundreds of these predators (see also Williams, Chapter 15 in this volume). As such, it is not necessary to advocate control of whale populations by killer whales to suggest that whaling could have dramatically altered the total food resources available to them and thus generated changes in feeding behaviors, including increased consumption of alternative prey species.

Critics of this hypothesis have also pointed out what they perceive as a paucity of direct observations of killer whale attacks on great whales. As Reeves et al. point out (Chapter 14 in this volume), scientifically observed attacks do appear to be uncommon, although this impression has also been reinforced by the attitude of past authors that all observations of attacks must be mere flukes. One aspect of this problem that has not been well addressed is the actual likelihood of observing attacks on whales, given the number of whales, numbers of attacks, and observation effort. This issue, stated bluntly, is that most whales are not seen most of the time and thus it is probably not surprising that attacks are rarely observed, even if the attacks themselves are common. To illustrate just how unlikely it is that scientific observers will see killer whale attacks on great whales, we used data from the National Marine Fisheries Service's Oregon, California, and Washington Line-transect Experiment (ORCAWALE) project off the west coast of North America in 1996 (Von Saunder and Barlow 1999). This project included a total of 114 days of work at sea, resulting in 14,592 km of line-transect observations, with visibility in a strip up to 12 nautical miles wide. The great whales seen totaled 84 sperms, 115 fins, 9 minke, 137 humpbacks, and 147 blues. To estimate the chances of seeing killer whale attacks on these sighted whales, we used the mean estimates of current North Pacific whale numbers to calculate the median estimated deaths per year from our different current models. Next, we used the highest attack rate that we modeled above (50% of deaths of all ages are due to killer whale attacks) and the highest observation efficiency imaginable (that each whale seen on the surveys was observed for an entire day; the ship speed during the surveys was 10 knots). With the assumption that attacks occur independently of one another (assuming that clumped attacks will reduce the likelihood of observing any attacks), we can then estimate the probability of seeing even one (i.e., >zero) attack on each species on this cruise. The computed probabilities are 0.0076 for sperms, 0.0105 for fins, 0.0013 for minkes, 0.0168 for humpbacks, and 0.0098 for blues. Even using the extraordinarily liberal assumptions we have outlined, these numbers make clear that attacks are extremely unlikely to be observed, even if they are actually quite common events. This suggests that observations of killer whale attacks, such as those detailed by Pitman et al. (2001), are likely to be unusual more because people happen to see them than because they are actually uncommon.

By using a combination of demographic and energetic methods, we have attempted not only to test the plausibility of the predation dynamics that could have preceded declines of SSLs and sea otters but also to clarify the pieces of information needed to evaluate better the role of killer whales and their effects on and responses to prey populations. Unfortunately, there are no parameters governing the process of attack and consumption that are well enough known that they do not warrant further research. Even the most basic data on how much a killer whale can consume at one time are known only for a single—captive—animal. Some parameters, including this one, could be greatly improved through careful observations of attacks in areas where they are common and predictable. One reason to formulate a model that demands quantitative estimates for each part of the predation process is to begin to understand the range of parameters we still must obtain to begin to understand the interactions of killer whales with great whales or any other of their prey species.

Ideally, any effort to test the plausibility of one explanation should also contrast this possibility with other explanations (Hilborn and Mangel 1997; Burnham and Anderson 1998). In a sense we have done so here, defining combinations of population regulation processes and abundance estimates that would or would not generate large declines in great-whale deaths or make killer whales more or less sensitive to these changes. However, besides this null hypothesis, there are no clear alternative hypotheses for the series of changes in marine mammal populations that have occurred in the Aleutians and, perhaps more broadly, the North Pacific. While we believe that it is worth pursuing a better understanding of the interactions of killer whales with their potential mammalian prey, science usually proceeds best when multiple working hypotheses are in competition. As the study of other carnivores has shown, preconceptions can misdirect understanding of predator-prey effects, and even on land the difficulty of seeing predation can mislead researchers as to the true nature of predator-prey relationships (Stirling and Oritsland 1995; Reeves et al., Chapter 14 in this volume). We hope that new ideas, as well as new data, arise to further understanding of these still poorly understood marine mammal interactions and how human harvesting may have changed their structure and dynamics.

Acknowledgments

We thank R. Gisiner for support of the metabolic studies through the Office of Naval Research grant N00014-0010761-3 to T. M. Williams and NSF awards DEB-9816980 and DEB-0087078 to D. F. Doak. A Pew Fellowship to J. A. Estes provided partial funding for analysis and modeling. D. DeMaster and one anonymous reviewer provided valuable comments on the manuscript. K. Forney, J. Mead, and W. Perrin kindly helped us to locate necessary information for our analyses.

Literature Cited

Best, P.B. 1993. Increase rates in severely depleted stocks of baleen whales. *ICES Journal of Marine Science* 50: 169–186.

Best, P.B. and H. Kishino. 1998. Estimating natural mortality rate in reproductively active female southern right whales, *Eubalaena australis*. *Marine Mammal Science* 14: 738–749.

Boyd, I.L., C. Lockyer, and H.D. Marsh. 1999. Reproduction in marine mammals, in *Biology of marine mammals*. J.E. Reynolds III and S.A. Rommel, eds. Washington, DC: Smithsonian Institution Press, pp. 218–286.

Brault, S. and H. Caswell. 1993. Pod-specific demography of killer whales *(Orcinus orca)*. *Ecology* 74: 1444–1454.

Burnham, K.P. and D.R. Anderson. 1998. *Model selection and inference: a practical information-theoretic approach.* New York: Springer-Verlag.

Butterworth, D.S., A.E. Punt, H.F. Geromont, H. Kato, and Y. Fujise. 1999. Inferences on the dynamics of Southern Hemisphere minke whales from ADAPT analyses of catch-at-age information. *Journal of Cetacean Research and Management* 1: 11–32.

Caswell, H. 2001. *Matrix population models: construction, analysis and interpretation.* 2nd edition. Sunderland, MA: Sinauer Associates.

Cole, L.L.C. 1954. The population consequences of life history phenomena. *Quarterly Review of Biology* 19: 103–137.

Crouse, D., L. Crowder, and H. Caswell. 1987. A stage-based population model for loggerhead sea turtles and implications for conservation. *Ecology* 68: 1412–1423.

Crowder, L., D. Crouse, S. Heppell, and T. Martin. 1994. Predicting the impact of turtle excluder devices on loggerhead sea turtle populations. *Ecological Applications* 4: 437–445.

Dolphin, W.F. 1987. Observations of humpback whale, *Megaptera novaeangliae*–killer whale, *Orcinus orca*, interactions in Alaska: comparison with terrestrial predator-prey relationships. *Canadian Field-Naturalist* 101: 70–75.

Estes, J.A., M.T. Tinker, T.M. Williams, and D.F. Doak. 1998. Killer whale predation on sea otters linking oceanic and nearshore ecosystems. *Science* 282: 473–476.

Harding, E.K., D.F. Doak, and J.D. Albertson. 2001. Evaluating the effectiveness of predator control for the non-native red fox. *Conservation Biology* 15: 1114–1122.

Hilborn, R. and M. Mangel. 1997. *The ecological detective: confronting models with data.* Princeton, NJ: Princeton University Press.

Holmes, E.E. and A.E. York. 2003. Using age structure to detect impacts on threatened populations: a case study with Steller sea lions. *Conservation Biology* 17: 1794–1806.

Horwood, J. 1987. The sei whale: population biology, ecology, and management. London: Croom Helm.

Jefferson, T.A., P.J. Stacey, and R.W. Baird. 1991. A review of killer whale interactions with other marine mammals: predation to co-existence. *Mammal Review* 21: 151–180.

Lande, R. 1988. Demographic models of the Northern spotted owl. *Oecologia* 75: 601–607.

Lockyer, C. 1976. Body weights of some species of large whales. *Journal du Conseil International pour L'Exploration de la Mer* 36: 259–273.

Ludwig, D. and C.J. Walters. 1985. Are age-structured models appropriate for catch-effort data? *Canadian Journal of Fisheries and Aquatic Sciences* 42: 1066–1072.

Mizroch, S.A. and A.E. York. 1984. Have pregnancy rates of Southern hemisphere fin whales, *Balaenoptera physalus*, increased? *Report of the International Whaling Commission* (Special Issue) 6: 401–410.

Morris, W.F. and D.F. Doak. 2002. *Quantitative conservation biology: the theory and practice of population viability analysis.* Sunderland, MA: Sinauer Associates.

National Research Council. 2003. *The decline of the Steller sea lion in Alaskan waters: untangling food webs and fishing nets.* Washington, DC: National Academies Press.

Ohsumi, S. 1979. Interspecific relationships among some biological parameters in cetaceans and estimation of natural mortality coefficient of the Southern Hemisphere minke whale. *Report of the International Whaling Commission* 29: 397–407.

Olesiuk, P.F., M.A. Bigg, and G.M. Ellis. 1990. Life history and population dynamics of resident killer whales *(Orcinus orca)* in the coastal waters of British Columbia and Washington State. *Report of the International Whaling Commission* (Special Issue) 12: 209–244.

Omura, H., S. Ohsumi, T. Nemoto, K. Nasu, and R. Kasuya. 1969. Black right whales in the North Pacific. *Scientific Reports of the Whales Research Institute* 21: 1–78.

Perrin, W.F., B. Würsig, and J.G.M. Thewissen (editors). 2002. *Encyclopedia of marine mammals.* San Diego: Academic Press.

Perry, S.L., D.P. DeMaster, and G.K. Silber. 1999. The great whales: history and status of six species listed as endangered under the U.S. Endangered Species Act of 1973. *Marine Fisheries Review* 61: 1–74.

Pitman, R.L., L.T. Ballance, S.L. Mesnick, and S.J. Chivers. 2001. Killer whale predation on sperm whales: observations and implications. *Marine Mammal Science* 17: 494–507.

Reilly, S.B. 1984. Observed and maximum rates of increase in gray whales, *Eschrichtius robustus.* *Report of the International Whaling Commission* (Special Issue) 6: 389–399.

Schaller, G.B. 1972. *The Serengeti lion: a study of predator-prey relations.* Chicago: University of Chicago Press.

Siniff, D.B. and J.L. Bengtson. 1977. Observations and hypotheses concerning interactions among crabeater seals, leopard seals, and killer whales. *Journal of Mammalogy* 58: 414–416.

Springer, A.M., J.A. Estes, G.B. van Vliet, T.M. Williams, D.F. Doak, E.M. Danner, K.A. Forney, and B. Pfister. 2003. Sequential megafaunal collapse in the North Pacific Ocean: an ongoing legacy of industrial whaling? *Proceedings of the National Academy of Sciences* 100: 12223–12228.

Stirling, I. and N.A. Oritsland. 1995. Relationships between estimates of ringed seal *(Phoca hispida)* and polar bear *(Ursus maritimus)* populations in the Canadian Arctic. *Canadian Journal of Fisheries and Aquatic Sciences* 52: 2594–2612.

Tomilin, A.G. 1967. *Mammals of the USSR and adjacent countries. Mammals of eastern Europe and northern Asia.* Vol. IX. *Cetacea.* Israel Program for Scientific Translations, trans. Jerusalem: Israel Program for Scientific Translations.

Trites, A.W. and C.P. Donnelly. 2003. The declines of Steller sea lions *Eumetopias jubatus* in Alaska: a review of the nutrional stress hypothesis. *Mammal Review* 33: 3–28.

Von Saunder, A. and J. Barlow. 1999. A report of the Oregon, California, and Washington Line-transect Experiment (ORCAWALE) conducted in West Coast waters during summer/fall 1996. NOAA Technical Memorandum NMFS-SWFSC-264. La Jolla, CA: Southwest Fisheries Science Center.

Weller, D.W. 2002. Predation on marine mammals, in *Encyclopedia of marine mammals.* W.F. Perrin, B. Würsig, and J.G.M. Thewissen, eds. San Diego: Academic Press, pp. 985–994.

Williams, T.M., J.A. Estes, D.F. Doak, and A.M. Springer. 2004. Killer appetites: assessing the role of predators in ecological communities. *Ecology* 85: 3373–3384.

NINETEEN

Whales and Whaling in the North Pacific Ocean and Bering Sea

Oceanographic Insights and Ecosystem Impacts

ALAN M. SPRINGER, GUS B. VAN VLIET, JOHN F. PIATT,
AND ERIC M. DANNER

It is not clear whether or not there are enough fin whales in the
North Pacific to allow the continuation of operations on such a
scale. It may be wise to take measures to prevent any abrupt
expansion of whaling operations. The situation should be watched
closely from the viewpoint of the conservation of whale stocks.

OMURA (1955)

Hideo Omura's concern over the killing of fin whales in the
early 1950s, just a few years after the resumption of whaling
following World War II, foreshadowed the demise of most of
the remaining great whales in the following two decades as
the slaughter expanded across the North Pacific (Springer
et al. 2003). His concern was warranted because of the development of an unprecedented human capability for large-scale
harvests of even the fastest whales using high-speed catcher
boats and mechanized factory ships designed specifically for
this purpose. In 1955, the nominal harvest of fin whales was
about 2,100, and it doubled over the next 10 years to a peak
of some 4,000 in 1965 before declining through 1975 when
the harvest was ended. Other species were similarly exploited
and depleted until pelagic hunting was successively halted
for humpback and blue whales in 1965, for sei whales in
1975, and for sperm and Bryde's whales in 1979. Populations

of most species were severely depleted by the mid-1980s
(Stewart et al. 1987; Rice and Wolman 1982). Exceptions
were humpback whales in the northeastern Pacific, which
had begun to recover slowly after protection was enacted in
1965 (Calambokidis et al. 2001), and bowhead whales and
gray whales that had been ravaged 100 years earlier (Raferty
et al. 1995; Rugh et al. 1999).

Reports of larger numbers of some species in the southeastern Bering Sea in the late 1980s in comparison with the
mid- to late 1970s (Baretta and Hunt 1994) are difficult to
interpret because changes in distribution rather than abundance could explain the difference in numbers between
decades (Tynan 2004). Fin whales were possibly showing
signs of slow recovery by the late 1990s in the northern Gulf
of Alaska and Bering Sea, although sperm whales remained
scarce (Moore et al. 2000; Tynan 2004).

While the rapid depletion of great whales in the North Pacific during the postwar decades was catastrophic to each species, an important ancillary issue is whether the removal of great whales had significant community-level effects (National Research Council 1996; Trites et al. 1999; Springer et al. 2003). Did the removal of megatons of upper-trophic-level consumers significantly alter food-web dynamics by (1) removing significant levels of predatory controls over prey populations, (2) removing an important prey resource for predator populations (i.e., killer whales), and (3) changing the sensitivity of the ecosystem to physical forcing because of new predator-prey functional relationships?

In order to address these questions, it is necessary to understand where and when whales were harvested in the North Pacific Ocean, and how this ultimately affected whale distribution. Whales were not uniformly distributed across this broad region, and the roles they played were concentrated in relatively small areas. Here we show where great whales formerly were found in abundance in the North Pacific, relate those distributions to oceanography, and briefly explore some examples of the magnitude of change that might have resulted from the loss of great whales in the Aleutian Islands and Bering Sea.

Data Sources and Caveats

The geographic focus of the following accounts of whales and whaling is primarily the northern and eastern North Pacific Ocean and the Bering Sea. We do not include information from the western North Pacific or its marginal seas, except for the Japanese shore-based fishery and the region including the Kurile Islands, western Aleutian Archipelago, and eastern Kamchatka Peninsula. Information on whale harvests in the 1800s and first half of the 1900s was taken from various published documents cited in appropriate locations throughout this chapter. We have plotted these data along with harvest data compiled by the International Whaling Commission (IWC) for all series where they exist. In some analyses, we have excluded information provided by the former Soviet Union (USSR), as it is known that they falsified data on the number of whales harvested, the species composition of their catches, and the locations of catches (Brownell et al. 2000a; R.L. Brownell, personal communication). The USSR data are included in the figures of harvest time series for comparative purposes, because although the USSR data were underreported and misrepresented in some cases, a better sense of the magnitude of the total harvest is achieved when these data are included than when they are excluded. Most harvest location data submitted by Japan for the pelagic harvest were reported at a resolution of 1° latitude by 2° longitude and are considered to be accurate in terms of numbers, locations, and species compositions (R.L. Brownell, personal communication).

The maps of whale distributions used in the following sections show where whales were harvested by shore-based and pelagic fisheries (excluding USSR data) since 1946. The maps thus reflect the principal summer foraging grounds of those species in the northern North Pacific, except that many of the whales taken in Japanese and Canadian coastal waters by the shore-based fisheries were actually migrating to more northerly feeding grounds (Gregr et al. 2000; Kasuya and Miyashita 1988; Nishiwaki 1966). The maps do not show the full summer range of any species, as hunting was focused on areas of greatest concentration. For example, in 1941 Japanese whalers prospected in the Chukchi Sea and killed 74 fin whales and 101 humpback whales along the Chukotka coast (Nemoto 1959), but these whales are not included in the data from which the maps were drawn. Also, because Soviet data have been excluded, harvests, and thus distributions, in the Sea of Okhotsk are poorly rendered.

Additional bias was introduced by the assignment of harvest quotas by the IWC. For example, humpback whales were apparently much more numerous in the Aleutian Islands than harvest data indicate, as the small allotment for humpbacks made it unprofitable to pursue them there (Nishiwaki 1966). The same can be said for sei whales, at least through the early 1960s; they were more abundant in the Aleutian Archipelago than it appears from harvest data.

History of Whaling in the North Pacific

Nineteenth Century

Intense commercial whaling in the northern North Pacific began in the early 1840s with the discovery of right whaling grounds in the Gulf of Alaska and off the Kamchatka Peninsula and Kurile Islands. The number of American whaling ships operating north of 50° N increased rapidly from just a few in 1840 to 108 by 1843, 292 in 1846, and 300–400 off the Kodiak Grounds between 1846 and 1851 (Scarff 1991; Gilmore 1978). In the first 10 years, between 1840 and 1850, some 11,000 right whales were taken by the fleet (Figure 19.1). Only about 3,000 were taken in the second decade from 1850 to 1860, reflecting both the depletion of the stock and the discovery of bowheads in the northern Bering Sea.

Whaling for bowheads grew equally rapidly, from one ship in the Bering Strait in 1848, to 50 in 1849, and 220 in 1852 (Bockstoce and Botkin 1983). The first few years of the fishery proved to be disastrous for bowheads, as it had for right whales, with a third of the total pelagic catch taken by 1852 and half by 1865 (Figure 19.1). The population plummeted from about 18,000 to about 3,000 by the end of the century (Woodby and Botkin 1993).

Shortly after the initial slaughter of right and bowhead whales in the northern North Pacific, gray whale calving grounds in Baja California were discovered and a commercial harvest began there (Scammon 1874). Nearly 6,600 gray whales were killed during the peak years, 1855–1870 (Figure 19.1). The population in 1846 was estimated to have been about 12,000, down from an historical high of about 24,000 because of an aboriginal take of around 600 y^{-1} just prior to the 1800s (Reilly 1981). By the 1880s the population had collapsed to about 2,000.

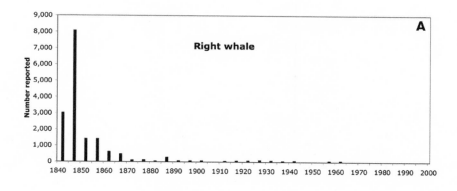

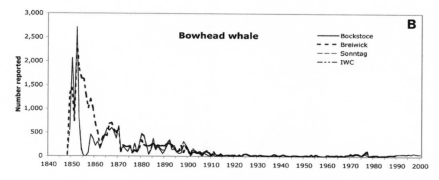

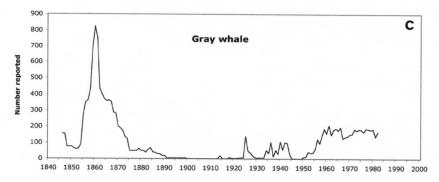

FIGURE 19.1. Annual harvests of right, bowhead, and gray whales in the North Pacific. Data from Bockstoce and Botkin (1983), Breiwick et al. (1981), Sonntag and Broadhead (1989), Best (1987), Reilly (1981), and IWC (All nations; unpublished data). Harvests of right whales compiled by 5-year intervals.

Twentieth Century Prior to World War II

Whaling in the first half of the twentieth century continued to be episodic by region and time. Modern whaling using catcher boats with mounted harpoon cannons began in Korea in 1889, in Japan in 1899, and in British Columbia and southeastern Alaska in 1905 and 1907 (Rice 1978). By the early 1900s, Japan had a flourishing coastal fishery, taking as many as 1,000 fin whales, 700 sei whales, and 250 blue whales each year. The fin whale harvest was excessive, and the population fell by an estimated 35% in just the eight years between 1910 and 1917 (Ohsumi et al. 1971). Likewise, the take of blue whales fell rapidly between 1910 and 1920 as the stock declined. The sei whale population was able to accommodate the harvest and changed little until after

World War II. Right whales were afforded worldwide protection in 1935, but continued illegal hunting, particularly by the Soviet Union, after 1949 and through the early 1960s drove them nearly to extinction (Rice 1974; Wada 1979; Gambell 1976; Brownell et al. 2001).

Whalers in the eastern North Pacific in the early 1900s were hunting primarily humpback whales, which were abundant along the coast from Washington to southeastern Alaska (Figure 19.2). Catch statistics for earlier years are not available, but it has been estimated that 4,000–5,000 humpbacks were killed in Alaska and British Columbia between 1905 and 1910 (Rice 1978). Between 1920 and 1930 attention shifted south to humpbacks off Baja California and California, where during that time nearly 4,000 were killed. By 1930

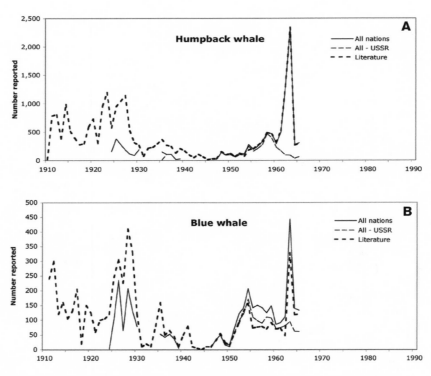

FIGURE 19.2. Annual harvests of humpback and blue whales in the North Pacific. Data from Omura (1955), Rice (1978), and IWC (All nations and All minus USSR; unpublished data).

some 18,000 humpbacks had been killed in the North Pacific, and the population had fallen from an estimated 15,000 to about 6,000 (Rice 1978).

The blue whale population in the North Pacific was historically small, and few were taken in the eastern North Pacific until the mid-1920s, when the animals congregating off Baja California in spring were targeted (Figure 19.2). The combined harvest of blue whales off Baja California and from the northern summering grounds during 1925–1930 was approximately 1,600.

POSTWAR WHALING AND THE END OF AN ERA

With the end of World War II, Japanese and Soviet pelagic whaling expanded in the North Pacific. Intensive hunting of sperm whales to the east of Kamchatka and around the Commander Islands began in 1954. Harvests increased rapidly in the years following the conversion of the Japanese fleet from solid fuel to liquid fuel in 1957. By 1963 there were three Japanese and four Russian fleets operating regularly (Rice 1978).

Humpback and blue whales were harvested heavily in Alaska in the early 1960s just prior to protection. The numbers of fin, sei, and sperm whales taken each year grew rapidly to peak levels in the middle to late 1960s (Figures 19.3 and 19.4). Bryde's whales were not hunted until the 1970s, following the depletion of larger whales in more northern waters (Figure 19.3). Pelagic fleets did not target the small minke whales; only about 12,000 were reported killed by all

nations during the period 1947–1987, primarily (about 10,500) by the Japanese shore-based fishery. Other species of whales were taken incidentally between 1947 and 1987, including Baird's beaked whale (618 reported, all nations), Cuvier's beaked whale (2 reported, all nations), pilot whales (482 reported, all nations), and killer whales (319 reported, all nations).

The abrupt, intense harvest of the larger whales beginning in the early 1950s reduced to very small numbers species already depleted before the war. By 1965, when humpback and blue whales were given protection, there were approximately 1,000–1,500 of each remaining (Rice 1978, Mizroch et al. 1984). For other species with much larger initial populations—the fin, sei, and sperm whales—estimates of abundance before and after the slaughter are less reliable.

There is little doubt, however, that the overall abundance of most species declined dramatically, particularly on the northern grounds (Cook 1985; Ohsumi et al. 1971; Ohsumi 1980; Kasuya 1991). By the end of the 1950s, for example, there was a pronounced shift in size of male sperm whales taken in the Bering Sea, to smaller (younger) animals, which forewarned of the collapse of the stock (Berzin 1964). Although sperm whales were not protected until 1979, hunting of them in the Bering Sea ended in 1972 because they were so scarce (Kasuya 1991). A whale census in the northern Gulf of Alaska in summer 1980 concluded that all species of great whales were severely depleted—in an area of approximately 2.2×10^5 km², which formerly supported thousands

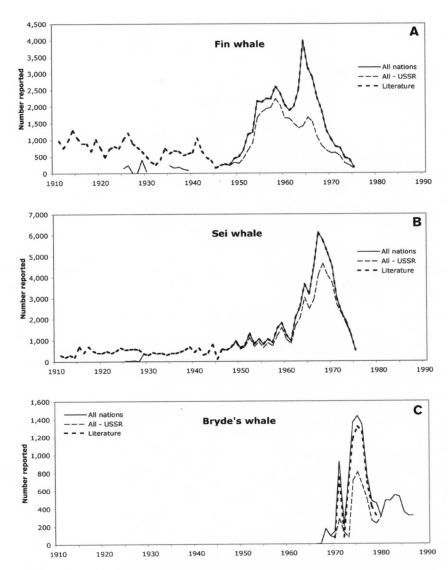

FIGURE 19.3. Annual harvests of fin, sei, and Bryde's whales in the North Pacific. Data from Omura (1955), Ohsumi et al. (1971), IWS (1930, 1937, 1948), and IWC (All nations and All minus USSR; unpublished data).

of whales, population estimates were fin, 159; humpback, 364; sperm, blue, sei, and right, few (only 36 sperm whales and none of the other species were sighted) (Rice and Wolman 1982).

Whales and Oceanography

Years ago, numerous scientists produced maps of summer distributions of whales in the North Pacific and described the patterns in relation to habitat and ocean productivity (Nasu 1966; Nemoto 1959, 1963; Nishiwaki 1966; Omura 1955; Uda 1962). We repeat that approach here, and we improve on their excellent earlier work only by broadening the horizon with locations of large numbers of whales killed in the 1960s and 1970s and with some additional insights gained from the

great amount of information on oceanography obtained since those earlier studies.

Right Whales

Right whales summered in the Sea of Okhotsk south to Japan and the East China Sea, in the southeastern Bering Sea, and in the northern Gulf of Alaska south to British Columbia (Townsend 1935; Omura et al. 1969; Braham and Rice 1984; Clapham et al. 2004). In the northeastern North Pacific they were concentrated on the southeastern Bering Sea outer shelf and slope and along the shelf edge in the western Gulf of Alaska from Kodiak Island to the eastern Aleutian Islands. Today the remnant population is known to occur only in the middle shelf domain of the eastern Bering Sea, as far north

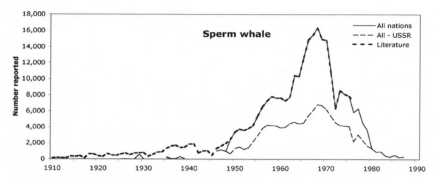

FIGURE 19.4. Annual harvests of sperm whales in the North Pacific. Data from Ohsumi (1980) and IWC (All nations and All minus USSR; unpublished data).

as St. Matthew Island (J. F. Piatt, unpublished data), where they number in the tens of individuals (Goddard and Rugh 1998; Tynan et al. 2001).

Historically, the diets of right whales consisted primarily of the copepods *Neocalanus cristatus* and *N. plumchrus*, as indicated by a small sample of whales taken during the 1950s–1970s (Nemoto and Kawamura 1977, cited in Kawamura 1980; Omura et al. 1969). These are the dominant species of copepods in the oceanic and outer shelf regions (Cooney 1981). Right whales also are known to have fed occasionally on larvae of *Euphausia pacifica*, and possibly other species of euphausiids. Recently, right whales on the middle shelf of the eastern Bering Sea are thought to be feeding on *Calanus marshallae* (Tynan et al. 2001), the large calanoid that replaced oceanic *N. cristatus* and *N. plumchrus* on the shelf (Cooney 1981). The whales may now also be feeding on euphausiids, which are abundant there (Cooney 1981; Smith and Vidal 1984).

Tynan et al. (2001) believe the present distribution of right whales on the eastern Bering Sea middle shelf is toward the periphery of their former feeding grounds, perhaps because of a change in the productivity of the different regions. They note that the abundance of *C. marshallae* in the middle shelf in the late 1990s was much higher than in the 1980s. An alternative hypothesis is that these particular animals exist where they always have, at the fringe of their range, which served as a refuge for them during the whaling days. Even in the 1980s and early 1990s right whales apparently were relatively abundant in the inner shelf domain near Bristol Bay (Vladimirov 1994). A similar situation developed with bowhead whales, where today they summer primarily only in regions where they found refuge during their era of exploitation—the ice-covered waters of the Beaufort and Chuckchi Seas, inaccessible to whaling ships.

Bowhead Whales

The historic summer range of bowheads includes the Bering, Ohkotsk, Chukchi, and Beaufort Seas (Townsend 1935; Bockstoce and Botkin 1983; Braham 1984a). In the Bering Sea, they formerly summered off Cape Navarin, south along the edge of the Kamchatka shelf in the Kamchatka Current, and north across the shallow Bering-Chukchi shelf. Today nearly all bowheads in the western Arctic still summer in the Chukchi and Beaufort seas, their refuge from whaling.

Diets of bowhead whales in the Bering Sea are not known. Contemporary diet samples have come mostly from Barrow, Alaska, in the western Beaufort Sea, and have contained approximately equal amounts of copepods and euphausiids and insignificant amounts of mysids and other invertebrates. Farther east in the Beaufort Sea at Kaktovik, Alaska, copepods contribute somewhat more, and euphausiids somewhat less, to diets (Lowry 1993, Lowry et al. 2004).

Considering the former distribution of bowheads in the northern and western Bering Sea relative to the distribution and biomass of zooplankton, it is likely that diets there consisted primarily of copepods and euphausiids as well. Bowheads lived in the Anadyr Current, the northern branch of the Bering Sea Green Belt (Springer et al. 1996), and in the headwaters of the Kamchatka Current that carries the Green Belt around the western side of the Bering Sea. Both currents originate at depth along the shelf break in the northwestern Bering Sea (Coachman et al. 1975), and the Anadyr Current transports vast amounts of nutrients and zooplankton biomass across the shallow Bering-Chukchi shelf, transforming it into one of the most highly productive marine pelagic regions in the world (Springer et al. 1989; Springer and McRoy 1993).

Gray Whales

The eastern North Pacific (ENP) stock of gray whales summers primarily in the northern Bering Sea and Chukchi Sea (Braham 1984b; Omura 1984; Rice et al. 1984). The remaining small population of western gray whales summers in the northern Sea of Okhotsk, mainly off the northwestern coast of Sakhalin Island (Rice et al. 1984; Weller et al. 1999).

ENP gray whales feed for the most part on the northern Bering-Chukchi continental shelf on benthic invertebrates, primarily ampeliscid amphipods (review by Nerini 1984;

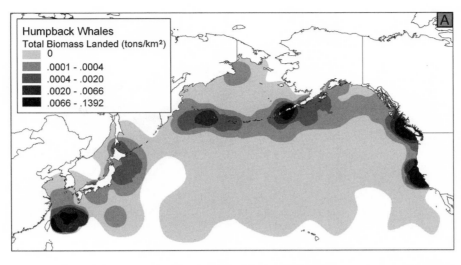

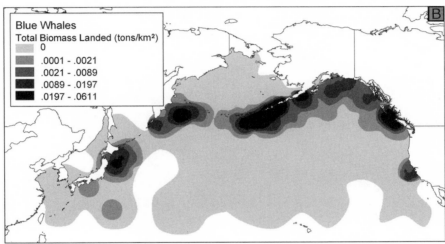

FIGURE 19.5. Summer distribution of humpback and blue whales in the North Pacific. May–September harvest locations after 1947 reported to the IWC (unpublished data) by all nations except the USSR.

Highsmith et al., Chapter 23 in this volume). Productivity of amphipods and other benthic invertebrates is extremely high in this region for the same reasons that pelagic production is high—the Anadyr Current. Nutrients supplied in the flow lead to annual primary production in the order of 500 g C m^{-2} y^{-1} (Springer and McRoy 1993), most of which falls to the sea floor and fuels the prolific benthic communities (Grebmeier et al. 1988). Although gray whales were decimated on their wintering grounds in Baja California, their loss from the Bering-Chukchi shelf undoubtedly altered benthic community structure and productivity.

Humpback Whales

Humpbacks are distributed widely in the North Pacific (Johnson and Wolman 1984). After World War II, most were killed in the eastern North Pacific (Figure 19.5). As noted previously, they were more abundant in the Aleutian Islands, as well as in the Bering Sea, than harvest records indicate.

Indeed, humpbacks were, and again are, numerous in the Bering and Chukchi Seas (Moore et al. 2000; Nemoto 1959; Sleptsov 1961; Tynan 2004; Votrogov and Ivashin 1980).

Humpback whales eat a mixture of fish and euphausiids. Copepods do not appear to be important in their diet. They require dense concentrations of prey and commonly feed on schooling species of forage fishes, such as capelin, sand lance, herring, Atka mackerel, and cods, as well as on dense swarms of euphausiids (Nemoto 1959; Piatt and Methven 1992).

Because of their prey preferences, humpbacks feed closer to shore than most of the other great whales. They are presently the most abundant species of large whale in the inshore waters of the Gulf of Alaska and eastern Bering Sea.

Blue Whales

Blue whales are found around the rim of the North Pacific from Japan to California (Mizroch et al. 1984). In summer they concentrated along the edge of the continental shelf

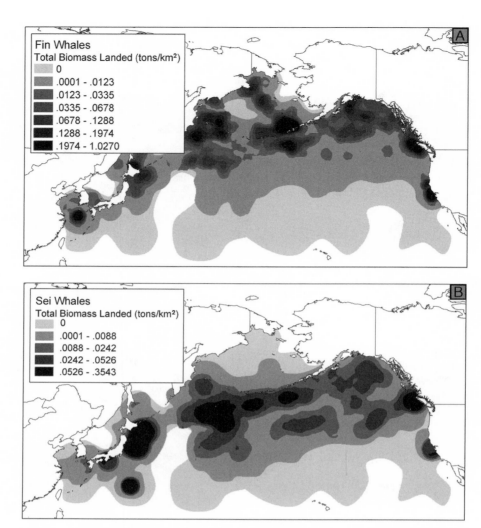

FIGURE 19.6. Summer distribution of fin, sei, and Bryde's whales in the North Pacific. May–September harvest locations after 1947 reported to the IWC (unpublished data) by all nations except the USSR.

from California to the Gulf of Alaska and along the south side of the Aleutian Archipelago (Figure 19.5). They penetrated into the Bering Sea in small numbers, and a few even ventured into the western Chukchi Sea (Sleptsov 1961), undoubtedly following the plume of the Anadyr Current northward.

Blue whales feed nearly exclusively on euphausiids in the North Pacific (Nemoto 1959; Kawamura 1980).

Fin Whales

Fin whales were very widely distributed in summer but were concentrated on their feeding grounds in a small part of the overall range (Figure 19.6). Most were found in particular locations around the rim of the North Pacific from California to Japan, including the Pacific Northwest (Washington-British Columbia), where a distinct subpopulation was exploited (Gregr et al. 2000). Also exploited in this region were whales migrating to more northern feeding grounds along the outer shelf and slope of the southeastern Bering Sea and shelf edge

to the northwest, where the greatest densities of fin whales in the North Pacific occurred, and in the northwestern North Pacific south of the Aleutian Islands. Fin whales ranged into the western Chukchi Sea in substantial numbers that are not apparent in the Japanese harvest data, as noted previously. The northern distribution can be seen in the Soviet harvest data, and it has been reported by Nemoto (1959) and Sleptsov (1961).

The diet of fin whales was geographically diverse in the North Pacific (Kawamura 1980, 1982; Nemoto 1959; Nemoto and Kasuya 1965). A variety of euphausiid species provided perhaps the bulk of the diet overall. However, Nemoto (1963) saw a strong correlation between the main distribution of fin whales on their southeastern Bering Sea feeding grounds and the main concentrations of *Neocalanus cristatus*. Along the shelf edge to the northwest and in the western Bering Sea, fishes replaced zooplankton as the dominant part of the diet. Different species dominated in different areas: pollock along the shelf edge, capelin downstream off Cape Navarin and in the Gulf of Anadyr, and herring along the shelf edge southwest

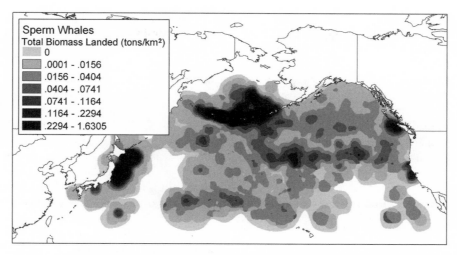

FIGURE 19.7. Summer distribution of sperm whales in the North Pacific. May–September harvest locations after 1947 reported to the IWC (unpublished data) by all nations except the USSR.

of Cape Navarin and off Cape Olyutorski. Even more so than humpbacks, fin whales require very high densities of schooling fish for successful foraging (Piatt and Methven 1992).

Diets south of the Aleutian Islands consisted primarily of euphausiids and copepods, with comparatively few fishes. *Neocalanus cristatus* was particularly important and *N. plumchrus* was common. Atka mackerel was the predominant species of fish taken in this region.

Sei Whales

Sei whales were most abundant in the western North Pacific off the coast of Japan, and south of the Aleutian Islands in the Alaska Stream (Figure 19.6). Although sei whales generally were not common north of the Aleutians, and few were taken by the fishery in the Bering Sea, they were formerly abundant on the northwestern shelf in July and August (Masaki 1976). As noted earlier, sei whales were more numerous in the Aleutian Islands than Japanese harvest records indicate because of management and economic reasons.

Sei whales are one of the smaller of the commercially exploited species and eat the smallest prey. The sei whale distribution in the northern North Pacific corresponded with the distribution of its main prey, *Neocalanus plumchrus*, although *N. cristatus* also was common prey, particularly at and beyond the shelf break in the eastern Gulf of Alaska. Euphausiids and fishes were of little importance (Kawamura 1980; Nemoto 1963; Nemoto and Kasuya 1965). The abundance of sei whales in the northwestern Bering was likely explained by the advection of the huge biomass of zooplankton, particularly *N. plumchrus*, from the basin in the flow of the Anadyr Current.

Bryde's Whales

Bryde's whales are found in tropical and warm-temperate regions of the Pacific. Their northern limit is defined generally by the 40° N parallel, although they occur somewhat north

of there, particularly in the central portion of their range (Omura 1959; Ohsumi 1977; Privalikhin and Berzin 1978). They are rare in the northern part of the North Pacific but are included here because they were targeted by pelagic fleets after more valuable species were depleted on the northern grounds, and protections were enacted for most of them. Most whales taken by the fishery were in central and western temperate and subtropical regions (Figure 19.6).

Bryde's whales taken on their more southern whaling grounds fed primarily on gonostomatid fish, and secondarily on euphausiids (Kawamura 1982). Off Japan, the migration of Bryde's whales seems to be keyed to the seasonal abundance of anchovy (Nemoto 1959).

Minke Whales

Minke whales are widely distributed in the North Pacific (Brownell et al. 2000b). However, because pelagic fisheries did not target them, there are no harvest records showing whether they concentrated in particular areas. They tend to be solitary or occasionally in pairs (Buckland et al. 1992; Moore et al. 2000; Tynan 2004), indicating that they maintain a low density over broad areas. Minke whales in Japanese waters feed on euphausiids and schooling forage fishes, such as pollock, herring, sand lance, and sardines (Kasamatsu and Hata 1985).

Sperm Whales

The majority of sperm whales in the northern North Pacific were males (Figure 19.7). Most females remained in more southerly waters throughout summer, although some did migrate to the Aleutian Islands and Gulf of Alaska, particularly in El Niño years. The greatest concentration of sperm whales in summer was found in the vicinity of the Aleutian Islands, over Brower's Ridge, which extends north off the central arc, and along the edge of the continental shelf in the

Bering Sea. They were particularly dense on the north side of the Aleutians, where squid were abundant (Uda 1962). These regions are part of the highly productive Green Belt, which is so important to other cetaceans.

Sperm whales eat both squid and fish throughout their range (Whitehead, Chapter 25, this volume). In the early 1960s, squids dominated in diets of sperm whales from the Aleutians, were of similar importance as fish along the shelf edge in the Bering Sea, and were less important than fish at the shelf edge in the Gulf of Alaska (Okutani and Nemoto 1964).

Whales and Whaling in an Ecosystem Context

Great whales played roles as both consumers and prey, and the loss of both functions is thought to be consequential to ecosystem structure in the northern North Pacific (National Research Council 1996; Springer et al. 2003; Croll et al., Chapter 16 in this volume). Dead whales falling to the deep-sea floor also provide detrital oases that support complex food webs in the abyss (Smith and Baco 2003; Smith, Chapter 22 in this volume).

Estimates of their quantitative importance in the ecosystem, and the effects their loss might have had are difficult to derive. Pre-whaling and post-whaling population sizes are not well known for most species, especially for any given region, nor is it known how many whale-days each year were spent by the various species on their feeding grounds. Estimates of historic population sizes of great whales vary widely and are hotly debated (Roman and Palumbi 2003). Nonetheless, we have quantitative data on the number and lengths (and therefore biomass) of great whales harvested in well-defined areas and over relatively short periods of time after the mid-1900s, and the nutritional requirements of these large predators can be estimated reasonably well.

Whales as Consumers

Omura et al. (1969) suggested that initial depletions of bowhead and right whales in the mid-1800s could have had a beneficial effect on populations of fin and sei whales in the Bering Sea and Gulf of Alaska. Right whales likely competed for copepods with fin whales in the eastern Aleutian Islands and southeastern Bering Sea, and with fins and sei whales in the Gulf of Alaska. Bowheads would have competed for copepods and euphausiids with fins and seis in the northern Bering Sea. Rights and bowheads both would have competed for prey with several species of planktivorous forage fishes, including pollock, capelin, and herring, which in turn could have had implications for fin whales. ENP gray whales are critical to community structure and productivity on their northern feeding grounds as both consumers and habitat architects (Oliver and Slattery 1985; Highsmith et al., Chapter 23 in this volume), and their reduction and subsequent recovery must have had important effects on benthic ecology. Worm et al. (Chapter 26, this volume) hypothesize that functional dominance in the Bering Sea shifted from marine mammals to fishes with the loss of whales.

Evidence of substantial increases in prey during the modern whaling era is provided indirectly by the dramatic density-dependent responses exhibited by some whales as their populations fell. During the period of intensive whaling, from the early 1950s until 1975, the average age at sexual maturity of female fin whales declined by half, from 12 years to 6 years, and of males by nearly 65%, from 11 years to 4 years (Ohsumi 1986). Male sperm whales grew faster after modern whaling began, particularly in the 1970s when over 80% of the total post-war take occurred (Kasuya 1991). Both of these examples indicate that as whale populations fell, individuals responded to improving feeding opportunities. While part of the improvement may have resulted from diminished interference competition, much of it likely resulted from both relative and absolute increases in prey availability as total consumption declined.

A sense of the magnitude of prey released by the loss of whales, and the potential implications for them and the system, can be seen by calculating daily and seasonal consumption budgets for three important species—bowhead, fin, and sperm whales (Table 19.1).

BOWHEAD WHALES

The exact number of bowheads that formerly summered in the northwestern Bering Sea and Bering Strait region is unknown, but some 7,000–9,000 were killed there in just the first six years following their discovery in 1848 (Bockstoce and Botkin 1983; Breiwick et al. 1981; Sonntag and Broadhead 1989). The slow reproductive rate of bowheads certainly precluded substantial replacement during such a short interval, and the population likely fell by approximately the number harvested. Further evidence that supports this case is the fact that by 1856 the Bering Strait grounds were virtually deserted of whales and whalers (Bockstoce 1986).

An average bowhead is assumed here to weigh 31 t (Pfister and DeMaster, Chapter 10 in this volume), and thus would consume 0.59–0.88 t d^{-1} of zooplankton biomass consisting of a mixture primarily of large calanoid copepods and euphausiids. If 6,000 bowheads were present on any given day in summer in the region from the Gulf of Anadyr to the Bering Strait (some bowheads were southwest of Cape Navarin), they would have consumed 3.5–5.3 × 10^3 t d^{-1}, or, with a diet of 50% copepods and 50% euphausiids, in the order of 5–8% of the daily advective input to the Bering Strait region of *Neocalanus cristatus* and *N. plumchrus* via the Anadyr Current in early to mid-summer, and 8–13% in late summer (calculated from data in Springer et al. 1989).

Although consumption by bowheads was not a great proportion of the daily supply of zooplankton to the Bering Strait region, it may still have been an important competitor of other baleen whales, as noted previously, as well as planktivorous auklets in the region. Least and crested auklets nest on all of the islands on the northern shelf, and in aggregate, number several million individuals, making this the richest region in the world for these species (Springer et al. 1993).

TABLE 19.1

Consumption by Bowhead, Fin, and Sperm Whales in the North Pacific

	Individual Mass	Individual Consumption (Wet Weight)			
	t	t d^{-1}, low	t d^{-1}, high	t y^{-1}, low	t y^{-1}, high
Bowhead whales, N Bering	31	0.59	0.88	71	106
Fin whales, SE Bering	38	0.69	1.03	82	123
Sperm whales, Aleutian Is.	27	0.53	0.80	64	96

NOTE: Masses assigned to fin and sperm whales are the mean weights of animals reported in the Japanese harvest data for the Bering Sea and Aleutian archipelago. The mass of bowhead whales is taken to be 31 t from Pfister and DeMaster (this volume). Daily intake rates to meet resting, or basal, metabolic requirements were calculated from Hain et al. (1985) as consumption (g wet weight d^{-1}) = 70 × (body weight in kg)$^{0.75}$. A best estimate range of daily consumption was arrived at by applying correction factors to resting rate of 3.6 × (active metabolism = 3 × resting; assimilation efficiency = 84%) and 5.4 × (food storage requirements = 1.5 × active metabolism + assimilation efficiency). Time spent by whales on their summer feeding grounds (the effective year length) was taken to be 4 months based on information in Masaki (1976) and Ohsumi (1966), as well as the length of the principal hunting season from June to September.

Auklets specialize on *Neocalanus* and euphausiids, and their otherwise paradoxical abundance on the shallow northern shelf is made possible by the Anadyr Current conveyor belt. Despite their immense numbers, auklets are small and constitute in biomass only the equivalent of about 2–5 bowheads per million birds, depending on species, and their combined consumption of zooplankton (ca. 2.0×10^2 t d^{-1}; Piatt and Springer 2003) is more than an order of magnitude less than that of bowheads. Thus, they have an even smaller impact on the vast zooplankton stocks in the region. However, the opposite effect is plausible, that auklets benefited from the release of several thousand tonnes per day of zooplankton biomass entrained in a marine river flowing past their nesting colonies, particularly since bowheads likely targeted the same dense concentrations of zooplankton as auklets did (Hunt and Harrison 1990).

FIN WHALES

The greatest concentration of fin whales in the North Pacific in summer was over the broad slope in the eastern Aleutian Basin of the southeastern Bering Sea. At a mass of 38 t (average of fin whales harvested in the Bering Sea; IWC, unpublished data), one whale would have consumed about 0.69–1.03 t d^{-1} of zooplankton (Table 19.1). The nominal all nations harvest of fin whales from the region totaled 8,144. The instantaneous standing stock of whales is not known, so assuming two scenarios, that (1) the number of whales present in the region at any time in summer was equal to the total harvest, or (2) equal to half the total harvest, they would have consumed at least 2.8×10^3 t d^{-1} of zooplankton biomass (4,072 whales × 0.69 t d^{-1} whale^{-1}), and at most 8.4×10^3 t d^{-1} (8,144 whales × 1.03 t d^{-1} whale^{-1}).

The main distribution of fin whales in the southeastern Bering Sea corresponded with the center of the spawning distribution of the offshore segment of the pollock stock of the eastern Bering Sea, located in the Bogoslof Island area (Management area 515: Hinckley 1987; Wespestad et al. 1990; D. Arciprete in Napp et al. 2000). Diets of fin whales and pollock overlap extensively, as pollock also prey predominantly on *Neocalanus* copepods, euphausiids, and small fishes, particularly young-of-year pollock (Bailey and Dunn 1979; Dwyer et al. 1987; Livingston 1989; Takahashi and Yamaguchi 1972; Yoshida 1994). The proportion of fishes in pollock diets off the shelf is low compared to on the shelf, so assuming pollock in the Bogoslof area consume predominantly copepods and euphausiids, as do fin whales in this region, the amount of zooplankton released by the depletion of whales is equivalent to the amount consumed daily by $1.4–4.2 \times 10^5$ t of pollock age 1–4, or $2.8–8.4 \times 10^5$ t of pollock >4 years old (assuming daily consumption is 2% × body weight for 1–4 year old fish and 1% × body weight for >4 year old fish) (from Springer 1992).

Estimates of trends in biomass of pollock in the eastern Aleutian Basin have not been made. In 1989 a standing stock of about 1×10^6 t was estimated (Wespestad et al. 1990). The harvest there grew rapidly in the late 1980s, peaking at 3.8×10^5 t in 1987, but declined rapidly because of dwindling abundance. Much of the harvest in the 1980s was of fish from the unusually strong 1978 year class. The biomass of prey released by the loss of fin whales was thus of the same order as the requirement of pollock in the Bogoslof area, even at its highest.

SPERM WHALES

The center of distribution of sperm whales in the northern North Pacific was the Aleutian Arc from the Near Islands to Unimak Pass and around the eastern perimeter of the Aleutian Basin. The nominal all-nations catch from this area, including waters 100 nautical miles south of the Aleutians, was 40,850 whales, of which 29,766 were taken within

100 nautical miles on either side of the Aleutians. This is a minimum estimate, as the USSR underreported sperm whale harvests by as much as 60% (Brownell et al. 2000a).

An average male sperm whale of 27 t (mean mass of whales caught in this area; IWC, unpublished data) consumes about 0.53–0.80 t d^{-1} (Table 19.1). For the Aleutians, where sperm whales were most highly concentrated and assuming between 14,883 and 29,766 whales were present at any time in summer, they would have consumed in the order of 0.79–2.4 × 10^4 t d^{-1} of prey biomass. In the southeastern Bering Sea, the nominal harvest was 16,279 sperm whales, and using the same assumptions as above, the whales there would have consumed in the order of 0.43–1.3 × 10^4 t d^{-1}.

Squids dominated diets of sperm whales in the Aleutian Islands and Bering Sea. Squids consume a variety of prey from zooplankton to fishes to other squids depending on species and size, although larger squids of the size commonly eaten by sperm whales were likely piscivorous or teuthivorous. Although there might have been other predators available to take up the surplus biomass released by the removal of sperm whales, a more immediate result of an increase in the abundance of second- and third-order squids might have been their effect as predators on prey populations. Over the course of a 120-day season in the Aleutians absent sperm whales, 0.95–2.9 × 10^6 t of additional biomass would have been available as predators or prey. In the southeastern Bering Sea absent sperm whales, an additional 0.52–1.6 × 10^6 t would have been available to eat and be eaten and to participate in the rebalancing of food web dynamics brought on by the removal of sperm whales and fin whales.

Whales as Prey

The only significant predators of great whales, other than people, are killer whales. Notably, the commercial fishery in the North Pacific did not target killer whales, and the majority of the 391 reported to the IWC by all nations between 1949 and 1964 were taken in the northwestern Pacific off the Kamchatka Peninsula and Kurile Islands (IWC, unpublished data).

Killer whales are known to prey on all of the great whales (Jefferson et al. 1991; Reeves et al., Chapter 14 in this volume). Highly choreographed defensive formations and evasion tactics of great whales are well documented (Finley 1990; Pitman et al. 2001; Whitehead 2003) and it is argued (George et al. 1994; Corkeron and Connor 1999; but see Clapham 2001) that long seasonal migrations of some species are made primarily to avoid killer whales. Skilled, cooperative attacks on individuals and groups of great whales by pods of killer whales are obviously learned behaviors to effectively subdue large prey, and include ramming, exsanguination, and drowning (Jonsgard 1968; Jefferson et al. 1991; Pitman et al. 2001).

Sheer size does not appear to confer immunity to great whales. Nor do killer whales need to kill their prey in order to obtain benefits from an attack. In one case without confirmed mortality, a large pod of killer whales, in a highly coordinated attack, stripped long pieces of blubber from a 20-m blue whale (Tarpy 1979). In another case, killer whales bit large chunks from humpback whales without apparently killing them (Whitehead and Glass 1985).

The significance of killer whale predation to great whale abundances was probably low in general, but not necessarily always. In the eastern Canadian Arctic, bowheads are preyed upon by killer whales to the extent that their recovery from over exploitation might have been retarded (Mitchell and Reeves 1982). Finley (1990) reported that approximately 30% of bowheads in Isabella Bay, Baffin Islands, had scars from killer whales. Finley et al. suggest that although killer whales in the eastern Canadian Arctic can meet their dietary needs by feeding on other, more abundant prey, they nevertheless target bowheads when they are available. Branch and Williams (Chapter 20, this volume) have speculated that in the Southern Ocean killer whales may have reduced minke whale abundance following decimation of great whales there. Evidence that killer whales prey on great whales elsewhere also exists. Gray whales are taken across a major portion of their range—in Alaska they are killed as they migrate through the western Gulf of Alaska and southeastern Bering Sea, and while on their feeding grounds on the northern Bering-Chukchi-Beaufort shelf. In the past in southwestern Alaska, where grays are available only in spring and fall, some killer whales in summer logically could have targeted other great whales that were so highly concentrated in the Aleutian Islands and Bering Sea prior to depletion.

Other species of great whales are also attacked in Alaska. There are numerous reports, summarized by Jefferson et al. (1991), of attacks on humpback whales in southeastern Alaska; Spalding (1999) reported that some 15% of humpbacks in the Gulf of Alaska bore scars from killer whales; and a vigorous, bloody attack on humpbacks in southeast Alaska was recently observed from an Alaska ferry (G. Kruse, personal communication). George et al. (1994) examined bowhead whales taken by hunters at Barrow, Alaska, (western Beaufort Sea) for scarring by killer whales, 81 in 1976–1979 and 114 more in 1980–1992. The incidence of scarring in 1976–1979 was just 2.5%, whereas in 1980–1992 it rose to 7.9%. All whales were considered to have been "confidently examined," and it seems plausible that the difference was due to redirected killer whale predation following the demise of great whales farther south. Transient killer whales move long distances (Goley and Straley 1994) and relocation of transients from the depauperate Aleutian Islands and southern Bering Sea to areas with higher densities of marine mammals would be expected. Increases in the abundance of killer whales in the vicinity of the Pribilof Islands and in Bristol Bay in the late 1980s (Frost et al. 1992; Baretta and Hunt 1994), following the collapse of pinnipeds populations in the Aleutians, were accompanied by a resumption of the decline of fur seals following a brief interval of stability on St. Paul Island, and by numerous observations of attacks on several species of marine mammals in Bristol Bay. Residents of villages in the Bering Strait region (Russia and Alaska) are reporting unusually high numbers of killer whales in recent years

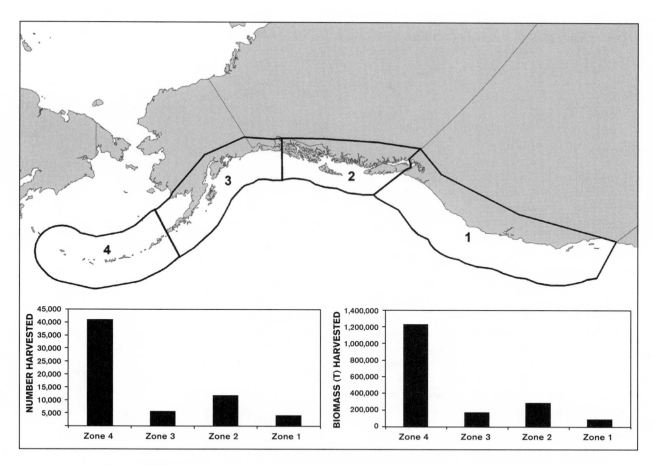

FIGURE 19.8. Numbers and biomass of whales harvested within 100 nautical miles of the coast in four regions of the North Pacific since 1947 reported to the IWC (unpublished data) by all nations except the USSR.

that are preying on bowheads, gray whales, walruses, and seals (G. Sheffield, personal communication; C. George, personal communication). Bowheads and gray whales are now the most abundant and concentrated large whales from the Aleutian Islands to Bering Strait. Minke whales also are commonly eaten by killer whales and may be particularly vulnerable because of their small size, broad distribution, and relative abundance following depletion of the larger species. There are no reliable estimates of abundance or trends in abundance of minkes, but nowadays they apparently are less numerous than many of the larger species (Moore et al. 2000; Tynan 2004), which raises the possibility that they did decline in the North Pacific, as they may have in sections of the Southern Ocean (Branch and Williams, Chapter 20, this volume).

Conclusions

Most whales were killed in the North Pacific during May–September while highly concentrated on their summer feeding grounds. These grounds are oceanographic hot spots where physical processes lead to enhanced production at numerous trophic levels (Uda 1962; Nasu 1966; Springer et al. 1996), and where prey must be concentrated for feeding to be efficient

(Nemoto 1972; Piatt et al. 1989; Piatt and Methven 1992). Thus, hot spots are found along the Aleutian Arc, the slope and shelf edge of the Bering Sea and Gulf of Alaska, the northern Bering-Chukchi shelf, the mixing zone of the Kuroshio and Oyashio currents, and the Western Subarctic Gyre.

Whalers since the mid-1800s were able to quickly deplete whales in given regions of the North Pacific, generally within spans of about 10 years (Danner et al., Chapter 11, this volume). Whales served as predators and prey, and the abrupt, extreme reductions of great whales from small areas likely focused the effects of the loss of these functions on ecosystems. The great whales' chief predators, killer whales, were taken in very small numbers by the fishery and likely included all three ecotypes (resident, transient, and offshore). Unfortunately, information necessary to evaluate the extent to which killer whales were and are dependent on large whales as prey remains to be collected.

We may never be certain of the magnitude or extent of the effects of commercial whaling. Cascades of response in communities, food webs, and ecosystems likely varied depending on local and regional characteristics, including basic oceanography and production regimes, the magnitude of whale biomass removed, and the status of other species in the matrix of interactions. For example, more than three fold more whales were killed in the modern fishery in the Aleutian Islands than in southeastern Alaska and British Columbia (Figure 19.8). By the time most whales were depleted on these grounds, the early 1970s, pinnipeds in the Pacific Northwest also were reduced to low numbers because of bounty programs and commercial hunting in the 1950s and 1960s (Bigg 1985; Olesiuk et al. 1990). But in the Aleutians, as well as the western Gulf of Alaska and Bering Sea, pinnipeds were still very abundant in the mid-1970s and did not begin to decline until about that time. Predator-prey relationships among marine mammals that existed then and that evolved in succeeding years were undoubtedly different in the two regions, and it is not surprising that conditions remain different today.

Great whale populations are recovering in the North Pacific, perhaps even the right whale stock (NMFS 2004). As population numbers grow, so too will the roles they play in the ecosystem. Whether food webs and communities return to their former condition remains to be seen, as much has changed in the intervening years. The mean climate state over the northern North Pacific has undergone three major shifts since the end of the modern whaling era (Hare and Mantua 2000; Mantua et al. 1997; Bond et al. 2003), and pinniped and sea otter populations throughout the Aleutian Islands and western Gulf of Alaska have collapsed (Estes et al. 1998; Doroff et al. 2003; Springer et al. 2003). The fundamental rules governing rates and pathways of energy exchange in the ocean are likely still the same (but see Jackson, Chapter 4 in this volume), but the constraints are certainly different now than they were in the hierarchy of the mature ecosystem 50 to150 years ago. Attention should be focused now on ways to improve our understanding of top-down oceanography (predator-prey interactions at all trophic levels, particularly high levels); how marine community structure and dynamics are influenced by those processes; and how ecosystems in their dramatically altered condition today behave in response to environmental change.

Acknowledgments

We thank D. DeMaster and R. Brownell, who read and improved an earlier draft of this paper; C. Allison (IWC), who provided updated summaries of the IWC catch data; and J. Wetzel (USGS), who produced Figures 19.5–19.7. The North Pacific Universities Marine Mammal Research Consortium and the Pew Charitable Trust provided support.

Literature Cited

Bailey, K. and J. Dunn. 1979. Spring and summer foods of walleye pollock, *Theragra chalcogramma*, in the eastern Bering Sea. *Fishery Bulletin* 77: 304–308.

Baretta, L. and G.L. Hunt, Jr. 1994. Changes in the numbers of cetaceans near the Pribilof Islands, Bering Sea, between 1975–78 and 1987–89. *Arctic* 47: 321–326.

Berzin, A.A. 1964. Determination of age composition of the sperm whale stock of the Bering Sea and adjacent parts of the Pacific, in *Soviet fisheries investigations in the northeast Pacific, Vol. 3*. P.A. Moiseev ed. Moscow: Pishchevaya Promyshlennost, pp. 263–266.

Best, P.B. 1987. Estimates of the landed catch of right (and other whalebone) whales in the American fishery, 1805–1909. *Fishery Bulletin* 85: 403–418.

Bigg, M.A. 1985. Status of the Steller sea lion *(Eumetopias jubatus)* and California sea lion *(Zalophus californianus)* in British Columbia. *Canadian Special Publication of Fisheries and Aquatic Sciences* 77: 1–20.

Bockstoce, J.R. 1986. *Whales, ice, and men: the history of whaling in the western Arctic*. Seattle: University of Washington Press.

Bockstoce, J.R. and D.B. Botkin. 1983. The historical status and reduction of the Western Arctic bowhead whale *(Balaena mysticetus)* population by the pelagic whaling industry, 1848–1914. *Report of the International Whaling Commission* (Special Issue) 5: 107–142.

Bond, N.A., J.E. Overland, M. Spillane, and P. Stabeno. 2003. Recent shifts in the state of the North Pacific. *Geophysical Research Letters* 30: 2183–2186.

Braham, H.W. 1984a. The bowhead whale, *Balaena mysticetes*. *Marine Fisheries Review* 46: 45–53.

———. 1984b. Distribution and migration of gray whales in Alaska, in *The gray whale Eschrichtius robustus*. M.L. Jones, S.L. Swartz, and S. Leatherwood, eds. New York: Academic Press, pp. 249–266.

Braham, H.W. and D.W. Rice. 1984. The right whale, *Balaena glacialis*. *Marine Fisheries Review* 46:38–44.

Breiwick, J.M., E.D. Mitchell, and D.G. Chapman. 1981. Estimated initial population size of the Bering Sea stock of bowhead whale, *Balaena mysticetus*: an iterative approach. *Fishery Bulletin* 78: 843–853.

Brownell, R.L. Jr., P.J. Clapham, T. Miyashita, and T. Kasuya. 2001. Conservation status of North Pacific right whales. *Journal of Cetacean Research and Management* Special Issue 2: 269–286.

Brownell, R.L. Jr., W.F. Perrin, L.A. Pastene, P.J. Palsbøll, J.G. Mead, A.N. Zerbini, T. Kasuya, and D.D. Tormosov. 2000b. Worldwide taxonomic status and geographic distribution of minke whales *(Balaenoptera acutorostrata* and *B. bonaerensis)*. Paper SC/52/027 presented to the International Whaling Commission Scientific Committee.

Brownell, R.L., Jr., A.V. Yablokov, and V.A. Zemsky. 2000a. USSR pelagic catches of North Pacific sperm whales, 1949–1979: conservation implications, in *Soviet whaling data (1949–1979)*. A.V. Yablokov and V.A. Zemsky, eds. Moscow: Center for Environmental Policy, pp. 123–130.

Buckland, S.T. and K.L. Cattanach. 1992. Minke whale abundance in the Northwest Pacific and the Okhotsk Sea, estimated from 1989 an 1990 sighting surveys. *Report of the International Whaling Commission* 42: 387–392.

Calambokidis, J., G.H. Steiger, J.M. Straley, L.M. Herman, S. Cerchio, D.R. Salden, J. Urbán-Ramirez, J.K. Jacobsen, O. von Ziegesar, K.C. Balcomb, C.M. Gabriele, M.E. Dahlheim, S.Uchida, G. Ellis, Y. Miyamura, P. Ladrón de Guevara-Porras, M. Yamaguchi, F. Sata, S.A. Mizroch, L. Schlender, K. Rasmussen, J. Barlow, and T.J. Quinn II. 2001. Movements and population structure of humpback whales in the North Pacific. *Marine Mammal Science* 17: 769–794.

Clapham, P.J. 2001. Why do baleen whales migrate? A response to Corkeron and Connor. *Marine Mammal Science* 17: 432–436.

Clapham, P.J., C. Good, S.E. Quinn, R.R. Reeves, J.E. Scarff, and R.L. Brownell, Jr. 2004. Distribution of North Pacific right whales *(Eubalaenus japonica)* as shown by 19th and 20th century whaling catch and sighting records. *Journal of Cetacean Research* 6: 1–6.

Coachman, L.K., K. Aagaard, and R.B. Tripp. 1975. *Bering Strait: regional physical oceanography.* Seattle: University of Washington Press.

Cook, J.G. 1985. Trends in the abundance of sperm whales in the western North Pacific. *Report of the International Whaling Commission* 35: 205–208.

Cooney, R.T. 1981. Bering Sea zooplankton and micronekton communities with emphasis on annual production, in *The eastern Bering Sea shelf: oceanography and resources*, Vol. 2. D.W. Hood and J.A. Calder, eds. Juneau, AK: NOAA, pp. 947–974.

Corkeron, P.J. and R.C. Connor. 1999. Why do baleen whales migrate? *Marine Mammal Science* 15: 1228–1245.

Doroff, A.M., J.A. Estes, M.T. Tinker, D.M. Burn, and T.J. Evans. 2003. Sea otter population declines in the Aleutian Archipelago. *Journal of Mammalogy* 84(1): 55–64.

Dwyer, D.A., K.M. Bailey, and P.A. Livingston. 1987. Feeding habits and daily ration of walleye pollock *(Theragra chalcogramma)* in the eastern Bering Sea, with special reference to cannibalism. *Canadian Journal of Fisheries and Aquatic Sciences* 44: 1972–1984.

Estes, J.A., M.T. Tinker, T.M. Williams, and D.F. Doak. 1998. Killer whale predation on sea otters linking oceanic and nearshore ecosystems. *Science* 282: 473–476.

Finley, K.J. 1990. Isabella Bay, Baffin Island: an important historical and present-day concentration area for the endangered bowhead whale *(Balaena mysticetus)* of the eastern Canadian Arctic. *Arctic* 43: 137–152.

Frost, K.J., R.B. Russell, and L.F. Lowry. 1992. Killer whales, *Orcinus orca*, in the southeastern Bering Sea: recent sightings and predation on other marine mammals. *Marine Mammal Science* 8: 110–119.

Gambell, R. 1976. World whale stocks. *Mammal Review* 6: 41–53.

George, J.C., L.M. Philo, K. Hazard, D. Withrow, G.M. Carroll, and R. Suydam. 1994. Frequency of killer whale *(Orcinus orca)* attacks and ship collisions based on scarring on bowhead whales *(Balaena mysticetus)* of the Bering-Chukchi-Beaufort seas stock. *Arctic* 47: 247–255.

Gilmore, R.M. 1978. Right whale, in *Marine mammals of eastern North Pacific and Arctic waters*. D. Haley, ed. Seattle: Pacific Search Press, pp. 62–69.

Goddard, P. and D.J. Rugh. 1998. A group of right whales seen in the Bering Sea in July 1996. *Marine Mammal Science* 14: 344–349.

Goley, P.D. and J.M. Straley. 1994. Attack on gray whales in Monterey Bay, California, by killer whales previously identified in Glacier Bay, Alaska. *Canadian Journal of Zoology* 72: 1528–1530.

Grebmeier, J.M., C.P. McRoy, and H.M. Feder. 1988. Pelagic-benthic coupling on the shelf of the northern Bering and Chukchi Seas. I. Food supply source and benthic biomass. *Marine Ecology Progress Series* 48: 57–67.

Gregr, E.J., N. Nichol, J.K.B. Ford, G. Ellis, and A.W. Trites. 2000. Migration and population structure of northeastern Pacific whales off coastal British Columbia: an analysis of commercial whaling records from 1908–1967. *Marine Mammal Science* 16: 699–727.

Hain, J.H.W., M.A.M. Hyman, R.D. Kenney, and H.E. Winn. 1985. The role of cetaceans in the shelf-edge region of the northeastern United States. *Marine Fisheries Review* 47: 13–17.

Hare, S.R. and N.J. Mantua. 2000. Empirical evidence for North Pacific regime shifts in 1997 and 1989. *Progress in Oceanography* 47: 103–145.

Hinckley, S. 1987. The reproductive biology of walleye pollock, *Theragra chalcogramma*, in the Bering Sea, with reference to spawning stock structure. *Fishery Bulletin* 85: 481–498.

Hunt, G.L., Jr. and N.M. Harrison. 1990. Foraging habitat and prey taken by least auklets at King Island, Alaska. *Marine Ecology Progress Series* 65: 141–150.

IWS. 1930. *International Whaling Statistics*, Vol. 2. Oslo: Commission for Whaling Statistics.

———. 1937. *International Whaling Statistics*, Vol. 9. Oslo: Commission for Whaling Statistics.

———. 1948. *International Whaling Statistics*, Vol. 19. Oslo: Commission for Whaling Statistics.

Jefferson, T.A., P.J. Stacey, and R.W. Baird. 1991. A review of killer whale interactions with other marine mammals: predation to co-existence. *Mammal Review* 21: 151–180.

Johnson, J.H. and A.A. Wolman. 1984. The humpback whale, *Megaptera novaeangliae*. *Marine Fisheries Review* 46(4): 1–100.

Jonsgård, Å. 1968. Another note on the attacking behavior of killer whale *(Orcinus orca)*. *Norsk Hvalfangst-Tidende* 6: 175–176.

Kasamatsu, F. and T. Hata. 1985. Notes on the minke whales in the Okhotsk Sea–west Pacific area. *Report of the International Whaling Commission* 35: 299–304.

Kasuya, T. 1991. Density dependent growth in North Pacific sperm whales. *Marine Mammal Science* 7: 230–257.

Kasuya, T. and T. Miyashita. 1988. Distribution of sperm whale stocks in the North Pacific. *Scientific Report of the Whales Research Institute* 39: 31–75.

Kawamura, A. 1980. A review of food of balaenopterid whales. *Scientific Report of the Whales Research Institute* 32: 155–197.

———. 1982. Food habits and prey distributions of three rorqual species in the North Pacific Ocean. *Scientific Report of the Whales Research Institute* 34: 59–91.

Livingston, P.A. 1989. Interannual trends in walleye pollock, *Theragra chalcogramma*, cannibalism in the eastern Bering Sea. *International Symposium on the Biology and Management of Walleye Pollock, Alaska Sea Grant*. Fairbanks: University of Alaska, pp. 275–296.

Lowry, L.F. 1993. Foods and feeding ecology, in *The bowhead whale*. Special Publication No. 2, Society for Marine Mammalogy. J.J. Burns, J.J. Montague, and C.J. Cowles, eds. Lawrence, KS: Allen Press, pp. 201–238.

Lowry, L.F., G. Sheffield, and J.C. George. 2004. Bowhead whale feeding in the Alaskan Beaufort Sea, based on stomach contents analyses. *Journal of Cetacean Research and Management* 6: 215–223.

Mantua, N.J., S.R. Hare, Y. Zhang, J.M. Wallace, and R.C. Francis. 1997. A Pacific interdecadal climate oscillation with impacts on salmon production. *Bulletin of the American Meteorological Society* 78: 1069–1079.

Masaki, Y. 1976. Biological studies on the North Pacific sei whale. *Bulletin of Far Seas Fisheries Research Laboratory (Shimizu)* 14: 1–104.

Mitchell, E., and R.R. Reeves. 1982. Factors affecting abundance of bowhead whales *Balaena mysticetus* in the eastern Arctic of North America, 1915–1980. *Biological Conservation* 22: 59–78.

Mizroch, S.A., D.W. Rice, and J.M. Breiwick. 1984a. The blue whale, *Balaenoptera musculus*. *Marine Fisheries Review* 46(4): 15–19.

Moore, S.E., J.M. Waite, L.L. Mazzuca, and R.C. Hobbs. 2000. Mysticete whale abundance and observations of prey associations on the central Bering Sea shelf. *Journal of Cetacean Research and Management* 2: 227–234.

Napp, J.M., A.W. Kendall, Jr., and J. Schumacher. 2000. A synthesis of biological and physical processes affecting the feeding environment of larval walleye pollock *(Theragra chalcogramma)* in the eastern Bering Sea. *Fisheries Oceanography* 9: 147–162.

Nasu, K. 1966. Fishery oceanographic study on the baleen whaling grounds. *Scientific Report of the Whales Research Institute* 20: 157–210.

Nemoto, T. 1959. Food of baleen whales with reference to whale movements. *Scientific Report of the Whales Research Institute* 14: 149–291.

———. 1963. Some aspects of the distribution of *Calanus cristatus* and *C. plumchrus* in the Bering and its neighboring waters, with reference to the feeding of baleen whales. *Scientific Report of the Whales Research Institute* 17: 157–170.

———. 1972. Feeding pattern of baleen whales in the ocean, in *Marine food chains*. J.H. Steele, ed. Edinburgh: Oliver and Boyd, pp. 241–252.

Nemoto, T. and T. Kasuya. 1965. Foods of baleen whales in the Gulf of Alaska of the North Pacific. *Scientific Report of the Whales Research Institute* 19: 45–51.

Nerini, M. 1984. A review of gray whale feeding ecology, in *The gray whale Eschrichtius robustus*. M.L. Jones, S.L. Swartz, and S. Leatherwood, eds. New York: Academic Press, pp. 423–463.

Nishiwaki, M. 1966. Distribution and migration of the larger cetaceans in the North Pacific as shown by Japanese whaling results, in *Whales, dolphins, and porpoises*. K.S. Norris, ed. Berkeley: University of California Press, pp. 171–191.

NMFS. 2004. Scientists double tally of known right whales. http://www.fakr.noaa.gov/newsreleases/rightwhale100104.htm. (Accessed October 1, 2004.)

National Research Council. 1996. *The Bering Sea Ecosystem.* Washington, DC: National Academies Press, p. 307.

Ohsumi, S. 1966. Sexual segregation of the sperm whale in the North Pacific. *Scientific Report of the Whales Research Institute* 20: 1–16.

———. 1977. Bryde's whales in the pelagic ground of the North Pacific. *Report of the International Whaling Commission* (Special Issue 1): 140–150.

———. 1980. Catches of sperm whales by modern whaling in the North Pacific. *Report of the International Whaling Commission* (Special Issue) 2: 11–18.

———. 1986. Yearly change in age and body length at sexual maturity of a fin whale stock in the eastern North Pacific. *Scientific Report of the Whales Research Institute* 37: 1–16.

Ohsumi, S., Y. Shimadzu, and T. Doi. 1971. The seventh memorandum on the results of Japanese stock assessment of whales in the North Pacific. *Report of the International Whaling Commission* 21: 76–89.

Okutani, T. and T. Nemoto. 1964. Squids as the food of sperm whales in the Bering Sea and Alaskan Gulf. *Scientific Report of the Whales Research Institute* 18: 111–121.

Olesiuk, P.F., M.A. Bigg, and G.M. Ellis. 1990. Life history and population dynamics of resident killer whales *(Orcinus orca)* in the coastal waters of British Columbia and Washington State. *Report of the International Whaling Commission* (Special Issue) 12: 209–244.

Oliver, J.S. and P.N. Slattery. 1985. Destruction and opportunity on the sea floor: effects of gray whale feeding. *Ecology* 66: 1965–1975.

Omura, H. 1955. Whales in the northern part of the North Pacific. *Norsk Hvalfangst-Tidende* 44(6): 323–345.

———. 1959. Bryde's whale from the coast of Japan. *Scientific Reports of the Whales Research Institute* 14: 1–33.

———. 1984. History of gray whales in Japan, in *The gray whale Eschrichtius robustus*. M.L. Jones, S.L. Swartz, and S. Leatherwood, eds. Orlando, FL: Academic Press, pp. 57–77.

Omura, H., S. Ohsumi, T. Nemoto, K. Nasu, and T. Kasuya. 1969. Black right whales in the north Pacific. *Scientific Report of the Whales Research Institute* 21: 1–78.

Piatt, J.F. and D.A. Methven. 1992. Threshold foraging behavior of baleen whales. *Marine Ecology Progress Series* 84: 205–210.

Piatt, J.F. and A.M. Springer. 2003. Advection, pelagic food webs and the biogeography of seabirds in Beringia. *Marine Ornithology* 31: 141–154.

Piatt, J.F., D.A. Methven, and A.E. Burger. 1989. Baleen whales and their prey in a coastal environment. *Canadian Journal of Zoology* 67: 1523–1530.

Pitman, R.L., L.T. Ballance, S.L. Mesnick, and S.J. Chivers. 2001. Killer whale predation on sperm whales: observations and implications. *Marine Mammal Science* 17: 494–507.

Privalikhin, V.I. and A.A. Berzin. 1978. Abundance and distribution of Bryde's whales *(Balaenoptera edeni)* in the Pacific Ocean. *Report of the International Whaling Commission* 28: 301–302.

Raferty, A., J. Zeh, and G. Givens. 1995. Revised estimate of bowhead rate of increase. *Report of the International Whaling Commission* 45: 158.

Reilly, S.B. 1981. *Gray whale population history: an age structured simulation.* Seattle: National Marine Mammal Laboratory.

Rice, D.W. 1974. Whales and whale research in the eastern North Pacific, in *The whale problem: a status report*. W.E. Schevill, ed. Cambridge, MA: Harvard University Press, pp. 170–195.

———. 1978. The humpback whale in the North Pacific: distribution, exploitation, and numbers, in *Report on a workshop on problems related to humpback whales (Megaptera novaeangliae) in Hawaii*. K.S. Norris and R.R. Reeves, eds. Report No. MMC-77/03. Washington, DC. Marine Mammal Commission, pp. 29–44.

Rice, D.W. and A.A. Wolman. 1982. Whale census in the Gulf of Alaska June to August 1980. *Report of the International Whaling Commission* 32: 491–497.

Rice, D.W., A.A. Wolman, and H.W. Braham. 1984. The gray whale, *Eschrichtius robustus*. *Marine Fisheries Review* 46: 7–14.

Roman, J. and S.R. Palumbi. 2003. Whales before whaling in the North Atlantic. *Science* 301: 508–510.

Rugh, D.J., M.M. Muto, S.E. Moore, and D.P. DeMaster. 1999. *Status review of the eastern north Pacific stock of gray whales.* U.S. Department of Commerce, NOAA Technical Memo NMFS-AFSC-103. Seattle: Alaska Fisheries Science Center.

Scammon, C.M. 1874. *The marine mammals of the northwestern coast of North America.* San Francisco: John H. Carmany and Co. Reprint, New York: Dover Publications, 1968. Reprint, Riverside, CA: Manessier, 1969.

Scarff, J.E. 1991. Historic distribution and abundance of the right whale, *Eubalaena glacialis*, in the North Pacific, Bering Sea, Sea of Okhotsk and Sea of Japan from the Maury Whale Charts. *Report of the International Whaling Commission* 41: 467–487.

Sleptsov, M.M. 1961. Fluctuations in the number of whales in the Chukchi Sea in various years. *Trudy Instituta Morfologii Zhivotnykh* 34: 54–64.

Smith, C.R. and A.R. Baco. 2003. Ecology of whale falls at the deep-sea floor. *Oceanography and Marine Biology: an Annual Review* 41: 311–354.

Smith, S.L. and J. Vidal. 1984. Spatial and temporal effects of salinity, temperature and chlorophyll on the communities of zooplankton in the southeastern Bering Sea. *Journal of Marine Research* 42: 221–257.

Sonntag, R.M. and G.C. Broadhead. 1989. Documentation for the revised bowhead whale catch (1848–1987). *Report of the International Whaling Commission* 39: 114–115.

Spalding, D.A.E. 1999. *Whales of the West Coast*. Madeira Park, British Columbia: Harbour Publishing.

Springer, A.M. 1992. A review: walleye pollock in the North Pacific—how much difference do they really make? *Fisheries Oceanography* 1: 80–96.

Springer, A.M., J.A. Estes, G.B. van Vliet, T.M. Williams, D.F. Doak, E.M. Danner, K.A. Forney, and B. Pfister. 2003. Sequential megafaunal collapse in the North Pacific Ocean: an ongoing legacy of industrial whaling? *Proceedings of the National Academy of Sciences* 100: 12223–12228.

Springer, A.M., A.Y. Kondratyev, H. Ogi, Y.V. Shibaev, and G.B. van Vliet. 1993. Status, ecology, and conservation of *Synthliboramphus* murrelets and auklets, in *The status, ecology, and conservation of marine birds of the North Pacific*. K. Vermeer, K.T. Briggs, K.H. Morgan, and D. Siegel-Causey, eds. Ottawa: Canadian Wildlife Service, pp. 187–201.

Springer, A.M. and C.P. McRoy. 1993. The paradox of pelagic food webs in the northern Bering Sea-III. Patterns of primary production. *Continental Shelf Research* 13: 575–599.

Springer, A.M., C.P. McRoy, and M.V. Flint. 1996. The Bering Sea Green Belt: shelf edge processes and ecosystem production. *Fisheries Oceanography* 5: 205–223.

Springer, A.M., C.P. McRoy, and K.R. Turco. 1989. The paradox of pelagic food webs in the northern Bering Sea—II. Zooplankton communities. *Continental Shelf Research* 9: 359–386.

Stewart, B.S., S.A. Karl, P.K. Yochem, S. Leatherwood, and J.L. Laake. 1987. Aerial surveys for cetaceans in the former Akutan, Alaska, whaling grounds. *Arctic* 40: 33–42.

Takahashi, Y. and H. Yamaguchi. 1972. Stock of the Alaska pollock in the eastern Bering Sea. *Bulletin of the Japanese Society of Scientific Fisheries* 38: 389–399.

Tarpy, C. 1979. Killer whale attack. *National Geographic* 155: 542–545.

Townsend, C.H. 1935. The distribution of certain whales as shown by logbook records of American whaleships. *Zoologica* 19: 1–50.

Trites, A.W., P.A. Livingston, S. Mackinson, M.C. Vasconcellos, A.M. Springer, and D. Pauly. 1999. *Ecosystem change and the decline of marine mammals in the eastern Bering Sea: testing the ecosystem shift and commercial whaling hypothesis*. Fisheries Centre Research Report 1999, 7(1). Vancouver: University of British Columbia.

Tynan, C.T. 2004. Cetacean populations on the SE Bering Sea shelf during the late 1990s: implications for decadal changes in ecosystem structure and carbon flow. *Marine Ecology Progress Series* 272: 281–300.

Tynan, C.T., D.P. DeMaster, and W.T. Peterson. 2001. Endangered right whales on the southeastern Bering Sea shelf. *Nature* 294: 1894.

Uda, M. 1962. Subarctic oceanography in relation to whaling and salmon fisheries. *Scientific Report of the Whales Research Institute* 16: 105–119.

Vladimirov, V.L. 1994. Recent distribution and abundance level of whales in Russian Far-Eastern seas. *Russian Journal of Marine Biology* 20: 1–9.

Votrogov, L.M. and M.V. Ivashin. 1980. Sightings of fin and humpback whales in the Bering and Chukchi seas. *Report of the International Whaling Commission* 30: 247–248.

Wada, S. 1979. Indices of abundance of large-sized whales in the North Pacific in the 1977 whaling season. *Report of the International Whaling Commission* 29: 253–264.

Weller, D.W., B. Würsig, A.L. Bradford, A.M. Burdin, S.A. Blokhin, H. Minakuchi, and R.L. Brownell, Jr. 1999. Gray whales (*Eschrichtius robustus*) off Sakhalin Island, Russia: seasonal and annual patterns of occurrence. *Marine Mammal Science* 15: 1208–1227.

Wespestad, V.G., R.G. Bakkala, and P. Dawson. 1990. Walleye pollock, in *Stock assessment and fishery evaluation document for groundfish resources in the Bering Sea-Aleutian Islands region as projected for 1991*. Compiled by the Plan Team for groundfish fisheries of the Bering Sea/Aleutian Islands of the North Pacific Fishery Management Council. Seattle: Alaska Fisheries Science Center, pp. 27–50.

Whitehead, H. 2003. *Sperm whales: social evolution in the ocean*. Chicago: University of Chicago Press.

Whitehead, H. and C. Glass. 1985. Orcas (killer whales) attack humpback whales. *Journal of Mammalogy* 66: 183–185.

Woodby, D.A. and D.B. Botkin. 1993. Stock sizes prior to commercial whaling, in *The bowhead whale*. J.J. Burns, J.J. Montaque, and C.J. Cowles, eds. Lawrence, KS: Society for Marine Mammalogy, pp. 387–407.

Yoshida, H. 1994. Food and feeding habits of pelagic pollock in the central Bering Sea in summer, 1976–1980. *Scientific Report of the Hokkaido Fisheries Experimental Station* 45: 1–35.

TWENTY

Legacy of Industrial Whaling

Could Killer Whales Be Responsible for Declines of Sea Lions, Elephant Seals, and Minke Whales in the Southern Hemisphere?

TREVOR A. BRANCH AND TERRIE M. WILLIAMS

Industrial whaling decimated populations of the great whales in the Southern Hemisphere and undoubtedly had a marked effect on the ecosystems in which they existed. In the Southern Hemisphere, industrial whaling fleets focused on a succession of less and less valuable species, from humpback *(Megaptera novaeangliae)*, blue *(Balaenoptera musculus)*, fin *(B. physalus)*, sperm *(Physeter macrocephalus)*, and sei whales *(B. borealis)* to, finally, Antarctic minke whales *(B. acutorostrata)* (Hilborn et al. 2003). This pattern occurred after the decimation of southern right whales *(Eubalaena glacialis australis)* in the nineteenth century. As a result, by the early 1970s, only Antarctic minke whales remained abundant in the Southern Hemisphere.

Such an unprecedented alteration in biomass has undoubtedly had significant effects on the Southern Hemisphere marine ecosystem. Large whales may be considered both predator and prey in these systems, and their carcasses provide life for unique ocean-bottom communities (e.g., Smith et al. 1989; Smith and Baco 2003; Smith, Chapter 22, this volume). Despite these varied roles, much of the speculation has focused on the role of great whales as predators, with a commonly accepted theory (e.g., Laws 1977) being that their removal resulted in a huge "krill surplus," which in turn

allowed other krill predators to increase in abundance (e.g., Laws 1977; Ballance et al., Chapter 17, this volume).

In comparison, little attention has been paid to the potential role of large cetaceans as prey. Killer whales *(Orcinus orca)* in particular are known to feed on great whales (Jefferson et al. 1991), and they account for the vast majority of predatory attacks on these large cetaceans. Even if large cetaceans were only a small component of the annual diet of killer whales, it is still worthwhile to question whether changes occurred in the diet of killer whales following the decimation of large-whale populations.

An intriguing hypothesis has recently been raised for the North Pacific Ocean and southern Bering Sea: that the reduction in large-cetacean biomass from whaling caused killer whales to prey on other species (Springer et al. 2003). The resulting increase in predation then initiated a sequential decline in populations of harbor seals *(Phoca vitulina)*, northern fur seals *(Callhorinus ursinus)*, Steller sea lions *(Eumetopias jubata)*, and sea otters *(Enhydra lutris)* in the area (Estes et al. 1998; Springer et al. 2003; Williams et al. 2004) (Figure 20.1). In turn, the loss of sea otters caused marked changes in kelp forest ecosystems (Estes et al. 2004). If the Springer et al. (2003) hypothesis is true in regards to the North Pacific

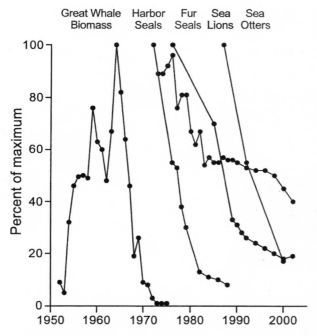

FIGURE 20.1. The sequential collapse of marine mammals in the North Pacific Ocean and southern Bering Sea, all shown as proportions of annual maxima. Great whales: International Whaling Commission reported landings (in biomass) within 370 km of the Aleutian archipelago and coast of the western Gulf of Alaska. Harbor seals: counts and modeled estimate (1972) of Tugidak Island. Fur seals: average pup production on St. Paul and St. George islands, Pribilof Islands. Steller sea lions: estimated abundance of the Alaska western stock. Sea otters: counts of Aleutian Islands. For fur seals and harbor seals, 100% represents population sizes at the time effects of excessive harvesting ended and "unexplained" declines began. Caption and figure reproduced with permission from Springer et al. (2003). Original sources cited in Springer et al. (2003).

Ocean and the southern Bering Sea, then similar patterns should be evident in the Southern Hemisphere. Indeed, more than 25 years ago Barrat and Mougin (1978) proposed an identical explanation for the decline of southern elephant seals in the Crozet Archipelago with figures analogous to those in Springer et al. (2003).

Here we examine the plausibility of the Springer et al. (2003) hypothesis for the Southern Hemisphere, focusing on marine mammal populations that occur year-round or seasonally south of about 40°S. We specifically looked for evidence of the following corollaries of Springer et al. (2003):

1. Marine mammal species that are frequently preyed on by killer whales should have declined in the 20th century.

2. Marine mammal species not often preyed on by killer whales should have increased (if depleted) or remained stable (if not depleted).

3. Any declines should be consistent with increased killer whale predation.

If these three corollaries hold, we can conclude that it is possible to reconcile the Springer et al. (2003) hypothesis with the situation in the Southern Ocean. We regard this as the first step toward developing an integrative model that explicitly takes into account sources of uncertainty and broader ecological links between trophic groups.

For this analysis we reviewed the evidence for a dietary shift by killer whales in the Southern Hemisphere following the depletion of the great whales. We focused on marine mammal populations south of 40°S corresponding to the location of industrial whaling. Primary prey of killer whales included southern elephant seals (*Mirounga leonina*), southern sea lions (*Otaria flavescens*), and Antarctic minke whales. These three species are considered "primary prey" because of the high number of predatory interactions with killer whales (Jefferson et al. [1991] recorded >200 predatory interactions for southern elephant seals and southern sea lions, and no more than 68+ for any other species), and because stomach content data indicate that minke whales are the single most important prey species for mammal-eating killer whales in the Southern Hemisphere (this chapter). Available evidence indicates that localized populations of these three species may have declined, whereas other marine mammal species have remained stable or increased.

Our energetic and demographic models suggest that it is plausible that these declines in southern elephant seals and southern sea lions were caused by increased predation by killer whales, but that the decrease in estimates for minke whales is too precipitous for predation to be the sole cause. Admittedly, these findings are speculative since direct evidence for or against this hypothesis is difficult to obtain. However, these results raise important questions regarding the historical importance of great whales in the diet of killer whales, and the long-term effects of industrial whaling on these predators in the Southern Hemisphere.

Depletion of Large-Cetacean Populations by Industrial Whaling

More than two million great whales were killed in the Southern Hemisphere during the twentieth century, 1.6 million in the Antarctic alone (90 million metric tons), compared to "just" 800,000 in the entire Northern Hemisphere. More than 54 million metric tons (60% of the total) were caught between 1938 and the mid-1960s (Figure 20.2). Today, humpback, southern right, and Antarctic blue whales (*Balaenoptrera musculus intermedia*) remain at small fractions of their pre-exploitation levels (IWC 2001; Johnson and Butterworth 2002; Zerbini 2004; Branch et al. 2004), and it is likely that sperm whales (Whitehead 2002), fin whales, and possibly sei whales are also still well below pre-exploitation levels. We have not attempted to combine abundance trajectories for each species, since these are available for only a few intensively-studied species (i.e., southern right whales, humpback whales, Antarctic blue whales) and because updated catch series that include illegal Soviet Union whaling have only recently become available. Major gaps in knowledge, particularly for important components

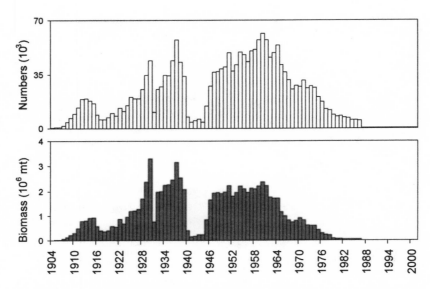

FIGURE 20.2. Total numbers (in thousands) and estimated biomass (in millions of metric tons) of large cetaceans killed each year in the Southern Hemisphere during the industrial whaling era. Source: C. Allison, International Whaling Commission, December 2002. Numbers were converted into biomass using the mean masses in Laws (1977). These catches include the most important species: blue, fin, sperm, humpback, sei and minke whales; catches for other species were relatively small in the twentieth century. Catches declined in the early 1940s because whaling vessels were diverted to other purposes during World War II.

such as sperm whales and fin whales, are likely to overwhelm available information from other species. Where previous authors have attempted this task, they have suggested that industrial whaling reduced the overall biomass of large cetaceans (excluding minke whales) to 16–20% of original levels (Laws 1977; Knox 1994). Given the great uncertainty in these estimates, however, we assume in this paper that current total biomass is 10–33% of the original biomass. This implies that whale biomass before 1938 was 64–73% of original levels (i.e., 2.2–6.4 times current biomass), since 40% of the catches by weight were caught before 1938 (Figure 20.2).

Rise and Fall of Megafaunal Populations in the Southern Hemisphere

Trends are not available for most marine mammal populations that reside year-round or seasonally south of 40°S, except for charismatic megafauna that can be easily counted. Of the well-studied megafauna, two fur seal species and three whale species have increased exponentially since exploitation of the great whales halted. In contrast, southern sea lions and southern elephant seals have shown marked declines in portions of their ranges; likewise, some evidence suggests that Antarctic minke whales may have declined during the latter half of the twentieth century.

Pack Ice Seals

Four species of seals breed in the Antarctic pack ice. Available circumpolar estimates of 7–12 million crabeater seals

(Lobodon carcinophagus) suggest that this phocid may be one of the most abundant seals in the world (Erickson and Hanson 1990), although these estimates are known to be imprecise. Numbers of Weddell seals (Leptonychotes weddellii), leopard seals (Hydrurga leptonyx), and Ross seals (Ommatophoca rossii) are even less precisely known, but each species probably numbers at least several hundred thousand (Erickson and Hanson 1990). More recent estimates are anticipated from the multinational APIS (Antarctic Pack Ice Seal) program (e.g., Ackley et al. 2003). Without long time series of abundance estimates it is difficult to infer trends in these species, although the estimated density of crabeater seals in the Weddell Sea dropped from 11.38 to 4.28 per square nautical mile between 1968–1969 and 1983, and in the Pacific Sector they dropped from 4.93 to 1.95 per square nautical mile between 1973–1974 and 1983 (Erickson and Hanson 1990). Reasons for these declines point away from killer whale predation. First, the apparent decrease may be partially an artifact associated with surveys being conducted at different times of their molt (Green et al. 1995). Second, scarring patterns indicate that leopard seals, not killer whales, are the major predators of crabeater seals (Siniff and Bengtson 1977).

Fur Seals

Antarctic and sub-Antarctic fur seals (Arctocephalus gazella and A. tropicalis) were both depleted to the point of extinction by the nineteenth century. Their subsequent recovery has been astonishing. Ninety-seven percent of Antarctic fur seals are located at South Georgia, but they have been increasing at 10–20% per annum, and have recolonized most of their former range to number 1.6 million in 1997 (Hofmeyr et al. 1997). Declines due to leopard seal predation have been reported at one location in the South Shetland Islands (Boveng et al. 1998). Everywhere else, fur seal numbers have increased. Similarly, all populations of sub-Antarctic fur seals are increasing rapidly, at rates of 0.4–23.8% p.a., with numbers >320,000 in the early 1990s (Hofmeyr et al. 1997).

Southern Sea Lions

Southern sea lions are currently found around South America. We collated abundance estimates for the Falkland Islands and three regions of Argentina. The Falkland Islands were historically the biggest southern sea lion component, but pup production at the Falkland Islands fell from 80,550 (1938) to 6,000 (1965), decreasing to a minimum between 1965 and 1990 before increasing to 2,050 in 1995, and then 2,747 in 2003 (Thompson et al. 2005). In Northern Patagonia, total counts dropped from 137,500 in 1938 to 18,400 in 1949 before stabilizing at pup counts of 3,187 (1975), 3,344 (1981), and 2,968 (1982) (Crespo and Pedraza 1991). The numbers then increased to an estimated total of 44,482 animals in 2002 (Dans et al. 2004). In Central and Southern Chubut, total counts declined from 33,000 (1947–49) to 8,800 (1974) before increasing to 10,500 in 1989 and then 14,900 in 1995 (Reyes et al. 1999). In Santa Cruz and Tierra del Fuego, total counts declined from 84,714 during 1946–49 to 12,310 in 1992–97.

To combine the abundance estimates from these sources, we applied the following methods:

1. The Central/Southern Chubut and Santa Cruz/Tierra del Fuego total counts were multiplied by 1.8 to correct for animals in the water (Reyes et al. 1999).

2. Pup counts were multiplied by 4.7 to obtain total individuals, using the ratio in Thompson et al. (2005).

3. Abundance was considered constant before the first survey in a region (potentially under-estimating the magnitude of the declines).

4. A linear change in population between surveys was assumed.

5. Where available, recent annual rates of increase were used to predict population sizes around the year of the most recent surveys as follows: Northern Patagonia (5.7%, 1983–2002; Dans et al. 2004), Central and Southern Chubut (3%, 1995 onwards; Schiavini et al. 2004), Falklands (8.5%, 1990–95, and 3.8%, 1995–2003; Thompson et al. 2005).

6. For Santa Cruz and Tierra del Fuego, we assumed a 4.2% annual increase after the 1992–1997 surveys (mean of other three regions during this period), as future increases were likely in that region (Schiavini et al. 2004).

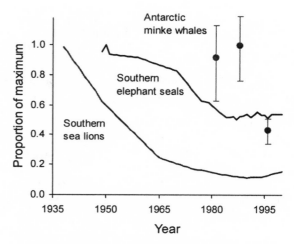

FIGURE 20.3. Estimated declines for southern sea lions, southern elephant seals, and Antarctic minke whales. The numbers for southern sea lions are a combination of population estimates from the Falklands Islands, and Argentina: Northern Patagonia, Central and Southern Chubut, and Santa Cruz and Tierra del Fuego. Southern elephant seals represent totals from Kerguelen, Macquarie, Heard, Marion, Crozet, Campbell, and Gough Islands; total declines may be underestimated because the Kerguelen population (the largest) was assumed to be stable before the first survey in 1970. Survey estimates were sporadic on most islands, requiring interpolation to estimate abundance in the intervening years, as discussed in the text.

When combined, the total southern sea lion abundance for the Falklands and Argentina declined from 729,000 in 1937 to 213,000 in 1963 and then declined further to 84,000 in 1990 before increasing to 129,000 in 2003 (Figure 20.3). However, given the sporadic timing of the surveys, it is difficult to pinpoint exactly when the declines ended and recovery began. For example, minimum abundances were likely reached between 1965 and 1990 for the Falklands and between 1946–1949 and 1992–1997 for Santa Cruz/Tierra del Fuego. The great majority of the earlier losses were clearly due to hunting, with catches totalling 432,000 from 1937 to 1962, but estimated declines of 499,000 between 1937 and 1962 were still greater than expected from sealing alone, and overall declines apparently continued after sealing halted (Thompson et al. 2005).

Southern Elephant Seals

Southern elephant seals are found in three distinct populations: South Georgia, Kerguelen, and Macquarie (Laws 1994). The South Georgia population (including South Georgia itself, the South Orkneys, the South Shetlands, the Falklands, Gough, and Peninsula Valdez) showed an overall increasing trend since the 1950s to 397,000 in 1990, except for the declining subpopulation at Gough Island, which had only 11 births in 1997 (Bester et al. 2001). Increases in the South Georgia population have been driven by a population explosion at

TABLE 20.1

Rates of Change for Subpopulations of Southern Elephant Seals in the Kerguelen and Macquarie Populations, and the
Periods Over Which They Were Estimated

Location	Population in 1990	Earlier Rate of Change (p.a.)	Subsequent Rate of Change (p.a.)	Reference
Isles Kerguelen	143,500	–3.6% (1970–1987)	+1.1% (1987–1997)	Guinet et al. (1999)
Macquarie Island	77,791	–2.1% (1959–1985) –1.5% (1985–1994)	No change (1994–2002)	McMahon et al. (1999); Burton, pers. comm.
Heard Island	40,355	–1.9% (1949–1985)	No change (1985–1992)	Slip and Burton (1999)
Isles Crozet	2,023	–5.35% (1970–1990)	No change (1990–1997)	Guinet et al. (1999)
Marion Island	2,009	–5.8% (1951–1994)	Increase (1993–2001)	Pistorius and Bester (2002); Bester, pers. comm.
Campbell Island	20	–7% (1940s–1988)	Not known	Laws (1994)

Peninsula Valdez from 1,000 in 1951 to 33,726 in 1990 and by an approximate 15% increase at South Georgia from 310,000 in 1951 to 357,000 in 1990 as a result of recovery of males after exploitation (Laws 1994). Note that pup counts at South Georgia itself remained stable during this latter period.

In contrast, in the Kerguelen population (Isles Kerguelen, Heard Island, Isles Crozets, and Marion Island; numbering 189,000 in 1990) and the Macquarie population (Macquarie Island, Campbell Island; numbering 78,000 in 1990), southern elephant seal abundances at various locations declined by 44–96% in the second half of the twentieth century (Table 20.1; Laws 1994). The rate of change in these locations was significantly correlated with 1960 abundance (estimated from linear interpolation between surveys), that is, declines were greater in small rookeries than in large rookeries (Figure 20.4). Declines in the Kerguelen and Macquarie populations halted or reversed in the late 1980s to mid-1990s (Table 20.1). Although these declines may have started in the 1940s or 1950s, Isles Kerguelen (the largest of the declining rookeries) was first surveyed in 1970, and, therefore, total declines may be underestimated. From 1970 to the mid-1980s, total southern elephant seal numbers in the Kerguelen and Macquarie populations declined from 450,000 to 280,000 at an annual rate of –3.2% before stabilizing or slightly increasing through to the present (Figure 20.3).

Cetaceans (Except Antarctic Minke Whales)

Abundance estimates for most large and small cetaceans are not available or do not have sufficient precision to detect trends (Branch and Butterworth 2001a), but most formerly heavily exploited cetacean species are probably recovering. There is statistical evidence of increases in three populations of southern right whales at annual rates of 6–9% (Bannister 2001; Best et al. 2001; Cooke et al. 2001), for two populations

of humpback whales at 10–11% p.a. (Bannister 1994; Paterson et al. 2001), and for Antarctic blue whales at 7.3% p.a. (Branch et al. 2004).

Antarctic Minke Whales

Antarctic minke whales are the only known Southern Hemisphere cetacean species that may be declining. This species was never heavily exploited and was thought to be at, or even above, pre-exploitation levels. Comparable estimates from the IDCR/SOWER surveys show significant decreases in minke whale numbers from 931,000 in the 1985/86–1990/91 circumpolar set of surveys, to 402,000 for the 1991/92–2000/01 set (Branch and Butterworth 2001b; Branch 2003), representing –10% per year. The International Whaling Commission (IWC) is currently investigating whether this decrease reflects a real decline in abundance or whether it can be explained by changes in survey methodologies (e.g., IWC 2003; Mori et al. 2003). For the purposes of this paper, we set aside the question of whether the decrease is real, and merely ask whether killer whale predation alone could have caused such a rapid decrease in minke abundance.

Killer Whale Ecotypes and Diets

Several ecotypes and dietary preferences have been identified for killer whales in the Southern Hemisphere. The following sections describe the abundance and diets of the three different ecotypes of killer whales identified in Antarctic waters.

Ecotypes and Abundance

Killer whales have a circumpolar distribution reaching into temperate waters and occasionally into subtropical or tropical waters (Forney and Wade, Chapter 12, this volume). They are more common in southerly waters and most abundant in and around the Antarctic pack ice (Budylenko 1981;

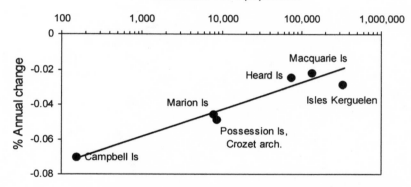

FIGURE 20.4. Relationship between the annual rate of change during declines (see Table 20.1) and 1960 abundance estimates of southern elephant seals (extrapolated or interpolated from surveys) for different locations within the Kerguelen and Macquarie populations. The correlation is $r^2 = 0.92$ ($P < 0.01$).

Dahlheim 1981). Three different ecotypes (Type A, B, and C), or possibly species, of killer whales occur in Antarctic waters (Berzin and Vladimirov 1983; Pitman and Ensor 2003). Type A killer whales are the familiar worldwide form of *Orcinus orca*, found throughout the Southern Hemisphere in the austral winter (Pitman and Ensor 2003). In summer, Type A killer whales migrate to the Antarctic, north of the pack ice, and prey primarily on minke whales (Pitman and Ensor 2003). Type B killer whales appear to prey primarily on seals in the Antarctic, although minke whales and humpback whales may also be targeted. Type B may migrate into the tropics but are typically found in loose Antarctic sea ice year-round (Pitman and Ensor 2003). Type C killer whales (putative *O. glacialis*; Berzin and Vladimirov 1983), penetrate further into the fast ice and may be resident year-round in the Antarctic (Pitman and Ensor 2003). This third type of killer whale specializes on fish, notably Antarctic toothfish *(Dissostichus mawsoni)*. Pods of the three killer whale types never intermingle and are readily distinguishable in the field by the shape and size of the white eyepatches and dorsal capes (Pitman and Ensor 2003).

Recent estimates of killer whales have been made from three circumpolar sets of surveys in the IDCR/SOWER program (described in Matsuoka et al. 2003). Estimates from the most recent two circumpolar sets were 27,000 (CV = 0.26, 95% interval 17,000–44,000) and 25,000 (CV = 0.23, 95% interval 16,000–39,000), and apply to killer whales outside the Antarctic pack ice south of 60°S during the austral summer (Branch and Butterworth 2001a). The first set of surveys in the early 1980s provided estimates of 91,000 and 180,000 killer whales (Hammond 1984, Branch and Butterworth 2001a), but were likely biased because one survey vessel remained close to the ice edge, where killer whales concentrate.

A combined estimate of 26,000 (mean of the most recent two sets of surveys) is a reasonable minimum point value, but is negatively biased because: (1) the second set did not extend northwards to 60°S, (2) the third set had not yet completely encircled the Antarctic, (3) some killer whales on the trackline

were missed, (4) some killer whales remained north of the survey area, and (5) some killer whales penetrated into the pack ice, where survey vessels cannot follow.

Unfortunately, the three types of killer whales were not distinguished in the IDCR/SOWER surveys. Therefore, additional assumptions must be made to obtain separate estimates for each type. School sizes provide one crude method of discrimination: Type A have a mean school size of 13.6 (range 1–38), similar to the 11.8 (range 2–31) for Type B schools, but Type C schools are considerably larger with a mean of 46.1 (range 10–200) (Berzin and Vladimirov 1983; Pitman and Ensor 2003). We assume that schools ≥29 are Type C, and schools ≤28 are Type A or B, based on normal likelihoods associated with the means and standard deviations of school sizes identified in Pitman and Ensor (2003). When separated in this manner, 87% of schools and 48% of killer whales sighted on the IDCR/SOWER surveys were Type A or B, equaling 12,500 whales. To separate Type A and B, it is worth noting that 85% of the marine mammal prey in killer whale stomach contents were minke whales, and the other 15% were pinnipeds (derived from Table 20.2, see below). If we assume that in the Antarctic, Type A killer whales eat only minke whales, and Type B eat only pinnipeds, then these data suggest that there are 10,600 Type A and 1,900 Type B killer whales. However, the stomach content data were likely non-randomly collected, and could therefore bias these rough estimates. An alternative approach is to add up the number of whales of each type summarized in Pitman and Ensor (2003), obtaining the following proportions: 27% Type A, 29% Type B, and 44% Type C, which translates into 7,020 Type A, 7,540 Type B, and 11,440 Type C whales. Unfortunately Pitman and Ensor's compilation is biased toward more frequented Antarctic regions (e.g., the Antarctic Peninsula). The uncertainties in this rough separation analysis overwhelm the available data, but there are likely to be thousands, or even tens of thousands of Type A killer whales. For this paper, we will assume there are 9,000 Type A killer whales (the rounded mean of the two estimates above).

TABLE 20.2

Summary of Published Southern Hemisphere Killer Whale Stomach Contents

Number Sampled	Area	Latitude	Month	Stomach Contents	Cited Reference	Notes
?	Temperate/warm waters	N of 50°S	?	Mainly dolphins/pinnipeds, one basking shark, also sperm whales	Mikhalev et al. (1981)	Original reference: Yukhov et al. (1975)
1	Rio Grande, Brazil	N of 50°S	?	Electric rays, sharks, eagle ray	Mikhalev et al. (1981)	Original reference: Castello (1977), stranding
7	E coast Argentina	N of 50°S	Nov	100% minke	Mikhalev et al. (1981)	7 stomachs full
9	Indian Ocean	N of 50°S	Mar	44% pinniped	Mikhalev et al. (1981)	5 empty
5	New Zealand waters	N of 50°S	Mar	40% other whales	Mikhalev et al. (1981)	3 empty
30	Temperate/warm waters	30-50°S	Mar-Apr	47.3% dolphins, 20.1% pinnipeds, 42.1% fish	Budylenko (1981)	Original reference: Shevchenko (1975)
1	NW Bouvet Is	50°S-60°S	Apr	100% dolphin	Budylenko (1981)	
49	Cold waters	S of 50°S	Nov-Mar	84.2% minke, 6.8% fish, 2.3% squid, number sampled may refer only to minkes.	Budylenko (1981)	Original reference: Shevchenko (1975)
?	Cold waters	S of 50°S	?	70-85% minke	Budylenko (1981)	Original reference: Doroshenko (1978)
4	Indian Ocean	S of 50°S	Nov	60% minke	Mikhalev et al. (1981)	2 stomachs full, 2 low
28	Indian Ocean	S of 50°S	Dec	64% minke, 25% pinniped	Mikhalev et al. (1981)	46.4% minke, 35.7% mixed minke/pinniped, 7.1% pinniped
?	Antarctic?	S of 60°S?	?	Meat, tongues from blue, fin, humpback, sei, leopard seals, Weddell seals	Mikhalev et al. (1981)	Original reference: Sleptsov (1965), cetacean remains from whaling vessels?
38	Antarctic Area I	S of 60°S	Jan	58% minke, 5% pinniped	Mikhalev et al. (1981)	14 stomachs empty, 21 minke, 2 minke+pinniped, 1 pinniped
9	Weddell Sea	S of 60°S	Jan	6% minke, 67% pinniped, 11% fish	Mikhalev et al. (1981)	5 pinniped, 1 minke/pinniped, 2 pinniped/fish
3?	Ross Sea	S of 60°S	Jan	25% minke, 25% pinniped, 25% fish, 25% squid	Mikhalev et al. (1981)	Minke/pinniped and fish/squid were common
55	Antarctic Area I	S of 60°S	Feb	15% minke, 5% pinniped, 1% fish, 16% squid	Mikhalev et al. (1981)	14 empty, 9 squid, 6 minke, 4 pinniped/minke, 1 pinniped/fish
3	Ross Sea	S of 60°S	Mar	67% minke	Mikhalev et al. (1981)	1 empty
1	Near Enderby Land	S of 60°S	Feb	45.4% pinnipeds	Budylenko (1981)	Original reference: Shevchenko (1975)
156	Antarctic	S of 60°S	?	89.7% marine mammals, 3.2% fish, 7.1% squid	Berzin and Vladimirov (1983)	Orcinus orca = Type A (possibly also Type B)[a]
629	Antarctic	S of 60°S	?	0.4% marine mammals, 98.5% fish, 1.1% squid	Berzin and Vladimirov (1983)	Orcinus glacialis = Type C[a]
362	Antarctic	S of 60°S	?	31% minke, 5% pinniped, 60% fish, 5% squid	Mikhalev et al. (1981)	Original reference: Ivashin (1981)

[a] Ivashin (1981) likely presented combined data that were later separated by Berzin and Vladimirov (1983) into Type A (89.7% marine mammal) and Type C (98.5% fish) stomach contents.

TABLE 20.3
Calculation of the Year-Round Diet of Type A Killer Whales in the Southern Hemisphere, 1975–1983

Number Sampled	North or South of 50°S	Large Cetaceans	Minke Whales	Pinnipeds	Dolphins	Fish	Squid	Other
1	North							100%
7	North		100%					
9	North			44%				
5	North	40%						
30	North			20.1%	47.3%	42.1%		
1	South				100%			
49	South		82.4%			6.8%	2.3%	
4	South		60%					
28	South		64%	25%				
38	South		58%	5%				
9	South		6%	67%		11%		
55	South		15%	5%		1%	16%	
3	South		67%					
1	South			45.4%				
156	South		76.2%	13.5%		3.2%	7.1%	
52	North total	4.3%	15.0%	21.3%	30.3%	27.0%	0.0%	2.1%
344	South total	0.0%	74.9%	13.8%	0.4%	3.5%	7.4%	0.0%
396	Year-round	2.8%	34.9%	18.8%	20.3%	19.1%	2.5%	1.4%

NOTE: Sample-weighted dietary proportions were calculated for stomach contents collected north and south of 50°S using the data in Table 20.2 (after excluding some rows, see text for details). Year-round diet was estimated by assuming that Type A killer whales spent eight months north of 50°S and four months south of 50°S.

We focused on Type A killer whales because they prey on Antarctic minke whales (Mikhalev et al. 1981), are found outside the Antarctic during the austral winter where southern sea lions and southern elephant seals reside (R. Pitman, pers. comm.), and were the most likely ecotype to be affected by industrial whaling. Therefore, Type A killer whales are most likely to be implicated in the declines of those species. The other two types of killer whales rarely prey on large cetaceans—although Pitman and Ensor (2003) note instances of Type B and Type C killer whales apparently attacking humpback whales—and also occur inside the pack ice or fast ice, where encounters with large cetaceans other than minke whales are unlikely. Presumably, therefore, the decimation of large cetaceans would have had little impact on their diet.

Assessing the Diet of Type A Killer Whales

Killer whales have been seen attacking or harassing a wide variety of large whales that are present in the Southern Hemisphere, including blue, fin, sei, sperm, minke, southern right, humpback, and Bryde's whales (B. edeni) (Jefferson et al. 1991). These predators also take pinnipeds, dolphins, sharks, fish, and squid (e.g., Yukhov et al. 1975; Siniff and Bengtson 1977; Jefferson et al. 1991). To determine the preferred year-round diet of Type A killer whales, we examined published accounts (summarized in Mikhalev et al. 1981;

Budylenko 1981; and Berzin and Vladimirov 1983) of the stomach contents of killer whales from 1975–83 (Table 20.2). We assumed that samples collected south of 50°S applied to four months of the year, and that samples collected north of 50°S were applicable to the remaining eight months. For year-round diet analysis, the data from Ivashin (1981) were omitted since they were likely a subset of the data in Berzin and Vladimirov (1983). The data for Type C killer whales in Berzin and Vladimirov (1983) were also omitted, and the unspecified "marine mammals" in the stomachs of probable Type A (or B) killer whales in that paper were assumed to be 85% minke whales and 15% pinnipeds—the sample-weighted ratio from the other Antarctic studies in Table 20.2. Overall averages were weighted by the number of stomachs measured (rows were omitted if the sample size was not recorded).

The stomach contents indicated that Type A killer whales primarily ate minke whales south of 50°S, but that pinnipeds, dolphins, and fish were important prey in warmer waters (Table 20.3). Their estimated annual diet consisted primarily of minke whales (36%) and pinnipeds (19%) but also included surprisingly large proportions of dolphins (20%) and fish (19%), although estimates for the latter two groups are disproportionately influenced by the 30 samples collected by Shevchenko (1975).

Large cetaceans constituted 2.8% of the year-round diet, although this proportion is entirely dependent on the results

TABLE 20.4

Historical Potential Year-Round Dietary Proportions for Type A Killer Whales

Year	Multiple	Large Cetaceans	Minke Whales	Pinnipeds	Dolphins	Fish	Squid	Other
1900	3×	8.5%	32.9%	17.7%	19.1%	18.0%	2.3%	1.3%
1900	10×	28.5%	25.7%	13.9%	15.0%	14.1%	1.8%	1.0%
1938	2.2×	6.3%	33.7%	18.2%	19.6%	18.5%	2.4%	1.4%
1938	6.4×	18.2%	29.4%	15.9%	17.1%	16.1%	2.1%	1.2%
1975–83	×	2.8%	34.9%	18.8%	20.3%	19.1%	2.5%	1.4%

NOTE: It is assumed that the killer whales' dietary intake of abundance of large cetaceans was 3 or 10 times greater in 1900 (before industrial whaling) than in 1975–1983. The proportion of other dietary groups in their diet is prorated accordingly. Estimates for 1938 assume that 40% of the declines in large cetaceans occurred between 1900 and 1937 (40% of catch weight was taken during that period), and hence that the proportion of large cetaceans in their diet in 1938 was 2.2 to 6.4 times greater than between 1975 and 1983.

of Mikhalev et al. (1981), who reported "other whales" in two out of five stomachs collected in New Zealand waters. Two additional citations (Yukhov et al. 1975; Sleptsov 1965) also reported large cetaceans in killer whale stomachs, but their sample size was not recorded in Mikhalev et al. (1981), and hence their results were excluded from the calculations of year-round diet. If Type A current diet includes 2.8% large cetaceans, then their diet in 1900 may have included 8.5–28.5% large cetaceans, assuming large cetaceans were 3–10 times more abundant in 1900 (Table 20.4). Furthermore, their 1938 diet may have included 6.3–18.2% large cetaceans, assuming that large cetaceans then were 2.2–6.4 times more abundant than in 1975–83. Higher proportions of large cetaceans in their diet imply lower percentages of pinnipeds: 13.9–17.7% in 1900, and 15.9–18.2% in 1938, compared to 18.8% in 1975–1983; and minke whales: 25.7–32.9% in 1900, and 29.4–33.7% in 1938, compared to 34.9% in 1975–83 (Table 20.4). In obtaining these estimates, we make the assumption that changes in killer whale diet were linearly related to the abundance of great whales.

Killer Whale Predation on Pinnipeds

Killer whales are known to prey on pinnipeds, including southern sea lions, with >200 predatory interactions in a review by Jefferson et al. (1991). In South America the presence of killer whales coincides with the availability of pups. Thirty individual killer whales have been identified off Northern Patagonia, 17 of which returned annually from 1985 to 1997; their mean school size was 5.6 animals (Iñíguez 2001). These killer whales were sighted most frequently in December and in March–May, coinciding with southern sea lion breeding season and the first entry of young pups into the water (Iñíguez 2001). Several killer whales off the coast of Argentina are well known for intentionally stranding themselves in the surf zone in their attempts to catch southern sea lions, which constitute 86% of their diet (Lopez and Lopez 1985; Hoelzel 1991).

At Marion Island, 25–30 killer whales (Keith et al. 2001) start arriving in September. A consistent annual peak in sightings during October–December is significantly correlated with the number of southern elephant seal pups and adult females in the area, and most (34–63%) of the killer whales remain extremely close to the shore (within 5 m) (Condy et al. 1978; Keith et al. 2001). At the Crozet Archipelago, predation on southern elephant seals is commonly observed (Guinet and Jouventin 1990; Guinet 1991, 1992; Guinet et al. 2000).

We have also found that killer whale abundance in discrete regions appears to be coupled with the status of southern elephant seal populations. When sightings data from the JSV database (1967/68–1987/88) in the 40°S–55°S latitude band (December–January) were examined, only 52 killer whales (0.091/day) were sighted around stable or increasing populations of elephant seals (20°W–70°W, corresponding to the South Georgia population). In comparison, 557 killer whales (0.285/day) and 398 killer whales (0.410/day) were sighted between 25°E and 95°E (Kerguelen population) and between 140°E and 180°E (Macquarie population), respectively. Both are regions where southern elephant seals were declining. A randomization test suggested that the differences in sighting rates were statistically significant in both cases ($P = 0.006$, $P = 0.0003$). This test involved drawing repeated (10,000 times) samples of the number of days in each region from the pooled data, and testing the means for differences in sightings rates equal to or greater than that observed. However, these results should be treated with caution, since the JSV sightings were collected on an opportunistic basis (they are not systematic surveys), and the relative density estimates may therefore be biased.

Killer Whale Predation on Antarctic Minke Whales

During the austral summer, killer whales migrate into Antarctic waters following minke whales, and there are numerous accounts of predation on minke whales (Budylenko 1981; Mikhalev et al. 1981; Jefferson et al. 1991; Guinet et al. 2000).

Most sightings at this time are close to the pack ice in areas where minke whales and seals are most common (Kasamatsu and Joyce 1995; Kasamatsu et al. 1996; Branch and Butterworth 2001a). The migration timing of killer whales and minke whales to the Antarctic is almost identical (Kasamatsu and Joyce 1995; Kasamatsu et al. 1996), and there is also a significant correlation between densities of minke whales and killer whales in survey strata ($r = 0.39$, $n = 102$, $P < 0.0001$) (Kasamatsu et al. 2000; Branch and Butterworth 2001a, b).

Dietary Switching Model

It is striking that the declines in southern sea lions and southern elephant seals, and possibly also Antarctic minke whales in the Southern Hemisphere (Figure 20.3), are analogous to declines in pinnipeds and sea otters in southeast Alaska following industrial whaling (Figure 20.1). In view of this, we investigated whether predation by killer whales could explain the decline of marine mammals in the Southern Oceans as has been suggested for North Pacific marine mammals (Estes et al. 1998; Springer et al. 2003). In other words, could increased predation from 9,000 Type A killer whales plausibly explain the declines? To assess this feasibility we used an energetics-demographic model similar to Williams et al. (2004), where the number of additional deaths were estimated using Leslie matrix models (Leslie 1945) and compared to the energetic requirements of adult killer whales.

Killer Whale Metabolism

Assuming an 82% assimilation efficiency (Kriete 1995), adult male killer whales require 287,331 kcal/day, and females require 193,211 kcal/day to meet basic metabolic needs (Williams et al. 2004). These are considered to be conservative estimates, as neither growth nor reproductive costs are included. For this analysis, we assumed that subadults eat approximately half as much as adult females and that calves obtain all their nutritional needs from their mothers.

MODEL STRUCTURE

The model accounts for only the female portion of the population and requires age-specific survival rates (S_a), catch mortality rates (F_a), and fecundity rates (R_a, female births per female). The numbers (N) at each age (a) in the initial population in each year (t) are then given by:

$$N_{a=0,t} = \sum_{a=1}^{A} N_{a,t-1} R_a$$
$$N_{a,t} = N_{a-1,t-1} S_{a-1} F_{a-1,t-1} \qquad 1 \le a < A$$
$$N_{A,t} = \frac{N_{A-1,t-1} S_{A-1} F_{A-1,t-1}}{1 - S_A F_{A-1,t-1}} \qquad a = A, \text{ plus group}$$

(20.1)

Stable age distributions were first obtained, and then survival rates were reduced (S^*) for either juveniles, adults, or for all ages, to match the observed rates of decline for each species. Average additional deaths over the period of the decline (from T_{start} year to T_{end} year) were calculated by:

$$\frac{1}{T_{\text{end}} - T_{\text{start}}} \sum_{t=T_{\text{start}}}^{t=T_{\text{end}}-1} \left(\sum_{a=0}^{A} N_{a,t} \left[S_a - S_a^* \right] \right)$$

(20.2)

Southern Sea Lions

For southern sea lions, we examined two time periods: 1937–1963 (including the entire catch history detailed in Thompson et al. 2005), and 1963–1990 (to the estimated end of the declines). Including the catch series is problematic because it starts in 1937, but the earliest estimates for each region vary from 1937 to 1949, and we assumed pre-survey abundance to be equal to the abundance estimates in the first surveys. The actual total abundance may therefore have been larger in 1937 than assumed in this model. For 1937–1963, the catch mortality rate was assumed to be zero for ages 0 and 1, and constant for older ages regardless of sex (Thompson et al. 2005). Age-specific parameters were obtained from Thompson et al. (2005): $S_0 = 0.6$, $S_{1-4} = 0.86$, $S_{5+} = 0.88$, $R_4 = 0.33R_{7+}$, $R_5 = 0.85R_{7+}$, $R_6 = 0.9R_{7+}$, $R_{7+} = 0.315$; except that for our model structure, R_{7+} was too small, and was replaced by 0.359 to obtain a stable age distribution of 729,000 with 64,210 female births. When catches were added, model abundance declined from 729,000 in 1937 to 268,000 in 1963 (similar to that year's estimated abundance of 213,000 sea lions).

We assumed that southern sea lions have the same body composition as Steller sea lions and that edible portions comprise 36% meat and 23% blubber. Average masses were 300 kg for adult males, 135 kg for adult females, and 25 kg for weaned pups (based on Vaz-Ferreira 1982).

To explain declines to 213,000 in 1963 and to 84,000 in 1990 from additional pup mortality alone, would require $S_0 = 0.554$ for 1937–1963, and 0.359 for 1963–1990, or 2,720 and 6,690 additional pup deaths annually in the two periods. Based on the energetic demands of the free-ranging animal and the caloric value of sea lion meat and blubber (Williams et al. 2004), an adult male killer whale would require six pups per day to meet basic nutritional needs; a female killer whale requires four pups per day. Killer whales are present around Northern Patagonia for about three months a year (Iñíguez 2001). According to our assumptions, eight killer whales (3 male, 3 females, 1 subadult, 1 calf; i.e., typical sex and age compositions recorded in Iñíguez 2001 and Keith et al. 2001) could eat 2,720 pups in three months; and 18 killer whales (7 males, 7 females, 2 subadults, 2 calves) could eat 6,690 pups in three months. Alternatively, 9,000 Type A killer whales could increase their year-round diet of sea lion pups by only 0.022% (1937–1963) or 0.049% (1963–1990) to account for the observed population declines.

If predation was solely on adult sea lion females, then their survival would have declined by 0.007 in the earlier period (to 0.853 and 0.873 for ages 1–4 and 5+, respectively), and by 0.036 in the later period (to 0.824 and 0.844 for ages 1–4 and 5+, respectively), with an additional 970 (earlier period) and 2,130 (later period) unexplained adult female deaths per year. A male killer whale requires 1.1 female sea lions per day; a female killer whale needs 0.75 sea lions per day. A school of 15 killer whales (5 males, 6 females, 2 subadults, and 2 calves) could account for the additional deaths in the earlier period, and 31 killer whales (12 adult males, 13 adult females, 3 subadults, and 3 calves) could account for additional deaths in the later period if their diet were composed solely of adult female southern sea lions. If adult male sea lions accounted for half the dietary intake of killer whales, these estimated killer whale numbers would need to double (although losses of male southern sea lions would not, in theory, cause an additional population decline). However, even a doubling of these numbers requires a year-round killer whale dietary increase of only 0.08% for the earlier decline and 0.17% for the later decline.

Southern Elephant Seals

A stable initial age distribution of 450,000 individuals was obtained with $R_3 = 0.248$, $R_4 = 0.416$, $R_5 = 0.441$, $R_{6+} = 0.5$, $S_0 = 0.6$, $S_{1,2,3} = 0.85$, $S_{4+} = 0.804$ (estimated), and 50,800 female pups (Pistorius et al. 1999; Pistorius et al. 2001; Pistorius and Bester 2002; McMahon et al. 2003). We assumed a female southern elephant seal would weigh 500 kg and a weaned pup 115 kg. Caloric values were based on those for phocid seals (Williams et al. 2004), assuming that 60% of both blubber and meat would be eaten.

For a decline from 450,000 to 280,000 (1970–1985) due solely to pup predation, first-year survival would need to decrease from 0.60 to 0.449, averaging 13,400 additional pup deaths annually. Based on our energetics model, a male killer whale requires 0.7 southern elephant seal pups per day, and a female 0.5 pups per day. For killer whales around the Marion Island southern elephant seal colonies, 29% were adult males, 46% adult females, 13% subadults, and 13% calves (Keith et al. 2001). If killer whales fed solely on elephant seal pups for three months, 325 killer whales would be needed to explain the declines (93 adult males, 148 adult females, 42 subadults, 42 calves). Alternatively, the year-round diet of Type A killer whales could increase by 0.89%.

If predation were solely on adult females (ages 4+), then a decline in survival from 0.804 to 0.744 would explain the decline in abundance, with an average of 4,440 additional female deaths annually. Just 311 killer whales (89 adult males, 142 adult females, 40 subadults, 40 calves) could cause the declines over three months. Alternatively, all Type A killer whales could increase their year-round diet by 0.85%.

Predation on the enormous adult male southern elephant seals (mass 2,000 kg) is probably not common, given the small school size (mean 2.8–2.9 killer whales, range 1–28) around Marion Island (Keith et al. 2001), and given that killer whale sightings are most frequent in months when southern elephant seal pups (and not adult males) are most abundant (Condy et al. 1978). In addition, no attacks have been observed on adult male southern elephant seals. Because of the harem structure in this species, an overwhelming majority of males would need to be killed (in theory) to cause population-level declines. We therefore did not model predation on male elephant seals.

Antarctic Minke Whales

Stable age distributions were difficult to obtain for Antarctic minke whales. The best available estimates of the input parameters are $S_{1+} = 0.932$ or 0.955 (Butterworth et al. 2002) and $R_{8+} = 0.392$ (Best 1982; Thomson et al. 1999). Therefore, S_0 would need to be implausibly low, much lower than for humpbacks (>0.8) and southern right whales (>0.9) (Best et al. 2001; Gabriele et al. 2001). Based on this, we assumed $S_{1+} = 0.932$ and $R_{9+} = 0.392$—a later age at maturity and minimum S_{1+} estimate. A stable age distribution of 931,000 was obtained with $S_0 = 0.305$ which was still considered extremely low, and 84,900 female calves born every year. In our modeling, calf predation alone, even with $S_0 = 0$, could not explain the estimated declines of Antarctic minke whales from 931,000 to 402,000 (1988–1996). As a result, we spread predation uniformly over all age groups for minke whales. Altered survivals of $S_0 = 0.215$ and $S_{1+} = 0.842$ replicated the estimated decline and required an additional 59,700 deaths of both sexes per year.

A 10-metric-ton adult minke would provide approximately 3,600 kg of meat and 1,000 kg of blubber depending on the condition of the whale. Killer whales need to eat 115 kg (for a male) or 77 kg (for a female) of whale meat per day. Alternatively, males would need 78 kg of blubber and meat (females, 48 kg). Gut capacity (Williams et al. 2001) limits adult killer whales to enough meat to last for two days or enough meat and blubber for three days. Because the mean school size of Type A killer whales in the Antarctic is 13.6 animals, there would be substantial wastage of adult carcasses, even if all school members satiated themselves (see also Doak et al., Chapter 18, this volume). We assumed that other killer whale schools are not drawn toward minke kills (attacks are usually brief, and minke whales rapidly sink), although this certainly occurs in attacks on other whale species (e.g., Pitman et al. 2001). Accounting for all of these limitations and assumptions, each killer whale school would require a minke whale every 2–3 days from mid-December to mid-February. With 9,000 Type A killer whales, there would be about 660 schools, which could eat only 13,000–20,000 minke whales in total in 60 days, just 23–35% of the 59,700 additional deaths required to account for the observed difference in minke estimates. If killer whale predation on minke whales is assumed to be year-round, then the proportion of minke whales in their diet would need to increase from 36% to 84–107%.

Discussion

The hypothesis outlined in this paper relies on the following key points:

1. Killer whales prey upon large cetaceans.

2. Large cetaceans were extensively whaled, reducing their abundance and biomass to a small fraction of former levels.

3. In response, killer whales increased predation on other marine mammals.

4. Those marine mammals subsequently declined (while others increased).

5. The magnitude of these declines can be explained by dietary changes in killer whales.

We also look at the anomalies that have been uncovered and discuss alternative hypotheses for the marine mammal declines.

Killer Whales Prey on Large Cetaceans

There is evidence for predatory activity by killer whales on most large cetacean species in the Southern Hemisphere (Jefferson et al. 1991), supported by the presence of sperm, fin, blue, and sei whale remains in killer whale stomachs (Sleptsov 1965; Yukhov et al. 1975; Mikhalev et al. 1981) and by characteristic scarring patterns on sperm (65% of individuals), fin (54.4%), sei (24.4%), humpback (17%), and minke whales (3.1–6.4%) (Shevchenko 1975; Best 1982; Naessig and Lanyon 2004). Scarring patterns in photo-identified humpback whales suggest that killer whale attacks are primarily on calves, with adults only rarely incurring fresh scars (Clapham 2001; Naessig and Lanyon 2004), and attacks on calves are also more common for other large cetaceans (Reeves et al., Chapter 14, this volume). Predation by killer whales is hypothesized to have led to the evolution of the migratory patterns of great whales from polar feeding grounds to temperate breeding grounds in the winter, in order to lessen killer whale attacks on calves (Corkeron and Connor 1999; Connor and Corkeron 2001), although this has been disputed by Clapham (2001).

The exact percentage of large cetaceans in the diet of killer whales is difficult to infer, even though there are considerably more stomach content analyses available for the Southern Hemisphere than for other regions. One problem is that there are many samples from the Antarctic during the austral summer, but fewer samples from temperate waters during the remainder of the year. In addition, stomach contents obtained during whaling operations may be compromised, since killer whales routinely scavenged on whale carcasses awaiting processing (e.g., Chrisp 1958; Mikhalev et al. 1981). The proportion of large cetaceans in Type A killer whale diet during 1975–1983, although estimated at 2.8%, is therefore best summarized as "a small proportion."

Large-Cetacean Populations Were Substantially Reduced by Whaling

Some cetacean species were reduced to 1–2% of their original levels from 1910 to 1970 due to industrial whaling (e.g., Branch et al. 2004). In total, 60% of the removals occurred after 1937. Estimates of the overall biomass depletion to 16–20% of original levels (Laws 1977; Knox 1994) are based on outdated assessments, exclude large-scale illegal Soviet Union whaling (e.g., Zemsky et al. 1995), and rely in particular on much guesswork about the depletion of sperm whales and fin whales. An overall reassessment of total biomass depletion is long overdue. Despite these caveats, it is unquestioned that industrial whaling severely depleted overall cetacean biomass levels—around South Georgia, the center of the industrial whaling fleets, sightings of large cetaceans are still rare (Moore et al. 1999). It therefore seems reasonable to assume that the original biomass of large cetaceans was perhaps 3–10 times greater than it is now.

Killer Whales Increased Predation on Alternative Prey

There are no long-term series of records for killer whale predation before, during, and after the declines in marine mammals in the Southern Oceans. In the absence of such data, we assumed that the proportion of large cetaceans in the diet of killer whales was 3–10 times greater than in 1975–1983 (note that this is true whether they fed on the adults, calves, or both adults and calves of large cetaceans). One of the major problems in collecting data on predation is that the events occur relatively quickly, are infrequent and widely spaced geographically, and occur in areas where human observers are rare. The rarity of observing predation events has been discussed for several marine mammals including killer whales. For example, only 2.7–14.3% of leopard seal predation events on Antarctic fur seals were observed, despite an intensive study by Hiruki et al. (1999). In 22 years of IDCR/SOWER surveys, there was only one well-documented account of killer whales eating a minke whale—their primary prey in the Antarctic. Thus, detecting an increase in predation from available data is unlikely, although a significant increase in killer whale predation on sea otters was inferred from only a small number of observed predation events (Estes et al. 1998; Hatfield et al. 1998).

Our hypothesis does not distinguish between a gradual increase in killer whale predation over time and periodic pulse predation. In one bay in the Crozet Archipelago, a quarter of all weaned southern elephant seal pups were eaten by killer whales (Guinet and Jouventin 1990). At Chiswell Island, Alaska, a single killer whale was responsible for depredating 13–24% of the Steller sea lion pups at that rookery in 2001 (Maldini et al. 2002). If pulse predation by killer whales occurs in Antarctic waters, it might explain why first-year survival of Macquarie Island southern elephant seal pups decreased linearly from 48% over 1951–1959 to just 2% in 1965 (Hindell 1991). This decreased survival preceded the overall population decline in this area. Conversely, at Marion Island, increased adult female mortality was considered to be the cause of the declines (Pistorius and Bester 2002). Other authors have hypothesized that killer whale predation can explain the patterns of declines in southern elephant seals between the Crozet Archipelago and Isles Kerguelen (Barrat and Mougin 1978; Guinet 1991; Guinet et al. 1999). At Isles Kerguelen (which has few sightings of killer whales), the elephant seal population is larger and the rate of decline less than at the Crozet Archipelago, where killer whales are sighted more frequently (Guinet et al. 1999; McMahon et al. 2003).

Alternative Prey Declined While Other Species Increased

Population trends for marine mammals in the Southern Oceans indicate that some populations of species that are primary prey for killer whales declined following industrial whaling, while those species rarely taken by killer whales flourished during the same time period. Declines in Argentine southern sea lions during the 1930s to 1980s, and elephant seals in the Indian Ocean sector during the 1940s to 1990s are well established. There was also an appreciable drop in the population estimates for Antarctic minke whales from the 1980s to the 1990s. All three of these species are primary prey for killer whales. In contrast, marine mammal species that are known to be increasing rapidly throughout the Southern Hemisphere (southern right whales, humpback whales, Antarctic fur seals, and sub-Antarctic fur seals), are rarely eaten by killer whales.

It is instructive to examine locations where an increasing species occurs together with a decreasing one. Marion Island offers this opportunity, with rookeries of southern elephant seals and sub-Antarctic fur seals. Local killer whales are sighted almost exclusively around the elephant seal rookeries, and only occasionally near the fur seal rookeries (Pistorius et al. 2002). In fact, although killer whale predation on Antarctic and sub-Antarctic fur seals (which are increasing rapidly throughout their range) is often presumed, such events have yet to be recorded (Jefferson et al. 1991).

Declines in Relation to Killer Whale Dietary Needs

The simple energetic/demographic models shown here demonstrate that the observed declines in southern sea lions can be explained by an additional 8–62 killer whales specializing on them for three months, or by a year-round increase in Type A killer whale diets of only 0.02–0.17%. Similarly, southern elephant seal declines can be explained by 311–325 additional killer whales in the area over three months, or a year-round dietary increase of 0.85–0.89%. If so, our results suggest that the proportion of pinnipeds in Type A killer whale diets would have increased from 15.9–18.2% in 1938 to 18.8% in the 1990s (i.e., by 0.6–3.9%), so these dietary increases seem plausible.

The situation is less plausible for (potential) minke whale declines. Even if all Type A killer whales fed solely on minke

whales of all ages for two months while in the Antarctic, the high number of additional deaths could not be explained. Killer whales would need to increase the proportion of minke whales in their *year-round* diet from 34.9% to 84–107%. This seems impossible, unless the abundance of Type A killer whales was grossly underestimated, or there was substantially greater wastage associated with a minke whale kill (and hence more frequent kills). As mentioned earlier, it is also possible that the estimated decline in Antarctic minke whales was an artifact associated with changes in the design and methodology of the surveys. Until better data are available, the conclusions of this part of the study should be interpreted with caution.

Anomalies

Why did some populations of southern elephant seals decline, while others remained stable or even increased? At Punta Norte, Argentina, southern elephant seals increased from 1,000 in 1951 to 33,700 in 1990 (Laws 1994), and in South Georgia, numbers have remained around 350,000 for decades (Laws 1994). There are several possible explanations. First, only 4% of the killer whale diet in Argentina's waters is composed of southern elephant seals, while southern sea lions, whose populations did decline, constitute 86% of the killer whale's diet (Iñíguez et al. 2002). Second, the population of elephant seals at South Georgia may be too large to be affected by a small, but consistent, level of killer whale predation. The smaller the initial population, the greater would be the measurable decline. Third, the whaling fleet in South Georgia once shot killer whales on sight because of their scavenging habits, wounding up to 12 in a single day (Chrisp 1958: 158). This may have substantially reduced killer whale numbers, such that killer whale sightings remain rare in this area (Moore et al. 1999). This notion is supported by JSV sighting rates of killer whales from 1968–1986, which were just 22–32% of those in waters around declining southern elephant seal populations.

Why did declines in both southern sea lions and southern elephant seals halt or reverse in the 1980s–1990s? If our hypothesis is true, when large cetaceans were removed, the balance between predator and prey was disturbed, but the killer whales did not starve—they merely caught different prey. One possibility is that this imbalance may have continued until killer whale numbers declined sufficiently through natural mortality for the equilibrium between predator and prey to return.

Alternative Hypotheses for the Declines

We have focused on a single hypothesis in this manuscript, but it is obvious that multiple factors may have caused the changes that we observed. A few of the alternative hypotheses are outlined in this section.

Declines in southern sea lions were primarily due to sealing (Thompson et al. 2005; Reyes et al. 1999), and substantial underreporting of catches would explain the discrepancies between total reported catches of 432,000 (1937–1962), and minimum declines of 645,000 (1937–1990). Nevertheless, initial declines were greater than can be explained by sealing alone, and unexplained declines continued even after the cessation of sealing (Thompson et al. 2005). It has been suggested that a density-independent factor, such as killer whale predation or pup deaths on the colony edges caused by conspecific adult males, is needed to explain why recovery did not occur earlier (Thompson et al. 2005).

For southern elephant seals, nutritional limitation has been the favored hypothesis to explain declines (e.g., Burton et al. 1997; McMahon et al. 2003), although population overshoot after reaching carrying capacity has also been proposed (Hindell et al. 1994). Nutritional stress has been inferred from the reduced size of breeding females in declining populations, although this might also be a consequence of increased predation. A second problem with the nutritional limitation hypothesis is that oceanwide estimates of resource availability do not exist. Consequently, there is no direct comparison between resources when the populations were declining and when they were stable or increasing. The fact that other marine mammal species were not limited by food availability, and increased spectacularly over the same period, also runs counter to this hypothesis. Finally, the declines in southern elephant seals were inversely proportional to their original population size, suggesting that a density-independent factor (e.g., predation) is responsible, not a factor that is density dependent (as would be associated with nutritional limitation and overshoot). Declines in some regions may be due to emigration, but if there was limited food availability, we would expect migration from large rookeries to small rookeries, not the reverse.

The differences in estimates for Antarctic minke whales may not reflect a true difference in abundance, as mentioned earlier. Many factors relating to survey design changed between the two sets of surveys and are being intensively studied as part of the IWC's comprehensive assessment of Antarctic minke whales. A full explanation of these factors is beyond the scope of this document, but interested readers can refer to Branch and Butterworth (2001b) and to IWC (2003) for further details.

Lastly, multidecadal regime shifts in climate patterns have been offered as an explanation of broad changes in the North Pacific, resulting from changes in the Pacific Decadal Oscillation, or PDO (Francis and Hare 1994; Mantua et al. 1997; Francis et al. 1998). Regime shifts have not yet been invoked as explanations of changes in the Southern Ocean; however, it is intriguing that the analog of the PDO in the Southern Oceans, the Antarctic Oscillation Index (AOI), was low during the 1960s and 1970s, and higher in the 1980s and 1990s (Gong and Wang 1999). These broad changes may have contributed to the reversal of declines in southern sea lions and elephant seals in the 1980s and 1990s.

We also do not discard the possibility that multiple factors may have played a role in the observed differences in estimates, and that teasing out the influence of a single factor or hypothesis on such widely-dispersed populations may be difficult or impossible.

Conclusions

Our study suggests that it is possible that killer whales increased the proportion of southern sea lions and southern elephant seals in their diet after large cetaceans were decimated by industrial whaling. Models estimating the required additional predation, and estimates of killer whale diets, suggest that it is plausible that they were responsible for subsequent declines in populations of southern sea lions and elephant seals. However, declines in Antarctic minke whales (if real) were too precipitous to be fully explained by increased killer whale predation. Nonetheless, there are many uncertainties in the available data from which our conclusions were drawn, and it is hoped that future research will be able to shed further light on this hypothesis, developed independently by Barrat and Mougin (1978) and Springer et al. (2003).

Acknowledgments

Many thanks to J. A. Estes for sparking T. A. B.'s interest in this project with a presentation at the 2002 Mote Symposium. Critical comments on the ideas presented here were received from P. B. Best, D. S. Butterworth, J. A. Estes, and from the reviewers R. L. Brownell and D. P. DeMaster, who suggested looking at the Antarctic Oscillation Index. Many thanks to C. Allison for providing updated summaries of the IWC catch data, and to T. Miyashita, National Research Institute of Far Seas Fisheries, Japan, for permission to use their JSV data. This work benefited immensely from presentations and discussions at the 2003 Symposium on Whales, Whaling and Ocean Ecosystems at Santa Cruz and at the 55th annual meeting of the Scientific Committee of the International Whaling Commission in Berlin. Acknowledging the input of these sources does not imply their endorsement of this chapter.

Literature Cited

Ackley, S.F., J.L. Bengtson, P. Boveng, M. Castellini, K.L. Daly, S. Jacobs, G.L. Kooyman, J. Laake, L. Quetin, R. Ross, D.B. Siniff, B.S. Stewart, I. Stirling, J. Torres, and P.K. Yochem. 2003. A top-down, multidisciplinary study of the structure and function of the pack-ice ecosystem in the eastern Ross Sea, Antarctica. *Polar Record* 39: 219–230.

Bannister, J.L. 1994. Continued increase in humpback whales off Western Australia. *Report of the International Whaling Commission* 44: 309–310.

_____. 2001. Status of southern right whales (*Eubalaena australis*) off Australia. *Journal of Cetacean Research and Management* (Special Issue) 2: 103–110.

Barrat, A., and J.L. Mougin. 1978. L'elephant de mer *Mirounga leonina* de l'ile de la Possession, archipel Crozet (46°25′S, 51°45′E). *Mammalia* 42: 143–174.

Berzin, A.A. and V.L. Vladimirov. 1983. A new species of killer whale *(Cetacea, Delphinidae)* from Antarctic waters. *Zoologicheskii Zhurnal* 62: 287–295 (in Russian).

Best, P.B. 1982. Seasonal abundance, feeding, reproduction, age and growth in minke whales off Durban (with incidental observations from the Antarctic). *Report of the International Whaling Commission* 32: 759–786.

Best, P.B., A. Brandão, and D.S. Butterworth. 2001. Demographic parameters of southern right whales off South Africa. *Journal of Cetacean Research and Management* (Special Issue) 2: 161–169.

Bester, M.N., H. Moller, J. Wium, and B. Enslin. 2001. An update on the status of southern elephant seals at Gough Island. *South African Journal of Wildlife Research* 31: 68–71.

Boveng, P.L., L.M. Hiruki, M.K. Schwartz, and J.L. Bengtson. 1998. Population growth of Antarctic fur seals: limitation by a top predator, the leopard seal? *Ecology* 79: 2863–2877.

Branch, T.A. 2003. Updated circumpolar abundance estimates for Southern Hemisphere minke whales including results from the 1998/99 to 2000/01 IDCR-SOWER surveys. *Journal of Cetacean Research and Management* (Suppl.) 5: 271–275.

Branch, T.A. and D.S. Butterworth. 2001a. Estimates of abundance south of 60 degrees S for cetacean species sighted frequently on the 1978/79 to 1997/98 IWC/IDCR-SOWER sighting surveys. *Journal of Cetacean Research and Management* 3: 251–270.

———. 2001b. Southern Hemisphere minke whales: standardised abundance estimates from the 1978/79 to 1997/98 IDCR/SOWER surveys. *Journal of Cetacean Research and Management* 3: 143–174.

Branch, T.A., K. Matsuoka, and T. Miyashita. 2004. Evidence for increases in Antarctic blue whales based on Bayesian modelling. *Marine Mammal Science* 20: 726–754.

Budylenko, G.A. 1981. Distribution and some aspects of the biology of killer whales in the South Atlantic. *Report of the International Whaling Commission* 31: 523–525.

Burton, H.R., T. Arnbom, I.L. Boyd, M. Bester, D. Vergani, and I. Wilkinson. 1997. Significant differences in weaning mass of southern elephant seals from five sub-Antarctic islands in relation to population declines, in *Antarctic communities: species, structure and survival*. D.W.H. Walton, ed. Cambridge, UK: Cambridge University Press, pp. 335–338.

Butterworth, D.S., A.E. Punt, T.A. Branch, Y. Fujise, R. Zenitani, and H. Kato. 2002. Updated ADAPT VPA recruitment and abundance trend estimates for Southern Hemisphere minke whales in areas IV and V. Paper SC/54/IA25 presented to the International Whaling Commission Scientific Committee.

Castello, H.P. 1977. Food of a killer whale: eagle sting-ray *Myliobatis* found in the stomach of a stranded *Orcinus orca*. *Scientific Reports of the Whales Research Institute* 29: 107–111.

Chrisp, J. 1958. *South of Cape Horn: a story of Antarctic whaling*. London: R. Hale.

Clapham, P.J. 2001. Why do baleen whales migrate? A response to Corkeron and Connor. *Marine Mammal Science* 17: 432–436.

Condy, P.R., R.J. van Aarde, and M.N. Bester. 1978. The seasonal occurrence and behaviour of killer whales *Orcinus orca*, at Marion Island. *Journal of Zoology* 184: 449–464.

Connor, R.C. and P.J. Corkeron. 2001. Predation past and present: killer whales and baleen whale migration. *Marine Mammal Science* 17: 436–439.

Cooke, J.G., V.J. Rowntree, and R. Payne. 2001. Estimates of demographic parameters for southern right whales *(Eubalaena australis)* observed off Península Valdéz, Argentina. *Journal of Cetacean Research and Management* (Special Issue) 2: 125–132.

Corkeron, P.J. and R.C. Connor. 1999. Why do baleen whales migrate? *Marine Mammal Science* 15: 1228–1245.

Crespo, E.A. and S.N. Pedraza. 1991. Estado actual y tendencia de la poblacion de lobos marinos de un pelo *(Otaria flavescens)* en el litoral norpatagonico. *Ecologia Austral* 1: 87–95.

Dahlheim, M.E. 1981. A review of the biology and exploitation of the killer whale, *Orcinus orca*, with comments on recent sightings from Antarctica. *Report of the International Whaling Commission* 31: 541–546.

Dans, S.L., E.A. Crespo, S.N. Pedraza, and M.K. Alonso. 2004. Recovery of the South American sea lion *(Otaria flavescens)* population in northern Patagonia. *Canadian Journal of Fisheries and Aquatic Sciences* 61: 1681–1690.

Doroshenko, N.V. 1978. On interrelationship between killer whales (predator-prey) in the Antarctic [in Russian], in *Marine mammals. Abstracts of reports to the 7th all-union meeting*. Moscow, pp. 107–109.

Erickson, A.W., and M.B. Hanson. 1990. Continental estimates and population trends of Antarctic ice seals, in *Antarctic ecosystems. Ecological change and conservation*. K.R. Kerry and G. Hempel, eds. Berlin: Springer-Verlag, pp. 253–264.

Estes, J.A., E.M. Danner, D.F. Doak, B. Konar, A.M. Springer, P.D. Steinberg, M.T. Tinker, and T.M. Williams. 2004. Complex trophic interactions in kelp forest ecosystems. *Bulletin of Marine Sciences* 74: 621–638.

Estes, J.A., M.T. Tinker, T.M. Williams, and D.F. Doak. 1998. Killer whale predation on sea otters linking oceanic and nearshore ecosystems. *Science* 282: 473–476.

Francis, R.C. and S.R. Hare. 1994. Decadal-scale regime shifts in the large marine ecosystems of the North-east Pacific: a case for historical science. *Fisheries Oceanography* 3(4): 279–291.

Francis, R.C., S.R. Hare, A.B. Hollowed, and W.S. Wooster. 1998. Effects of interdecadal climate variability on the oceanic ecosystems of the northeast Pacific. *Fisheries Oceanography* 7: 1–20.

Gabriele, C.M., J.M. Straley, S.A. Mizroch, C. Scott Baker, A.S. Craig, L.M. Herman, D. Glockner-Ferrari, M.J. Ferrari, S. Cerchio, O. von Ziegesar, J. Darling, D. McSweeney, T.J.I. Quinn, and J.K. Jacobsen. 2001. Estimating the mortality rate of humpback whale calves in the central North Pacific Ocean. *Canadian Journal of Zoology* 79: 589–600.

Gong, D. and S. Wang. 1999. Definition of Antarctic oscillation index. *Geophysical Research Letters* 26: 459–462.

Green, K., H.R. Burton, V. Wong, R.A. McFarlane, A.A. Flaherty, and S.A. Haigh. 1995. Difficulties in assessing population status of ice seals. *Wildlife Research* 22: 193–199.

Guinet, C. 1991. L'orque *(Orcinus orca)* autour de l'archipel Crozet: Comparaison avec d'autres localités. *Revue d'Ecologie (La Terre et la Vie)* 46: 321–337.

———. 1992. Comportement de chasse des orques *(Orcinus orca)* autour des îles Crozet. *Canadian Journal of Zoology* 70: 1656–1667.

Guinet, C., L.G. Barrett-Lennard, and B. Loyer. 2000. Co-ordinated attack behavior and prey sharing by killer whales at Crozet Archipelago: strategies for feeding on negatively buoyant prey. *Marine Mammal Science* 16: 829–834.

Guinet, C. and P. Jouventin. 1990. La vie sociale des "baleine tueuses". *La Recherche* 21: 508–510.

Guinet, C., P. Jouventin, and H. Weimerskirch. 1999. Recent population change of the southern elephant seal at Îles Crozet and Îles Kerguelen: the end of the decrease? *Antarctic Science* 11: 193–197.

Hammond, P.S. 1984. Abundance of killer whales in Antarctic Areas II, III, IV, and V. *Report of the International Whaling Commission* 34: 543–548.

Hatfield, B.B., D. Marks, M.T. Tinker, K. Nolan, and J. Peirce. 1998. Attacks on sea otters by killer whales. *Marine Mammal Science* 14: 888–894.

Hilborn, R., T.A. Branch, B. Ernst, A. Magnusson, C.V. Minte-Vera, M.D. Scheuerell, and J.L. Valero. 2003. State of the world's fisheries. *Annual Review of Environment and Resources* 28: 359–399.

Hindell, M.A. 1991. Some life-history parameters of a declining population of southern elephant seals, *Mirounga leonina*. *Journal of Animal Ecology* 60: 119–134.

Hindell, M.A., D.J. Slip, and H.R. Burton. 1994. Possible causes of the decline of southern elephant seal populations in the southern Pacific and southern Indian Oceans, in *Elephant seals: population ecology, behavior and physiology*. B.J. Le Beouf and R.M. Laws, eds. Berkeley: University of California Press, pp. 66–84.

Hiruki, L.M., M.K. Schwartz, and P.L. Boveng. 1999. Hunting and social behaviour of leopard seals *(Hydrurga leptonyx)* at Seal Island, South Shetland Islands, *Antarctica. Journal of Zoology* 249: 97–109.

Hoelzel, A.R. 1991. Killer whale predation on marine mammals at Punta Norte, Argentina: food sharing, provisioning and foraging strategy. *Behavioral Ecology and Sociobiology* 29: 197–204.

Hofmeyr, G.J.G., M.N. Bester, and F.C. Jonker. 1997. Changes in population sizes and distribution of fur seals at Marion Island. *Polar Biology* 17: 150–158.

Iñíguez, M.A. 2001. Seasonal distribution of killer whales *(Orcinus orca)* in Northern Patagonia, Argentina. *Aquatic Mammals* 27.2: 154–161.

Iñíguez, M.A., V.P. Tossenberger, and C. Gasparrou. 2002. Cooperative hunting and prey handling of killer whales in Punta Norte, Patagonia, Argentina, in *Fourth International Orca Symposium and Workshops*, Villiers en Bois, France, p. 85. http://www.cebc.cnrs.fr/Fr_collo/ORCA.pdf.

Ivashin, M.V. 1981. USSR. Progress report on cetacean research June 1979–May 1980. *Report of the International Whaling Commission* 31: 221–226.

IWC. 2001. Report of the Workshop on the Comprehensive Assessment of Right Whales: a worldwide comparison. *Journal of Cetacean Research and Management* (Special Issue) 2: 1–60.

———. 2003. Hypotheses that may explain why the estimates of abundance for the third circumpolar set of surveys (CP) using the "standard methods" of Branch and Butterworth (2001) are appreciably lower than estimates for the second CP. Appendix 10, Annexe G. *Journal of Cetacean Research and Management* (Suppl.) 5: 286–290.

Jefferson, T.A., P.J. Stacey, and R.W. Baird. 1991. A review of killer whale interactions with other marine mammals: predation to co-existence. *Mammal Review* 21: 151–180.

Johnson, S.J. and D.S. Butterworth. 2002. An assessment of the west and east Australian breeding stocks of Southern Hemisphere humpback whales using a model that allows for mixing in the feeding grounds. Paper SC/54/H17 presented to the International Whaling Commission Scientific Committe.

Kasamatsu, F. and G.G. Joyce. 1995. Current status of odontocetes in the Antarctic. *Antarctic Science* 7: 365–379.

Kasamatsu, F., G.G. Joyce, P. Ensor, and J. Mermoz. 1996. Current occurrence of baleen whales in the Antarctic. *Report of the International Whaling Commission* 46: 293–304.

Kasamatsu, F., K. Matsuoka, and T. Hakamada. 2000. Interspecific relationships in density among the whale community in the Antarctic. *Polar Biology* 23: 466–473.

Keith, M., M.N. Bester, P.A. Bartlett, and D. Baker. 2001. Killer whales *(Orcinus orca)* at Marion Island, Southern Ocean. *African Zoology* 36(2): 163–175.

Knox, G.A. 1994. *The biology of the Southern Ocean.* Poole, Dorset, UK: Blandford Press.

Kreite, B. 1995. Bioenergetics of the killer whale, *Orcinus orca.* Ph.D. dissertation, University of British Columbia, Vancouver.

Laws, R.M. 1977. Seals and whales of the Southern Ocean. *Philosophical Transactions of the Royal Society of London Series B: Biological Sciences* 279: 81–96.

Laws, R.M. 1994. History and present status of southern elephant seal populations, in *Elephant seals: population ecology, behavior and physiology*. R.M. Laws, ed. Berkeley: University of California Press, pp. 49–65.

Leslie, P.H. 1945. On the use of matrices in certain population mathematics. *Biometrika* 33: 183–212.

Lopez, J.C. and D. Lopez. 1985. Killer whales of Patagonia and their behavior of intentional stranding while hunting nearshore. *Journal of Mammalogy* 66: 181–183.

Maldini, D., J. Maniscalco, and A. Burdin. 2002. Predatory activity of a single killer whale, *Orcinus orca*, at a Steller sea lion, *Eumotopias jubatus*, rookery in Alaska, in *Fourth International Orca Symposium and Workshops*, Villiers en Bois, France, pp. 92–94. http://www.cebc.cnrs.fr/Fr_collo/ORCA.pdf.

Mantua, N.J., S.R. Hare, Y. Zhang, J.M. Wallace, and R.C. Francis. 1997. A Pacific interdecadal climate oscillation with impacts on salmon production. *Bulletin of the American Meteorological Society* 78: 1069–1079.

Matsuoka, K., P. Ensor, T. Hakamada, H. Shimada, S. Nishiwaki, F. Kasamatsu, and H. Kato. 2003. Overview of minke whale sightings surveys conducted on IWC/IDCR and SOWER Antarctic cruises from 1978/79 to 2000/01. *Journal of Cetacean Research and Management* 5: 173–201.

McMahon, C.R., H.R. Burton, and M.N. Bester. 1999. First-year survival of southern elephant seals, *Mirounga leonina*, at sub-Antarctic Macquarie Island. *Polar Biology* 21: 279–284.

———. 2003. A demographic comparison of two southern elephant seal populations. *Journal of Animal Ecology* 72: 61–74.

Mikhalev, Y.A., M.V. Ivashin, V.P. Savusin, and F.E. Zelenaya. 1981. The distribution and biology of killer whales in the Southern Hemisphere. *Report of the International Whaling Commission* 31: 551–566.

Moore, M.J., S.D. Berrow, B.A. Jensen, P. Carr, R. Sears, V.J. Rowntree, R. Payne, and P.K. Hamilton. 1999. Relative abundance of large whales around South Georgia (1979–1998). *Marine Mammal Science* 15: 1287–1302.

Mori, M., D.S. Butterworth, A. Brandão, R.A. Rademeyer, H. Okamura, and H. Matsuda. 2003. Observer experience and Antarctic minke whale sighting ability in IWC/IDCR-SOWER surveys. *Journal of Cetacean Research and Management* 5: 1–11.

Naessig, P.J. and J.M. Lanyon. 2004. Levels and probable origin of predatory scarring on humpback whales *(Megaptera novaeangliae)* in east Australian waters. *Wildlife Research* 31: 163–170.

Paterson, R., P. Paterson, and D.H. Cato. 2001. Status of humpback whales, *Megaptera novaeangliae*, in East Australia at the end of the 20th century. *Memoirs of the Queensland Museum* 47: 579–586.

Pistorius, P.A. and M.N. Bester. 2002. Juvenile survival and population regulation in southern elephant seals at Marion Island. *African Zoology* 37: 35–41.

Pistorius, P.A., M.N. Bester, and S.P. Kirkman. 1999. Dynamic age-distributions in a declining population of southern elephant seals. *Antarctic Science* 11: 445–450.

Pistorius, P.A., M.N. Bester, S.P. Kirkman, and F.E. Taylor. 2001. Temporal changes in fecundity and age at sexual maturity of southern elephant seals at Marion Island. *Polar Biology* 24: 343–348.

Pistorius, P.A., F.E. Taylor, C. Louw, B. Hanise, M.N. Bester, C. De Wet, A. du Plooy, N. Green, S. Klasen, P. Podile, and J. Schoeman. 2002. Distribution, movement and estimated populations of killer whales at Marion Island, December 2000. *South African Journal of Wildlife Research* 32: 86–92.

Pitman, R.L., L.T. Ballance, S.L. Mesnick, and S.J. Chivers. 2001. Killer whale predation on sperm whales: observations and implications. *Marine Mammal Science* 17: 494–507.

Pitman, R.L. and P. Ensor. 2003. Three forms of killer whales *(Orcinus orca)* in Antarctic waters. *Journal of Cetacean Research and Management* 5(2): 131–139.

Reyes, L.M., E.A. Crespo, and V. Szapkievich. 1999. Distribution and population size of the southern sea lion *(Otaria flavescens)* in Central and Southern Chubut, Patagonia, Argentina. *Marine Mammal Science* 15: 478–493.

Schiavini, A.C.M., E.A. Crespo, and V. Szapkievich. 2004. Status of the population of South American sea lion *(Otaria flavescens* Shaw, 1800) in southern Argentina. *Mammalian Biology* 69: 108–118.

Shevchenko, V.I. 1975. The nature of the interrelationships between killer whale and other cetaceans. *Morsk Mlekopitayushchie Chast'* 2: 173–174 (in Russian).

Siniff, D.B. and J.L. Bengtson. 1977. Observations and hypotheses concerning interactions among crabeater seals, leopard seals, and killer whales. *Journal of Mammalogy* 58: 414–416.

Sleptsov, M.M. 1965. The killer whales of the Southern Hemisphere, in *Marine Mammals, a collection of articles*. Izdatelstvo Nauka, pp. 60–64.

Slip, D.J. and H.R. Burton. 1999. Population status and seasonal haulout patterns of the southern elephant seal *(Mirounga leonina)* at Heard Island. *Antarctic Science* 11: 38–47.

Smith, C.R. and A.R. Baco. 2003. Ecology of whale falls at the deep-sea floor. *Oceanography and Marine Biology: an Annual Review* 41:311–354.

Smith, C.R., H. Kukert, R.A. Wheatcroft, P.A. Jumars, and J.W. Deming. 1989. Vent fauna on whale remains. *Nature* 341: 27–28.

Springer, A.M., J.A. Estes, G.B. van Vliet, T.M. Williams, D.F. Doak, E.M. Danner, K.A. Forney, and B. Pfister. 2003. Sequential megafaunal collapse in the North Pacific Ocean: an ongoing legacy of industrial whaling? *Proceedings of the National Academy of Sciences* 100: 12223–12228.

Thompson, D., I. Strange, M. Riddy, and C.D. Duck. 2005. The size and status of the population of southern sea lions *Otaria flavescens* in the Falkland Islands. *Biological Conservation* 121: 357–367.

Thomson, R.B., D.S. Butterworth, and H. Kato. 1999. Has the age at transition of Southern Hemisphere minke whales declined over recent decade? *Marine Mammal Science* 15: 661–682.

Vaz-Ferreira, R. 1982. *Otaria flavescens* (Shaw), South American sea lion, in *Mammals in the Seas. Volume IV. Small Cetaceans, Seals, Sirenians and Otters*. Rome: Food and Agriculture Organization of the United Nations, pp. 477–495.

Whitehead, H. 2002. Estimates of the current global population size and historical trajectory for sperm whales. *Marine Ecology Progress Series* 242: 295–304.

Williams, T.M., J.A. Estes, D.F. Doak, and A.M. Springer. 2004. Killer appetites: assessing the role of predators in ecological communities. *Ecology* 85: 3373–3384.

Williams, T.M., J. Haun, R.W. Davis, L.A. Fuiman, and S. Kohin. 2001. A killer appetite: Metabolic consequences of carnivory in marine mammals. *Comparative Biochemistry and Physiology A* 129: 785–796.

Yukhov, V.L., E.K. Vinogradova, and L.I. Medvedev. 1975. The diet of killer whales *(Orcinus orca* L.) in the Antarctic and adjacent waters. *Translation Series Fisheries And Marine Service* 3844.

Zemsky, V.A., A.A. Berzin, Y.A. Mikhaliev, and D.D. Tormosov. 1995. Soviet Antarctic pelagic whaling after WWII: review of actual catch data. *Report of the International Whaling Commission* 45: 131–135.

Zerbini, A.N. 2004. Status of the Southern Hemisphere humpback whale breeding stock A: preliminary results from a Bayesian assessment. Paper SC/56/SH7 presented to the International Whaling Commission Scientific Committee.

TWENTY-ONE

Predator Diet Breadth and Prey Population Dynamics
Mechanism and Modeling

MARC MANGEL AND NICHOLAS WOLF

The notion that the sequential megafaunal collapse of several subpopulations of harbor seal, Steller sea lion, and sea otter (sensu Springer et al. 2003) is due to killer whale predation relies on the assumption of expanding diet breadth for orcas. That is, the putative mechanism for the decline is a preference of killer whale for large whale species, but an inclusion of certain populations of harbor seal, fur seal, Steller sea lion, and sea otter as the preferred prey types become less available. When considering the observed sequential declines, a number of features need to be explained (Figure 2 of Springer et al. 2003): (1) the ordering of the declines (harbor seals first, Steller sea lions second, sea otters third), and (2) the apparent lack of inclusion of one prey type (as measured by a lack of decline) until the other prey types are nearly exhausted. For simplicity, we focus on harbor seals, Steller sea lions, and sea otters.

In this chapter, we use a combination of models from behavioral and population ecology to address these questions, and in doing so, elucidate these two points and certain other aspects of the community dynamics. Our goal is to show how the synthesis of two different scientific traditions (the literature on prey choice by predators and the literature on population dynamics) can lead to new understandings and new empirical challenges. Indeed, in anticipation of the meeting that lead to this chapter, one of us (MM) asked a colleague about this question and was told, "There is actually quite a large literature on body composition of pinnipeds although I can't recall anybody looking at them from the perspective of the prey" (I. Boyd, personal communication to MM, February 25, 2003). Each of these scientific traditions has an enormous literature and in order to keep the paper of reasonable length, we provide minimal citation to that literature, but an interested reader will have no trouble finding entry points, some of which we point out at relevant moments in the chapter. An evolutionary understanding of feeding behavior turns out to be a good starting point for combining the traditions of prey choice by predators and population dynamics.

The Four Questions of Tinbergen and the Diet Choice Model

Niko Tinbergen (1963) observed that, regarding any behavior, four questions need to be considered:

1. How does behavior develop within an individual (ontogeny)?

2. What is the evolutionary history of the behavior (phylogeny)?

3. What is the physiological mechanism of the behavior (proximate mechanism)?

4. What is the survival and/or reproductive value of the behavior (ultimate mechanism)?

The classic diet choice model (Stephens and Krebs 1986; Mangel and Clark 1988; Clark and Mangel 2000) begins with the assumption that Darwinian fitness (the long-term number of descendants, often approximated by the expected lifetime reproductive success) is maximized when the rate at which energy is obtained is maximized. As we note below, killer whales often hunt in groups. Group hunting is a complicating factor, but it does not necessarily affect the validity of the assumption of rate maximization. For example, energy obtained from the prey may simply be divided among the individual predators in the group. The current formulation of our model assumes that group size is fixed with respect to prey type, and that food is divided equally among members of the group, at least on average. The numerical values of energy gain that we report correspond to the share of an individual. We recognize that the strategy killer whales use in hunting may depart from this assumption, but it seems a reasonable starting point in this paper. Subsequent research should investigate the robustness of our conclusions to this hypothesis (see Discussion).

In this classic model of diet choice (Stephens and Krebs 1986), prey are characterized by three values. These are (1) the energy, E_i (units, kcal), obtained by chasing, killing, and consuming a single individual of prey type i; (2) the time, h_i (units, hours), needed to "handle" (chase, kill, and consume) a single individual of prey type i; and (3) the encounter rate, λ_i (units, 1/hours), with prey type i. In the standard model of diet choice, the encounter rate is treated as fixed; in the analysis that we conduct next the encounter rate will vary according to the population dynamics. In particular, the encounter rates with a prey species will decline as its population declines.

We next assume that the only activities of the predator consist of searching for prey and handling prey. Thus, any long period of time T is divided into a total time spent searching S and a total time spent handling H. These assumptions are sufficient to compute the long-term rate of energy gain for any diet selection rule that the predator uses. We then choose the diet selection rule that maximizes the rate of energy return.

The key results of this theory are that items are ranked by profitability, E_i/h_i, that a prey species is either completely included in the diet or not (there are no partial preferences), and that only the most profitable prey species ($i = 1$) is included as long as

$$\lambda_1 > \frac{E_2}{E_1 h_2 - E_2 h_1} \qquad (21.1)$$

where $i = 2$ corresponds to the next most profitable prey species. This equation may be derived in another manner,

assuming that at each moment the predators are maximizing the rate of energy gain from an encountered prey. If an individual of prey species type 2 is encountered and consumed, the rate of energy flow to the predator from this prey is E_2/h_2. On average, taking into account both handling time and search time ($= 1/\lambda_1$), the rate of energy flow (the profitability) to the predator from the preferred prey species is $E_1/(h_1 + 1/\lambda_1)$, and we predict that the predator will take only the most profitable prey as long as $E_1/(h_1 + 1/\lambda_1) > E_2/h_2$, which can be rearranged to give Equation 21.1.

Equation 21.1 defines a *switching value* for the encounter rate with the most profitable prey species. Note that the right-hand side of this equation does not contain the encounter rate with the second most profitable prey species. That is, as long as the most profitable prey species is sufficiently abundant, we predict that it will be the only species selected by the predator, regardless of the abundance of other prey types. Thus, a predator foraging optimally, according to this description, on a high initial density of its most profitable prey species is predicted to eat only that species until it is sufficiently depleted. Once the most profitable prey species is sufficiently depleted, the predator is predicted to expand diet breadth and include the second most profitable prey species.

Equation 21.1 is generalized by considering the rate of energy return, R_k, when the most profitable k prey types are included in the diet:

$$R_k = \frac{\displaystyle\sum_{i=1}^{k} E_i \lambda_i}{1 + \displaystyle\sum_{i=1}^{k} h_i \lambda_i} \qquad (21.2)$$

The first link between predator diet breadth and prey population dynamics is the assumption that the encounter rates are determined by prey abundance. That is, if $N_i(t)$ is the abundance of prey type i at the start of year t, we assume that

$$\lambda_i = q_i N_i(t) \qquad (21.3)$$

where q_i is a measure of the search effectiveness of the predator for prey species i. It has units of rate per prey individual. We recognize that transient killer whales do not hunt as individuals but as pods of two to seven animals when preying on relatively small prey such as pinnipeds. Even so, this assumption is a reasonable starting point since any additions make the analysis more complicated (e.g., Mangel and Clark 1988, especially Chapter 3).

There is another interpretation of q_i, and it is this interpretation that allows us to couple the diet choice and prey population dynamics. For simplicity, consider a single, focal prey individual and a single predator. Equation 21.3 is equivalent to the assumption (Hilborn and Mangel 1997) that in a small interval of time, dt, the probability of the prey encountering the predator and being killed is approximately

$q_i dt$ (it is actually $q_i dt + o(dt)$, where $o(dt)$ represents terms that are higher-order powers of dt; see Hilborn and Mangel 1997). This can be integrated (Hilborn and Mangel 1997) over a year to show that the probability that an individual of the focal prey avoids predation by a single orca is $\exp(-q_i)$. If there are $O(t)$ orcas present at the start of year t, and the predators do not interfere with each other, the probability that a focal prey escapes predation is $[\exp(-q_i)]^{O(t)} = \exp[-q_i O(t)]$, and we thus conclude that the fraction of prey experiencing predation will be $1 - \exp[-q_i O(t)]$. This assumes, of course, that the prey species we are discussing is included in the diet and that the predators are insatiable (we discuss the latter assumption later in the paper). As described above, the hunting of killer whales in pods may require reinterpretation of this derivation; in such a case $O(t)$ would represent the number of hunting groups and q_i would represent the hunting effectiveness of a group. (Once again, see Mangel and Clark 1988, Chapter 3.)

Coupling Diet Choice and Population Dynamics

We now couple prey dynamics and predator diet choice. To do this, we explicitly separate out the orca predation from other sources of regulation of the prey at the level of the population. In the absence of predation by orcas, we assume that the ith prey species grows according to the a-logistic model

$$N_i(t+1) = N_i(t) + r_i N_i(t)\left[1 - \left(\frac{N_i(t)}{K_i}\right)^a\right] \tag{21.4}$$

where we set the parameter $a = 2.4$ (Taylor et al. 2000), although the actual range may be 1–8 and 2.4 is used out of tradition (D. DeMaster, personal communication). In addition, r_i is the maximum per capita growth rate (corresponding to low population sizes), and K_i is the carrying capacity of species i.

Suppose now that predator diet breadth is fixed for the course of a year and we define an acceptance variable according to $S_i(t) = 1$ if the ith prey species is included in the diet and $S_i(t) = 0$ if the prey species is not included in the diet. Combining the result from the end of the last section with Equation 21.4, the prey population dynamics become

$$N_i(t+1) = N_i(t) + r_i N_i(t)\left[1 - \left(\frac{N_i(t)}{K_i}\right)^a\right]$$
$$- S_i(1 - e^{-qO(t)})N_i(t) \tag{21.5}$$

and these are the equations that will allow us to predict the population dynamics of the prey, as a function of the predator's diet choice behavior. Note that in deriving this equation we have assumed that all killer whales (or killer whale pods more specifically) behave similarly. This is likely not true in the sense that pods develop expertise for different prey

species and there is inertia when trying to learn how to forage on new species. But, once again, these are elaborations awaiting future work.

To begin, let us hold the orca population constant, so that $O(t)$ does not depend on time. We may then ask for the steady states of the prey population assuming that they are included in the diet, which are population sizes for which

$$r_i \overline{N}(t)\left[1 - \left(\frac{\overline{N}(t)}{K_i}\right)^a\right] = (1 - e^{-qO(t)})\overline{N}(t) \tag{21.6}$$

and for which one solution always is $\overline{N}_i = 0$, corresponding to extinction of the prey. The other solution can be envisioned as the intersection of the line on the right-hand side with the parabola-like curve on the left-hand side (Figure 21.1). At this point of intersection, removals ("harvest") by the predators balance population growth ("production") by the prey.

Parameter Estimates and Sample Calculations

In the Chapter Appendix, we use the theory developed so far to determine the conditions of parameters in which a cascade is predicted. There we show that there are two sets of conditions, one involving 9 parameters and the other involving 14 parameters. Rather than conducting such a full analysis here to examine the plausibility of the diet breadth hypothesis, we will explore the numerical iteration of Equation 21.5. In order to use Equation 21.5 to explore prey population dynamics as a function of predator diet breadth, we need estimates of the various parameters. Some of these have simply not been measured, or can only be estimated for some of the species. Even so, it is instructive to choose parameters and estimate the dynamics.

Matkin et al. (2002) estimate about 200–300 transient killer whales in Alaska, so that we hold $O(t)$ constant at 200; and, lacking any information, we set $q_i = 0.002$ for each species. Note that it is the product $q_i O(t)$ that matters. In this case, the product is 0.4, implying that killer whales, once they have added a prey species to the diet, can take about 40% of the population per year. Although this appears to be a large fraction, the reader should recognize that there are no other sources of mortality in the model.

Regarding the information that is needed for diet choice calculation, Barrett-Lennard et al. (1995, p. 12) report the following data, based on multiple sources: killer whales take about 30 minutes to kill and consume harbor seals and about 1–2 hours to kill and consume Steller sea lions. Hence we set $h_1 = 0.5$ hr, $h_2 = 2.0$ hr. We lacked data for sea otters, so we set $h_3 = 0.25$ hr.

Barrett-Lennard et al. (1995) also report that when preying on harbor seals and southern sea lions, killer whales predominantly attack and take pups and juveniles (50–80% of the kills or attacks). We lack specific data on energy content of the prey species, which is required for the parameter E_i.

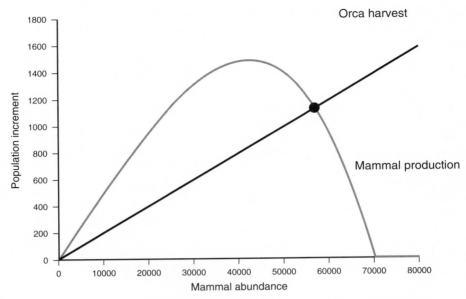

FIGURE 21.1. If the number of orcas is held constant, the steady-state population size of the prey species is determined by the balance between biological production (the curve) and the harvest taken by orcas (the line). Since the slope of that line is proportional to q_iO, increasing either the number of orcas or the efficiency with which they can find prey will rotate the line counterclockwise and will thus decrease the steady-state population size of the prey species. Furthermore, if the slope of the line is greater than the slope of the growth curve at the origin, the only potential steady state is that which corresponds to extinction of the prey. Here we have explicitly shown the case in which take by orcas leads to a steady state in which the prey species persist (at the intersection of the curve and the line). However, for the megafaunal collapse to occur, the parameters should be such that the only steady state is the origin.

However, we assume as a first approximation that body mass is a reasonable measure of energy content. Harbor seals are about 10 kg at birth and about 150 kg (males 170 kg, females 130 kg) at maturity; Steller sea lions about 20 kg at birth and 350 (female)–1100 (male) kg at maturity; and sea otters about 2 kg at birth and 40 kg (males 45 kg, females 33 kg) at maturity (Reeves et al. 2002). We do not know the mixture of pups, juveniles, and adults of any of the species taken by killer whales; this is another empirical question.

Lacking such information, we proceed as follows. We assume that E_i in Equations 21.1 and 21.2 is a fraction f of the mass of mature animals, where f is the assimilation efficiency divided by the killer whale group size. Note, however, that if this fraction is the same for all the prey species, then the particular value of the fraction has no effect on the diet choice rule in Equation 21.1 as it appears in both the numerator and the denominator. Furthermore, the relative variation in mass at maturity of harbor seals and sea otters is small, so we stipulate that $E_1 = 150$ kg and $E_3 = 40$ kg, respectively. On the other hand, Steller sea lions show a much greater size dimorphism, so we will treat E_2 as a parameter in our investigations.

Following Peters (1983), we assume that maximum per capita growth rate of species i, r_i, is related to female mass at maturity, W_i, according to $r_i = bW_i^{-0.26}$, where we picked the

constant b so that the maximum per capita growth rate of otters was 0.1/yr. The allometry then gives 0.07 and 0.05 for seals and sea lions, respectively. These may be a bit low; Taylor et al. (2000) indicate a value of 0.12 for harbor seals, and 0.2 and 0.09 may be more appropriate for otters and sea lions, respectively (D. DeMaster, personal communication). In the conditions that characterize the cascade (given in the Chapter Appendix), we show that these parameters are confounded with the number of orcas and the search efficiency of the orcas, so that determining the precise values of one set of parameters without knowing the others provides little advantage. Estimates of carrying capacities for any population of marine mammals are difficult to measure, and the carrying capacities themselves will surely change on a number of different time scales. For purposes of computation and illustration, we set the carrying capacities to be the maximum observed population sizes (about 20,500 in 1964, 70,000 in 1979, and 52,000 in 1959 for seals, sea lions, and otters, respectively).

We start each population at its carrying capacity, and we use Equation 21.2 to determine the diet breadth of the killer whales, and Equation 21.5 to iterate the population dynamics forward. The key result is shown in Figure 21.2, in which we have fixed all parameters in the model except E_2 (which is now a proxy for energy content of a sea lion). Because all

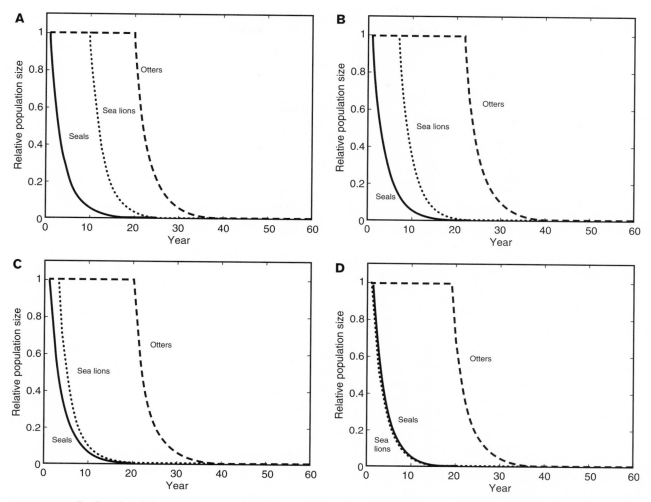

FIGURE 21.2. Predicted population dynamics of harbor seals, Steller sea lions, and sea otters, assuming the dynamics given by Equation 21.5, and diet breadth determined by Equation 21.2. We have held all parameters except E_2 constant ($h_1 = 0.5$ hr, $h_2 = 2$ hr, $h_3 = 0.25$ hr, $E_1 = 150$ kg, $E_3 = 40$ kg), for $E_2 = 350$ kg (panel a), 450 kg (panel b), 550 kg (panel c), or 650 kg (panel d).

parameters are fixed except this one, the profitability of sea lions changes as we vary it. If E_2 is not too large, our simple model captures the qualitative dynamics of the sequential collapse; but if E_2 is sufficiently large, then the prediction is that both seals and sea lions will be taken by the killer whales from the outset, or that sea lions would be depleted before harbor seals were included in the diet. Thus, to reproduce, at least qualitatively, the observed results, we required that E_2 is no more than about 450–500 kg. A similar calculation can be done to determine a minimum value for sea lions. That is, as E_2 becomes smaller, we reach a point (about 325 kg) at which sea lions and otters are predicted to be included in the diet simultaneously, rather than sequentially.

In summary, by coupling a model for predator diet choice and simple population dynamics for the prey, we have shown that predator diet breadth is a plausible mechanism for the megafaunaul collapse. The question of how likely it is to be the cause requires an assessment of the weight of evidence for all potential mechanisms simultaneously (Wolf and Mangel 2006), but that is a different topic.

Summary

The modeling effort described here thus makes the notion of expansion of diet breadth a more plausible mechanism for the observed sequential decline, and leads to clear predictions and suggestions for empirical work. For example, three predictions are the following:

Prediction #1: The profitability of potential killer whale prey is ranked: harbor seals > Steller sea lions > sea otters; note that this refers to the profitability E_i/h_i of a single encounter.

Prediction #2. The energetic content of Steller sea lions taken by killer whales is no more than about three times greater than the energetic content of harbor seals taken by killer whales.

Prediction #3: All else being equal, at those Steller sea lion rookeries where harbor seal numbers are smaller, sea lions are more likely to decline than they are at sea lion rookeries where the harbor seal numbers are larger.

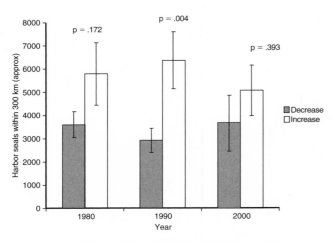

FIGURE 21.3. Estimates (±1 SE) of the harbor seal populations within 300 km of rookeries with declining and growing Steller sea lion populations in 1980, 1990, and 2000. The significance test is a two-tailed *t*-test. See Wolf and Mangel (2006) for further details.

As a rudimentary test of this prediction, we estimated the number of harbor seals within the 300 km foraging radius (Gerber and Van Blaricom 2001) of each Steller sea lion rookery, and compared harbor seal numbers between rookeries where the sea lion population was declining and where the sea lion population was growing (Wolf and Mangel 2006). Harbor seal population time series for various large and small regions of Alaska were constructed by interpolating between published censuses and scaling the regional counts in proportion to "trend" data (counts from sub-areas containing an unknown fraction of the regional total) in years when only trend data were available. Estimates of the number of animals at a rookery before the earliest count or after the latest count were assumed to be fixed at the level of the earliest or latest count, respectively. The number of seals near each sea lion rookery was calculated as the sum of the counts from all regions multiplied by the fraction of their respective area (or linear extent) falling within 300 km of the rookery. The results (Figure 21.3) support the prediction: In all three of the years considered, rookeries with declining sea lion populations tended to have fewer seals within 300 km. The difference is statistically significant in the 1990 data (two-tailed *t*-test, *p* = 0.0043). The figure lends qualitative support to the idea that there is a threshold harbor seal density below which Steller sea lions become acceptable prey. The threshold is apparently somewhere around 4,500 seals per 283,000 km² (the area of a circle with a radius of 300 km), or about 0.016 seals per km².

Our work also suggests three important empirical tasks:

Empirical task #1: Determine the profitabilities of harbor seals, northern fur seals, Steller sea lions, and sea otters for killer whales.

Empirical task #2: Conduct studies to determine encounter rates of killer whales with different prey species.

Empirical Task #3: Determine the mixture of pups, juveniles, and adults taken by killer whales. Each of these will contribute further details in the investigation of the diet breadth hypothesis.

A Variety of Caveats

There is little chance of "validating" this model (for a fuller discussion of this point, see Stamps et al. 1998; Mangel et al. 2001). Clearly, with enough tunable parameters, one can make any model match a series of data points. Furthermore, in a stochastic world (which happens to be the kind in which we live) even if the model is exactly correct, it is possible for the model and the observed trajectory to differ considerably because of chance fluctuations. So what should we do? First, one should think about testing the assumptions that go into the model. Second, one should recognize that models are always approximations and that we use them to understand nature; so the question is, "what is the alternative model to explain the pattern of decline of seals, sea lions, and otters?"

Regarding the first point, there are key assumptions that we already know are wrong, but could be incorporated directly by using dynamic state variable models based on stochastic dynamic programming (Mangel and Clark 1988; Clark and Mangel 2000). We have ignored physiological state, both energetic reserves and gut content, of the killer whales and these can often affect the predictions of diet breadth in subtle ways. We know that killer whales hunt in groups and that there is individual variation in group formation and social interactions (Baird and Whitehead 2000). The classic diet choice model can be extended to include each of these, but such extensions are beyond the scope of this paper, and will not change the qualitative ideas developed here. To object that a model is not realistic because nature is more complicated than the model is a specious argument. Nature will always be more complicated than any model. We fully concur that physiology and social interactions are important, but also that theory becomes even more important in those circumstances.

Conclusion

Our goal in this paper has been to raise a variety of interlocking points. Understanding behavior is essential, not an add-on, for understanding population dynamics. In this area, theory plays a key and informative role. We have modeled a plausible mechanism and identified through that model a range of values for relevant parameters that allows us to provide a quantitative basis for the qualitative pattern of decline of several subpopulations of harbor seal, sea lion, and sea otter. The process of modeling, in addition to generally sharpening our thinking about the issues, helps identify what needs to be measured in the field and allows us to recognize the key assumptions that underlie the plausible mechanism of diet breadth expansion. To be sure, the model described here is a simple one, but it demonstrates that the tools allowing us to connect behavioral and population ecology to determine insights into community ecology are available. We now need to use them.

Appendix: The General Conditions for a Cascade

In this appendix, we derive the general conditions for a cascade, under the assumptions of diet choice and population dynamics given in the text. The sequence of the cascade from harbor seal (prey species 1) to sea lion (prey species 2) to sea otter (prey species 3) requires that $E_1/h_1 > E_2/h_2$ and that $E_2/h_2 > E_3/h_3$. The cascade will proceed from harbor seal to sea lion if the harbor seal steady-state population size is sufficiently low so that Equation 21.1 of the text is violated. That is

$$q_1 K_1 \left(1 - \frac{1 - e^{-qO(t)}}{r_1}\right)^{\frac{1}{a}} \leq \frac{E_2}{E_1 h_2 - E_2 h_1} \quad (21.A1)$$

The cascade proceeds from sea lion to sea otter if both the harbor seal and sea lion steady-state population sizes are low enough that sea otter is predicted to be included in the diet. This condition is

$$q_2 K_2 \left(1 - \frac{1 - e^{-q_2 O(t)}}{r_2}\right)^{\frac{1}{a}}$$

$$\leq \frac{E_3 + q_1 K_1 \left(1 - \frac{1 - e^{-qO(t)}}{r_1}\right)^{\frac{1}{a}} (E_3 h_1 - E_1 h_3)}{E_2 h_3 - E_3 h_2} \quad (21.A2)$$

Equation 21.A1 involves 9 parameters, and Equation 21.A2 involves 14 parameters. Our results show that the cascade occurs for certain values of the parameters, so exploration of Equations 21.A1 and 21.A2 will allow us to determine the subsets of parameter space in which the cascade is predicted to occur. But that exploration is beyond the scope of this chapter.

Acknowledgments

For comments on the work, we thank members of the Mangel Research Group, and for very helpful comments on a previous version of the manuscript, an anonymous referee, Doug DeMaster, and Dan Doak.

Literature Cited

Baird, R.W. and H. Whitehead. 2000. Social organization of mammal-eating killer whales: group stability and dispersal patterns. *Canadian Journal of Zoology* 78: 2096–2105.

Barrett-Lennard, L.G., K.A. Heise, E. Saulitis, G. Ellis, and C. Matkin. 1995. *The impact of killer whale predation on Steller sea lion populations in British Columbia and Alaska*. Vancouver, BC: North Pacific Universities Marine Mammal Research Consortium.

Clark, C.W. and M. Mangel. 2000. *Dynamic state variable models in ecology. Methods and applications*. New York: Oxford University Press.

Gerber, L.R. and G. R.Van Blaricom. 2001. Implications of three viability models for the conservation status of the western population of Steller sea lions *(Eumetopias jubatus)*. *Biological Conservation* 102: 261–269.

Hilborn, R. and M. Mangel. 1997. *The ecological detective. Confronting models with data*. Princeton, NJ: Princeton University Press.

Mangel, M. and C.W. Clark. 1988. *Dynamic modeling in behavioral ecology*. Princeton, NJ: Princeton University Press.

Mangel, M., O. Fiksen, and J. Giske. 2001. Theoretical and statistical models in natural resource management and research, in *Modeling in natural resource management. Development, interpretation, and application*. T.M. Shenk and A.B. Franklin, eds. Washington, DC: Island Press, pp. 57–72.

Matkin, C.O., L.G. Barrett-Lennard, and G. Ellis. 2002. Killer whales and predation on Steller sea lions, in *Steller sea lion decline: Is it food II*. D. DeMaster and S. Atkinson, eds. Fairbanks: University of Alaska Sea Grant College Program, pp. 61–66.

Peters, R.H. 1983. *The ecological implications of body size*. Cambridge, UK: Cambridge University Press.

Reeves, R.R., B.S. Stewart, P.J. Clapham, and J.A. Powell, eds. 2002. *Guide to Marine Mammals of the World*. New York: Knopf.

Springer, A.M., J.A. Estes, G.B. van Vliet, T.M. Williams, D.F. Doak, E.M. Danner, K.A. Forney, and B. Pfister. 2003. Sequential megafaunal collapse in the North Pacific Ocean: An ongoing legacy of industrial whaling. *Proceedings of the National Academy of Science* 100: 12223–12228.

Stamps, J.A., M. Mangel, and J.A. Phillips. 1998. A new look at relationships between size at maturity and asymptotic size. *American Naturalist* 152: 470–479.

Stephens, D.W., and J.R. Krebs. 1986. *Foraging theory*. Princeton, N.J.: Princeton University Press.

Taylor, B.R., P.R. Wade, D.P. De Master, and J. Barlow. 2000. Incorporating uncertainty into management models for marine mammals. *Conservation Biology* 14: 1243–1252.

Tinbergen, N. 1963. On aims and methods of ethology. *Zeitschrift für Tierpsychologie* 20: 410–433.

Wolf, N., J. Melbourne, and M. Mangel. 2006. The method of multiple hypothese and the decline of Stellar sea lions in western Alaska, in *Top predators in marine ecosystems. their role in monitoring and management*. I. Boyd, S. Wanless, and C. J. Camphusen, eds. Cambridge: Cambridge University Press, pp. 275–293.

TWENTY-TWO

Bigger Is Better

The Role of Whales as Detritus in Marine Ecosystems

CRAIG R. SMITH

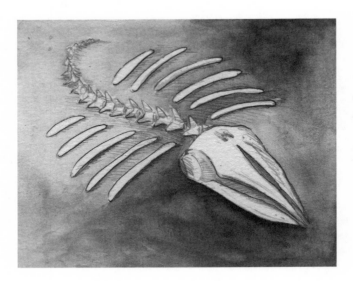

Organic detritus (i.e., nonliving organic matter) plays fundamental roles in the structure and dynamics of all marine ecosystems. The importance of a particular type of organic detritus in an ecosystem depends on several key characteristics of the material, including (1) the size of the detrital particles; (2) the nature of organic materials contained within the particles (e.g., the presence of nutritious lipids and proteins); (3) the flux of organic carbon, or limiting nutrient, entering the ecosystem in the form of the detritus (especially relative to fluxes in other forms); (4) the frequency of occurrence of the detrital particles (essentially flux divided by particle size). These characteristics constrain the use of a particular detrital type by detritivores and ultimately control the ecological and evolutionary opportunities (and selective milieu) provided by the detritus.

Dead-whale detritus has remarkable characteristics and thus may play unusual roles in marine ecosystems. Cetaceans are by far the largest parcels of organic matter formed in the ocean, with adult body masses of the nine largest species, or the "great whales," ranging from 5 to >160 metric tons (e.g., Lockyer 1976). The enormous size of adult great whales provides a refuge from most predators, with the consequence that much of the natural whale mortality may occur from nutritional or disease stresses sustained during migrations (see, e.g., Gaskin 1982; Corkeron and Conner 1999; Moore et al. 2003). Based on relative population production rates,

even the successful whale predators, the killer whales (Orcinus orca), appear to utilize only a small proportion of adult great-whale production. When predation events do occur, the disproportionate mass of a great whale and the sinking of the carcass often prevent predators from consuming most of it (e.g., Silber et al. 1990; Guinet et al. 2000). Thus, in contrast to most other marine animals, great-whale biomass typically enters the marine food web as fresh carrion parcels many metric tons in size (Britton and Morton 1994).

A fresh detrital whale consists mostly of soft tissues (87%–92% by weight; Robineau and de Buffrénil, 1993), with a 40-metric-ton carcass containing 1.6×10^6 g C in labile organic compounds such as lipids and proteins. As a consequence, dead whales are among the most nutrient-rich of all types of detritus on both a weight- and particle-specific basis. The cetacean skeleton is also laden with organic material, with large bones often exceeding 60% lipid by weight (Deming et al. 1997, Smith and Baco 2003; Schuller et al. 2004). Thus, the ossified skeleton of a 40-metric-ton whale may harbor 2,000–3,000 kg of lipid (Smith and Baco 2003), potentially providing substantial nutritional resources, as well as habitat, for a variety of organisms.

Because of their large body size, whale populations have low production rates compared to most other organisms in the ocean (Katona and Whitehead 1988); thus, when averaged over large areas, the flux of carbon through whale detritus is small relative to total detrital flux, even in the most organic-poor ecosystems such as the abyssal seafloor (Jelmert and Oppen-Bernsten 1996; see calculations in this chapter). Nonetheless, end-member characteristics in particle size and quality potentially allow whale detritus to play disproportionate roles in the structure and evolution of marine ecosystems.

Below I discuss current ecosystem responses to the input of whale detritus. I then estimate the effects of industrial whaling on the production of dead whales and speculate on the consequences of these changes for marine ecosystems. Finally, I propose an experimental approach to test some of these speculations.

Current Ecosystem Responses to Whale Detritus

Production and Initial Fate of Whale Detritus

Great whales suffering natural mortality are typically in poor nutritional condition and negatively buoyant upon death; as a consequence, most whale carcasses initially sink toward the seafloor (Ashley 1926; Schafer 1972; Gaskin 1982; Guinet et al. 2000; Smith and Baco 2003; D. W. Rice, personal communication). Because there appear to be few scavengers on whale sized particles in midwater (Britton and Morton 1994), and because whale carcasses will sink rapidly following lung deflation from hydrostatic pressure, it is extremely likely that relatively little tissue removal will occur during a dead whale's descent to the seafloor. If a whale carcass sinks in deep enough water, hydrostatic pressure will limit the generation of buoyant decompositional gases through reduction of gas volume and increased gas solubility (Allison et al. 1991). At depths greater than 1,000 m, the amount of microbial tissue decay required to generate carcass buoyancy (e.g., >67 % of carcass mass through fermentation) is prohibitive; the soft tissue of a carcass will be removed by scavengers or disintegrate from microbial decomposition long before positive buoyancy can be generated, and the carcass will remain on the seafloor (Allison et al. 1991). At shallower depths, there is some probability that gas generation could refloat a whale carcass, although this will depend in part on the outcome of competition for soft tissue between scavengers and microbes. It is interesting to note that an essentially intact gray whale carcass has been found at 150 m depth in Alaskan waters, suggesting that, in cold waters where whales often abound, even 15 atm of pressure may prevent decompositional flotation of a whale carcass (Smith and Baco 2003).

Based on this reasoning, most great-whale "detritus" will be rapidly deposited onto the seafloor. Because 88% of the ocean is underlain by ocean bottom deeper than 1000 m, the vast bulk of great-whale detritus is very likely to begin recycling at the deep-sea floor. In contrast, although whale strandings receive prominent play in the mass media, a relatively small proportion of great-whale detritus appears to reach the intertidal zone. For example, out of ~1,600 gray whales (*Eschrichtius robustus*) dying annually in the Northeast Pacific (Smith and Baco 2003, and references therein), only 50 or so become stranded along the shoreline in a typical year, and the record 273 strandings in 1999 represented less than 20% of average annual gray whale mortality (Rugh et al. 1999). Because most whale detritus likely ends up at the deep-sea floor, the deep-sea ecosystem response to great-whale detritus is discussed first.

Deep-Sea Effects of Whale Detritus

When averaged over the entire deep-sea floor, the flux of particulate organic carbon (POC) in the form of great-whale carcasses is modest. The flux of small POC to the deep-sea floor ranges between ~0.3 and 10 g C_{org} m^{-2} y^{-1} (e.g., Smith and Demopoulos 2003). For comparison, Smith and Baco (2003) estimated that approximately 69,000 great whales die each year. If we assume that whale biomass is 5% organic carbon, that the average weight of a dying great whale is 40 metric tons, and that the ocean covers 3.6×10^8 km^2, the flux of organic carbon to the seafloor from whale falls averages 3.8×10^{-4} g C_{org} m^{-2} y^{-1} (see Jelmert and Oppen-Bernsten 1996 for similar calculations). This is only about 0.1% of the background POC flux to the deep-sea floor under the most oligotrophic central gyre waters. Even if whale mortality and flux is tenfold greater along migration corridors or in whale feeding grounds, background POC flux will also be higher in these regions because they typically occur along ocean margins or oceanographic fronts. Thus, it is difficult to imagine that the flux of great-whale detritus would exceed 0.3% of seafloor POC flux anywhere in the deep sea.

However, whales do not sink as an even veneer of organic matter but rather as giant, organic-rich lumps (e.g., Butman et al. 1995). The ~50 m^2 of sediments immediately underlying a fresh whale fall sustains, in a single pulse, the equivalent of about 2000 yr of background POC flux at abyssal depths (Smith and Baco 2003). In addition, these massive enrichment events can be common on regional scales. For example, Smith and Baco (2003) estimated conservatively that within the North Pacific gray whale range, whale falls occur annually with an average nearest neighbor distance of <16 km. If whale falls produced organic-rich "islands" at the food-poor deep-sea floor for extended time periods (see, e.g., Stockton and DeLaca 1982), they could support archipelagos of specialized communities, much as hydrothermal vents and cold seeps do (Van Dover 2000).

How do deep-sea ecosystems respond to the massive flux event of a whale fall? Although the deep-sea floor is remote and relatively poorly studied, there is now substantial evidence that sunken whales create persistent, ecologically significant habitats. Most information concerning the seafloor fate and impacts of whale detritus comes from the California slope, beneath the migration corridor of the northeast Pacific gray whale. I will review the California slope data first, and then summarize knowledge from other deep-sea regions.

The first natural whale-fall community was discovered on the California slope in 1987 (Smith et al. 1989). Study of this assemblage led to the hypothesis that deep-sea whale falls pass through four successional stages (Bennett et al. 1994):

1. A *mobile-scavenger stage,* during which necrophagous fish and invertebrates rapidly remove whale soft tissue

2. An *enrichment-opportunist stage,* during which dense assemblages of heterotrophic bacteria and invertebrates colonize the lipid-laden skeleton and surrounding sediments enriched by whale tissue "fallout"

3. A *sulfophilic stage,* during which chemoautotrophic assemblages colonize the skeleton as it emits sulfide from anaerobic decomposition of internal lipids

4. A *reef stage,* during which the hard, elevated skeletal remains are colonized by suspension feeders exploiting flow enhancement

Experimental, time-series studies of whale falls at depths between 1,000 and 2,000 m on the California slope provide strong evidence for the first three successional stages; these data are reviewed in the following sections.

THE MOBILE-SCAVENGER STAGE

Whale carcasses (*n* = 2, wet weights of 5 and 35 t) studied at 0.25 and 1.5 months after arrival at the seafloor exhibited community patterns consistent with a mobile-scavenger stage (Figure 22.1A). Within this time frame, carcasses were largely intact, with the predominant scavengers including hundreds of hagfish *(mostly Eptatretus deani)* and several sleeper sharks *(Somniosus pacificus)* 1.5 to 3.5 m in length (Smith et al. 2002). Other important scavengers included many thousands of small (~0.5 cm long) lysianassid amphipods on one carcass, and large lithodid crabs, possibly *Paralomis multispina,* on the other (Smith and Baco 2003). During this stage, hagfish were drawn from minimum distances of 0.6 to 0.8 km (Smith and Baco 2003), and the stage lasted approximately 0.3 to 1.5 yr, depending on carcass size (5 t or 35 t). Time-lapse photography and *in situ* sampling suggested that most of the soft tissue was directly removed by necrophages, especially *S. pacificus,* even though putrefaction was occurring within the whale flesh. The resultant tissue removal rates estimated for the scavenger assemblages (40–60 kg d^{-1}) imply that a 160 t blue whale *(Balaenoptera musculus)* carcass might support a mobile scavenger stage for as long as 7–11 yr. A total of 38 species of megafauna and macrofauna have been identified from whale falls in the mobile scavenger stage (Baco-Taylor, 2002; Smith and Baco 2003), with most species apparently being generalized scavengers. Calculations combining whale-fall spacing (for *Eschrichtius robustus* in the northeast Pacific) with scavenger foraging rates and fasting times indicate that large mobile scavengers such as rattails, hagfish, and lysianassids are unable to specialize on whale falls, given current stock sizes

of great whales in the northeast Pacific (Smith and Baco 2003). Nonetheless, the scavenger assemblages on the California slope are well adapted to recycle the soft tissue of whale carcasses over surprisingly short time scales (i.e., months to years).

THE ENRICHMENT-OPPORTUNIST STAGE

Communities consistent with an enrichment-opportunist stage were documented on carcasses (*n* = 3) ranging in size from 5 to 35 t at the seafloor for 0.3–4.5 yr. During this stage, sediments within 1–3 m of the skeleton were heavily enriched in organic matter (in some cases exceeding 10% organic carbon by weight) from tissue particles dispersed by scavengers. Organic-rich bones and sediments during this time were colonized by extremely high densities of heterotrophic macrobenthic polychaetes, mollusks, and crustaceans (Figure 22.1B–D)(Smith et al. 2002; Smith and Baco 2003). In some areas, bacterial mats also covered sediments nearby the skeleton. Macrofaunal densities in the sediments within 1–3 m of the carcass attained 20,000–45,000 individuals m^{-2} in as little as 4 months (Figure 22.2A); these densities exceeded background levels by an order of magnitude and are the highest ever reported for macrofauna below 1,000 m depths (Smith and Baco 2003). A number of the most abundant species in organic-rich sediments and on whale bones are new to science (e.g., two dorvilleid polychaetes, a chrysopetallid polychaete, and a gastropod) and could be whale-fall specialists; other species abundant on the whale falls during this stage have been collected at other types of organic enrichment (e.g., fish falls; Smith 1986) and are likely to be generalized opportunists. Despite high macrofaunal densities near the whale carcasses, species diversity adjacent to the skeletons was low (e.g., only 18 macrofaunal species) (Figure 22.2B). This rapid colonization by a high-density, low-diversity assemblage is strongly reminiscent of shallow-water opportunistic communities around sewage outfalls and beneath salmon pens (e.g., Pearson and Rosenberg 1978; Weston 1990; Zmarzly et al. 1994) and indicates that intense pulses of organic enrichment (whale falls, large kelp falls, etc.) are common enough at slope depths off California to have allowed the evolution of enrichment opportunists (Smith and Baco 2003). The duration of the *enrichment-opportunist stage* is likely to vary substantially with whale carcass size and ranges from <2 yr for a 5–10 t carcass, to at least 4.5 yr for a 35 t carcass.

THE SULFUPHILIC STAGE

Following scavenger removal of soft tissue from great-whale carcasses on the California slope, the recycling of lipids trapped within the skeleton (5–8% of total body mass) appears to be dominated by anaerobic microbial decomposition (Smith 1992; Deming et al. 1997; Smith and Baco 2003). Sulfate reduction is particularly important, providing a sustained efflux of sulfides, which can support sulfide-based chemoautotrophic bacteria, both free-living and endosymbiotic within

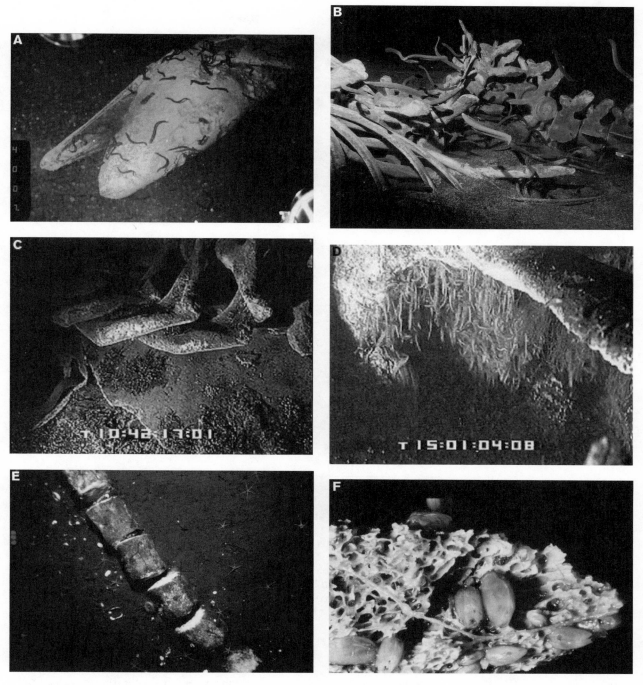

FIGURE 22.1. Photographs of whale falls at the seafloor on the California slope illustrating three successional stages. (A) A ~35 t gray whale carcass on the seafloor for 1.5 months at 1,675 m in the Santa Cruz Basin in the *mobile-scavenger stage*. Dozens of hagfish *(Eptatretus deani)*, each ~30-cm long, are feeding on the white carcass. Large bite marks formed by sleeper sharks *(Somniosus pacificus)* are also visible. (B–D) The Santa Cruz carcass after 18 months on the seafloor, now in the *enrichment-opportunist stage*. The whale soft tissue has been almost completely removed by scavengers, exposing vertebrae and ribs. The sediments around the skeleton (B) are colonized by a dense assemblage of gastropods, juvenile bivalves, cumacean crustaceans, and dorvilleid polychaetes (visible as white dots). The organic-rich bones, including the scapula (C) and ribs (D), harbor high densities of polychaetes, including a new species of chrysopetalid *(Vigtorniella fokati)* that forms grasslike patches (C) and hanging curtains (D) on some areas of the skeleton. For scale, the polychaetes are 1–2 cm long. (E–F) The 21-m long skeleton of a balaenopterid at 1,240 m in the Santa Catalina Basin, illustrating the *sulfophilic stage*. This skeleton has been on the seafloor for several decades. (E) Visible on the bones *in situ* are white bacterial mats covering the ends of vertebrae, and the shells of vesicomyid clams (~10 cm long). (F) A bone recovered from the carcass harboring large numbers of the mussel *Idas washingtonia* nestled into bone crevices to exploit effluxing hydrogen sulfide (for scale, mussels are 0.5–0.8 cm long).

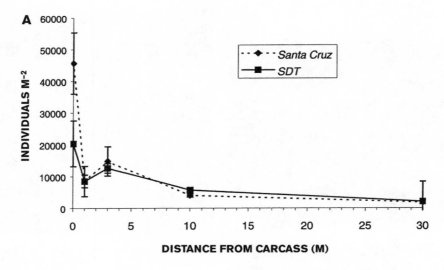

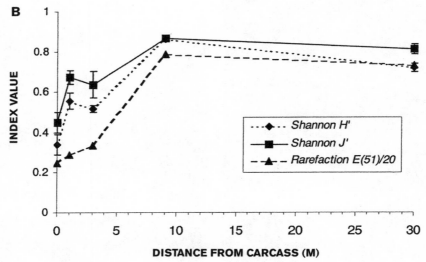

FIGURE 22.2. Macrofaunal community patterns around implanted whale falls in the San Diego Trough (*t* = 4 months) and the Santa Cruz Basin (*t* = 18 months) during the *enrichment-opportunist stage*. (A) Sediment macrofaunal densities around experimentally implanted whale falls in the San Diego Trough at 4 months, and in the Santa Cruz Basin at 18 months. Means ± one standard error are given. (B) Macrofaunal species diversity versus distance for the Santa Cruz Basin carcass.

the tissues of mussels, clams, and vestimentiferan polychaetes. Such a sulfophilic stage, composed of chemoautotrophs and other sulfide-tolerant species, has been documented on four California-slope whale skeletons that had been on the seafloor for >2 yr (Bennett et al. 1994; Smith and Baco 2003). This stage is characterized by several key components including (Smith and Baco 2003):

1. Mats of heterotrophic and chemoautotrophic bacteria growing on bone surfaces and within bone sutures and trabaculae (Figure 22.1E)

2. Large populations (>10,000 individuals per skeleton) of the mussel *Idas washingtonia* (Figure 22.1F), which harbors chemoautotrophic endosymbionts

3. Rich macrofaunal communities (>30,000 individuals) composed of bivalves, isopods, amphipods, polychaetes, limpets, and snails constituting at least three trophic levels

Whale-fall communities in the sulfophilic stage are remarkably species-rich, with an average of 185 species per skeleton; they appear to have the highest local species richness of any known deep-sea, hard-substrate community (Baco and Smith 2003). Many of the species from the sulfophilic stage are extremely abundant on whale skeletons but have rarely, if ever, been collected in surrounding habitats; they thus may be specialists that have evolved in sulfide-rich, whale-skeleton habitats (see subsequent discussion). The sulfophilic stage also exhibits faunal overlap with other deep-sea chemosynthetic communities, sharing 11 species (including vesicomyid clams, bathymodiolin mussels, and a vestimentiferan polychaete) with hydothermal vents, and 20 species with cold seeps (Baco et al. 1999; Smith and Baco 2003).

Large whale skeletons on the California slope sustain rich sulfophilic communities for extended time periods. Schuller et al. (2004) used $^{210}Pb/^{226}Ra$ disequilibrium and lipid degradation rates in whale bones to show that large whale skeletons may support sulfophilic communities for 40–80 yrs. The skeletons of juvenile gray whales appear to support the sulfophilic stage for much shorter periods of time (e.g., several years) because the poorly calcified bones disintegrate, releasing the lipid reservoir, much more rapidly (Baco-Taylor 2002, Smith and Baco 2003).

WHALE-FALL SUCCESSION IN OTHER REGIONS

Considerably less is known about deep-sea community response to whale falls beyond the California slope, but there is evidence that a succession of scavengers, enrichment opportunists, and sulfophiles will also colonize carcasses in other regions. For example, numerous studies suggest that mobile scavengers will feed voraciously on fresh whale falls throughout the well-oxygenated deep sea (Isaacs and Schwartzlose 1975; Hessler et al. 1978; Jones et al. 1998; Smith and Baco 2003). Furthermore, organic-rich sediments with an abundant microbial assemblage are documented beneath a whale fall in the western Pacific (Naganuma et al. 2001), and enrichment opportunists are known from sites of organic loading in a range of deep-sea settings (e.g., Turner 1977; Grassle and Morse-Porteus 1987; Desbruyeres and Laubier 1988; Levin et al. 1994; Snelgrove et al. 1994; Kitazato and Shirayama 1996; Snelgrove and Smith 2002). Finally, sulfophilic assemblages appear to be widespread on whale carcasses in the deep sea, because bathymodiolin mussels with chemoautotrophic endosymbionts have been recovered from whale bones in the North and South Atlantic and in the northwestern and southwestern Pacific, at depths ranging from 220 to 4,037 m (Wada et al. 1994; Naganuma et al. 1996, 2001; Baco-Taylor 2002; Smith and Baco 2003). Sulfophilic assemblages have also been found on fossil deep-sea whale skeletons as old as 30 Ma (Squires et al. 1991; Goedert et al. 1995), indicating that whale skeletons have supported chemoautotrophic communities over evolutionary time (Distel et al. 2000). Thus, succession on whale falls in the deep sea in general is likely to be functionally similar to that on the California slope, and this successional process, including colonization by sulfophiles, is likely to have occurred for at least 30 million years. Nonetheless, species structure and rates of successional change may differ dramatically in other parts of the deep sea, and patterns of succession are likely to have varied following the radiation of large whales since the Miocene (Gaskin 1982; Distel et al. 2000). In particular, in the modern ocean in regions such as the North Pacific central gyre, where whale falls should be much less common and seafloor communities are much more depauperate (e.g., Smith and Demopoulos 2003), whale-fall succession is expected to be extremely protracted (potentially lasting >100 yr) and species-poor compared to the California slope.

BIODIVERSITY AND WHALE-FALL SPECIALISTS

Deep-sea whale-fall communities, in particular those in the sulfophilic stage, may harbor remarkable levels of both local and global species richness. Whale falls are perhaps the least-studied chemosynthetic habitats in the deep sea, having been intensively sampled only along the California slope. Nonetheless, 407 animal species are known from whale falls, with 91% coming from California-slope whale falls alone (Baco and Smith 2003). This rivals the global species richness (469) known for far more intensively studied hydrothermal vents (Tunnicliffe et al. 1998) and substantially exceeds the number (~230) known from cold seeps (Sibuet and Olu 1998; Baco and Smith 2003). The relatively high species richness on lipid-rich whale skeletons likely results from the broad array of nutritional modes sustained by whale falls. A whale skeleton supports sulfophiles (e.g., species with chemoautotrophic endosymbionts), bone-matrix feeders, saprophages, generalized organic-enrichment respondents, and typical deep-sea deposit feeders and suspension feeders, all in close proximity (Baco and Smith 2003). Clearly, whale falls are heavily exploited habitat islands at the deep-sea floor.

There is increasing evidence that whale falls provide habitat for a specialized fauna: a suite of species that is specifically adapted to live on whale remains. Bennett et al. (1994) first noted a bimodal pattern in the frequency distribution of species abundances on whale skeletons, suggesting the presence of core species particularly adapted to whale-bone niches (qualitatively similar patterns are observed in dung and carrion assemblages in terrestrial environments). To date, 36 macrofaunal species were first collected on whale falls, and 28 of these have not been found in any other habitat (Table 22.1). A number of the species thus far unique to whale carcasses are extremely abundant, indicating that they are well adapted to whale falls and can attain substantial population sizes given suitable conditions. The absence of these species in samples from other related habitats (e.g., wood falls, algal falls, enriched sediment trays, hydrothermal vents, and cold seeps) suggests that they may indeed be endemic to whale falls.

In addition to the 28 potential whale-fall endemics, there are at least five other species that may be dependent on whale falls (Table 22.2). These are species that attain extraordinary abundance on whale carcasses but occur only as isolated individuals in other habitats. It is likely that a large proportion of the total individuals within these species live on whale falls, essentially making them whale-fall specialists; that is, their evolution has been largely shaped by selective pressures at whale falls (Bennett et al. 1994). This brings the total

TABLE 22.1
Species First Recorded at Large Whale Falls

Higher Taxon	Species	Known only at whale falls[a]	Estimated pop. Size[b]	Location[c]	Reference
Mollusca					
Archaegastropoda	*Pyropelta wakefieldi*	×	>100	California	McLean 1992
	Cocculina craigsmithi		300–1,100	California	McLean 1992
	Paracocculina cervae			New Zealand	Marshall 1994
	Osteopelta praeceps	×	>200	New Zealand	Marshall 1994
	Osteopelta ceticola			Iceland	Warén 1989
	Osteopelta mirabilis	×		New Zealand	Marshall 1987
	Protolira thorvaldsoni			Iceland	Warén 1996
Gastropoda	*Bruciella laevigata*	×		New Zealand	Marshall 1994
	Bruciella pruinosa	×		New Zealand	Marshall 1994
	Xylodiscula osteophila	×		New Zealand	Marshall 1994
	Hyalogyrina n.sp.			California	J. H. McLean and A. Warén, pers. comm.
Bivalvia					
Bathymodiolinae	*Adipicola pelagica*	×		South Atlantic	Dell 1987
	Myrina (Adipicola) Pacifica	×		Japan, Hawai'i	Dell 1987
	Adipicola (Idas) arcuatilis			New Zealand	Dell 1995
	Adipicola osseocola			New Zealand	Dell 1995
	Idas pelagica	×		North Atlantic	Warén 1993
	Idas ghisottii			North Atlantic	Warén 1993
Vesicomyid	New species?	×		California	Baco et al. 1999
Thyasiridae	*Axinodon* sp. nov.	×		California	P. Scott, pers. comm.
Aplacophora	New genus	×		California	A. Scheltema in prep.
Arthropoda					
Anomura	*Paralomis manningi*	×		California	Williams et al. 2000
Annelida					
Polychaeta					
Polynoidae	*Harmathoe craigsmithi*	×		California	Pettibone 1993
	Peinaleopolynoe santacatalina	×		California	Pettibone 1993
Chrysopetalidae	*Vigtorniella flokati*	×	1,000–100,000	California	Smith et al. 2002, Dahlgren et al. 2004
Ampharetidae	New genus	×	>10	California	B. Hilbig, pers. comm.
	Asabellides sp. nov.	×	>10	California	B. Hilbig, pers. comm.
	Anobothrus sp. nov.	×		California	B. Hilbig, pers. comm.
Siboglinidae	*Osedax frankpressi*	×	>1,000	California	Rouse et al. 2004
	Osedax rubiplumus	×	>1,000	California	Rouse et al. 2004
	Osedax, 3 sp. nov.	×	>1,000	California	Pers. obs.
	Osedax mucofloris	×	>1,000	Sweden	Glover et al. 2005
Dorvilleidae[d]	*Palpiphitime* sp.nov.	×	>10,000	California	B. Hilbig pers. comm.
	Dorvilleid sp. nov.	×		California	B. Hilbig pers. comm.
Sipuncula	*Phascolosoma saprophagicum*	×	>20–>200	New Zealand	Gibbs 1987

NOTE: Modified from Smith and Baco (2003).

[a] The 28 species marked as "known only at whale falls" have been reported from no other habitat.

[b] Where available, estimated population sizes on whale falls are given.

[c] Note that more than half of these species have been collected from southern California whale falls, suggesting that whale-fall habitats in other regions may be grossly undersampled.

[d] In addition to *Palpiphitime* sp. nov., an estimated 38 unidentified species of dorvilleids, with population sizes ranging from 10s to 1,000s of individuals per whale fall, have been collected from whale falls in the Santa Catalina Basin, San Diego Trough, San Clemente Basin, and Santa Cruz Basin (Baco and Smith 2003; C. Smith and I. Altamira, unpublished data). Many of these species appear to be new to science.

TABLE 22.2
Macrofaunal Species That Appear to Be Overwhelmingly More Abundant on Whale Skeletons
Than in Any Other Known Habitat

Species	Population Size on Whale Skeletons[a]	Number Collected in Other Habitat(s)
Bivalvia		
Idas washingtonia	>10,000– >20,000	1–10 (wood, vents, seeps)
Gastropoda		
Cocculina craigsmithi	300–1,100	1–10 (vents)
Pyropelta corymba	>1,000	1–10 (vents)
Pyropelta musaica	>250	1–10 (vents)
Crustacea		
Ilyarachna profunda	500–1,800	1–90 (sediments, seeps)

[a] Estimated population sizes on whale skeletons, and the total number of specimens collected in other habitats, are indicated for each species. Data from Bennett et al. (1994), Smith et al. (1998), Baco-Taylor (2002), Smith et al. (2002), Smith and Baco (1998), Baco and Smith (2003), D. Poehls et al. (in preparation), and J. H. McLean (personal communication). Table modified from Smith and Baco (2003).

number of potential whale-fall specialists to 33. This number is likely to rise substantially as the diverse dorvilleid (estimated to be >40 species), amphipod, and copepod components of the California-slope whale-fall fauna are rigorously examined by taxonomists and as whale-fall communities are more intensively sampled throughout the world ocean.

It should be noted that potential whale-fall specialists span a broad range of taxonomic and functional groups. These "specialists" come from five different phyla and appear to include whale-bone feeders (*Osedax*, a sipunculid, and some limpets), bacterial grazers (some limpets and *Ilyarachna profunda*), species utilizing chemoautotrophic endosymbionts (the bathymodiolins, thyasirid, vesicomyid, and siboglinid), deposit feeders (the ampharetids), facultative suspension feeders (the bathymodiolins), and predators (the polynoids and *Paralomis manningi*). This diversity suggests that a variety of taxa and trophic types have become specifically adapted to whale-fall niches and depend (in aggregate) on a variety of resources provided by the whale-fall habitat.

Whale Detritus at Shelf Depths

Remarkably little is know about the ecosystem response to whale falls at shelf depths. Because seafloor POC flux rates are typically much higher on the shelf than in the deep sea, the flux of organic carbon to the shelf floor in the form of whale detritus likely makes an insignificant contribution to the nutrient budgets of the continental shelf (e.g., Katona and Whitehead 1988). Exceptions to this generalization might occur in calving lagoons, such as the Ojo de Liebre and the San Ignacio Lagoons in Mexico, where gray whale strandings, and mortality in general, are likely to be concentrated in unusually small areas (e.g., Rugh et al. 1999).

Whale falls are certain to attract scavenger aggregations and undergo community succession on the continental shelf floor, but only very limited, anecdotal information concerning such shelf processes is available. At 150 m depths off Alaska, a gray whale carcass with substantial remaining soft tissue had attracted dense clouds of scavenging lysianassid amphipods (T. Shirley, personnal communication). At 90 m depths in the Strait of Juan de Fuca near San Juan Island, a 30 t fin whale *(Balaenoptera physalus)* placed at the seafloor for three months attracted a moderate diversity of facultative fish and shrimp scavengers, although little tissue removal had occurred (A. Shepard, D. Duggins, and C. Smith, unpublished data). In this relatively high-flow setting, no bacteria mats were visible on the carcass, possibly because of disruption by currents. After 28 months at the seafloor, the fin whale carcass had been stripped of soft tissue (D. Duggins, personal communication).

There are only a few data to indicate whether lipid-rich whale bones support a specialized fauna at shelf depths. The bone-eating siboglinid *Osedax mucofloris* occurs on whale bones at 40–125 m off the coast of Sweden (Glover et al. 2005). The mussel *Myrina pacifica*, which is thus far known only from whale bones, has been collected at 220 m on the Japanese slope (Baco-Taylor 2002, Smith and Baco 2003), but this may reflect the upper end of a bathyal (deep-sea) depth distribution. In addition, a new species of Polyplacophora *(Lepidozona balaenophila)* has been collected on whale bones from 240 m off Concepción, Chile (Schwabe and Sellanes 2004), but once again it is unclear whether this is predominantly a shelf or bathyal species. It is conceivable that whale falls, like hydrothermal vents and cold seeps (Van Dover 2000; Tarasov et al. 2005), primarily support a speciose endemic fauna in the deep sea, below depths of a few hundred meters. If true, this contrasts with other organic-rich substrates, in particular wood falls, which support a substantial number of highly specialized (albeit nonoverlapping) species in both the deep sea and shallow water (e.g., Turner 1973, 1977; Coan et al. 2000). Clearly, the dynamics and biogeography of whale falls at shelf depths merit substantial further study, especially in regions sustaining coastal migrations, or feeding and breeding aggregations, of great whales (e.g., the west coast of North America and the margins of Chile, New Zealand, and Antarctica).

Whale Detritus in the Intertidal

Only a small percentage of great-whale mortalities result in strandings in the intertidal, even for essentially coastal species such as *E. robustus* (Jones et al. 1984). For example, roughly 50 gray whales in the northeast Pacific, comprising <5% of annual mortality, come ashore in a typical year (Rugh et al. 1999; Smith and Baco 2003). This represents approximately one gray whale stranding per year per ~200 km of coastline along the ~10,000-km gray whale migration route, or a flux of organic carbon from whale detritus of ~10 g C_{org} $m^{-1}y^{-1}$ (assuming each whale carcass weighs 30 t and is 5% organic carbon). The flux of drift carrion from other sources (e.g., jellyfishes, fishes, turtles, seabirds, and other marine mammals) to beaches in the northeast Pacific, based on very limited measurements, appears to be roughly an order of magnitude higher (Columbini and Chelazzi 2003). Thus, whale detritus (if left undisturbed on the beach) appears to be a relatively minor source of carrion for intertidal scavengers (e.g., Rose and Polis 1998). However, it has been suggested that cetacean carcasses are important in the diet of some highly mobile terrestrial scavengers such as polar bears *(Ursus maritima)* and Arctic fox *(Alopex lagopus),* and that whale carrion may have helped coastal populations of California condors *(Gymnogyps californianus)* to survive following the extinction of the Pleistocene terrestrial megafauna (Katona and Whitehead 1988). Reliance on whale carrion by local populations of terrestrial scavengers seems especially likely around calving lagoons, such as the Ojo de Liebre and San Ignacio Lagoons in Mexico, where whale strandings are especially frequent (Rugh et al. 1999).

Very few data appear to be available on the natural recycling of stranded whale carcasses, but some generalizations appear possible. Although scavengers, such as seabirds, shorebirds, polar bears, foxes and vultures, may remove some of the soft tissue from whale carcasses (e.g., Schafer 1972; Katona and Whitehead 1988; Columbini and Chelazzi 2003), stranded cetaceans appear to be recycled primarily by microbes and terrestrial arthropods (e.g., flies, ants, and trogid, dermestid, and silphid beetles) (Columbini and Chelazzi 2003). Carcass reduction may take many months or even many years if mummification occurs (Schafer 1972) and involves a variety of decompositional stages (e.g., bloat, internal liquification, and dry-tissue stages) with successional patterns resembling those for large carcasses in fully terrestrial habitats (Schafer 1972; Cornaby 1974; Columbini and Chelazzi 2003). Thus, whale detritus stranded on beaches appears to be largely removed from marine food webs, with very little direct impact on marine ecosystems. In essence, whale strandings constitute a small, natural detrital flux from the ocean to land.

Whale Detritus in the Pelagic Realm

Whales that die in shallow water become inflated with decomposition gases within days, becoming buoyant detrital particles that may drift at the sea surface for weeks. The total flux of organic material in the form of whale detritus clearly is very small compared to other pelagic detrital sources (whale detrital carbon flux is <0.0005% of primary production rates even in oligotrophic regions), suggesting the energy input from whale detritus is not significant. Very limited observations indicate that during the early stages of decomposition, floating whale carcasses may be scavenged by sharks (e.g., blue sharks, *Prionace glauca,* and tiger sharks, *Galeocerdo cuvier*) and seabirds, although massive tissue removal is not usually observed (C. Smith, personal observations). Over periods of weeks, microbial decay weakens the cetacean connective tissues, and large skeletal components with tissue attached (e.g., the jaw, skull, sections of vertebrae) break off the carcass and sink to the seafloor (Schafer 1972). Thus, even for whales that die and initially float at the sea surface, much of the organic matter contained in the carcass ultimately becomes recycled at the seafloor. The small flux and short residence time of whale detritus at the sea surface suggests that there is little opportunity for whale carcasses to support a specialized community in pelagic ecosystems.

Impacts of Whaling on the Roles of Whale Detritus

Hunting by humans has caused massive reductions of great whale populations throughout the world oceans. The patterns of whale population depletion, carcass utilization, and, in some cases, whale population recovery, have differed substantially over time, among cetacean species, and among ocean basins, with the consequence that whaling has had complex effects on the availability of great-whale detritus to marine ecosystems. Below, I attempt to reconstruct patterns of whale detritus depletion resulting from commercial whaling and speculate on some of the consequences, particularly for deep-sea whale-fall communities.

Effects of Whaling on the Production of Whale Detritus

To evaluate the impacts of whaling on the production of whale detritus, it would be extremely useful to reconstruct the population trajectories of exploited cetaceans in each ocean. Despite the efforts of the International Whaling Commission (IWC) and numerous scientists, such reconstructions are generally not possible, and estimates of prewhaling cetacean population levels have remained controversial and politically charged (e.g., Roman and Palumbi 2003). It does seem clear that great whales, especially coastal species such as the Atlantic gray whale, began to be intensively exploited in the North Atlantic in the early nineteenth century (Tønnessen and Johnsen 1982; Whitehead 2002). Whaling efforts then intensified in the tropical and temperate Pacific in the midnineteenth century, in Antarctic waters after 1910, and in

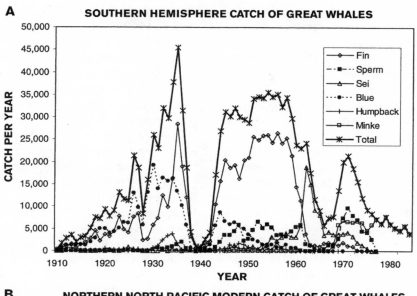

A SOUTHERN HEMISPHERE CATCH OF GREAT WHALES

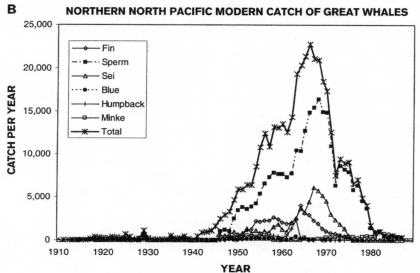

B NORTHERN NORTH PACIFIC MODERN CATCH OF GREAT WHALES

FIGURE 22.3. Annual catches of great whales in the southern hemisphere (A) and in the northern North Pacific (B) by whalers, between 1910 and 1985. Data are from the International Whaling Commission, compiled by Eric Danner in 2003.

higher latitudes of the North Pacific and in the Bering Sea as late as the 1950s (IWC 1993; Whitehead 2002; Springer et al. 2003; E. Danner, personal communication based on IWC catch statistics; Figure. 22.3). However, for many large species (e.g., fin, sperm, sei, blue, humpback, and minke whales), the bulk of the worldwide take occurred between approximately 1920 and 1980 (i.e., during 1–2 generations of a great whale), with the IWC estimating that roughly 2 million great whales were harvested from the oceans over this period (Figure 22.3; E. Danner, personal communication). For all but the sperm whale, it appears safe to say that great-whale population sizes were reduced an order of magnitude or more by whaling (e.g., IWC 1993; Best 1993; Roman and Palumbi 2003); for sperm whales, a reduction to ~30% of prewhaling values appears to be the best estimate (Whitehead 2002). Thus, as a

rule of thumb for estimating whaling effects on the production of whale detritus, I will assume a 10-fold reduction in great-whale standing stock. In many cases, especially in the North Atlantic, boreal North Pacific, and Southern Oceans, great-whale populations remain at only 10%–20% of prewhaling levels (e.g., Best 1993; Springer et al. 2003; Roman and Palumbi 2003), with a few notable exceptions (e.g., the northeast Pacific gray whale; Rugh et al. 1999).

Pelagic/Shelf/Intertidal Effects of Whaling

During both the open-boat and modern era of whaling (i.e., pre- and post-1900, respectively; Whitehead 2002), the net effect of whaling must have been a straightforward reduction of whale detrital inputs to pelagic, shelf, and intertidal ecosystems. This is because, during the open-boat era, whale carcasses taken in the high seas typically were stripped of blubber and released to sink to the deep-sea floor (Tønnessen and Johnsen 1982), whereas whales caught near shore (e.g., over the continental shelf) were likely to have been towed ashore for processing. Modern whaling leaves little detritus anywhere for the marine ecosystem because entire carcasses are processed on factory ships or on shore (e.g., Tønnessen and Johnsen 1982). Thus, with very localized exceptions (e.g., the inter- and subtidal in the immediate vicinity of whaling stations), harvested carcasses were essentially removed from pelagic, shelf, and intertidal ecosystems.

By reducing great-whale populations by an order of magnitude, whaling must have forced a roughly 10-fold decline in the flux and availability of great-whale detritus in pelagic, shelf, and intertidal ecosystems. For coastal populations of the California condor (Gymnogyps californianus), the loss of whale carrion may have been significant and could have caused dramatic population declines. It is also conceivable that coastal populations of other wide-ranging scavengers that fed on stranded whales, such as polar bears, arctic foxes, and grizzly bears (Ursus arctos horribilis), declined as a consequence of commercial whaling. It is interesting to note that, although gray whale populations have rebounded in the northeast Pacific, most stranded whale carcasses are still removed from beaches (C. Smith, personal observations), yielding on ongoing depletion of whale carrion. From a community-level perspective, the current energetic contribution of great-whale detritus to pelagic and shelf ecosystems appears to be so small that even if great-whale detrital fluxes were restored to prewhaling levels, the ecosystem consequences would be modest. However, in some intertidal areas, a 10-fold increase in the frequency of whale strandings could yield a carbon flux approaching that from other sources of marine carrion (see discussion in the preceding sections), suggesting that, prior to commercial whaling, stranded whales could have been a significant source of carrion to mobile-scavenger assemblages along coastlines. This conclusion must remain tentative until the intertidal flux of carrion from all sources is more intensively studied in various ocean regions.

The Impacts of Whaling on Deep-Sea Ecosystems

As in the case of shallow marine systems, whaling must ultimately have led to a dramatic decline in whale-fall habitats at the deep-sea floor, potentially leading to extinction of whale-fall specialists and limiting the dispersal of species dependent on sulfide-rich whale skeletons as habitat steppingstones (Butman et al. 1995, 1996; Committee on Biological Diversity in Marine Systems 1995; Smith and Baco 2003). However, the effects of whaling on whale-fall abundance in the deep sea were not necessarily monotonic, because of two opposing factors.

First, prior to ~1900, the initiation of whaling in a region increased flux of whale carcasses to the seafloor because whale carcasses were discarded to sink after removal of blubber,

baleen, spermaceti, and minor components (Tønnessen and Johnsen 1982; Butman et al. 1995). The flux of carcasses to the deep-sea floor must also have been redistributed by carcass discards relative to natural whale mortaility, because early whaling was initiated near home ports and then moved further afield as local populations became depleted (e.g., Butman et al. 1995; Springer et al. 2003). In contrast, natural whale mortality is likely to have been distributed along migration routes, in calving grounds, or in regions where whales spend substantial portions of their life cycles (Butman et al. 1995; Rugh et al. 1999; Smith and Baco 2003).

Second, ultimately (and immediately in the modern era), whaling decreased the flux of carcasses to the deep-sea floor because whale populations were driven downward, leaving far fewer whales to suffer natural mortality and sink to the seafloor.

The initial increase and subsequent decrease in carcass production resulting from whaling, combined with spatial and temporal variations in the activities of whalers, have caused historical patterns of dead-whale flux to vary among ocean basins. In addition, whale-fall communities pass through successional stages with different persistence times (ranging from months to decades), yielding time lags between a reduction in whale-carcass flux and the decline of particular community types at the seafloor. Without accurate population trajectories for all great whales in all basins, a detailed reconstruction of whale-fall habitat loss and likely patterns of species extinctions is not possible. However, the limited data available on whale-population trajectories do provide some insights into the historical biogeography of whale-fall habitat loss.

The most comprehensive population trajectory available in the refereed literature for any great-whale species was developed by Whitehead (2002) for the sperm whale (Physeter macrocephalus), with reconstruction of the global population size since 1800. Using a few reasonable assumptions, the abundance over time of sperm whale carcasses supporting mobile-scavenger, enrichment-opportunist, and sulfophilic communities in the deep sea can be estimated using the sperm whale population trajectory (Figure 22.4A). The largest number of potential whale-fall specialists are found in sulfophilic communities (Tables 22.1 and 22.2) (Smith and Baco 2003), so the dynamics of this community type are perhaps most relevant to species extinction. Several points emerge from modeling the abundance of sperm whale falls at the deep-sea floor over time since 1800. First, the discard of whale carcasses only modestly enhanced the number of whale-fall communities; for example, the number of sulfophilic communities increased by ~20% over natural processes in 1850 (Figure 22.4A). Second, because of short residence times, the abundances of mobile-scavenger and enrichment-opportunist communities respond rapidly to changes in whale-fall abundance, while sulfophilic communities respond with a 40-year time lag to whale depletion. Thus, the number of sulfophilic communities on sperm whale skeletons is estimated to be declining now, even though sperm whale abundance passed a minimum in 1981 and is currently about 40% of pre-exploitation levels (Figure 22.4A). Based on the global sperm-

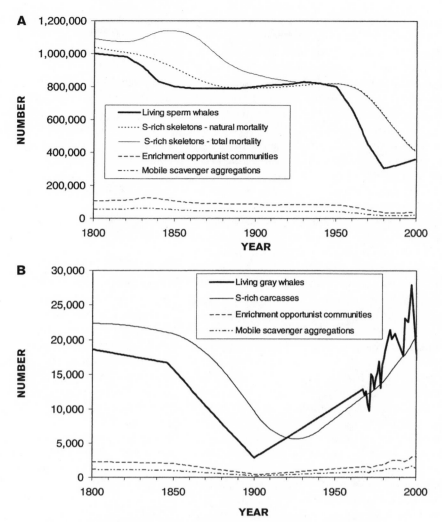

FIGURE 22.4. (A) "Population" trajectories for living sperm whales *(Physeter macro-cephalus)*, and the number of sperm whale falls in various successional stages (mobile-scavenger stage, enrichment-opportunist stage, and sulfophilic stage) at the deep-sea floor since 1800. Living sperm whale trajectory is from Whitehead (2002). The estimates of the number of whale-fall communities in the various successional stages depend on the following assumptions: (1) natural sperm whale mortality rate is 0.05 y^{-1}; (2) 90% of discarded carcasses and 50% of carcasses resulting from natural mortality sink to the deep-sea floor; (3) persistence times are 1 yr, 4 yr, and 40 yr for the mobile-scavenger stage, enrichment-opportunist stage, and sulfophilic stage, respectively (Smith and Baco, 2003; Schuller et al. 2004). (B) Similar trajectories, based on similar assumptions as in part A, for gray whales *(Eschrichtius robustus)* in the northeast Pacific. The trajectory for living gray whales is a combination of data from IWC (1993) and Rugh et al., in preparation (as communicated by J. Breiwick, 2003).

whale trajectory, whale-fall specialists may be only now approaching their greatest habitat loss, potentially causing species extinctions to be occurring at their highest historical rates. Species-area relationships suggest that loss of 60% of the area of a habitat, such has occurred for sperm whale falls, should yield extinction of 20% or more of its endemic fauna (e.g., Pimm and Askins 1995; Ney-Nilfe and Mangel 2000).

Regional asynchrony in the extermination of great whales suggests that ocean basins may be in different phases of whale-fall habitat loss and species extinction. Whale populations

were first reduced in the North Atlantic in the 1800s and even now may remain at <25% of pre-whaling levels (Roman and Palumbi 2003). Because whale abundance has remained low in the North Atlantic for approximately 150 years (i.e., much longer than the lag time resulting from sulfophilic community persistence), the number of whale-fall habitats have long since adjusted to low whale abundance, and species extinction driven by habitat loss is likely to be well advanced (e.g., Brooks et al. 1999). The loss of species may be substantial in the North Atlantic because whale-fall habitat abundance has been held at 10%–25% of pre-exploitation levels for an extended time; species-area relationships (e.g., Pimm and Askins 1995; Ney-Nilfe and Mangel 2000) suggest that such habitat reduction will extinguish 30%–50% of the specialized whale-fall fauna. In contrast, southern-hemisphere great whales were heavily exploited much later, between 1920 and 1965 (Figure 22.3), with their populations remaining low to the present (e.g., Best 1993; Clapham et al. 1999; Young 2000). As a consequence, sulfophilic communities in the Southern Ocean are likely only now to be approaching their historic lows, with extinction of whale-fall specialists in the acceleration stage. Species extinctions are probably least advanced in the northeast Pacific, where the greatest depletion of most large whales did not occur until the 1970s (Figure 22.3; Springer et al. 2003). In addition, some species, such the gray whale, had substantially recovered by 1970 from depredation suffered in the 1800s (Figure 22.4B), with the consequence that whale-fall habitats in the northeast Pacific may never have reached the relative lows experienced in the North Atlantic. Thus, one can predict that species extinctions and diversity loss in whale-fall communities have been greatest in the North Atlantic, have been substantial and are likely accelerating in the Southern Ocean, and have been least intense in the northeast Pacific. If species extinction due to whaling has dramatically altered the biodiversity of whale-fall communities, one would predict that current biodiversity levels are lowest in the North Atlantic and highest in the northeast Pacific.

Can we rigorously test this prediction to determine whether patterns of whale-fall biodiversity are consistent with whaling-induced species extinctions? A reasonable experimental approach would be to emplace uniform packages of lipid-rich whale bones at similar depths in the North Atlantic, Southern Ocean, and northeast Pacific and then, after a sufficient time period (i.e., 2–3 years), compare biodiversity levels of bone-colonizing assemblages across basins. This experimental approach is quite feasible, because bone implantations have fostered sulfophilic community development on the California slope, and similar experimental approaches (i.e., using standardized colonization substrates) have been used to assess regional variations in the biodiversity of fouling assemblages in shallow-water communities (e.g., Ruiz et al. 2000).

Conclusions

Whale carcasses are end members in the spectrum of marine detritus, constituting the largest, most energy-rich organic particles in the ocean. Most great-whale carcasses sink essentially intact to the deep-sea floor, where they are recycled by a succession of scavenger, enrichment-opportunist, and sulfophilic assemblages. Although the flux of organic carbon in whale falls is small compared to total detrital flux, the massive energy concentrated in a whale fall can support a diverse deep-sea community (~370 species in the northeast Pacific) for decades, including a significant number of potential whale-fall specialists (>33 species). The ecosystem impacts of detrital whales in epipelagic, shelf, and intertidal ecosystems

is poorly known but appears to be small, although some highly mobile intertidal scavengers (e.g., polar bears) could obtain important nutritional inputs from whale carrion.

Commercial whaling has drastically reduced the flux of whale detritus to all marine ecosystems. In intertidal habitats, this may have caused population declines in some scavenging species (e.g., the California condor) dependent on whale carrion. At the deep-sea floor, whaling led to substantial habitat loss to whale-fall communities and likely caused the first anthropogenic extinctions of marine invertebrates in the 1800s in the North Atlantic. Extinctions of whale-fall specialists are probably ongoing and, to date, are likely to have been most severe in North Atlantic, intermediate in the Southern Ocean, and least intense in northeast Pacific whale-fall communities.

Acknowledgments

I especially thank Jim Estes for conceiving and organizing the fascinating and important symposium that has led to this volume on *Whales, Whaling, and Ocean Ecosystems*. I also warmly thank the many people who have assisted at sea and on land during our whale-fall studies. The whale-fall work has been generously supported by the National Undersea Research Center Alaska (now the West Coast and Polar Regions Undersea Research Center), the U.S. National Science Foundation, the National Geographic Society, the British Broadcasting Corporation, and the University of Hawaii Research Council. This is contribution no. 6819 from the School of Ocean and Earth Science and Technology, University of Hawai'i at Manoa.

Literature Cited

Allison, P.A., C.R. Smith, H. Kukert, J.W. Deming, and B.A. Bennett. 1991. Deep-water taphonomy of vertebrate carcasses: a whale skeleton in the bathyal Santa Catalina Basin. *Paleobiology* 17: 78–89.

Ashley, C.W. 1926. *The Yankee whaler.* Boston: Houghton Mifflin.

Baco, A.R. and C.R. Smith. 2003. High biodiversity levels on deep-sea whale skeletons. *Marine Ecology Progress Series* 260: 109–114.

Baco, A.R., C.R. Smith, A.S. Peek, G.K. Roderick, and R.C. Vrijenhoek. 1999. The phylogenetic relationships of whale-fall vesicomyid clams based on mitochondrial COI DNA sequences. *Marine Ecology Progress Series* 182: 137–147.

Baco-Taylor, A.R. 2002. Food-web structure, succession and phylogenetics on deep-sea whale skeletons. Ph.D. thesis, University of Hawai'i.

Bennett, B.A., C.R. Smith, B. Glaser, and H.L. Maybaum. 1994. Faunal community structure of a chemoautotrophic assemblage on whale bones in the deep Northeast Pacific Ocean. *Marine Ecology Progress Series* 108: 205–223.

Best, P.B. 1993. Increase rates in severely depleted stocks of baleen whales. *ICES Journal of Marine Science* 50: 169–186.

Britton, J.C. and B. Morton. 1994. Marine carrion and scavengers. *Oceanography and Marine Biology: An Annual Review* 32: 369–434.

Brooks, T.M., S.L. Pimm, and J.O. Oyugi. 1999. Time lage between deforestation and bird extinction in tropical forests. *Conservation Biology* 13: 1140–1150.

Butman, C.A., J.T. Carlton, and S.R. Palumbi. 1995. Whaling effects on deep-sea biodiversity. *Conservation Biology* 9: 462–464.

———. 1996. Whales don't fall like snow: reply to Jelmert. *Conservation Biology* 10: 655–657.

Clapham, P.J., S.B. Young, and R.L. Brownell, Jr. 1999. Baleen whales: conservation issues and the status of the most endangered populations. *Mammal Review* 29: 35–60.

Coan, E.V., P.V. Scott, and F.R. Bernard. 2000. Bivalve shells of western North America: marine bivalve mollusks from Arctic Alaska to Baja California. *Santa Barbara Museum of Natural History Monographs* 2: 1–764.

Columbini, L. and L. Chelazzi. 2003. Influence of marine allochthonous input on sandy beach communities. *Oceanography and Marine Biology: an Annual Review* 41: 115–160.

Committee on Biological Diversity in Marine Systems. 1995. *Understanding marine biodiversity: a research agenda for the nation.* Washington, DC: National Academies Press.

Corkeron, P.J. and R.C. Connor. 1999. Why do baleen whales migrate? *Marine Mammal Science* 15: 1228–1245.

Cornaby, B.W. 1974. Carrion reduction by animals in contrasting environments. *Biotropica* 6: 51–63.

Dahlgren, T.G., A.G. Glover, and C.R. Smith. 2004. Fauna of whale falls: systematics and ecology of a new polychaetes (Annelida: Chrysopetalidae) from the deep Pacific Ocean. *Deep-Sea Research I* 51: 1873–1887.

Dell, R.K. 1987. Mollusca of the family Mytilidae (Bivalvia) associated with organic remains from deep water off New Zealand, with revisions of the genera *Adipicola* Dautzenborg, 1927 and *Idasola* Iredale, 1915. *National Museum of New Zealand Records* 3: 17–36.

———. 1995. New species and records of deep-water mollusca from off New Zealand. *National Museum of New Zealand Records* 2: 1–26.

Deming, J., A.L. Reysenbach, S.A. Macko, and C.R. Smith. 1997. The microbial diversity at a whale fall on the seafloor: bone-colonizing mats and animal-associated symbionts. *Microscopy Research and Technique* 37: 162–170.

Desbruyeres, D. and L. Laubier. 1988. Exploitation d'une source de matière organique concentrée dans l'océan profond: intervention d'une annelide polychete nouvelle. *Comptes Rendus de L'Academie des Sciences* 307: 329–335.

Distel, D.L., A.R. Baco, E. Chuang, W. Morril, C. Cavanaugh, and C.R. Smith. 2000. Do mussels take wooden steps to deep-sea vents? *Nature* 403: 725–726.

Gaskin, D.E. 1982. *Ecology of whales and dolphins.* Portsmouth: Heinemann.

Gibbs, P.E. 1987. A new species of *Phascolosoma* (Sipuncula) associated with a decaying whale's skull trawled at 880 m depth in the southwest Pacific. *New Zealand Journal of Zoology* 14: 135–137.

Glover, A.G., B. Källström, C.R. Smith, and T.G. Dahlgren. 2005. World-wide whale worms? A new species of *Osedax* from the shallow north Atlantic. *Proceedings of the Royal Society of London* B 3275: 1–4.

Goedert, J.L., R.L. Squires, and L.G. Barnes. 1995. Paleoecology of whale-fall habitats from deep-water Oligocene rocks, Olympic Peninsula, Washington state. *Palaeogeography, Palaeoclimatology, Palaeoecology* 118: 151–158.

Grassle, J.F. and L. Morse-Porteous. 1987. Macrofaunal colonization of disturbed deep-sea environments and the structure of deep-sea benthic communities. *Deep-Sea Research I* 34: 1911–1950.

Guinet, C., L.G. Barrett-Lennard, and B. Loyer. 2000. Co-ordinated attack behavior and prey sharing by killer whales at Crozet Archipelago: strategies for feeding on negatively buoyant prey. *Marine Mammal Science* 16: 829–834.

Hessler, R.R., C.L. Ingram, A.A. Yayanos, and B.R. Burnett. 1978. Scavenging amphipods from the floor of the Philippine Trench. *Deep-Sea Research I* 25: 1029–1047.

Isaacs, J.D. and R.A. Schwartzlose. 1975. Active animals of the deep-sea floor. *Scientific American* 233: 85–91.

IWC. 1993. Report of the Special Meeting of the Scientific Committee on the Assessment of Gray Whales. *Report of the International Whaling Commission* 43: 241–259.

Jelmert, A, and D.O. Oppen-Bernsten. 1996. Whaling and deep-sea biodiversity. *Conservation Biology* 10: 653–654.

Jones, E.G., M.A. Collins, P.M. Bagley, S. Addison, and I.G. Priede. 1998. The fate of cetacean carcasses in the deep sea: observations on consumption rates and succession of scavenging species in the abyssal Northeast Atlantic Ocean. *Proceedings of the Royal Society of London Series B: Biological Sciences* 265: 1119–1127.

Jones, M.L., S.L. Swartz, and S. Leatherwood (editors). 1984. *The gray whale* Eschrichtius robustus. San Diego: Academic Press.

Katona, S. and H. Whitehead. 1988. Are cetaceans ecologically important? *Oceanography and Marine Biology Annual Reviews* 26: 553–568.

Kitazato, H. and Y. Shirayama. 1996. Rapid creation of a reduced environment and an early stage of a chemosynthetic community on cattle bones at the deep-sea bottom in Sagami Bay, central Japan. *Vie et Milieu* 46: 1–5.

Levin, L.A., G.R. Plaia, and C.L. Huggett. 1994. The influence of natural organic enhancement on life histories and community structure of bathyal polychaetes, in *Reproduction, larval biology and recruitment of the deep-sea benthos.* C. Young and K. Eckelbarger, eds. New York: Columbia University Press, pp. 261–283.

Lockyer, C. 1976. Body weights of some species of large whales. *Journal du Conseil International pour L'Exploration de la Mer* 36: 259–273.

Marshall, B.A. 1987. Osteopeltidae (Mollusca: Gastropoda): a new family of limpets associated with whale bone in the deep sea. *Journal of Molluscan Studies* 53: 121–127.

———. 1994. Deep-sea gastropods from the New Zealand region associated with recent whale bones and an Eocene turtle. *Nautilus* 108: 1–8.

McLean, J.H. 1992. Cocculiniform limpets (Cocculinidae and Pyropeltidae) living on whale bone in the deep sea off California. *Journal of Molluscan Studies* 58: 401–414.

Moore, S.E., J.M. Grebmeier, and J.R. Davis. 2003. Gray whale distribution relative to forage habitat in the northern Bering Sea: current conditions and retrospective summary. *Canadian Journal of Zoology* 81: 734–742

Naganuma, R., H. Wada, K. Fujioka, K. 1996. Biological community and sediment fatty acids associated with the deep-sea whale skeleton at the Torishima seamount. *Journal of Oceanography* 52: 1–15.

Naganuma, T., M. Hattori, K. Akimoto, J. Hashimoto, H. Momma, and C.J. Meisel. 2001. Apparent microfloral response to organic degradation on bathyal seafloor: an analysis based on sediment fatty acids. *Pubblicazioni della Stazione Zoologica di Napoli—Marine Ecology* 22: 1–16.

Ney-Nilfe, M. and M. Mangel. 2000. Habitat loss and changes in the speciea-area relationship. *Conservation Biology* 14: 893–898.

Pearson, T.H. and R. Rosenberg. 1978. Macrobenthic succession in relation to organic enrichment and pollution of the marine environment. *Oceanography and Marine Biology: an Annual Review* 16: 229–311.

Pettibone, M.H. 1993. Polynoid polychaetes associated with a whole skeleton in the bathyal Santa Catalina Basin. *Proceedings of the Biological Society of Washington* 106: 678–688.

Pimm, S.L. and R. Askins. 1995. Forest losses predict bird extinctions in eastern North America. *Proceedings of the National Academy of Sciences* 92: 9343–9347.

Robineau, D. and V. de Buffrénil, V. 1993. Nouvelles données sur la masse du squelette chez les grands cétacés (Mammalia, Cetacea). *Canadian Journal of Zoology* 71: 828–834.

Roman, R. and S.R. Palumbi. 2003. Whales before whaling in the North Atlantic. *Science* 301: 508–510.

Rose, M.D. and G.A. Polis. 1998. The distribution and abundance of coyotes: the effects of allochthonous food subsidies from the sea. *Ecology* 79: 998–1007.

Rouse, G.W., S.K. Goffredi, and R. Vrijenhoek. 2004. *Osedax*: bone-eating marine worms with dwarf males. *Science* 305: 668–671.

Rugh, D.J., M. Muto, S.E. Moore, and D.P. DeMaster. 1999. *Status review of the eastern North Pacific stock of gray whales.* U.S. Department of Commerce, NOAA Technical Memorandum NMFS-AFSC-103. Seattle: Alaska Fisheries Service Center.

Ruiz, G.M., P. Fofonoff, J.T. Carlton, M.J. Wonham, and A.H. Hines. 2000. Invasions of coastal marine communities in North America: apparent patterns, processes, and biases. *Annual Review of Ecology and Systematics* 31: 481–531.

Schafer, W. 1972. *Ecology and palaeoecology of marine environments.* Chicago: University of Chicago Press.

Schuller, D., D. Kadko, and C.R. Smith. 2004. Use of $^{210}Pb/^{226}Ra$ disequilibria in the dating of deep-sea whale falls. *Earth and Planetary Science Letters* 218: 277–289.

Schwabe, E. and J. Sellanes. 2004. A new species of *Lepidozona* (Mollusca: Polyplacophora: Ischnochitonidae), found on whale bones off the coast of Chile. *Iberus* 22: 147–153.

Sibuet, M. and K. Olu. 1998. Biogeography, biodiversity, and fluid dependence of deep-sea cold seep communities at active and passive margins. *Deep-Sea Research II* 45: 517–567.

Silber, G.K., M.W. Newcomer, and M.H. Perez-Cortes. 1990. Killer whales *(Orcinus orca)* attack and eat a Bryde's whale *(Balaenoptera edeni). Canadian Journal of Zoology* 68: 1603–1606.

Smith, C.R. 1986. Nekton falls, low-density disturbance, and community structure of infaunal benthos in the deep-sea. *Journal of Marine Research* 44: 567–600.

———. 1992. Whale falls: chemosynthesis on the deep-sea floor. *Oceanus* 36: 74–78.

Smith, C.R. and A.R. Baco. 1998. Phylogenetic and functional affinities between whale-fall, seep, and vent communities. *Cahiers de Biologie Marine* 39: 345–346.

———. 2003. Ecology of whale falls at the deep-sea floor. *Oceanography and Marine Biology: an Annual Review* 41: 311–354.

Smith, C.R., A.R. Baco, and A. Glover. 2002. Faunal succession on replicate deep-sea whale falls: time scales and vent-seep affinities. *Cahiers de Biologie Marine* 43: 293–297.

Smith, C.R. and A.W.J. Demopoulos. 2003. The deep Pacific Ocean floor, in *Ecosystems of the World,* Vol. 28: *Ecosystems of the Deep Ocean.* P.A. Tyler, ed. Amsterdam: Elsevier, pp. 179 – 218.

Smith, C.R., H. Kukert, R.A. Wheatcroft, P.A. Jumars, and J.W. Deming. 1989. Vent fauna on whale remains. *Nature* 34: 27–128.

Snelgrove, P.V.R., J.F. Grassle, and R.F. Petrecca. 1994. Macrofaunal response to artificial enrichments and depressions in a deep-sea habitat. *Journal of Marine Research* 52: 345–369.

Snelgrove, P.V.R. and C.R. Smith. 2002. A riot of species in an environmental calm: the paradox of the species-rich deep-sea floor. *Oceanography and Marine Biology: an Annual Review* 40: 311–342.

Springer, A.M., J.A. Estes, G.B. van Vliet, T.M. Williams, D.F. Doak, E.M. Danner, K.A. Forney, and B. Pfister. 2003. Sequential megafaunal collapse in the North Pacific Ocean: an ongoing legacy of industrial whaling? *Proceedings of the National Academy of Sciences* 100: 12223–12228.

Squires, R.L., J.L. Goedert, and L.G. Barnes. 1991. Whale carcasses. *Nature* 349: 574.

Stockton, W.L. and T.E. DeLaca. 1982. Food falls in the deep sea: occurrence, quality, and significance. *Deep-Sea Research I* 29: 157–169.

Tarasov, V.G., A.V. Gebruk, A.V. Mironov, and L.I. Moskalev. 2005. Deep-sea and shallow-water hydrothermal vent communities: Two different phenomena? *Chemical Geology* 224: 5–39.

Tønnessen, J.N. and A.O. Johnsen. 1982. *The history of modern whaling.* Berkeley: University of California Press.

Tunnicliffe, V., A.G. McArthur, and D. McHugh. 1998. A biogeographical perspective of deep-sea hydrothermal vent fauna. *Advances in Marine Biology* 34: 353–442.

Turner, R.D. 1973. Wood-boring bivalves, opportunistic species in the deep sea. *Science* 180: 1377–1379.

———. 1977. Wood, mollusks, and deep-sea food chains. *Bulletin of the American Malacology Union* 213: 13–19.

Van Dover, C.L. 2000. *The ecology of deep-sea hydrothermal vents.* Princeton, NJ: Princeton University Press.

Wada, H., T. Naganuma, K. Fujioka, H. Ditazato, K. Kawamura, K., and Y. Akazawa. 1994. The discovery of the Torishima whale bone animal community and its meaning: the results of revisit dives by the "Shinkai 6500". *Japan Marine Science and Technology Center/Deep-Sea Research* 10: 38–47.

Warén, A. 1989. New and little known Mollusca from Iceland. *Sarsia* 74: 1–28.

———. 1993. New and little known Mollusca from Iceland and Scandinavia. Part 2. *Sarsia* 78: 159–201.

———. 1996. New and little known Mollusca from Iceland and Scandinavia. Part 3. *Sarsia* 81: 197–245.

Weston, D.P. 1990. Quantitative examination of macrobenthic community changes along an organic enrichment gradient. *Marine Ecology Progress Series* 61: 233–244.

Whitehead, H. 2002. Estimates of the current global population size and historical trajectory for sperm whales. *Marine Ecology Progress Series* 242: 295–304.

Williams, A.B., C.R. Smith, and A.R. Baco. 2000. New species of *Paralomis* (Decapoda, Anomura, Lithodidae) from a sunken whale carcass in the San Clemente Basin of southern California. *Journal of Crustacean Biology* 20 (Special Issue 2): 282–285.

Young, J.W. 2000. Do large whales have an impact on commercial fishing in the South Pacific Ocean? *Journal of International Wildlife Conservation Law and Policy* 3: 1–32.

Zmarzly, D.L., Stebbins, T.D., Pasko, D., Duggan, R.M., and Barwick, K.L. 1994. Spatial patterns and temporal succession in soft-bottom macroinvertebrate assemblages surrounding an ocean outfall on the southern San Diego shelf: relation to anthropogenic and natural event. *Marine Biology* 118: 293–307.

CASE STUDIES

TWENTY-THREE

Gray Whales in the Bering and Chukchi Seas

RAYMOND C. HIGHSMITH, KENNETH O. COYLE,
BODIL A. BLUHM, AND BRENDA KONAR

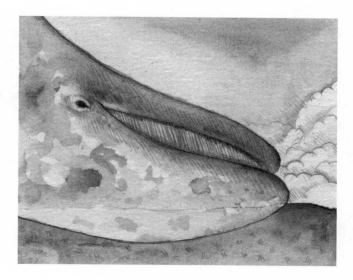

Among the large cetaceans, gray whales *(Eschrichtius robustus)* are unique in three important ways: They are benthic feeders; they undertake one of the longest migrations of any mammal; and they may be fully recovered (at least the eastern Pacific stock) from overharvesting by commercial whaling. The eastern (Chukotka-California) gray whales migrate annually between the mating regions and calving lagoons on the west coast of Baja California, Mexico, to summer feeding grounds in the northern Bering Sea and the Chukchi Sea (Rice and Wolman 1971; Marquette and Braham 1982; Findley and Vidal 2002). The purpose of this chapter is to explore the arctic ecosystem dynamics that justify such a migration and the impacts of the whales upon the system. We will conclude the chapter with an oceanographic production model that both explains the current location of major gray whale feeding sites and can be used for predictive purposes.

Commercial whaling is thought to have reduced the eastern North Pacific gray whale population from 15,000–20,000 to a few thousand animals by the late nineteenth or early twentieth century (Rice and Wolman 1971; Reilly 1984; Henderson 1984). Recovery to prewhaling levels had occurred by about 1980 (Reilly 1984), although numbers continued to increase through the 1990s and reached an estimated maximum of 28,000 by 1998 (Rugh et al. 2003). Based on studies of the gray whale prey community, composed of benthic ampeliscid amphipods, in a major feeding site (the Chirikov Basin, B5 in Figure 23.1), it was predicted that the growing gray whale population would not be supportable at the site (Highsmith and Coyle 1992; Coyle and Highsmith 1994), and indeed recent survey trends suggest that the population may be approaching carrying capacity (LeBoeuf et al. 2000; Moore et al. 2001; Perryman et al. 2002; Rugh et al. 2003).

Northern Feeding Sites

The distribution and relative abundance of gray whales in the Bering and Chukchi seas can be inferred from the results of various aerial and shipboard surveys conducted by Russian and U.S. scientists over the past several decades. The whales are widely dispersed and are usually observed singly or in small groups (Zimushko 1970; Berzin 1984; Table 23.1), not a surprising distribution given the large benthic area needed to support the energy requirements of each whale (Highsmith and Coyle 1992). In the west, consistently large numbers have been seen between Cape Dezhnev (northeastern tip of Chukotka) in Bering Strait to about Cape Schmidt (Table 23.1; C1–C3 in Figure 23.1), with the greatest concentrations off Cape Serdtse Kamen (C1 in Figure 23.1; Berzin 1984). Relatively few whales were observed in or south of the

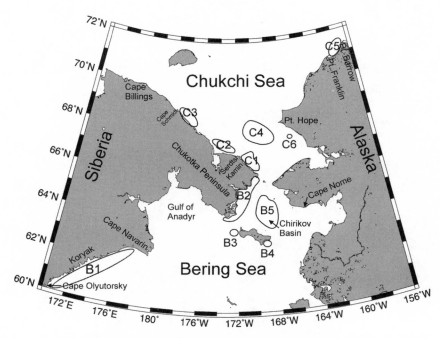

FIGURE 23.1. Map of the northern Bering and Chukchi Seas showing locations of significant gray whale feeding sites. See Table 23.1 for references.

Gulf of Anadyr. The westernmost sighting of gray whales in the Chukchi is 178°30′E (between Cape Schmidt and Cape Billings; Figure 23.1; Berzin and Rovnin 1966; Doroshenko 1981; Berzin 1984). There have been reports of some whales in the C4 area on both sides of the Convention Line (Maher 1960; Wilke and Fiscus 1961; Johnson et al. 1981; Marquette et al. 1982; Braham 1984; Moore and Ljungblad 1984). In the east, gray whales occasionally venture into the western Beaufort Sea (Maher 1960; Rugh and Fraker 1981; Marquette and Braham 1982).

Berzin (1984) suggested that the western (~179°W) and eastern (~157°W) distribution limits of feeding gray whales in the Chukchi Sea are set by the point at which the whales encounter the ice pack. Specific locations vary annually, and with the retreat of the ice pack in recent decades it will be interesting to see whether the whales extend their foraging range. An alternative hypothesis (see the last section of this chapter) is that the western and eastern foraging limits are determined by high benthic production derived from the variable location of the broad (60 km) frontal zone formed off the north Chukotka coast by the intersection of the Siberian Coastal Current and Bering Sea Water (Figure 23.2; Weingartner et al. 1999). This front extends northeastward and intersects the Alaskan coast near the easternmost distribution of significant numbers of feeding gray whales near Point Barrow (C5 in Figure 23.1). The lack of concentrated food may also explain the general absence of gray whales in the north-central Chukchi Sea (Berzin 1984). The southern limit of gray whale foraging in the eastern Bering Sea is off the southeastern coast of St. Lawrence Island at approximately 63°N (B4; Figure 23.1). In the western Bering Sea,

after the end of commercial whaling, the southern extent of Chukotka-California gray whale feeding aggregations was at Glubokaya Bay (61°12′N; B1 in Figure 23.1), but as the protected population grew, sightings extended to Cape Olyutorsky (60°N) by the early 1980s (Berzin 1984). These southern distribution limits are presumably determined by the location of soft-bottom depositional habitats with high prey biomass.

Most eastern North Pacific gray whales feed each summer on benthic organisms, especially infaunal amphipods, on the broad continental shelves of the northern Bering Sea and the Chukchi Sea. The majority of foraging activity occurs in water depths of approximately 50 m or less (Rice and Wolman 1971; Berzin 1984; Braham 1984; Nerini 1984; Yablokov and Bogoslovskaya 1984; Highsmith and Coyle 1990, 1992; Moore et al. 2000; Clarke and Moore 2002) and where benthic biomass is very high (Berzin and Rovnin 1966). A variety of foraging sites have been determined (Figure 23.1; Table 23.1). Because gray whales are submerged most of the time (Wursig et al. 1986), the number of whales counted (Table 23.1) is sometimes multiplied by a factor of up to three to obtain an estimate of the actual number of whales at a site (Berzin 1984).

The Chirikov Basin has been identified repeatedly as a major gray whale feeding site (B5 in Figure 23.1; Braham 1984; Moore and Ljungblad 1984). Concentrations of foraging animals occur at both ends of St. Lawrence Island (B3 and B4 in Figure 23.1). Early-arriving whales probably feed in the region until ice conditions allow dispersal farther north (Braham 1984). Thomson and Martin (1986) estimated that 15% of the total whale population remained in the Chirikov

TABLE 23.1
Gray Whale Population Counts and Estimates at Feeding Sites in the Northern Bering and Chukchi Seas

Area	Density	Counted (Estimated)	Survey Type	Survey Date	Reference
B1	Low		Aerial	July 1968	Berzin 1984
	Low	5	Aerial	Aug., Oct. 1973	Berzin 1984
	High		Ship	Sept., Oct. 1974	Berzin 1984
	Low		Aerial	Sept., Oct. 1975	Berzin 1984
	High		Ship	Sept., Oct. 1982	Berzin 1984
		200	Ship	Sept. 1936	Zenkovich 1937a
B2	Few		Aerial	July 1968	Berzin 1984
	High		Ship	Sept., Oct. 1974	Berzin 1984
		4/km	Aerial	Sept., Oct. 1974	Berzin 1984
	High		Aerial	Sept., Oct. 1975	Berzin 1984
	High		Ship	Sept., Oct. 1982	Berzin 1984
		1197	Ship	1942	Vadivasov 1947
		1154	Ship	Aug.–Sept. 1942	Yablokov and Bogoslovskaya 1984
		1033	Ship	1962	Berzin & Rovnin 1966
B3	High		Aerial	1976–1978	Berzin 1984
B4	High		Aerial	1976–1978	Berzin 1984
B5	High		Aerial	Sept., Oct. 1975	Berzin 1984
	High		Ship	Sept., Oct. 1979	Berzin 1984
	High	>500/yr	Aerial	1979–1978	Berzin 1984
	High	299	Aerial	May–June 1981	Moore and Ljungblad 1984
		(3,300–3,500)		1982	Berzin 1984
	High		Aerial	1982–1991	Moore et al. 2000
	Low		Aerial	July 2002	Moore et al. 2003
C1		17	Aerial	July 1968	Berzin 1984
	High	0.11/km²	Aerial	Sept., Oct. 1973	Berzin 1984
	High		Aerial	Sept., Oct. 1975	Berzin 1984
	High	3,000	Ship	July–Aug. 1982	Berzin 1984
	High	180	Ship	Aug. 1982	Berzin 1984
	High		Ship	Sept., Oct. 1982	Berzin 1984
C2	Low		Aerial	July 1968	Berzin 1984
	High		Aerial	Sept., Oct. 1975	Berzin 1984
	High		Ship	July–Aug. 1982	Berzin 1984
C3	Low		Aerial	July 1968	Berzin 1984
	High		Ship	Sept.–Nov. 1980	Berzin 1984
	Medium		Ship	Sept.–Oct. 1982	Berzin 1984
C4	High	44; 132	Aerial	Aug., Oct. 1973	Berzin 1984
	High		Aerial	Sept., Oct. 1975	Berzin 1984
	High	260	Ship	Sept., Oct. 1979	Berzin 1984
	High	588 (2,000)	Ship	Sept.–Nov. 1980	Berzin 1984
	High	1,021	Ship	Sept.–Nov. 1980	Berzin 1984
		190	Ship	Aug. 1982	Berzin 1984
		588 (2,000)	Ship	Sept. 1979	Doroshenko 1981
		1,021		Sept. 1980	
		127	Aerial	Oct. 1979	Johnson et al. 1981
		125	Aerial	Sept. 1980	Marquette et al. 1982
		>1,000 (3,000)	Ship	June, Sept. 2003	This chapter
		128	Aerial	Aug.–Nov. 1980–1989	Clarke and Moore 2002
C5		40	Ship	Aug. 1982	Berzin 1984
		60	Ship	Aug. 1982	Berzin 1984
		200	Ship	Aug. 1982	Berzin 1984
	High		Aerial	1982–1991	Moore et al. 2000
C6		(200)	Ship	June, Sept. 2002	This chapter

NOTE: For areas referenced, see Figure 23.1. Total for B1 and B2 was 1,800–2,000; total for C1 and C4 was 873.

FIGURE 23.2. Map showing the East Siberian Current and region of wide frontal zones established when the northward-flowing Bering Sea Water is encountered. From Weingartner et al. 1999.

area to feed during the summer. Soviet scientists estimated that 3,300–3,500 animals were Chirikov residents in the summer of 1982 (Berzin 1984), approximately 19% of the total population (Breiwick et al. 1988). An estimated 70% of the total population (Highsmith and Coyle 1992) feed in the Chirikov while transiting through the area during the spring and fall migration.

In 2003, we made two cruises (June and September) to the C4 region east of the Convention Line. Within an area bounded by 67°00′N to 68°10′N and 168°00′W to 169°00′W (eastern fourth of C4 in Figure 23.1), we counted more than 1,000 gray whales in 4–5 days. Each cruise used a point transect survey. Of those whales observed closely, about half produced mud plumes, an indication of bottom feeding. The whales were arrayed along a broad east-west oceanographic front created by the confluence of the eastward-flowing Siberian Coastal Current (cold, dilute) and northward-flowing Bering Sea Water, which is warmer and saltier (Figure 23.2; Weingartner et al. 1999). It should be noted, however, that the Bering Sea Water includes cold, dense nutrient-laden Anadyr Water, which may underlie the warmer water for tens of kilometers. The highest whale concentrations were observed along the Bering Sea Water (southern) side of the frontal area. We posit that whales observed feeding west of the Convention Line were also orienting to the same frontal system (Figure 23.2).

Animals that are as large and abundant as gray whales and that concentrate most of their feeding during approximately half of the year (Rice and Wolman 1971; Oliver et al. 1983; Perryman and Lynn 2002), need to forage where there is either sustained high primary production or a concentration of production from elsewhere (Moore and DeMaster 1997). Various studies (Walsh et al. 1989; Springer et al. 1996; Weingartner 1999; Figure 23.2) and satellite images (Figure 23.3) show that a plume of production extends from the Gulf of Anadyr through the Bering Strait to the gray whale feeding grounds in the Chukchi Sea (Figure 23.1). The Anadyr Water is upwelled and delivers nutrients as well as phytoplankton to the Chukchi

sites. As has been suggested for the Anadyr-fed Chirikov Basin (B5 in Figure 23.1; Hansell 1989; Walsh et al. 1989; Springer et al. 1996), there must be poor coupling between phytoplankton production and zooplankton grazers in areas C1–C5. Thus, much of the diatom biomass settles to the shallow (<50 m) bottom (Grebmeier and McRoy 1989; Fukuchi et al. 1993; Feder et al. 1994) and is consumed by the benthos, which in turn is preyed upon by gray whales and other benthic feeders.

Although the northern feeding grounds are clearly important influences on the distribution and abundance of gray whales, it has become apparent that feeding may occur during the northern migration (Braham 1984) and that tens to a few hundred whales may remain at places, such as Vancouver Island and Kodiak Island, along the migration route during the summer (Darling 1984; Murison et al. 1984; Oliver et al. 1984; Kvitek and Oliver 1986; Calambokidis et al. 2002; Dunham and Duffus 2002; Stelle 2002; Moore et al. 2003). The feeding methods utilized during migration tend to vary, and may include skimming, filter feeding, and benthic feeding in soft-bottom locations. These observations are important for present-day understanding of gray whale ecology and also provide interesting insights into gray whale survival over the millennia. As recently as 10–12 thousand years ago, sea level was roughly 75 m lower than it is today (Hopkins 1967), one consequence being that the areas that are currently the major northern feeding sites utilized by gray whales were emergent. This raises questions about how these animals survived in the absence of required food resources and feeding sites as we presently know them. The ability to utilize alternative feeding modes and locations may have been critical to survival of the species during glacial periods when the continental shelf areas of the northern seas were above sea level.

Benthic Feeding in the North

Gray whale stomach contents collected from the Chirikov Basin and Russian coast indicate that the primary food is infaunal amphipods, primarily the large *Ampelisca macrocephala*

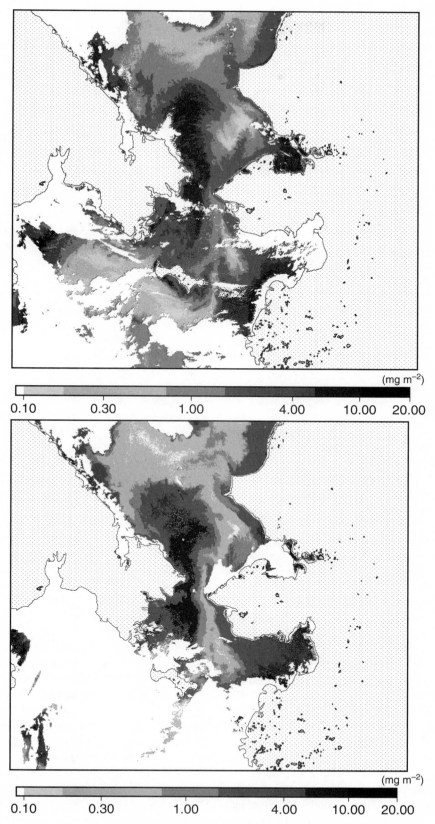

FIGURE 23.3. Satellite images showing the highly productive Anadyr Water entering the Chukchi Sea through the Bering Strait. The images are 3 days apart in June 2003 and illustrate variability in the front location. The authors would like to thank the SeaWifs Project (Code 970.2) and the Goddard Earth Sciences Data and Information Services Center/Distributed Active Archive Center (Code 902) at the Goddard Space Flight Center, Greenbelt MD 20771, for the production and distribution of these data, respectively. These activities are sponsored by NASA's Earth Sciences Enterprise.

(Nerini 1984; Yablokov and Bogoslovskaya 1984; Highsmith and Coyle 1990, and references therein). The whales prey on the amphipods by placing one side of their mouth on the sandy-mud bottom and sucking out large pits as they swim (Ray and Schevill 1974; Nerini and Oliver 1983; Nerini 1984; Swartz and Jones 1987). Many gray whales produce a mud plume when they surface, which is interpreted to indicate that the whale just made a feeding dive (Scammon 1874; Moore and Ljungblad 1984; Nerini 1984). The distribution of barnacles *(Cryptolepas rhachianecti)* on the whales' heads suggests that they tend to roll on their right side to feed as they work the bottom sediments (Kasuya and Rice 1970; Blohkin 1984). Also, baleen on the right side tends to show greater wear.

In contrast to the Chirikov Basin (B5 in Figure 23.1) and the northern Russian sites (B2 and C1–C3 in Figure 23.1), whales feeding along the Koryak Coast (B1 in Figure 23.1) and on the northeast shore of the Gulf of Anadyr tend to feed on amphipods in the genus *Pontoporeia* (Haustoriidae), particularly *P. femorata* (Nerini 1984; Yablokov and Bogoslovskaya 1984). Stomach contents also included the amphipods *Anonyx* (Lysianassidae) and *Atylus* (Demaxinidae) and polychaetes in the family Oweniidae (Blokhin 1984). Adults of the various amphipod species reach lengths of 13 to 27 mm (Nerini 1984).

The gray whale feeding site in which the prey community has been studied most extensively is the Chirikov Basin (B5 in Figure 23.1; Nerini and Oliver 1983; Braham 1984; Nerini 1984; Nelson and Johnson 1987; Highsmith and Coyle 1990, 1991, 1992; Moore et al. 2003). More than 90% of the stomach contents from this area were *A. macrocephala* (Bogoslovskaya et al. 1981), which requires a soft-sediment habitat. The Chirikov seafloor is covered by an approximate 2-m-thick layer of fine sand (0.125 mm), deposited at the end of the last glacial period, about 10,000 years ago (Johnson and Nelson 1984; Nelson and Johnson 1987).

Ampelisca macrocephala is the largest and most abundant of the eight ampeliscid species present in the Chirikov Basin, reaching lengths of 3 cm and mean densities of about 2,500/m² (Coyle and Highsmith 1989; Highsmith and Coyle 1990, 1992). This species constructs mucous tubes that extend a few centimeters into the loose sediment. *A. macrocephala* feed by positioning themselves ventral side up at the opening to their tubes and creating a mouthward feeding current with their abdominal appendages (Highsmith and Coyle 1991). Diatoms in the water column near the seafloor are the primary food source for the amphipods, but they are also capable of sweeping the sand surface around their tubes to a distance of about 1 cm with their long second antennae and wafting any food items detected into the feeding current. The latter feeding method was utilized in experiments in which settling larvae and newly metamorphosed juveniles of a potential space competitor, the sand dollar *Echinarachnius parma*, were eaten, suggesting that *A. macrocephala* is also a facultative predator (Highsmith and Coyle 1991). *A. macrocephala* is long-lived with low fecundity (~60 eggs/brood) and a high age (4–5 years) at first reproduction (Highsmith

and Coyle 1991; Coyle and Highsmith 1994). *A. macrocephala* appears to dominate smaller amphipod species with higher reproductive rates and shorter life cycles by outcompeting them for space (Coyle and Highsmith 1994).

The population dynamics and life history features of gray whale prey items at other feeding sites have not been studied. Also, there appears to be little stomach content information from other northern areas. The stomach of a first-year male near Cape Lisbourne contained the benthic isopod *Tecticeps alascensis* (Kim and Oliver 1989).

Nerini (1984) proposed that when gray whales forage on the seafloor, they consume the benthic invertebrates susceptible to the sucking mode of feeding in proportion to their abundance in the community, as the whales lack the means to sort prey. Approximately 60 benthic amphipod species and 80–90 other benthic invertebrate species have been recorded in the diets of gray whales (Blokhin 1980, 1984; Zimushko and Ivashin 1980; Bogoslavskaya et al. 1982; Nerini 1984; Yabolokov and Bogoslovskaya 1984), but most of these species are a small proportion of the stomach contents, commensurate with field abundances. In contrast, the pelagic mode of feeding usually reported along the migration route (Nerini 1984; Dunham and Duffus 2002; Stelle 2002) appears to target swarms or schools of a particular species (Kim and Oliver 1989).

Our 2003 work in the Chukchi Sea (the eastern portion of C4 in Figure 23.1) indicates that gray whales feeding there utilize prey other than infaunal amphipods. We collected 105 grab samples in the area where the whales were feeding and found no *A. macrocephala* and few other infaunal amphipods. The surface sediments were light brown and very soft, but at 2–3 cm deep they turned dark and claylike in texture and appeared to be anoxic. Confirming that the surface sediments were unconsolidated, otter trawls collected large numbers of shallow burrowers such as the small sea cucumbers *Cucumaria japonica* and *Chiridota* sp.; the protobranch bivalves *Ennucula bellotii, Nuculana pernula,* and *Yoldia hyperborea;* and the moonsnail *Cryptonatica affinis.* Epifauna was quite abundant as judged by 10-minute otter trawls that typically included numerous snow crabs *(Chionoecetes opilio);* lyre crabs *(Hyas coarctatus);* hippolytid, pandalid, and crangonid shrimp; small seastars *(Leptasterias* sp.); brittle stars *(Ophiura sarsi);* basket stars *(Gorgonocephalus caryi);* and occasional sea anemones, compound ascidians, flatfish, and sculpins. Acoustic scattering layers over the bottom consisted of krill, *Thysanoessa raschii,* and arctic cod, *Boreogadus saida.* Most of the listed invertebrates previously have been found in gray whale stomachs near the Chukotka Peninsula, where they appear to have been incidentally caught while the whales were targeting the dominant (>90%) infaunal amphipods (Blokhin 1980; Bogoslovskaya et al. 1982; Nerini 1984; Yablokov and Bogoslovskaya 1984). With infaunal amphipods absent at C4, one or more of the locally abundant shallow-infaunal or epifaunal species, or the dense near-bottom crustacean and fish accumulations, or some combination of these groups, must constitute the gray whale's principal prey.

Of the approximately 70 gray whales near the ship that we observed closely, about half produced light-colored mud plumes, similar to the upper 2–3 cm of surface sediment brought up in van Veen grabs. These findings further suggest that the whales are feeding on near-bottom epifauna or skimming shallow-dwelling infauna from the bottom.

Impacts of Gray Whale Feeding

A large, abundant predator is expected to have a strong role in structuring the biological community in which it feeds (Paine, Chapter 2 of this volume; Jackson, Chapter 4 of this volume; Williams, Chapter 15 of this volume). Gray whales in the northern feeding grounds probably exert their strongest impacts through sediment disruption and removal of the spatially dominant prey species (Nerini and Oliver 1983; Oliver and Slattery 1985; Moore and DeMaster 1997).

Biological Impacts

Daily consumption of benthic organisms by large adult gray whales has been estimated at 379–2,496 kg (Tomilin 1946) and at 1,200 kg (Zimushko and Lenskaya 1970). The latter authors also estimated that the gray whale population at that time consumed 773,000 metric tons of food per year on the northern feeding grounds. This estimate is based upon estimated daily food intake, number of feeding days, and population size. For example, using 1,200 kg day^{-1} and 180 feeding days (May–October) translates into a population of about 3,600 whales needed to consume 773,000 mt yr^{-1}. This estimate is probably too low, because the 1980 population estimate of 15,500 whales (Reilly 1984; Rugh 1984) extrapolates to a 1970 population of about 11,000 individuals, based on a growth rate of 3% per year (Reilly et al. 1983). Eleven thousand gray whales would consume about 2,462,400 mt yr^{-1}. Estimates of this type are subject to substantial errors, but the inescapable conclusion must be that a huge invertebrate liveweight biomass is removed from the benthic communities of the northern Bering and Chukchi Seas each year. This underscores the requirement of gray whales for feeding sites with very high secondary production.

In the well-studied Chirikov Basin (B5 in Figure 23.1), ampeliscid production has been estimated at 170–230 kcal m^{-2} yr^{-1}, more than entire community production at other highly productive locations such as Georges Bank and Long Island Sound (Highsmith and Coyle 1990, 1992). Indeed, the amphipod community inhabiting approximately 40,000 km^2 of the Chirikov Basin in the late 1980s was one of the most productive benthic communities in the world. Based on whale energy requirements and population growth and benthic amphipod production, Highsmith and Coyle (1992) and Coyle and Highsmith (1994) predicted that gray whales were at or near the carrying capacity of the Basin and that by 2000 the amphipod community would not be able to support continued whale predation. Research cruises in late June through early July and September of 2002 and 2003, and a National Oceanic and Atmospheric Administration (NOAA) survey in 2002 (Moore et al. 2003), yielded very few sightings of gray whales in the central Chirikov region, supporting that conclusion. Preliminary results of our 2002–2003 field work indicate that amphipods are still present in the Chirikov Basin, but they have a patchier distribution and overall lower densities than in the 1980s.

Physical Impacts

The major physical impact of gray whale feeding is the reworking of surface sediments (Johnson and Nelson 1984; Nerini 1984; Nelson and Johnson 1987; Nelson et al. 1987). A variety of information indicates that this is an important ecosystem-level process. Using side-scan sonar, Johnson and Nelson (1984) and Nerini (1984) detected numerous pits in the sandy floor of the Chirikov Basin that they attributed to the benthic feeding activities of gray whales. In shallow, nearshore areas off the Southeast Cape of St. Lawrence Island, pits were also observed by divers (Nerini 1984). Comparisons of fauna in recently made pits (same season) with adjacent areas revealed a reduction in infaunal abundance of up to 50% and a shift in species from tube-building polychaetes and amphipods to free-living amphipods, especially scavengers (Klaus et al. 1990). Feeding pits cover an estimated 1,200 km^2 or 5.6% of the Chirikov feeding area each year. Associated with this process, an estimated 172 million cubic meters of sand is resuspended, with the clay fraction (4.3×10^6 mt) advected by currents to the Chukchi Sea (Nelson and Johnson 1987; Nelson et al. 1987).

Gray Whale Distribution Model: Bering and Chukchi Seas

The data reviewed in this chapter and in recent oceanographic publications suggest a unifying model that incorporates the major ocean currents, pelagic primary production, and subsequent benthic secondary production to explain the distribution of gray whale feeding sites in the Bering and Chukchi Seas (Figure 23.4). Large marine mammals, especially those that migrate annually, are good indicators of ecosystem productivity, because they are forced to feed efficiently and therefore aggregate in areas of plentiful prey (Moore and DeMaster 1997). The most important gray whale feeding locations are B5 in the Chirikov Basin (at least until recently) and C4 in the Hope Basin (Figure 23.4). The next most important feeding locations appear to be B1, B3, B4, C2, and especially C1, where numerous whales are typically observed. The most important foraging location along the Alaskan coast is C5 between Point Franklin and Point Barrow. Areas C3 and C6 each appear to be utilized by 100–200 whales (Berzin 1984; personal observation, respectively). The Chirikov Basin (B5) was not utilized by many whales in 2002 or 2003, but we include it in our model because it has been the most commonly noted gray whale feeding site over the last century and probably has been a major feeding location since the last ice age.

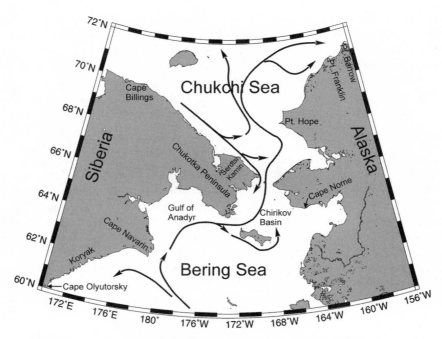

FIGURE 23.4. Model of gray whale distribution in the Bering and Chukchi Seas. Cold, nutrient-rich water upwells near Cape Navarin and the northern Gulf of Anadyr and delivers nutrients and primary production biomass at frontal zones to benthic communities with high secondary production utilized as feeding sites by gray whales.

The Green Belt in the Bering Sea consists of waters running from southeast to northwest along the Bering shelf slope (Hansell et al. 1989; Springer et al. 1996). Deep, nutrient-rich waters of the Green Belt upwell near Cape Navarin and in the Gulf of Anadyr (Figure 23.4). Some of the nutrient-laden water deflects southward along the Koryak coast and fuels high production in feeding area B1. Water deflecting north transits the Gulf of Anadyr, carrying high nutrient concentrations that generate high primary production (Sambrotto et al. 1984), especially by diatoms (Whitledge et al. 1988). The Anadyr Water passes between the Russian coast and St. Lawrence Island to Bering Strait (Walsh et al. 1989), thus enriching feeding area B2 and much of B5 (Grebmeier et al. 1988; Highsmith and Coyle 1990, 1992; Fukuchi et al. 1993). Some of the Anadyr Water also transits around the south side of St. Lawrence Island and then crosses the Chirikov Basin to Bering Strait. The Chirikov Basin appears to be a depositional area as northbound water slows because of the constriction at the Bering Strait (Coachman et al. 1975). Also, the cold, dense Anadyr Water extends along the seafloor many kilometers eastward beneath the warmer Alaska Coastal Water.

Up to one-third of local and advected production, both of which are dominated by diatoms (Whitledge et al. 1988; Highsmith and Coyle 1990, 1992; Fukuchi et al. 1993), settles to the seafloor in the Chirikov Basin and provides food for the ampeliscid amphipods in feeding area B5 (Figure 23.4).

Thus, in the Bering Sea this upwelling system provides nutrients and food for benthic amphipods, such as *Pontoporeia femorata* along the Koryak coast (B1) and southern Gulf of Anadyr, and *Ampelisca macrocephala* along the Chukotka Peninsula (B2), the western and eastern ends of St. Lawrence Island (B3, B4; Grebmeier and Cooper 1995), and the Chirikov Basin (B5).

Once through the Bering Strait, the plume of productive Bering Sea Water (Figure 23.3) slows and moves north and west until it encounters the East Siberian Current and forms a front approximately 60 km wide (Figure 23.2), which may occur anywhere from Serdtse Kamen west to the Cape Schmidt area (C1–C3; Figure 23.5; Weingartner et al. 1999). The resulting meander of the front provides an abundant food supply to the benthic communities in the region, especially areas C1, C2, and C4 (Blackburn 1987; Grebmeier et al. 1988; Grebmeier 1993), and to some extent C3. The front is wide in area C4 because the dense Anadyr Water, bearing high phytoplankton concentrations from the Bering Sea, underlies warmer, less productive surface waters for tens of kilometers, and thus delivers considerable production to the benthos. The extension of the bottom front northward along the Hope Valley then swerves northeastward (Weingartner et al. 1999) to Point Franklin on the Alaska coast and provides an energy source for gray whale prey in area C5 (Feder et al. 1994). The only feeding site not presently accounted for by this model is the small feeding area C6, directly east of C4. It is possible that the front with productive Anadyr water on occasion extends that far east to deposit diatoms, but we have no data at present indicating that this occurs.

As gray whales appear to feed on shallow infauna and epifauna, soft sediments are a necessary condition for gray whale bottom feeding. However, sediment grain size may not be as restrictive (Feder et al. 1994) as sometimes has been suggested (Johnson and Nelson 1984; Nelson and Johnson 1987; Moore et al. 2003). For example, fine sand occurs in the Chirikov Basin (Johnson and Nelson 1984), but the bottom in the Hope Basin (C4; Figure 23.4) is dense black silt and clay overlain by 2–3 cm of loose, fine sediments (unpublished observation). We have also observed ampeliscid amphipods in coarser sediments than those in the central Chirikov Basin. Thus, extensive gray whale feeding areas in the Bering and Chukchi seas are soft-sediment habitats with very abundant infaunal or epifaunal prey species supported by high deposition rates of phytoplankton produced by upwelled waters of the Green Belt and Anadyr stream, termed Bering Sea Water in the Chukchi Sea.

The apparent mortality of as many as 11,000 gray whales over a recent four-year period, indicating a food shortage, and the apparent nonuse of much of the Chirikov Basin in 2002 and 2003, suggest that the cause of the population decline is associated with reduced primary or secondary production in area B5 (Figure 23.4). This decline in production could be caused by a reduced flow of nutrient-rich, productive Anadyr Water through the Chirikov Basin or by reduced production of ampeliscid amphipods as a result of overgrazing by the whales. Indeed, if the Green Belt ceases to function, resulting in reduction or loss of upwelled water moving through the northern Bering Sea and setting up a major frontal system in the Chukchi Sea, the known feeding sites would probably not be able to support current levels of gray whale predation.

Gray whales have undoubtedly had to respond to changing food supplies over evolutionary time and are known to feed on a variety of pelagic and benthic prey along the northern migration route between Baja California and Alaska (Dunham and Duffus 2002). These observations raise the intriguing question of where and how eastern gray whales fed during the Pleistocene glacial advances, the most recent of which ended just 10,000–12,000 YBP. At that time, all of the present foraging locations in the Bering and Chukchi seas were emergent, and marine access to the Arctic was blocked by the Bering land bridge. Perhaps eastern gray whales in these earlier times mingled with the western stock and fed in the Sea of Okhotsk or spread out in small numbers at various locations along the present migratory route. We may never have a complete understanding of long-past events, but the model we propose accounts for the current feeding ecology of the gray whale. Changes in the upwelled currents and formation of fronts delivering particulate organic carbon to the seafloor have the potential to expand, reduce, or eliminate feeding sites and to create new sites.

Concluding Remarks

Gray whales offer a unique opportunity to advance our knowledge of large whale ecology because they follow well-known, coastal migratory routes (making them countable), feed largely on benthic organisms at well-defined locations, and have profoundly responded to whaling and to management. Indeed, eastern North Pacific gray whales are the only whale stock/species to have recovered to estimated prewhaling numbers following the cessation of industrial whaling. The coastal migration route makes it possible to estimate changes in population size and calving rates. The shallow feeding sites make it possible to study the foraging ecology of the whales and their impacts on the prey community and habitat on a systemwide basis. Gray whale management thus can be viewed as a large experiment that produced interesting and informative results. We have made great progress in our understanding of gray whales, but the species and its associated ecosystem have much more to tell us. The challenge, in our view, is to conduct multidisciplinary research on appropriate scales of space and time.

Literature Cited

Berzin, A.A. 1984. Soviet studies of the distribution and numbers of the gray whale in the Bering and Chukchi Seas, from 1968–1982, in *The gray whale* Eschrichtius robustus. M.L. Jones, S.L. Swartz, and S. Leatherwood, eds. Orlando, FL: Academic Press, pp. 409–419.

Berzin, A.A. and A.A. Rovnin. 1966. The distribution and present condition of the number of migrating whales in the northwest part of the Pacific Ocean, in the Bering and Chukchi Seas. *Izvestia TINRO* 58: 179–207.

Blackburn, T.H. 1987. Microbial food webs in sediments, in *Microbes in the seas*. M.A. Sleigh, ed. Chichester, UK: Ellis Horwood, pp. 39–58.

Blokhin, S.A. 1980. Investigations on gray whales taken off Chukotka in 1980. *Report of the International Whaling Commission* 32: 375–380.

———. 1984. Investigations of gray whales taken in the Chukchi Coastal Waters, USSR, in *The gray whale* Eschrichtius robustus. M.L. Jones, S.L. Swartz, and S. Leatherwood, eds. Orlando, FL: Academic Press, pp. 487–509.

Bogoslovskaya, L.S., L.M. Votrogov, and T.N. Seminova. 1981. Feeding habits of the gray whale off Chukotka. *Report of the International Whaling Commission* 31: 507–510.

———. 1982. Distribution and feeding of gray whales off Chukotka in the summer and autumn of 1980. *Report of the International Whaling Commission* 32: 385–389.

Braham, H.W. 1984. Distribution and migration of gray whales in Alaska, in *The gray whale* Eschrichtius robustus. Jones, M.L., S.L. Swartz, and S. Leatherwood, eds. Orlando, FL: Academic Press, pp. 249–266.

Breiwick, J.M., D.J. Rugh, D.E. Withrow, M.E. Dahlheim, and S.T. Buckland. 1988. Preliminary population estimate of gray whales during the 1987/88 southward migration. Paper SC/40/PS12 presented to the International Whaling Commission Scientific Committee.

Calambokidis, J., J.D. Darling, V. Deecke, P. Gearin, M. Gosho, W. Megill, C.M. Tombach, D. Goley, C. Toropove, and B. Gisborne. 2002. Abundance, range and movements of a feeding aggregation of gray whales *(Eschrichtius robustus)* from California to southeastern Alaska in 1998. *Journal of Cetacean Research and Management* 4: 267–276.

Clarke, J.T. and S.E. Moore. 2002. A note on observations of gray whales in the southern Chukchi and northern Bering Seas, August–November, 1980–89. *Journal of Cetacean Research and Management* 4: 283–288.

Coachman, L.K., K. Aagaard, and R.B. Tripp. 1975. *Bering Strait: the regional physical oceanography*. Seattle: University of Washington Press.

Coyle, K.O. and R.C. Highsmith. 1989. Arctic ampeliscid amphipods: three new species. *Journal of Crustacean Biology* 9: 157–175.

———. 1994. Benthic amphipod community in the northern Bering Sea: analysis of potential structuring mechanisms. *Marine Ecology Progress Series* 107: 233–244.

Darling, J.D. 1984. Gray whales off Vancouver Island, British Columbia, in *The gray whale* Eschrichtius robustus. M.L. Jones, S.L. Swartz, and S. Leatherwood, eds. Orlando, FL: Academic Press, pp. 267–287.

Doroshenko, N.V. 1981. Concise results of cetacean research during the cruise of R/V *Razyashchy* in the Bering, Chukchi and East Siberian Seas, in *A collection of research works on marine mammals of the northern Pacific Ocean in 1980–1981*. [Proj. 02.06-61] Moscow: VNIRO, pp. 13–16.

Dunham, J.S. and D.A. Duffus. 2002. Diet of gray whales *(Eschrichtius robustus)* in Clayoquot Sound, British Columbia, Canada. *Marine Mammal Science* 18: 419–427.

Feder, H.M., A.S. Naidu, S.C. Jewett, J.M. Hameedi, W.R. Johnson, and T.E. Whitledge. 1994. The northeastern Chukchi Sea: benthos-environmental interactions. *Marine Ecology Progress Series* 111: 171–190.

Findley, L.T. and O. Vidal. 2002. Gray whale *(Eschrichtius robustus)* at calving sites in the Gulf of California, Mexico. *Journal of Cetacean Research and Management* 4: 27–40.

Fukuchi, M., H. Sasaki, H. Hatori, O. Matsuda, A. Tanimura, N. Handa, and C.P. McRoy. 1993. Temporal variability of particulate flux in the northern Bering Sea. *Continental Shelf Research* 13: 693–704.

Grebmeier, J.M. 1993. Studies on pelagic-benthic coupling extended onto the Russian continental shelf in the Bering and Chukchi Seas. *Continental Shelf Research* 13: 653–668.

Grebmeier, J.M. and L.W. Cooper. 1995. Influence of the St. Lawrence Island polynya on the Bering Sea benthos. *Journal of Geophysical Research* 100: 4439–4460.

Grebmeier, J.M. and C.P. McRoy. 1989. Pelagic-benthic coupling on the shelf of the Bering and Chukchi Seas. III. Benthic food supply and carbon cycling. *Marine Ecology Progress Series* 53: 79–91.

Grebmeier, J.M., C.P. McRoy, and H.M. Feder. 1988. Pelagic-benthic coupling on the shelf of the northern Bering and Chukchi Seas. I. Food supply source and benthic biomass. *Marine Ecology Progress Series* 48: 57–67.

Hansell, D.A., J.J. Goering, J.J. Walsh, C.P. McRoy, L.K. Coachman, and T.E. Whitledge. 1989. Summer phytoplankton production and transport along the shelf break in the Bering Sea. *Continental Shelf Research* 9: 1085–1104.

Henderson, D.A. 1984. Nineteenth century gray whaling: grounds, catches and kills, practices and depletion of the whale population, in *The gray whale* Eschrichtius robustus. M.L. Jones, S.L. Swartz, and S. Leatherwood, eds. Orlando, FL: Academic Press, 159–186.

Highsmith, R.C. and K.O. Coyle. 1990. High productivity of northern Bering Sea benthic amphipods. *Nature* 344: 862–864.

———. 1991. Amphipod life histories: community structure, impact of temperature on decoupled growth and maturation rates, productivity, and P:B ratios. *American Zoologist* 31: 861–873.

———. 1992. Productivity of arctic amphipods relative to gray whale energy requirements. *Marine Ecology Progress Series* 83: 141–150.

Hopkins, D.M. 1967. *The Bering land bridge*. Stanford, CA: Stanford University Press.

Johnson, J., H. Braham, B. Krogman, W. Marquette, R. Sonntag, and D. Rugh. 1981. Research conducted on bowhead whales, June 1979 to June 1980. *Report of the International Whaling Commission* 31: 461–475.

Johnson, K.R. and C.H. Nelson. 1984. Side-scan sonar assessment of gray whale feeding in the Bering Sea. *Science* 225: 1150–1152.

Kasuya, T. and D.W. Rice. 1970. Notes on baleen plates and on arrangement of parasitic barnacles of gray whale. *Scientific Reports of the Whales Research Institute* 22: 39–43.

Kellog, R. 1929. What is known of the migration of some of the whalebone whales, in *Smithsonian Institution Annual Report*, Washington, DC: Smithsonian Institution, pp. 467–494.

Kim, S.L. and J.S. Oliver. 1989. Swarming benthic crustaceans in the Bering and Chukchi Seas and their relation to geographic patterns in gray whale feeding. *Canadian Journal of Zoology* 67: 1531–1542.

Klaus, A.D., J.S. Oliver, and R.G. Kvitek. 1990. The effects of gray whale, walrus and ice gouging disturbance on benthic communities in the Bering Sea and Chukchi Sea, Alaska. *National Geographic Research* 6: 470–484.

Kvitek, R.G. and J.S. Oliver. 1986. Side-scan sonar estimates of the utilization of gray whale feeding grounds along Vancouver Island, Canada. *Continental Shelf Research* 6: 639–654.

LeBoeuf, B.J., H. Perez-Cortes Moreno, J. Urbán Ramirez, B.R. Mate, and F. Ollervides U. 2000. High gray whale mortality and low recruitment in 1999: potential causes and implications. *Journal of Cetacean Research and Management* 2: 85–99.

Maher, W.J. 1960. Recent records of the California gray whale *(Eschrichtius glaucus)* along the north coast of Alaska. *Arctic* 13: 257–265.

Marquette, W.M. and H.W. Braham. 1982. Gray whale distribution and catch by Alaskan Eskimos: a replacement for the bowhead whale? *Arctic* 35: 386–394.

Marquette, W.M., H.W. Braham, M.K. Nerini, and R.V. Miller. 1982. Bowhead whale studies, autumn 1980–spring 1981: harvest, biology and distribution. *Report of the International Whaling Commission* 32: 357–370.

Moore, S.E. and D.P. DeMaster. 1997. Cetacean habitats in the Alaskan Arctic. *Journal of Northwest Atlantic Fisheries Science* 22: 55–69.

Moore, S.E., D.P. DeMaster, and P.K. Dayton. 2000. Cetacean habitat selection in the Alaskan Arctic during summer and autumn. *Arctic* 53: 432–447.

Moore, S.E., J.M. Grebmeier, and J.R. Davies. 2003. Gray whale distribution relative to habitat in the northern Bering Sea:

current conditions and retrospective summary. *Canadian Journal of Zoology* 81: 734–742.

Moore, S.E. and D.K. Ljungblad. 1984. Gray whales in the Beaufort, Chukchi and Bering Seas: distribution and sound production, in *The gray whale* Eschrichtius robustus. M.L. Jones, S.L. Swartz, and S. Leatherwood, eds. Orlando, FL: Academic Press, pp. 543–559.

Moore, S.E., J. Urbán Ramirez, W.L. Perryman, F. Gulland, H. Perez-Cortes Moreno, P.R. Wade, L. Rojas Bracho, and T. Rowles. 2001. Are gray whales hitting "K" hard? *Marine Mammal Science* 17: 970–974.

Murison, L., D. Murie, K. Morin, J. da Silva Curiel. 1984. Foraging of the gray whale along the west coast of Vancouver Island, British Columbia, in *The gray whale* Eschrichtius robustus. M.L. Jones, S.L. Swartz, and S. Leatherwood, eds. Orlando, FL: Academic Press, pp. 451–463.

Nelson, C.H. and K.R. Johnson. 1987. Whales and walruses as tillers of the sea floor. *Scientific American* 256: 112–117.

Nelson, C.H., K.R. Johnson, and J.H. Barber, Jr. 1987. Gray whale and walrus feeding excavation on the Bering shelf, Alaska. *Journal of Sedimentary Petrology* 57: 419–430.

Nerini, M. 1984. A review of gray whale feeding ecology, in *The gray whale* Eschrichtius robustus. M.L. Jones, S.L. Swartz, and S. Leatherwood, eds. Orlando, FL: Academic Press, pp. 423–450.

Nerini, M.K. and J.S. Oliver. 1983. Gray whales and the structure of the Bering Sea benthos. *Oecologia* 59: 224–225.

Oliver, J.S. and P.N. Slattery. 1985. Destruction and opportunity on the seafloor: effects of gray whale feeding. *Ecology* 66: 1965–1975.

Oliver, J.S., P.N. Slattery, M.A. Silberstein, and E.F. O'Connor. 1983. A comparison of gray whale feeding in the Bering Sea and Baja California. *Fishery Bulletin* 81: 513–522.

———. 1984. Gray whale feeding on dense ampeliscid communities near Bamfield, British Columbia. *Canadian Journal of Zoology* 62: 41–49.

Perryman, W.L., M.A. Donahue, P.C. Perkins, and S.B. Reilly. 2002. Gray whale calf production 1994–2000: are observed fluctuations related to changes in seasonal ice cover? *Marine Mammal Science* 18: 121–144.

Perryman, W.L. and M.S. Lynn. 2002. Evaluation of nutritive condition and reproductive status of migrating gray whales *(Eschrichtius robustus)* based on analysis of photogrammetric data. *Journal of Cetacean Research and Management* 4: 155–164.

Reilly, S.B. 1984. Assessing gray whale abundance: a review, in *The gray whale* Eschrichtius robustus. M.L. Jones, S.L. Swartz, and S. Leatherwood, eds. Orlando, FL: Academic Press, pp. 203–223.

Reilly, S.B., D.W. Rice, and A.A. Wolman. 1983. Population assessment of the gray whale, *Eschrichtius robustus*, from California shore censuses, 1967–1980. *Fishery Bulletin* 81: 267–281.

Rice, D.W. and A.A. Wolman. 1971. *The life history and ecology of the gray whale* (Eschrichtius robustus). American Society of Mammalogists, Special Publication No. 3. Lawrence, KS: Allen Press.

Rugh, D.J. 1984. Census of gray whales at Unimak Pass, Alaska, November–December 1977–1979, in *The gray whale* Eschrichtius robustus. M.L. Jones, S.L. Swartz, and S. Leatherwood, eds. Orlando, FL: Academic Press, pp. 225–248.

Rugh, D.J. and M.A. Fraker. 1981. Gray whale *(Eschrichtius robustus)* sightings in the eastern Beaufort Sea. *Arctic* 34: 186–187.

Rugh, D.J., R.C. Hobbs, J.A. Lerczak, and J.M. Breiwick. 2003. Estimates of abundance of the eastern north Pacific stock of gray whales 1997 to 2002. Paper SC/55/BRG13 presented to the International Whaling Commission Scientific Committee.

Sambrotto, R.N., J.J. Goering, and C.P. McRoy. 1984. Large yearly production of phytoplankton in the western Bering Sea. *Science* 225: 1147–1150.

Scammon, C.M. 1874. *The marine mammals of the northwest coast of North America.* San Francisco: John H. Carmany and Co.

Springer, A.M., C.P. McRoy, and M.V. Flint. 1996. The Bering Sea Green Belt: shelf edge processes and ecosystem production. *Fisheries Oceanography* 5: 205–223.

Stelle, L.L. 2002. Behavioral ecology of summer resident gray whales *(Eschrichtius robustus)* feeding on mysids in British Columbia, Canada. *Dissertation Abstracts International, Part B—Science and Engineering* 62: 3886.

Swartz, S.L. and M.L. Jones. 1987. Gray whales make a comeback. *National Geographic* 171: 754–771.

Thomson, D.H. and L.R. Martin. 1986. Feeding ecology of gray whales *(Eschrichtius robustus)* in the Chirikof Basin, summer 1982. *U.S. Department of Commerce, NOAA, Outer Continental Shelf Environmental Assessment Program Final Report* 43: 377–460.

Tomilin, A.G. 1946. Problem of lactation and nourishment in Cetacea. *Byulleten' Moskofskoe obshchestvo ispytatelei prirody, Biologicheskii Otdel* 51: 44–57.

Walsh, J., C.P. McRoy, L.K. Coachman, J.J. Goering, J.J. Nihoul, T.E. Whitledge, T.H. Blackburn, P.L. Parker, C.D. Wirick, P.G. Shuert, J.M. Grebmeier, A.M. Springer, R.D. Tripp, D.A. Hansell, S. Djenidi, E. Deleersnijder, K. Henriksen, B.A. Lund, P. Andersen, F.E. Müller-Karger, and K. Dean. 1989. Carbon and nitrogen cycling within the Bering/Chukchi Seas: source regions for organic matter affecting AOU demands of the Arctic Ocean. *Progress in Oceanography* 22: 277–359.

Weingartner, T.J., S. Danielson, Y. Sasaki, V. Pavlov, and M. Kulakov. 1999. The Siberian Coastal Current: a wind- and buoyancy-forced Arctic coastal current. *Journal of Geophysical Research* 104: 29697–29713.

Whitledge, T.E., R.R. Bidigare, S.I. Zeeman, R.N. Sambrotto, P.F. Roscigno, P.R. Jensen, J.M. Brooks, C. Treest, and D.M. Veidt. 1988. Biological measurements and related chemical features in Soviet and United States regions of the Bering Sea. *Continental Shelf Research* 8: 1299–1319.

Wilke, F. and C.H. Fiscus. 1961. Gray whale observations. *Journal of Mammology* 42: 108–109.

Würsig, B., R.S. Wells, and D.A. Croll. 1986. Behavior of gray whales summering near St. Lawrence Island, Bering Sea. *Canadian Journal of Zoology* 64: 611–621.

Yablokov, A.V. and L.S. Bogoslovskaya. 1984. A review of Russian research on the biology and commercial whaling of the gray whale, in *The gray whale* Eschrichtius robustus. M.L. Jones, S.L. Swartz, and S.L. Leathergood, eds. Orlando, FL: Academic Press, pp. 465–485.

Zimushko, V.V. 1970. An aerial-visual count of the numbers and observations of the distribution of gray whales in the coastal waters of the Chukshi. *Izvestiya TINRO* 71: 289–294.

Zimushko, V.V. and M.V. Ivashin. 1980. Some results of Soviet investigations and whaling of gray whales *(Eschrichtius robustus,* Lilljeborg, 1861). *Report of the International Whaling Commission* 30: 237–246.

Zimushko, V.V. and S.A. Lenskaya. 1970. On the diet of the gray whale on its feeding grounds. *Ekologiya (Sverdlovsk)* 3: 26–35.

TWENTY-FOUR

Whales, Whaling, and Ecosystems in the North Atlantic Ocean

PHILLIP J. CLAPHAM AND JASON S. LINK

Although whaling for subsistence purposes has occurred in various locations worldwide from prehistoric times, commercial whaling—in the sense of a continuous, directed effort pursued for profit—likely had its origin in the North Atlantic Ocean (Figure 24.1). The earliest known commercial ventures were those conducted by the Basques of northern Spain and southern France, who operated a well-organized hunt in the Bay of Biscay beginning in or before the eleventh century (Aguilar 1986; Reeves and Smith, Chapter 8 of this volume). Various coastal whale hunts existed in Europe in medieval times, and by the sixteenth century whalers were already pursuing their quarry much further afield. This expansion was also led by the Basques, who reached the New World shortly after (and perhaps even before) its discovery by Columbus.

The colonization of eastern North America brought with it additional ventures, culminating in the extensive pelagic whaling industry based in New England, which was marked by great wealth and a truly global reach. By the time the whalers of Nantucket and New Bedford had reached other oceans in their pursuit of oil and whalebone, several populations of whales in the North Atlantic had already been seriously reduced in size, and in a few cases this depletion continued well into the twentieth century.

Because of the long history and intensive nature of whaling in the North Atlantic, and the magnitude of exploitation of other resources (notably fish) as well as natural population and environmental cycles, the ecosystem of this ocean basin is certainly different today from what it was in centuries past. The effects of whaling on whale populations are reasonably well documented and understood. However, assessing the impact of whaling on other aspects of the North Atlantic ecosystem presents novel challenges. Many questions can be posed in this regard: What might historic ecosystems have looked like? Did whaling-induced changes to whale populations directly or indirectly alter other biota? Did exploitation of or other changes to other biota influence whale populations (directly or indirectly)? Has climate change altered the abundance of these organisms? Given the general dearth of contemporary information, let alone information extending back through past centuries, addressing these issues regarding the changing role of whales in the North Atlantic ecosystem remains difficult.

To explore further the past, current, and changing roles of large whales in the North Atlantic ecosystem, we briefly review the history of whaling in this region and the status of the exploited species. We present case studies to contrast the exploitation history and subsequent recovery status of three

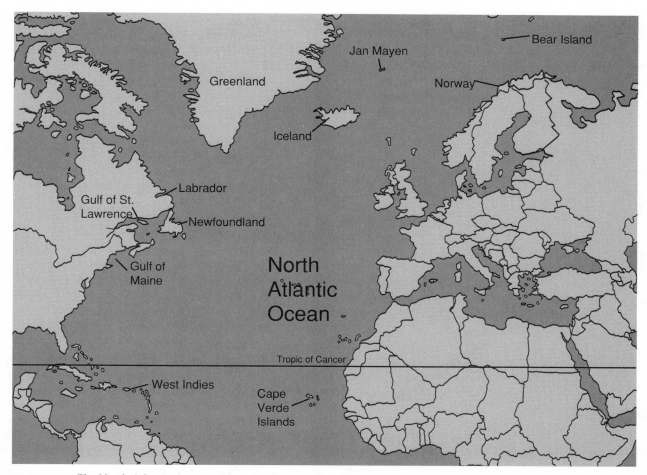

FIGURE 24.1. The North Atlantic Ocean, with major features identified as used in the text.

species with differing ecological roles and life histories. Our basic tenet is that, even if the functional role of whales and their interaction with other species does not fundamentally change, changing the abundance of whales does alter the relative magnitude of these roles and interactions, thus ultimately altering the relative importance whales may have in an ecosystem (Figures 24.2a and 24.2b).

Whaling in the North Atlantic: A Brief History and Current Population Status

Detailed reviews of aspects of North Atlantic whaling can be found in several papers (Brown 1976; Aguilar 1986; Reeves and Mitchell 1986a, 1986b; Reeves and Smith, Chapter 8, this volume) as well as in more general whaling history texts (notably Tønnessen and Johnsen 1982). A multicentury perspective, particularly with regard to the life history and life span of large whales, is important to grasp fully the changing ecological role of these organisms in the North Atlantic ecosystem. It is not so much that we need a historical perspective for the sake of history, rather that the long life spans of these organisms effectively requires such a view.

As noted previously, the first known commercial whaling in the North Atlantic was conducted by Basques in the Bay of Biscay (Aguilar 1986). The principal target of this hunt was the North Atlantic right whale *(Eubalaena glacialis)*, a species that likely possesses the dubious distinction of having the longest hunting history of any of the great whales. Medieval market and tax records indicate that the Basque right whale hunt began no later than 1059 and appears to have been a cooperative effort among coastal towns. There is some evidence that mothers and calves were disproportionately targeted, a characteristic that likely accelerated the eventual decline of the population (Aguilar 1986).

It is not clear when right whales began to be scarce in southern European waters, although it is known that the Basque whalers began to pursue pelagic whaling well before the coastal hunt was abandoned for lack of whales. By the summer of 1530, Basque whaling ships were working in Red Bay, Labrador (Cumbaa 1986), as well as in various other locations in the general vicinity of the Strait of Belle Isle. Recent genetic analysis of bones from the Red Bay site (McLeod et al. 2003) suggests that the bowhead whale *(Balaena mysticetus)* was the main species taken in these Basque hunts, although it is not clear whether the sample used in the

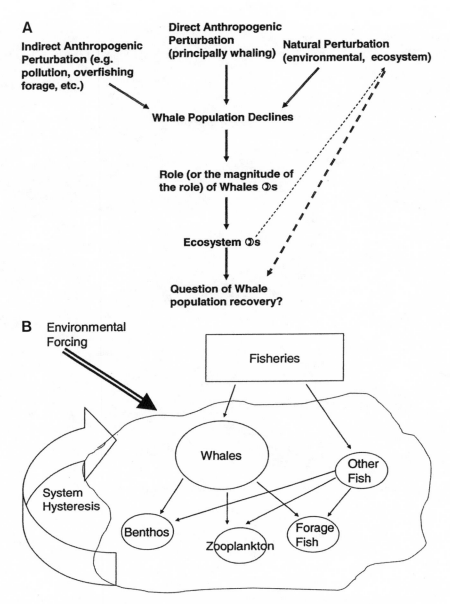

FIGURE 24.2. (A) Schematic of the role of large whales in the North Atlantic ecosystem and (B) how it potentially changes as a result of anthropogenic or environmental perturbations. The term *hysteresis* refers to an effect lagging behind its cause.

genetic study was representative of contemporary whaling effort. The genetic analysis contradicts earlier work by Cumbaa (1986), who incorrectly identified many of the same bones as coming from right whales. That the Basque hunt occurred during the so-called "Little Ice Age" likely explains why the distribution of bowhead whales in the early sixteenth century extended much further south than it does today. Furthermore, given the minimal overlap between right whales and bowheads today, it is quite possible that right whales were not common off the Labrador coast at that time and thus were not a substantial component of the Basque hunt in that region.

In the late sixteenth century, several European nations followed the lead of the Basques and entered into extensive pelagic operations. Beginning about 1570, the British established an Arctic hunt based largely on bowhead whales. The British were quickly followed in this enterprise by whalers from Holland, France, Denmark, Norway, and Germany (Reeves and Smith, Chapter 8, this volume).

Meanwhile, the colonization of the New World (notably New England) brought additional whaling effort and the subsequent exploitation of pristine populations. When the *Mayflower* Pilgrims arrived at Provincetown (Cape Cod, Massachusetts) in November 1620, they remarked on the abundance of right whales in local waters and regretted that they had not brought with them implements for whaling. As the seventeenth century progressed, the colonists' occasional opportunistic whale hunts evolved into directed whaling, with right whales again being the primary species taken.

The great American ("Yankee") pelagic whaling era began about 1730, attaining its peak in the nineteenth century (Reeves and Smith, Chapter 8 of this volume). Although the majority of whalers from Nantucket, New Bedford, and other New England ports plied their trade far from home in other oceans, there remained a substantial American presence on the North Atlantic whaling grounds, where right, humpback, and sperm whales were the principal targets. Catches were made over a wide range, from the tropics to high latitudes. The Yankee pelagic era was already in decline during the second half of the nineteenth century and came to a close in 1925.

The invention of the explosive harpoon and of the steam engine revolutionized whaling in the North Atlantic and elsewhere. What Reeves and Smith (Chapter 8 of this volume) term the "Norwegian coastal era" began about 1854, and by 1900 shore-based whalers were routinely employing bow-mounted harpoon cannons on small steam-powered catcher boats. In addition to further depleting already-hunted populations, the new technology provided a means to catch faster species that had previously remained largely unexploited, including fin (Balaenoptera physalus), sei (B. borealis), and blue whales (B. musculus). This essentially modern-type whaling was concentrated in coastal areas in higher latitudes, from Europe to North America.

It is probably accurate to state that by 1900 there was not a single area of the North Atlantic that had not been subject to whaling effort at some point in time. Furthermore, because exploitation occurred over a much longer duration in this region than in other oceans, North Atlantic populations of the great whales were often depleted or even extirpated earlier than whale populations elsewhere in the world.

As of 2006, there were five active whaling operations in the North Atlantic Ocean; of these, one consists of shore-based commercial whaling (for minke whales, Balaenoptera acutorostrata) conducted by Norway; one consists of whaling under special permit ("scientific whaling") by Iceland (minke whales), and three others are native subsistence hunts. The latter include Inuit whaling off Greenland (fin and minke whales) and eastern Canada (a small hunt for bowhead whales), as well as a small hunt for humpback whales by St. Vincent and the Grenadines, off the island of Bequia in the southeastern Caribbean. The impact of these operations varies from insignificant (in the case of Bequia humpbacks and, probably, eastern Arctic bowheads) to potentially damaging in the case of Greenland fin whales (IWC 2004).

The current status of whale populations that were (or still are) subject to exploitation in the North Atlantic ranges from abundant to critically endangered (Clapham et al. 1999). One population, the Atlantic gray whale (Eschrichtius robustus), became

extinct in relatively recent times (see below). In the critically endangered category is the western population of the North Atlantic right whale, while the eastern Atlantic population of this species is functionally extinct. The Spitsbergen/Barents Sea population of the bowhead whale is also close to extinction, with only a few sightings in recent years. Blue whales remain endangered, with the population that summers in the Gulf of St. Lawrence likely being in the low hundreds (Sears et al. 1990). Although the eastern North Atlantic blue whale population is likely larger (Sigurjónsson 1995), there are no reliable estimates of current abundance. Conversely, humpback whales (Megaptera novaeangliae) are recovering strongly in most areas. Fin, sei, and minke whales are also believed to be generally abundant. The differential status of these populations provides a reasonable contrast for evaluating the changing roles of large whales in the North Atlantic ecosystem. We next examine three case studies representative of such differing populations.

Case Studies

The three case studies presented here include species with considerable differences in hunting history, recovery status, life history, and ecology. Of the three, the gray whale is extinct in the Atlantic, the right whale remains critically endangered, and the humpback whale is recovering strongly from past exploitation. In terms of foraging ecology, these three whales are, respectively, a species that primarily (at least elsewhere) consumes benthic invertebrates, a zooplankton specialist, and a generalist feeding on small fish and euphausiids (krill). The role of these whales in the ecosystem, the change in the magnitude of the effects of that role as whale population sizes declined, and the hypothesized causes for the observed declines, lack of recovery, or increases in population sizes are discussed.

The Atlantic Gray Whale

The gray whale today exists only in the North Pacific, where the status of the two extant populations is starkly different. The eastern North Pacific population (also called the "California" stock) is so abundant that it was removed from the U.S. list of endangered species in 1994 (Angliss and Lodge 2003). In contrast, the western North Pacific (or "Korean") stock consists of approximately 100 animals and suffers from poor reproduction, low calf survival, nutritional stress, and potential habitat exclusion as a result of oil and gas development (Clapham et al. 1999; Weller et al. 2002).

The extinction of the Atlantic population remains a mystery. Subfossil specimens have been recovered from both North America and Europe, and some of these have been dated to the seventeenth century (Mead and Mitchell 1984; Bryant 1995). A variety of records from medieval times, and in one case from the American colonial era (Dudley 1725), have been interpreted as referring to gray whales. Mead and Mitchell (1984) deemed credible only three observations of living gray whales in the Atlantic: from Iceland (sixteenth century), Spitsbergen (seventeenth century), and the eastern coast of North America (eighteenth century). Lindquist (2000) assigns credibility to a rather larger number of medieval reports, though with dubious justification. Regardless of how many of the various descriptions are indeed of gray whales, records of this species were few and far between in the several hundred years preceding its extinction.

Accordingly, it is difficult to believe that whaling was solely responsible for the depletion and eventual extinction of the Atlantic gray whale, given the general lack of widespread whaling prior to that era. The fate of this population was driven by four possible causes, and others have been similarly hypothesized:

1. That substantial unrecorded whaling in medieval times was the primary cause of extinction. Again, given the abundance of market and related records concerning the exploitation of other species (notably right whales), it seems inconceivable that a hunt large enough to have rendered a population extinct could have passed undocumented.

2. That the Atlantic gray whale became extinct from natural causes. In this scenario, a remnant population, which perhaps had failed to adapt to major post-glacial climatic and habitat changes, would have slowly slid into extinction, perhaps aided by occasional unrecorded opportunistic catches. It is possible that alternate ecosystem states caused by natural predator-prey cycles or North Atlantic Oscillation (NAO) shifts could have fundamentally changed the Atlantic ecosystem such that gray whale populations could not persist there. However, we know nothing about the habitat and foraging requirements of Atlantic gray whales.

3. That the population was greatly depleted by exploitation in prehistoric times. One can speculate that the historic distribution of the Atlantic gray whale was similar to that of its modern eastern North Pacific counterpart, being characterized by a strong dependence upon coastal waters and the use of sheltered lagoons where whales aggregate to mate and calve (Jones et al. 1984). Such a distribution would have made the Atlantic stock highly vulnerable to exploitation by primitive hunters. However, no archaeological evidence is available that points to the existence of a prehistoric hunt.

4. Gray whales might have permanently migrated to the Pacific basin. Alternatively, the Atlantic population may have been a small spillover from the species' primary range in the Pacific during periods when the Northwest Passage was passable.

That gray whales were once distributed from Europe to North America is apparent, but other information about the population—its abundance, structure, habitat use, foraging ecology, and life history—is entirely absent.

Little is certain about any aspect of the gray whale's existence in the North Atlantic, but we can infer some aspects of the role of the gray whale in the Atlantic from its Pacific analogs. If the Atlantic gray whale fed primarily on benthic invertebrates, as occurs in the Pacific (Nerini 1984; Highsmith et al., Chapter 23 of this volume), then the loss of this population could have cascaded into a predatory release on some benthic communities; it could have altered the bentho-pelagic coupling in the entire North Atlantic; it could have provided a competitive release for localized demersal fish populations that also fed on benthos; or it could simply be indicative of a broader response to a systemic flip (Spencer 1997; Link 1999), whereby the principal source of energy changed from a vertical orientation to a horizontal (i.e., more pelagic) one. The exact nature and magnitude of the role of gray whales in the North Atlantic is difficult to ascertain.

It is similarly difficult to assess the ecosystem responses that occurred as this population was shrinking. However, one thing is certain: Rather obviously, whatever the role of this species may have been, it was no longer an active factor in any Atlantic ecosystem following its extinction.

North Atlantic Right Whale

The North Atlantic right whale is a long-lived planktivorous species that today migrates from feeding grounds in temperate or high latitudes to a single known calving area off the southeastern United States (Winn et al. 1986). As noted earlier in this chapter, right whales were probably the first species to be regularly taken on a commercial basis in the North Atlantic, beginning with the Basque hunts in Europe in the eleventh century. Subsequent exploitation proceeded in the following order, with approximate starting dates noted in parentheses (see Aguilar 1986; Brown 1986; Cumbaa 1986; Reeves and Mitchell 1986a, 1986b): Newfoundland/Labrador (early sixteenth century); New England and adjacent colonial states (seventeenth century); Cintra Bay (Africa) and Cape Farewell region (Greenland) (both nineteenth century); and ending with catches from a remnant population off northwestern Scotland, Ireland, Iceland, and Norway (late nineteenth century to 1924). The European population was probably depleted by about 1700.

By the time of the last recorded catch, in 1935, several populations were effectively extinct. These included the eastern North Atlantic stock, which may have been slowly recovering when it was all but extirpated by a burst of Norwegian coastal whaling beginning in 1881 off Scotland and elsewhere (Brown 1976). The eastern stock's demise was likely hastened by nineteenth-century American pelagic whaling at Cintra Bay (western Africa), which may have been the population's principal calving ground (Reeves and Mitchell 1986b).

Today, the range of the remaining right whale population is largely confined to waters off the eastern coast of North America. These include the only known calving area, off the Florida/Georgia coast, as well as a variety of foraging habitats off New England and Nova Scotia (Winn et al. 1986). Occasional observations of identified individuals in the central or eastern North Atlantic (Knowlton et al. 1992) suggest the existence of a subpopulation in what were historically important habitats. Clearly the historical decline of right whales was due to over-exploitation.

The current population is estimated at 300–350 animals and may be in decline (Caswell et al. 1999; IWC 2001). Although this decline is primarily due to anthropogenic impacts (notably mortalities from fishing gear entanglements and ship collisions), the population's vulnerable status has been exacerbated by highly variable reproductive rates, which are approximately half of those observed from healthy populations of the congeneric southern right whale *(E. australis)* (IWC 2001). The mean calving interval of North Atlantic right whales has increased since 1992 from 3.67 years to more than 5 years, a significant trend; by contrast, the interval for southern rights is typically about 3 years (Kraus et al. 2001).

Because the North Atlantic right whale had been reduced to a remnant population by the twentieth century, any historical assessment of its ecological role in this ocean basin is difficult. Any ecosystem changes brought about by the large-scale removal of right whales would have occurred well before 1900, would be difficult to document, and would have been overlain by major ecosystem perturbations since. Currently right whales feed exclusively on zooplankton, notably copepods of the genus *Calanus* (Mayo and Marx 1989). As such, right whales forage at a lower trophic level than most rorquals. It is probable that as the abundance of the right whale population decreased, other zooplanktivorous competitors could have experienced a release in competitive pressure (see also Worm et al., Chapter 26 of this volume), but it is unclear what historical changes in the zooplankton community may have occurred or whether the zooplankton were even a limiting factor for other planktivorous species. Conversely, as the right whale population declined, it may have become competitively inferior. The idea that anthropogenic perturbations in the marine ecosystem have led to increased abundance of piscine taxa that may compete with right whales for copepods is a significant issue (Worm et al., Chapter 26 of this volume), even if only theoretical.

There are several hypotheses for the nonrecovery of this species, and many relate to how the system may be influencing whale population dynamics via its ecological role. In particular, several hypotheses have been considered to explain the observed low reproductive rates (Reeves et al. 2001), including infectious disease, environmental contaminants/endocrine disruptors, marine biotoxins, low genetic diversity, and food limitation/nutritional stress. Of these, the last appears to be the best-supported possibility . This raises what is termed here the "*Calanus* paradox": That a remnant population feeding upon the most abundant zooplankter in the North Atlantic should be food-limited would seem to be an implausible proposition. However, it is possible that the absolute (oceanic-scale) abundance of *Calanus* is not the key factor for right whales, and that efforts focusing on this feature mistake the scale of the problem. Right whale foraging is known to be triggered by a minimum copepod density threshold (Mayo and Marx 1989), and thus the primary effector may be the distribution and occurrence of suitably dense prey patches in right whale feeding habitats. Whether the small size of the population is also a factor (i.e., whether right whales need other right whales to locate food resources more effectively) is unknown.

Although examining the potential role of ecosystem effects in the decline of right whales is inevitably difficult, it is perhaps more practicable in the Gulf of Maine and adjacent regions than elsewhere. Although much of the North Atlantic Ocean has been subject to intensive fishing pressure, nowhere has this overexploitation and its effect on the ecosystem been better documented than off the northeastern United States. The northeast U.S. continental shelf ecosystem represents an area encompassing formerly rich fishing areas including the Gulf of Maine/Georges Bank and the waters off the mid-Atlantic states (New York to North Carolina).

The Northeast Fisheries Science Center (NEFSC) and its collaborating institutions have conducted long-term monitoring of this ecosystem since the 1960s, and this research effort represents one of the longest time series of such data anywhere in the world. Information—much of it from standardized surveys and some dating back to 1963—has been collected on fish abundance and fish catches, chlorophyll α", ^{14}C primary production, zooplankton and ichthyoplankton abundance, and inorganic nutrients. Thus, although we of course have no data on the historical state of this ecosystem prior to the inception of large-scale fishing efforts, some of the changes occurring in this region over the past four decades have been documented with some precision.

Substantial changes in the Northeast U.S. shelf ecosystem occurred in the late 1970s and early 1980s, with many abiotic, biotic, and human metrics exhibiting coincident increases or decreases (see review by Link and Brodziak 2002; Link et al. 2002). It is generally accepted that in recent years the ecosystem has become "horizontal" rather than "vertical" in terms of energy flow and is now dominated by migratory small pelagic fish, such as herring and mackerel species, rather than by more resident demersal taxa (Link 1999). This is reflected in biomass, energy fluxes, and community structure throughout the region. These changes are likely attributable to a combination of top-down forcing from fishing pressure and bottom-influences resulting from inherent natural variation in ecosystem processes (Link and Brodziak 2002; Link et al. 2002).

The large-scale depletion of Atlantic herring *(Clupea harengus)* by the distant water fleet in the 1960s, along with the overexploitation of mackerel *(Scomber scombrus)*, apparently vacated a major zooplanktivorous role in the Georges Bank/Gulf of Maine ecosystem. In the late 1970s and early 1980s this niche was at least partly filled by sand lance (*Ammodytes* sp.), the abundance of which increased dramatically but not to the level of mackerel or herring (Smith et al. 1980; Brown and Halliday 1983). It was apparent that sand lance were responding to some combination of lower mortalities (i.e., no directed fishing) and competitive release from the decline in mackerel and herring (Fogarty et al. 1991). After a cessation of exploitation on mackerel and herring, those populations began a recovery trajectory in the early 1990s. These populations are now at the highest levels of abundance ever recorded (Overholtz and Friedland 2002). Although not conclusively demonstrated, the hypothesis remains viable that the overall decline in abundance of zooplanktivores in the late 1970s and early 1980s did not last long enough for right whales to consume additional food, and more recently the unprecedented resurgence of herring and mackerel may be a limiting factor for right whales in terms of food competition. A recent alternative hypothesis to explain low reproduction in right whales is that climate variation and related oceanographic changes have resulted in food limitation (Greene and Pershing 2004).

The Humpback Whale

Humpback whales are widely distributed throughout the North Atlantic and elsewhere. The humpback is a highly migratory species that moves seasonally between feeding areas in temperate and high latitudes to breeding and calving grounds in tropical or subtropical waters (Clapham and Mead 1999).

Humpback whales were hunted in the North Atlantic from the seventeenth century or earlier, although extensive takes probably did not begin until the nineteenth century (Mitchell and Reeves 1983; Reeves and Smith 2002). Catch data are relatively complete for the twentieth century but are "fragmented and incomplete for earlier times" (Reeves and Smith 2002). Smith and Reeves (2003) provided an updated summary of known and extrapolated catch data from 1665 to 2001; the total catch over this period is estimated at approximately 29,000 whales. The two most intensive humpback whale hunts were a mechanized operation conducted off Iceland at the turn of the twentieth century and the Yankee sail-based pelagic hunt in the West Indies (Reeves and Smith 2002).

Humpback whales are currently known to be abundant and increasing; the most recent estimates of abundance and population growth for the Gulf of Maine are 902 (CV = 0.41) and 4%, respectively (Clapham et al. 2003). The best estimate for the abundance of North Atlantic humpback whales derives from the mark-recapture-based study of Stevick et al. (2003), which gave 11,570 (95% CI = 10,290–13,390) animals. However, this is certainly an underestimate of current abundance; it is more than a decade old, may be subject to negative bias from heterogeneity, and there is strong evidence for population growth of between 3.5% and 6.5% in most areas (Barlow and Clapham 1997, Stevick et al. 2003). Today, continued population growth and possible range expansion (e.g., to Irish waters; S. Berrow, personal communication) may indicate that the North Atlantic humpback whale population remains below carrying capacity. However, the estimated catch total (approximately 29,000) suggests that the number of humpbacks in the North Atlantic prior to the onset of significant whaling for this species was probably not greatly larger than current abundance (likely underestimated at some 11,000 animals; see above).

A genetics-based estimate of abundance of 240,000 humpback whales by Roman and Palumbi (2003) represents a harmonic mean over evolutionary time and cannot be tied to a specific period immediately prior to whaling. Given the constraints of carrying capacity and suitable habitat, it is very difficult to imagine a quarter-million humpback whales (together with large pre-exploitation populations of other species) in the North Atlantic at any time in prehistory. The genetic estimate is not reconcilable with catch and abundance data, even when an unrealistically generous allowance is made for gaps in the historical catch record (Baker and Clapham 2004).

It is apparent that humpbacks are recovering strongly from commercial whaling in most areas, with the apparent exception of West Greenland (IWC 2002, Larsen and Hammond 2004). That this is very different from the situation with right whales may be due to one or more of several factors. These potentially include larger population size (and greater range) after whaling, a shorter interbirth interval (2–3 years vs. 3–5 years in right whales; Clapham and Mayo 1990), earlier attainment of sexual maturity (mean of 5 years vs. 8–9 years; Clapham 1992, Kraus et al. 2001), greater genetic variability (Palsbøll et al. 1997, Schaeff et al. 1997), and a lower rate of mortality from ship collisions and entanglements.

Humpbacks have a catholic diet, feeding upon a variety of small schooling fish as well as euphausiids. Given the continued persistence of this species, it is probable that the magnitude of the impact of this predator upon the marine ecosystem may not have changed substantially since historical times. It may be that the role played by humpback whales, and their trophic interactions with other taxa, are subtly different today as a result of fundamental anthropogenic changes in the ecosystem itself.

Whether the broader diet of humpback whales, or their preference for small pelagic fish over plankton, has also given them greater ability to respond to changing ecological conditions is not clear, but that it has is certainly a viable hypothesis. For instance, as a result of the drastic population declines in herring and mackerel observed in the 1960s and early 1970s, sand lance became the primary prey for humpback whales, which presumably had previously relied upon herring as a major diet item (Overholtz and Nicolas 1979). This change was reflected in a distribution shift; specifically, the local density of humpback whales increased in those areas characterized by abundant sand lance, notably the southwestern Gulf of Maine. Now that herring has returned to historic levels of abundance, it would be useful to undertake an analysis of the current and past distribution of humpback whales to investigate patterns of occurrence relative to these shifts in the availability of prey.

Synthesis

These three case studies show a contrast in exploitation and demonstrate a range of recovery (or nonrecovery) trajectories.

If a species is extinct, it no longer has a role in an ecosystem. What is less clear is the relative difference in the role of extant whale populations relative to their historical roles. The specifics of the ecological role for these whales change in response to changes in the ecosystem (e.g., switching from herring to sand lance to herring as prey), but the generalized roles likely have remained consistent. The magnitude or effects of the ecological role played by these whales in the ecosystem is most likely coupled tightly with their abundance. Based on densities alone, it is probable that the critically endangered right whale has a much smaller role in the ecosystem than it did in historical times. Conversely, the (recovering or recovered) humpback whale likely has at least a similar role to what it once had prior to extensive whaling.

Relationships among the four major whale species (the two described above as well as fin and sei whales) in the Gulf of Maine and their prey were explored by Payne et al. (1990), who analyzed baleen whale distribution and abundance data surrounding the major shift that occurred in the mid-1980s. Right whales are exclusively planktivorous, while sei whales eat both zooplankton and small fish. Humpback and fin whales consume herring, sand lance, capelin, and sometimes other small fish; in more northerly parts of the region, euphausiids (notably *Meganyctiphanes norvegica*) are often a significant component of the diet.

As documented by Payne et al. (1990), in the summer of 1986 the population of sand lance in Massachusetts Bay (southwestern Gulf of Maine) crashed, and *Calanus* became unusually abundant for the time of year. Following the sand lance crash, humpback and fin whales abandoned the area, foraging instead in the Great South Channel (east of Cape Cod) where sand lance remained abundant. Right whales, which are usually not found in Massachusetts Bay after late spring, became resident in the area for much of the summer (Hamilton and Mayo 1990, Payne et al. 1990). Sei whales, which had been essentially absent from this area in daily summertime observations over ten years, also moved in for several weeks to feed upon the abundant copepods.

Payne et al. (1990) suggested that the shifts in distribution and local abundance of whales in the 1980s reflected top-down ecosystem impacts precipitated by overfishing of herring and mackerel (described above, see also Worm et al., Chapter 26 of this volume). Payne et al. further speculated that the removal of the fin-fish predators of sand lance resulted in increased exploitative or interference competition for copepods, with negative impacts upon planktivorous whales, notably the right whale. Although speculative, this hypothesis is worth considering in light of the right whale population's failure to recover, and the evidence tentatively linking low reproduction with nutritional stress (Reeves et al. 2001). While it is possible that the change in right whale (and other mysticete) distribution in the mid-1980s was in part related to local sand lance abundance, the population explosion of the latter was relatively short-lived and was followed by a resurgence in numbers of herring and mackerel. Thus, sand lance population dynamics alone cannot explain the phenomenon documented by Payne et al. (1990) and subsequent changes in the local abundance and distribution of whales.

As noted in the previous discussion of the "*Calanus* paradox," the key factor determining foraging success in right whales may not be abundance per se, but the availability of suitably dense prey patches. If *Calanus* patches are frequently reduced in size and density by predation from any other planktivore (which would include sand lance but also herring or mackerel), conceivably right whales could suffer a significant negative impact as a result. At this point, little more can be said without better data on several variables, including right whale feeding thresholds, the abundance and density of copepod patches, and patterns and rates of copepod consumption by sand lance, herring, mackerel, and similar predators.

If such interspecific competition is real, sei whales may have been less affected because of their ability to exploit a wider range of prey, including fish. Similarly, the broad diets of humpback and fin whales would presumably provide some insulation from major ecosystem shifts, including the well-documented changes in the abundance of herring and sand lance in the Gulf of Maine.

Summary and Approaches to Future Work

Whaling has a long history in the North Atlantic, and its impact on whale populations is still evident today. One species (the gray whale) is extinct in this ocean, although whether whaling pressure was the primary cause remains unclear. The balaenid whales (rights and bowheads) remain depleted despite decades of protection from whaling, and in some areas they may no longer represent significant parts of the local ecosystem. In contrast, most of the balaenopterids (notably humpback and fin whales) are recovering strongly from exploitation, and occupy major niches in the ecosystem.

Even with well-studied whale populations in well-studied ecosystems, major gaps exist in our knowledge. However, there is some evidence that top-down changes from overexploiting competing fish or forage fish populations have precipitated shifts in the distribution and local abundance of baleen whales in the Gulf of Maine. It is possible that changes in community structure and a consequent increase in competition for copepods between North Atlantic right whales and sand lance and/or herring and mackerel has been a contributing factor in the former species' failure to recover.

If we are to understand better the roles played by large whales in the marine ecosystem, it is essential that future marine mammal surveys be integrated with multidisciplinary ecosystem studies. This will be particularly fruitful in ecosystems that are already well studied, such as that of the northeast U.S. continental shelf. Another approach would be to conduct such integrated surveys in simpler ecosystems where whales have to date failed to recover from whaling (Clapham and Hatch 2000, Baker and Clapham 2004). For example, humpback, fin, and blue whales were all virtually extirpated by whaling at South Georgia, where the ecosystem is (compared to many areas of the North Atlantic) comparatively simple and based largely upon euphausiids. In such areas, long-term monitoring of the ecosystem might provide insights into how community structure and dynamics change as whales recover. However, the scope and expense of such a long-term study might well be prohibitive unless the cetacean component could be piggybacked onto existing projects.

All oceanic ecosystems have been greatly perturbed as a result of human activities, and the North Atlantic is among the best examples. In many fundamental ways, the North Atlantic today presents a very different ecosystem from what existed a thousand years ago before the onset of whaling. Human fisheries induced major changes in the structure and community composition of the ecosystem. Because of this, and the absence of relevant historical data, it is difficult to assess the impacts of the large-scale removal of so many mysticete predators from this ocean basin. More germane is how the ecological role of these whales may assist or hinder their long-term recovery.

Literature Cited

Aguilar, A. 1986. A review of old Basque whaling and its effect on the right whales (Eubalaena glacialis) of the North Atlantic. Report of the International Whaling Commission (Special Issue) 10: 191–199.

Angliss, R. P. and K. L. Lodge. 2003. Gray whale (Eschrichtius robustus): eastern North Pacific stock, in Alaska marine mammal stock assessments, 2003. U.S. Department of Commerce, NOAA Technical Memorandum NMFS-AFSC-144. Seattle: Alaska Fisheries Science Center, pp. 138–146.

Baker, C. S. and P. J. Clapham. 2004. Modeling the past and future of whales and whaling. Trends in Ecology and Evolution 19: 365–371.

Barlow, J. and P. J. Clapham. 1997. A new birth-interval approach to estimating demographic parameters of humpback whales. Ecology 78: 535–546.

Brown, B. E., V. C. Anthony, E. D. Anderson, R. C. Hennemuth, and K. Sherman. 1983. The dynamics of pelagic fishery resources off the northeastern coast of the United States under conditions of extreme fishing perturbations, in Proceedings of the expert consultation to examine changes in abundance and species composition of neritic fish resources. FAO Fisheries Report 291, Vol. 2. Rome: UN Food and Agricultural Organization, Fisheries Department, pp. 427–468. Available at ftp://ftp.fao.org/docrep/fao/005/x6850b/x6850b18.pdf. Accessed June 17, 2006.

Brown, S. G. 1976. Modern whaling in Britain and the north-east Atlantic Ocean. Mammal Review 6: 25–36.

———. 1986. Twentieth century records of right whales (Eubalaena glacialis) in the northeast Atlantic Ocean. Reports of the International Whaling Commission (Special Issue) 10: 121–127.

Bryant, P. J. 1995. Dating remains of gray whales from the eastern North Atlantic. Journal of Mammalogy 76: 857–861.

Caswell, H., S. Brault, and M. Fujiwara. 1999. Declining survival probability threatens the North Atlantic right whale. Proceedings of the National Academy of Sciences USA 96: 3308–3313.

Clapham, P. J. 1992. The attainment of sexual maturity in humpback whales. Canadian Journal of Zoology 70: 1470–1472.

Clapham, P. J., J. Barlow, M. Bessinger, T. Cole, D. Mattila, R. Pace, D. Palka, J. Robbins, and R. Seton. 2003. Abundance and demographic parameters of humpback whales from the Gulf of Maine, and stock definition relative to the Scotian Shelf. Journal of Cetacean Research and Management 5: 13–22.

Clapham, P. J. and L. T. Hatch. 2000. Determining spatial and temporal scales for population management units: lessons from whaling. Paper SC/52/SD2 submitted to the International Whaling Commission Scientific Committee.

Clapham, P. J. and C. A. Mayo. 1990. Reproduction of humpback whales, Megaptera novaeangliae, observed in the Gulf of Maine. Report of the International Whaling Commission (Special Issue) 12: 171–175.

Clapham, P. J. and J. G. Mead. 1999. Megaptera novaeangliae. Mammalian Species 604: 1–9.

Clapham, P. J., S. B. Young, and R. L. Brownell, Jr. 1999. Baleen whales: conservation issues and the status of the most endangered populations. Mammal Review 29: 35–60.

Cumbaa, S. L. 1986. Archaeological evidence of the 16th century Basque right whale fishery in Labrador. Report of the International Whaling Commission (Special Issue) 10: 187–190.

Dudley, P. 1725. An essay upon the natural history of whales. Philosophical Transactions of the Royal Society of London 33: 256–269.

Fogarty, M. J., E. B. Cohen, W. L. Michaels, and W. W. Morse. 1991. Predation and the regulation of sand lance populations: an exploratory analysis. ICES Marine Science Symposium 193: 120–124.

Greene, C. H. and A. J. Pershing. 2004. Climate and the conservation biology of North Atlantic right whales: the right whale at the wrong time? Frontiers in Ecology and the Environment 2: 29–34.

Hamilton, P. K. and C. A. Mayo. 1990. Population characteristics of right whales (Eubalaena glacialis) observed in Cape Cod and Massachusetts Bays, 1978–1986. Report of the International Whaling Commission (Special Issue) 12: 203–208.

IWC. 2001. Report of the workshop on status and trends in North Atlantic right whales. Journal of Cetacean Research and Management (Special Issue) 2: 61–87.

———. 2002. Report of the Sub-committee on the Comprehensive Assessment of North Atlantic humpback whales. Report of the Scientific Committee, Annex H. Journal of Cetacean Research and Management 4 (Supplement): 230–260.

———. 2004. Report of the Standing Working Group (SWG) on the Development of an Aboriginal Subsistence Whaling Management procedure (AWMP). Journal of Cetacean Research and Management (Supplement) 185–210.

Jones, M. L., S. L. Swartz, and S. Leatherwood, editors. 1984. The gray whale Eschrichtius robustus. San Diego: Academic Press.

Knowlton, A. R., J. Sigurjónsson, J. N. Ciano, and S. D. Kraus. 1992. Long-distance movements of North Atlantic right whales (Eubalaena glacialis). Marine Mammal Science 8: 397–405.

Kraus, S. D., P. K. Hamilton, R. D. Kenney, A. R. Knowlton, and C. K. Slay. 2001. Reproductive parameters of the North Atlantic right whale. Journal of Cetacean Research and Management (Special Issue) 2: 231–236.

Larsen, F. and P. S. Hammond. 2004. Distribution and abundance of West Greenland humpback whales. *Journal of Zoology*, London 263: 343–358.

Lindquist, O. 2000. *The North Atlantic gray whale* (Eschrichtius robustus)*: a historical outline based on Icelandic, Danish-Icelandic, English and Swedish sources dating from ca 1000 AD to 1792*. St. Andrews, UK: Universities of St. Andrews and Stirling, The Centre for Environmental History and Policy.

Link, J. 1999. (Re)constructing food webs and managing fisheries, in *Proceedings of the 16th Lowell Wakefield Fisheries Symposium—Ecosystem Considerations in Fisheries Management*. AK-SG-99-01. Fairbanks, AK: University of Alaska Sea Grant, pp. 571–588.

Link, J. S. and J. K. T. Brodziak. 2002. *Status of the Northeast U.S. Continental Shelf Ecosystem*. NEFSC Reference Document 02-11. Woods Hole, MA: Northeast Fisheries Science Center.

Link, J., J. K. T. Brodziak, S. F. Edwards, W. J. Overholtz, D. Mountain, J. W. Jossi, T. D. Smith, and M. J. Fogarty. 2002. Marine ecosystem assessment in a fisheries management context. *Canadian Journal of Fisheries and Aquatic Science* 59: 1429–1440.

Mayo, C. A. and Marx, M. K. 1989. Surface foraging behavior of the North Atlantic right whale, *Eubalaena glacialis*, and associated zooplankton characteristics. *Canadian Journal of Zoology* 68: 2214–2220.

McLeod, B. A., T. Rastogi, T. R. Frasier, M. W. Brown, R. Grenier, S. L. Cumbaa, J. Nadarajah, and B. N. White. 2003. DNA analysis of 58 16th century whale bones from Basque sites in Labrador. North Atlantic Right Whale Consortium annual meeting, November 4–5 (abstract).

Mead, J. G. and E. D. Mitchell. 1984. Atlantic gray whales, in *The gray whale* Eschrichtius robustus. M. L. Jones, S. L. Swartz, and S. Leatherwood, eds. New York: Academic Press, pp. 33–53.

Mitchell, E. and R. R. Reeves. 1983. Catch history, abundance and present status of northwest Atlantic humpback whales. *Report of the International Whaling Commission* (Special Issue) 5:153–212.

Nerini, M. 1984. A review of gray whale feeding ecology, in *The gray whale* Eschrichtius robustus. M. L. Jones, S. L. Swartz, and S. Leatherwood, eds. San Diego: Academic Press, pp. 423–450.

Overholtz, W. J. and K. D. Friedland. 2002. Recovery of the Gulf of Maine-Georges Bank Atlantic herring *(Clupea harengus)* complex: perspectives based on bottom trawl survey data. *Fishery Bulletin* 100: 593–608.

Overholtz, W. J. and J. R. Nicholas. 1979. Apparent feeding by the fin whale, *Balaenoptera physalus*, and the humpback whale, *Megaptera novaeangliae*, on the American sand lance, *Ammodytes americanus*, in the northwest Atlantic. *Fishery Bulletin* 77: 285–287.

Palsbøll, P. J., J. Allen, M. Bérubé, P. J. Clapham, T. P. Feddersen, P. Hammond, H. Jørgensen, S. Katona, A. H. Larsen, F. Larsen, J. Lien, D. K. Mattila, J. Sigurjónsson, R. Sears, T. Smith, R. Sponer, P. Stevick, and N. Øien. 1997. Genetic tagging of humpback whales. *Nature* 388: 767–769.

Payne, P. M., D. Wiley, S. Young, S. Pittman, P. J. Clapham, and J. W. Jossi. 1990. Recent fluctuations in the abundance of baleen whales in the southern Gulf of Maine in relation to changes in selected prey. *Fishery Bulletin* 88: 687–696.

Reeves, R. R. and E. D. Mitchell. 1986a. The Long Island, New York, right whale fishery: 1650–1924. *Report of the International Whaling Commission* (Special Issue) 10: 201–220.

———. 1986b. American pelagic whaling for right whales in the North Atlantic. *Report of the International Whaling Commission* (Special Issue) 10: 221–254.

Reeves, R. R., R. Rolland, and P. J. Clapham, eds. 2001. *Causes of reproductive failure in North Atlantic right whales: new avenues of research*. NEFSC Reference Document 01-16. Woods Hole, MA: Northeast Fisheries Science Center.

Reeves, R. R. and T. D. Smith. 2002. Historical catches of humpback whales in the North Atlantic Ocean: an overview of sources. *Journal of Cetacean Research and Management* 4: 219–234.

Roman, J. and S. R. Palumbi. 2003. Whales before whaling in the North Atlantic. *Science* 301: 508–510.

Schaeff, C. M., S. D. Kraus, M. W. Brown, J. Perkins, R. Payne, and B. N. White. 1997. Comparison of genetic variability of North and South Atlantic right whales (*Eubalaena*) using DNA fingerprinting. *Canadian Journal of Zoology* 75: 1073–1080.

Sears, R., M. J. Williamson, F. W. Wenzel, M. Bérubé, D. Gendron, and P. Jones. 1990. Photographic identification of the blue whale *(Balaenoptera musculus)* in the Gulf of St. Lawrence, Canada. *Report of the International Whaling Commission* (Special Issue) 12: 335–342.

Sigurjónsson, J. 1995. On the life history and autecology of North Atlantic rorquals, in *Whales, seals, fish and man*. A. S. Blix, L. Walløe, and Ø. Ulltang, eds. New York: Elsevier Science, pp. 425–441.

Smith, T. D. and R. R. Reeves. 2003. Estimating historical humpback whale removals from the North Atlantic: an update. *Journal of Cetacean Research and Management* 5 (Supplement): 301–311.

Smith, W. D., D. McMillan, P. Rosenberg, M. Silverman, and A. Wells. 1980. *Seasonal and annual changes in the distribution, abundance, and species composition of fish eggs and larvae off the northeastern United States, 1977–1979*. ICES CM 1978/L:30. Copenhagen: International Council for Exploration of the Sea.

Spencer, P. D. 1997. Optimal harvesting of fish populations with nonlinear rates of predation and autocorrelated environmental variability. *Canadian Journal of Fisheries and Aquatic Science* 54: 59–74.

Stevick, P. T., J. Allen, P. J. Clapham, N. Friday, S. K. Katona, F. Larsen, J. Lien, D. K. Mattila, P. J. Palsbøll, J. Sigurjónsson, T. D. Smith, N. Øien, and P. S. Hammond. 2003. North Atlantic humpback whale abundance and rate of increase four decades after protection from whaling. *Marine Ecology Progress Series* 258: 263–273.

Tønnessen, J. N. and A. O. Johnsen. 1982. *The history of modern whaling*. Berkeley: University of California Press.

Weller, D. W., A. M. Burdin, B. Würsig, B. L. Taylor, and R. L. Brownell, Jr. 2002. The western gray whale: a review of past exploitation, current status and potential threats. *Journal of Cetacean Research and Management* 4: 7–12.

Winn, H. E., C. A. Price, and P. W. Sorensen. 1986. The distributional biology of the right whale *(Eubalaena glacialis)* in the western North Atlantic. *Report of the International Whaling Commission* (Special Issue) 10: 129–138.

TWENTY-FIVE

Sperm Whales in Ocean Ecosystems

HAL WHITEHEAD

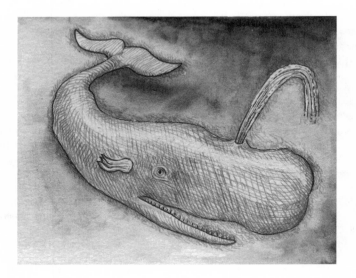

For Herman Melville the sperm whale, *Physeter macrocephalus*, was "The Whale," and many supposedly generic illustrations of whales portray sperm whales. But the sperm whale is not a generic whale; in fact, from almost any perspective, it is the most atypical of species. One class of attributes by which the sperm whale has no close analogs, or homologs, is ecology. The ecology of the sperm whale forms the principal subject of this chapter. I will describe the trophic web surrounding the sperm whale with the goal of addressing three questions: How are sperm whale populations regulated? How do sperm whales affect other elements of the marine environment? How has sperm whaling affected marine ecosystems?

The ecological role of an organism is a function of its other biological attributes. Thus I will begin this chapter by summarizing some elements of the biology of the sperm whale. In describing sperm whale ecology, I will summarize what we know of sperm whale prey, competitors, life history, and predators and try to evaluate both the bottom-up (through competition for prey) and top-down (through predation by orcas) hypotheses for the natural regulation of populations. Whaling certainly affected sperm whale populations, and I summarize this huge enterprise and consider the potential effects of the killing of so many sperm whales on other elements of the ecosystem. I conclude the chapter by suggesting ways in which we may reduce our

huge uncertainty about the ecological role of this spectacular species.

Evolution to Extremes

Viewed from almost any perspective, the sperm whale looks odd, but oddest of all is its head. Square and huge, the head of the sperm whale forms roughly 35% of the animal. Most of this is the spermaceti organ, a large and complex oil-filled structure (see Cranford 1999). The function of the spermaceti organ has been debated from before Melville's time (e.g., Beale 1839) to the present (e.g., Carrier et al. 2002). In the view of most, but not all, sperm whale scientists, available data overwhelmingly support the acoustic hypothesis of Norris and Harvey (1972): that the spermaceti organ and surrounding structures have evolved primarily to produce clicks (see Cranford 1999). These are very loud, highly directional clicks; the spermaceti organ complex seems to be the most powerful sonar system in the natural world (Møhl et al. 2000). I have argued that the development of the spermaceti organ complex may have been key to the evolution of the other extreme characteristics of the sperm whale (Whitehead 2003).

One such characteristic is size. Adult sperm whales reach 11–18 m, larger than any other toothed whale and exceeded only by the large baleen whales. Large animals must eat a lot, and the largest have developed particularly efficient methods of extracting nutrition from their environment. Bulk feeding allows massive size in the baleen whales, and so does the powerful sonar of the sperm whale.

As absolute size of an animal increases, so do the sizes of constituent body parts, and the sperm whale contains other extreme organs in addition to its sonar. One is its intestine (Williams, Chapter 15 in this volume), and another is its brain. The sperm whale has, at 7.8 kg, the largest brain on Earth (Kojima 1951). This may mean little if cognitive abilities correlate with brain size relative to body size, by which measure the sperm whale is far from exceptional. However, recent studies suggest that, among primate species, absolute brain size better predicts complex cognitive function (e.g., Beran et al. 1999). If this result can be extrapolated to a very different type of animal, then the sperm whale may have some cognitive surprises for us, and these could include unusual ways of interacting with the environment.

An adult female sperm whale, at about 11 m and 15 t, is large, but the male can reach 17 m and 45 t or more. There are other differences between the sexes. For instance, males have proportionally larger spermaceti organs (Cranford 1999). Females can be found almost anywhere where waters are warmer than 15 °C at the surface and deeper than 1,000 m (Rice 1989), but mature males have even wider distributions. They are usually in cool waters, including those off the edge of pack ice in both hemispheres, and will sometimes use areas less than 100 m deep (e.g., Scott and Sadove 1997). The evolutionary forces that produced these differences between the sexes are unclear (see Best 1979; Whitehead 2003 for some speculations), but they have led to a species that has the largest geographical sexual segregation of any organism and a sexual size dimorphism greater than that of any other mammal that copulates in the ocean. Extreme male-biased sexual size dimorphism is usually related to defense of either females or territory, but both are hard or impossible in a three-dimensional medium such as the ocean (e.g., Connor et al. 2000).

There is nothing much like a sperm whale, except other sperm whales. The common ancestor of the sperm and its nearest phylogenetic relatives (the dwarf sperm whale, *Kogia simus,* and pygmy sperm whale, *K. breviceps*) lived at least 8 million years ago (Berta and Sumich 1999), and the kogiids are in a separate family as well as much smaller (2–4 m) than the sperm. Physeteridae is the only oceanic odontocete family containing just one species. The closest ecological analog is sometimes considered to be the northern bottlenose whale, *Hyperoodon ampullatus* (Mitchell 1977), a fairly large (7–9 m), sexually-dimorphic, deep-diving odontocete. But the northern bottlenose has a much more restricted distribution, a totally different social system, and considerably more localized movement patterns than the sperm whale (Whitehead 2003). In some ways a closer ecological analog of the sperm whale is the 6 m Cuvier's beaked whale, *Ziphius cavirostris* (this comparison was suggested by an anonymous reviewer of the manuscript). Cuvier's has the widest distribution of any of the beaked whales (Ziphiidae), a distribution that closely parallels that of female sperm whales, but Cuvier's

does not seem to use the cooler waters where only male sperm whales are found (Heyning 1989). Like sperm whales, Cuvier's beaked whale also eats mid- and deep-water squid (MacLeod et al. 2003); although there is considerable overlap between the diets of the two species (MacLeod et al. 2003; Whitehead 2003), sperm whales eat a considerably larger proportion of large squid and have a wider niche breadth (see subsequent sections). Furthermore, Cuvier's beaked whale shows little or no sexual dimorphism in size (Heyning 1989).

So how did the sperm whale evolve into such a distinct area of biological space? I believe that the spermaceti organ complex was key (Whitehead 2003). With a more powerful sonar system than other members of its guild, the sperm whale was able to outcompete most other cetacean predators as well as possibly other competitor groups, for the considerable resources of the mesopelagic ocean (see subsequent sections). Animals with similar habits to the sperm whale went extinct or, like the northern bottlenose whale, adopted specialized niches where they could survive in a world dominated by the great generalist.

This suggests that intraspecific competition for food regulates sperm whale populations in a bottom-up fashion and that sperm whales may at least partially regulate the populations of their prey. However, predators may also play a role in sperm whale population dynamics; one predator—*Homo sapiens*—certainly has.

Ecology

The Trophic Scene Beneath the Sperm Whale

Crude calculations suggest that, globally, sperm whales remove 75 million metric tons of food per year from the ocean, roughly the same as all human marine fisheries (Whitehead 2003; see also Williams, Chapter 15, and Croll et al., Chapter 16, in this volume). With a few exceptions, however (such as the jumbo squid, *Dosidicus gigas,* and some large deep-water fish species of higher latitudes), sperm whale diet overlaps little with the species caught by humans. So there is a huge, but little known, proliferation of life in the deeper waters of the oceans—sperm whales seem to feed mostly in waters of depths between about 300 and 1,200 m (Lockyer 1977; Clarke 1980; Papastavrou et al. 1989; Watkins et al. 2002).

Sperm whales mainly eat cephalopods, but these vary enormously. The modal prey of a female sperm whale may be a 400 g histioteuthid, which is approximately equivalent, in percent body mass, to a human eating a walnut (Clarke et al. 1993). The females' cephalopod prey ranges, however, from chiroteuthids of <100 g (equivalent to a pea for a human) to a 100 kg architeuthid giant squid, equivalent to a "half-pound steak" in the words of Clarke et al. (1993). Male sperm whales can eat even larger architeuthids of up to 400 kg. These cephalopods vary in activity from the fierce jumbo squid, which *National Geographic* once called "a living horror from the deep" (Duncan 1941) and that must be a challenging quarry, to the poorly muscled histioteuthids,

TABLE 25.1

Most Trophically Distinct Marine Mammals[a]

Common Name	Scientific Name	Maximum Niche Overlap (%) with Another Marine Mammal
Leopard seal	*Hydrurga leptonyx*	65.0
Sperm whale	*Physeter macrocephalus*	70.0
Orca	*Orcinus orca*	70.0
Bryde's whale	*Balaenoptera edeni*	75.0
Antarctic fur seal	*Arctocephalus gazella*	75.0
Minke whale	*Balaenoptera acutorostrata*	75.0
Humpback whale	*Megaptera novaeangliae*	75.0

[a] As indicated by smallest maximum niche overlap (%) with other marine mammals of 97 species listed by Pauly et al. (1998).

which seem to move little (Clarke et al. 1993). Sperm whales, and male sperm whales in particular, also eat a wide range of noncephalopod prey, especially large demersal fish including sharks, rays, and teleosts (see, e.g., Rice 1989).

All this points to a highly successful, opportunistic, and generalist predator that can make a living in a wide range of environments with diverse sets of competitors. Squid biomass generally varies enormously over quite short timescales (Arnold 1979), as do indicators of sperm whale feeding success (Whitehead 1996), so, in most areas, the resources on which sperm whales subsist are anything but predictable. Thus one can visualize the trophic scene beneath the sperm whale as rich, but diverse and fickle, cloudlike formations at a variety of depths beneath the surface.

Competitors

Predators of deep-water cephalopods come from a variety of families. Cephalopods derive much of their sustenance from other cephalopods, and cannibalism is a prominent feature of this order (O'Dor 1998), so there are undoubtedly strong cephalopod-to-cephalopod links in the trophic webs of the deep ocean. Thus, sperm whales compete with larger cephalopods for smaller ones, but, because the large squid are also sperm whale prey, their existence may have little effect, or even a positive effect, on sperm whale subsistence.

Some large pelagic fish, such as swordfish *(Xiphius gladius)*, feed on similar organisms as the sperm whales do and at similar depths (see, e.g., Hernández-García 1995). Unlike the large cephalopods, these are not themselves important sperm whale food, and their biomass is of a similar order to that of the sperm whales (see Whitehead 2003), so they may be significant competitors, although they generally take smaller

TABLE 25.2

Niche Breadth Indices for Teuthivorous Mammals

Species	Niche Breadth Index
Sperm whale	1.25
Southern elephant seal	0.90
Cuvier's beaked whale	0.68
Northern bottlenose whale	0.21

food than the sperm whales do. Dissolved oxygen generally reaches a minimum at the depths where sperm whales forage (Nybakken 2001), hampering the energetics of any animal that filters its oxygen from the water. As a mammal, the sperm whale thus possesses a potential advantage over both its prey and its non-air-breathing competitors, although it must return periodically to the surface to breathe.

Other marine mammals possess the same advantages and disadvantages and would seem more likely to be direct competitors. Using the coarse diet classifications of Pauly et al. (1998), the sperm whale has the greatest diet overlap (70%) with the pygmy sperm whale and the southern elephant seal *(Mirounga leonina)*, and diet overlaps of 65% with nine other species, mostly beaked whales (Whitehead 2003). When these 11 potential competitors (those with a diet overlap >65% with sperm whales) are compared with one another, each has a diet overlap of at least 90% with one of the other species on the list, emphasizing the sperm whale's trophic distinctiveness among mammalian teuthivores. As shown in Table 25.1, the sperm whale is one of the three most trophically distinct marine mammals, as measured by the percent overlap index on the data presented by Pauly et al. (1998), the other two being high-level predators, the leopard seal *(Hydrurga leptonyx)* and orca *(Orcinus orca)*. The proportional use of large (>0.5 m mantle length) squid principally distinguishes the sperm whale from other marine mammals. Pauly et al. (1998) estimate that the large squid constitute about 60% of the sperm whale's diet, but no more than 40% for any other marine mammal species, except True's beaked whale *(Mesoplodon mirus)*, about which little is known. In the Antarctic this is illustrated by the sperm whale's predominant use of the large (2 m mantle length), deep-living *Mesonychoteuthis hamiltoni*, a species that has yet to be associated with any potential marine mammal competitor (Whitehead 2003). Similarly in the southeast Pacific, the majority of the sperm whales' diet consists of the large squid *Dosidicus gigas* (Clarke et al. 1976).

The sperm whale's diet is not only different from that of its competitors; it also seems to be generally wider. For the four species of mammalian mesopelagic teuthivores for which sufficient data existed, I calculated measures of niche breadth based on the slope of a logistic curve relating the generic diversity of cephalopod beaks found in a stomach to the number of beaks examined (see Whitehead et al. 2003 for methodology). As shown in Table 25.2, the measure was greatest for the sperm whale, at 1.25. Thus, probably principally because of the inclusion of larger squid in the diet, the sperm whale's niche is wider than those of these competitors, and especially so in the case of the northern bottlenose whale, whose diet is dominated by squid of one genus, *Gonatus* (Hooker et al. 2001).

Sperm whales usually seem to exceed potential mammalian competitors in terms of biomass. In five ocean areas I found, or calculated, estimates of the biomass of the principal groups of mammalian mesopelagic teuthivore. In the Antarctic, sperm whales constitute an estimated 67% of this biomass; off California, 60%; in the eastern tropical Pacific, 66%; in the northwestern Gulf of Mexico, 94%; and in the shelf-edge region off the northeast coast of the United States, 42% (for numbers and references see Whitehead 2003). Only in the last of these cases does any other species, in this instance the long-finned pilot whale *(Globicephala melas)*, eclipse the sperm whale.

In ecological terms, then, the sperm whale seems to dominate its ecosystem. Partly this is because it regularly eats the larger deep-water squid, which seem to be beyond the reach of most mammals, but it can also make a very good living from the staples of the mesopelagic, such as the histioteuthids, as well as from very different prey types, including large demersal fish. Success over such a wide range points to some key advantage, and my guess is that this is the spermaceti organ complex.

Life History

Before "K-selected" fell from grace as an ecological adjective (Stearns 1992), the sperm whale would have been seen as an epitome of such a species. Females first give birth at about age 10 following a 14–16 month gestation and then produce single offspring about every five years, suckling each for approximately two years (Best et al. 1984; Rice 1989). A female's fecundity is reduced during her forties, and there may be menopause, in the sense of a substantial part of a female's life span being lived following her last parturition (Best et al. 1984). Females live at least into their seventies and probably, on occasion, much longer. Males become sexually mature in their teens but, being geographically segregated from the females, do not seem to take much of a part in breeding until their late twenties, when they start to migrate from their cold-water feeding grounds to the lower latitudes inhabited by females (Best 1979). Estimates of sperm whale mortality (IWC 1980) are derived from estimated age distributions in catch data and have little reliability, but adult natural mortality is likely to be low, at most a few percent per year.

All these characteristics make for a species with poor ability to recover from depletion. The set of life history parameters last used by the International Whaling Commission (IWC) (IWC 1980) gives a maximal rate of increase of 0.9%/yr. An age-specific fecundity rate calculated from the data of Best et al. (1984) and the age-specific mortality rates calculated for

orcas by Olesiuk et al. (1990) give a maximal rate of increase of 1.1%/yr (Whitehead 2003).

This result indicates that during recent evolutionary history the sperm whale has existed in an environment in which the ability to reproduce fast gave little fitness advantage; an environment in which sperm whales filled their niche fairly tightly and populations were close to carrying capacity.

Natural Regulation of Sperm Whale Populations: The Bottom-Up Scenario and Density Dependence

Phylogenetic remoteness, a distinct niche, high biomass, and slow life history traits all suggest that the sperm whale, perhaps as a result of the power of the spermaceti organ complex's sonar, dominates a large and important part of the marine ecosystem. Under this hypothesis, related animals, such as the other physeterids that diverged about 15 Ma, are extirpated; most competitors, such as many of the beaked whales, are pushed into peripheral, specialized habitats; and sperm whales compete for food largely with each other. A prediction of this bottom-up hypothesis for the regulation of sperm whale populations is that when sperm whale populations were decimated by whaling, life history parameters should have responded in a density-dependent fashion.

Several factors complicate our ability to identify density-dependent responses in sperm whales. Measures of natural mortality are not precise enough to detect important changes with whaling; we do not understand enough about the effects of open-boat whaling to estimate with any certainty the relative depletion of the population when modern sperm whaling hit its stride in about 1950 (see subsequent sections); and the expected density-dependent change in fecundity— a higher birth rate with fewer animals—is potentially confounded by two socially mediated factors. First, during modern sperm whaling there was an emphasis on killing large males. The adult sex ratio was changed sufficiently that there was concern (May and Beddington 1980), and some data (Clarke et al. 1980), that the female pregnancy rate had been compromised. Second, female sperm whales have a complex social structure, within which females assist one another in raising their offspring (Whitehead 2003). In a very congruent social structure—that of the African elephant, *Loxodonta africana* (Weilgart et al. 1996)—the killing of females reduced the fecundity of surviving females within their social groups (Poole and Thomsen 1989; McComb et al. 2001). Either of these factors would counteract the expected increase in fecundity at lower densities. Together with the technical problems of defining stock boundaries and data quality, these have been suggested as factors that may have led several studies to fail to detect density dependence in sperm whale life history characteristics during the course of modern whaling (see Kasuya 1991 for a summary of attempts to detect density dependence in sperm whales).

Despite these difficulties and failures, there are two pieces of evidence that, as sperm whale densities were reduced, life history parameters responded. Off South Africa, following 10 years of exploitation, fecundity rates increased by about 10% (Best et al. 1984); and in the North Pacific during the 1970s the proportion of male sperm whales longer than 16.8 m increased from 0% to greater than 20%, a change that Kasuya (1991) relates to better feeding conditions for growing males in the 1950s and 1960s as the population was reduced. Kasuya's (1991) study was performed before the extent of data falsification in the North Pacific sperm whale fishery was made public (e.g., Kasuya 1999).

These indications of density dependence support the bottom-up hypothesis for sperm whale population regulation. But there is an alternative: that even before humans began their slaughter, predators had a major impact on sperm whales and their populations.

FIGURE 25.1. Orca tooth-marks on the flukes of a sperm whale (H. Whitehead laboratory).

Natural Predation on Sperm Whales

The sperm whale may be large and look fearsome, but it has natural enemies. Foremost among these is the orca, but sharks, pilot whales (*Globicephala* spp.), and false killer whales (*Pseudorca crassidens*) have also been implicated as attackers of sperm whales.

Bite marks from sharks have been found on the flukes of some sperm whales (Dufault 1994) as well as on the carcasses of calves (Best et al. 1984), and opened shark stomachs have contained remains of sperm whales (Rice 1989). However, in no case is it clear that sharks have killed a sperm whale or even badly injured one; bite marks on live animals indicate nonlethal injuries, those on carcasses were probably produced after death, and the sperm whale parts in shark stomachs likely represent scavenging (Rice 1989).

Pilot whales and false killer whales are sometimes seen closely interacting with sperm whales (Palacios and Mate 1996; Weller et al. 1996); from the evident discomfort of the sperm whales, the smaller odontocetes appear to be at least harassing the sperm whales, and perhaps attacking them. In one incident involving false killer whales, pieces of flesh were observed in the water afterwards (Palacios and Mate 1996). It is not clear what the smaller whales were doing in these incidents; suggestions include "snacking" on flukes and fins, play, trying to induce vomiting in the sperm whales, or attempting to drive sperm whales away from shared food resources (Palacios and Mate 1996; Weller et al. 1996). But it does seem improbable that any significant sperm whale mortality results.

Orcas are another matter, much more powerful than pilot and false killer whales and, unlike sharks, possessing sophisticated cooperative foraging techniques (Baird 2000). Furthermore, orcas have definitely killed sperm whales (Pitman et al. 2001). But how important are these interactions?

As a rough indicator of the incidence of interactions between sperm whales and orcas, during 294 days spent tracking groups of female sperm whales in the South Pacific, we have observed orcas with or near the sperm whales on five occasions. Assuming that interactions between the species are equally likely to occur during day or night (we observed them only in the daytime), this suggests that a female sperm whale in this area encounters orcas about 12 times per year, or on 745 occasions over a 60-year lifespan. This is a high encounter rate, but most of these events are demonstrably benign; of the five encounters we observed (summarized in Whitehead 2003), for one there was no sign of behavioral reaction at all by either species; in three others the sperm whales appeared to react, but there was no sign of close approach by the orcas; and in a final case (see Arnbom et al. 1987) the orcas attacked the sperm whales, causing superficial wounds but no observable serious injury. Similarly, in their literature review, Jefferson et al. (1991) note 33 "non-predatory interactions" between sperm whales and orcas and six "attacks." Thus, the number of orca attacks suffered by a female sperm whale during her lifetime might be of the order of 150 (20% of 745), the remaining encounters perhaps being either with orcas of non-sperm-whale-eating ecotypes (see Barrett-Lennard and Heise, Chapter 13, this volume; Reeves et al., Chapter 14, this volume) or potentially dangerous, but not currently interested, orcas. These attacks will be on the group as a whole and may not be directed towards the particular female.

One hundred and fifty is still a large number of attacks for an animal to face in its lifetime, and it is not surprising that sperm whales frequently bear scars from orca teeth (Figure 25.1). But are any of these attacks fatal, or sufficiently serious that the fitness of the victims is compromised? There are several lines of evidence that some are:

- That the orcas continue to attack sperm whales suggests that, at least sometimes, they are successful. If they always failed, evolutionary theory predicts that they would not bother, although it is possible to develop scenarios other than predation for the apparent attacks, such as those suggested above for the encounters of sperm whales with pilot and false killer whales.

- Although nine of the ten reported attacks of orcas on sperm whales (in Jefferson et al. 1991; Brennan and Rodriguez 1994; Pitman et al. 2001; Whitehead 2003) were apparently unsuccessful, one clearly was. Pitman et al. (2001) graphically describe a prolonged attack off California in which one sperm whale was certainly killed, and several other members of its group were likely fatally injured.

- Less directly, several aspects of the complex social system of female sperm whales, including clustering at the surface, babysitting, and bonds between unrelated females, suggest social evolution in the presence of a dangerous predator: the orca (Best 1979; Whitehead 2003).

Natural Regulation of Sperm Whale Populations: the Top-Down Scenario

In the orca, sperm whales have a predator that at least occasionally kills, and is strongly implicated in their evolution. But did, or do, orcas regulate sperm whale populations? If female sperm whales are attacked by orcas 150 times during their lives, only a very small proportion of those attacks would need to be successful for orca predation to be a major contributor to sperm whale mortality. If orcas kill one female out of a group of 20 during 10% of the attacks (a 0.5% mortality rate for each individual per attack), this indicates that orcas account for about $[1 - 0.995^{150}] \times 100\% = 53\%$ of sperm whale mortality. As noted in a preceding section, the very slow life history processes of the sperm whale are expected from the bottom-up scenario of a species whose population is near carrying capacity, but these apparently "K-selected" features could also be consistent with a situation in which young animals are very vulnerable to predation and need to receive great care if they are to have much chance of survival, and great care is what they seem to get (Whitehead 2003).

However, two of the other attributes of sperm whales previously summarized do, I think, support the theory that sperm whale populations naturally remained near carrying capacity and so were not substantially regulated by orca predation. First, the observation of density-dependent effects following whaling indicates a significant increase in the resources available after population reduction, and thus a prewhaling population near carrying capacity. Second, the lack of close trophic homologs or analogs indicates that sperm whales were taking a large proportion of the available resources in a large niche, squeezing out potential competitors.

Thus, the evidence seems to support bottom-up regulation of sperm whale numbers through prey limitation. However, the case for control from above by orcas is far from closed.

Sperm Whaling

Unlike some of the baleen whales and smaller odontocetes (see Reeves and Smith, Chapter 8, this volume), sperm whales were generally not prey of preindustrial humans. The only known exception is the sperm whaling at Lamalera, Indonesia, which seems to have begun before contact with Western whalers (Barnes 1991). Sperm whales' deep-water habitat usually separated them from the ranges of early whalers, except in cases, such as Lamalera, where deep water is found very close to shore.

In 1712, a Nantucket whaling ship cruising for right whales *(Eubalaena glacialis)* was blown south of Nantucket island into deeper waters, found a sperm whale, killed it using a rowed whaleboat, and towed it back to the island. Industrial sperm whaling had begun. During the next 150 years, killing sperm whales became one of the major economic enterprises of the Industrial Revolution. The following are some of the key events of this "open-boat" hunt (see Starbuck 1878):

- The invention of a clean-burning candle made from the oil in the spermaceti organ in the mid-eighteenth century, making sperm whales a major source of light in the Western world, and sperm whaling extremely lucrative

- The invention of the onboard tryworks at about the same time, allowing sperm whales to be processed entirely at sea, resulting in a truly pelagic industry

- The spreading of pelagic sperm whaling out of the Atlantic and into the Pacific and Indian Oceans during the early nineteenth century

- A decline in the industry starting in about 1840, as sperm whales became increasingly hard to find, petroleum began to substitute for spermaceti oil, and external events (including the American Civil War and the loss of a large number of whaleships in the ice off Alaska in 1871) affected the profitability of the enterprise

- A relict industry persisting into the twentieth century, with the final American sailing voyage being made in 1925 (Rice 1989)

- The transfer of sperm whaling technology to various locations around the world, including the Azores and West Indies, where it persisted until recently

Open-boat whaling for sperm whales overlapped with the modern whaling industry, in which steam- or diesel-powered catcher boats replaced rowed whaleboats, large guns were used in place of muscle for propelling harpoons, and the whaleship itself became a huge modern factory. In the early part of the twentieth century, modern whalers concentrated on the larger baleen whales, but by about 1950 the populations of all the larger baleen whales were severely reduced and the sperm whale became a major target. Sperm whaling peaked in the 1960s, with tens of thousands of animals being killed each year, and then declined as the IWC and other bodies increasingly began to control the industry. Modern sperm whaling virtually ended with the IWC moratorium in 1985, although Japan has taken a few animals in some years since then.

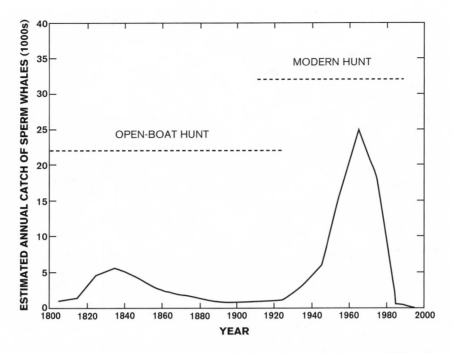

FIGURE 25.2. Estimated numbers of sperm whales caught per year between 1800 and 2000. This likely underestimates the true catches because of missing data and falsification. (Figure reproduced from Whitehead 2003; data from from Best 1983 and FAO Fisheries Department, Fishery Information, Data and Statistics Unit. FISHSTAT Plus: Universal software for fishery statistical time series. Version 2.3. 2000.)

An estimated global sperm whale catch history from 1800 to 2000 is shown in Figure 25.2 based on reported landings of oil and whales. The catch is likely to be underestimated because of unreported or falsified catches (see, e.g., Best 1983; Zemsky et al. 1995).

Human Regulation of Sperm Whale Populations

In trying to reconstruct the effects of whaling on sperm whale populations, we are faced with a range of difficulties. These include uncertainties about catch numbers and life history parameters (noted previously) as well as how to deal with possible population segregation. I estimated the current global population extrapolating from visual surveys and then back-calculated a population trajectory using the estimated catches (Figure 25.2), an estimate of the maximum rate of increase and the form of density dependence (Whitehead 2002). The best-estimate trajectory is shown in Figure 25.3 along with trajectories using randomly-selected but reasonable values of the model's parameters. This analysis suggests that starting with a prewhaling population of a little over a million sperm whales (95% confidence interval: 672,000–1,512,000), the open-boat whalers had reduced numbers to about 71% of the initial population in 1880 (95% confidence interval 52%–100%), and modern whalers to about 32% of the initial population in 1999 (95% confidence interval 19%–62%). The projected increase in population during the late twentieth century shown in Figure 25.3 is purely theoretical, based on

a lack of whaling and a population below carrying capacity. There is no empirical evidence for an increase in sperm whale populations anywhere in the world over the last 20 years, and there are concerns that the population might fail to recover or continue to fall following the cessation of whaling because of the impact of the killing on sperm whale social systems (Whitehead 2003).

Ecological Effects of Sperm Whales and Sperm Whaling

Effects on Prey

If sperm whales remove 75 million metric tons from the ocean annually, it seems likely that they have an impact on the populations of some of their prey species. However, there is no direct evidence of this, as we know virtually nothing of the population biology of any major sperm whale prey. The best indirect evidence comes from the density-dependent responses already noted, with sperm whales growing faster or becoming more fecund with fewer other sperm whales around. This strongly suggests that sperm whales do depress prey populations.

Such a predator might then be expected to exert a top-down influence on the trophic system beneath. So little is known of mesopelagic squid that it is hard to even suggest the strength and nature of such an effect, and the task is further complicated by the sperm whale taking prey from at

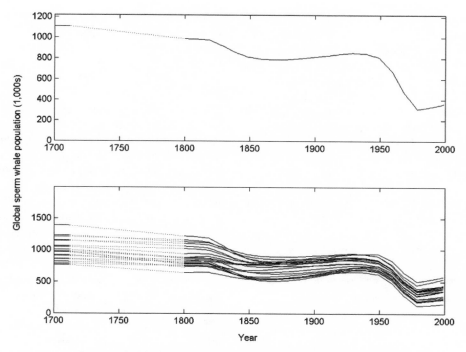

FIGURE 25.3. Estimated global sperm whale population from 1700 to 1999 (reproduced from Whitehead 2003 based on methodology in Whitehead 2002). The upper plot shows the best estimate of the population trajectory; the lower plot shows 20 trajectories calculated using parameter values randomly chosen from within reasonable ranges. Information is very limited for the period 1712–1800.

least two trophic levels. If sperm whales disproportionately target the smaller squid of lower trophic levels, we might expect that sperm whale predation will have a double negative effect on the larger squid: directly when the larger squid are eaten, and indirectly through competition. If, on the other hand, the larger squid are preferred, then smaller squid may actually benefit from the presence of sperm whales, the direct sperm whale predation being offset by the reduction in numbers of their cephalopod predators. That larger squid often constitute a substantial part of sperm whale diet (e.g., Clarke 1980; Kawakami 1980) perhaps suggests the latter scenario.

Sperm whaling will have changed this picture. It seems likely that populations of the larger cephalopods will have become larger with reduced predation pressure, but the effects on the smaller species could have gone either way (Croll et al., Chapter 16 in this volume; Lindberg and Pyenson, Chapter 7 in this volume).

Effects on Competitors

The lack of close phylogenetic relatives, morphologically similar species, or close ecological analogs indicates that the sperm whale has an important effect on potential competitors, close competitors either being extirpated or evolving into specialized niches (as suggested earlier in this chapter). The removal of sperm whales should have benefited potential competitors, but, once again, there is no direct evidence for this. Populations of northern elephant seals (*Mirounga angustirostris*), important mesopelagic predators, increased dramatically during the second half of the twentieth century as sperm whale numbers on their North Pacific feeding grounds were reduced (Le Boeuf and Laws 1994). However, correlation does not prove causation, and some southern elephant seal populations declined over the period of Antarctic sperm whaling (Le Boeuf and Laws 1994).

Effects on the Orca

I have suggested earlier in this chapter that the little available evidence, principally from density-dependent effects, indicates that orca predation has and had little role in regulating sperm whale populations, but this does not necessarily mean that the interaction was also insignificant for orca population dynamics. Even if only a very small proportion of the encounters between orcas and sperm whales result in kills, they could be important for orcas, because there are many sperm whales and they are large. If the orcas kill just sick or disabled animals, which have little reproductive value, or even if they simply scavenge sperm carcasses, there could be a significant energy flow with little or no effect on sperm whale population biology.

To give some idea of the scale of the potential effect, a prewhaling global population of 1,100,000 sperm whales with an average mass of 20 metric tons suggests a standing biomass of 22 Mt. If 0.1% of this was eaten by orcas each year through predation or scavenging, this is 22,000 t/yr, enough to support about 300 orcas (assuming an orca needs an average of about 73 t of sperm whale per year, following the methodology in Doak et al., Chapter 18 in this volume). Thus, what is from the sperm whales' perspective a relatively small amount of predation or scavenging could be an important food source for orcas. However, this calculation includes all waters of the globe—a 300-member sperm whale–eating orca ecotype would have an extremely, and perhaps unsustainably, low density. It is perhaps more likely that sperm whales are only part of the diet of a carnivorous ecotype. Alternatively, if orcas were significant predators on sperm whales, removing perhaps 1% of biomass per year (mean natural mortality rates of sperm whales are perhaps in the range 1%–5%/yr; see Whitehead 2003), then this could support 3,000 orcas worldwide, and more if sperm whales formed only part of the diet.

If there were a set of orcas that depended to a substantial extent on sperm whales as prey, then the whalers' depletion of the sperm whale populations (Figure 25.3) removed a large and dependable source of food. For a consideration of the consequences, see other chapters in this volume.

Other Ecological Effects

There are other animals for which sperm whale carcasses may be important, such as sharks, but it is when a sperm whale carcass descends far beneath the rich photic zone that it becomes a real bonanza (Smith and Baco 2003). Due to their abundance, size, and widespread distribution over deep water, sperm whales may be particularly important in the strange and significant ecology of whale falls (see Croll et al., Chapter 16 in this volume; Smith, Chapter 22 in this volume).

Sperm whales are clearly important for the organisms that live on or in their bodies (see Rice 1989 for a review), but there may be other species with a commensal relationship with sperm whales. Off the Galápagos Islands, the stomachs of waved albatross (Diomedea irrorata) contain remains of squid species that live in far deeper waters than the birds can dive (Harris 1973). Perhaps the birds are eating sperm whale feces or vomitus.

Sperm whales feed at depth and generally defecate at the surface, thus recycling nutrients into the photic zone (Clarke 1977). If ecological processes are constant through space and time, then the significance of the nutrients in sperm whale feces to primary production will be small (see calculations in Katona and Whitehead 1988). However, in dynamic systems, sperm whale feces may be significant to the fundamental processes of biological oceanography. They could prolong a plankton bloom or cause a substantial rise of productivity in nutrient-poor waters to which deep-water squid migrate for mating or other purposes (Katona and Whitehead 1988).

All of these putative ecological services will have diminished as sperm whale populations were reduced by whaling (Figure 25.3).

Conclusion

The sperm whale is an extraordinary animal, strange from almost any perspective, extreme in many. Attributes that contribute to its ecological significance include size, numbers, biomass, and extremely wide distribution—other large whales are generally restricted to continental margins and highly productive, high latitude waters, whereas sperm whales are found almost everywhere. Given its prominence in the ocean, it is remarkable how little we know of the sperm whale's food sources—ocean ecosystems are poorly understood, but the trophic world of mesopelagic is particularly opaque. This lack makes it difficult to address my questions on the regulation of sperm whale populations, the effects of sperm whales on other elements of the environment, and the ecosystem consequences of sperm whaling.

Indirect evidence, particularly from density-dependent effects and the apparent exclusion of competitors, suggest bottom-up regulation of sperm whale populations and that sperm whales have considerable impact on the populations of their prey and competitors. They could also be important in the population biology of their principal natural predator, the orca, and in fundamental ecosystem processes. If I am right in drawing these conclusions, then the whalers' reduction of sperm whale populations by perhaps 68% must have had important consequences for the ecosystems and animals of the deep ocean.

These conclusions are both tentative and broad. There will have been variation with time, space, and other factors. Sperm whales probably reduced squid populations heavily in some places at some times but had little effect at others. Orcas may have regulated sperm whale populations in some areas but not generally. Also, we are now realizing that, as with orcas, sperm whale populations have cultural substructure (Rendell and Whitehead 2003). Within areas, sperm whale groups from different cultural clans use habitat and forage in distinct ways (Whitehead and Rendell 2004). It seems likely that the clans have characteristic ways of interacting with the trophic environment and thus affecting prey, competitors, and predators.

To get a better handle on the principal elements of the ecology of the sperm whale, we need to keep collecting data using the tried tracking and observational methods, but we also need new research techniques: to examine, on the one hand, the small-scale behavior of sperm whales, orcas, and mesopelagic squid, and to monitor, on the other hand, the large-scale distributional dynamics of these ecosystem players. Tags may be key for small-scale studies, satellite imagery for the large-scale, with acoustic arrays and biochemical techniques (such as fatty-acid and isotope ratio methods) providing valuable data on both small-scale interactions and large-scale distributions.

This research-to-come will help address the specific relationships between sperm whales and other species. It will also shed light on some of the broader biological issues raised by this remarkable animal. Has one anatomical innovation (the spermaceti organ complex) changed a whole ecosystem? How does a species with exceedingly low fecundity survive on a food supply that is enormously variable? What is the significance of the mesopelagic to other oceanic ecosystems?

Summary

The sperm whale is extreme or unusual in several characteristics. The development of an uniquely powerful sonar in the spermaceti organ complex may have been key in this evolution to extremes. Although the sperm whale has a catholic diet, its partiality for large squid and its dominant biomass among teuthivores distinguish the sperm whale from most potential competitors. The sperm whale's phylogenetic and morphological distinctiveness, slow life history traits, and some evidence of density-dependent responses during modern whaling, indicate a prewhaling population close to carrying capacity and primarily regulated by intraspecific competition for food. It is thought that predation by orcas was important in the evolution of the sperm whale social system, but the significance of interactions between the two species on the population dynamics of either is unknown. Industrial whaling for sperm whales was an important industry in the eighteenth, nineteenth, and twentieth centuries. Sperm whale populations were certainly reduced, and, consequently, populations of sperm whale prey, competitors, and predators were probably affected. However, the scale, and even the direction, of these effects are unknown.

Literature Cited

Arnbom, T., V. Papastavrou, L.S. Weilgart, and H. Whitehead. 1987. Sperm whales react to an attack by killer whales. *Journal of Mammalogy* 68: 450–453.

Arnold, G.P. 1979. *Squid: a review of their biology and fisheries.* Laboratory Leaflet 48. Lowestoft, UK: MAFF Directorate of Fisheries Research.

Baird, R.W. 2000. The killer whale: foraging specializations and group hunting, in *Cetacean societies: field studies of dolphins and whales.* J. Mann, R.C. Connor, P.L. Tyack, and H. Whitehead, eds. Chicago: University of Chicago Press, pp. 127–153.

Barnes, R.H. 1991. Indigenous whaling and porpoise hunting in Indonesia, in *Cetaceans and cetacean research in the Indian Ocean Sanctuary.* S. Leatherwood and G.P. Donovan, eds. Marine Mammal Technical Report 3. Nairobi: United Nations Environment Programme, pp. 99–106.

Beale, T. 1839. *The natural history of the sperm whale.* London: John van Voorst.

Beran, M.J., K.R. Gibson, and D.M. Rumbaugh. 1999. Predicted hominid performance on the transfer index: body size and cranial capacity as predictors of transfer ability, in *The descent of mind.* M. Corballis and S.E.G. Lea, eds. Oxford: Oxford University Press, pp. 87–97.

Berta, A. and J.L. Sumich. 1999. *Marine mammals: evolutionary biology.* San Diego, CA: Academic Press.

Best, P.B. 1979. Social organization in sperm whales, *Physeter macrocephalus,* in *Behavior of marine animals.* H.E. Winn and B.L. Olla, eds. New York: Plenum Press, pp. 227–289.

———. 1983. Sperm whale stock assessments and the relevance of historical whaling records. *Report of the International Whaling Commission* (Special Issue) 5: 41–55.

Best, P.B., P.A.S. Canham, and N. MacLeod. 1984. Patterns of reproduction in sperm whales, *Physeter macrocephalus. Report of the International Whaling Commission* (Special Issue) 6: 51–79.

Brennan, B. and P. Rodriguez. 1994. Report of two orca attacks on cetaceans in Galápagos. *Noticias de Galápagos* 54: 28–29.

Carrier, D.R., S.M. Deban, and J. Otterstrom. 2002. The face that sank the *Essex:* potential function of the spermaceti organ in aggression. *Journal of Experimental Biology* 205: 1755–1763.

Clarke, M.R. 1977. Beaks, nets and numbers. *Symposia of the Zoological Society, London* 38: 89–126.

———. 1980. Cephalopoda in the diet of sperm whales of the southern hemisphere and their bearing on sperm whale biology. *Discovery Reports* 37: 1–324.

Clarke, M.R., N. MacLeod, and O. Paliza. 1976. Cephalopod remains from the stomachs of sperm whales caught off Peru and Chile. *Journal of Zoology (London)* 180: 477–493.

Clarke, M.R., H.R. Martins, and P. Pascoe. 1993. The diet of sperm whales (*Physeter macrocephalus* Linnaeus 1758) off the Azores. *Philosophical Transactions of the Royal Society of London. (Series B: Biological Sciences)* 339: 67–82.

Clarke, R., A. Aguayo, and O. Paliza. 1980. Pregnancy rates of sperm whales in the southeast Pacific between 1959 and 1962 and a comparison with those from Paita, Peru between 1975 and 1977. *Report of the International Whaling Commission* (Special Issue) 2: 151–158.

Connor, R.C., A.J. Read, and R. Wrangham. 2000. Male reproductive strategies and social bonds, in *Cetacean societies.* J. Mann, R.C. Connor, P.L. Tyack, and H. Whitehead, eds. Chicago: University of Chicago Press, pp. 247–269.

Cranford, T.W. 1999. The sperm whale's nose: sexual selection on a grand scale? *Marine Mammal Science* 15: 1133–1157.

Dufault, S. 1994. A photoidentification study of the geographic stock structure of sperm whales *(Physeter macrocephalus)* in the South Pacific. M.Sc. dissertation, Dalhousie University, Halifax, Nova Scotia, Canada.

Duncan, D.D. 1941. Fighting giants of the Humboldt. *National Geographic* 79: 373–400.

Harris, M.P. 1973. The biology of the waved albatross *Diomedea irrorata* of Hood Island, Galápagos. *Ibis* 115: 483–510.

Hernández-García, V. 1995. The diet of the swordfish *Xiphias gladius* Linnaeus, 1758, in the central east Atlantic, with emphasis on the role of cephalopods. *Fishery Bulletin* 93: 403–411.

Heyning, J.E. 1989. Cuvier's beaked whale *Ziphius cavirostris* G. Cuvier, 1823, in *Handbook of marine mammals*, Vol. 4, S.H. Ridgway and R. Harrison, eds. London: Academic Press. pp. 289–308.

Hooker, S.K., S.J. Iverson, P. Ostrom, and S.C. Smith. 2001. Diet of northern bottlenose whales as inferred from fatty acid and stable isotope analyses of biopsy samples. *Canadian Journal of Zoology* 79: 1442–1454.

IWC. 1980. Report of the Special Meeting on Sperm Whale Assessments, La Jolla, 27 November to 8 December 1978. *Report of the International Whaling Commission* (Special Issue) 2: 107–136.

Jefferson, T.A., P.J. Stacey, and R.W. Baird. 1991. A review of killer whale interactions with other marine mammals: predation to co-existence. *Mammal Review* 21: 151–180.

Kasuya, T. 1991. Density-dependent growth in North Pacific sperm whales. *Marine Mammal Science* 7: 230–257.

Kasuya, T. 1999. Examination of the reliability of catch statistics in the Japanese coastal sperm whale fishery. *Journal of Cetacean Research and Management* 1: 109–122.

Katona, S. and H. Whitehead. 1988. Are Cetacea ecologically important? *Oceanography and Marine Biology Annual Reviews* 26: 553–568.

Kawakami, T. 1980. A review of sperm whale food. *Scientific Reports of the Whales Research Institute* 32: 199–218.

Kojima, T. 1951. On the brain of the sperm whale (*Physeter catodon* L.). *Scientific Reports of the Whales Research Institute* 6: 49–72.

Le Boeuf, B.J. and R.M. Laws. 1994. *Elephant seals: population ecology, behavior, and physiology*. Berkeley: University of California Press.

Lockyer, C. 1977. Observations on diving behaviour of the sperm whale, in *A voyage of discovery*. M. Angel, ed. Oxford: Pergamon, pp. 591–609.

MacLeod, C.D., M.B. Santos, and G.J. Pierce. 2003. Review of data on diets of beaked whales: evidence of niche separation and geographic segregation. *Journal of the Marine Biological Association of the United Kingdom* 83: 651–665.

May, R.M., and J.R. Beddington. 1980. The effect of adult sex ratio and density on the fecundity of sperm whales. *Report of the International Whaling Commission* (Special Issue) 2: 213–217.

McComb, K., C. Moss, S.M. Durant, L. Baker, and S. Sayialel. 2001. Matriarchs as repositories of social knowledge in African elephants. *Science* 292: 491–494.

Mitchell, E. 1977. Evidence that the northern bottlenose whale is depleted. *Report of the International Whaling Commission* 27: 195–203.

Møhl, B., M. Wahlberg, P.T. Madsen, L.A. Miller, and A. Surlykke. 2000. Sperm whales clicks: directionality and source level revisited. *Journal of the Acoustical Society of America* 107: 638–648.

Norris, K.S. and G.W. Harvey. 1972. A theory for the function of the spermaceti organ of the sperm whale (*Physeter catodon* L.), in *Animal orientation and navigation*. S.R. Galler, K. Schmidt-Koenig, G.J. Jacobs, and R.E. Belleville, eds. Washington, DC: NASA Special Publications, pp. 397–417.

Nybakken, J.W. 2001. *Marine biology: an ecological approach*, 5th edition. San Francisco, CA: Benjamin-Cummings.

O'Dor, R.K. 1998. Can understanding squid life-history strategies and recruitment improve management? *South African Journal of Marine Science* 20: 193–206.

Olesiuk, P.F., M.A. Bigg, and G.M. Ellis. 1990. Life history and population dynamics of resident killer whales *(Orcinus orca)* in the coastal waters of British Columbia and Washington State. *Report of the International Whaling Commission* (Special Issue) 12: 209–244.

Palacios, D.M. and B.R. Mate. 1996. Attack by false killer whales *(Pseudorca crassidens)* on sperm whales *(Physeter macrocephalus)* in the Galápagos Islands. *Marine Mammal Science* 12: 582–587.

Papastavrou, V., S.C. Smith, and H. Whitehead. 1989. Diving behaviour of the sperm whale, *Physeter macrocephalus*, off the Galápagos Islands. *Canadian Journal of Zoology* 67: 839–846.

Pauly, D., A.W. Trites, E. Capuli, and V. Christensen. 1998. Diet composition and trophic levels of marine mammals. *ICES Journal of Marine Science* 55: 467–481.

Pitman, R.L., L.T. Ballance, S.L. Mesnick, and S.J. Chivers. 2001. Killer whale predation on sperm whales: observations and implications. *Marine Mammal Science* 17: 494–507.

Poole, J.H. and J. Thomsen. 1989. Elephants are not beetles: implications of the ivory trade for the survival of the African elephant. *Oryx* 23: 188–198.

Rendell, L. and H. Whitehead. 2003. Vocal clans in sperm whales (*Physeter macrocephalus*). *Proceedings of the Royal Society of London Series B: Biological Sciences* 270: 225–231.

Rice, D.W. 1989. Sperm whale. *Physeter macrocephalus* Linnaeus, 1758, in *Handbook of marine mammals*, Vol. 4. S.H. Ridgway and R. Harrison, eds. London: Academic Press, pp. 177–233.

Scott, T.M. and S.S. Sadove. 1997. Sperm whale, *Physeter macrocephalus*, sightings in the shallow shelf waters off Long Island, New York. *Marine Mammal Science* 13: 317–321.

Smith, C.R. and A.R. Baco. 2003. Ecology of whale falls at the deep-sea floor. *Oceanography and Marine Biology: an Annual Review* 41:311–354.

Starbuck, A. 1878. History of the American whale fishery from its earliest inception to the year 1876, in *United States Commission on Fish and Fisheries, Report of the Commissioner for 1875–1876*. Washington, DC: Government Printing Office, Appendix A.

Stearns, S.C. 1992. *The evolution of life histories*. Oxford: Oxford University Press.

Watkins, W.A., M.A. Daher, N.A. DiMarzio, A. Samuels, D. Wartzok, K.M. Fristrup, P.W. Howey, and R.R. Maiefski. 2002. Sperm whale dives tracked by radio tag telemetry. *Marine Mammal Science* 18: 55–68.

Weilgart, L., H. Whitehead, and K. Payne. 1996. A colossal convergence. *American Scientist* 84: 278–287.

Weller, D.W., B. Würsig, H. Whitehead, J.C. Norris, S.K. Lynn, R.W. Davis, N. Clauss, and P. Brown. 1996. Observations of an interaction between sperm whales and short-finned pilot whales in the Gulf of Mexico. *Marine Mammal Science* 12: 588–594.

Whitehead, H. 1996. Variation in the feeding success of sperm whales: temporal scale, spatial scale and relationship to migrations. *Journal of Animal Ecology* 65: 429–438.

———. 2002. Estimates of the current global population size and historical trajectory for sperm whales. *Marine Ecology Progress Series* 242: 295–304.

———. 2003. *Sperm whales: social evolution in the ocean*. Chicago: University of Chicago Press.

Whitehead, H., C.D. MacLeod, and P. Rodhouse. 2003. Differences in niche breadth among some teuthivorous mesopelagic marine mammals. *Marine Mammal Science* 19: 153–159.

Whitehead, H. and L. Rendell. 2004. Movements, habitat use and feeding success of cultural clans of South Pacific sperm whales. *Journal of Animal Ecology* 73: 190–196.

Zemsky, V.A., A.A. Berzin, Yu. A. Mikhaliev, and D.D. Tormosov. 1995. Soviet Antarctic pelagic whaling after WW II: review of actual catch data. *Report of the International Whaling Commission* 45: 131–135.

TWENTY-SIX

Ecosystem Effects of Fishing and Whaling in the North Pacific and Atlantic Oceans

BORIS WORM, HEIKE K. LOTZE, AND RANSOM A. MYERS

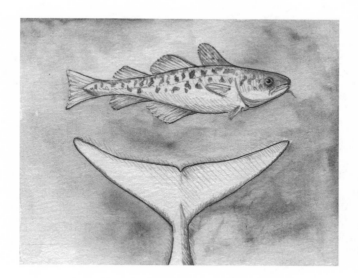

Human alterations of marine ecosystems have occurred throughout history, but only over the last century have these reached global proportions. Three major types of changes have been described: (1) the changing of nutrient cycles and climate, which may affect ecosystem structure from the bottom up, (2) fishing, which may affect ecosystems from the top down, and (3) habitat alteration and pollution, which affect all trophic levels and therefore were recently termed side-in impacts (Lotze and Milewski 2004). Although the large-scale consequences of these changes for marine food webs and ecosystems are only beginning to be understood (Pauly et al. 1998; Micheli 1999; Jackson et al. 2001; Beaugrand et al. 2002; Worm et al. 2002; Worm and Myers 2003; Lotze and Milewski 2004), the implications for management are often profound (Lotze 2004).

Fishing and whaling were arguably the first massive human-induced alteration of the marine environment, preceding other impacts such as pollution and climate change (Jackson et al. 2001). What were the ecosystem impacts of removing millions of large whales from the ocean (Katona and Whitehead 1988)? Because much of the changes occurred in past centuries (Reeves and Smith, Chapter 8 in this volume), we may never know with absolute certainty, but we can formulate hypotheses based on (1) what we know

about the role of whales in the food web and (2) what has been observed in other species playing a similar role. Then we may explore whether the available evidence supports these hypotheses. Experiments and detailed observations in lakes, streams, and coastal and shelf ecosystems have shown that the removal of large predatory fishes or marine mammals almost always causes release of prey populations, which often set off ecological chain reactions such as trophic cascades (Estes and Duggins 1995; Micheli 1999; Pace et al. 1999; Shurin et al. 2002; Worm and Myers 2003; Frank et al. 2005). Another important interaction is competitive release, in which formerly suppressed species replace formerly dominant ones that were reduced by fishing (Fogarty and Murawski 1998; Myers and Worm 2003). Although both prey release and competitive release appear to be general ecosystem effects of fishing, it is unclear how these processes apply to whales and whaling.

In this paper we use time-series analysis to explore whether prey release and competitive release have occurred during or shortly after the period of industrial whaling (from 1950 to 1980) in the Northwest Atlantic and Northeast Pacific Oceans, respectively. We focus on these two regions because they are well known, food webs are relatively simple, and there are data available on whale abundance, diets, and the abundance

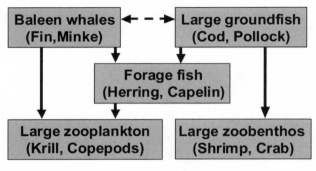

FIGURE 26.1. Predator-prey (solid arrows) and potential competitive (broken arrows) relationships between whales, groundfish, forage fish, benthic and pelagic invertebrates. Species names represent the focal species discussed in this chapter.

of prey and competitor populations. Unfortunately, time-series abundance data predating the 1950s are very scarce. Therefore, at this point, we were not able to explore the long-term (century-scale) effects of whaling, which began as early as the eleventh century in the Atlantic (Clapham and Link, Chapter 24 in this volume). Also, we caution that short time series such as those that are available to us bear a number of statistical problems, such as low sample size and temporal autocorrelation.

The focal species of this chapter and their interactions are shown in Figure 26.1. Diet data indicate that baleen whales feed primarily on small schooling fish (often called forage fish), such as herring *(Clupea* spp.), capelin *(Mallotus villosus)*, or sand lance *(Ammodytes* spp.), and on zooplankton such as krill or calanoid copepods (Kawamura 1980). Forage fish and benthic invertebrates (such as northern shrimp, *Pandalus borealis,* or snow crab, *Chionoecetes opilio*) represent the primary diet of large groundfish such as Atlantic cod *(Gadus morhua)* and wall-eye pollock *(Theragra chalcogramma)* (Lilly 1984; Dwyer et al. 1987; Link and Garrison 2002). Because of diet overlap with respect to forage fish, groundfish and baleen whales may potentially compete for food in regions where both reach high abundance (Figure 26.1). Exploitation competition for food could be exacerbated by interference competition, in which whales (for example, minke whales; Lindstrøm et al. 1998) feed on groundfish. This may be particularly important in the Bering Sea, where school-forming juvenile walleye pollock represent an important forage fish species for marine mammals (Brodeur et al. 1996; Merrick 1997).

Recent studies have shown that the biomasses of groundfish and their prey generally show countervailing abundance trends and strong negative correlations in the Northwest Atlantic and Northeast Pacific, respectively (Figure 26.2). This was true particularly for the Northern shrimp (Figure 26.2A, B), which is an important prey species of groundfish such as cod, but trends were also evident for other crustaceans such as snow crab (Figure 26.2D) and forage fishes such as herring and capelin (Figure 26.2C). Strong predation or top-down effects should result in such a negative correlation between predator and prey, because predators suppress prey

abundance (McQueen et al. 1989). Most predator-prey models, other than donor-controlled models (Pimm 1991), predict such a negative relationship if strong predator-prey linkages are assumed. Although trends in the North Atlantic were clearly indicative of a predation effect (Worm and Myers 2003), large concurrent changes in climate in the North Pacific may render the situation there more complex (Anderson and Piatt 1999). Note, however, that community changes shown in Figure 26.2 were already well under way when the main climate shift occurred in 1977. An interesting contrast between the two regions is that groundfish abundance has increased and reached high levels in the North Pacific, whereas it has collapsed to low levels in most of the North Atlantic due to overfishing (Myers et al. 1996). In both cases, prey species such as benthic invertebrates and forage fishes showed the opposite pattern to groundfish, which supports the hypothesis of a general top-down linkage (Figure 26.2).

Based on this top-down hypothesis, we predict that there may be negative correlations between whale and forage fish abundance (prey release), between groundfish and forage fish abundance (prey release), and possibly between whale and groundfish abundance (competitive release) during or shortly after the period of industrial whaling in the Northwest Atlantic and Northeast Pacific.

Methods

We assembled time-series data of selected forage fish, groundfish, and baleen whale species abundances for the Southern Grand Banks (NW Atlantic, NAFO region 3 NO) and the Bering Sea (NE Pacific, PICES region BSC) from 1950 to 1980, when industrial whaling for large baleen whales occurred (Table 26.1). Inclusion of the period before 1950 would be desirable but would require a different approach, because time-series data are not readily available.

For whales, we focused on fin *(Balaenoptera physalus)* and minke whales *(B. acutorostrata)* because fin and minke whales represented the majority of catches between 1950 and 1975 in the NW Atlantic, and fin whales dominated the catch of baleen whales in the NE Pacific. In contrast to other baleen whales that were harvested concurrently (such as blue and sei whales), fin and minke are considered important consumers of forage fishes in these regions (Kenney et al. 1997; Lindstrøm et al. 1998). We used current fin and minke whale abundances for the Bering Sea/Aleutian region as given by Pfister and DeMaster (Chapter 10 in this volume). These estimates are preliminary but the best available data at this point (see Pfister and DeMaster, Chapter 10 in this volume, for a detailed discussion of caveats). Minke whale abundance was assumed stable over time in the Pacific (Pfister and DeMaster, Chapter 10 in this volume), and fin whale abundance was assumed to have declined substantially as a result of whaling (Danner et al., Chapter 11 in this volume). For the Atlantic, we used the published estimates for the NW Atlantic fin and Canadian East Coast

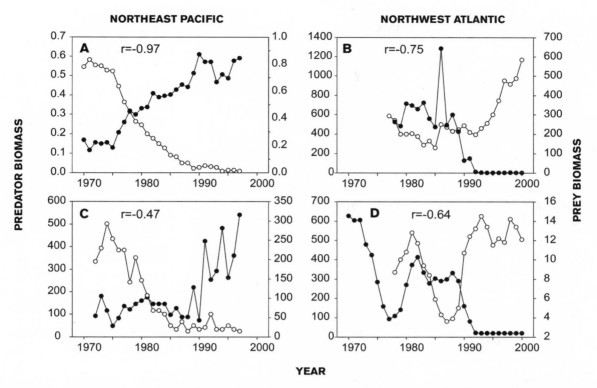

FIGURE 26.2. Strongly inverse abundance trends of predator (black circles) and prey populations (white circles) in the North Atlantic (Newfoundland Shelf) and North Pacific (Gulf of Alaska). Time series represent biomass estimates for total groundfish and Northern shrimp *(Pandalus borealis)* (A), Atlantic cod *(Gadus morhua)* and Northern shrimp (B), total groundfish and forage fish (herring, *Clupea pallasi,* and capelin, *Mallotus villosus*) (C) and groundfish and snow crab *(Chionoecetes opilio)* (D), respectively. Strong negative correlations may be interpreted as evidence for general top-down effects of large groundfish on prey populations. All data except for snow crab are based on research surveys. Units are: survey biomass as kg per tow (A, C), total biomass in thousands of metric tons (B, D: groundfish), and catch per unit of effort (CPUE; catch per trap) (D: snow crab). Modified after data from Anderson and Piatt (1999), Worm and Myers (2003), and R. A. Myers, B. Worm, and W. Blanchard (unpublished manuscript).

TABLE 26.1

Data Used for Analysis

	Species	Method	Unit	Source
NE Pacific				
Whales	*Balaenoptera physalus, B. acutorostrata*	Reconstruction from catches	Individuals	IWC
Groundfish	*Theragra chalcogramma*	Biomass (SPA)	1000 mt	Myers et al. 1995
Forage fish	*Clupea pallasi*	Biomass (SPA)	1000 mt	Myers et al. 1995
NW Atlantic				
Whales	*Balaenoptera physalus, B. acutorostrata*	Reconstruction from catches	Individuals	IWC
Groundfish	*Gadus morhua, Melanogrammus aeglefinus*[a]	Biomass (Survey)	1000 mt	Casey 2000
Forage Fish	*Ammodytes* sp., *Mallotus villosus*	Biomass (Survey)	Proportion of tows	Casey 2000

[a] NW Atlantic groundfish also included several other Gadidae, Pleuronectidae, Rajidae, Scorpaenidae, and Anarhichadidae.

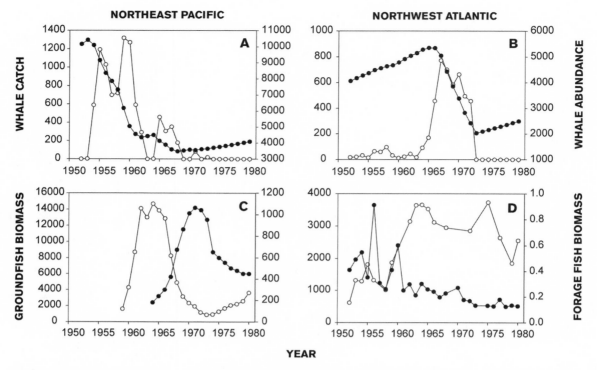

FIGURE 26.3. Trajectories of reconstructed fin and minke whale abundance (A, B, black circles), and catches (A, B, white circles), groundfish abundance (C: walleye pollock, D: Atlantic cod and haddock, black circles) and forage fish abundance (C: Pacific herring, D: capelin and sand lance, white circles) in the Bering Sea, Northeast Pacific and the Newfoundland shelf, Northwest Atlantic, 1950–1980. For data description refer to Table 26.1; for analysis, refer to Table 26.2. Units are numbers of whales (A, B), total biomass as thousands of metric tons (C, D: groundfish, herring), and survey catch per unit effort (CPUE) as proportions of tows (D: capelin and sand lance).

minke whale populations (Waring et al. 2001). These estimates are mostly based on direct ship and flight transect survey counts and should also be considered preliminary. Because no time-series abundance data were available for large whales in these regions, we back-calculated probable trends in whale numbers from current abundances and catches using minimal assumptions. Whale abundance X at time t was reconstructed from recent estimates using

$$X_t = (X_{t+1} + C_t)(1-r), \qquad (26.1)$$

where we assumed that the catch C was known without error, and the specific rate of population growth r was derived from published values for Northern Hemisphere baleen whales, as compiled by the IWC (http://www.iwcoffice.org/estimate.htm). We assumed that the population was sufficiently low during this period that density dependence was not important. This assumption is consistent with the strong convexity hypothesis: that density dependence for cetaceans occurs only at or close to carrying capacity (Fowler 1981; Best 1993). Catches were compiled from the International Whaling Commission (IWC) database for the Bering Sea (Japanese whaling operations) and the East Canadian Shelf (Newfoundland and Nova Scotian operations) regions, respectively. The resulting time series likely underestimate total declines, at least

in the Bering Sea, as only Japanese catch data are available at this point, but other nations are known to have operated there as well. The resulting time series are not meant to represent accurate representations of historic trajectories but rather a simple representation of how catches may have affected trends in whale abundance over the period of industrial whaling.

For selected forage and groundfish species, abundance time series were compiled from published sources (Table 26.1). For the Grand Banks, data on the abundance of large groundfish (mostly Atlantic cod and haddock), and forage fish (capelin, sand lance) from standardized research trawl surveys were used (Casey 2000). For the Bering Sea, abundance estimates from published stock assessments for walleye pollock and Pacific herring were used. These stock assessments used sequential population analysis (SPA) of catch and survey data to estimate changes in abundance over time. Data and assessment details are available from Ransom Myers's Stock-Recruit Database (Myers et al. 1995), which is available online at http:// fish.dal.ca.

Relationships between time series were analyzed using standard techniques for partial correlation. We used partial correlation in order to control for the confounding effects of changes in groundfish abundance when testing effects of whales on forage fish and vice versa. All time series data were log-transformed. Time series showed intermediate to strong

TABLE 26.2
Partial Correlation Analysis

	Slope	r	t	P	P'^a	Model r	Adjusted r^2
NE Pacific							
Dependent: Forage fish biomass							
Whales	−0.694	−0.577	−2.55	0.0243	0.4430	0.862	0.704
Groundfish	−1.379	−0.815	−5.07	0.0002	0.1850		
Dependent: Groundfish biomass							
Whales	−0.622	−0.876	−6.55	<0.0001	0.3201	0.954	0.897
Forage fish	−0.481	−0.815	−5.06	0.0002	0.1850		
NW Atlantic							
Dependent: Forage fish biomass							
Whales	0.421	0.416	1.71	0.1090	0.7269	0.675	0.378
Groundfish	−0.837	−0.673	−3.40	0.0043	0.3280		
Dependent: Groundfish biomass							
Whales	0.555	0.682	3.49	0.0036	0.3180	0.805	0.598
Forage fish	−0.541	−0.673	−3.40	0.0043	0.3280		

[a] Note that P' values were corrected for autocorrelation.

autocorrelation ($r = 0.6 - 0.96$ at 1-year lag) and correction for autocorrelation was performed using the modified Chelton method as described by Pyper and Peterman (1998).

Results

Whale catches increased sharply through the 1950s and 1960s in the Bering Sea (Figure 26.3A) and NW Atlantic (Figure 26.3B) but then ended abruptly in 1972 and 1975, respectively, because of legal protection from commercial whaling. While catches were high, whale abundance likely declined steeply in the 1950s and 1960s and may have started to recover in the 1970s, after commercial whaling had ceased (Figure 26.3A, B). Total reported removals of fin and minke whales were 9,114 whales from the Bering Sea (Japanese whaling) and 5,768 whales from the East Canadian Shelf (Newfoundland and Nova Scotia whaling).

While catches were high and whale abundance declined in the Bering Sea, first herring and then walleye pollock abundance increased by an order of magnitude (Figure 26.3C). As pollock spawning stock biomass reached its historic peak of 14 million metric tons, herring declined to low levels but started to recover when pollock biomass declined again in the 1970s. Partial correlations revealed strong negative relationships between whales and herring ($r_{partial} = -0.58$), pollock and herring ($r_{partial} = -0.82$), and whales and pollock ($r_{partial} = -0.88$), respectively (Table 26.2). These relationships were statistically significant ($P < 0.05$) only if autocorrelation of time series was ignored. If we corrected for autocorrelation,

true sample sizes declined to $n \approx 4$, and significance levels increased to $P > 0.05$ (Table 26.2).

In the NW Atlantic, groundfish biomass was high in the 1950s but then declined gradually from 1950 to 1980, while capelin and sand lance abundance showed an inverse pattern in abundance (Figure 26.3D). Forage fish abundance was weakly positively correlated with whale abundance ($r_{partial} = 0.42$) and strongly negatively correlated with groundfish biomass ($r_{partial} = -0.67$). Groundfish abundance was positively related to whale abundance ($r_{partial} = 0.68$). Again, formal significance of the latter two relationships broke down when temporal autocorrelation was accounted for (Table 26.2).

Overall, partial correlations among whales, groundfish, and forage fish could explain 83%–91% of total variance in the NE Pacific but only 38%–60% of variance in the NW Atlantic. A sensitivity analysis revealed that negative correlations in both data sets remained stable or became stronger when only years with whale removals (1952–1973) were analyzed. Positive correlations between whales and groundfish in the NW Atlantic weakened considerably in this alternative analysis ($r_{partial} = 0.45$).

Discussion

The evidence discussed in this chapter is in accordance with a general top-down effect of groundfish on forage species (schooling fish and benthic invertebrates), but the effects of whales on forage species appeared more ambiguous and

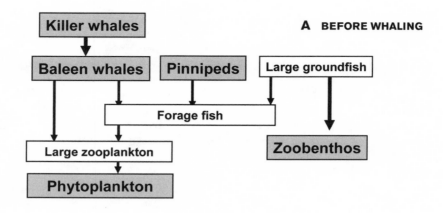

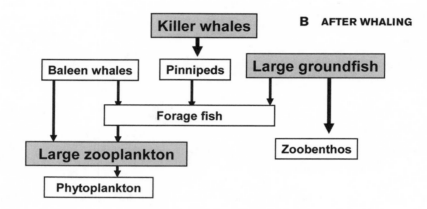

FIGURE 26.4. Hypothetical effects of industrial whaling on some major components of the Bering Sea ecosystem. Functionally dominant groups are shaded. Prewhaling (before 1950) the ecosystem is dominated by marine mammals; postwhaling (after 1972) it is dominated by groundfish.

differed markedly between the NW Atlantic and NE Pacific. This difference is likely explained by the contrasting exploitation history of the two regions: Massive whaling operations had occurred before 1950 in the NW Atlantic but not to the same extent in the NE Pacific.

In the NE Pacific, the available data indicate large declines of whale biomass in the 1960s, which may have triggered subsequent increases in herring and walleye pollock, the latter reducing herring again to low levels. Pollock increased more slowly than herring, possibly because of its lower intrinsic rate of increase and higher age at maturity. All species showed strong negative correlations, the whale-pollock relationship being the strongest, explaining 77% of the variance in pollock abundance. Thus, whaling may have contributed to a long-term shift from marine mammal to groundfish dominance in this ecosystem (Figure 26.4). It is important to note that the removal of whales coincided with massive removals of pinnipeds such as fur seals and Steller sea lions from the same region, which may have accentuated the shift from marine mammals to fish in this ecosystem (Merrick 1997). Killer whales, switching from whales to

alternative prey species (Figure 26.4), may have further escalated the decline of pinnipeds and, later, sea otters in this region (Estes et al. 1998; Springer et al. 2003). Finally, if top-down effects were general, we hypothesize that the decline of whales may have increased large-zooplankton biomass, possibly causing phytoplankton abundance to decline, and the increase in groundfish may have reduced zoobenthic biomass (Figure 26.4).

In the NW Atlantic a gradual depletion of both whales and groundfish occurred from the 1950 to the 1970s, and no dominance shift was observed. Capelin and sand lance increased in the 1960s, but the partial correlation analysis suggested that these increases were explained by release from groundfish (not whale) predation (Table 26.2). Indeed, whales showed weak positive correlations with forage fish and groundfish, which does not support the top-down hypothesis. Positive linkages could indicate a weak bottom-up effect (such as similar response to climate variation) or that humans exploited all groups at the same time, and trajectories co-vary for that reason. Ultimately, the striking differences observed between the two regions (NE Pacific,

NW Atlantic) may be explained by their different exploitation history. The NW Atlantic has been exploited heavily for many decades, and intense industrial whaling and a large-scale groundfishery were in place there long before 1950 (Lear 1997). In contrast, the Bering Sea experienced fewer removals of baleen whales before 1950 (but note that right, bowhead, humpback, and gray whales were hunted before 1950), and an industrial-scale fishery for walleye pollock emerged only in the 1960s and 1970s. Moreover, this latter fishery was managed relatively conservatively, avoiding the depletion seen in the NW Atlantic and its associated ecosystem effects.

Of course, a correlative historical analysis like this has many caveats. First, reconstructions of past whale abundances rely heavily on simplifying assumptions and can most likely only reproduce broad trends. There are also unresolved questions with respect to current abundance estimates and the completeness of whaling records, among others (Danner et al., Chapter 11 in this volume; Pfister and DeMaster, Chapter 10 in this volume). Secondly, correlations are one way to summarize patterns in nature, but we cannot use them to infer causality. Third, short time series offer only limited information, particularly if they are heavily autocorrelated. Still, they may allow us to examine trends and to formulate interesting hypotheses, such as the dominance shift hypothesis outlined in Figure 26.4. Finally, we are presenting data from only two regions, which represent an effective sample size of $n = 2$. A broader analysis would combine results from a larger set of species from many regions, allowing for a proper meta-analysis of species interactions, as presented by Micheli (1999) and Worm and Myers (2003). Unfortunately, long-term time-series data are not readily available for forage fish and whales in particular. Carefully constructing such data sets from historical records represents an important challenge (Reeves and Smith, Chapter 8 in this volume).

A largely open question is how the effects of fishing and whaling propagate further through the food web. Increases in forage fishes due to removal of whales (NE Pacific) and groundfish (NW Atlantic) could have had strong effects on zooplankton and phytoplankton. Evidence for such pelagic trophic cascades, for example in the NW Atlantic (Carscadden et al. 2001; Frank et al. 2005) and NE Pacific (Shiomoto et al. 1997) is accumulating, but arguments are often based on weak evidence. In a comprehensive meta-analysis of 20 pelagic food webs, Micheli (1999) found strong evidence for a fish-zooplankton link but no general effect on phytoplankton. Another potential effect of the increased forage fish biomass could be that humans switched from whales to forage fish, as the whales become scarce. Indeed, large purse-seine fisheries for herring and capelin developed in the NW Atlantic and NE Pacific in the 1960s and 1970s, partly replacing the role of whales as consumers of forage fishes. This pattern of serial depletion of predators and their prey has been documented as a general pathology of global fisheries (Pauly et al. 1998).

Conclusions

This paper is about hypotheses. Although we will likely never know for sure what the precise ecosystem effects of past fishing and whaling were, a careful exploration of the topic may teach us some important rules about managing marine ecosystems. Here, we suggest that exploitation can cause trophic cascades from humans to fish and invertebrates and that evidence for strong top-down linkages is often seen, where data exist and where they are carefully analyzed. Although many of the underlying data sets undoubtedly have serious problems, it is the collective weight of the evidence that points toward a more general phenomenon. Marine ecosystems can change rapidly in response to perturbations and may shift to new stable states, as seen in the Bering Sea example. They also may completely reorganize, leading to outbreaks of some species and near-extirpation of others (Scheffer et al. 2001). These shifts could be due to whaling, fishing, climate change, or a combination of these factors, all of which are partly or entirely anthropogenic and therefore within our responsibility. From the present analysis it appears at least conceivable that industrial whaling in the Bering Sea, 1950–1975, has contributed to, and possibly initiated, a shift in food web structure from marine mammals to groundfish, which may have been further emphasized by fisheries management and climate instability, as suggested by Merrick (1997). The lack of such clear responses during the same time period in the NW Atlantic could be interpreted as an indication that baleen whales were ecologically less important in this ecosystem. However, the North Atlantic has the longest whaling record of any ocean, going back to the eleventh century (Clapham and Link, Chapter 24 in this volume). At that point in time, whales may have well been dominant, and perhaps only their decline over the following centuries gave rise to the great abundance of fish that made this region world-famous. However, industrial fishing again transformed that situation. The recent collapses of cod and other gadoids in the NW Atlantic had clear ecosystem effects, such as massive increases in forage species (Fogarty and Murawski 1998; Worm and Myers 2003) and possible shifts to new stable states (Swain and Sinclair 2000). The lesson may be that the effects of whaling can be understood only as part of a bigger, emerging picture: the destabilizing effects of removing most large predators, such as whales, seals, turtles, bony fishes, sharks, and rays, effectively terminating their roles as key functional components of marine ecosystems.

Acknowledgments

We thank J. Casey, E. Danner, N. Friday, M. Kaufmann, and S. Moore for providing data, and W. de la Mare, J. Estes, and the participants of the "Ecosystem Effects of Whaling" workshop for discussion and insight. We acknowledge support from the Deutsche Forschungsgemeinschaft (B.W.), and the

Natural Sciences and Engineering Research Council of Canada (R.A.M.).

Literature Cited

Anderson, P.J. and J.F. Piatt. 1999. Community reorganization in the Gulf of Alaska following ocean climate regime shift. *Marine Ecology Progress Series* 189: 117–123.

Beaugrand, G., P.C. Reid, F. Ibañez, J.A. Lindley, and M. Edwards. 2002. Reorganization of North Atlantic marine copepod diversity and climate. *Science* 296: 1692–1694.

Best, P.B. 1993. Increase rates in severely depleted stocks of baleen whales. *ICES Journal of Marine Science* 50: 169–186.

Brodeur, R.D., P.A. Livingston, T.R. Loughlin, and A.B. Hollowed. 1996. *Ecology of juvenile walleye pollock.* NOAA Technical Report 126. Washington, DC: National Marine Fisheries Service.

Carscadden, J.E., K.T. Frank, and W.C. Leggett. 2001. Ecosystem changes and the effects on capelin *(Mallotus villosus)*, a major forage species. *Canadian Journal of Fisheries and Aquatic Sciences* 58: 73–85.

Casey, J.M. 2000. Fish assemblages on the Grand Banks of Newfoundland. M.Sc. thesis, Memorial University of Newfoundland, St. John's, Newfoundland, Canada.

Dwyer, D.A., K.M. Bailey, and P.A. Livingston. 1987. Feeding habits and daily ration of walleye pollock *(Theragra chalcogramma)* in the eastern Bering Sea, with special reference to cannibalism. *Canadian Journal of Fisheries and Aquatic Sciences* 44: 1972–1984.

Estes, J.A. and D.O. Duggins. 1995. Sea otters and kelp forests in Alaska: generality and variation in a community ecological paradigm. *Ecological Monographs* 65: 75–100.

Estes, J.A., M.T. Tinker, T.M. Williams, and D.F. Doak. 1998. Killer whale predation on sea otters linking oceanic and nearshore ecosystems. *Science* 282: 473–476.

Fogarty, M.J. and S.A. Murawski. 1998. Large-scale disturbance and the structure of marine systems: fisheries impacts on Georges Bank. *Ecological Applications* 8 (Supplemental): 175–192.

Fowler, C.W. 1981. Density dependence as related to life history strategies. *Ecology* 62: 602–610.

Frank, K.T., B. Pietrie, J.S. Choi, and W.C. Leggett. 2005. Trophic cascades in a formerly cod-dominated ecosystem. *Science* 308: 1621–1623.

Jackson, J.B.C., M.X. Kirby, W.H. Berger, K.A. Bjorndal, L.W. Botsford, B.J. Bourque, R. Bradbury, R. Cooke, J. Erlandson, J.A. Estes, T.P. Hughes, S. Kidwell, C.B. Lange, H.S. Lenihan, J.M. Pandolfi, C.H. Peterson, R.S. Steneck, M.J. Tegner, and R. Warner. 2001. Historical overfishing and the recent collapse of coastal ecosystems. *Science* 293: 629–638.

Katona, S. and H. Whitehead. 1988. Are Cetacea ecologically important? *Oceanography and Marine Biology Annual Reviews* 26: 553–568.

Kawamura, A. 1980. A review of food of balaenopterid whales. *Scientific Reports of the Whales Research Institute* 32: 155–197.

Kenney, R.D., G.P. Scott, T.J. Thompson, and H.E. Winn. 1997. Estimates of prey consumption and trophic impacts of cetaceans in the USA Northeast continental shelf ecosystem. *Journal of Northwest Atlantic Fishery Science* 22: 155–171.

Lear, W.H. 1997. History of fisheries in the Northwest Atlantic: the 500-year perspective. *Journal of Northwest Atlantic Fishery Science* 23: 41–73.

Lilly, G.R. 1984. *Predation by Atlantic cod on shrimps and crabs off northeastern Newfoundland in autumn of 1977–82.* ICES CM 1984/G:53. Copenhagen: International Council for the Exploration of the Sea.

Lindstrøm, U., Y. Fujise, T. Haug, and T. Tamura. 1998. Feeding habits of western North Pacific minke whales, *Balaenoptera acutorostrata*, as observed in July–September 1996. *Report of the International Whaling Commission* 48: 463–469.

Link, J.S. and L.P. Garrison. 2002. Trophic ecology of Atlantic cod *Gadus morhua* on the northeast US continental shelf. *Marine Ecology Progress Series* 227: 109–123.

Lotze, H.K. 2004. Repetitive history of resource depletion and mismanagement: the need for a shift in perspective. *Marine Ecology Progress Series* 274: 282–285.

Lotze, H.K. and I. Milewski. 2004. Two centuries of multiple human impacts and successive changes in a North Atlantic food web. *Ecological Applications* 14: 1428–1447.

McQueen, D.J., M.R.S. Johannes, J.R. Post, D.J. Stewart, and D.R.S. Lean. 1989. Bottom-up and top-down impacts on freshwater pelagic community structure. *Ecological Monographs* 59: 289–309.

Merrick, R.L. 1997. Current and historical roles of apex predators in the Bering Sea ecosystem. *Journal of Northwest Atlantic Fisheries Science* 22: 343–355.

Micheli, F. 1999. Eutrophication, fisheries, and consumer-resource dynamics in marine pelagic ecosystems. *Science* 285: 1396–1398.

Myers, R.A., J. Bridson, and N.J. Barrowman. 1995. *Summary of worldwide stock and recruitment data.* Canadian Technical Report of Fisheries and Aquatic Sciences 2024. Ottawa: Fisheries and Oceans Canada.

Myers, R.A., J.A. Hutchings, and N.J. Barrowman. 1996. Hypotheses for the decline of cod in the North Atlantic. *Marine Ecology Progress Series* 138: 293–308.

Myers, R.A. and B. Worm. 2003. Rapid worldwide depletion of predatory fish communities. *Nature* 423: 280–283.

Pace, M.L., J.J. Cole, S.R. Carpenter, and J.F. Kitchell. 1999. Trophic cascades revealed in diverse ecosystems. *Trends in Ecology and Evolution* 14: 483–488.

Pauly, D., V. Christensen, J. Dalsgaard, R. Froese, and F.C. Torres, Jr. 1998. Fishing down marine food webs. *Science* 279: 860–863.

Pimm, S.L. 1991. *The balance of nature?* Chicago: University of Chicago Press.

Pyper, B.J. and R.M. Peterman. 1998. Comparison of methods to account for autocorrelation in correlation analysis of fish data. *Canadian Journal of Fisheries and Aquatic Sciences* 55: 2127–2140.

Scheffer, M., S. Carpenter, J.A. Foley, C. Folke, and B. Walker. 2001. Catastrophic shifts in ecosystems. *Nature* 413: 591–596

Shiomoto, A., K. Tadokoro, K. Nagasawa, and Y. Ishida. 1997. Trophic relations in the subarctic North Pacific ecosystem: possible feeding effects from pink salmon. *Marine Ecology Progress Series* 150: 75–85.

Shurin, J.B., E.T. Borer, E.W. Seabloom, K. Anderson, C.A. Blanchette, B. Broitman, S.D. Cooper, and B.S. Halpern. 2002. A cross-ecosystem comparison of the strength of trophic cascades. *Ecology Letters* 5: 785–791.

Springer, A.M., J.A. Estes, G.B. van Vliet, T.M. Williams, D.F. Doak, E.M. Danner, K.A. Forney, and B. Pfister. 2003.

Sequential megafaunal collapse in the North Pacific Ocean: an ongoing legacy of industrial whaling? *Proceedings of the National Academy of Sciences* 100: 12223–12228.

Swain, D.P., and A.F. Sinclair. 2000. Pelagic fishes and the cod recruitment dilemma in the Northwest Atlantic. *Canadian Journal of Fisheries and Aquatic Sciences* 57: 1321–1325.

Waring, G.T., J.M. Quintal, and S.L. Swartz. 2001. *U.S. Atlantic and Gulf of Mexico marine mammal stock assessments 2001.* NOAA Technical Memorandum NMFS-NE-168. Woods Hole, MA: Northeast Fisheries Science Center.

Worm, B., H.K. Lotze, H. Hillebrand, and U. Sommer. 2002. Consumer versus resource control of species diversity and ecosystem functioning. *Nature* 417: 848–851.

Worm, B. and R.A. Myers. 2003. Meta-analysis of cod-shrimp interactions reveals top-down control in oceanic food webs. *Ecology* 84: 162–173.

TWENTY-SEVEN

Potential Influences of Whaling on the Status and Trends of Pinniped Populations

DANIEL P. COSTA, MICHAEL J. WEISE, AND JOHN P. Y. ARNOULD

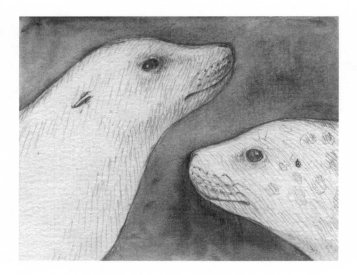

Although this volume focuses on whales and whaling, the depletion of great whales over the last 50 to 150 years perturbed the marine interaction web, thus influencing many other species and ecosystem processes (Estes, Chapter 1 of this volume; Paine, Chapter 2 of this volume). Such interaction web effects have been hypothesized for several pinniped species. For example, the reduction of great whales in the Southern Ocean may have caused seal and penguin populations to increase because of reduced competition for their shared prey, krill (Laws 1977; Ballance et al., Chapter 17 of this volume). In addition, pinnipeds share some of the same predators, especially killer whales, as large whales do. Declines in whale populations may thus have caused the decline of certain pinniped populations because of redirected predation by killer whales (Springer et al. 2003; Branch and Williams, Chapter 20 of this volume). These purported indirect effects of whales on pinnipeds are poorly documented and controversial. Since most of the arguments are area- or species-specific, a global overview of the known patterns and causes for pinniped population change is topical and relevant.

Because of differences in body size and life history, pinnipeds are both easier to study, and possibly more sensitive to environmental fluctuations, than most cetaceans

are. Pinnipeds are 1 to 2 orders of magnitude smaller in mass than whales, which result in greater mass-specific rates of food consumption. Thus the pinnipeds have physiological and environmental scaling functions that must be considerably different from those of the great whales. For example, although some pinnipeds have remarkable abilities to fast, even the most extreme durations of fasting in pinnipeds fall easily within the abilities of large cetaceans. The relatively small size of pinnipeds compared with cetaceans results in a much higher mass-specific metabolism and thus a shorter fasting duration. These differences should constrain pinnipeds to operate at smaller spatial and temporal scales than the large cetaceans, thus making pinnipeds more sensitive to variations in prey abundance and distribution. Smaller size is also linked to a shorter generation time in pinnipeds, which makes their populations more vulnerable to environmental disturbances but also affords them a greater potential for population growth. All of these characteristics suggest that pinniped populations should be more responsive to changes in their environment than the large whales are.

Pinnipeds have a nearly cosmopolitan distribution in the world oceans, although most species occur in temperate to polar regions. Abundances range across species from a few hundred to tens of millions of individuals. Estimates of abundance or trends in population numbers are the most useful indicators of population status. Most populations were severely depleted by commercial harvesting. However, species distributions and population abundances before sealing are often unknown, because sealing ships did not keep adequate records. Furthermore, reliable modern abundance estimates are lacking for many species. Despite these problems, the history and trends in abundance of the majority of pinnipeds is reasonably well known.

In this chapter we review the current status and trends of pinniped populations worldwide, and, where possible, we summarize the known or suspected reasons for recent declines. Trends in pinniped populations attributed to natural biological processes are evaluated in terms of reproductive strategies, physiological limitations, and the resultant susceptibility to disturbance in prey resources and predation brought about by these factors.

Pinniped Population Trends

The present-day abundances of species do not always reflect their pre-exploitation numbers. Some species that were decimated to near-extinction are now very abundant, whereas others have either not recovered or have recovered and subsequently declined. Population abundance in pinnipeds ranges over four orders of magnitude across species from the Mediterranean and Hawaiian monk seals, which number in the hundreds of individuals, to the crabeater seal with an estimated abundance of 10 to 15 million individuals (Table 27.1). Phocids are generally more abundant than otariids. Fifteen of the 19 phocid species number greater than 100,000 individuals, whereas only 8 of the 17 otariid species number greater than 100,000 individuals.

Pinnipeds range throughout the world oceans. Although the preponderance of species occurs in the northern hemisphere (Figures 27.1 and 27.2), the southern hemisphere contains far more individuals. The abundance of crabeater and Antarctic fur seals alone exceeds the combined abundance of all northern hemisphere species. The lesser number of species in the southern hemisphere may reflect a northern hemisphere center of origin for otariids and phocids (Costa 1993; Demere 1994; Demere et al. 2003). The larger numbers of individuals in the southern hemisphere likely result from highly productive Antarctic and sub-Antarctic waters coupled with an abundance of predator-free islands. The relative scarcity of human settlements (which invariably lead to habitat loss, direct and indirect pinniped/fisheries interactions, and hunting pressure) may also contribute to the larger sizes of southern hemisphere pinnipeds. The relative abundance of phocids is likely due to their generally inhabiting the highly productive polar and subpolar regions (Bowen 1997). Similarly, the three most abundant otariid species—the northern, Antarctic,

and Cape fur seals—all forage in seasonally productive, high-latitude ecosystems.

Phocid Population Trends

ARCTIC SPECIES

There are six species of ice-breeding phocids in the northern hemisphere (harp, hooded, bearded, ringed, spotted, and ribbon seals), many of which annually migrate between sub-arctic and arctic regions. Because of the difficulty in conducting surveys in this harsh environment, the abundance of many of these species is not well known.

Harp and hooded seals are both divided into three stocks (eastern Canada, White Sea, and West Ice), each identified with a specific breeding site. Recent modeling efforts indicate that a harvest of 460,000 young harp seals per year is holding the eastern Canada stock stable at about 5.2 million individuals (Healey and Stenson 2000). The other harp seal stocks are smaller—approximately 1.5 to 2.0 million in the White Sea and 286,000 on the West Ice. The best current population estimate for hooded seals is 400,000 to 450,000 animals (Stenson et al. 1993). Marked increases in the number of harp and hooded seals occurred on Sable Island in the mid-1990s (Lucas and Daoust 2002).

Populations of bearded seals were decimated by early commercial sealing. Russia continued a commercial harvest of bearded seals, with catches exceeding 10,000 animals yr^{-1} during the 1950s and 1960s. In the 1970s and 1980s quotas were introduced to limit harvests on declining populations to a few thousand animals annually (Kovacs 2002). Today, bearded seals are an important subsistence resource to arctic peoples, with a few thousand animals taken annually for use as human food, dog food, and clothing. Reliable estimates of the total population of bearded seals do not exist. Early estimates of just the Bering-Chukchi Sea population ranged from 250,000 to 300,000. Discrepancies in recent survey efforts in 1999 and 2000 have precluded an updated estimate, but the abundance may be much greater than previously described (Waring et al. 2002).

Currently, five distinct subspecies of ringed seals are recognized. Population estimates for most of these are outdated, and there are many uncertainties in the estimation and sampling methods. Nonetheless, Bychkov (in Miyazaki 2002) estimated that there were 2.5 million in the Arctic Ocean and 800,000 to 1 million in the Sea of Okhotsk in 1971. The Baltic ringed seal population decreased from 190,000 to 220,000 animals at the beginning of the twentieth century to approximately 5,000 during the 1970s. In the mid-1960s, the remaining seals were afflicted by sterility, likely caused by organochlorides (Harding and Harkonen 1999; Reijnders and Aguilar 2002), which inhibited natural population growth during the subsequent 25-year period. Ringed seals are hunted in many regions (Miyazaki 2002). Thus, the decrease in seal numbers was a consequence of excessive hunting in combination

TABLE 27.1
Pinniped Population Numbers and Trends Worldwide

Common Name	Species	Population Size	Trend
Northern Hemisphere			
Eared Seals	Otariidae		
Guadalupe fur seal (GFS)	*Arctocephalus townsendi*	7,000	Increasing
California sea lion (CSL)	*Zalophus californianus*	237,000–244,000	Increasing
Northern fur seal (NFS)	*Callorhinus ursinus*	1,400,000	Decreasing
Steller sea lion (SSL)	*Eumatopias jubatus*	<75,000	Decreasing[a]
Galápagos sea lion (GSL)	*Zalophus wollebaeki*	5,000	Fluctuating
Galápagos fur seal (GAFS)	*Arctocephalus galapagoensis*	12,000	Fluctuating
Japanese sea lion	*Zalophus japonicus*	Extinct	Extinct
Walruses	Odobenidae		
Pacific walrus	*Odobenus rosmarus divergens*	200,000	Decreasing
Atlantic walrus	*Odobenus rosmarus rosmarus*	>14,000	Unknown
Earless Seals	Phocidae		
Hooded seal (HOS)	*Cystophora cristata*	>400,000	Increasing
Gray seal (GS)	*Halichoerus grypus*	Unknown	Increasing
Ribbon seal (RIS)	*Histriophoca fasciata*	490,000	Increasing
Northern elephant seal (NES)	*Mirounga angustirostris*	127,000	Increasing
Harp seal (HAS)	*Pagophilus groenlandicus*	7,486,000	Increasing
Western Atlantic harbor seal (HS)	*Phoca vitulina concolor*	40,000–100,0000	Increasing
Western Pacific harbor seal (HS)	*Phoca vitulina richardsi*	146,900	Stable
Mediterranean monk seal (MMS)	*Monachus monachus*	250–500	Decreasing
Hawaiian monk seal (HMS)	*Monachus schauinslandi*	1,400	Decreasing
Ungava harbor seal (HS)	*Phoca vitulina mellonae*	120–600	Decreasing
Caspian seal (CS)	*Phoca caspica*	<100,000	Decreasing
Baikal seal (BS)	*Phoca sibirica*	5,000–6,000	Decreasing
Eastern Atlantic harbor seal (HS)	*Phoca vitulina vitulina*	98,000	Fluctuating
Bearded seal (BS)	*Erignathus barbatus*	100,000s	Unknown
Eastern Pacific harbor seal (HS)	*Phoca vitulina stejnegeri*	7,300	Unknown
Spotted (Largha) seal (SS)	*Phoca largha*	335,000–450,000	Unknown
Ringed seal (RS)	*Pusa hispida hispida*	2,500,000	Unknown
Baltic seal (RS)	*Pusa hispida botnica*	5,000	Unknown
Ladoga seal (RS)	*Pusa hispida ladogensis*	5,000	Unknown
Sea of Okhotsk ringed seal (RS)	*Pusa hispida ochotensis*	800,000–1,000,000	Unknown
Saimaa seal (RS)	*Pusa hispida saimensis*	2,000–5,000	Unknown
Caribbean monk seal	*Monachus tropicalis*	Extinct	Extinct
Southern Hemisphere			
Eared Seals	Otariidae		
South American fur seal (SAFS)	*Arctocephalus australis*	235,000–285,000	Increasing
New Zealand fur seal (NZFS)	*Arctocephalus forsteri*	135,000	Increasing
Juan Fernandez fur seal (JFS)	*Arctocephalus philippii*	18,000	Increasing

TABLE 27.1 (CONTINUED)
Pinniped Population Numbers and Trends Worldwide

Common Name	Species	Population Size	Trend
Southern Hemisphere			
Eared Seals	Otariidae		
Australian fur seal (AFS)	*Arctocephalus pusillus doriferus*	60,000	Increasing
Cape fur seal (CFS)	*Arctocephalus pusillus pusillus*	1,700,000	Increasing
Subantarctic fur seal (SFS)	*Arctocephalus tropicalis*	>310,000	Increasing
Antarctic fur seal (ANFS)	*Arctocephalus gazella*	1,600,000	Increasing
Australian sea lion (ASL)	*Neophoca cinerea*	9,300–11,700	Stable/decreasing
New Zealand sea lion (NZSL)	*Phocarctos hookeri*	13,000	Stable
South American sea lion (SASL)	*Otaria flavenscens*	275,000	Decreasing
Earless Seals	Phocidae		
Leopard seal (LS)	*Hydruga leptonyx*	220,000–440,000	Stable
Weddell seal (WS)	*Leptonychotes weddelli*	500,000–1,000,000	Stable
Crabeater seal (CE)	*Lobodon carcinophaga*	10,000,000–15,000,000	Stable
Southern elephant seal (SES)	*Mirounga leonina*	640,000	Stable/decreasing
Ross seal (ROS)	*Ommatophoca rossi*	100,000–650,000	Unknown

a Stock specific.

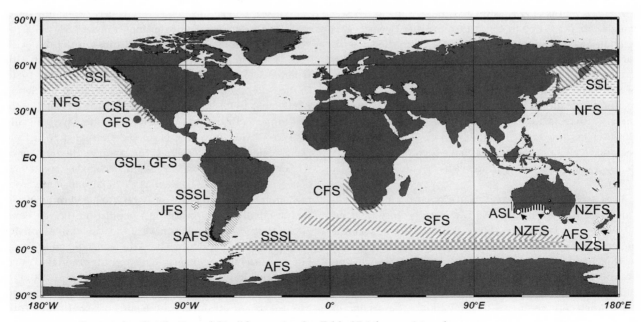

FIGURE 27.1. Present day distribution of *Otariidae* species. See Table 27.1 for species codes.

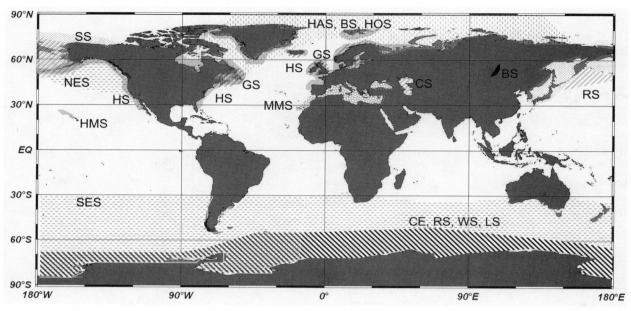

FIGURE 27.2. Present day distribution of Phocidae species. See Table 27.1 for species codes.

with lowered fertility rates after 1965 (Harding and Harkonen 1999).

The best estimate of ribbon seal abundance in the Bering Sea is 120,000 to 140,000 animals, recorded in 1987 (Angliss and Lodge 2002). Two additional populations occur in the Okhotsk Sea. The estimated total abundance for the species is 370,000 animals (Fedoseev 2000). An average of 9,000 to 11,000 ribbon seals were harvested annually from the 1950s through the 1960s.

In 1973, Burns (1973) estimated the world spotted seal population at 335,000 to 450,000 animals. Fedoseev (1971) estimated a population of 168,000 in the Okhotsk Sea. Aerial surveys of spotted seals hauled-out on the Bering Sea pack ice and along the western Alaskan coast produced an estimate of 59,214 for this region (Angliss and Lodge 2002). Because of the uncertainties in the population estimates for this species, there is no information of population trends. However, because spotted seals rely on ice, they are likely to respond to climate changes that have been observed in the Bering Sea over the last 10 to 15 years (Tynan and DeMaster 1997).

The Caspian seal declined from an estimated 1 million individuals at the beginning of the twentieth century to about 70,000 by the late 1980s (Miyazaki 2002) This species is presently considered to be one of the twenty most threatened marine mammals in the world. The Caspian seal population decline was largely a consequence of overexploitation—115,000 to 174,000 have been harvested annually since the early nineteenth century (Miyazaki 2002). The decline of Caspian seals was exacerbated by a mass mortality event of epizootic origin in 1997, which killed several thousand animals. The Baikal seal has also declined steadily in recent years, from about 70,000 animals in the 1970s to about 5,000 animals

currently. Mass mortalities from morbillivirus (Likhoshway et al. 1989) occurred in 1987–1988 and in 1998.

TEMPERATE AND TROPICAL SPECIES

Harbor seals occur widely in coastal, estuarine, and occasionally freshwater habitats across the North Atlantic and Pacific oceans. The nonmigratory nature of this species apparently has resulted in considerable regional genetic differentiation. Harbor seal population trends vary widely depending upon area and habitat. Populations are increasing at 3.5% to 7% per year in California, Oregon, and Washington (Jeffries et al. 1997; Carretta et al. 2001). These increases contrast with reported declines of 65% to 85% during the 1970s and 1980s in the Gulf of Alaska, Prince William Sound, and the Bering Sea (Pitcher 1990; Small et al. 2003; see Figure 27.3). Harbor seal numbers in the western North Atlantic have generally increased during the past several decades, although there have been significant local declines. For example, the number of harbor seal pups born on Sable Island declined by about 95% between 1989 and 1997 (Bowen et al. 2003), apparently from predation by sharks (Lucas and Stobo 2000) and competition with a rapidly growing gray seal population (Bowen et al. 2003). Harbor seal numbers have also declined substantially in the Eastern Atlantic, but here the apparent cause was a phocine distemper epidemic (Heide-Jorgensen and Harkonen 1992; Heide-Jorgensen et al. 1992; Thompson et al. 2002).

Once-abundant northern elephant seals were exploited extensively for oil during the eighteenth and nineteenth centuries. By 1900 the species had been reduced to 20 to 30 individuals (Hoelzel et al. 1993; Hoelzel 1999). Despite the resulting reduction in genetic diversity (Hoelzel 1999), northern elephant seals have recovered at an estimated

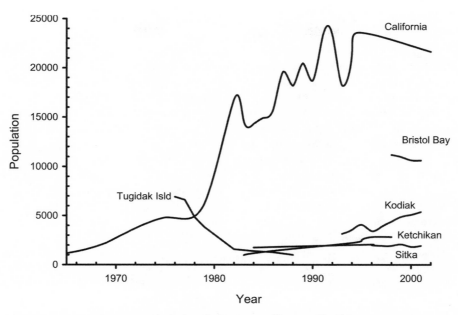

FIGURE 27.3. Population trends for Pacific harbor seal populations off the California coast (Carretta et al. 2002), the Gulf of Alaska (Kodiak Island), Southeastern Alaska (Sitka and Ketchikan) (Small et al. 2003), and Tugidak Island (Pitcher 1990).

8.3% yr⁻¹ throughout the species' range (Cooper and Stewart 1983; Stewart et al. 1994). Northern elephant seals in California were estimated at 101,000 individuals in 2001 (Carretta et al. 2002).

Monk seals are the only tropical/subtropical phocid. Populations of these species may never have been numerous because of the generally low productivity of tropical oceans. Because of their small population size, monk seals are vulnerable to die-offs resulting from disease, inbreeding and low genetic variability, and human disturbance. Of the three recent species, the Hawaiian and Mediterranean monk seals are endangered, and the Caribbean monk seal is considered extinct (Kenyon 1977). Although overall numbers are currently stable, some Hawaiian monk seal colonies are increasing while others are in decline. Reasons for the declines include human disturbance, habitat loss, disease, competition with fisheries, shark predation, intraspecific aggression, and entanglement with fisheries gear. Mediterranean monk seals are presently estimated at 250 to 500 individuals. The largest known aggregation of this species, which occurs at the peninsula of Cape Blanc in the Western Sahara, suffered a mass mortality event of unknown origin in 1997 that reduced its numbers from 317 to 109 individuals (Forcada et al. 1999). A hitherto unknown colony was recently discovered in the Cilician Bay off Turkey (Gucu et al. 2004).

SOUTHERN HEMISPHERE SPECIES

Like the northern elephant seal, the once-abundant southern elephant seal was heavily exploited during the eighteenth and nineteenth centuries. Populations were severely reduced at all major breeding sites. Although controlled harvests were reinstituted early in the twentieth century, southern elephant seal populations increased in most areas. The current world population is estimated at 640,000 individuals, with major breeding colonies located on South Georgia (113,000 pups per year), the Kerguelen Archipelago (43,000 pups per year), Macquarie Island (19,000 pups per year), Heard Island (17,000 pups per year), and Peninsula Valdes (14,500 pups per year). Paradoxically, despite their recovery following commercial sealing and a lack of subsequent hunting pressure, southern elephant seal populations in the Pacific and Indian oceans declined remarkably (50% to 80%) between the 1950s and 1990s (McMahon et al. 2003). The proximate causes of theses decreases are poorly understood. Some authors have attributed them to long-term environmental change leading to resource limitation (Burton et al. 1997; McMahon et al. 2003), whereas others have proposed the cause to be increased killer whale predation (Barrat and Mougin 1978; Guinet et al. 1992; Branch and Williams, Chapter 20 of this volume).

The remaining southern hemisphere phocids (Weddell, Ross, crabeater, and leopard seals) are restricted to the Antarctic continent and its surrounding pack ice. Although these species have been harvested irregularly in past years, they were never so extensively depleted as pinnipeds elsewhere in the world. Current harvest levels are regulated under the Convention on the Conservation of Antarctic Seals (CCAS).

For various logistical reasons, estimates of abundance for the Antarctic phocids have been difficult to obtain. Current estimates are 220,000 to 440,000 for leopard seals; 500,000 to 1 million for Weddell seals; 10 to 15 million for crabeater seals; and 100,000 to 650,000 for Ross seals. Given the absence of historical estimates and the large uncertainties associated with modern-day population estimates, it is not possible to ascertain current population trends for any of the Antarctic phocids. Nonetheless, the relatively small number of animals harvested (ca. 39,000 since 1892) over a wide geographical range is unlikely to have adversely affected populations of any of these species; thus, changes in distribution and abundance are likely due to other factors. Testa et al. (1991) documented cyclic patterns in the age structure and cohort strength of crabeater, Weddell, and leopard seals related to the Southern Oscillation Index (SOI). Furthermore, Bengtson and Laws (1985) and Ballance et al. (Chapter 17 of this volume) suggested that a reported decline in the age of first reproduction of crabeater seals was caused by an increased availability of krill, which ostensibly resulted from the depletion of baleen whales. The age of first reproduction increased again between 1963 through 1976, a presumed physiological response to reduced prey availability.

Otariid Population Trends

Like temperate-latitude phocids, most otariid species were nearly extirpated by overharvesting by the end of the nineteenth century. With the cessation of harvest, most stocks recovered to varying degrees. Overall, fur seal populations appear to have recovered more rapidly than sea lion populations. Fur seals also outnumber sea lions by an order of magnitude (nearly 8 million fur seals worldwide, in contrast to just over 600,000 sea lions). Variation in diet, foraging strategies, and regional productivity may explain the differences in abundance and recovery rates between the two groups.

Antarctic fur seals were thought to be extinct until a remnant population of 1,000 to 1,200 animals was discovered at Bouvetøya in 1928 (Fevoden and Sømme 1976), and another 100 animals were discovered off Bird Island in the 1930s (Laws 1973). Since that time the Antarctic fur seal population has grown at about 10% yr^{-1}, to a 1990 estimate of 1.6 million animals (Boyd 1993). Although Antarctic fur seals have increased throughout their range, Boveng et al. (1998) suggested that the slower recovery rate on the South Shetland Islands is a result of leopard seal predation.

Northern fur seals inhabit the North Pacific Ocean and the Bering Sea, with primary breeding rookeries on the Pribilof and Commander Islands and smaller colonies in the Sea of Okhotsk and Kurile Islands. An outlying colony also occurs on San Miguel Island off the coast of southern California. Following the overharvest and depletion of Northern fur seals during the eighteenth and nineteenth centuries, they were protected in 1911 and recovered significantly during the first half of the twentieth century, However, in contrast with Antarctic fur seals, northern fur seal numbers began to decline again in the 1950s, a trend that has continued to the present day (Angliss and Lodge 2002; Carretta et al. 2002). The cause of this ongoing decline is unknown.

The closely related Cape and Australian fur seals were both greatly depleted by commercial sealing. However, their populations have since taken rather different trajectories. Cape fur seal populations, while subjected to controlled hunts,

have increased to an estimated 1.7 million individuals and are growing at 3% per year (Butterworth et al. 1995). In contrast, Australian fur seal populations, which have not been hunted since 1923, have increased only to an estimated 60,000 individuals, well below the presealing estimate of 175,000 to 225,000 individuals. These different population growth rates and abundances have been attributed to variation in food availability and differing foraging strategies (Arnould and Warneke 2002).

Other fur seal populations were also nearly hunted to extinction during the commercial sealing era but have similarly rebounded. The sub-Antarctic fur seal, which breeds just north of the Antarctic polar front on sub-Antarctic and subtemperate islands, numbers more than 235,000 to 285,000 individuals and is increasing by as much as 9% to 14% yr^{-1} in a few colonies (Wickens and York 1997). The South American fur seal, which occurs along the Pacific coast of South America, in the islands west of Tierra del Fuego, in the Falkland Islands, and in Uruguay and along the southern coast of Brazil, was hunted during the sealing era and, more recently, in regular small, controlled harvests in Uruguay until the 1990s. While numbers of South American fur seals are increasing, they remain diminished, apparently due to the recent commercial harvest. The New Zealand fur seal, an estimated 135,000 animals, is protected and increasing throughout its range in New Zealand and Australia (Wickens and York 1997). The Guadalupe fur seal, now the rarest of all fur seal species (about 7,000 individuals), was presumed extinct until a small breeding group was discovered at Isla de Guadalupe in 1928 (Townsend 1928). This small colony was then nearly exterminated by museum collectors (Bartholomew 1950; Hubbs 1956). The population bottleneck resulted in a substantial loss of genetic variability (Weber et al. 2004). The Galápagos fur seal, which is limited to the Galápagos Islands and was greatly depleted during the sealing era, recovered to its estimated presealing population level of about 40,000 animals by 1977–1978. However, population size in this species fluctuates considerably in response to El Niño events (Trillmich et al. 1991). For example, 90% pup mortality and 45% overall mortality reduced the population to an estimated 6,000–8,000 animals following the 1997–1998 El Niño event (Salazar 2002). The Juan Fernandez fur seal, which is confined to the islands off the coast of Chile, contained an estimated four million animals before the sealing era. This species, thought to be extinct until a small population was rediscovered in 1966, is currently estimated at approximately 18,000 animals and growing (Wickens and York 1997).

In contrast to the fur seals, most sea lion populations are in decline, and one species (the Japanese sea lion) is probably extinct. Galápagos sea lions have continued to decline from anthropogenic impacts, El Niño events, and disease outbreaks, with a most recent count in 2002 of between 14,000 and 16,000 animals (Salazar 2002). The New Zealand (or Hooker's) sea lion, which once occurred all around New Zealand, was depleted by subsistence and commercial hunting. Presently, its breeding range is restricted to a few sub-Antarctic islands (Gales and Fletcher 1999). Australian sea lions also were depleted by sealing, even though the larger populations of sympatric fur seals were the primary targets. This species is now one of the world's rarest and most unusual sea lions, on account of its 17.5-month breeding cycle (Higgins 1993), in contrast to the 12-month cycle typical of this group. Populations of both New Zealand and Australian sea lions, while far below their estimated pre-exploitation levels of about 13,000 and 10,000 (Gales et al. 1994; P. D. Shaughnessy et al. 2005) respectively, are presently stable. However, unlike the New Zealand sea lion, the Australian sea lion is widely distributed among 67 scattered colonies in southern and Western Australia, and may in fact be in decline.

The South American sea lion, which ranges from Brazil to Peru, was decimated in the hunt for oil. Today's populations stand at approximately 20% of their historical numbers (Cappozzo 2002). Although populations along the Pacific coast of South America are poorly known, many animals apparently abandon this region during severe El Niño events. Thompson et al. (2005) reported population declines in the Falkland Islands to <1.5% of the 1937 abundance estimate between 1965 and 1990, for reasons that remain unclear. Pup production in the Falklands since 1990 has increased at 3.8% to 8.5% per year, thus putting this population on a similar growth trajectory to the adjacent Argentinean population.

The northern or Steller sea lion (SSL), which ranges from California to Japan, was hunted for various reasons until the early 1970s. There is still a small subsistence take by Alaska natives. Two stocks are currently recognized in U.S. waters, with Cape Suckling, Alaska (144°W) the demarcation point between the eastern and western stocks. Despite the cessation of hunting and protection from other disturbances, the western stock began a precipitous decline in the late 1970s or early 1980s (Loughlin et al. 1992; NRC 2003). The western stock is currently listed as Endangered and the eastern stock as Threatened under the U.S. Endangered Species Act. The known or suspected causes of SSL mortality include incidental losses in fisheries gear, entanglement in marine debris, shooting, competition with fisheries for food, ocean climate change, and predation by killer whales. SSL and their associated ecosystem have been the objects of intensive research, yet there is still no widely accepted explanation for sea lions' recent decline (NRC 2003).

The California sea lion, which ranges from Mexico to British Columbia, is the most abundant of all sea lion species. Culling, largely because of perceived damage to commercial catches and competition for salmonid fishery resources (Everitt and Beach 1982), reduced the abundance of California sea lions in southern California and Mexico to approximately 1,500 individuals by the 1920s. The species has increased steadily at 5% to 6.2% yr^{-1} through the latter part of the twentieth century (NMFS 1997). Presently, there are an estimated 204,000 to 214,000 California sea lions in U.S. waters (Carretta et al.

2002), and an additional 44,000 to 53,000 animals in Mexico (Aurioles-Gamboa and Zavala-Gonzalez 1994).

Potential Causes of Population Declines

In the preceding sections of this chapter we have provided a broad overview of the population status and trends of pinnipeds worldwide. Most species were reduced substantially during the era of commercial sealing. The explanation for these early declines is clear and certain—humans killed them. However, populations have not all responded to the cessation of commercial sealing in the same way. Some species or local populations increased, often rapidly. Here again there is little mystery as to why—reduced mortality, together with resource surpluses created by the earlier population reduction, probably fueled population growth in a manner expected from simple demographic and ecological theory. However, other species and populations either did not recover from sealing or did recover but have subsequently again declined. These latter cases are more difficult to understand. In the final sections of this chapter we attempt to shed light on these perplexing trends by mapping known or suspected patterns of variation in life history, behavior, and environment with reported population trajectories. Our synthesis focuses on three key patterns and processes: differing reproductive strategies between phocids and otariids; physiological limitations associated with particular foraging and diving behavior; and the resulting susceptibility to disturbance in prey resources and predation.

Life History and Behavioral Correlates

Phocids and otariids have solved the conflicting demands of terrestrial parturition and marine feeding in different ways (Bartholomew 1970; Costa 1993). Most phocids are capital breeders, storing, prior to parturition, sufficient energy for the entire lactation period. Otariids, in contrast, are income breeders, feeding more or less continuously during lactation (Costa 1991a,b, 1993; Boyd 2000). These different strategies confer differing benefits and costs to phocids and otariids. Capital breeding disassociates reproductive success from local food availability. The nutritional provisioning of pups by phocid mothers is thus largely unconstrained by traveling time to and from the foraging grounds, thereby allowing them to utilize prey that are more dispersed, patchy, unpredictable, or distant from the rookery. The necessity of feeding during lactation constrains otariid females to forage closer to the rookery, thus linking reproductive success and local prey abundance (Costa 1993) and thereby potentially connecting population status to localized environmental changes such as El Niño/Southern Oscillation events (Trillmich et al. 1991; see Figure 27.4 for the California sea lion). Significant alterations in trip duration, female condition, fecundity, pup growth rate, and survival in response to reductions in prey availability caused by changing oceanographic conditions are common in otariids (Costa et al. 1989,

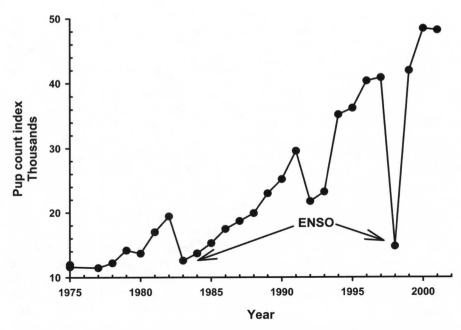

FIGURE 27.4. Pup production of California sea lions off the California coast (Carretta et al. 2002). Notice the different effect of the 1983 and 1998 strong El Niño/Southern Oscillation (ENSO) events.

Testa et al. 1991; Trillmich et al. 1991; Boyd et al. 1994; Boyd and Murray 2001). On the other hand, fasting or reduced feeding during lactation limits the total amount of energy and protein that can be invested in phocid young, resulting in a smaller relative pup mass at nutritional independence. Phocid weaning mass reflects the mother's foraging success over the previous year; postweaning survival is related to both weaning mass (energy reserves provided by the mother) and postweaning resource availability (Stewart and Lavigne 1984). Furthermore, weaning in phocids is abrupt, thus preventing pups from learning from their mothers how to forage—a potential disadvantage in times of food shortage. Weaning is often synchronous within species, which means large numbers of inexperienced individuals will be simultaneously testing the waters near breeding colonies, thus making them susceptible to predation. The typically short breeding period in phocids also allows them to utilize unstable breeding substrates, such as pack ice (Stirling 1975, 1983; Costa 1993).

Implications of the aforementioned differences in reproduction and foraging between phocids and otariids are potentially exemplified by the striking differences in recovery rates of fur seals, elephant seals, and sea lions on Guadalupe Island, Mexico. While fur seals and elephant seals both increased following the cessation of hunting, the elephant seal recovery has been far more dramatic (Figure 27.5). During this same period, sea lion numbers at Guadalupe Island have remained low but relatively stable, whereas populations elsewhere have increased rapidly. As income breeders, California sea lions and Guadalupe fur seals must remain with their pups for almost a year, thus constraining them to feed near Guadalupe

Island. As a capital breeder, the northern elephant seal can forage almost anywhere in the North Pacific Ocean (Stewart and DeLong 1993; LeBoeuf et al. 2000), thus providing this species with greater access to prey resources.

Variation in dive behavior can also influence foraging efficiency and, thus, the potential for prey limitation in population regulation. That is, pinnipeds that operate at or near their physiological limits should have little capacity to adjust foraging effort in response to food availability, whereas those that operate below their physiological limits should not be so constrained. Thus, one would expect stronger covariation between population trends and food availability for species in the former than in the latter group (Costa et al. 2001; Costa and Gales 2003; Costa et al. 2004). This proposed relationship should be particularly evident between benthic and water column foragers, because benthic foragers may be working at levels closer to their maximum physiological diving capacity. Benthic foraging requires longer transit times, thus reducing the time beneath the surface that is available to search for prey (Costa and Williams 1999). Because adults of benthic foraging species are working at or near their physiological limit, the smaller juveniles, with their reduced physiological capabilities and oxygen stores, should be particularly vulnerable to resource limitation. Survival of juveniles in benthic foraging species might thus be a major determinant of demographic trends. Furthermore, benthic foraging species might be particularly sensitive to bottom trawlers, which disrupt the habitat and remove the larger size-classes of fishes upon which they often depend (Thrush et al. 1998). On the other hand, the benthos may be a more predictable source of prey than the water column, and thus benthic foraging species may be less

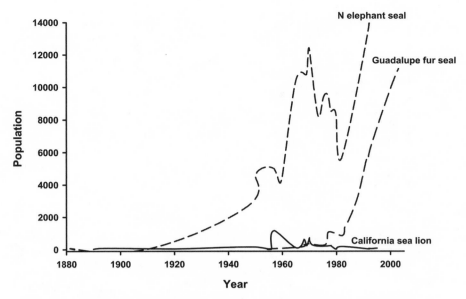

FIGURE 27.5. Population trends from the three species of pinniped that are found on Guadalupe Island, Mexico: California sea lion, Guadalupe fur seal, and Northern elephant seal (data from J. Pablo-Gallo, unpublished).

affected by oceanographic perturbations such as El Niños than water column feeders (Miller and Sydeman 2004).

Observed relationships between aerobic dive limit (ADL), which is a measure of physiological dive capability, and foraging behavior across five otariid species (Antarctic and Australian fur seals; Australian, California, and New Zealand sea lions) provides an initial test of these predictions (Costa et al. 2004). The Antarctic fur seal makes short, shallow dives, while the Australian and New Zealand sea lions and the Australian fur seal make deep prolonged dives to the benthos (Boyd et al. 1991; Costa and Gales 2000, 2003). California sea lions are especially interesting because they (at least the females and juveniles) dive epipelagically off the coast of southern California (Feldkamp et al. 1989), whereas they forage much deeper on mesopelagic prey in the Sea of Cortez (C. E. Kuhn, D. Aurioles-Gamboa, and D. P. Costa, unpublished). The near-surface-feeding Antarctic fur seal and California sea lion in southern California forage well within their calculated ADL (cADL), whereas the three benthically foraging species/populations routinely exceed their cADL (Figure 27.6). For both fur seals and sea lions, the benthic foragers spend >40% of their time at sea diving, whereas the epipelagic foragers spend <30% of their time at sea diving (Figure 27.7). These findings may explain why populations of many fur seal species (and the pelagic foraging California sea lion) have increased whereas most sea lions (many of which occupy the same area as near-surface-feeding fur seal species; Costa and Gales 2003) have remained stable or declined (Boyd et al. 1995; Sydeman and Allen 1999; Gales and Fletcher 1999; Gales et al. 1994).

Risk of Predation

Pinnipeds are preyed on by a variety of species, including other pinnipeds, humans, polar bears, wolves, foxes, coyotes, hyenas and jackals, eagles, sharks, and killer whales (Weller 2002). In the northern hemisphere, land- or ice-based predators (people, bears, foxes) are particularly important, whereas in the southern hemisphere, ice seals are free from terrestrial predators but are subjected to several aquatic predators. The clearly divergent predator avoidance tactics between Arctic and Antarctic pinnipeds, with Arctic species fleeing into the water to escape predation and Antarctic species seeking refuge on the ice (Stirling 1975, 1983; Weller 2002), attest to the strength and importance of these predator-prey interactions.

Killer whales (Orcinus orca) are probably the most important aquatic predator of marine mammals. Harbor seals are the most commonly reported prey of killer whales in the northern hemisphere (Jefferson et al. 1991). However, killer whales are also known to prey on many other species of pinnipeds, including the crabeater seal (Smith et al. 1981), southern sea lion and southern elephant seal (López and López 1985; Guinet et al. 2000), northern fur seal (M. Goebel, personal communication), and California sea lion. In some areas, killer whales intentionally strand themselves in order to seize pinniped prey, such as southern sea lions and southern elephant seals, on the beach (López and López 1985).

Although predator control of pinniped populations is difficult to verify, several studies either demonstrate or suggest significant population level impacts of predation. Harbor seal pup production at Sable Island declined from 600 in 1989 to just 32 in 1997 (Lucas and Stobo 2000; Bowen et al. 2003). An estimated 45% of the total pup production was killed by sharks during 1996. The increasing grey seal population may have indirectly influenced harbor seals by attracting sharks to the region and competing with harbor seals for prey (Bowen et al. 2003).

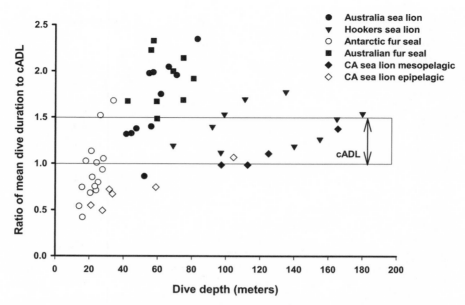

FIGURE 27.6. Dive performance, defined as the ratio of average dive duration to the calculated aerobic dive limit (cADL), as a function of dive depth in five pinniped species. Range of cADL outlined by the box is the cADL plus 50% to account for the variability in FMR estimates. Notice that both the dive duration and tendency to exceed the cADL are greater in benthic foraging species.

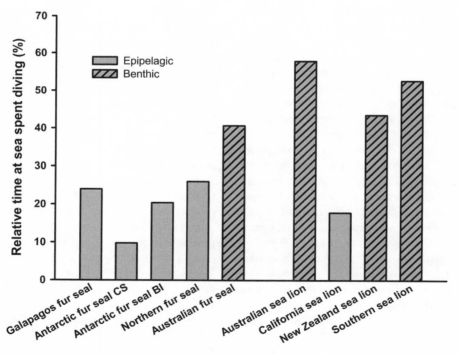

FIGURE 27.7. The relative time spent foraging while at sea, compared across eight species of otariids. The group to the left consists of fur seals and the group to the right consists of sea lions. Of the sea lions only the California sea lion forages epipelagically, whereas only one fur seal forages benthically. Data for Antarctic fur seals comes from Cape Shirreff (CS) and Bird Island (BI) (Costa and Gales 2003).

Springer et al. (2003) recently speculated that killer whale predation was principally responsible for widespread population declines of sea otters and pinnipeds in southwest Alaska. Their argument was based largely on feasibility analyses from demographic and energetic modeling (Williams et al. 2004), strong evidence that killer whale predation caused the sea otter decline (Estes et al. 1998), and various inconsistencies in the available information with other purported explanations for the declines (NRC 2003). Branch and Williams (Chapter 20 of this volume), on the basis of similar evidence and analyses, have concluded that killer whale predation might also have figured prominently in the decline of various Southern Ocean pinniped populations. In support of this latter suggestion, killer whales were observed taking up to 25% of the southern elephant seal weanlings from one beach at Crozet Island and have been implicated in the decline of that population (Guinet 1992; Guinet et al. 1992).

Pinnipeds also prey on one another. A single Hooker's sea lion on Macquarie Island killed 43% of the 130 Antarctic and sub-Antarctic fur seal pups born over a two year period (Robinson et al. 1999). Similarly, leopard seals killed an estimated 34% of the Antarctic fur seal pup production at Seal Island (Boveng et al. 1998). In Punta San Juan, Peru, up to 8.3% of South American fur seal pups are reportedly killed by Southern sea lions (Harcourt 1992). Northern fur seal pups are killed and eaten by adult male Steller sea lions (Gentry and Johnson 1981).

Summary

In this chapter we have reviewed the current status and trends of pinniped populations worldwide, with a focus on the anthropogenic and natural biological processes that might be responsible for these trends. Although little or no data are available for some species, useful time series exist for many others. Increasing populations include most of the southern hemisphere fur seals, the California sea lion, harbor seal populations off the west coast of the United States, and the northern elephant seal. Populations in decline include northern and southern sea lions, the northern fur seal, the southern elephant seal in parts of the Southern Ocean, and the harbor seal in southwest Alaska. The tropical monk seals are either stable at low levels or in decline. Population trends for polar species are poorly known, although by and large these species appear to be both abundant and fairly stable.

Recovery failures or recent population declines are most commonly attributed to interactions with fisheries and environmental change. Pinniped interactions with fisheries include both operational and ecological effects (Harvey 1987; Mate and Harvey 1987). Ecological effects largely result from direct competition, thus ostensibly reducing both potential fishery yields and the environmental carrying capacity for pinnipeds. Operational effects occur when pinnipeds and fishery operations come into direct contact. For example, pinnipeds can be incidentally entangled and damage fishing gear and remove or damage fish caught in nets or on fishing lines. A currently expanding and largely unregulated trade in seal products (Reeves 2002), poor or misguided population management, and continued overexploitation of some populations are contributing factors. For example, although commercial sealing has declined considerably since the 1960s, native hunters kill more than 100,000 ringed, bearded, ribbon, harp, hooded, and spotted seals annually (Reeves 2002). In addition, after a reduction in takes of harp seals in the 1980s, government subsidies have reinvigorated the Canadian commercial hunt, with approximately 350,000 harp seals being taken in eastern Canada and Greenland in 1998 (Lavigne 1999). Norwegian and Russian ships also take tens of thousands of harp and hooded seals annually in the Greenland and Barents seas. In the southern hemisphere, South American fur seals were harvested in Uruguay until the 1990s. and the centuries-old hunt of Cape fur seals in southwestern Africa continues to take thousands of fur seals annually.

Pinniped populations also have suffered adverse impacts by human modifications of coastal and marine environments, thus resulting in disturbance, loss of breeding and resting sites, and alteration of foraging grounds (Reeves 2002). Exposure to various pollutants is a problem in some areas, and disease outbreaks are often related to immune-response suppression caused by a variety of pollutants (Reijnders and Aguilar 2002).

In addition to environmental forcing, pinniped population trends are influenced by variation in life history, behavior, and physiological capacity. Capital breeders or species that have life history patterns that allow them to forage across ocean basins have a greater capability of recovery from exploitation, whereas income breeders are more sensitive to local oceanographic variations and associated limitations in prey resources may recover more slowly. Similarly, differences in the foraging strategy of otariids may also be a factor in their ability to respond to environmental fluctuations. Benthic diving otariids (e.g., Steller, Australian, southern, and New Zealand sea lions) have a lower reproductive output than epipelagic species because they spend more time at sea diving and they push their physiological limits. This is further compounded by a potential for reduced juvenile survival in benthic foraging species because of the reduced diving capability of juveniles. Differences in the foraging strategies and reproductive patterns also may make certain pinnipeds more susceptible than others to predation. With both reproductive strategies the young are exposed to predation, but adult female otariids will be more susceptible to increased predation because income breeders must make multiple visits to and from the rookery.

Although our review provides an overview of pinniped population trends worldwide and a synthetic effort to understand reasons for recent population declines and the failure of various other populations to even recover from overexploitation, in truth these patterns are poorly understood. Although capital breeding would seem to convey an advantage to phocids over otariids, it cannot explain why both harbor seals and Steller sea lions have declined at the same rate and magnitude over largely the same regions of southwest Alaska or why northern elephant seals have recovered so spectacularly whereas many southern elephant seal populations have collapsed. Similarly, although diving behavior and physiological limitation would seem to convey a relative disadvantage to benthic foraging over epipelagic foraging otariids, this explanation alone cannot account for the spectacular population collapse of the benthically foraging Steller sea lion. On the other hand, although there is compelling evidence that predators have driven the declines of small isolated pinniped colonies and there have been arguments for the importance of killer whale predation in the large scale population declines of seals, sea lions, and otters of the North Pacific, conclusive evidence is lacking.

Given the extremely dynamic nature of pinniped populations and the high degree of uncertainty over ultimate causes of pinniped population irruptions and declines, it is not unreasonable to imagine that these dynamics were influenced at least to some degree by the effects of whaling on ocean ecosystems. As Croll et al. (Chapter 16 of this volume) point out, the great whales co-opted a significant proportion of the world's marine production before whaling, and if Roman and Palumbi's (2003) prewhaling abundance estimates are even close to being accurate, the magnitude of this effect would have been even greater. In a simplistic sense, less production being co-opted by whales means potentially more for other ocean consumers. Such effects have been proposed in the southern ocean, where the removal of krill-eating whales ostensibly led to increases in other krill consumers, including some of the pinnipeds (Ballance et al., Chapter 17 of this volume). However, as Paine (Chapter 2 of this volume) points out, food web dynamics are rarely so simple. Although complex food web dynamics of this sort are poorly known for ocean ecosystems, the very fact that pinniped population dynamics fit so poorly into traditional explanatory molds raises the distinct possibility that the ecological influences of whaling are associated with some of this uncertainty.

Literature Cited

Angliss, R.P. and K.L. Lodge. 2002. *Alaska marine mammal stock assessments, 2002.* U.S. Department of Commerce, NOAA Technical Memorandum NMFS-AFSC-133. Seattle: Alaska Fisheries Science Center.

Arnould, J.P.Y., and R.M. Warneke. 2002. Growth and condition in Australian fur seals *(Arctocepha pusillus doriferus)* (Carnivora: Pinnipedia). *Australian Journal of Zoology* 50: 53–66.

Aurioles-Gamboa, D. and A. Zavala-Gonzalez. 1994. Ecological factors that determine distribution and abundance of the California sea lion *Zalophus californianus* in the Gulf of California. *Ciencias Marinas* 20: 535–553.

Barrat, A. and J.L. Mougin. 1978. The Southern elephant seal, *Mirounga leonina,* of Possession Island, Crozet Archipelago, 46°25′ south, 51°45′ east. *Mammalia* 42: 143–174 (in French).

Bartholomew, G.A. 1950. A male GFS on San Nicolas Island, California. *Journal of Mammalogy* 31:175–180.

———. 1970. A model for the evolution of pinniped polygyny. *Evolution* 24: 546–559.

Bengtson, J.L. and R.M. Laws. 1985. Trends in crabeater seal age *Lobodon carcinophagus* at sexual maturity: an insight into Antarctic marine interactions, in *Antarctic nutrient cycles and food webs.* W.R. Siegfried, P.R. Condy, and R.M. Laws, eds. Berlin: Springer-Verlag, pp. 667–675.

Boveng, P.L., L.M. Hiruki, M.K. Schwartz, and J.L. Bengtson. 1998. Population growth of Antarctic fur seals: limitation by a top predator, the leopard seal? *Ecology* 79: 2863–2877.

Bowen, W.D. 1997. Role of marine mammals in aquatic ecosystems. *Marine Ecology Progress Series* 158: 267–274.

Bowen, W.D., S.L. Ellis, S.J. Iverson, and D.J. Boness. 2003. Maternal and newborn life-history traits during periods of contrasting population trends: Implications for explaining the decline of harbour seals *(Phoca vitulina),* on Sable Island. *Journal of Zoology* 261: 155–163.

Boyd, I.L. 1993. Pup production and distribution of breeding Antarctic fur seals *(Arctocephalus gazella)* at South Georgia. *Antarctic Science* 5(1): 17–24.

———. 2000. State-dependent fertility in pinnipeds: contrasting capital and income breeders. *Functional Ecology* 14: 623–630.

Boyd, I.L., J.P.Y. Arnould, T. Barton, and J.P. Croxall. 1994. Foraging behaviour of Antarctic fur seals during periods of contrasting prey abundance. *Journal of Animal Ecology* 63: 703–713.

Boyd, I.L., J.P. Croxall, N.J. Lunn, and K. Reid. 1995. Population demography of Antarctic fur seals: the costs of reproduction and implications for life-histories. *Journal of Animal Ecology* 64: 505–518.

Boyd, I.L., N.J. Lunn, and T. Barton. 1991. Time budgets and foraging characteristics of lactating Antarctic fur seals. *Journal of Animal Ecology* 60: 577–592.

Boyd, I.L. and A.W.A. Murray. 2001. Monitoring a marine ecosystem using responses of upper trophic level predators. *Journal of Animal Ecology* 70: 747–760.

Boyd, I.L., T.R. Walker, and J. Poncet. 1996. Status of southern elephant seals at South Georgia. *Antarctic Science* 8: 237–244.

Burns, J.J. 1973. *Marine mammal report.* Pittman-Robertson Project Report W-17-3, W-17-4, and W-17-5. Juneau: Alaska Department of Fish and Game.

Burton, H.R., T. Arnbom, I.L. Boyd, M. Bester, D. Vergani, and I. Wilkinson. 1997. Significant differences in weaning mass of southern elephant seals from five sub-Antarctic islands in relation to population declines, in *Antarctic communities: species, structure and survival.* D.W.H. Walton, ed. Cambridge, UK: Cambridge University Press, pp. 335–338.

Butterworth, D.S., A.E. Punt, W.H. Oosthuizen, and P.A. Wickens. 1995. The effects of future consumption by the Cape fur seal on catches and catch rates of the Cape hakes. 3. Modelling the dynamics of the Cape fur seal *Arctocephalus pusillus pusillus. South African Journal of Marine Science* 16: 161–183.

Cappozzo, H.L. 2002. South American sea lion, in *Encyclopedia of marine mammals.* W.F. Perrin, B. Würsig, and J.G.M. Thewissen, eds. San Diego: Academic Press, pp. 348–351.

Carretta, J.V., J. Barlow, K.A. Forney, M.M. Muto, and J. Baker. 2001. *U.S. Pacific marine mammal stock assessments: 2001.* NOAA Technical Memorandum NMFS-SWFSC-317. La Jolla, CA: Southwest Fisheries Science Center.

Carretta, J.V., M.M. Muto, J. Barlow, J. Baker, K.A. Forney, and M. Lowry. 2002. *U.S. Pacific marine mammal stock assessments:*

2002. NOAA Technical Memorandum NMFS-SWFSC-346. La Jolla, CA: Southwest Fisheries Science Center.

Cooper, C.F. and B.S. Stewart. 1983. Demography of northern elephant seals *Mirounga angustirostris* 1911–1982. *Science* 219: 969–971.

Costa, D.P. 1991a. Reproductive and foraging energetics of high-latitude penguins, albatrosses, and pinnipeds: implications for life history patterns. *American Zoologist* 31: 111–130.

———. 1991b. Reproductive and foraging energetics of pinnipeds: implications for life history patterns, in *Behaviour of pinnipeds.* D. Renouf, ed. London: Chapman and Hall, pp. 300–344.

———. 1993. The relationship between reproductive and foraging energetics and the evolution of the Pinnipedia, in *Marine mammals: advances in behavioural and population biology.* I.L. Boyd, ed. Oxford: Oxford University Press, pp. 293–314.

Costa D.P., J.P. Croxall, and C. Duck. 1989. Foraging energetics of Antarctic fur seals, *Arctocephalus gazella*, in relation to changes in prey availability. *Ecology* 70: 596–606.

Costa D.P. and N.J. Gales. 2000. Foraging energetics and diving behaviour of lactating New Zealand seal lions, *Phocarctos hookeri. Journal of Experimental Biology* 203: 3655–3665.

———. 2003. Energetics of a benthic diver: seasonal foraging ecology of the Australian sea lion, *Neophoca cinerea. Ecological Monographs* 73: 27–43.

Costa, D.P., N.J. Gales, and M.E. Goebel. 2001. Aerobic dive limit: How often does it occur in nature? *Comparative Biochemistry and Physiology Part A: Molecular and Integrative Physiology* 129A: 771–783.

Costa, D.P., C.E. Kuhn, M.J. Weise, S.A. Shaffer, and J.P.Y. Arnould. 2004. When does physiology limit the foraging behaviour of freely diving mammals? in *Animals and environments: proceedings of the 3rd International Conference on Comparative Physiology and Biochemistry, KwaZulu/Natal, South Africa, 7–13 August 2004.* S. Morris and A. Vosloo, eds. International Congress Series 1275. Amsterdam: Elsevier, pp. 359–366.

Costa, D.P. and T.M. Williams. 1999. Marine mammal energetics, in *Biology of marine mammals.* J.E. Reynolds and S.A. Rommel, eds. Washington, DC: Smithsonian Institution Press, pp. 176–217.

Demere, T.A. 1994. The family Odobenidae: A phylogenetic analysis of fossil and living taxa. *Proceedings of the San Diego Society of Natural History* 29: 99–123.

Demere, T.A., A. Berta, and P.J. Adam. 2003. Pinnipedimorph evolutionary biogeography. *Bulletin of the American Museum of Natural History* 279: 32–76.

Estes, J.A., M.T. Tinker, T.M. Williams, and D.F. Doak. 1998. Killer whale predation on sea otters linking oceanic and nearshore ecosystems. *Science* 282: 473–476.

Everitt, R.D. and R.J. Beach. 1982. Marine mammal-fisheries interactions in Oregon and Washington: an overview, in *Marine mammals: conflicts with fisheries, other management problems, and research needs.* D.G. Chapman and L.L. Eberhardt, eds. Transactions of the 47th North American Wildlife and Natural Resources Conference. Washington, DC: Wildlife Management Institute, pp. 253–263.

Fedoseev, G.A. 1971. Distribution and population of seals at pup and molting rookeries in the Sea of Okhotsk. *Trudy Atlanticheskogo Nauchno-Issledovatel'Skogo Instituta Rybnogo Khozyaistva i Okeanografii* 39: 87–99 (in Russian).

———. 2000. *Population biology of ice-associated forms of seals and their role in the Northern Pacific ecosystems.* Moscow: Center for Russian Environmental Policy.

Feldkamp, S.D., R.L. DeLong, and G.A. Antonelis. 1989. Diving patterns of California sea lions, *Zalophus californianus. Canadian Journal of Zoology* 67: 872–883.

Fevoden, S.E. and L. Sømme. 1976. Observations on birds and seals at Bouvetøya. *Norsk Polarinstitutt Årbok* 1976: 367–371.

Forcada, J., P.S. Hammond, and A. Aguilar. 1999. Status of the Mediterranean monk seal *Monachus monachus* in the Western Sahara and the implications of a mass mortality event. *Marine Ecology-Progress Series* 188: 249–261.

Gales, N.J. and D.J. Fletcher. 1999. Abundance, distribution and status of the New Zealand sea lion, *Phocarctos hookeri. Wildlife Research* 26: 35–52.

Gales, N.J., P.D. Shaughnessy, and T.E. Dennis. 1994. Distribution, abundance and breeding cycle of the Australian sea lion *Neophoca cinerea* (Mammalia: Pinnipedia). *Journal of Zoology (London)* 234: 353–370.

Gentry, R.L. and J.H. Johnson. 1981. Predation by sea lions on northern fur seal neonates. *Mammalia* 45: 423–430.

Gucu, A.C., G. Gucu, and H. Orek. 2004. Habitat use and preliminary demographic evaluation of the critically endangered Mediterranean monk seal *(Monachus monachus)* in the Cilician Basin (Eastern Mediterranean). *Biological Conservation* 116: 417–431.

Guinet, C. 1992. Hunting behavior of killer whales *(Orcinus orca)* around the Crozet Islands. *Canadian Journal of Zoology* 70: 1656–1667 (in French).

Guinet, C., P. Jouventin, and H. Weimerskirch. 1992. Population changes, movements of southern elephant seals on Crozet and Kerguelen Archipelagos in the last decades. *Polar Biology* 12: 349–356.

Guinet, C., L.G. Barrett-Lennard, and B. Loyer. 2000. Co-ordinated attack behavior and prey sharing by killer whales at Crozet Archipelago: strategies for feeding on negatively buoyant prey. *Marine Mammal Science* 16: 829–834.

Harcourt, R. 1992. Factors affecting early mortality in the South American fur seal *Arctocephalus australis* in Peru: density-related effects and predation. *Journal of Zoology* 226: 259–270.

Harding, K.C. and Harkonen, T.J. 1999. Development in the Baltic grey seal *(Halichoerus grypus)* and ringed seal *(Phoca hispida)* populations during the 20th century. *Ambio* 28(7): 619–627.

Harvey, J.T. 1987. Population dynamics, annual food consumption, movements, and dive behavior of harbor seals, *Phoca vitulina*, in Oregon. Ph.D. thesis, Oregon State University, Corvallis.

Healey, B.P., and G.B. Stenson. 2000. *Estimating pup production and population size of the northwest Atlantic harp seal* (Phoca groenlandica). CSAS Research Document 2000/081. Ottawa: Canadian Science Advisory Secretariat.

Heide-Jorgensen, M.P., and T. Harkonen. 1992. Epizootiology of the seal disease in the eastern North Sea. *Journal of Applied Ecology* 29: 99–107.

Heide-Jorgensen, M.P., T. Harkonen, R. Dietz, and P.M. Thompson. 1992. Retrospective of the 1988 European seal epizootic. *Diseases of Aquatic Organisms* 13: 37–62.

Higgins, L.V. 1993. The nonannual, nonseasonal breeding cycle of the Australian sea lion, *Neophoca cinerea. Journal of Mammalogy* 74: 270–274.

Hoelzel, A.R. 1999. Impact of population bottlenecks on genetic variation and the importance of life-history: a case study of the northern elephant seal. *Biological Journal of the Linnean Society* 68: 23–39.

Hoelzel, A.R., J. Halley, S.J. O'Brien, C. Campagna, T. Arnbom, B. Le Boeuf, K. Ralls, and G.A. Dover. 1993. Elephant seal genetic variation and the use of simulation models to investigate historical population bottlenecks. *Journal of Heredity* 84: 443–449.

Hubbs, C.L. 1956. Back from oblivion, Guadalupe fur seal: still a living species. *Pacific Discovery* 9: 14–21.

Jefferson, T.A., P.J. Stacey, and R.W. Baird. 1991. A review of killer whale interactions with other marine mammals: predation to co-existence. *Mammal Review* 21: 151–180.

Jeffries, S.J., R.F. Brown, H.R. Huber, and R.L. DeLong. 1997. Assessment of harbor seals in Washington and Oregon, 1996, in *Marine Mammal Protection Act and Endangered Species Act implementation program, 1996*. P.S. Hill and D.P. DeMaster, eds. AFSC Processed Report 97-10. Seattle: Alaska Fisheries Science Center, pp. 83–94.

Kenyon, K.W. 1977. Caribbean monk seal extinct. *Journal of Mammalogy* 58: 97–98.

Kovacs, K.M. 2002. Bearded seal, in *Encyclopedia of marine mammals*. W.F. Perrin, B. Würsig, and J.G.M. Thewissen, eds. San Diego: Academic Press, pp. 84–87.

Lavigne, D.M. 1999. Estimating total kill of Northwest Atlantic harp seals, 1994–1998. *Marine Mammal Science* 15: 871–878.

Laws, R.M. 1973. Population increases of fur seals at South Georgia. *Polar Record* 16: 856–858.

———. 1977. Seals and whales of the Southern Ocean. *Philosophical Transactions of the Royal Society of London Series B: Biological Sciences* 279: 81–96.

LeBoeuf, B.J., D.E. Crocker, D.P. Costa, S.B. Blackwell, P.M. Webb, and D.S. Houser. 2000. Foraging ecology of northern elephant seals. *Ecological Monographs* 70: 353–382.

Likhoshway, Y.V., M.A. Grachev, V.P. Kumarev, Y.V. Solodun, O.A. Goldberg, O.I. Belykh, F.G. Nagiev, V.G. Nikulina, and B.S. Kolesnik. 1989. Baikal seal virus. *Nature* 339(6222): 266.

López, J.C. and D. López. 1985. Killer whales *(Orcinus orca)* of Patagonia and their behavior of intentional stranding while hunting nearshore. *Journal of Mammalogy* 66: 181–183.

Loughlin, T.R., A.S. Perlov, and V.A. Vladimirov. 1992. Range-wide survey and estimation of total number of Steller sea lions in 1989. *Marine Mammal Science* 8: 220–239.

Lucas, Z. and P.Y. Daoust. 2002. Large increases of harp seals *(Phoca groenlandica)* and hooded seals *(Cystophora cristata)* on Sable Island, Nova Scotia, since 1995. *Polar Biology* 25: 562–568.

Lucas, Z. and W.T. Stobo. 2000. Shark-inflicted mortality on a population of harbour seals *(Phoca vitulina)* at Sable Island, Nova Scotia. *Journal of Zoology* 252: 405–414.

Mate, B.R. and J.T. Harvey (editors). 1987. *Acoustical deterrents in marine mammal conflicts with fisheries: a workshop held February 17–18, 1986, at Newport, Oregon*. Publ. No. ORESU-W-86-001. Corvallis: Oregon State University.

McMahon, C.R., H.R. Burton, and M.N. Bester. 2003. A demographic comparison of two southern elephant seal populations. *Journal of Animal Ecology* 72: 61–74.

Miller, A.K. and W.J. Sydeman. 2004. Rockfish response to low-frequency ocean climate change as revealed by the diet of a marine bird over multiple time scales. *Marine Ecology Progress Series* 281: 207–216.

Miyazaki, N. 2002. Ringed, Caspian, and Baikal seals, in *Encyclopedia of marine mammals*. W.F. Perrin, B. Würsig, and J.G.M. Thewissen, eds. San Diego: Academic Press, pp. 1033–1037.

NMFS. 1997. *Impacts of California sea lions and Pacific harbor seals on salmonids and the coastal ecosystems of Washington, Oregon, and California*. NOAA Technical Memorandum NMFS-NWFSC-28. Seattle: Northwest Fisheries Science Center.

NRC. 2003. *The decline of the Steller sea lion in Alaskan waters: untangling food webs and fishing nets*. Washington, DC: National Academies Press.

Pitcher, K.W. 1990. Major decline in number of harbor seals *Phoca vitulina richardsi* on Tugidak Island, Gulf of Alaska, North Pacific Ocean. *Marine Mammal Science* 6: 121–134.

Reeves, R.R. 2002. Conservation efforts, in *Encyclopedia of marine mammals*. W.F. Perrin, B. Würsig, and J.G.M. Thewissen eds. San Diego: Academic Press, pp. 276–297

Reijnders, P.J.H. and A. Aguilar. 2002. Pollution and marine mammals, in *Encyclopedia of marine mammals*. W.F. Perrin, B. Würsig, and J.G.M. Thewissen, eds. San Diego: Academic Press, pp. 948–957.

Robinson, S., L. Wynen, and S. Goldsworthy. 1999. Predation by a Hooker's sea lion *(Phocarctos hookeri)* on a small population of fur seals *(Arctocephalus spp.)* at Macquarie Island. *Marine Mammal Science* 15: 888–893.

Roman, R. and S.R. Palumbi. 2003. Whales before whaling in the North Atlantic. *Science* 301: 508–510.

Salazar, S.K. 2002. Lobo marino y lobo peletero, in *Reserva marina de Galápagos: línea base de la biodiversidad*. E. Danulat and G.J. Edgar, eds. Puerto Ayora, Santa Cruz, Galápagos, Ecuador: Charles Darwin Foundation, pp. 267–289.

Shaughnessy, P.D., T.E. Dennis, and P.G. Seager. 2005. Status of Australian sea lions, *Neophoca cinerea*, and New Zealand fur seals, *Arctocephalus forsteri*, on Eyre Peninsula and the far west coast of South Australia. *Wildlife Research* 32(1): 85–101.

Small, R.J., G.W. Pendleton, and K.W. Pitcher. 2003. Trends in abundance of Alaska harbor seals, 1983-2001. *Marine Mammal Science* 19: 344–362.

Smith, T.G., D.B. Siniff, R. Reichle, S. Stone. 1981. Coordinated behavior of killer whales, *Orca orcinus* hunting crab-eater seal, *Lobodon carcinophagus*. *Canadian Journal of Zoology* 59(6): 1185–1189.

Springer, A.M., J.A. Estes, G.B. van Vliet, T.M. Williams, D.F. Doak, E.M. Danner, K.A. Forney, and B. Pfister. 2003. Sequential megafaunal collapse in the North Pacific Ocean: an ongoing legacy of industrial whaling? *Proceedings of the National Academy of Sciences* 100: 12223–12228.

Stenson, G.B., R.A. Myers, M.O. Hammill, I.H. Ni, W.G. Warren, and M.C.S. Kingsley. 1993. Pup production of harp seals, *Phoca groenlandica*, in the northwest Atlantic. *Canadian Journal of Fisheries and Aquatic Sciences* 50: 2429–2439.

Stewart, B.S. and R.L. De Long. 1993. Seasonal dispersion and habitat use of foraging northern elephant seals. *Symposia of the Zoological Society of London* 66: 179–194.

Stewart, B.S., P.K. Yochem, H.R. Huber, R.L. DeLong, R.J. Jameson, W.J. Sydeman, S.G. Allen, and B.J. Le Boeuf. 1994. History and present status of the northern elephant seal population, in *Elephant seals: population ecology, behavior, and physiology*. B.J. Le

Boeuf and R.M. Laws, eds. Berkeley: University of California Press, pp. 29–48.

Stewart, R.E.A. and D.M. Lavigne. 1984. Energy transfer and female condition in nursing harp seals *Phoca groenlandica*. *Holarctic Ecology* 7: 182–194.

Stirling, I. 1975. Factors affecting the evolution of social behavior in the pinnipedia. *Rapports et Procès-verbaux des Réunions du Conseil International pour l'Exploration de la Mer* 169: 205–212.

———. 1983. The evolution of mating systems in pinnipeds, in *Advances in the study of mammalian behavior*. J.F. Eisenberg and D.G. Kleiman, eds. American Society of Mammalogists, Special Publication No. 7. Lawrence, KS: Allen Press, pp. 489–527.

Sydeman, W.J. and S.G. Allen. 1999. Pinniped population dynamics in central California: correlations with sea surface temperature and upwelling indices. *Marine Mammal Science* 15: 446–461.

Testa, J.W., G. Oehlert, D.G. Ainley, J.L. Bengtson, D.B. Siniff, R.M. Laws, and D. Rounsevell. 1991. Temporal variability in Antarctic marine ecosystems: periodic fluctuations in the phocid seals. *Canadian Journal of Fisheries and Aquatic Sciences* 48: 631–639.

Thompson, D., I. Strange, M. Riddy, and C.D. Duck. 2005. The size and status of the population of southern sea lions *Otaria flavescens* in the Falkland Islands. *Biological Conservation* 121: 357–367.

Thompson. P.M., H. Thompson, and A.J. Hall. 2002. Prevalence of morbillivirus antibodies in Scottish harbour seals. *Veterinary Record* 151(20): 609–610.

Thrush, S.F., J.E. Hewitt, V.J. Cummings, P.K. Dayton, M. Cryer, S.J. Turner, G.A. Funnell, R.G. Budd, C.J. Milburn, and M.R. Wilkinson. 1998. Disturbance of the marine benthic habitat by commercial fishing: impacts at the scale of the fishery. *Ecological Applications* 8: 866–879.

Townsend, C.H. 1928. Reappearance of the lower California fur seal. *Bulletin of the New York Zoological Society* 31: 173–174.

Trillmich, F., K.A. Ono, D.P. Costa, R.L. DeLong, S.D. Feldkamp, J.M. Francis, R.L. Gentry, C.B. Heath, B.J. LeBoeuf, P. Majluf, and A.E. York. 1991. The effects of El Niño on pinniped populations in the eastern Pacific, in *Pinnipeds and El Niño: responses to environmental stress*. F. Trillmich and K.A. Ono, eds., in *Ecological Studies*, vol. 88. Berlin: Springer-Verlag, pp. 247–270.

Tynan, C.T. and D.P. DeMaster. 1997. Observations and predictions of Arctic climatic change: potential effects on marine mammals. *Arctic* 50: 308–322.

Waring, G.T., J.M. Quintal, and C.P. Fairfield. 2002. *U.S. Atlantic and Gulf of Mexico marine mammal stock assessments 2002*. NOAA Technical Memorandum NMFS-NE-169. Woods Hole, MA: Northeast Fisheries Science Center.

Weber, D.S., B.S. Stewart, and N. Lehman. 2004. Genetic consequences of a severe population bottleneck in the Guadalupe fur seal *(Arctocephalus townsendi)*. *Journal of Heredity* 95: 144–153.

Weller, D.W. 2002. Predation on marine mammals, in *Encyclopedia of marine mammals*. W.F. Perrin, B. Würsig, and J.G.M. Thewissen, eds. San Diego: Academic Press, pp. 985–994.

Wickens, P., and A.E. York. 1997. Comparative population dynamics of fur seals. *Marine Mammal Science* 13: 241–292.

Williams, T.M., J.A. Estes, D.F. Doak, and A.M. Springer. 2004. Killer appetites: assessing the role of predators in ecological communities. *Ecology* 85: 3373–3384.

SOCIAL CONTEXT

TWENTY-EIGHT

The Dynamic Between Social Systems and Ocean Ecosystems

Are There Lessons from Commercial Whaling?

DANIEL W. BROMLEY

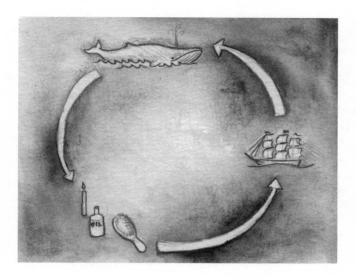

The general issue to be explored here concerns the dynamic nexus between the gradual intensification of human inter-action with—and extraction from—nature and the inevitable biological response to that activity. We may think of this dynamic as an instance of co-evolutionary adaptation in which (1) biological processes undergo transformation in the face of extensive human exploitation, and (2) human processes (and associations) in turn undergo transformation in the face of biological feedback onto human communities that have been organized and structured around this very interaction. Much of the literature in this general area concerns the first phenomena—humans bringing about (generally destructive) effects on natural systems. The feed-back from natural systems to human systems is much less explored. I shall argue here that both aspects of this nexus must be understood if we are to gain a plausible under-standing of why many natural resources are so often overex-ploited. With that understanding in hand, we may gain insights into the ways in which human interaction with nature might be modified and mediated such that timely adjustment of exploitation trajectories is both possible and feasible. The pertinence of this evolutionary perspective is found in its illumination of important aspects of the history of commercial whaling, and in alerting us to emerging

problems before they lead to the sort of destructive history we see in commercial whaling.

A General Model of Co-Evolutionary Elaboration

The general co-evolutionary model envisions four phases:

1. Emergence of the *idea* of a resource
2. *Elaboration* (reconstitution) of the use of that natural resource
3. *Naturalization* of that elaborated activity; and
4. *Apologetics* concerning the naturalized and elaborated enterprises now dependent upon nature.

I will discuss the general co-evolutionary model before turning to the history (and possible future) of commercial whaling.

The Idea of a Resource

A resource represents something of value—a resource is useful, instrumental, and it contributes to something. A resource can be thought of as a means whereby some end might be achieved. Human resources are the particular bundles of skills and attributes embodied in people. One invests in human resources—one's own, through further education, or those of others, by engaging their services—and thereby becomes more valuable in the marketplace. A natural resource is something found in nature, and while one can invest effort in a natural resource—by selectively thinning and pruning a forest or by adding fertilizer to a meager soil—much of our interaction with natural resources consists in taking them as we find them. Natural resources are thought of as free gifts from nature.

This idea of a *natural* resource relates to the matter of how humans come to regard nature. In somewhat different terms, this brings us to the collective idea of the *purposes of nature*—what, exactly, *nature is for*. Humans see nature as being *for* something when we come to understand precisely how it can be instrumentalized. That is, what specific purposes, presently imagined to be important, can be served by a particular pattern of use and extraction focused on a particular physical (tangible) aspect of the natural world? The end result of this mental process is that a particular part of the natural world *becomes a resource*. Notice that this identification of what is a resource can play out at the level of the physical object itself or at the level of the constituent parts (the attributes) of the item.

Teak forests are an example of the first idea. In this case, wood of a very particular kind became the object of human interest, and there were (are) no near substitutes for the "look and the feel" of teak. When teak became scarce, that scarcity induced a search for feasible, though much inferior, substitutes. However, forests also offer resources of a less specific nature, such as firewood. If one particularly desirable form of firewood disappears, there are others that, while not necessarily perfect substitutes, will provide the services (fire) associated with the preferred species.

We see that a resource "becomes"—comes into being—when some part of nature is found to have usefulness to humans. Before a particular object (or attribute of nature) acquires specific human-defined utility, it remains simply an invisible and undifferentiated part of "the environment" (or "the ecosystem"). But once a particular part of the natural environment becomes commoditized in this way, we tend to see nature in the light of this specific object. In a sense, the object of our interest comes to represent for us the prevailing purpose of nature. The philosopher John Locke was an important contributor to this general idea. Locke embellished a prevailing "creation story" in which a Calvinist God gave the earth to humans *in common* and admonished them to take dominion over that commons by, among other things, mixing their labor with it (Kreuckeberg 1999). On the Lockean presumption that humans have "natural rights" in their own labor power, it was then easy to believe that the only way for humans to have an incentive to mix labor with something that, in the beginning, belonged to no one was to reward them, both with the fruits of their labor and with the thing on which they have so assiduously labored. On land

this shows up in the prevailing "property rights" dogma, which motivates many to insist that the primary role of government is to protect property rights (Bromley 1993, 1997). In ocean fisheries we see a version of this Lockean ideology in the rush to bestow (free of charge) individual fishing quotas (IFQs) to all who can demonstrate some credible catch history in particular fisheries. These free gifts of the public wealth to an extractive industry represent billions of dollars of annual income and wealth. For example, the total value of U.S. commercial fish landings in 2002 was $3.1 billion. If this is a plausible prediction of annual future landings, then the present value of this income stream into perpetuity discounted at 3% is approximately $95 billion. The Lockean myth persists in the assertion that making fishers "owners" of a share of a total allowable catch will thereby make them good stewards of nature. Such claims are not supported by economic theory (Bromley 1991; Clark 1973; Macinko and Bromley 2002, 2004; Page 1977; Smith 1968).

Notice that this first stage of the co-evolutionary model builds on an idea about the purpose of some specific part of nature and then ratifies this idea by insisting that this new use is precisely *what nature is for*.

Elaboration and Reconstitution

In the second phase of this co-evolutionary process, a growing dependence on specific parts of the natural world becomes regularized. This regularization leads to the extractive activity becoming both elaborated and reconstituted. Specifically, as the extraction of some particular part of nature evolves, those individuals associated with its extraction will become important and integral parts of the specific "economy" defined by the resource. Examples include loggers, oilmen, coal miners, whalers, and fishers.

It is here that we see the "professionalization" of those engaged in natural resource extraction. That is, an *activity* becomes transformed into an *industry*. The process becomes "industrial" to the extent that there is a particular emergent and evolved production technology with associated patterns of hierarchy, control, and social and economic advantage. Those engaged in the newly industrialized exploitation of nature come to be recognized as participants in a legitimate *going concern*. With this recognition, the emergent industry becomes the justifiable focus of government attention, and therefore the rightful (deserving) recipient of official sanction and financial largess. Something else occurs in this phase. An uncoordinated aggregation of individuals, engaged in what may be random, periodic, and opportunistic economic behavior, become transformed—reconstituted—as professionals. This transition into a proper and recognized "industrial category" becomes the necessary grounds for reconstituting those involved in that endeavor. Notice that this reconstitution legitimizes the participants—by now becoming identified with the resource, they come to be socially and economically differentiated. This differentiation has great value to them and to their "place" in the evolved economic geography. Place becomes an essential part of their reconstitution. Certain towns are ranching towns (Elko, Nevada), other places are oil towns (Midland, Texas), others are (were) whaling towns (New Bedford, Massachusetts), while other places are lumber towns (Medford, Oregon). The evolved industry, based on the extraction of a natural resource, comes to constitute the meaning of particular places that have become dependent on that extraction.

Naturalization

In the third phase the elaborated and reconstituted industrial activity becomes *naturalized*. By "naturalized" I mean that the specific use of some natural resource comes to be seen as right, inevitable, good—even necessary. Notice that this naturalization follows from the first phase, in which the social definition of the purpose of nature (what nature is for) renders it quite inevitable that those who then implement this vision come to be regarded as serving society by doing what has come to be thought as "natural" to do. It is only natural that cattle graze on the open range—it is too dry for agriculture. It is only natural that trees be harvested, otherwise they will simply grow old and die. It is quite natural that the "surplus" production of fish be harvested, for if not they will ultimately die and thus be "wasted" (never mind that dead fish surely contribute *something* to the chemical and physical composition of their habitat, and in that sense, there are no available surpluses in nature).

Once this particular form of human interaction with nature has become naturalized, it is inevitable that members of society will come to see it as a rationalization—a justification—of those practices central to the viability of the naturalized process. In some environmental disputes, efforts to alter the use of nature will be opposed on quite different grounds than might have been present when the interaction began. Fish landings that threaten sustainable stock levels will be defended on the grounds that particular communities depend on the jobs and income. As part of this process, the efforts of scientists to monitor stock levels may be discredited and disputed. Efforts to protect old-growth timber as habitat for spotted owls will be opposed on the grounds that small rural towns will suffer without logging. The loss of logging jobs will threaten the social and economic viability of these places that have come to be defined by the harvesting of trees. Efforts to scale back grazing on public lands will be opposed on the grounds that some ranchers will not be able to survive with smaller herds of cattle. Even pollution control efforts will be resisted on the grounds that stopping pollution will threaten the economic viability of particular factories and towns.

Efforts to limit or redefine the extent of the "naturalized" uses of nature are inevitably opposed on the grounds that doing so will be detrimental to jobs and economic prosperity in the immediate vicinity. Notice that as the resource-based industry becomes elaborated and naturalized, the purpose of nature undergoes a transformation. While the original idea of a natural resource arises in terms of its potential usefulness to those who undertake its exploitation (or to those who purchase it from the exploiters), when problems arise, it is not uncommon to see the defense of the naturalized industry shift to other ground. Suddenly the use of some specific part of nature is justified in terms of protecting the livelihoods of those engaged in the exploitation of the resource. It is rare to find that the *original* purpose of nature was concerned with jobs and income growth in a particular place. Rather, resource extraction results in the reification of place, and then the importance of place comes to be a central and validating aspect of natural resource policy. This observation brings us to the fourth and final stage of evolution.

Apologetics

The final phase in the co-evolutionary model concerns the politics of victimization once it is realized that existing patterns of resource use cannot be sustained. It is here that participants in the human enterprise (industry) suddenly associated with deleterious impacts on the biological system will often characterize themselves as the "innocent victims" of a new force (environmentalism) that (1) fails to understand the defining purpose of the industry (what nature is "really" for); (2) does not appreciate the extent to which participants in the industry voluntarily take actions to protect nature; and (3) tends to value nature more than it seems to value the families who depend on the jobs and products of the industry now under scrutiny. The full measure of this rationalization emerges when the industry is able to mobilize political support for financial indemnification to mitigate the costs of necessary adjustment in their interactions with nature. Indemnification not only eases the financial burden on those whose economic pursuits are now judged to be antisocial; it also puts the citizenry (taxpayers) on notice that it will not be costless to challenge the industrial economic system now found menacing to nature.

I now turn to a discussion of this co-evolutionary model in the context of commercial whaling.

The Rise and Fall of Commercial Whaling

In their book *In Pursuit of Leviathan*, economic historians Lance Davis, Robert Gallman, and Karin Gleiter open with the observation that:

> Whaling, today, is pursued by small fleets of Norwegian and Japanese vessels, and by Inuit living beside the Arctic Ocean. In the context of the world economy, the industry is minute. . . . One hundred and fifty years ago, the world was different. Whaling was a major economic activity, and it was centered on the United States. In value of output, whaling was fifth among U.S. industries. . . . New Bedford, Massachusetts, the leading whaling port, was said to be the richest town in the country, and Hetty Green, later called the Witch of Wall Street, would shortly inherit two New Bedford whaling fortunes and become the wealthiest woman in America.

How the U.S. industry achieved this distinction and why it declined so rapidly, disappearing before the end of the 1920s, are questions of substance, not grist for antiquarian mills. There are lessons to be learned from the history of American whaling that are germane to modern interests—indeed modern preoccupations (Davis et al. 1997: 4).

As discussed in the first section, we are interested in the delicate balancing that is the inevitable accompaniment of the intense and often large-scale human extraction of a particular natural resource. Davis et al. are prescient when they observe that American commercial whaling informs "modern interests—indeed modern preoccupations." It would seem that the past and the future of commercial whaling offer a unique opportunity to explore the larger lessons pertinent to contemporary relations between ecosystems and social systems. This issue is of concern in two important respects. First, some species of whale are now said to be "recovered" and thus "sustained harvesting" poses no threat to that recovery. Second, a number of other disputes over human interaction with nature can be illuminated by a careful assessment of the rise and decline of the commercial whaling industry.

That history entails the massive and fairly rapid mobilization of labor and capital dedicated to the acquisition of a natural resource with very specific market attributes—in this case, baleen, lighting, and lubrication. The same story exists for beaver and other furs when America was still a vast and wild frontier. A contemporary account might concern genetic resources constituting the varied biomes of the tropics. Returning to the sea, recent accounts seem to suggest that the bulk of the world's fisheries are in an advanced state of exploitation and degradation (FAO 2002), and many of the large predatory fish in the world's oceans have been seriously depleted (Meyers and Worm 2003; Pauly et al. 1998, 2002).

The lesson here concerns how human avarice becomes systematized in evolving patterns of interaction—such patterns then impelling the elaboration and expansion of these emergent systems that are soon revealed to be biologically unsustainable. Standard accounts of resource degradation focus attention on insecure property rights, or on what is known as a "faulty telescopic faculty"—the inability to see the future correctly and the correlated failure to use (extract) parts of nature at a rate consistent with its ability to reproduce and thereby be sustainable. These standard accounts often pay too little attention to the context within which human systems evolve as they become defined by the natural systems with which they interact—and on which they become dependent.

In the context of the co-evolutionary model spelled out in the foregoing paragraphs, whaling had emerged at a subsistence level because of particular attributes of whale blubber found useful for lighting and other local needs. Over time these quite specific benefits became widely understood beyond their origins in subsistence communities, and it was not long until whale oil dominated the lighting market. Whale blubber also played a role in lubrication before the advent of petroleum products. We see that resources "become" (and recede) depending upon human needs and wants for the attributes of those resources, and their general availability to humans.

The co-evolutionary pathways under discussion concern the stages through which human processes, institutional arrangements, and organizations pass and are transformed by the very act of extracting particular aspects of nature. Once the initial idea takes hold—that some particular piece of the natural world holds important and useful attributes—the basic idea is developed and elaborated. In whaling, early practice was confined to gathering so-called drift whales. From this small beginning there emerged an "artisanal," inshore boat-whaling phase, which then evolved into commercial offshore whaling as we came to know it. This evolution required larger vessels, skilled captains, and, most certainly, skilled hunters. As this process evolved, there was a transition from part-time and opportunistic whaling to serious and persistent commercial whaling over both time and space. In essence, the process of elaboration and reconstitution entailed the *professionalization* of whaling. Professionalization implies that an economic *activity* is transformed into an *industry*. The process becomes "industrial" to the extent that there is a particular emergent and evolved *production technology* with associated patterns of hierarchy, control, and social and economic advantage.

One indicator of this is seen with the bounty system introduced by the Massachusetts legislature following the Revolutionary War. The war had been hard on the industry, so the state of Massachusetts offered a financial reward for whale products returned by vessels owned and operated by state residents (Davis et al. 1997). This new involvement by government agencies reflects the official recognition of an organized *industry* worthy of legislative attention. With this recognition, the emergent industry becomes the legitimate focus of government attention and, therefore, the rightful (deserving) recipient of official sanction and financial largess.

As we saw above, something else occurs in this phase. An uncoordinated aggregation of individuals engaged in random, periodic, and opportunistic economic behavior became transformed—reconstituted—as *whalers*. This transition into a proper and recognized "industrial category" then became the grounds for reconstituting those involved in that endeavor. Notice that this reconstitution legitimizes the participants—by now becoming a *whaler* they come to be socially and economically differentiated and this differentiation has great value to them and to their "place" in the evolved economic geography. Notice that *place* becomes an essential part of their reconstitution. New Bedford was to whalers as northern Michigan, Minnesota, and Wisconsin were to lumberjacks and what Cheyenne was to cowboys. Today we see similar connections between place and particular industrial activities in Alaskan fisheries.

In whaling, full reconstitution—the transition into a "modern industry"—probably occurred at the end of the 1860s with the introduction of steam-powered catcher

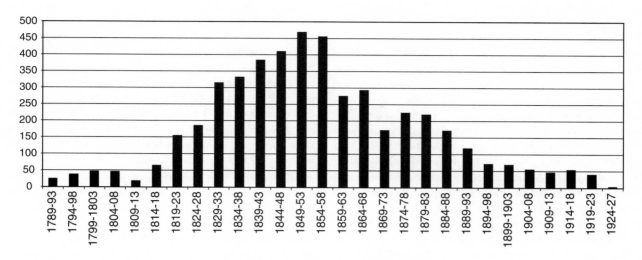

FIGURE 28.1. Whaling voyages from New Bedford, Massachusetts (Davis et al. 1997).

vessels carrying high-powered harpoon guns (Davis et al. 1997). These technical innovations allowed more lethal pursuit of traditional targets (right, bowhead, and sperm whales) but also permitted whalers to pursue the faster rorquals (sei, fin, humpback, and blue whales and, later, Bryde's and minke whales). Figure 28.1 reveals the rapid growth in voyages from New Bedford that, interestingly, predated these technical innovations. Indeed, these innovations appear to have missed the peak period in the American industry; this is not at all surprising in the exploitation of natural resources.

The rapid growth of any industry will always trigger interest in the technology and practices of that industry. Given the lags and inevitable development times, technical innovation often comes along behind the original expansionary peaks. The important difference in commercial whaling—and many other natural resource activities—is that this induced phase of technical change often comes *after* the natural resource has begun to decline in response to the original flurry of extractive activity. As new technology tends to prolong the profitable extraction of the resource beyond that possible under the original technology, this new technology can, if not controlled, accelerate resource decline. After all, the driving impetus for technical change is to lower the costs of resource extraction, so it is inevitable that new technology allows the new industry to be more efficacious in its extraction of economic value from nature.

With full industrial reconstitution, the new going concern becomes of such scope and importance that we see the emergence of a range of formal associations whose purpose is to provide a forum for collective activities associated with the going concern. These can be trade associations, advocacy groups, or quality assurance endeavors. Given the global (as opposed to merely national) reach of the commercial whaling industry, it came into its own with the creation of the International Whaling Commission (IWC) in 1946. The League of

Nations attempted, without much success, to regulate whaling as early as 1924. The 1931 Convention for the Regulation of Whaling committed the signatory nations to protect right whales, immature whales, and females with calves (Davis et al. 1997: 52). But with Japan and Germany refusing to sign, the effort failed. It was not until after World War II that regulation of the industry became an international concern. In particular,

> The International Whaling Commission (IWC) was set up under the International Convention for the Regulation of Whaling which was signed in Washington DC on 2nd December 1946. The purpose of the Convention is to provide for the proper conservation of whale stocks and thus make possible the orderly development of the whaling industry.
>
> The main duty of the IWC is to keep under review and revise as necessary the measures laid down in the Schedule to the Convention which govern the conduct of whaling throughout the world. These measures, among other things, provide for the complete protection of certain species; designate specified areas as whale sanctuaries; set limits on the numbers and size of whales which may be taken; prescribe open and closed seasons and areas for whaling; and prohibit the capture of suckling calves and female whales accompanied by calves. The compilation of catch reports and other statistical and biological records is also required.
>
> In addition, the Commission encourages, co-ordinates and funds whale research, publishes the results of scientific research and promotes studies into related matters such as the humaneness of the killing operations. (IWC 2006) [www.iwcoffice.org].

Notice that a key component of the IWC is not to protect whales *per se* but rather to "make possible the orderly development of the whaling industry." That is, a legitimate and "reconstituted" industry now finds it necessary and valuable to organize and mobilize for the "orderly development" of that industry. Very soon, commercial whaling became "naturalized." As above, naturalization is the process of coming to see the reconstituted process as natural, right, necessary, good, and as a higher form of social and economic organization. At the beginning of the twentieth century it would not be an exaggeration to suggest that whaling was seen as natural, necessary, good, and right. After all, if whaling were to collapse, it would have meant a loss of many jobs, New Bedford would have suffered serious economic losses, and the ripple effect throughout the New England whaling economy would have been severe. We see that the emerged industrial process had become too important—politically, economically, socially, and culturally—to entertain the prospects of decline.

The creation of the IWC was an essential aspect of this process. At last the industry could point to the IWC charter as proof of the seriousness of the industry in protecting its asset base (the source of its wealth), and it could show that it was intent on ensuring the orderly "development" of the industry. Whalers, in other words, were no longer unorganized opportunists but were the very heart of an international effort to ensure the sustainability of the industry and, incidentally, the whales on which that industry was dependent. When, in the middle of the twentieth century, public concern over whaling began to emerge, the industry was able to point to the IWC as proof of its commitment to sustainable harvests.

But, of course, this defense missed the point. By the time public sentiment had turned against whaling, the central issue was no longer a concern for *sustainable harvests*. The issue had become, rather, the existence of an industry that *pursued and killed mammals*—and not just any mammal, but the largest mammal on earth, a mammal that seems to communicate with its peers, an animal that can be viewed from coastal waters and charter boats, and an animal that seems to produce musical sounds. Here was an "intelligent" and empathetic animal—*charismatic megafauna* is the current term of art. With these realizations, whales were suddenly too much like humans for comfort, and thus *any* level of whaling came to be seen as unacceptable. Suddenly nature was no longer "for" sustainable harvests of creatures that seemed a bit too "human." That was not the "purpose" of nature. When the reigning purposes of nature suddenly are redefined in the face of an industry that has grown accustomed to the manifold political and economic benefits of social legitimacy, problems arise.

It cannot be a surprise, therefore, that the next phase in this evolutionary process finds the industry regarding itself as the innocent victim of a new force that (1) fails to understand its defining purpose; (2) does not appreciate the extent to which participants in the industry undertake actions to protect its asset base; and (3) tends to value animals more than it values the families who depend on the jobs and products of the industry under assault. This victimization is compounded when only a subset of the international community finds an industrial activity to be objectionable, while the remainder of the international community continues to see the contested activity as quite normal and therefore "right."

It is during this phase that we tend to see the emergence of a sense of *entitlement* on the part of the extractive industry. The political and economic legitimacy that comes from the creation of an organization such as the IWC reinforces this sense of entitlement. In some industries, demands will be made for a variety of compensation schemes. In the case of the American whaling industry, most of the reason for its slow decline was not external pressure from those offended by the activity, nor was there concern over the extinction or increased scarcity of whales. Rather, the decline seems to be attributable to the ready availability of superior substitutes (Davis et al. 1997). But of course international whaling continued—principally by Norway, Japan, and the former Soviet Union. By 1960 it was apparent that the Antarctic whales were in decline, and this raised general alarms about all whales.

On the Future of Commercial Whaling

Under the auspices of the International Whaling Commission, commercial whaling was suspended in 1986. By 1992, only Japan and Iceland were still engaged in commercial whaling, although it was in that year that Norway resumed commercial whaling. At present it seems that several populations of the two species of the northern right whale, the western North Pacific population of the gray whale, and four of the five populations of the bowhead whale are threatened with extinction in the near future; all other stocks are either said to be viable and capable of recovering, considered to be recovering, or classified as recovered. The three nations that currently practice commercial or scientific whaling (Iceland, Japan, and Norway) restrict their activities to fin, sei, minke, and sperm whales, along with a few of the smaller cetaceans. None of these species appears to be in any danger from current levels of harvesting (Davis et al. 1997).

As we consider the history of commercial whaling, the often hostile reaction to those few nations that continue to engage in scientific and commercial whaling, and the current management protocols advanced by the IWC, we can explore the scientific and ethical issues central to the past and possible future of whaling. For example, if some species of whales seem to be recovered—or appear to be on the road to recovery—we face two interesting questions: (1) Why is continued harvesting so contentious? (2) What is to be expected if there should emerge an interest among other nations in beginning to engage in commercial harvesting?

There are two related issues here. The first issue pertains to the claim that particular stocks are "now recovered" or "on the road to recovery." The second issue pertains to the claim that "a well-managed subsistence harvest" of a particular species is "consistent with recovery to healthy population levels." That is, one must ask what, exactly, is meant when scientists declare that whales have recovered (or are recovering), and that particular harvest protocols do not threaten that recovery?

Recovery

Notice that the assertion about a species being recovered is more robust (parsimonious) than is the derivative assertion concerning prospects of renewed harvests that do not pose a threat to recovery. By "parsimonious" I mean that the concept of recovery is predicated upon reasonably well-known mathematical structures and empirical claims, whereas the assertion about the innocence of renewed harvests requires additional strong assumptions about the levels of those harvests, the likely responses to evidence of population decline, the pressures to keep up unsustainable harvest levels, and other factors. In other words, respected biologists can write down their models, their assumptions, and their empirical evidence, and they can develop analytical processes to generate plausible time-dependent parameters of various whale populations. In short, they can produce plausible conjectures about past, current, and future whale populations given clearly stipulated scenarios concerning specific whale groups and the environments in which they live. When a number of these analyses have been carried out and reviewed and debated by the community of whale biologists, the community of whale "experts" will come to feel confident in asserting that a particular species—or a group of species—is now viable and either recovered, or on the road to recovery.

Such assertions are often described as statements about what the "science tells us" with respect to a particular issue. When people hear what the "science tells us," there is a sense that the matter has now reached closure. It may be said that we have reached decisive belief—or that we have produced decisive evidence—concerning the recovery of certain whale populations. Of course, we are never certain in scientific matters, and so there is no presumption here of absolute certainty. And most who listen to what the "science tells us" know very well that there will always remain some uncertainty. But, in general, when the community of whale experts speaks with one voice about "recovery," the rest of us may generally take this apparent consensus as quite decisive evidence that certain species of whales are now out of "danger."

However, further clarification is called for. In particular, just because those who study the population dynamics of whales have reached consensus on the alleged recovery of particular whale populations, there is no necessary assurance that these empirical claims will garner wide agreement within the broader biological community. Put somewhat differently, the assertions of a small subset of the natural science community is not necessarily decisive when subjected to the scrutiny of others who have related skills but a somewhat different perspective on individual populations embedded within a dynamic ecological setting. Most often this different perspective will come in the form of challenges to the axioms and assumptions underlying the analysis of whale populations. One does not usefully challenge the conclusions of

science—one challenges the axioms and assumptions of the analytical engines that produce particular theoretical or empirical claims (the "conclusions") advanced by scientists (Bromley 2004).

Perhaps oceanographers will have quite different ideas about currents, temperature profiles, seasonal patterns, and other factors critical to the empirical claims of the population experts. Physiologists may have quite different ideas about nutritional stress, reproductive success, life expectancy, sex ratios, or other matters that affect the empirical claims of the population experts. Pathologists may have quite different ideas about pathogens, about life expectancy, and about vulnerability to new and unanticipated threats to whales. Each of these branches of the natural sciences has legitimate reasons to challenge the empirical claims about the allegedly sustainable recovery of particular whale populations.

We see that adhering to what "the science says" may not be as straightforward as it might seem. *Warranted assertions* are those arising from settled beliefs emanating from a community of individuals thought to have special epistemic sanction to study, ponder, undertake research about, and then pronounce on, particular matters. However, not all assertions from a scientific discipline are warranted assertions. Specifically, only those assertions that enjoy widespread assent within the broader encompassing discipline earn the right to be regarded as warranted assertions—or warranted beliefs. This broader community goes beyond population biologists and would encompass, as above, oceanographers, physiologists, and pathologists. The purpose of warranted beliefs is to help the rest of us figure out what we ought to believe.

The arrival at knowledge that is sufficient to produce warranted assertions is not always straightforward. Indeed, the evolution of science has been toward increasing specialization, so that the highly particular knowledge is very much more elaborated than hitherto, but also more abstracted (isolated) from the complexities of the world in which that particularistic scientific understanding is embedded and dependent. We have a situation in which highly articulated scientific specialization compounds the curse of exogeneity. That is, the narrower we become in our analyses, the greater are the implied connections (dependencies and feedbacks) with the larger world that must, for tractability's sake, be increasingly characterized and parameterized by strong assumptions. It is for this reason that the settled deliberations (the confident pronouncements) of a particular subdiscipline (or a particular research field or specialization) must constantly be held accountable by those who themselves are specialists working *at the boundary* of the isolated analytical work out of which presumptively warranted assertions arise.

Moreover, this high degree of specialization also places a premium on the integrative insights of those who are working one level removed from (above) the narrow specializations. We may regard these individuals as the theorists, or the synthesizers, or those who can see and integrate the bigger picture emerging from a number of isolated specialties that constitute a scientific discipline or field. In practical terms, subdisciplines, and the highly specialized fields of science, must always answer to the integrating judgments of others in the parent discipline before they can regard their findings as warranted assertions. Once this broader community of scientists has accepted the empirical and conceptual claims of one or more of its constituent parts, then—and only then—can one say that particular empirical claims are indeed *warranted* (Bromley 2004). In the absence of that affirmation from the broader epistemic community, all empirical claims are merely provisional (conditional).

Our focus now shifts from the producers of scientific assertions to the consumers of those assertions. The reality of an educated and discerning citizenry implies that the rest of us are not mere passive vessels into which warranted assertions are poured and immediately acted upon. There was a time when that state of affairs gave free rein to presumptuous princes, kings, military leaders, charismatic dictators, charlatans, and religious leaders who invoked a higher authority to lend credence to their specialized wishes. However, those days are mostly gone. Now, as discerning agents *(homo sapiens)*, we have the obligation to consider pronouncements from all "experts" (or opinion makers) and decide for ourselves whether we will now hold to our prior belief or change it. Specifically, *valuable assertions* are those that can be justified to an audience of attentive and contemplative agents focused on a particular decision (or action). A valuable assertion (or belief) is one upon which I am now prepared to act. Do I believe what the doctors tell me about the reliability of screening for prostate cancer? Do I believe what the geneticists tell me about the innocence of genetically modified organisms? Do I believe the whale experts when they assure us that particular populations are now recovered or on the road to recovery?

The idea of *valuable belief* may be troublesome to some scientists. It will be troublesome precisely because we live in an age in which science is said to hold particular credibility, and scientists work hard to provide the public with the "truth" about specific matters. In the face of this quest and desire for clarity, there will be times when the warranted assertions of scientific experts will be found interesting but not necessarily compelling. That is, not all *warranted assertions are valuable assertions*. In practical terms, the rest of us are under no special obligation, upon hearing the warranted assertions of a particular epistemic (disciplinary) community, to stop what we are doing, immediately accept those assertions as "truth," and act on them. As sapient agents we have the right to demand clear and convincing justification for discarding what we now believe. If the proffered justification by the scientific experts is regarded as somehow deficient, it means that we have not yet been presented with valuable assertions. The choice of what is to be believed is ours, not the scientists'. In more general terms, democracy bestows on individual citizens the authority to decide which plausible futures they find the more compelling. And this requires that individuals retain the right to process authoritative assertions and to reach their own conclusions about what seems better to believe (Bromley 2006).

This democratic ideal is not always reflected in public discourse. Those who resist scientific assertions about particular aspects of public life are often accused of letting emotion get in the way of the objective facts or of failing to engage in rational thought predicated on scientific "truth." They may be accused of being romantic "environmentalists" incapable of clear thought about serious matters. When public discourse concerns the assertions of the latest science, it is often difficult to have a balanced conversation, because those who might disagree with what "the science says" are dismissed as hopelessly uninformed and in the grip of some odd emotional attachments.

The interesting issue for public policy concerns why, exactly, those beliefs contrary to the latest scientific "truth" are vigorously attacked and impugned? The problem centers on what passes for "truth." Those inclined to resist the latest scientific assertion may point out that scientific "truths" seem to change quite regularly. There was a time when the bounty of the world's oceans was said to be limitless. Why should we now eagerly believe what the "science" asserts about recovered whale stocks?

Perhaps the public has some experience with new "truths" appearing that suddenly rattle the collective confidence with which particular beliefs are held, and that therefore ratify particular actions as reasonable and prudent. Why should we now expect individuals, in the face of assertions from the biological community, to find comfort and assurance in the knowledge that the world's whales are safely out of danger of extinction? We see that valuable belief is predicated on the idea of *acquired decisiveness*. By "decisiveness" I mean the arrival at a point in the consideration of possible action that individuals or groups can finally and honestly declare, "This seems the better thing to do at this time." When we can say to ourselves (or to our colleagues in the parliaments, the legislature, the administrative agencies, or the court chambers) that we have reached a decisive point, it means that our settled deliberations have given us a new coherent belief. Moreover, a belief is that upon which we are prepared to act. In other words, we have now found *sufficient reasons* to reconsider the current ban on commercial whaling. Populations have recovered, and there is now compelling evidence that certain species of whales are out of danger.

Sustainable Resumption of Whaling

We now come to the second strand of the problem. Earlier I suggested that the recovery question was more parsimonious than the question about the long-run harmlessness of resumed whaling. Parsimony implies that it requires fewer *strong* assumptions. Reflecting on the evolution of American whaling—indeed, of commercial whaling on the international level—the proposition that sustainable harvests pose no threat to the continued recovery of particular populations of whales is fraught with assumptions. Recall that the exploitation of nature has a history of going from a useful idea, to an elaboration and reconstitution of the participants into an industry, to the politics of naturalization, and finally to apologetics and claims of victimhood.

We see that the resumption of whaling entails a number of important questions. What assurance do we have that these harvests will not recapitulate the earlier account in resource extraction? The strongest assumption is that whaling could somehow avoid the elaboration, the naturalization, and the the presumption of entitlement and sense of victimization that characterizes most human-nature interactions.

It is possible to acquire a sense of this evolutionary process from the development of the American fisheries industry since the establishment of the Exclusive Economic Zone in the late 1970s and continuing to the present time. The nature of the political commitment to an aggressive extraction of resources is apparent from a statement by Senator Warren Magnuson, the individual after whom the first fisheries "conservation and management" act was named. In a 1968 address to federal and state fishery scientists and managers, Senator Magnuson declared:

> You have no time to form study committees. You have no time for biologically researching the animal. . . . Your time must be devoted to determining how we can get out and catch fish. Every activity. . . . whether by the federal or state governments, should be primarily programmed to that goal. Let us not study our resources to death, let us harvest them. (Magnuson 1968: 7–8).

Perhaps those who worry about confident assertions regarding the innocence of renewed commercial harvests of whales have reasons for their apprehensions. Why, the skeptic might ask, should we assume that renewed commercial harvesting of whales would somehow manage to avoid the evolutionary pathway of other extractive industries?

In a larger sense, we have returned to the issue of what nature is "for." It was not so very long ago that extraction was the dominant idea about what nature is for. Those with financial means went off to darkest Africa to return with trophies large and small. Stories abound of hunters slaughtering hundreds of animals on a safari lasting several weeks. Most of them returned home to be acclaimed as local heroes. Today such behavior is rare. It would seem that nature is now *for* something else. Perhaps it is for viewing and for taking pictures.

What is nature for? Why should whales be harvested at all—even if it can be shown *beyond any doubt* that harvesting will not threaten them with extinction? It is not, after all, imperative that animals in the oceans *must* be harvested. And with that being the case, the matter then turns on various reasons that might be mustered concerning *why and why not?* Ought we to do what we can do, merely because we can? The resumption of commercial harvesting of whales entails the asking for and giving of reasons (Brandom 2000). Why, exactly, do you believe it is right for whales to be harvested?

The presumption of species protection via the Endangered Species Act is that bad (human) actions should be stopped to protect particularly valuable aspects of nature. But problems arise if one thing we like is possibly responsible for (or contributes to) the demise of another thing we also happen to like. What do we do when a large number of marine mammals are thought to be "valuable" and worthy of protection? More seriously, what do we do when the evolved purpose of nature is no longer to extract pieces of nature we find valuable but rather to leave it "alone" and then agonize over the evolving reality of the possible disappearance of particular pieces of nature that we have come to revere?

This will be difficult precisely because *we have not done that before*. All conversation about possible future actions with respect to whales and other valued sea life is a conversation involving contested "created imaginings" (Shackle 1961). These contending created imaginings are stories we tell ourselves and others about the future. Created imaginings are contending prescriptions, and all prescriptions are necessarily predictions: Do X, and Y will happen. Future policy about the oceans—and the unavoidable legacy of large-scale removals of biomass—cannot be worked out in the abstract. That policy can only be worked out as we come to grips with the data and the settled deliberations of the broader community of natural and social scientists. It will be worked out by what comes to be seen as the *better story to believe*. And if this story seems the right one to believe, then this story indeed represents the "truth"—at least for now.

Conclusions

Public policy—including environmental policy—is best understood as a process of collective action in liberation, and restraint of individual action (Bromley 1989, 2004). It is *collective* because policy occurs in bodies collectively constructed for precisely this purpose—legislature, courts, executive branch agencies. It is *individual* because it redefines choice sets for each of us—what we may and may not do, what we must and must not do; what we can do with the aid of the collective power, what we cannot expect the collective power to help us to do (Bromley 2006).

Public policy with respect to whaling, or with respect to other natural resources, will often be difficult, because the mental benchmark for anchoring many discussions about nature tends to consist of the sanctification of prehuman settings and circumstances, and this then becomes reified as some "ideal" state. While this account is not quite on a par with the "noble savage" of anthropology, it is clearly the *Edenic Ideal* that relates to what Leo Marx called the "garden" or the "pastoral ideal (Marx 1964)." In much of the literature on natural resource degradation we see this Edenic Ideal advanced as how nature *ought to be (and as it was)* before humans came along to ruin it. This is, appropriately, an example of the *naturalistic fallacy*. The idea here is that the way things *are* (or once were) constitutes the basis for how they *ought to be*.

However, the Edenic Ideal is not very helpful in serious debates about what is best to do with respect to contemporary resource problems. Resource policy issues are best understood as continual conversations, not about the past but about the future. Policy is a process of figuring out which among a number of plausible futures seems most compelling to us. We learn from our history, but history is informative, not determinative. When we consider the history of resource extraction in co-evolutionary terms, it permits us to ponder the dual transformations under way. Of course, human extraction alters natural systems. But the more profound alteration for policy purposes is what occurs within social (human) systems as they come to be defined and redefined by their interaction with natural systems.

This model of co-evolutionary pathways is offered as a plausible set of categories and transitional stages in the process of contemporary human interaction with nature. We see that as a promising and useful idea about a "resource" soon becomes elaborated, this elaboration reconstitutes the participants into an industrial category, these reconstituted participants and the "industry" they represent (and constitute) come to be seen as engaging in normal and right actions ("apologetics"), and then the stage is set for claims of victimhood when problems emerge from the interaction between this human system (this going concern) and the natural world on which it depends. The emergence of a new *human system* can be seen as the emergence of an *economy*— by which I mean a domain of economizing behavior, a realm of job and income creation, and thus a source of socially sanctioned wealth accumulation.

This evolutionary process throws light, I believe, on the inevitable tensions that arise when—for a variety reasons—the reigning purpose of nature that underpins this evolved economy begins to fall into disrepute. The history of commercial whaling certainly fits this category. Whether commercial whaling will again be regarded as a good idea cannot yet be foretold. I very much doubt that this will happen. But we can be sure that if it is to resume, there will be extraordinary and justified concerns that whaling must not be allowed to follow the evolutionary pathway that characterizes so many other industrial activities that draw on nature to create income and wealth.

Acknowledgments

I am grateful to Douglas DeMaster, Dan Doak, Jim Estes, and Terrie Williams for helpful comments on an earlier version of this manuscript.

Literature Cited

Brandom, R.B. 2000. *Articulating reasons*. Cambridge, MA: Harvard University Press.

Bromley, D.W. 1991. *Environment and economy: property rights and public policy*. Oxford: Blackwell.

———. 1993. Regulatory takings: coherent concept or logical contradiction? *Vermont Law Review* 17(3): 647–682.

———. 1997. Constitutional political economy: property claims in a dynamic world. *Contemporary Economic Policy* 15(4): 43–54.

———. 2004. Reconsidering environmental policy: prescriptive consequentialism and volitional pragmatism. *Environmental and Resource Economics* 28(1): 73–99.

———. 2006. *Sufficient reason: volitional pragmatism and the meaning of economic institutions*. Princeton, NJ: Princeton University Press.

Clark, C.W. 1973. Profit maximization and the extinction of animal species. *Journal of Political Economy* 81: 950–961.

Davis, L.E., R.E. Gallman, and K. Gleiter. 1997. *In pursuit of leviathan*. Chicago: University of Chicago Press.

FAO. 2002. *State of world fisheries and aquaculture*. Rome: UN Food and Agricultural Organization. Available at http://www.fao.org/sof/sofia/index_en.htm. Accessed June 20, 2006.

IWC. 2006. IWC information. Cambridge, UK: International Whaling Commission. Available at http://www.iwcoffice.org/commission/iwcmain.htm. Accessed June 20, 2006.

Krueckeberg, D.A. 1999. Private property in Africa: creation stories of economy, state, and culture. *Journal of Planning Education and Research* 19(Winter): 176–182.

Macinko, S. and D.W. Bromley. 2002. *Who owns America's fisheries?* Washington, DC: Island Press.

———. 2004. Property and fisheries for the twenty-first century: seeking coherence from legal and economic doctrine. *Vermont Law Review* 28(3): 623–661.

Magnuson, W.G. 1968. The opportunity is waiting . . . make the most of it, in *The Future of the fishing industry in the United States*. Publications in Fisheries, vol. 4. Seattle: University of Washington Press, pp. 7–8.

Marx, L. 1964. *The machine in the garden*. Oxford: Oxford University Press.

Meyers, R.A. and B. Worm. 2003. Rapid worldwide depletion of predatory fish communities. *Nature* 423(15 May): 280–283.

Page, T. 1977., *Conservation and economic efficiency*. Baltimore, MD: Johns Hopkins University Press.

Pauly, D., V. Christensen, J.D., R. Froese, and F. Torres, Jr. 1998. Fishing down marine food webs. *Science* 279(February 6): 860–863.

Pauly, D., V. Christensen, S. Guenette, T.J. Pitcher, U.R. Sumaila, C.J. Walters, R. Watson, and D. Zeller. 2002. Towards sustainability in world fisheries. *Nature* 418(8 August): 689–695.

Shackle, G.L.S. 1961. *Decision, order, and time in human affairs*. Cambridge: Cambridge University Press.

Smith, V.L. 1968. Economics of production from natural resources. *American Economic Review* 58(3): 409–431.

TWENTY-NINE

Whaling, Law, and Culture

MICHAEL K. ORBACH

All environmental law and policy, including policy governing whaling, involves trade-offs between the state of the world's *biophysical* ecology (i.e., all nonhuman elements of the world and their relationships with one another) and the state of the world's *human* ecology (i.e., humans and their relationships with one another, including human governance institutions). All rules of governance directly affect only human behavior; through that behavior they shape the biophysical world. The configuration of the biophysical environment, in turn, defines the form of the costs and benefits incurred and received by humans in the use of that environment. Furthermore, every decision regarding our relationship with the biophysical environment involves some form of trade-off. The more marine organisms humans extract, the fewer are left in the ocean, which in turn affects the trophic webs of which those organisms are a part. This is true for whales and all other biophysical entities.

All of the trade-offs we make in drafting and implementing our policies are guided by some set of human values. If we value a part of the biophysical ocean environment for *consumptive* purposes, we know there will be less of that item left in the ocean, but we judge that the value to humans of extraction outweighs the value of that item to the biophysical ecosystem. If we value a part of the biophysical environment for ecosystem function or for aesthetic purposes, sometimes we leave that part in its biophysical environment to be appreciated or function in that biophysical setting; this is

nonextractive or *nonconsumptive* use. When we make governance rules such as laws and administrative policies, we are expressing some particular form of human values regarding the state of both the human and the biophysical environments. We do not always do this with will full knowledge or forethought, but there is at least a presumptive trade-off made with each decision based on some human value structure.

In this chapter I will discuss the broad-brush history of human value–based governance with respect to whales and whaling. The human history of relationships with whales is long, and I will not dwell on all aspects of that history (see, for example, Reeves and Smith, Chapter 8 of this volume). I will rather focus on the principles of human values and behavior that have formed our recent laws and policies and thus reflected the human attitudes toward, and relationships with, whales. I will further focus on those attitudes and relationships that have been expressed through law and policy in the United States and the principal international organization involved in this area to which the United States has been a party: the International Whaling Commission.

Law and Culture

All law is an expression of culture. Legal statutes and rules are those elements of culture about which we feel strongly enough, and that we share enough with each other, that we write them down as rules of behavior and create some form of sanction for their transgression (Nader 1969). Culture varies throughout the world, and accordingly, so does law. Different local governments, different states, and different nations have different laws reflecting their common culture within their particular governance structure. Thus, different states have different fishing regulations, as do different nations. Law and regulation can vary regionally within one nation. For example, under U.S. law and policy some billfish species are not allowed to be caught or sold commercially on the U.S. East Coast but are considered commercial products on the West Coast. The biological conditions of the Atlantic and Pacific stocks of these species do not differ markedly, but Americans in different regions have decided as a matter of cultural preference to use them in different ways.

With respect to whaling, different nations of the world have widely differing laws with respect to whaling (Friedheim 2001; Gambell 1999). Some consider whales a commercial, consumptive product, while others consider whales a special category of wildlife, essentially to be fully protected from human predation and even from any human effect. These are expressions of cultural values, which these countries have translated into their law and policy.

The idea and activity of "conservation" itself is a culturally defined phenomenon. There is no single, most desirable state of the biophysical ecology—simply those states that are more or less desired by humans. "Pristine nature" apart from human influence does not really exist on the earth today; essentially every part of the biophysical ecology is anthropogenically affected in some way. When we seek to "conserve" a species or environment, we are adopting a set of human values with regard to a desired ratio of those species and environments to others and to the flow of costs and benefits of those species and environments to humans. Even if humans decide to place a value on a species or environment for biophysical environmental function, it is we humans who are doing the deciding, and defining the value of that function.

In general, when we set out to "conserve" a part of the biophysical environment, we mean to be able to sustain its use or enjoyment by humans over time. Some forms of this use or enjoyment may be extractive, others nonextractive. If we decide to enjoy part of the biophysical environment in a nonextractive way and develop rules accordingly, we often use the term "preservation" rather than "conservation," as in the terms "preservation ethic" and "conservation ethic" associated with John Muir and Gifford Pinchot, respectively (Miller et al. 1987). The form of law and policy we construct reflects our cultural values and our preferred "ethic" of interaction with the biophysical environment.

Whaling, Law, and Public Policy

There is much cultural, and therefore legal, discord in the world over cultural values with respect to whaling (Friedheim 2001). It is useful to trace these cultural values as they

have been expressed in law and policy in the last half century.

One of the first major attempts at public policymaking with respect to whaling occurred, somewhat paradoxically, at an international scale. After a decade of a series of intergovernmental agreements aimed primarily at the market economics of the whaling industry, the International Whaling Commission (IWC) was created in 1946 with the expressed purpose "to provide for the proper conservation of whale stocks and thus make possible the orderly development of the whaling industry" (Gambell 1999: 181). The cultural values expressed in the creation of this organization were (1) conservation of whale stocks in the classic sense of sustainable extractive use, and (2) the wish to develop and maintain an extractive industry based on the whale resource. At the point at which the IWC was created, many of the major whale stocks had been diminished due to overharvest, and many other industries had grown up to satisfy the demand for whale products using products from other sources. In a sense, by the time the IWC was created, whaling was already in decline, although whale harvest would continue and even increase on some stocks in the next few decades.

Throughout the 1950s and 1960s in the United States the environmental movement grew strong. One element of this movement was attention to ocean creatures, not just as a consumptive resource to be sustained, but as a special part of the ocean's biophysical ecology. Many cultural and behavioral trends, including depletion of marine mammal stocks and a focus on whales and other marine mammals in the popular media, led to the Marine Mammal Protection Act (MMPA) of 1972. Every statutory law has certain major defining characteristics, and the MMPA was a very unique law among those passed in the great flush of environmental legislation in the 1970s. The Magnuson-Stevens Fishery Conservation and Management Act (FCMA) of 1976, for example, was clearly a *consumptive* law, based on the principle of maximum sustainable yield (MSY). The Endangered Species Act (ESA) of 1973 was based on the principle that humans should not drive species to extinction and, further, that if a species becomes threatened or endangered, humans should not further "jeopardize" the habitat of that species. The ESA was clearly a *nonconsumptive* law, based on the principle of species "preservation" (Cicin-Sain and Knecht 2000).

The MMPA, however, had some very special qualities. The MMPA mandates that even if a marine mammal species is not threatened or endangered, no "take" of that species is allowed, with few intended exceptions. "Take" is defined as "to harass, hunt, capture or kill, or attempt to harass, hunt, capture or kill any marine mammal" (Baur et al. 1999: 56). Further, the MMPA is based on the principle of optimum sustainable population (OSP), defined as "with respect to any population stock, the number of animals which will result in the maximum productivity of the population of the species keeping in mind the carrying capacity of the habitat and the health of the ecosystem of which they form a constituent element" (Baur et al. 1999: 55). The MMPA is essentially a "species preference law"—one that places a different, and arguably higher, human value on marine mammals than on other parts of the biophysical ecosystem. Although the law did facilitate attention to biophysical ecosystem relationships (Baur et al. 1999), there is a clear priority given to marine mammal populations of the "optimal" size (i.e., the largest unharvested size given the constraints of their environment). (There are limited exceptions to the "no take" rule for certain fishing operations and Native American harvests; see Baur et al. 1999).

The passage of the MMPA created two sets of problems: one domestic and one international. The domestic problem is that the achievement of the objectives of the MMPA (complete protection of marine mammals) often conflicts with the objectives of other law and policy (the MSY of the FCMA, or other human uses of the ocean such as recreational boating or military uses) (Cicin-Sain and Knecht 2000). Even though there are provisions in the MMPA governing the "taking" of marine mammals in fishing operations, the resolution of such situations has not been easy or straightforward. How do we simultaneously maximize fish harvest and avoid "taking" marine mammals, given the very broad definition of that term? The same is true for such situations as a merchant or naval ship or recreational boater colliding with ("striking" and thus "taking") a marine mammal. How do we sustain maritime commerce and recreation in the face of marine mammal strikes? In a third case, when a marine mammal, such as a killer whale, protected under the MMPA begins consuming a threatened or endangered species such as salmon or sea otters that are protected under the ESA, which law takes precedence? Do humans intervene? In what way? Although various amendments to the MMPA and other applicable law over time have addressed some of these conflicts, many still remain (Cicin-Sain and Knecht 2000).

Internationally, a similar set of problems emerged. Not all nations share the cultural values of the United States with respect to marine mammals. In many countries, notably Japan and Norway and many small island nations, marine mammals are considered consumptive resources, as they were formerly considered in the United States before the MMPA. Much of the argument between the United States and its allies on the IWC regarding commercial whaling is not over the proper route to conservation but over the question of conservation (sustained consumptive use) versus preservation (complete nonconsumptive protection). Cultural attitudes in the United States toward whaling have evolved over time, as has our law and policy. In the 1970s, at the same time as the United States was pushing for a complete moratorium on commercial whaling in the IWC, a survey of U.S. residents revealed that a majority of three to one were in favor of allowing commercial whaling as long as the species was conserved and the products fully utilized. Lobbying by marine mammal interest groups and by U.S. government representatives was effective in the IWC, however, and the commercial moratorium was put into effect by the IWC in 1982 (Friedheim 2001). The ratio of public opinion had reversed by the early 1990s, with the surveyed population being opposed to commercial whaling by a ratio of three to one (Lavigne et al. 1999), thus bringing public opinion in line with official U.S. policy.

In the IWC, however, the cultural disagreement between the United States and its allies and the whaling nations tends to be mischaracterized as a "conservation" discussion, as opposed to an argument over conservation/extractive use (whaling nations) versus preservation/complete protection (United States and allies). Such mischaracterization hinders the resolution of the debate, because the different sides are operating under different cultural assumptions regarding the proper goal of the law and policy.

The Important Questions

The fact of the different—and changing—configurations of human values expressed in these laws and policies raises several important questions.

First, how do we reconcile the different cultural values reflected in conflicting public policies in the United States? This question becomes particularly important as the populations of many marine mammals recover and expand in numbers and range. What do we do when expanding seal populations literally take over beaches and piers, precluding their use by humans, as they have in parts of California? What do we do when protected killer whales consume threatened or endangered salmon or sea otters, as they do in Alaskan waters? What do we do when protected whale and other marine mammal populations are affected by commercial or recreational fishing and boating, as they are nationwide?

Second, how do we reconcile the cross-cultural differences in human values reflected in the IWC debate over the moratorium on commercial whaling? Even though the Scientific Committee of the IWC has expressed the opinion that certain populations of certain species of whales could bear a sustainable harvest, the Commission itself has refused to allow commercial harvest of those populations based on the predominant cultural value set of the Commission. If the IWC is going to reflect the cultural values of the U.S. MMPA in its decisions, should not its charter be changed to reflect those values, as opposed to the sustainable harvest-type values that charter now expresses? That is, should the IWC charter not be rewritten as a document requiring complete protection as opposed to "the orderly development of the whaling industry"?

Third, how do we address the larger trade-off questions such as the desired ratio of whales to other marine species and to other human uses? For example, if the moratorium continued and whales were to recover to their pre-exploitation levels, what other changes in the biophysical environment would this produce? Would there be fewer of other species of value to humans, such as mackerels and tunas, or of other species such as krill that serve important biophysical ecosystem functions? How would we judge significantly increased populations of whales against significantly reduced populations of tunas or krill? If the order of magnitude of increase in whale populations were large, how would we handle the question of interaction with ships and boats? Significant increases in marine mammals would likely produce significant increases in conflict with all manner of human ocean uses, from commercial and recreational fishing, to oil and gas development, to merchant and naval shipping. How would we view those trade-offs? Would we maintain our "species preference" for marine mammals if it meant significant curtailment of those other activities? It is clear that we do not currently have an adequate framework to deal with these larger-scale trade-offs, a framework many refer to as "integrated management" (Cicin-Sain and Knecht 1998).

Designing the Ocean Environment

The answer to all of these questions lies in two areas. The first is increased knowledge of the biophysical ecology of the world's oceans and of the human ecology of our use of the oceans, in order that the trade-off inherent in different rules of governance might be accurately estimated. The second is the ability to format the discussion in a reasonable and productive way and to use our increased knowledge to construct our law and policy rationally.

In making public policy regarding whales and whaling, we necessarily must choose among many different possible governance rules, all of which are interactive with human behavior and law and policy in related policy arenas such as fisheries, shipping, or oil and gas exploration. This is a case of what we might define as "ecosystem management," a term that is much bandied about but that is useful in the present context: managing human behavior is a way so as to prevent, for example, the inappropriate "taking" of whales while still allowing other human uses of the ocean, if that is our ultimate goal. The important point is that we need to know as much in a documented way about the human ecological configuration as we do about the biophysical ecological configuration, including the structure of human values that underlie our current, and potential future, law and policy.

Following from this point, all of the resulting governance rules will be based on some set of human cultural values. These values will reflect our perceptions and attitudes regarding the desired state of the biophysical and human environments. In this sense, we are engaged in a design exercise for both the biophysical and the human ecologies and the ways in which the two map onto one another. To manage human behavior effectively at the scale necessary for the world's whale populations, we will have to proceed toward a "policy enclosure" of the world's ocean, in order that the resulting governance rules for all parts of the ocean ecosystem can be effectively developed and applied (Orbach 2000). The future of whales and whaling will be guided by the design we select for whales and their companion ocean environments and resources and for the pattern of human uses of those environments and resources.

Literature Cited

Baur, D., M. Bean, and M. Gosliner. 1999. The laws governing marine mammal conservation in the United States, in *Conservation and management of marine mammals*. J. Twiss and R. Reeves, eds. Washington, DC: Smithsonian Institution Press, pp. 48–86.

Cicin-Sain, B. and R. Knecht. 1998. *Integrated coastal and ocean management*. Washington, DC: Island Press.

———. 2000. *The future of U.S. ocean policy*. Washington, DC: Island Press.

Friedheim, R. 2001. *Toward a sustainable whaling regime*. Seattle: University of Washington Press.

Gambell, R. 1999. The International Whaling Commission and the contemporary whaling debate, in *Conservation and management of marine mammals*. J. Twiss and R. Reeves, eds. Washington, DC: Smithsonian Institution Press, pp. 179–198.

Lavigne, D., V. Scheffer, and S. Kellert. 1999. The evolution of North American attitudes toward marine mammals, in *Conservation and management of marine mammals*. J. Twiss and R. Reeves, eds. Washington, DC: Smithsonian Institution Press, pp. 10–47.

Miller, M., R.P. Gale, and P.J. Brown. 1987. *Social science in natural resource management systems*. Boulder, CO: Westview Press.

Nader, L. 1969. *Law and Culture in Society*. Berkeley: University of California Press.

Orbach, M. 2002. *Beyond the freedom of the seas: ocean policy for the third millennium. Roger Revelle Memorial Lecture*. Washington, DC: National Research Council.

OVERVIEW AND SYNTHESIS

THIRTY

Whales Are Big and It Matters

PETER KAREIVA, CHRISTOPHER YUAN-FARRELL,
AND CASEY O'CONNOR

Whales have a unique place in conservation lore. Their plight is widely known, and the beaching of even a single whale is a major news event, typically drawing hundreds of spectators. In addition, as a marine mammal, whales are given favored legal protection above and beyond that given to the bulk of the planet's biodiversity. The irony is that, although whales have become a symbol of the human capacity for greedy overharvest and a rallying point for environmental activists, we know surprisingly little about their ecological role. Our ignorance regarding the ecology of whales is symptomatic of conservation: Too often conservation is about species as "symbols," while ecological processes and the functioning of ecosystems are ignored. Although conservation biology typically lacks fundamental ecological information, the knowledge gaps surrounding whales are especially surprising given their potential to be a keystone species in marine ecosystems. The contents of this volume reveal that even something as basic as who eats what is a matter of great controversy, and the implications of species interactions have only recently begun to receive the attention they deserve. Here we extrapolate from the scant information available on the ecological role of whales to explore what might be the broader consequences should whales recover to their historic abundance.

The Iconic Status of Whales in Conservation

The Marine Mammal Protection Act (MMPA) of 1972 affords special protection to all marine mammals. This act established a moratorium on the taking of marine mammals in U.S. waters and by U.S. citizens on the high seas, and generally prohibited the import of marine mammal products into the U.S. (U.S. Public Law 92–522). Unlike the Endangered Species Act, the protection afforded by the MMPA is provided for all species of marine mammals regardless of their risk of extinction. Academic researchers also fall under the spell of marine mammals and give them far more attention than would be expected on the basis of their total diversity, biomass, or numbers. For example, a survey of the journal *Conservation Biology* reveals a striking preponderance of articles regarding marine mammals compared with insects or even birds (see Figure 30.1).

Although all marine mammals enjoy favored treatment, whales in particular hold special status in the public eye. For example, popular culture has made whales the subject of enormously successful commercial movies, notably *Free Willy* and its two sequels, as well as the 2002 Oscar-nominated film *Whale Rider*. Whales are also a common subject in literature, most famously the white whale of *Moby Dick* (also adapted for

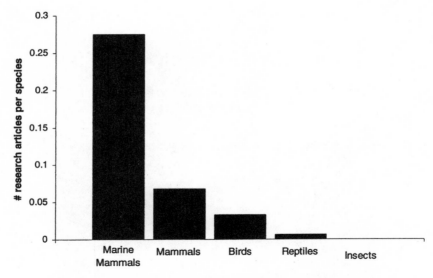

FIGURE 30.1. The number of research articles focusing on any species within five major taxonomic groups relative to the number of species within each taxon. Data were obtained by searching each issue of *Conservation Biology* from 1990 through 2003, but excluding letters, editorials, comments, and book reviews. If an article discussed multiple taxa, then that article would be counted once for each taxa. However, we did not count general discussions of "biodiversity" that only briefly mentioned some species as a contributed research article on that taxa. Altogether 711 articles were examined. The number of species per taxon was obtained from: Raven et al. 2005.

numerous films). One of the most renowned theme parks in the world, Sea World owes much of its success to Shamu, the killer whale who has fascinated children and adults alike for more than forty years. In addition, Greenpeace's interference with whaling ships provides some of the most vivid examples of environmental activism ever captured by the media (Greenpeace 2004). In fact, when just one whale gets in trouble, it often makes international news. As just one example, when a young right whale became entangled in fishing gear off the eastern coast of the United States in March 2004 (e.g., Holmes 2004), the rescue attempt triggered hundreds of newspaper articles during that week, as documented by a Google News search for "right whales" on March 30 (for an apparent happy ending, see NOAA 2005a; for a similar incident, see NOAA 2005b). It is hard to imagine such attention being given to any other group of organisms.

Much of the mysticism surrounding whales stems from what anthropologist Arne Kalland (1993) terms the "super-whale." The "super-whale" is a composite species fabricated by environmental and animal rights groups. This symbolic creature possesses the unique traits of several whale species; thus, it is the largest animal in existence (blue whale), one of the world's most severely endangered species (right whale), has beautiful songs (humpback whale), and is extremely friendly (gray whale). The fact that no one whale species combines all these attributes does not diminish the symbolic appeal. Of course, long before environmental groups adopted whales, Northwestern Native Americans made whales one of the animals most often depicted on the crests of their totem poles

(Stewart 1990). The symbolism and mysticism surrounding whales can be a useful motivator of goodwill, but can also be an obstacle to fact-based discussions and science-based policy.

Whales as Just Another Overharvested Natural Resource?

While whales may be special in terms of public appeal, the decimation of whales is not at all unusual compared to the slaughter of land-based vertebrates around the world. At the same time that whales were being harpooned in the oceans, terrestrial species in North America were being slaughtered with an almost unimaginable abandon. For example, passenger pigeons were totally extinguished by 1914—a formidable feat given that, only two hundred years earlier, four out of every ten birds in North America were passenger pigeons, making them the most abundant bird on the planet (Flannery 2001). The killing spree that was the downfall of the passenger pigeon was truly horrific—to win a pigeon shooting competition in the 1870s, an individual had to kill 30,000 birds in one day (Stephan 1932). Perhaps more analogous to the whale harvest, however, was the buffalo hunt on the Great Plains, which, in less than 100 years, managed to reduce the bison, once numbering between 30 and 60 million, to less than one thousand individuals (Flannery 2001). Although the bison were not driven entirely extinct, their ecological role on the North American grasslands is surely lost forever. Bison urine was a fertilizer that kept the prairies productive, and bison wallows harbored their own unique plant communities, which were in

TABLE 30.1
Whale Population Estimates for 2050

Species	Current Estimate	Annual Growth Rate (r) %	Carrying Capacity (K)	2050 Estimate[a]	% of K
Eastern North Pacific gray whale[b]	26,000	2.5	37,364	33,301	89.1
Sperm whale (global population)[c]	452,000	0.9	1,110,000	1,092,520	98.4
Bering Sea bowhead whale (conservative K)	8,000[d]	3.1[e]	10,000[f]	9,588	95.9
Bering Sea bowhead whale (liberal K)	8,000[d]	3.1[e]	23,000[f]	17,424	75.8
Humpback whale (conservative r)	33,600[d]	3.9[g]	115,000[h]	85,519	74.4
Humpback whale (liberal r)	33,600[d]	12.6[g]	115,000[h]	114,491	99.6

[a] For simplicity, all populations were assumed to grow logistically according to the equation $N_t = K/(1 + be^{-rt})$, where N_t = the estimated population in 2050, K = carrying capacity, $b = (K - N_0)/N_0$, N_0 = current population size, r = annual growth rate, and $t = (2050 - \text{year of current estimate})$.

[b] Rugh et al. 1999.

[c] Whitehead 2002.

[d] Gerber et al. 2000.

[e] Cited in Best 1993.

[f] Cited in Sheldon and Rugh 1995.

[g] Clapham et al. 2001

[h] Evans 1987.

turn favored by American antelope and other large herbivores (Knapp et al. 1999). The ecological extinction of bison and the total extinction of passenger pigeons have produced a radically different North American ecosystem, and any notion of returning to a natural pre-nineteenth-century state has to be discarded as hopelessly unrealistic. Never again will bison or passenger pigeons play an important ecological role.

In contrast, whales may again become dominant components of oceanic ecosystems. Even now, killer whales appear to be acting as keystone species in the Bering Sea (Estes, Chapter 1 of this volume), thus a major distinction between the stories of the bison and passenger pigeon and that of whales concerns the possibility of recovery. Although data are spotty and estimates have high uncertainty, it is clear that some populations of whales are currently rebounding. To demonstrate how quickly future populations of whales could rebound, we estimated future abundance for two global and two localized populations from well-studied stocks: the Eastern North Pacific gray whale, the Bering Sea bowhead whale, the global sperm whale, and the global humpback whale populations. Extrapolating recent estimates of population growth, we can expect all of these species to be nearing historical carrying capacity in less than fifty years (see Table 30.1). By touting the recovery of selected whale populations, we do not intend to diminish the peril of such whales as the North Atlantic right whale, which is on the brink of extinction (Caswell et al. 1999); our point is simply that many whales indeed have an excellent prospect of returning to historical abundances.

The key difference between the future for whales and the future for land-based megafauna such as the bison entails habitat loss and habitat conversion. Not only were bison (and numerous other large terrestrial vertebrates) hunted to near extinction, but their habitat was converted to agricultural and residential areas. In the ocean, humans have not established permanent communities or large-scale farming activities. Thus, if sources of whale mortality are reduced, there is a habitat in which the whales can rebuild to historical numbers. Unlike the terrestrial megafauna of North America, which will never reclaim a major ecological role, we might well see whales becoming dominant forces in our oceans once again.

Factors Limiting a Return to Cetacean Primacy

Superficially, it would seem that a cessation of harvest might ensure the recovery of whales to past abundances, and for many whale species or populations, recovery is proceeding. But whales are also imperiled by a nexus of human impacts more subtle than outright harvest. Global warming may cause serious problems for species that live in the ice (e.g., bowhead whales), and habitat degradation is a growing cause of concern. Most notably, toxic industrial and agricultural chemicals accumulate in whale blubber, which acts as a reservoir for lipophilic chemical contaminants. Elevated levels of polychlorinated biphenyls (PCBs) have been found in a number of whale species, including belugas (Becker et al.

2000; Hobbs et al. 2003) and orcas (Hayteas and Duffield 2000; Ross et al. 2000). DDT and a host of other organochlorine pollutants have also infiltrated marine waters. While there may be no reason to suspect that exposure to these chemicals is directly lethal to the large whale species, plenty of evidence implicates organochlorine exposure, including DDT and PCBs, in reducing reproductive potential and harming general health (Colborn and Smolen 1996).

Navy sonar and other sources of human-generated noise have been implicated in beaching events and death in several whale species (NRC 2003a; Frantzis 1998; Simmonds and Lopez-Jurado 1991), and several marine mammal species have been observed to change their vocalization and behavioral patterns in the presence of human-induced noise. These beaching events are still too poorly understood to know whether whale recovery might be significantly constrained by noise pollution. In addition, it is not clear whether or not human-generated noise in the oceans is still increasing; for example, although there are more commercial ships, they tend to be quieter (NRC 2003a).

Other potential constraints on whale recovery include entanglement in fishing gear and collisions with ships, which are implicated most clearly in right whales (Fujiwara and Caswell 2001). Some have argued that changes in the ocean environment related to global warming may reduce zooplankton productivity and thereby threaten whale recovery as well, although the data are not compelling (Greene et al. 2004). While direct mortality due to commercial harvest was obviously the cause of the original decline in whales throughout the world, factors other than harvest may now operate to slow or halt recovery. Nonetheless, the situation for the oceans' great whales is certainly more optimistic than for many terrestrial megafauna such as bison or grizzly bears.

Why So Much Numerology and So Little Ecology?

Although the majesty and charismatic appeal of whales evoke enormous interest in their conservation, it is troubling how little is known about their ecological roles. The vast majority of papers on whales in conservation journals discuss their abundance but overlook their impacts on other species or ecosystem processes. For example, a survey of all 106 articles published in the journal *Marine Ecology* from 1999 through 2003 failed to reveal a single article about whales. There have been a number of major articles (eight between 1990 and 2003) about "whale conservation" in the leading international journal *Conservation Biology*, but only one of these (Butman et al. 1995) discusses species interactions or ecological processes. Only 17 of the 94 whale-related papers published in *Marine Mammal Science* deal with species interactions and effects on ecosystem processes. In lieu of ecology, the recent scientific literature on whales is dominated by articles about whale numbers (e.g., population estimates, historical abundances, and population trends). This disproportionate attention given to whale numbers as opposed to

ecological interactions is striking but not surprising. Numbers loom large because there is a contentious debate about a possible resumption of commercial whaling, with whale numbers acting as the ultimate arbiter of whether or not whaling is resumed. In particular, Japan, Norway, and Iceland have argued since the enactment of the 1986 moratorium by the International Whaling Commission (IWC) that the numbers of several whale species are now high enough to allow limited commercial whaling (Nagasaki 1990). These nations argue that the IWC is for "the regulation of whaling, not the prohibition of whaling" (Norway IWC commissioner, quoted in Aldhous and Swinbanks 1991).

The two key numbers with respect to whaling regulations are current abundance and historical abundance. In 1975, the IWC set 0.54K (where K equals the pre-exploitation level of a given stock) as the threshold population size that had to be exceeded in order to allow commercial whaling. Although there is discussion of more sophisticated decision algorithms, the IWC Web site still emphasizes the 0.54K benchmark: "catches should not be allowed on stocks below 54% of the estimated carrying capacity" (IWC 2006). Such a policy obviously focuses attention on numbers and, especially, abundance estimates. The problem is that the uncertainty surrounding these numbers is huge. In fact, antiwhaling nations such as the United States and the United Kingdom argue that most abundance estimates are much too uncertain to allow a commercial whaling industry (Butterworth 1992). The most striking example of how little we may know about the historical condition comes from a genetic study by Palumbi and Roman (Chapter 9 of this volume; see also Roman and Palumbi 2003). By sampling the magnitude of genetic variation and assuming a range of mutation rates, Roman and Palumbi derived estimates for historical abundances of fin, humpback, and minke whales in the North Atlantic. In the case of fin and humpback whales, genetic-based population estimates are 7 and 12 times higher, respectively, than have ever before been proposed. Thus, even though skepticism about whale numbers is thought by many to be a smoke screen for different cultural values concerning the hunting of whales (Butterworth 1992; Nagasaki 1990), confidence in prewhaling population estimates is premature. Given Roman and Palumbi's recent work, the carrying capacity estimates provided in Table 30.1 might be underestimated by a factor of ten. Given the political pressures and technical challenges of counting whales, past and present, one has to wonder whether studies of whale ecology might yield far more useful and informative insights than yet another Bayesian estimation of whale population size.

Whales are Hugely Important to Ocean Ecosystem Functioning

Several lines of evidence point towards an ecologically significant role for whales in marine ecosystems. First, even if we assume low estimates of past abundance, simply by virtue of their energetics, whales must have been important

TABLE 30.2
Historic and Current Global Population Estimates for Six Whale Species

Species	Historic Estimate[a]	Current[b]	Current[c]	Current[d]	Current Average
Right	100,000	3,000 (3.0)	7,850 (7.9)	NA	5,425 (5.4)
Fin	548,000	120,000 (21.9)	139,050 (25.4)	47,300 (8.6)	102,117 (18.6)
Sei	256,000	54,000 (21.1)	25,110 (9.8)	NA	39,555 (15.5)
Minke	490,000	505,000 (103.1)	NA	935,000 (191)	720,000 (147)
Humpback	115,000	10,000 (8.7)	35,600 (31.0)	21,570 (18.8)	22,390 (19.5)
Sperm	1,110,000[e]	452,000[e] (40.7)	NA	NA	452,000 (40.7)
Total	3,909,000	1,144,000 (29.3)			1,341,487 (34.3)
Total except Minke	3,419,000	639,000 (18.7)			621,487 (18.2)

[a] For consistency, all historic estimates are taken from Evans (1987) with the exception of sperm whales, because extensive research has been conducted on that species. Because most available estimates refer to specific stocks, individual stock abundances were totaled for each species to yield global population numbers. The percentage of the historic abundance is given in parentheses after the current population. Current abundance reflects the most recent estimates and may not reflect actual 2004 population numbers. Current abundance was averaged by species and subsequently totaled to give global levels.

[b] Evans 1987.

[c] Gerber et al. 2000.

[d] IWC 2004.

[e] Whitehead 2002.

consumers (Williams, Chapter 15 of this volume). For example, off the Northeast Continental Shelf of the United States, six whale species (right, fin, sei, minke, humpback, and sperm) currently command roughly 8–18% of net primary productivity (based on Kenney et al. 1997). If these whale populations were to triple in magnitude, which is not unreasonable given that these six species average at only 34% of their global historic levels (Table 30.2), these whales could consume 24–54% of all primary productivity in the area. Croll et al. (Chapter 16 in this volume) estimate a similar scenario for the North Pacific, where large whale species currently consume 9% of the net primary production—nearly 7 times less than historic consumption rates of 62% in the same geographic area. Although it is unlikely that whale populations will grow homogenously worldwide, rising whale numbers will certainly alter energy flows and species composition in many marine communities.

In addition to food web effects due to consumption, whales also impact ecosystems in a variety of other ways, including substrate disturbance and sediment resuspension. For example, gray whales use suction to gather prey from the sea floor and in doing so create large disturbances (Highsmith et al., Chapter 23 of this volume), in some feeding areas plowing through over 30% of the floor (Oliver and Kvitek 1984). In the northeastern Bering Sea, gray whales resuspend 120 million m^3 of sediment annually, which is twice the amount introduced by the Yukon River (Johnson and Nelson 1984). Pilskaln and colleagues (1998) speculate that sediment mixing, in this case from biogenic perturbation, could shift organic matter between aerobic and anaerobic environments and greatly alter primary productivity. Sediment disturbance will release significant organic material as well as ammonium and nitrate into the water column (Pilskaln et al. 1998). The disturbances created by this feeding also generate open habitat

patches, which are then exploited by a diverse fauna of scavenger populations (Oliver and Slattery 1985).

Whale carcasses have been shown to support a diversity of marine life totaling more than 350 species in the northeast Pacific. They serve as important habitats for polychaetes, mollusks, crustaceans, and bottom-dwelling, chemoautotrophic organisms, as well as nutrient sources for scavenging fish and sharks. A large whale carcass (>30 tons) can support successional communities for up to 80 years or more before being totally decomposed (Smith, Chapter 22 of this volume). The combination of the mass and longevity of these whale carcasses makes them extremely important havens for biodiversity in deep-sea ecosystems (Jones et al. 1998).

Perhaps most compelling is the large number of experiments pointing out that predators commonly have a cascade of effects that reverberate through ecosystems as a result of trophic interactions (Paine, Chapter 2 of this volume). There is no reason to expect whales, especially top predators such as killer whales, to be any less influential than keystone predators in other systems. Unfortunately, in the absence of manipulative field experiments, inferences about trophic impacts of whales are not straightforward. One approach has been to take advantage of human-induced changes in marine food webs and then, through either statistical techniques or models, draw conclusions about the role of whales. For example, the food web model analyzed by Essington (Chapter 5 of this volume) suggests that the decline of sperm whales in the tropical and subtropical Pacific led to a shift toward pelagic systems dominated by squid and small tuna. Using analyses of population trend data as opposed to a food web model, Worm et al. (Chapter 26 of this volume) provide a plausible scenario by which the decline of whales in the 1950s and 1960s led to groundfish-dominated systems. The concordance of marine mammal declines in the

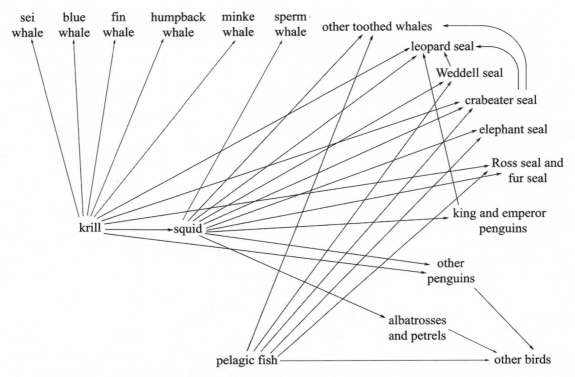

FIGURE 30.2. Primary and secondary pelagic consumers of krill and squid. Modeled after Laws 1985.

Bering Sea and possible prey switching by killer whales (Estes, Chapter 1 of this volume; Estes et al. 1998; Springer et al. 2003) is yet another line of evidence suggesting that hunting whales led to major shifts in species interactions with reverberations throughout marine ecosystems. Although more research is needed in this area, the weight of evidence to date suggests that whales profoundly alter the composition and functioning of oceanic ecosystems.

Some Whales are Coming Back—What Will This Mean for Our Oceans?

Any substantial increase in whale numbers is likely to have profound effects on ocean ecosystems. In fact, much of the motivation for this book is the spark provided by the controversy surrounding the National Marine Fisheries Service's (NMFS) conclusion that indirect fishing effects could be contributing to Steller sea lion declines, and therefore the pollock fishery needed to be limited (NRC 2003b). Contrary to NMFS's inference, several scientists have suggested that pollock may have no role or a negligible role in the sea lion declines; instead, recent declines in Steller sea lions could have been driven by a modest number of killer whales switching to the consumption of sea lions after having lost great whales as potential food sources through overharvest (Estes et al. 2004; NRC 2003b; Springer et al. 2003; Williams et al. 2004). This scenario is at best a hypothesis, but it is a plausible hypothesis and an idea that should be taken seriously. If true, then as the larger prey of killer whales recover, predation on the pinnipeds should be relaxed and pinniped

abundance recover. The expected recovery of great whales in the Pacific Ocean will provide a terrific opportunity to assess the killer whale hypothesis.

Certainly, as whale numbers increase, we can expect a reduction in the primary prey of many whales, such as krill; the effects of which will resonate throughout the food web (Butman et al. 1995; Laws 1985; May et al. 1979). For example, despite the enormous complexity of the Southern Ocean ecosystem and the difficulty in sorting out the myriad influences affecting the abundance of krill and their predators (Ballance et al., Chapter 17 of this volume), it is hard to imagine the recovery of baleen whales not having a dramatic effect on krill and the Southern Ocean food web. Krill in the Southern Ocean are eaten primarily and secondarily by squid, fish, birds, seals, and whales (Figure 30.2). Since 1900, whale contribution to total annual krill consumption has dropped from over 40% to less than 10%, which equates to 170 million and 35 million metric tons, respectively (Laws 1985). If whales increase in abundance and deplete krill populations, other consumers may suffer. Moreover, as whales recover, they will compete with the commercial fleets that harvest krill. The krill fishery now takes in an estimated 160,000 metric tons annually, and demand for commercially harvested krill is expected to rise in coming decades, both within current markets and with possible uses in biotechnology and pharmaceuticals (Nicol and Endo 1997).

In general, a return of cetacean primacy could impact the prey availability for many fishes currently harvested by commercial fisheries (Essington, Chapter 5 of this volume; Croll et al., Chapter 16 of this volume), with unanticipated consequences. When it comes to harvest and depletion of fish or squid populations, humans have had no rivals over the last century. It is interesting to speculate what it would mean for whales to resume historical abundances and themselves act as major consumers. Because of their energetic demands and adaptations for open-ocean foraging, whales could compete with the commercially important species for limited prey resources. According to Croll et al. (Chapter 16 of this volume), current whale populations in the North Pacific are already comparable with commercial fisheries in terms of consumption (55,155 metric tons day^{-1} and 75,468 metric tons day^{-1}, respectively) and trophic level (3.4 compared with 3.2, respectively). The regulation and management of fisheries has always suffered from a myopic single-species syndrome (Zabel et al. 2003); the recovery of whales could finally force fisheries managers to abandon the foolishness of managing fisheries as though harvested species exist in ecological vacuums.

Which Is More Important: Whale Abundance or Environmental Fluctuations?

There has been a long-standing debate in ecology between those who emphasize environmental factors and those who emphasize biotic interactions and trophic effects. These alternative perspectives about population regulation pervade such subjects as density dependence (Smith 1961), supply-side ecology (Lewin 1986), and the debate regarding bottom-up versus top-down control of ecosystems (Hunter and Price 1992; see also Matson and Hunter 1992). Some researchers, including contributors to this volume (e.g., Hunt, Chapter 6 of this volume), discount the hypothesis that whales can drive marine ecosystems, because they take the observation that fish populations often track climate variability as evidence against top-down control. It may be that the debate about top-down and bottom-up control is more a result of the nuances of language than fundamental scientific differences. For example, Hunt (Chapter 6 of this volume) concludes that the strong responses of fish, pinnipeds, and seabirds to climate variability are a "manifestation of bottom-up forcing." That is a reasonable inference. But "forcing" is different than control. The environment and food supply can drive fluctuations, while at the same time higher trophic levels constrain the average abundances around which populations fluctuate.

It is crucial to distinguish between those factors that best statistically account for variation in rates of population change and those factors that govern the persistence and long-run average abundance of a species. In the literature discussing whales and the food webs in which they are embedded, too much is made of casual correlations between population fluctuations and environmental variation. If one is to take a statistical approach when assessing top-down versus bottom-up factors, there must be an analysis that includes variation in the upper trophic levels and not simply variation in environmental conditions or primary productivity. A factor such as temperature could explain a great deal of the annual variation in population change, yet it could be predation that determines the persistence and average abundance of a species (Hassell et al. 1998). Ballance et al. (Chapter 17 of this volume) recognize this point when they remind us that all of our studies of pinnipeds, krill, and seabirds have taken place long after whales were severely depleted and that we might see an entirely different ecosystem if whales returned to historical levels.

One of the best statistical explorations of top-down versus bottom-up hypotheses in ocean systems does not concern whales but is nonetheless informative. In a study of cod-shrimp interactions in the North Atlantic Ocean, Worm and Myers (2003) found compelling evidence of top-down control, mediated somewhat by temperature. The decline of cod (due to overharvest) has clearly led to an increase in shrimp (a common prey of cod). In contrast to much of the discussion regarding correlations between climate and pinnipeds or krill, Worm and Myers adopted a formal hypothesis-testing approach entailing competing hypotheses. Without a formal contrast of competing hypotheses, spatial correlations between primary productivity and consumer abundances simply do not imply an absence of top-down control, as some have argued (Hunt, Chapter 6 of this volume). Of course, the only incontrovertible way of establishing which is more important—top-down or bottom-up forces—is an experiment that simultaneously manipulates both in a factorial manner.

There has been one notable meta-analysis of experimental manipulations performed in marine ecosystems, in which the effect of removing predators versus adding nutrients were compared (Micheli 1999). Responses were measured in terms of the log ratio of treatment (either predator removal or nutrient supplementation) to control biomass (more precisely as "log response ratios," which are the log-transformed ratios of the treatment mean divided by the control mean, weighted by sampling variances). Both nutrient supplementation and predator removal had effects in these marine experiments, but predator removal yielded effects on average 2–3 times larger than the effects of nutrient addition. Obviously none of the experiments reviewed by Micheli involved whales, but they were marine systems with similar numbers of trophic levels to those found in whale food webs.

Since experiments are impractical for whales, and correlations between population fluctuations and environmental conditions are by themselves suspect, what might be tools for testing ideas about whales as agents of top-down control of marine systems? We think the most fruitful approach is an examination of mechanistic models of trophic interaction, with estimates of energetic needs and consumption rates (e.g., Doak et al., Chapter 18 of this volume). These models lead to much more precise hypotheses than environmental correlations do, and these hypotheses can be incontrovertibly falsified by data on consumption rates, mortality, or responses to altered predator density. Spatial or temporal variation in whale abundance and in whale recovery can be used in conjunction with these models to test their merit.

Our Understanding of Ocean Ecology Should be Enhanced Due to the Recovery of Whales

The recovery of whales is a huge "experiment" at the scale of ocean subbasins. Because whale recovery will be uneven, we have an opportunity to make predictions about food-web impacts and community impacts that could then be tested over the next 50 years. But we will never learn anything from whale recovery if we neglect their ecological interactions and continue to focus solely on their numbers. Whereas intertidal and terrestrial ecologists have long recognized the importance of species interactions and the interplay of competition and predation, biological oceanography has focused on primary productivity and nutrients. When top-down effects have been looked for (Worm and Myers 2003; Micheli 1999; Paine, Chapter 2 of this volume), they have been found in marine systems. Studies of whale ecology and recovery need to examine the possibility of profound top-down impacts. Much could be gained by simultaneously sampling whale populations and key components of the biotic community in ocean subbasins where whales are increasing and in subbasins where they are not. Of course, this is not a real experiment because of lack of randomization and control; nonetheless, appropriate observational studies of contrasting ocean subbasins in conjunction with models such as those put forth by May et al. (1979) can yield strong inferences. The recovery of whales could provide the biological perturbation that changes the way we think about oceans—but only if we open our eyes to species interactions and trophic effects.

Literature Cited

Aldhous, P. and D. Swinbanks. 1991. Whaling ban versus science. *Nature* 351: 259.

Becker, P.R., M.M. Krahn, E.A. Mackey, R. Demiralp, M.M. Shantz, M.S. Epstein, M.K. Donais, B.J. Porter, D.C.G. Muir, and S.A. Wise. 2000. Concentrations of polychlorinated biphenyls (PCBs), chlorinated pesticides, and heavy metals and other elements in tissues of belugas, *Delphinapterus leucas*, from Cook Inlet, Alaska. *Marine Fisheries Review* 62: 81–98.

Best, P.B. 1993. Increase rates in severely depleted stocks of baleen whales. *ICES Journal of Marine Science* 50: 169–186.

Butman, C.A., J.T. Carlton, and S.R. Palumbi. 1995. Whaling effects on deep-sea biodiversity. *Conservation Biology* 9: 462–464.

Butterworth, D.S. 1992. Science and sentimentality. *Nature* 357: 532–534.

Caswell, H., M. Fujiwara, and S. Brault. 1999. Declining probability threatens the North Atlantic right whale. *Proceeding of the National Academy of Sciences* 96: 3308–3313.

Clapham, P., J. Robbins, M. Brown, P. Wade, and K. Findlay. 2001. A note on plausible rates of population growth for humpback whales. *Journal of Cetacean Research and Management*. 3(suppl.): 196–197.

Colburn, T. and M.J. Smolen. 1996. Epidemiological analysis of persistent orgaochlorine contaminants in cetaceans. *Review of Environmental Contamination Toxicology* 146: 91–172.

Estes, J.A., E.M. Danner, D.F. Doak, B. Konar, A.M. Springer, P.D. Steinberg, M.T. Tinker, and T.M. Williams. 2004. Complex trophic interactions in kelp forest ecosystems. *Bulletin of Marine Sciences* 74: 621–638.

Estes, J.A., M.T. Tinker, T.M. Williams, and D.F. Doak. 1998. Killer whale predation on sea otters linking oceanic and nearshore ecosystems. *Science* 282: 473–476.

Evans, P.G.H. 1987. *The natural history of whales and dolphins.* New York: Facts on File.

Flannery, T. 2001. *The eternal frontier.* New York: Grove Press.

Frantzis. A. 1998. Does acoustic testing strand whales? *Nature* 392: 29.

Fujiwara, M. and H. Caswell. 2001. Demography of the endangered north Atlantic right whale. *Nature* 414: 537–541.

Gerber, L.R., D.P. DeMaster, and S.P. Roberts. 2000. Measuring success in conservation. *Scientific American* 88: 316–324.

Greenpeace. 2004. Whales. Accessed May 19, 2004, at http://www.greenpeace.org/international_en/campaigns/intro?campaign_id=4017.

Greene, A.J. and C.H. Pershing. 2004. Climate and conservation biology of North Atlantic right whales: the right whale at the wrong time? *Frontiers in Ecology and Environment* 2: 29–34.

Hassell, M.P., M.J. Crawley, H.C.J. Godfray, and J.H. Lawton. 1998. Top-down versus bottom-up and the Ruritanian bean bug. *Proceedings of the National Academy of Sciences* 95: 10661–10664.

Hayteas, D.L., and D.A. Duffield. 2000. High levels of PCB and *p,p'*-DDE found in the blubber of killer whales *(Orcinus orca)*. *Marine Pollution Bulletin* 40: 558–561.

Hobbs, K.E., D.C.G. Muir, R. Michaud, P. Beland, R.J. Letcher, and R.J. Norstrom. 2003. PCBs and organochlorine pesticides in blubber biopsies from free-ranging St. Lawrence River Estuary beluga whales *(Delphinapterus leucas)*, 1994–1998. *Environmental Pollution* 122: 291–302.

Holmes, W.L. 2004. Rescue attempt for entangled whale is called off as it moves farther offshore. Associated Press: March 26, 2004. Available at http://www.findarticles.com/p/articles/mi_qn4188/is_20040326/ai_n11447897. Accessed June 21, 2006.

Hunter, M.D. and P.W. Price. 1992. Playing chutes and ladders: heterogeneity and the relative roles of bottom-up and top-down forces in natural communities. *Ecology* 73: 724–732.

IWC. 2004. Whale population estimates. Cambridge, UK: International Whaling Commission. Available at http://www.iwcoffice.org/conservation/estimate.htm. Updated May 5, 2004. Accessed June 21, 2006.

———. 2006. Revised management procedure. Cambridge, UK: International Whaling Commission. Available at http://www.iwcoffice.org/conservation/rmp.htm. Accessed June 21, 2006.

Johnson, K.R. and C.H. Nelson. 1984. Side-scan sonar assessment of gray whale feeding in Bering Sea. *Science* 225: 1150–1152.

Jones, E.G., M.A. Collins, P.M. Bagley, S. Addison, and I.G. Priede. 1998. The fate of cetacean carcasses in the deep sea: observations on consumption rates and succession of scavenging species in the abyssal north-east Atlantic Ocean. *Proceedings of the Royal Society of London* Series B 265: 1119–1127.

Kalland, A. 1993. Whale politics and green legitimacy: a critique of the anti-whaling campaign. *Anthropology Today* 9: 3–7.

Kenney, R.D., G.P. Scott, T.J. Thompson, and H.E. Winn. 1997. Estimates of prey consumption and trophic impacts of cetaceans in the USA Northeast continental shelf ecosystem. *Journal of Northwest Atlantic Fishery Science* 22: 155–171.

Knapp, A.K., J.M. Blair, J.M. Briggs, S.L. Collins, D.C. Hartnett, L.C. Johnson, and G. Towne. 1999. The keystone role of bison in North American tallgrass prairie. *Bioscience* 49: 39–50.

Laws, R.M. 1985. The ecology of the Southern Ocean. *American Scientist* 73: 26–40.

Lewin, R. 1986. Supply-side ecology. *Science* 234: 25–27.

Matson, P.A. and Hunter, M.D. Special feature: the relative contributions to top-down and bottom-up forces in population and community ecology. *Ecology* 73: 723.

May, R.M., J.R. Beddington, C.W. Clark, S.J. Holt, and R.M. Laws. 1979. Management of multi-species fisheries. *Science* 205: 267–277.

Micheli, F. 1999. Eutrophication, fisheries, and consumer-resource dynamics in marine pelagic ecosystems. *Science* 285: 1396–1398.

Nagasaki, F. 1990. The case for scientific whaling. *Nature* 344: 189–190.

National Research Council. 2003a. *Ocean Noise and Marine Mammals*. Washington, DC: The National Academy Press.

———. 2003b. *The decline of the Steller sea lion in Alaskan waters: untangling food webs and fishing nets*. Washington, DC: National Academies Press.

Nicol, S. and Y. Endo. 1997. Krill fisheries of the world. FAO: *Fisheries technical paper 367*.

NOAA. 2005a. Kingfisher the entangled right whale re-sighted off the Georgia coast. Washington, DC: U.S. Department of Commerce, January 13, 2005. Available at http://www.publicaffairs. noaa.gov/releases2005/jan05/noaa05-005.html. Accessed June 21, 2006.

———. 2005b. NOAA organizes rescue team to disentangle North Atlantic right whale. Washington, DC: U.S. Department of Commerce, December 6, 2005. Available at http://www.publicaffairs. noaa.gov/releases2005/dec05/noaa05-r136.html. Accessed June 21, 2006.

Oliver, J.S. and R.G. Kvitek. 1984. Side-scan sonar records and diver observations of the gray whale *(Eschrichtius robustus)* feeding grounds. *Biological Bulletin* 167: 264–269.

Oliver, J.S. and P.N. Slattery. 1985. Destruction and opportunity on the sea floor: effects of gray whale feeding. *Ecology* 66: 1965–1975.

Pilskaln, C.H., J.H. Churchill, and L.M. Mayer. 1998. Resuspension of sediment by bottom trawling in the Gulf of Maine and potential geochemical consequences. *Conservation Biology* 12: 1223–1229.

Raven, P.H., G.B. Johnson, J. Losos, and S. Singer. 2005. *Biology*. 7th edition. New York: McGraw-Hill.

Roman, R. and S.R. Palumbi. 2003. Whales before whaling in the North Atlantic. *Science* 301: 508–510.

Ross, P.S., G.M. Ellis, M.G. Ikonomou, L.G. Barrett-Lennard, and R.F. Addison. 2000. High PCB concentrations in free-ranging Pacific killer whales, *Orcinus orca*: effects of age, sex, and dietary preference. *Marine Pollution Bulletin* 40: 504–515.

Rugh, D.J., M.M. Muto, S.E. Moore, and D.P. DeMaster. 1999. *Status review of the eastern north Pacific stock of gray whales*. U.S. Department of Commerce, NOAA Technical Memo NMFS-AFSC-103. Seattle: Alaska Fisheries Science Center.

Sheldon, K.E.W. and D.J. Rugh. 1995. The bowhead whale, *Balaena myticetus*: its historic and current status. *Marine Fisheries Review* 57: 1–20.

Simmonds, M.P. and L.F. Lopez-Jurado. 1991. Whales and the military. *Nature* 351: 448.

Smith, F.E. 1961. Density dependence in the Australian thrips. *Ecology* 42: 403–407.

Springer, A.M., J.A. Estes, G.B. van Vliet, T.M. Williams, D.F. Doak, E.M. Danner, K.A. Forney, and B. Pfister. 2003. Sequential megafaunal collapse in the North Pacific Ocean: an ongoing legacy of industrial whaling? *Proceedings of the National Academy of Sciences* 100: 12223–12228.

Stephan, S.A. 1932. The passing of the passenger pigeon. *Game Stories* 18019.

Marine Mammal Protection Act of 1972. U.S. Public Law 92–522, December 21, 1972 (16 U.S.C. 1361 et seq.; 86 Stat. 1027).

Stewart, H. 1990. *Totem poles*. Seattle: University of Washington Press.

Whitehead, H. 2002. Estimates of the current global population size and historical trajectory for sperm whales. *Marine Ecology Progress Series* 242: 295–304.

Williams, T.M., J.A. Estes, D.F. Doak, and A.M. Springer. 2004. Killer appetites: assessing the role of predators in ecological communities. *Ecology* 85: 3373–3384.

Worm, B. and R.A. Myers. 2003. Meta-analysis of cod-shrimp interactions reveals top-down control in oceanic food webs. *Ecology* 84: 162–173.

Zabel, R.W., C.J. Harvey, S.L. Katz, T.P. Good, and P.S. Levin. 2003. Ecologically sustainable yield. *American Scientist* 91:150–157.

Retrospection and Review

J.A. ESTES, D.P. DEMASTER, R.L. BROWNELL, JR., D.F. DOAK,
AND T.M. WILLIAMS

This final chapter is a review of the volume's content, written with the intent of revisiting our initial goals (Chapter 1), assessing the degree to which we have succeeded in achieving these goals, and providing some direction for future research on the central question underlying this volume: How did whales and whaling influence the dynamics of ocean ecosystems? Our synopsis also highlights the significant findings and ideas in each of the preceding chapters.

We begin by reiterating both the reasons for wondering about the influences of whales and whaling, and our approach to addressing the question. Given the diversity of species, habitats, and conservation problems in today's world, why should whales and whaling deserve special attention? The answer at one level is simply a matter of opinion and choice. Because of our various backgrounds, each of the authors of this volume was interested in larger issues that involved these topics. Different personal histories might have led one or more of us in different directions, so there is an element of serendipity in what has been produced.

Our reasons for asking about the ecological effects of whales and whaling were logical and multifaceted. However, the best way of addressing these questions was far less clear and required two key elements: a conceptual structure and the right people to implement it. The conceptual structure emerged in two forms. One of these arose from the ways we imagined that whales influenced ecosystem-level processes: as predators, as prey, and as

detritus. The other arose from the different approaches people take in practicing science: as theory and broad concepts, as the interface across academic disciplines, and as case studies. Recognizing these differences, we decided to invite two kinds of people: those who were familiar with whales and whaling and those who were familiar with the relevant theories or scientific disciplines. These groups were not mutually exclusive, because most people who study whales also have broader disciplinary interests. However, there were several participants who were invited because of their disciplinary skills and had little or no prior knowledge of either whales or whaling.

Why Are Whales Interesting?

The editors of this volume were drawn to questions concerning whales and whaling by two forces. One derives from the whales themselves. As the largest animals on earth, their biology is inherently interesting. As oceanic mammals primarily hidden from man's view, their role in ecosystems remains a mystery. The important organismal-level features of whales are discussed by Williams in Chapter 15, who explores the problems of being extremely large, and the physiological and ecological consequences of large body size. This unique biology established an *a priori* case for the ecological importance of whales and the subsequent impact of their disappearance through whaling.

Whaling, a human endeavor, substantially changed the distribution and abundance of most whale species across the world's oceans (chronicled in many of the chapters). The same is true of many other large vertebrates, but these changes seemed especially important for whales because of their great historical abundance on the one hand and their large size, high metabolic rates, and high trophic status on the other.

The other force that drew us together was a growing appreciation in ecology and conservation biology for the functional importance of top-down forcing processes initiated by high-trophic-status consumers, a point that is emphasized in various ways in Chapters 2, 3, and 4. Chapter 2, by Paine, provides an overview of both the relevant ecological concepts and theory, and from that view asks the question: What might we infer about the ecological effects of whales and whaling, based on what is known about food web dynamics from other, more tractable species and ecosystems? Paine concludes that if photosynthetic carbon in the world's oceans is largely or entirely utilized, the conclusion that whales and whaling were ecologically significant forces is inescapable. In Chapter 4, Donlan et al. extend Paine's conceptual argument in two dimensions: to the land and over deep time, with an emphasis on the functional importance of megafauna. The authors point out that on these expansive scales of space and time, ecological influences by the megafauna can be seen in two ways: through the loss of native species and by the introduction of exotics. Donlan et al. reinforce the longstanding argument that humans have played a significant role in the megafaunal demise, both recently and in the past. They further suggest that these losses left many species anachronistic on their landscapes, thus leading in some cases to dysfunctional ecosystems. Chapter 5 provides a similar view, but for the coastal oceans. In this chapter Jackson makes a plea for "retrospective ecology," arguing that quasi-pristine modern ecosystems are time machines for understanding how systems worked when their megafauna were intact. He explores this idea for the coastal oceans (emphasizing the Caribbean) and concludes that the loss of large vertebrates was of immense ecological consequence—among other things, leading to destabilizations that in turn may have subsequently contributed to reduced production and carrying capacity. Seen in this light, whales and their depletion by whaling provide a relatively recent and well-understood case of the repeated loss of megafauna around the world, allowing us a better chance to think about and analyze the consequences of such losses than is possible for many other systems.

Conceptual Structure and Approaches to the Question

The Question of Population Size

Central to understanding the interplay among whales, whaling and ocean ecosystems is the seemingly simple question: How many whales are there? No one would challenge the claim that the abundance of most large whale species was reduced by whaling. That point is made repeatedly throughout the volume and, indeed, is the fundamental event upon which the volume is based. However, the *magnitude* of these reductions is a matter of uncertainty and debate. Reeves and Smith (Chapter 8) provide a historical synopsis of modern whaling, including its chronology, methods, species, and geography, beginning with the Basques more than a thousand years ago and ending with the sudden curtailment of industrial-level commercial whaling in the 1970s. Logbook records from the various whale fisheries discussed by Reeves and Smith provide one basis for estimating the size of whale stocks before whaling, and thus the degree to which they have been depleted. An alternative approach, described and evaluated by Palumbi and Roman (Chapter 9), is based on theoretical relationships between genetic diversity and population size, combined with genetic information from living whales. When applied to fin and humpback whale populations in the North Atlantic, this approach provides population estimates greatly exceeding those obtained from logbook records, a conclusion that has generated a great deal of debate (IWC 2005). Palumbi and Roman argue that new analyses of logbook records are unlikely to provide much additional insight into the different conclusions of these methods. Conversely, genetic-based estimates will inevitably improve as more detailed information (i.e., more genes sequenced from more individuals of more species) becomes available and as the theoretical bases for these estimates are refined. On the other hand, those who question the veracity of the Palumbi and Roman estimates (IWC 2005) believe that a more thorough integration of the logbook records and whale process records and capability would likely lead to the conclusion that whale populations as large as those proposed by Palumbi and Roman (Chapter 9) were unlikely. Where the truth lies is not a mere academic curiosity, since estimates of prewhaling numbers will influence both recovery targets and the assessments of ecosystem-level impacts of whaling.

Other chapters in this volume provide additional information concerning the abundance and distribution of whales before and after whaling. By relating spatiotemporal patterns in the North Pacific harvest records to various oceanographic features of that region, Springer et al. (Chapter 19) identify potential oceanographic hot spots. This provides a mechanistic basis for the significant stock structure suggested by Danner et al. (Chapter 11) and for the related implication that local rates of whale depletion and ecosystem effects of whaling were much greater than might be inferred solely from an ocean basin–wide view. Likewise, Springer et al. point out that vastly more whales were taken from the southwestern region, whereas bounty and predator-control programs likely had stronger depletion effects on pinnipeds in southeast Alaska. Perhaps, then, it is not surprising that sea otter and pinniped populations have collapsed in southwest Alaska, whereas this generally has not occurred in southeast Alaska.

Pfister and DeMaster (Chapter 10) provide another perspective on the greater North Pacific ecosystem by contrasting historic and modern estimates of marine mammal biomass among species and species groups. Although these authors carefully qualify their results based on many uncertainties and the poor quality of some of the data, their findings demonstrate that marine mammal biomass in the North Pacific was and currently is dominated by the large whales, even though abundances are now much reduced across most taxa.

In addition to understanding the population size of great whales, there has been much recent interest in determining the abundance of killer whales in ocean ecosystems. This stems from claims that some killer whales obtained significant nutritional input by feeding on large whales (Springer et al. 2003) or large-whale carcasses (Whitehead and Reeves 2005). It has further been suggested that as these nutritional resources declined due to whaling, killer whales may have fed more intensively on pinnipeds and sea otters, thus driving their populations downward in some areas (Chapters 1, 14, 18, 20). The distribution and abundance of killer whales is an important component in any analysis of these hypotheses. Chapter 12, by Forney and Wade, provides such a worldwide assessment of the distribution and abundance of killer whales. These authors conclude that the world killer whale population presently exceeds 50,000 individuals, with the greatest population density occurring at high latitudes. Interestingly, modern-day patterns of killer whale distribution and abundance correlate well with global patterns of ocean productivity. Given the interest and controversy over the ecological importance of killer whale predation, it is appropriate that the Forney and Wade chapter is followed by a more detailed account by Barrett-Lennard and Heise (Chapter 13) of the natural history and behavior of killer whales. These authors focus on the well known dichotomy between fish eating (*resident* types) and marine mammal eating (*transient* type) whales, pointing out that this division has long been known, and that clear genetic differences exist between *resident* and *transient* whales. Different feeding behaviors or dietary preferences are the products of long periods of cultural evolution. However, a great deal of variation in diet and foraging behavior exists within each type. The authors also point out an apparent paradox in killer whale diet and foraging behavior—a strong inertia against change on the one hand, and the capacity for rapid change on the other.

Large Whales as Predators

But what evidence is there for ecologically important influences of whales, beyond their former abundance and currently reduced numbers? To answer that question, we sought contributors to address evidence for the role of large whales as important consumers. In this case there is evidence available from each of the three scientific approaches—theoretical/conceptual, disciplinary interfaces, and specific case studies. The most concrete and definitive evidence comes from studies on gray whales in the Bering and Chukchi seas (Chapter 23).

Highsmith et al. establish that gray whale feeding in the shallow soft-sediment benthos has had important effects on this ecosystem by consuming epi- and infaunal organisms and through the resuspension of sediments and nutrients from the benthos to the water column. These authors also point out, however, that such effects vary spatially, depending on such factors as sediment type and local oceanographic features.

The general conclusion that large whales are (or were) important consumers in ocean ecosystems is reinforced by several other chapters. Essington (Chapter 5) explored the larger issue of the effects of fishing and whaling on pelagic ecosystems by using known food web structures and dynamic food web models to ask how this system may have changed following population reductions of the large consumers, in this case the sperm whales, tunas, and billfishes. Essington concludes that population reductions of these species led to changes, but these appear to be modest compared with those observed in other ecosystems. Whitehead (Chapter 25) reached a similar conclusion for sperm whales, based on the great abundance and high trophic status of this species. Conversely, Essington reports that his modeling approach cannot capture qualitative shifts in process; thus, the possibility of more radical shifts in ecosystem organization cannot be discounted.

A different approach is taken by Croll et al. (Chapter 16). Here estimates of ocean production, whale abundance, whale trophic status, and trophic transfer efficiency are used to evaluate the potential impact of foraging whales in ocean ecosystems. Focusing on the North Pacific (although other regions would likely provide similar results), these authors suggest that, prior to whaling, the large whales coopted more than 50% of the total primary production, and that after whaling this was reduced to less than 10%. Croll et al. are careful to explain that these calculations do not necessarily transfer directly to the strength of food web interactions. However, based on similar measurements and analyses from other species and ecosystems in which consumer-mediated influences are known or thought to be important, Croll et al.'s conclusions fall closely in line with those made by other authors in this volume—that the great whales were important consumers in ocean ecosystems, and this interaction likely was substantially altered by the effects of whaling.

Focusing mainly on mysticete/krill interactions in the Southern Ocean, Ballance et al. (Chapter 17) also concluded that whaling must have exerted a significant force on this food web, although these influences cannot easily be separated from the confounding effects of oceanographic change. The latter in particular clearly affected krill population dynamics and, therefore, the population dynamics of krill consumers. Worm et al. (Chapter 26) approached the question of whaling and fishing effects on ocean ecosystems by exploring time series of whale abundance estimates and stock assessment data for fin fishes and forage fishes in the North Atlantic and North Pacific oceans. These authors concluded that there are significant trophic linkages among various forage fish, fin fishes, and great whales, and that whaling probably reduced the intensity of competition between some species of whales and fin fishes for a limiting forage fish resource. The result was increased fin fish populations, especially during the later part of the twentieth century in the North Pacific. Yet Worm et al. find no strong evidence for this effect in the North Atlantic. In contrast to this last result, Clapham and Link (Chapter 24) speculate that such relationships did occur in this area, thus suggesting that the lack of a significant finding in Worm et al.'s time series may to some degree be caused by a temporal mismatch between the periods of whale removal (earlier) and the available time series of fin fish and forage fish data.

Large Whales as Detritus

We turn next to the role of dead whales as detritus (Chapter 22). Here Smith summarizes the empirical evidence for the significance of this process that has been obtained from experimental and opportunistic studies of the colonization, growth, and extinction of detritivores on dead whale carcasses. The author emphasizes a remarkable possibility—that a single whale carcass may influence the nearby seafloor community for decades. Smith also concludes that this process probably is not of great significance to seafloor ecosystems on continental shelves but that it may be extremely important in the deep sea. Another interesting dimension to this process is that, because of the great longevity of whale carcasses on the seafloor, deep-sea ecosystems may just now be feeling the effects of the impoverishment of this resource that resulted from industrial whaling.

Large Whales as Prey

The ecological role of large whales as prey is the most controversial of the three potential food web pathways that we have identified. This is due in part to the nature of the evidence and in part to equally controversial indirect effects of the interaction on other important components of the food web (Springer et al. 2003). Several chapters of the volume bring a variety of interesting dimensions to questions regarding the nature and magnitude of killer whale predation on large whales. Reeves et al. (Chapter 14) summarize the evidence and arguments for and against the view of killer whales as consequential predators of large whales, although they draw no firm conclusions. In Chapter 18, Doak et al. address the plausibility of the most controversial aspect of Springer et al. (2003)—the question of the nutritional importance of great whales to killer whales and how this changed as a result of whaling. Not surprisingly, the results depend strongly on the assumptions, but they illustrate that large whales could have provided a significant food resource for *transient* killer whales, even if attacks are rarely seen and if only limited tissues or segments of large-whale populations were consumed. Readers interested in the Springer et al. hypothesis are referred to recent papers by Whitehead and Reeves (2005) and DeMaster et al. (2006).

There have been two proposed examples of the so-called whaling/cascade hypothesis, Springer et al.'s (2003) account for the North Pacific and southeastern Bering Sea, and a much earlier poorly known account by Barrat and Mougin (1978) from the Southern Ocean. Branch and Williams (Chapter 20) build on these two reports and explore the feasibility for the Springer et al. hypothesis for the Southern Ocean. Although evidence for causality is problematic (as emphasized by Ballance et al. in Chapter 17), these authors conclude that the whaling/cascade hypothesis is an easily feasible explanation for various southern ocean pinniped population collapses (southern elephant seals and southern sea lions in this case), and the hypothesis is consistent with much of the available information. Branch and Williams further conclude that recently reported minke whale declines are unlikely to have been driven by killer whale predation, although that conclusion is partly based upon the assumption that all or much of the minke whale carcasses are consumed. The authors also point out that any explanation of these declines raises a large number of unanswered questions, and hence, like the analyses by Doak et al. (Chapter 18), the contribution by Branch and Williams is not so much an argument as it is a feasibility analysis. Chapter 21 by Mangel and Wolf explores the Springer et al. hypothesis using optimal foraging theory and decision analysis, essentially asking the question: Does it make sense? These authors conclude that the hypothesis is indeed plausible, but they also identify the need for additional information. Plausibility, of course, does not establish factuality. It does, however, in the absence of empirical evidence, establish the ease or difficulty with which an ecological interaction might occur. This knowledge, in turn, provides a quantifiable sense of likelihood as well as an indication of whether or not the pursuit of additional empirical evidence is worth the effort.

The Broader View—Discipline, Timescale, Socioeconomics, and Process

The volume also includes a number of chapters that explore broader issues concerning whales and whaling in ocean ecosystems from the perspectives of deeper time, other interaction web processes, policy, and law. Although this volume is about whales and whaling, not fisheries and oceanography, it would be incorrect and unfair not to acknowledge the influence of these disciplines on the way we look at ocean ecology. Chapter 6, by Hunt, provides that perspective by discussing the importance of bottom-up control, largely seen to be driven by climate change and human exploitation. The author gives a synopsis of the evidence for bottom-up forcing effects and points out that these effects cannot be disregarded in exploring the consequences of predator removals and top-down effects. Further, Hunt notes that for marine mammals, the occurrence of top-down control appears to be rare and that bottom-up and top-down forcing are processes whose importance in a given marine ecosystem is scale-dependent. In Chapter 7, Lindberg and Pyenson explore the relationships between whales and ocean ecosystems in macroevolutionary time, from the Cretaceous-Tertiary (K-T) boundary through the period of the fossil record of cetaceans and their recent ancestors. These authors summarize a large body of paleontological information, from which they suggest that loss of marine reptiles at the K-T boundary had a strong influence on cetacean evolution, affecting such fundamental characteristics as diet, body size, and species interactions. This is a rich potential area for future research. Chapter 27 by Costa et al. would seemingly appear as an outlier to the volume, because it addresses pinnipeds rather than whales or whaling. However, these authors present a global analysis of the patterns and reported causes of pinniped population declines and, in doing so, establish the oceanic conditions facing the higher-trophic-level marine species, including whales. Costa et al. indicate that pinniped declines have occurred in various species and in many different areas. Despite many interesting patterns, especially relating to taxonomy, behavior, and physiology, the authors also recognize perplexing inconsistencies. Thus, even under the best of circumstances the actual causes of these declines are in most cases poorly understood.

Because whaling was a human endeavor with important social and economic dimensions, no volume on whales and whaling would be complete without some attention paid to the question of what they meant and continue to mean to people. The chapters by Bromley (Chapter 28) and Orbach (29) discuss whaling from social, economic, legal, and policy perspectives. Two of us (D.P.D. and R.L.B.) note that the discussion of the IWC is more complex than described by Orbach when one factors in the issue of the Revised Management Scheme, which is currently under development by the IWC.

Finally, in Chapter 30 Kareiva et al. deal with the immediate problem of whale conservation, casting the main approaches, findings, and arguments from various contributions in this volume into the broader context of the discipline of ecology and the more general issues and needs of conservation. From each perspective, the authors make the following important points. The first is that scientific culture has affected the manner in which we view the questions, and this very often leads us somewhat astray. Second, while the problem of whale conservation and recovery may seem dire, it is in fact much more hopeful than similar issues on the land, where species are now extinct and habitats have been grossly altered. Third, continued human exploitation is not the only threat to the recovery of whales. And lastly, part of what makes this entire problem interesting is the fact that whales are a human cultural phenomenon. Kareiva et al. go one step further to chastise the cetacean research and conservation community for placing too much emphasis on counting and not enough emphasis on ecology, and they raise the question of whether the larger community of ocean scien-

tists may have placed too much emphasis on bottom-up forcing in their efforts to understand ocean ecosystem dynamics. These authors also ask the most forward-looking question in this volume: If the whales recover, what will this mean to the oceans? Their conclusion is that it will probably mean quite a lot.

Have We Achieved Our Goals in This Volume?

There is no simple answer to this question. A careful look back through the chapters reveals that, in fact, there is little in the way of concrete empirical evidence for the ecological effects of whales and whaling. There can be little doubt that the oceans have changed, but the degree to which whales and whaling caused or contributed to these changes is wrought with uncertainty and no small amount of controversy. It is difficult to separate the influences of whaling from the influences of oceanographic and climate change, or from the effects of more recent or ongoing fisheries, especially given that so much attention has been paid to these latter influences and so little attention has been paid to the influences of whales and whaling. Viewed in this way, one might conclude that we have failed. But while the empirical evidence for the effects of whales and whaling is thin for each case, a collective view of the combined empirical evidence, theory-based approaches, and process-based evidence from other more tractable species and ecosystems yields a substantially different interpretation. This more synthetic approach leads to the inescapable conclusion that whales have had large and important effects on the structure, function, and evolution of ocean ecosystems and that whaling has altered the structure and function of ocean ecosystems over the last two centuries. Viewed in this way, we believe that we have succeeded. Although case-by-case arguments remain, no longer is it reasonable to think of whales simply as passengers in a changing ocean. The volume provides a clear view of the processes by which whales and whaling can influence the oceans as well as a road map for how these processes can be further studied and understood.

The Future

Considering what we know and don't know about whales, whaling and ocean ecosystems, and looking now to the future, what would we like to know and how should science proceed? If nothing more, this volume highlights an important point—that whales and whaling have influenced the oceans in significant ways. As decisions are being made about how future whale stocks should be managed, this perspective must be added to the ever-evolving algorithms used to advise managers, if not brought to the fore. Although collectively we agree with Kareiva et al. (Chapter 30) that in the past there has been too much attention to numerology and not enough attention paid to ecology or, for that matter, other biological disciplines, individually we are of differing minds on the importance of these measures. Three of us (J.A.E., D.F.D., and T.M.W.) believe that improved estimates of historical whale abundance and current carrying capacities will be necessary to properly inform management decisions because of their central importance in the assessment of the potential ecosystem impacts of whaling and in the establishment of targets for recovery. Two of us (D.P.D. and R.L.B.) do not support that view, instead believing that the central goal of research should be to obtain improved information on current abundance, stock structure, and factors that limit the growth rates of large whales and to distinguish between changes in large-whale abundance caused by anthropogenic impacts and those caused by environmental change. Achieving a capability for the implementation of either view will likely be problematic, but the beginnings of a path have been established in many of the chapters found in this volume.

Looking beyond numerology, what about ecology, physiology, behavior, and evolution? What sorts of research might be conducted in the future to understand better the influences of whales and whaling on ocean ecosystems? One obvious approach is simply to watch ecosystem change as whale stocks recover, while keeping the confounding influences of other factors such as fisheries and oceanographic change well in mind. This is a potentially powerful research tool that can be employed to test some of the current thinking that is both hypothetical and controversial. As suggested by Clapham and Link (Chapter 24), if we are to understand better the ecological roles played by whales, future marine mammal surveys must be integrated with multidisciplinary ecosystem research, notably in areas that are already well studied. They further note that a parallel approach would be to conduct such integrated surveys in simpler ecosystems where whales have to date failed to recover from whaling, in order to monitor system changes as recovery occurs.

A second approach to the problem is the increased use of theory and interdisciplinary synthesis. Physiological limitations of the animals comprised in ecosystems represent a powerful, yet sorely underutilized, avenue for predicting what can or cannot occur. Despite their size, whales can swim only so fast when foraging and can process only so much food in a day when feeding. Such biological limitations and traits define the limit of their influence and the resulting strength of interactions between individual predators and prey. Coupled with demographic and behavioral information regarding prey choice, such interdisciplinary approaches have changed how we view whales in ecosystems. The chapters by Doak et al. (18), Williams (15), and Branch and Williams (20) illustrate the power of this method, but they are only a beginning. Modeling approaches could be used to understand better such fundamental problems as whether the killer whale/marine mammal assemblages in different ocean basins are sustainable without historic abundance levels of large whales, and whether population change is more sensitive to bottom-up or to top-down forcing processes. As pointed out by Doak et al., these approaches usually do not provide final definitive answers, but they do help establish the degree of plausibility of hypothesized processes and thus are an important step in the quest for final answers.

Finally, we encourage future cetologists to build their conceptual visions and conduct their research in the company of interdisciplinary collaborators. There has been some effort already to join forces with oceanographers, and that of course should continue. But the dimensions to the interplay between whales, whaling, and ocean ecosystems are much bigger than this. We hope this volume has made that point. We also note that whales are not so unique that our understanding of them cannot inform and be informed by comparisons with and insights from other taxa and the rest of ecology.

In conclusion, we note with interest the parallel paths Western culture has taken when attempting to place the roles of humans and whales in an ecological context. Both have proven difficult, although perhaps for different reasons. It appears that it was easier for us to consider humans an important element of the marine ecosystem than it was for us to consider large whales important. We hope this volume has contributed to the position that whales and whaling are aspects of the marine ecosystem that have to be integrated into the more traditional fields of fishery, wildlife biology, and oceanography. In the absence of such integration, many of the mysteries regarding how the world's oceans function will remain enigmatic.

Literature Cited

Barrat, A. and J.L. Mougin. 1978. The Southern elephant seal, *Mirounga leonina*, of Possession Island, Crozet Archipelago, 46°25′ south, 51°45′ east. *Mammalia* 42: 143–174 (in French).

DeMaster, D.P., A.W. Trites, P. Clapham, S. Mizroch, P. Wade, and R.J. Small. 2006. The sequential megafaunal collapse hypothesis: testing with existing data. *Progress in Oceanography* 68: 329–342.

IWC. 2005. Report of the Scientific Committee. *Journal of Cetacean Research and Management* (Supplement) 7: 32–34.

Springer, A.M., J.A. Estes, G.B. van Vliet, T.M. Williams, D.F. Doak, E.M. Danner, K.A. Forney, and B. Pfister. 2003. Sequential megafaunal collapse in the North Pacific Ocean: an ongoing legacy of industrial whaling? *Proceedings of the National Academy of Sciences* 100: 12223–12228.

Whitehead, H. and R.R. Reeves. 2005. Killer whales and whaling: the scavenging hypothesis. *Biology Letters* 1(4): 415–418.

INDEX

ice seals, 219–220
Krill Surplus Hypothesis, 217–218
loss of prey for killer whales, 226
overview of, 215
penguins, 221–226
pinnipeds and seabirds, 220–221, 223–224
post-whaling trends, 218–226
Ross Sea Shelf and Slope areas, 226–227
as predators, 390–391
as prey, 391
Lascaux cavern, 29
law and culture, relationship between, 373–374
less-than-zero-sum ecology, 34
limpet (*Lottia alveus*), extinction of, 9
Lindberg, David R., 67
line-transect survey (LTR), 148**t**, 149
longline catch rates, decline of, 47
Lotze, Heike K., 335

MacArthur, Robert, 9
Magnuson, Senator Warner, 371
Magnuson-Stevens Fisher Conservation Act (FCMA) of 1976, 374
Mangel, Marc, 279
marine bird populations
at-sea distribution, 55
bottom-up control of, 54–55
and El Niño-Southern Oscillation (ENSO) events, 54
and food limitation, 54
foraging and prey availability, 55
penguins, 221–226
and removal of large whales from the Southern Ocean, 220–221
marine ecology
absence of large animals, implications of, 30
baseline community concept, 30
basic principles, origins of, 29
biological oceanographers vs. ecologists and fisheries biologists, 30
biomass pyramids in, 29–30
marine ecosystems. *See also* ocean ecosystems
anthropogenic influences on, 19–20, 23, 68
coral reef, 22, 30–31
effects of fishing and whaling on, 1, 296–298, 335–341
freshwater, 8–9
history of whale populations as backdrop for management of, 102–103
human alterations, overview of, 335–336. *See also* fishing; whaling
management, 376
methodical dismantling of, 33
Marine Mammal Protection Act (MMPA) of 1972, 374–375, 374
marine mammals. *See also* pinnipeds
bottom-up effects on, 55–56
decline of
classic diet choice model as explanation for, 279–281, 284

coupling diet choice and prey population dynamics, 281
in North Pacific, predation and, 232–234
parameter estimates and sample calculations, 281–283
predictions, 283–284
in Southern Hemisphere, 274–275
ecological roles of, 116
evaluating impacts of commercial harvest on, 123–124
heart mass in relation to body mass, 194**f**
international attitudes toward, 375
intestinal length in relation to body length, 195**f**
killer whale predation by, 168–169
metabolic rate in relation to body mass, 193**f**
migratory patterns, 117
most trophically distinct, 326**t**
public interest in, 379–380
sequential collapse of in North Pacific Ocean, 263**f**
mark-recapture (MR) method, population estimates, 148**t**, 149
Martin, Paul S., 14
mass extinctions, 28
maximum sustainable yield (MSY), principle of, 374
megafauna
in Africa, 29
climate changes and decline of, 28
defined, 27
ecological consequences of the loss of, 28
humans and the decline of, 16, 27–28, 29
megafaunal collapse, and killer whale predation, 279
Melville, Herman, 324
mesonychids, 68
microbial loop, 30, 34
MIGRATE program, 108
migration
age- or sex-specific patterns of, 117
estimates of, 108
seasonal variations in, 117
to avoid predation by killer whales, 170
minke whale (*Balaenoptera acutorostrata*)
Antarctic, 266, 270–271, 272
carrying capacity, 55
dwarf, 191
genetic diversity in, 107
migratory patterns of, 135
North Pacific Ocean, 253
population estimates, 111
population structure of, 135
whaling and, 88
monk seals, 33, 34
Muir, John, 374
multispecies relationships. *See* interaction webs
mutation rate
extreme rate heterogeneity, 106
genetic divergence, 104–105
hypermutation, 106–107
measuring, 104–107
variation, cautions concerning, 105–106

muted cascade, 10
Myers, Ransom A., 335

narwhals, whaling of, 82, 83
natural resource, defined, 364
neotropical anachronisms, 1
net whaling, 88
non-standardized survey (SURV), 148**t**, 149–150
North Atlantic Oscillation (NAO) atmospheric circulation pattern, 53
North Atlantic right whale (*Eubalaena glacialis*), 90, 182, 318–320
North Pacific gray whale (*Eschrichtius robustus*), 55
North Pacific Ocean
blue whales, 248**f**, 251–252
bowhead whales, 247**f**, 250, 254–255
Bryde's whales, 249**f**, 252**f**, 253
commercial whaling, effects of, 257–258
depletion of whales in, 245–246, 248
fin whales, 249**f**, 252–253, 255
gray whales, 250–251
history of whaling in, 246–249
nineteenth century, 246
postwar whaling, 248
twentieth century prior to World War II, 247–248
humpback whales, 248**f**, 251
killer whales, 256–257
marine mammals, sequential collapse of, 263**f**
minke whales, 253
right whales, 247**f**, 249–250
sei whales, 249**f**, 252**f**, 253
sperm whales, 250**f**, 253–254, 255–256
whale depletion in, 258
whales harvested within 100 nautical miles of the coast, 257**f**
North Pacific right whale (*Eubalaena japonica*), 88, 124
northern bottlenose whale (*Hyperoodon ampullatus*), 82, 92
northern elephant seal, 348
northern fur seal (*Callorhinus ursinus*), 116, 122, 350
Norwegian-style shore whaling, 84, 91–92
no take rule, 374, 375
nutritional limitation hypothesis, 2

observation and anecdotal information (OBS), 148**t**, 150
ocean ecosystems. *See also* marine ecosystems
biodiversity loss and, 202
bottom-up effects on, 2
climate change and, 51
El Niño-Southern Oscillation (ENSO) events, 2
evolution and history of, 2
fisheries removal and, 51
importance of whales to, 382–384
large predators in, 196–199
management of, 376
removal of marine mammals from, effects of, 116